Directed C—H Bond Functionalization

Directed C—H Bond Functionalization

Concepts and Applications

Edited by Debabrata Maiti and Supriya Rej

WILEY-VCH

Editors

Debabrata Maiti
Indian Institute of Technology Bombay
Main Gate Rd, IIT Area
Powai, Mumbai
IN, 400076

Supriya Rej
Indian Institute of Technology Dharwad
Permanent Campus
Chikka Malligwad, Karnataka
IN, 580007

Cover Design: Wiley
Cover Image: © Double Brain/Shutterstock, Courtesy of Prof. Supriya Rej

Typesetting: Straive, Chennai, India
Printing and Binding: CPI Group (UK) Ltd, Croydon, CR0 4YY

Library of Congress Card No: applied for

British Library Cataloguing-in-Publication Data A catalogue record for this book is available from the British Library.

Bibliographic information published by the Deutsche Nationalbibliothek
The Deutsche Nationalbibliothek lists this publication in the Deutsche Nationalbibliografie; detailed bibliographic data are available on the Internet at <http://dnb.d-nb.de>.

Print ISBN: 978-3-527-35419-1
ePDF ISBN: 978-3-527-84847-8
ePub ISBN: 978-3-527-84846-1
oBook ISBN: 978-3-527-84848-5

Contents

Preface

The development of site-selective C–H bond functionalization has revolutionized the field of organic synthesis. Among the most powerful strategies to achieve such selectivity is *directed C–H functionalization*, wherein a suitably directing group guides the metal catalyst to activate a specific C–H bond, often with remarkable regio- and stereocontrol. This approach has transformed how chemists build molecular complexity, enabling streamlined access to diverse scaffolds with unprecedented precision and efficiency. This book, titled *Directed C–H Bond Functionalization: Concepts and Applications*, aims to provide a comprehensive yet accessible resource that captures the foundational principles, mechanistic insights, and cutting-edge applications of this field. The past three decades have witnessed explosive growth in transition metal-catalyzed C–H functionalization, particularly with palladium, ruthenium, rhodium, cobalt, and iridium complexes, guided by a wide array of directing groups.

The first part of the book introduces the conceptual underpinnings of C–H activation and the role of directing groups in overcoming the inherent challenges of site selectivity. Detailed discussions cover mechanistic pathways. This is followed by a systematic survey of key evolutionary strategies for selective C–H functionalization reactions. The second part of the book is devoted to specific applications and recent advances. Case studies explore the use of directed C–H activation in natural product synthesis, pharmaceutical lead modification, and late-stage functionalization. Several chapters highlight various directing strategies, including in-built directing groups, traceless directing groups, removable auxiliary-directed approaches, external template-assisted distal C–H bond functionalization, cooperative catalysis, and transient directing strategies. Special attention is given to sustainable and atom-economical methodologies, such as redox catalysis, photochemical C–H bond activation, and industrially relevant flow techniques. In addition, recent progress in asymmetric C–H functionalization and metal-free strategies are discussed, showcasing the expanding scope and utility of this transformative methodology.

This book is intended for graduate students, researchers, academic scientists, and industrial chemists who are engaged in organic synthesis, catalysis, medicinal chemistry, and related areas. It may also serve as a reference for educators and instructors who wish to incorporate modern synthetic methods into advanced coursework. We hope that this book will inspire readers to explore the vast opportunities that directed C–H bond functionalization offers and contribute further to its evolution. The field continues to challenge conventions and redefine what is possible in synthesis. While this book captures a snapshot of its exciting evolution, we believe that the potential of C–H bond functionalization is far from fully realized, and it is just the tip of the iceberg.

October 2025

Debabrata Maiti
Supriya Rej

1

History of Directed C—H Bond Activation and its Discovery

Susmita Mondal[1,*], *Sumit Ghosh*[2,*], *Asim Kumar Ghosh*[2], *and Alakananda Hajra*[2]

[1] *Department of Chemistry, Central Ayurveda Research Institute, 4-CN Block, Bidhannagar, Kolkata, 700091, West Bengal, India*
[2] *Department of Chemistry, Visva-Bharati (A Central University), Santiniketan, 731235, West Bengal, India*

1.1 Introduction

C—H Functionalization is a groundbreaking technique in synthetic organic chemistry that allows the direct transformation of C—H bonds into C—C or C—X (X = heteroatom) bonds [1]. However, regioselective functionalization of C—H bonds is a highly challenging task due to its ubiquitous nature [2]. Thus, in past decades, pre-functionalization was performed on starting molecules for further functionalization [3]. Within this realm, transition metal-catalyzed C—H bond activation for the construction of C—C and C—X bonds is one of the most significant and challenging fields of research due to its more research-laboratory-friendly conditions, high efficiency, and selective C—H functionalization [4]. A molecular complex can be built from readily available simple hydrocarbon counterparts by using this strategy [5]. Therefore, functional-group-assisted site-selective C—H bond activation for functionalization shows great promise to researchers. The coordination of transition metal catalysts with the directing group controls the site selectivity of the reaction. The pioneering work in the field was reported by Lewis and Murai in 1986 who reported regioselective mono- and di-*ortho*-alkylations of phenol with ethylene, which proceed through the formation of ruthenium phosphite complex [6]. After that, various directing group-assisted chelation-controlled C—H activation strategies have been developed for the incorporation of various functional groups like amide, anilide, imine, heterocyclic, amine, carboxylic acid, ester, ketone, and hydroxyl [7]. There are mainly two types of directing group-assisted C—H functionalizations: one is the built-in directing group-assisted C—H bond functionalization and another is the removable directing group-assisted C—H bond functionalization. The built-in directing group is already present in the functionalized molecule; thus, it reduces the step economic issue, whereas the

*These authors have first authorship and have equal contribution.

Directed C–H Bond Functionalization: Concepts and Applications, First Edition.
Edited by Debabrata Maiti and Supriya Rej.

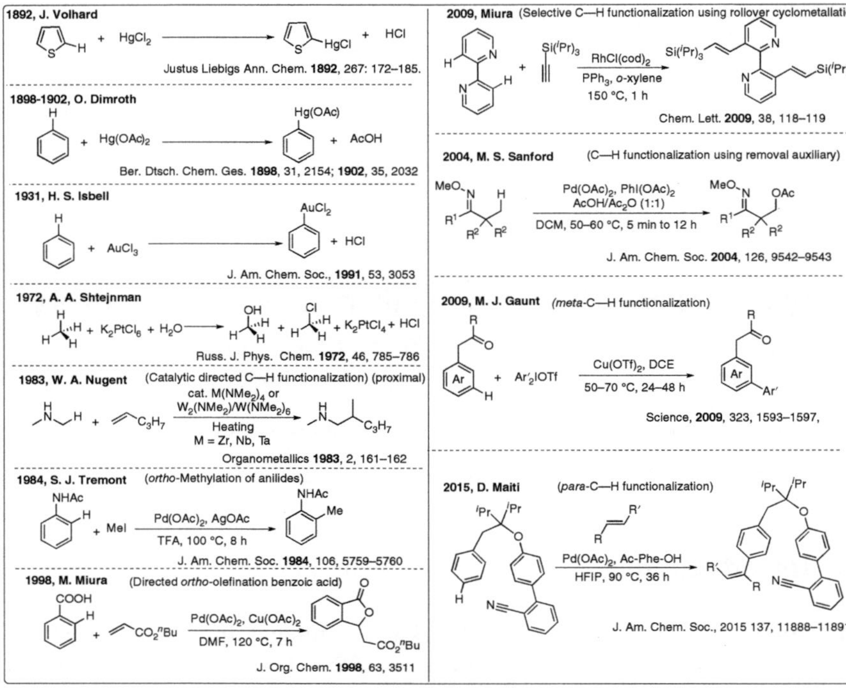

Figure 1.1 Schematic representation of the evolution of C–H activation.

removal directing group may be pre-installed or installed *in situ* in a molecule to achieve a favored coordination with the metal for selective C—H bond functionalization. In the former directing group, there is a possibility of unprecedented reactions due to the weak coordination ability to a metal, but the removal directing group has an optimum coordinating ability toward the metal catalyst and is widely used for the functionalization of a non-biased C—H bond.

Therefore, a brief overview on the history of directed C—H functionalization is much needed. Our group has also worked on directed C—H functionalization [8]. In this chapter, we have presented a strategic evaluation of directing group-assisted selective C—H bond functionalization via C–H activation. We have included early discoveries on stoichiometric metal-promoted proximal C—H bond activation, directing group (both built-in functional group and removal directing group)-assisted catalytic proximal C—H bond functionalization, and directing group-assisted distal (both *meta* & *para*) C—H bond activation in this chapter. Figure 1.1 represents the historical evolution of C—H bond activation.

1.2 Importance of C–H Activation

C—H Bond activation allows the synthesis of compounds without the requirement of pre-halogenated or pre-functionalized starting materials, which are typically required in traditional methods; hence, this approach is more environmentally friendly. On the other hand, C—H bond activation simplifies the synthetic process by reducing the number of steps needed. It also allows for the direct functionalization of hydrocarbons, which are the primary raw materials derived from oil and natural gas used in the chemical industry. Not only that, it has a significant interest in late-stage modification in the pharmaceutical and material industries.

1.3 Early Discoveries in Stoichiometric Metal-promoted Proximal C–H Bond Functionalization

In 1937, A. Farkas and L. Farkas first discovered the interaction between a C—H bond and a transition metal during the catalytic exchange of benzene and D_2 on a platinum foil [9]. After that, Garnett [10] and Parshall [11] also reported the metal atom interaction with either aromatic or aliphatic C—H bond. These discoveries inspired researchers to activate unreactive C—H bonds for functionalization purposes. In this regard, in 1955 and 1956, Murahashi and Horiie [12] carried out Co-promoted C—H carbonylation with carbon monoxide of a Schiff base and azobenzene, respectively (Scheme 1.1a). The following two reactions proceeded through the formation of a five-membered cobaltacycle. After that, a number of research groups were actively involved in the carbonylation of different moieties via C—H bond activation. In 1958, Rosenthal et al. [13] reported Co-promoted *ortho*-carbonylation of benzophenone oxime (Scheme 1.1b), and Bagga et al. [14] developed Fe-promoted carbonylation of a Schiff base (Scheme 1.1c). Other metallacycles like nickelacycle [15], ferrocycle [16], platinacycle [17], and palladacycle [17] (formed from corresponding metal complexes with the interaction of *N*-atom of organic core) have been isolated, which may serve as a directing group for C—H functionalization via the activation of a C—H bond. In this connection, Fahey group [18] used palladacycles formed from the reaction of azobenzene and $PdCl_2$ for halogenation (chlorination and bromination) of azobenzenes (Scheme 1.1d). Komiya group [19] used a ruthenium hydride complex for the deuteration of a C—H bond with DCl (Scheme 1.1e).

1.4 Directing Group-assisted Catalytic Proximal C–H Bond Functionalization

1.4.1 In-built Functional-Group-directed Proximal C–H Bond Functionalization

In the previous section, we have discussed transition-metal-complex-mediated C—H bond activation for functionalization. Meanwhile, catalytic-directed C—H activation was first done by Nugent et al. [20] in 1983, who reported C—H bond activation of dimethylamine for alkylation with alkenes using a metal dimethylamide precursor (Scheme 1.2).

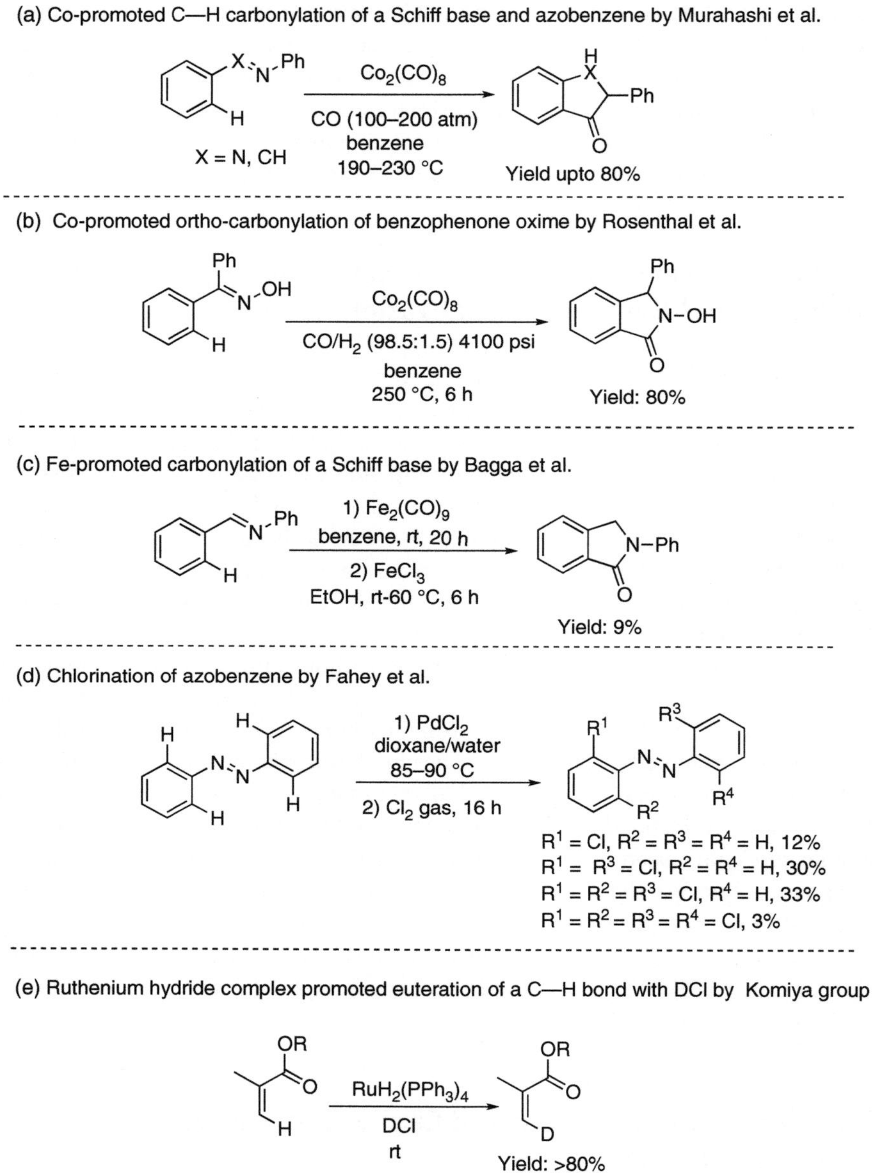

Scheme 1.1 Pioneering reports on directed C–H functionalization reactions.

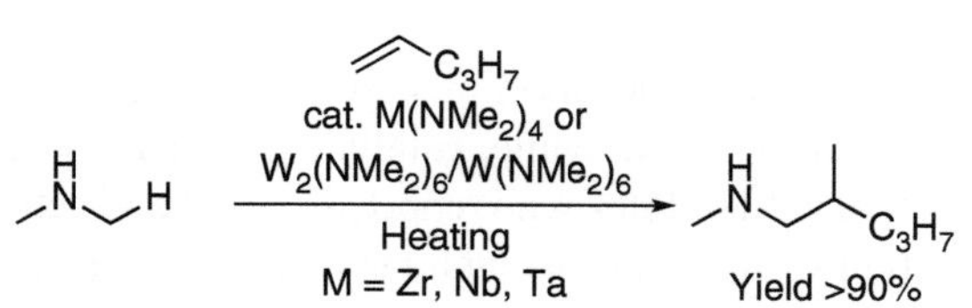

Scheme 1.2 C–H bond activation of dimethylamine for alkylation.

One of the essential catalytic C—H activation reactions was reported by Murai et al. [21] in 1993 (Scheme 1.3). They used a ketone as a directing group for the regioselective *ortho*-C—H alkylation of aromatic ketones with alkenes. The reaction proceeds through a metallacycle, which was the key intermediate of the reaction.

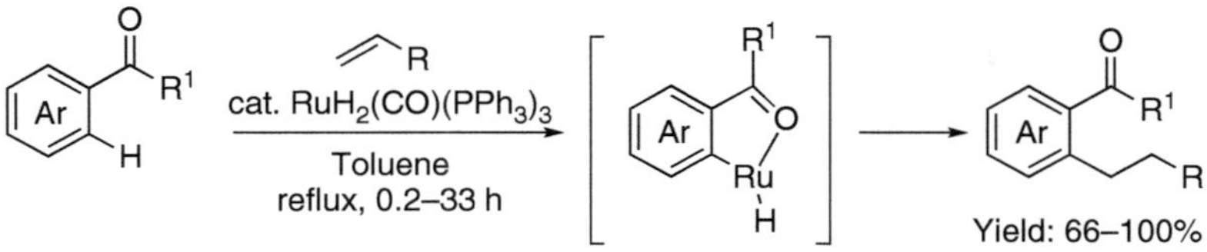

Scheme 1.3 Regioselective *ortho*-C—H alkylation of aromatic ketones.

After that, numerous research works have been published on this field using various functional groups as directing moiety, such as carbonyl, amine, amides, hydroxyl, carboxylic acid, and sulfonic acid derivatives. Daugulis and Zaitsev [22] reported palladium-catalyzed *ortho*-arylation (both mono- and di-) of anilides using amide as a directing group and aryl iodides as an arylating agent in 2005 (Scheme 1.4). The method is highly tolerant to functional groups showing up to 1000 turnovers for this reaction.

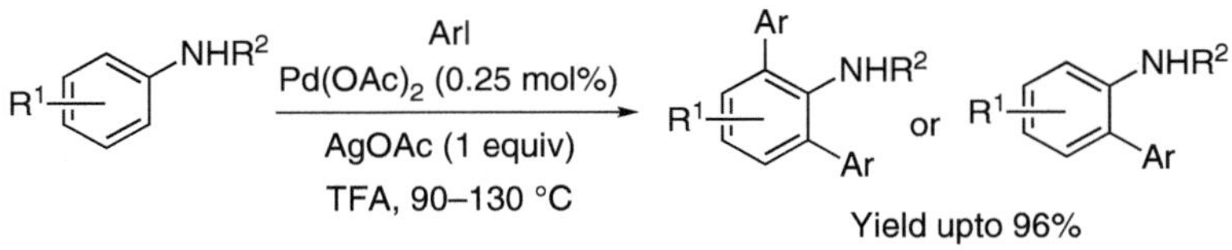

Scheme 1.4 Palladium-catalyzed *ortho*-arylation of anilides.

Kuninobu et al. [23] developed a method for direct aryl C—H addition to aldehyde using *N*-containing directing group in the presence of 5 mol% $[MnBr(CO)_5]$ in toluene (Scheme 1.5). This present protocol could be applied for asymmetric transformation using an aromatic compound with a chiral substituent.

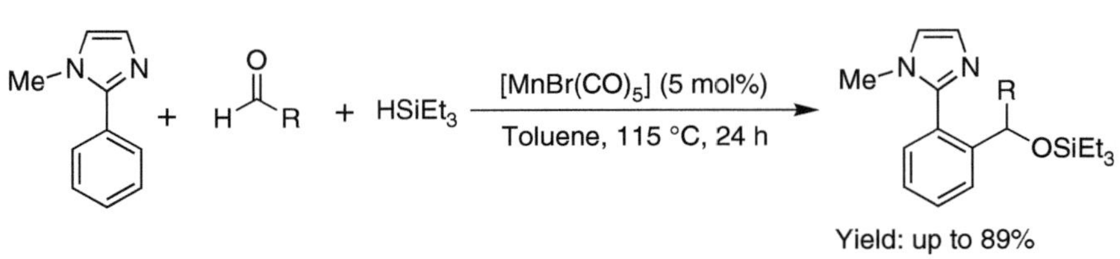

Scheme 1.5 Mn-catalyzed remote C—H addition to aldehyde.

In 2009, Zhang and Yu [24] developed a versatile method for *ortho*-hydroxylation of benzoic acids using Pd catalyst under O_2 atmosphere (Scheme 1.6). Mechanistic investigations suggested that direct oxygenation took the place of aryl-Pd species

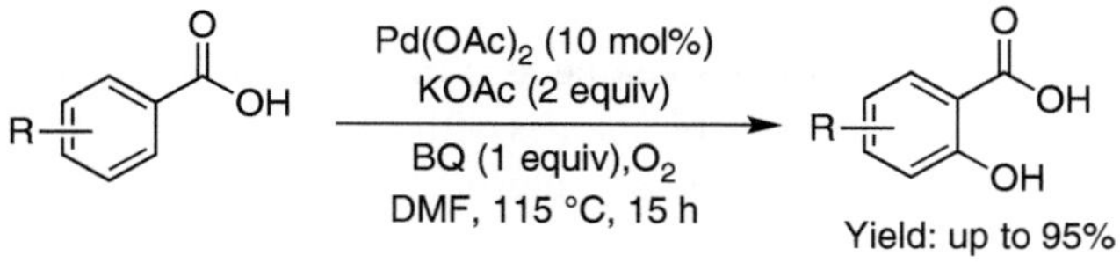

Scheme 1.6 Pd-catalyzed *ortho*-hydroxylation of acid derivatives.

from molecular O_2. A wide range of functional groups were well tolerable in the reaction.

Rhodium-catalyzed Grignard-type arylation of activated aldehyde was described by Li and co-workers [25] in 2011 (Scheme 1.7). Remote C—H addition of activated aldehyde occurred at the *ortho*-position of the phenyl ring of 2-phenyl pyridine derivatives assisted by pyridine ring in the presence of 5 mol% $[(CH_3CN)_3Cp^*Rh(SbF_6)_2]$.

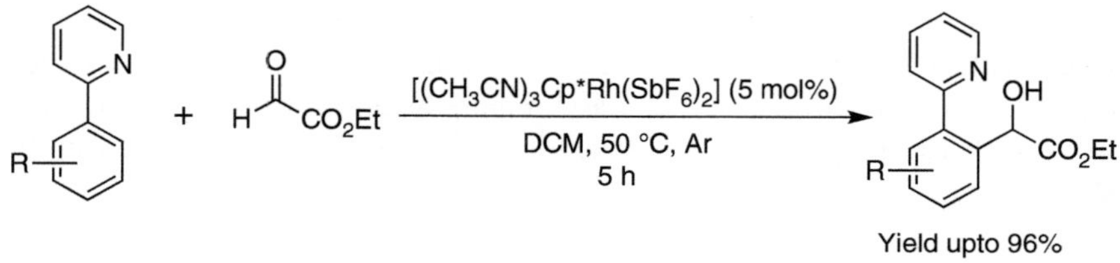

Scheme 1.7 Arylation of activated aldehyde.

The following year, Shi and co-workers [26] also reported *N*-directing group-assisted remote aryl C—H addition to activated aldehydes to produce biaryl methanols in the presence of Rh(III) catalyst (Scheme 1.8). The present reaction is compatible with water and air.

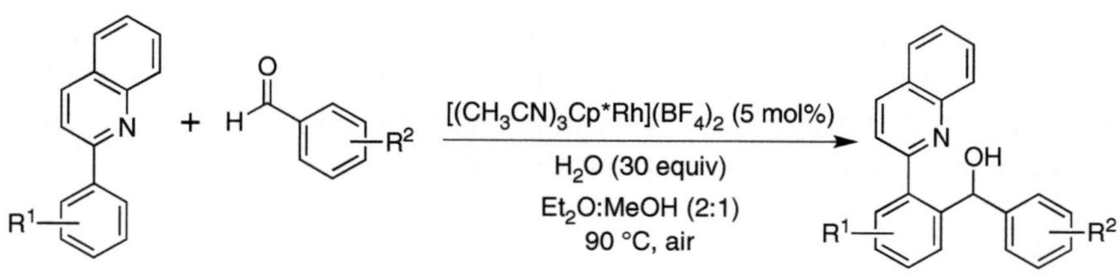

Scheme 1.8 C—H addition of 2-arylquinolines to activated aldehyde.

In the same year, Ackermann and co-workers [27] reported ruthenium-catalyzed remote C—H hydroxylation of benzamide derivatives using PIDA as an oxidant in trifluoroacetic acid/trifluoroacetic anhydride (TFA/TFAA) solvent mixture using Ru-catalyst (Scheme 1.9).

Shi and co-workers [28] described a Rh-catalyzed direct intermolecular aromatic C—H bond nucleophilic addition to ketones directed by quinoline (Scheme 1.10). A wide variety of quinoline derivatives were well tolerable in this transformation.

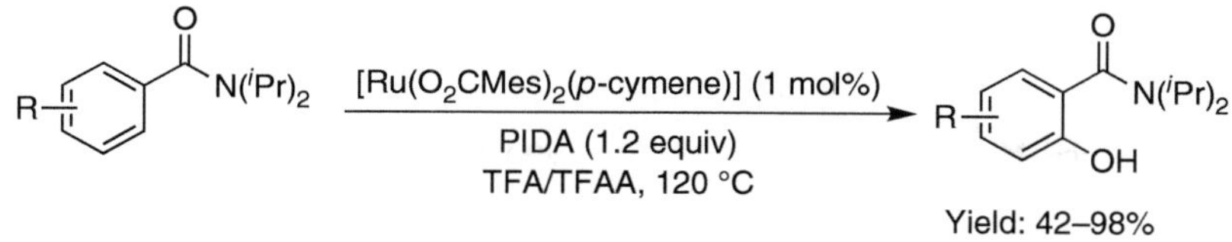

Scheme 1.9 Ruthenium-catalyzed hydroxylation of benzamides.

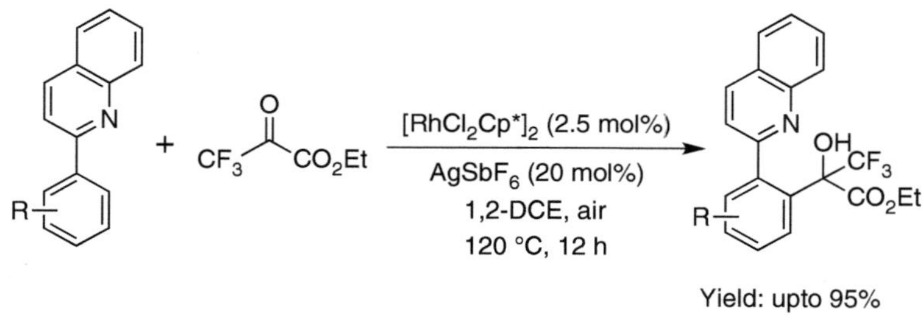

Scheme 1.10 Rh(III)-catalyzed remote C–H addition to ketones.

In 2013, Jiao and co-workers [29] described a direct C(sp^2)—H hydroxylation of 2-phenylpyridines using $PdCl_2$ as a catalyst and *N*-hydroxyphthalimide (NHPI) as a co-catalyst in toluene at 100 °C (Scheme 1.11). They used molecular O_2 as a green oxidant. Here, pyridine ring acted as a directing group for the C-2 hydroxylation of arenes. A variety of 2-phenylpyridines underwent reaction under optimized reaction conditions and produced desired products in moderate-to-good yields.

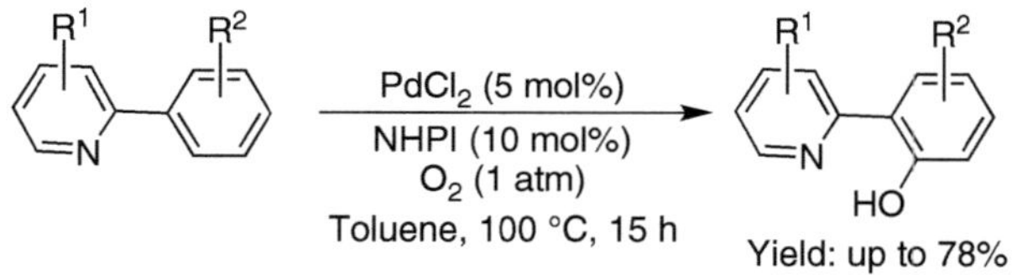

Scheme 1.11 Pyridine-directed hydroxylation of arenes.

The following year, Zhu and co-workers [30] described Rh(III)-catalyzed *N*-nitroso-directed C—H addition to ethyl-2-oxoacetate at the *ortho*-position of *N*-nitrosoaniline derivatives (Scheme 1.12). They also synthesized indazoles from the ethyl 2-hydroxy-2-(2-(methyl(nitroso)amino)phenyl)acetate derivatives via intramolecular cycloaddition.

Rh(III)-catalyzed divergent C—H activation of 2-phenylimidazo[1,2-*a*]pyridines with alkynes has been described by Li and co-workers [31] in 2015 (Scheme 1.13). Selective mono vs twofold C—H activation has been attained by using 4 mol% $[Cp^*RhCl_2]_2$. In the presence of AgOAc, 5,6-disubstituted naphtho[1′,2′:4,5]imidazo [1,2-*a*]pyridines were obtained by the chelation-assisted C—H activation at the benzene ring followed by rollover C—H activation. In addition, the reaction afforded a fused isoquinolinium via C—C and C—N coupling employing $AgBF_4$ as a co-oxidant.

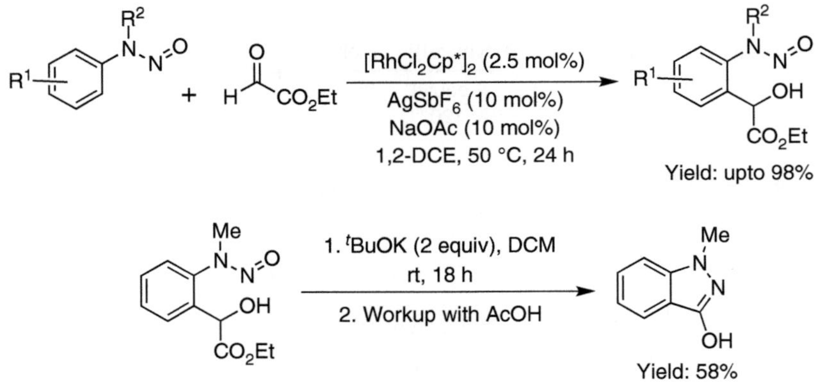

Scheme 1.12 Rh(III)-catalyzed remote C—H addition to ethyl 2-oxoacetate.

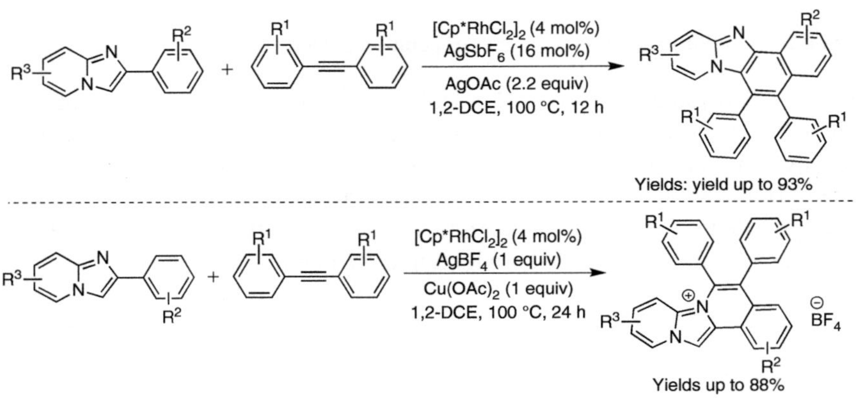

Scheme 1.13 Rh(III)-catalyzed C—H activation of 2-phenylimidazo[1,2-*a*]pyridines.

In 2015, Hajra and co-workers [32] reported a Pd-catalyzed direct dehydrogenative annulation of imidazo[1,2-*a*]pyridines with diarylalkynes through dual cleavage of C—H bonds (Scheme 1.14). This protocol provides a straightforward route for the synthesis of π-conjugated polyaromatic heterocycles from readily available imidazo[1,2-*a*]pyridines. A variety of 2,3,4-triarylphenyl-1,7*b*-diaza-cyclopenta[*cd*] indene derivatives were obtained in high yields by this protocol.

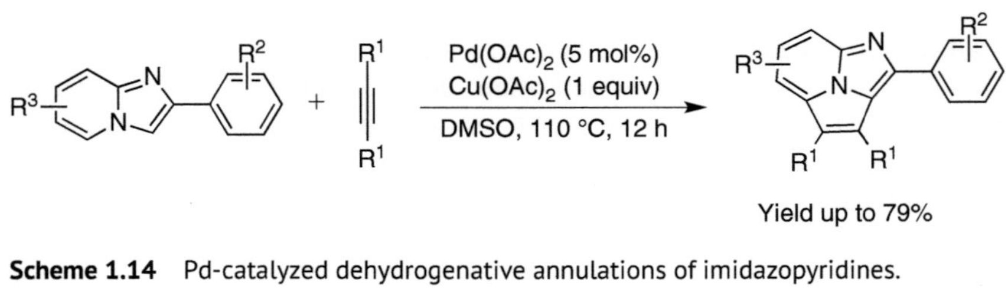

Scheme 1.14 Pd-catalyzed dehydrogenative annulations of imidazopyridines.

In 2015, Sawant's research group [33] described the Ru-catalyzed alkenylation of 2-phenylimidazo[1,2-*a*]pyridine with an alkene to give *ortho*-monoalkenylated products in good yields (Scheme 1.15). The reaction was performed with $[RuCl_2(p\text{-cymene})]_2$ in the presence of $AgSbF_6$ and $Cu(OAc)_2{\cdot}H_2O$ in 1,2-DCE at 100 °C for 24 hours. The reaction proceeded regioselectively through *ortho*-C—H bond activation via the formation of a five-membered ruthenacycle intermediate.

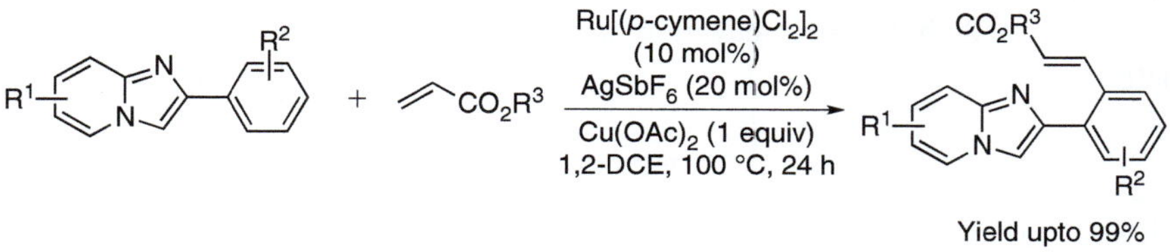

Scheme 1.15 Ruthenium-catalyzed alkenylation of 2-phenylimidazo[1,2-*a*]pyridine.

In 2016, Sakhuja and co-workers [34] developed a convergent and efficient methodology for the regioselective synthesis of *ortho*-amidated imidazoheterocycles via a facile $C(sp^2)$—H bond functionalization with aryl isocyanates employing $[Ru(p\text{-cymene})Cl_2]_2$ and KPF_6 in catalytic amounts (Scheme 1.16). A library of *ortho*-amidated 2-arylimidazo[1,2-*a*]pyridine derivatives with broad functionalities was synthesized in moderate-to-good yields. This present protocol could be applied for other imidazo-fused heterocycles such as imidazo[2,1-*b*]thiazole, benzo[*d*]imidazo[2,1-*b*]thiazole, and imidazo[1,2-*a*]pyrimidine.

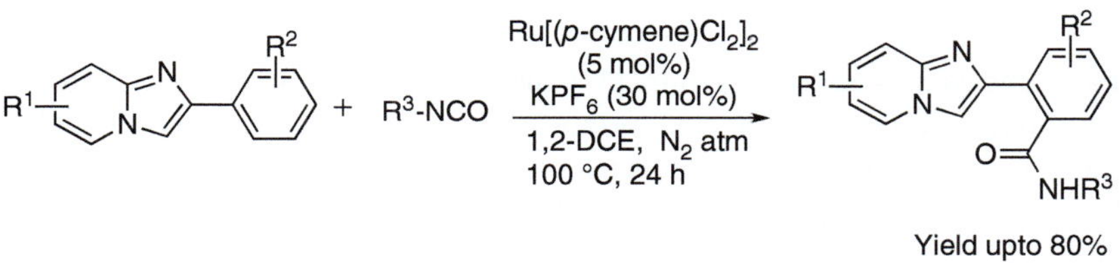

Scheme 1.16 Ru(II)-catalyzed *ortho*-amidation of imidazoheterocycles with isocyanates.

In 2019, Hajra and co-workers [35] reported a rhodium-catalyzed *ortho*-selective C—H amidation reaction of 2-arylimidazoheterocycles with dioxazolones to afford *N*-(2-(imidazo[1,2-*a*]pyridin-2-yl)phenyl)acetamide derivatives through C—H activation (Scheme 1.17). A variety of acetamide derivatives was formed in good-to-excellent yields.

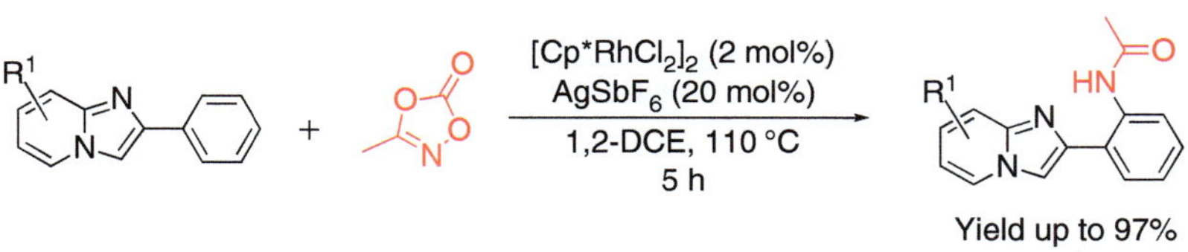

Scheme 1.17 Rhodium-catalyzed *ortho*-selective C—H amidation reaction of 2-arylimidazoheterocycles.

In 2020, Hajra and co-workers [36] reported $[Cp^*RhCl_2]_2$-catalyzed directed C—H functionalization of 2-arylindazoles with *N*-sulfonylformaldimines and activated aldehydes like ethyl glyoxalate and 2,2,2-trifluoroacetaldehyde (Scheme 1.18). A variety of *N*-benzylarylsulfonamide derivatives and hydroxyl derivatives were produced under optimized reaction conditions with good-to-excellent yields. The present protocol is also applicable in large-scale synthesis. To establish the mechanistic pathway of the reaction, they performed kinetic isotopic effect studies, which suggested that C—H activation step may be the rate-limiting step of the reaction.

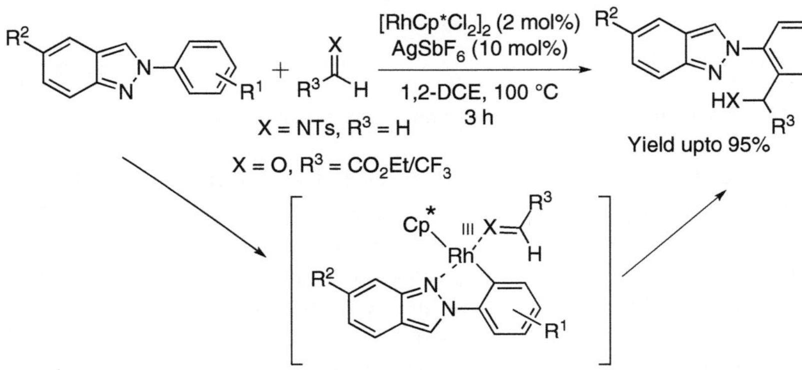

Scheme 1.18 Rhodium-catalyzed C–H functionalization of 2-arylindazoles with *N*-sulfonylformaldimines.

In the same year, the same group [37] also developed an efficient protocol for the rhodium-catalyzed directed *ortho*-C—H amidation of 2-arylindazoles (Scheme 1.19). A variety of 2-arylindazoles underwent reaction effectively with a wide range of dioxazolones having alkyl, aryl, and heteroaryl functionalities

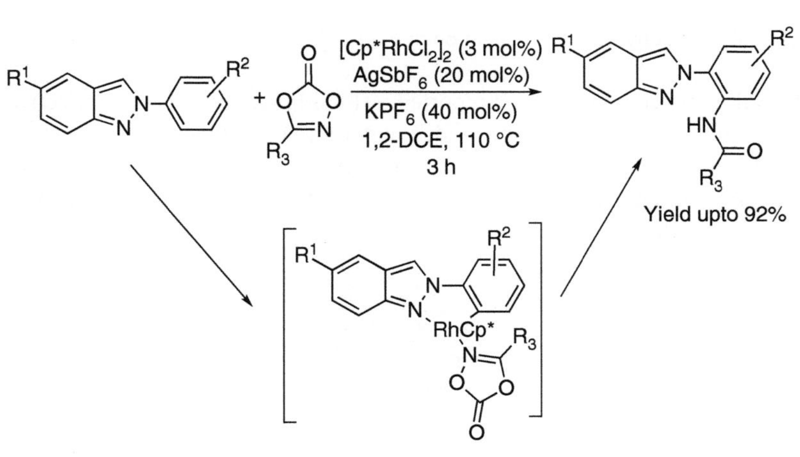

Scheme 1.19 Rhodium-catalyzed C–H amidation of 2-arylindazoles.

to form a series of *N*-(2-(2*H*-indazol-2-yl)phenyl)acetamide derivatives in excellent yields. To show the synthetic applicability of the reaction, they hydrolyzed the amide derivatives to amines by using 10 mol% KOH as a hydrolyzing agent. Mechanistic experiments suggested that C—H bond cleavage may be the rate-determining step of the reaction.

1.4.2 Removable Directing Group-assisted Proximal C—H Bond Functionalization

For removable directing group assisted C—H bond activation, a removable directing group must be installed in the substrate molecule, which involves chelation with the metal center activating one position of the substrate. After completing the reaction, the directing group can be removed without any difficulties. Now, this strategy has been extensively used by scientists due to its high efficacy. In this regard, in 1972, Breslow group [38] made selective C—H abstraction of 3α-cholestanol, using benzophenone moiety, connected through an ester linkage as a removal directing group. There are three main categories of removable directing groups: (i) pre-installed and post-removable directing groups, (ii) traceless directing group, and (iii) transient directing group. All these three categories are elaborately explained below.

1.4.2.1 Pre-installed and Post-removable Directing Groups-assisted C—H Bond Activation

In this activation process, two additional steps for installation and removal of directing group are required to achieve a selective C—H activation. The most explored designed bidentate directing groups are quinoline-, oxazole-, pyrazole-, pyridine-, triazole-based directing groups having an *N*,*N*-coordinating site for metal coordination; pyridine *N*-oxide-, oxalyl amide-, amino-acid-based directing groups having *N*,*O*-coordinating sites; and thiomethyl aniline-, sulfinyl aniline-based directing groups having *N*,*S*-coordinating sites, which are shown in Figure 1.2.

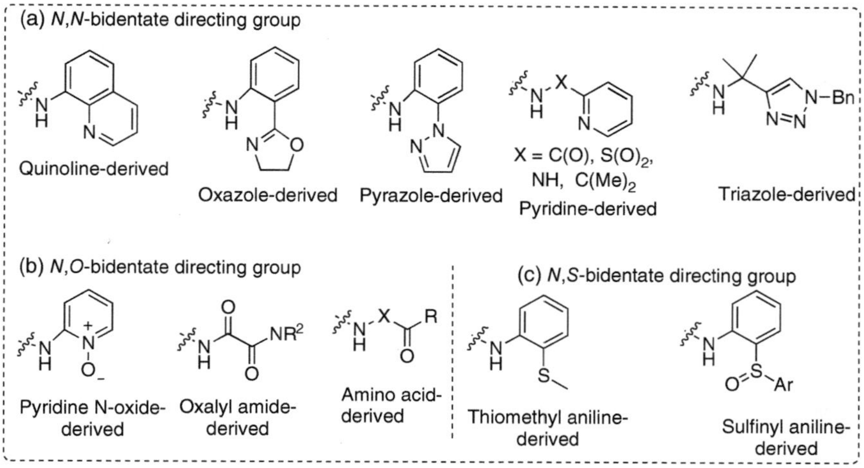

Figure 1.2 Some bidentate directing group.

One of the early examples on this topic was reported by Sanford and co-workers [39] in 2004. They carried out the Pd(II)-catalyzed acetoxylation of unactivated sp^3-C—H bond of ketone derivatives using an oxime moiety as a removable directing group (Scheme 1.20). In this method, $PhI(OAc)_2$ was used as a stoichiometric oxidant.

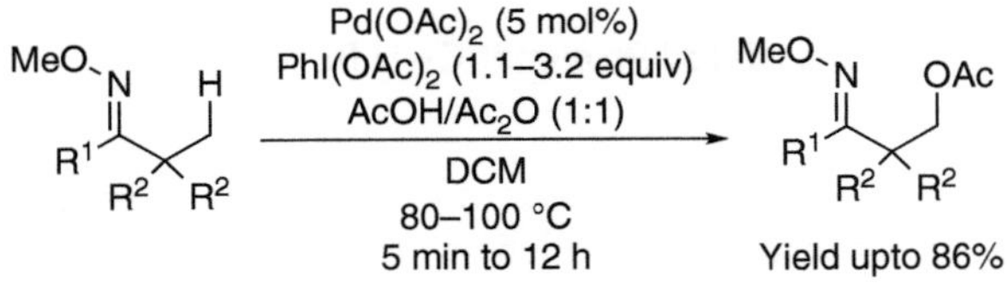

Scheme 1.20 Pd(II)-catalyzed acetoxylation of ketone derivatives.

In 2005, Yu and co-workers [40] reported that oxazole, which was prepared from the parent acid, can also serve as a directing group for the functionalization of sp^3-C—H bonds of acid derivatives. Considering the significance of removable directing groups, various removable directing groups have now been employed in C—H functionalization reactions of a variety of substrates. Later, Shi group [41] developed copper-mediated C—H hydroxylation of arenes and heteroarenes using a bidentate directing group derived from 2-(pyridine-2-yl)isopropylamine (PIP-amine) (Scheme 1.21). The optimized reaction condition was achieved using 1 equiv $Cu(OAc)_2{\cdot}H_2O$, 2 equiv Ag_2CO_3, and 2 equiv TBAI in DMF at 100 °C.

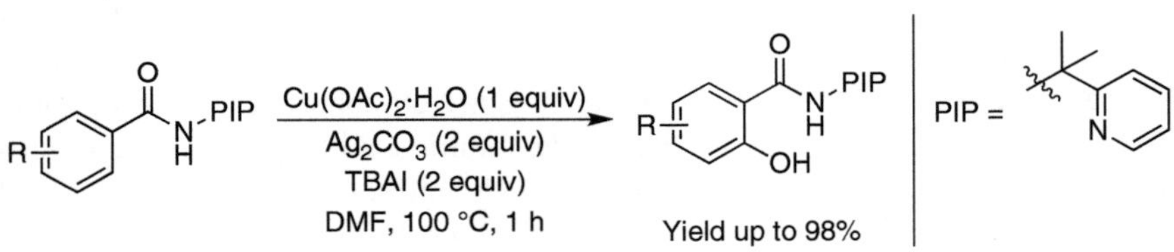

Scheme 1.21 Hydroxylation of arenes.

In 2015, Lin and co-workers [42] reported directing group-promoted *ortho*-C—H hydroxylation of aryl C—H bond using Cu(II) catalyst and molecular oxygen as the sole oxidant (Scheme 1.22). The reaction proceeded through the formation of an acetoxylated product.

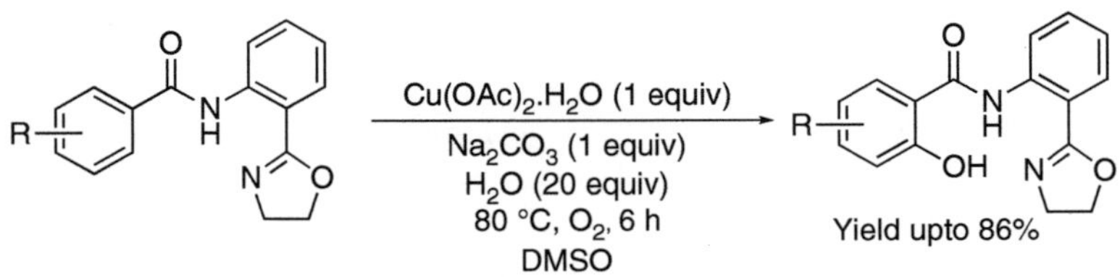

Scheme 1.22 Cu-catalyzed hydroxylation of arenes.

In 2015, Lu and co-workers [43] developed a novel method for arylation of unactivated at β- or γ-Csp^3—H bonds in aliphatic amides with aryl iodides using $Pd(OAc)_2$

as a catatist and pyridine *N*-oxides as a directing group (Scheme 1.23). Here, pyridine N—O⁻ and amide N—H group was found to play a key chelation-assisted role in activating Csp^3—H bond. The reaction proceeded through the formation of palladabicyclic intermediate, which was confirmed by HR-MS and ^1H NMR methods.

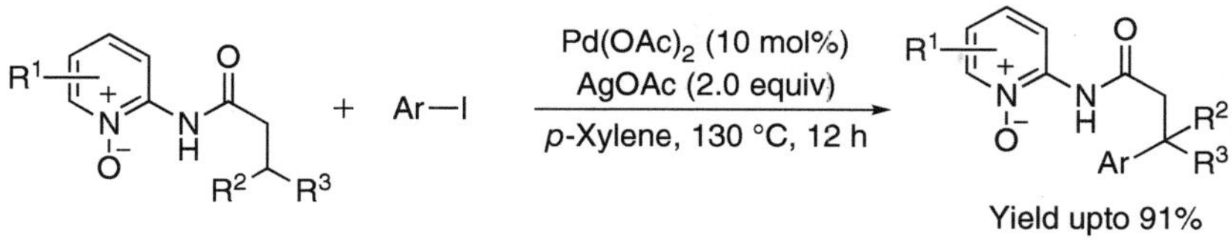

Scheme 1.23 Arylation of unactivated Csp^3–H bonds in aliphatic amides.

The following year, hydroxylation of arenes was also developed by Singh and Jana [44] using 8-aminoquinoline as a directing group, inexpensive copper(II)acetate monohydrate as a catalyst, and pyridine as a ligand (Scheme 1.24). A wide range of functional groups were well tolerable in this transformation.

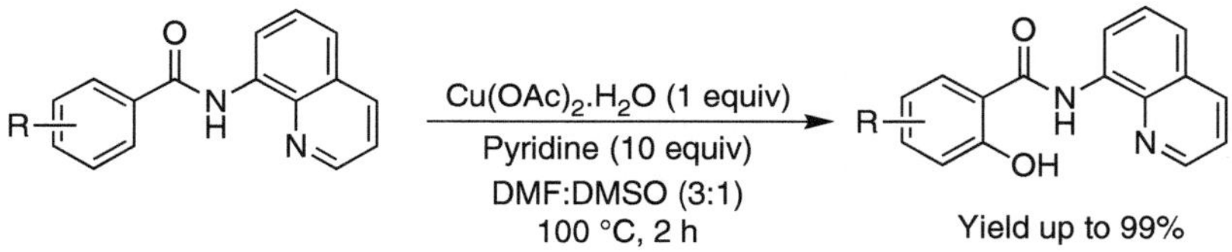

Scheme 1.24 Direct hydroxylation of arenes.

Iron is the most naturally abundant transition metal in comparison with other precious 5d and 4d transition metals, such as iridium, palladium, and rhodium. In 2019, Ackermann and co-workers [45] reported unprecedented iron-catalyzed C—H activation for arylation through oxidation-induced reductive elimination by electrocatalysis at room temperature (Scheme 1.25). They used a triazole-based removal directing group.

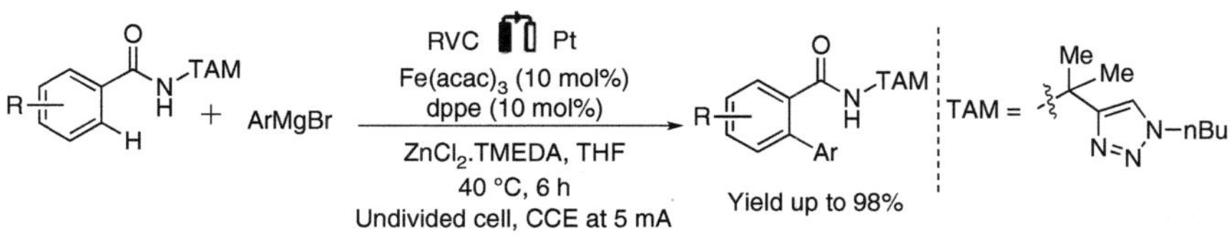

Scheme 1.25 Arylation of acid derivatives.

1.4.2.2 Traceless Directing Group-assisted C–H Bond Functionalization

A traceless directing group is either installed beforehand or is inherently present in naturally or commercially available compounds, which assists in selective C—H bond activation for functionalization and is spontaneously released after completion of C—H functionalization. In this section, we briefly discuss the discovery of traceless directing group-assisted C—H bond functionalization.

In 2008, Satoh and Miura group [46] first used carboxylic acid moiety as a traceless directing group for the C2- and C3-alkenylation of *N*-substituted indole derivatives using Pd/Cu catalytic system (Scheme 1.26). Protodecarboxylation occurred in one pot under the optimized conditions. Similarly, they also explored the reaction with pyrrole-, furan-, and thiophenecarboxylic acids, and all underwent decarboxylative vinylation efficiently.

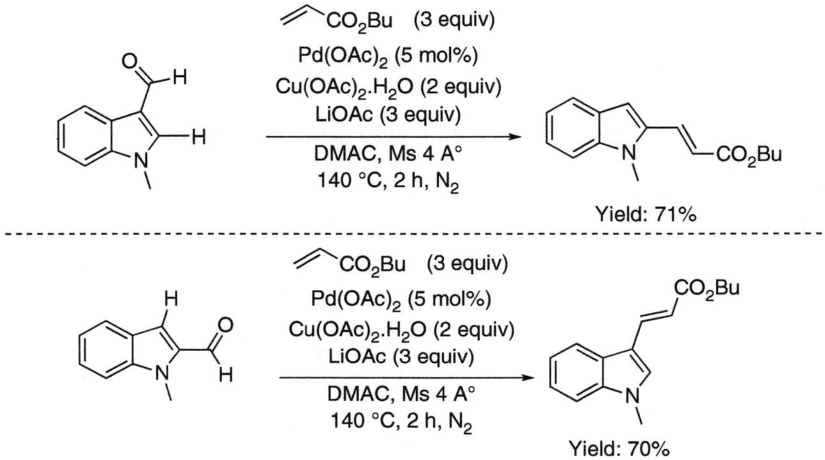

Scheme 1.26 C2- and C3-alkenylation of *N*-substituted indole derivatives.

Inspired by their result, a number of groups like carboxaldehyde, *N*-oxide, nitrones, *N*-nitroso, *N*—N bond-based, S=O bond-based, and SiR_3-based directing groups were used as traceless directing groups by various researchers [4c, 47]. In this section, we briefly discuss the discovery of traceless directing group-assisted C—H bond functionalization by different research groups.

In 2020, Shi and co-workers [48] reported Rh(III)-catalyzed regioselective *ortho*-alkylation of aromatic aldehydes using aldehyde as a traceless directing group. The reaction was performed in aerobic atmospheric conditions (Scheme 1.27). The addition of a trace amount of water was found to enhance the yield of the reaction. They have been used with various acrylates and acrylic acids as alkylating agents.

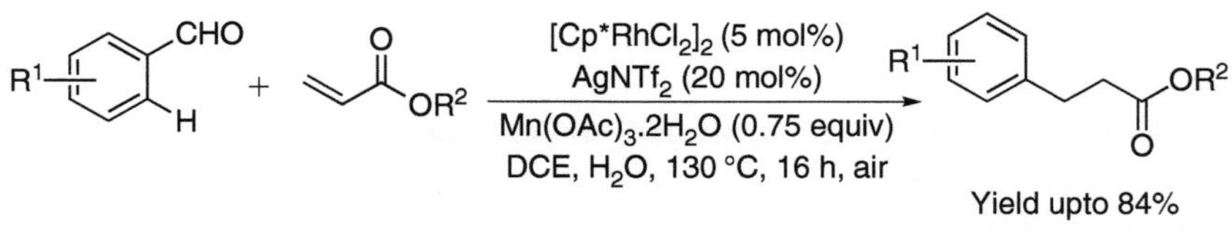

Scheme 1.27 Alkylation of aromatic aldehydes.

In 2018, Zeng and co-workers [49] reported an unprecedented method for the cyclization of *N*-nitrosoanilines with α-diazo-β-ketoesters for the formation of 3,3-disubstituted 2-oxindoles (Scheme 1.28). The reaction was catalyzed by cobalt(III) catalyst and proceeded through a combined C—H activation/Wolff rearrangement process. Density functional theory (DFT) studies were carried out by them, which suggested that the reaction proceeded through the trapping of ketene intermediates by cobalt metallocycles.

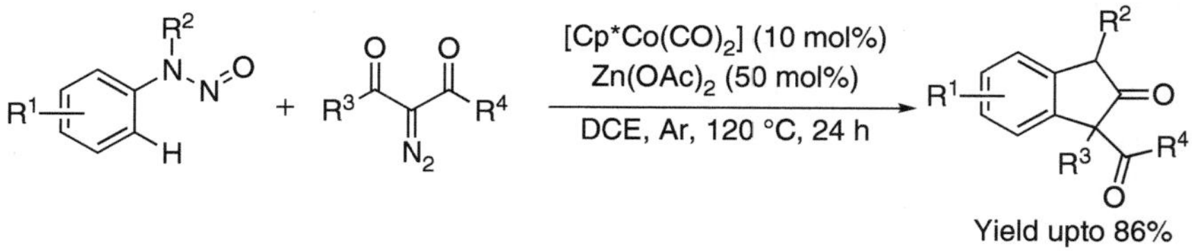

Scheme 1.28 Synthesis of 3,3-disubstituted 2-oxindoles.

Gevorgyan and co-workers [50] reported Pd(II)-catalyzed *ortho*-C—H alkenylation of phenols using di-*tert*-butylsilanol as a traceless directing group (Scheme 1.29). They synthesized a diverse range of alkenylated phenols in good-to-excellent yields. To show the synthetic applicability of the reaction, they also explored the alkenylation reaction of benzofuranone and estrone derivatives.

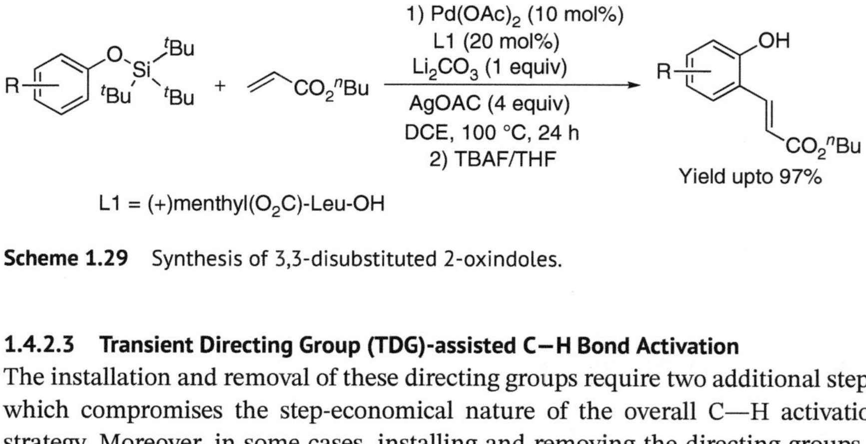

Scheme 1.29 Synthesis of 3,3-disubstituted 2-oxindoles.

1.4.2.3 Transient Directing Group (TDG)-assisted C–H Bond Activation

The installation and removal of these directing groups require two additional steps, which compromises the step-economical nature of the overall C—H activation strategy. Moreover, in some cases, installing and removing the directing groups is difficult. To overcome this drawback, researchers have developed the concept of a transient directing group, which can be easily incorporated into the substrate and easily removed from the product *in situ* during the functionalization; no extra additional step is required. The substrate scope of functionalization by using a transient directing group is limited to alcohols, carbonyl compounds, and amines. In this regard, phosphinites are used as directing groups for alcohol compounds, and imines for carbonyl and amine derivatives [51]. In this part, we have very briefly discussed the strategic evaluation of the transient directing group.

In 1969, Parshall et al. [52] first demonstrated that ruthenium, rhodium, and cobalt complexes of triphenylphosphite underwent cyclometallation at the *ortho*-position of one of the phenoxy groups of triphenylphosphite. They explored only stoichiometric C—H cyclometallations without further functionalizations. With the help of their concept, Lewis and Smith [6] successfully applied this transient phosphite DG to catalytic *ortho*-C—H alkylations of phenols by means of the first ruthenium-catalyzed alkene hydroarylation. In the year 2002, Bedford et al. [53] used phosphinite as a transient directing group for the Rh(I)-catalyzed arylation of phenol derivatives using aryl halides as a coupling partner (Scheme 1.30). The reaction proceeded through the formation of the rhodacycle with phosphinite. Inspired by their result, various groups like Bedford, Bergman, Ellman, Cole-Hamilton, Ye, Oi, and Inoue successfully explored phosphite or phosphinite as a transient directing group for *ortho*-selective alkylation and arylation of phenol derivatives [54].

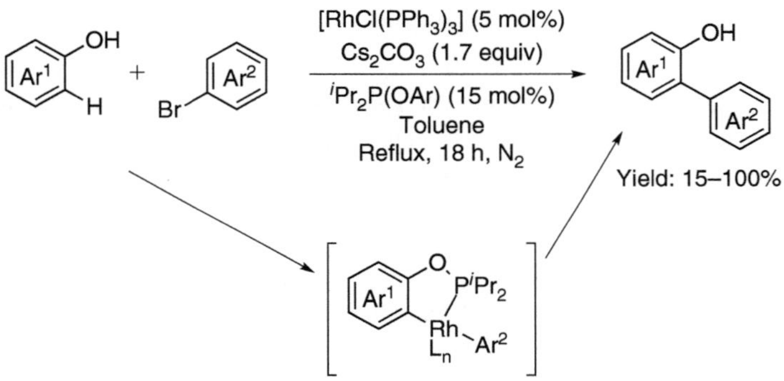

Scheme 1.30 Arylation of phenols.

Imines are broadly explored as transient directing groups for various functionalization through C—H of aldehydes, ketones, etc. Imines can be easily synthesized by the condensation of carbonyl compounds and amines; besides, they can be readily removed upon hydrolysis. In 1997, Jun et al. [55] first used an imine as a transient directing group for the hydroacylation of aldehydes with alkenes via the activation of the aldehyde C—H bond (Scheme 1.31). They used 2-amino-3-picolin for the generation of the transient directing group.

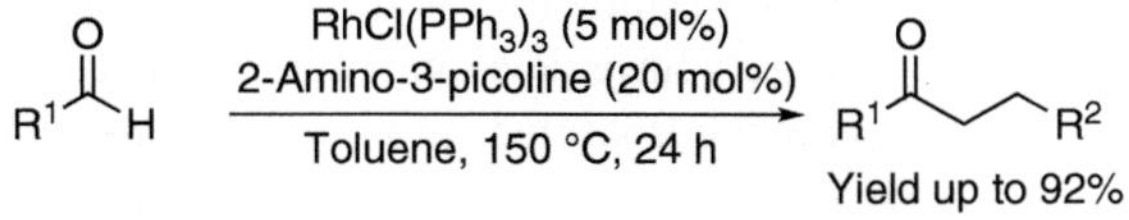

Scheme 1.31 Hydroacylation of aldehydes.

Subsequently, in 2000, Jun group [56] also explored Rh-catalyzed *ortho*-C—H alkylation reaction of the aldimine with tert-butylethylene using 2-amino-3-picoline as amine for formation of imine. Inspired by their work, many researchers successfully utilized imine as a transient directing group, for various functionalization like alkylation, arylation, alkenylation, amination, and halogenation of a variety of substrates: aldehyde C—H bonds, *ortho*-C—H bonds of aromatic aldehydes or ketones, the atroposelective C—H bond of biaryl carbonyl compounds, *ortho*-C—H bonds of ferrocene derivatives, benzylic C(sp^3)—H bonds of aromatic aldehydes, β- or γ-C(sp^3)—H bonds of aliphatic amines, and δ-C(sp^3)—H bond of aryl amines [57].

Some groups also explored enamine as a transient directing group for C—H functionalization. In 2012, Dong group [58] pioneered the use of an enamine as a transient directing group for Rh-catalyzed alkylation of the cyclic 1,2-diketone using 2-aminopyridine directing group (Scheme 1.32). The reaction proceeded through the coordination of pyridine nitrogen to the Rh-center. Subsequently, they also explored a similar strategy for the Rh-catalyzed α-alkylation of the ketone [59].

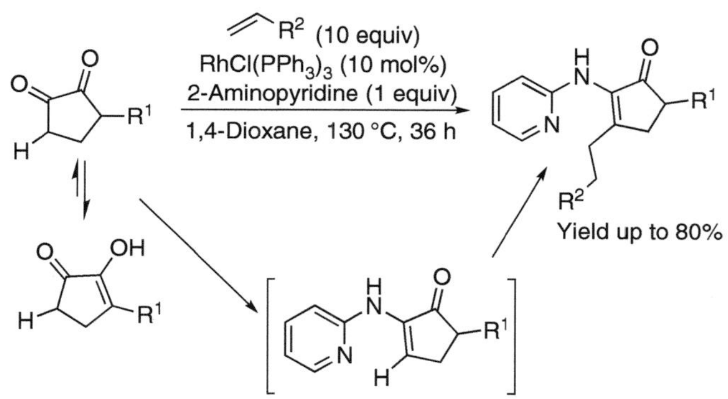

Scheme 1.32 Rh-catalyzed alkylation of the cyclic 1,2-diketone.

1.5 Directed Distal C—H Bond Functionalization

1.5.1 *meta*-C—H Bond Functionalization

In 2016, Yu and co-workers [60] first reported Pd(II)-catalyzed *meta*-C—H amination and alkynylation of aniline and phenol substrates using either *N*-benzoyloxyamines or alkynyl bromides as electrophilic reagents in presence of modified norbornene (methyl bicyclo[2.2.1]hept-2-ene-2-carboxylate) as a transient mediator and monoprotected 3-amino-2-hydroxypyridine ligands (Scheme 1.33). A variety of substrates, including different heterocycles like indole, indoline, and indazole, effectively reacted to produce the desired products in moderate-to-high yields.

Maiti and co-workers [61] developed a new protocol for selective β- and γ-C(sp^3)—H activation and arylation of aromatic and aliphatic carboxylic acid derivatives using a new traceless removal bidentate directing group, 3-amino-1-methyl-1*H*pyridin-2-one (AMP), using Pd catalyst (Scheme 1.34). To show the generality of the strategy, they also explored arylation at β- and γ-positions of C(sp^2)—H of the amide

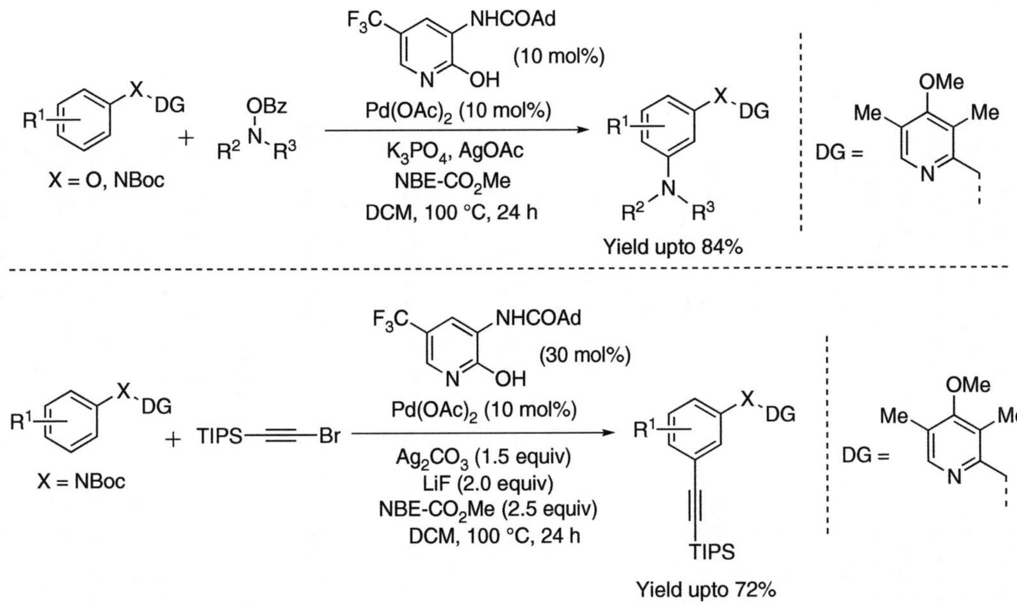

Scheme 1.33 *meta*-C–H amination and alkynylation of arenes.

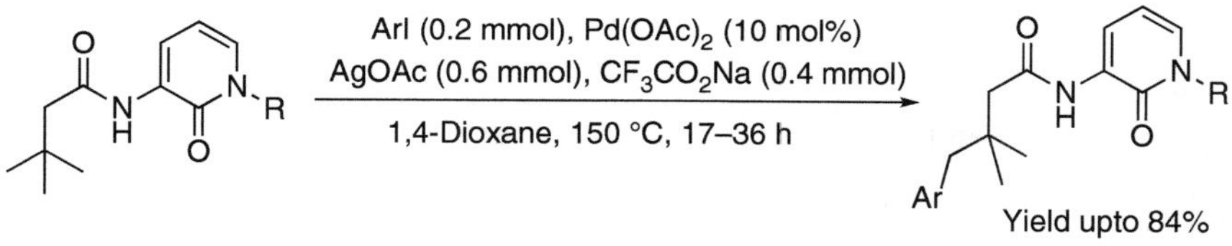

Scheme 1.34 γ-C(sp^3)–H arylation of carboxylic acid derivatives.

analogs. They also explored the γ-arylation of a chiral amino acid derivative like phthalimide-protected L-valine amide.

Yu group [62] used a pyridine-based U-shaped template for a diverse range of *meta*-C—H functionalizations of phenylacetic acid scaffolds using Pd catalyst. They explored *meta*-C—H olefination using ethyl/methyl acrylate derivatives, *meta*-C—H arylation with $ArBF_3K$, and *meta*-C—H iodination using 1,3-diiodo-5,5-dimethylhydantoin as the iodination reagent (Scheme 1.35). In all cases, the desired products were formed in moderate-to-good yields.

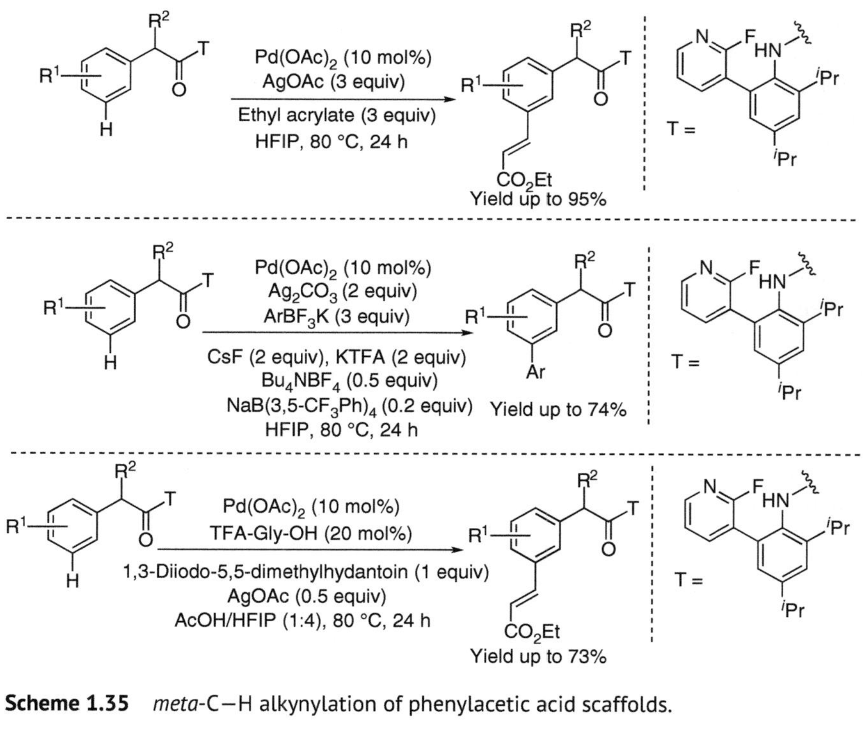

Scheme 1.35 *meta*-C–H alkynylation of phenylacetic acid scaffolds.

In 2019, Wang and co-workers [63] developed Pd(II)-catalyzed *meta*-C—H olefination of arene-tethered diols across different linker lengths attached to an easy removal well-designed pyrimidine template (Scheme 1.36). A variety of 3-phenylpropane-1,2-diol frameworks having electron-withdrawing and electron-donating groups efficiently underwent reaction under the optimized reaction

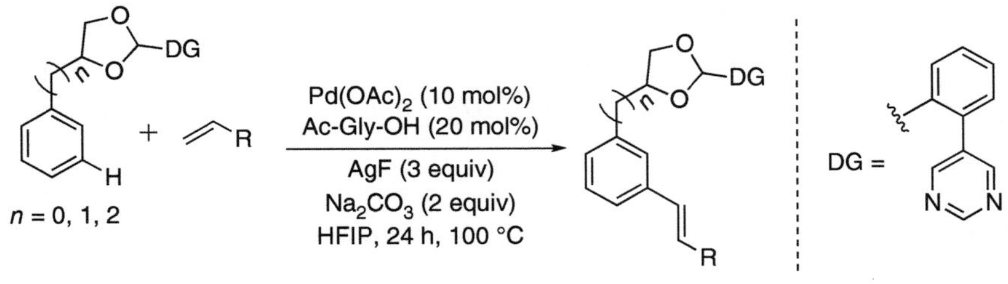

Scheme 1.36 Pd-catalyzed *meta*-C–H alkenylation of arenes.

conditions. To show the synthetic applicability of the reaction, they synthesized various diol-based natural products, such as coumarins, phenylpropanoids, stilbenes, and chalcones. They also performed the gram-scale reaction. Moreover, they also carried out different experimental and computational studies, including ^{1}H NMR, ESI-MS and IR, and DFT, to clarify the mechanistic complexity.

Introduction of different functionalities at distal positions in arenes in a regioselective manner is a challenging task. In 2020, Maiti and co-workers [64] reported pyrimidine-based auxiliary-directed palladium-catalyzed *meta*-C—H allylation of arenes using allyl phosphate as an allylating agent (Scheme 1.37). Here, the pyrimidine directing group of substrate coordinated to the palladium center, which enhances C—H activation. In this reaction, 1,1,1,3,3,3-hexafluoroisopropanol (HFIP) solvent played a critical role by lowering the energy of transition states and intermediates via its interactions with both catalysts and substrates. A variety of substrates like phenethyl ether, phenol, benzylsulfonyl ester, phenethylsulfonyl ester, phenylacetic acid, hydrocinnamic acid, and 2-phenylbenzoic acid derivatives were efficiently reacted under the optimized reaction conditions. They also synthesized eugenol derivative under the optimized reaction conditions. Various spectral analyses like ^{1}H NMR, ^{31}P NMR, ESI-MS, kinetic study, and DFT computations were performed to understand the mechanistic pathway of the reaction. The experimental results suggested that the reaction proceeded through a ligand-mediated *meta*-C—H activation followed by the addition of allyl to form a Pd-π-allyl complex, which finally had a turnover determining C—C bond formation step, leading to the formation of *meta*-allylated product.

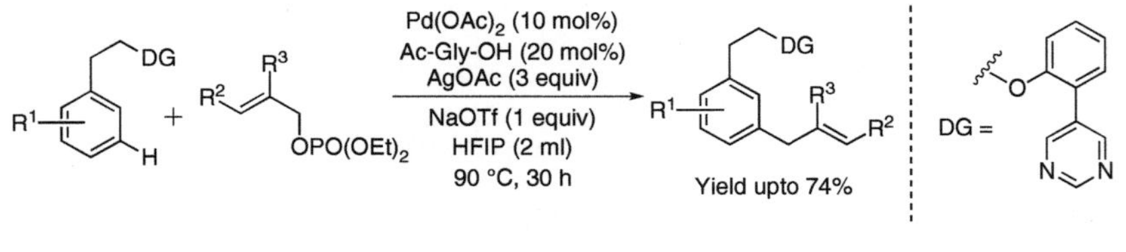

Scheme 1.37 *meta*-C–H allylation of arenes.

1.5.2 *para*-C–H Bond Functionalization

In 2014, Cong and Zeng [65] reported iron-catalyzed remote C-5 allylation of quinolines with allyl alcohol. The optimized reaction condition was achieved by using

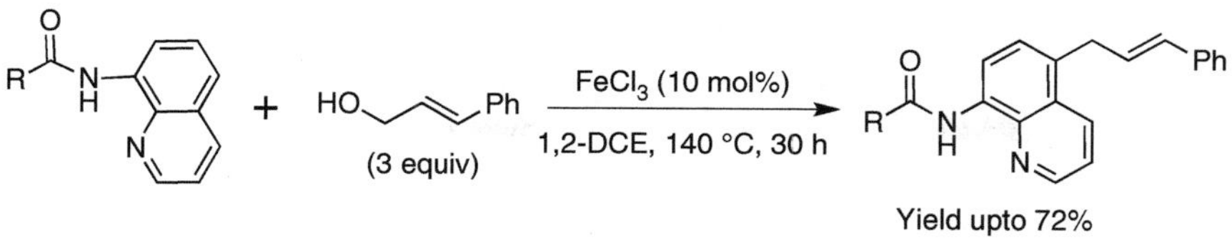

Scheme 1.38 Iron-catalyzed C-5 allylation of quinolones.

10 mol% $FeCl_3$ in 1,2-DCE at 140 °C (Scheme 1.38). Irrespective of the nature of the allyl alcohol, only the *E*-isomeric product was obtained under the optimized reaction condition.

In 2016, Ribas and co-workers [66] developed a facile and efficient approach for direct C-5 nitration of quinoline scaffolds using commercially available *tert*-butyl nitrite (TBN) as the nitrating reagent in the presence of Co catalyst (Scheme 1.39). The mechanistic pathway of this protocol involves the SET pathway. 5-Substituted-8-aminoquinoline derivatives undergo nitration at the C-7 position under the optimized reaction conditions.

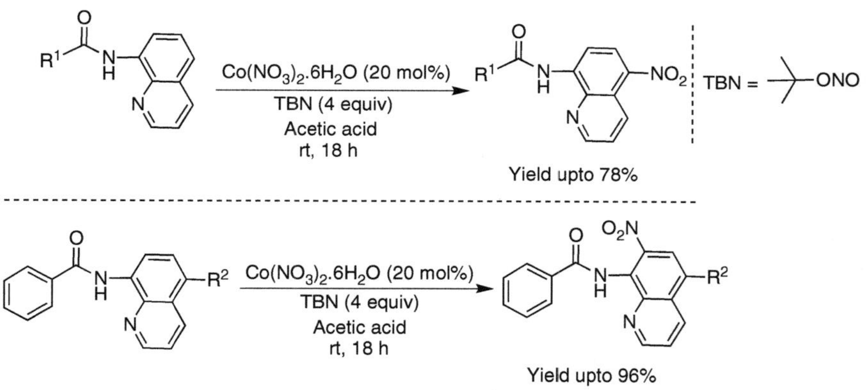

Scheme 1.39 Co-catalyzed nitration of quinolone.

Noncovalent interactions play a key role in controlling the chemical reactivity of similar types of C—H bonds within the substrate. In 2017, using noncovalent interaction between the substrate and an L-shaped ligand having a pendant noncovalent interacting site, Chattopadhyay and co-workers [67] developed an efficient method for Ir-catalyzed *para*-C—H borylation of aromatic esters (Scheme 1.40). They used an L-shaped ligand in which bipyridine was the core moiety, which provided an unprecedented controlling factor to activate selectively at the *para*-position of ester. A wide variety of substrates with different functional groups selectively borylated at the *para*-position and produced the desired products in good-to-excellent yields.

The development of regioselective C—Si bond formation reactions is extremely important due to the natural abundance and non-toxic nature of silicon [68]. Maiti and co-workers [69] developed a template-assisted regioselective *para*-silylation of toluene derivatives (Scheme 1.41). They designed the templates depending on their experimental and computational studies. They chose a dimethoxy-substituted

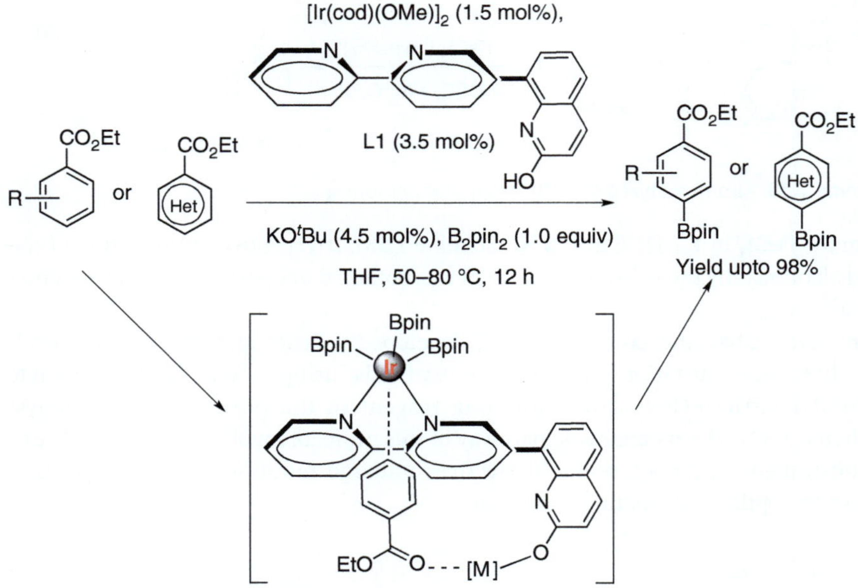

Scheme 1.40 Ir-catalyzed C-5 allylation of quinolones.

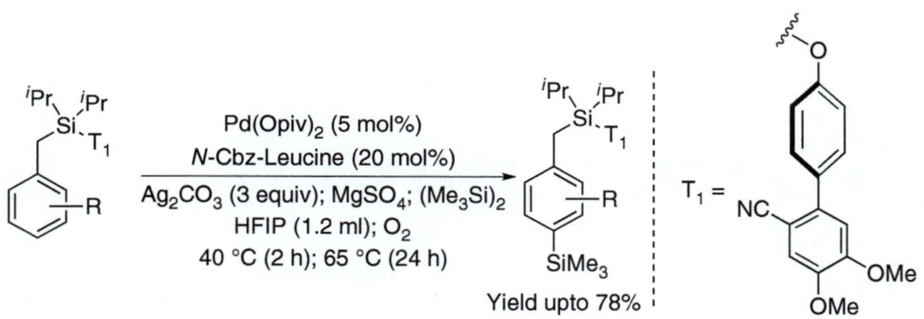

Scheme 1.41 Regioselective *para*-silylation of toluene derivatives.

cyano-based D-shaped template, which helped the *para*-C—H bond activation. Kinetic, spectroscopic, and computational studies shed light on the reaction mechanism. The reaction was also applicable in gram-scale synthesis. The synthetic advantage of this reaction is that the strategy was applied for the synthesis of potential lipophilic bioisostere of γ-aminobutyric acid (GABA).

Mondal and Hajra [70] developed a convenient protocol for the site-selective C—H addition of 8-aminoquinoline to ethyl glyoxalate via a ruthenium-catalyzed C—H activation under mild conditions (Scheme 1.42). This method describes selective functionalization at the C-5 position of quinoline ring. The reaction is not air- or moisture-sensitive.

Using the same template shown above, Maiti group [71] reported Pd(II)-catalyzed *para*-selective cyanation of arenes using CuCN as a cyanating agent (Scheme 1.43). This protocol holds its efficacy regardless of substrate electronic bias.

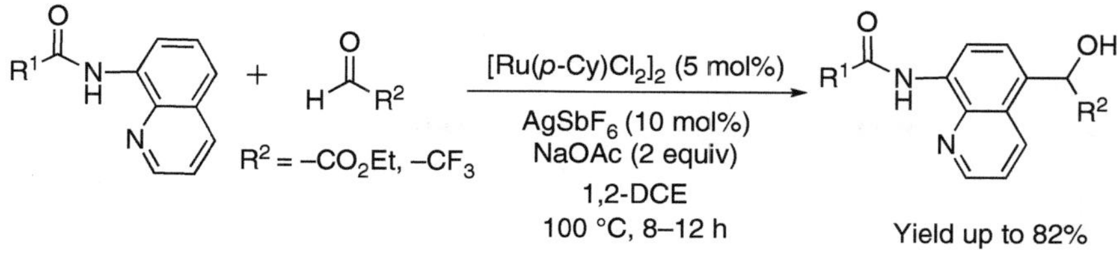

Scheme 1.42 Functionalization of 8-aminoquinoline amide.

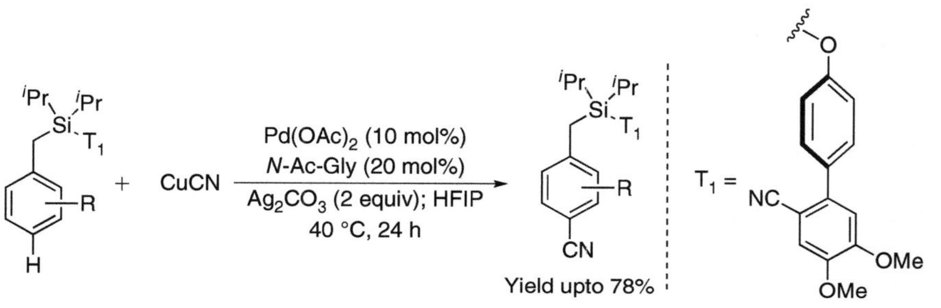

Scheme 1.43 H—Bonded template for regioselective *para*-cyanation of arenes.

Li and co-workers [72] reported an unprecedented Pd-catalyzed *para*-C—H alkenylation of arenes using a readily recycled pyridine-based *para*-directing template (DT) (Scheme 1.44). They explored the reaction by using three classes of arenes like phenylpropanoic acids, 2-phenyl benzoic acids, and benzyl alcohols as model substrates. A variety of alkenes, including perfluoroalkenes, efficiently reacted under the optimized reaction conditions. They have used air as the green terminal oxidant. They also scaled up their reaction into a 3 mmol scale.

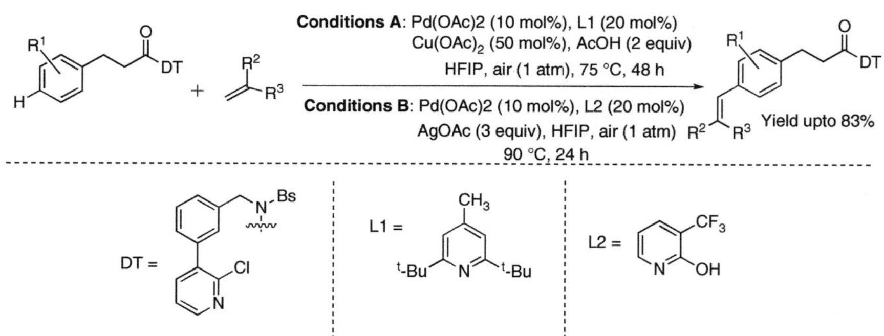

Scheme 1.44 Pd-catalyzed para-C—H alkenylation of arenes.

Regioselective C—H alkynylation of arenes via C—H activation is a challenging yet highly desirable transformation. A directing group having a suitable geometric orientation may help the transition metal catalyst to activate *para*-C—H bond. In this regard, Maiti and co-workers [73] developed a reusable dimethoxy-substituted cyano-based D-shaped template route for the unprecedented *para*-C—H

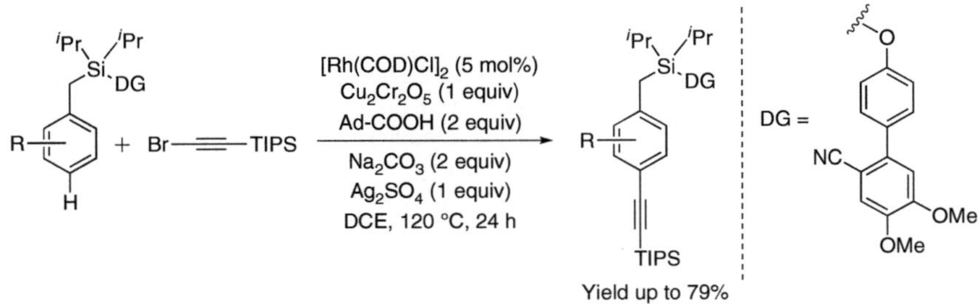

Scheme 1.45 Template-assisted *para*-C–H alkynylation of arenes.

alkynylation of arenes (Scheme 1.45). The reaction was catalyzed by Rh catalyst. This present protocol was well tolerated with a variety of substituents on the arene ring. Moreover, they also performed a late-stage modification of the product and functional group inter-conversion. To understand the mechanistic pathways, they performed computational studies, which suggested that the C—H activation step was not the overall rate-determining step; instead, migratory insertion of the alkyne to arenes was the overall rate-determining step.

1.5.3 Remote C–H Functionalization

Functionalization of remote C—H bonds, especially when the reaction site is distant from the directing group, is a major challenge in synthetic chemistry. This difficulty arises mainly because the reaction typically requires the formation of a macrocyclic metallacycle, which is a large, ring-shaped intermediate where a metal center coordinates with the substrate to guide the reaction. In this context, enzymes can also act as catalysts and can perform a reaction at a specific position of the bound substrate. In 1997, Breslow et al. [74] used manganese-porphyrin-carrying four β-cyclodextrin groups for selective hydroxylation of unactivated carbons in substrates, such as steroids, under the following optimized reaction conditions: 10 mol% of manganese catalyst (**Catalyst I**) and 5 equiv of iodosobenzene in water in the presence of 10 equiv of pyridine (Scheme 1.46). Molecular models suggested that two *tert*-butylphenyl groups of the substrate bound with the two *trans*-cyclodextrin rings of catalyst and placed the steroid ring B of steroid **A** directly above the metalloporphyrin ring and made it free for oxidation. 40% Conversion was found under the optimized reaction conditions.

In the same year (1997), the same group (Breslow group) [75] used manganese porphyrin, which had four β-cyclodextrin groups for site-selective hydroxylation of a substrate **A**. *Tert*-butylphenyl groups of substrate **A** could hydrophobically bind into the cyclodextrin cavities of the catalyst (Scheme 1.47). The turnover number of the catalyst was high. They also explored the reaction in a steroid derivative, and hydroxylation occurred regio- and stereoselectively at a single unactivated carbon atom with fewer turnovers. With a simple dihydrostilbene substrate, the hydroxylations occurred at the speed of corresponding cytochrome P450 enzymes and with hundreds of turnovers.

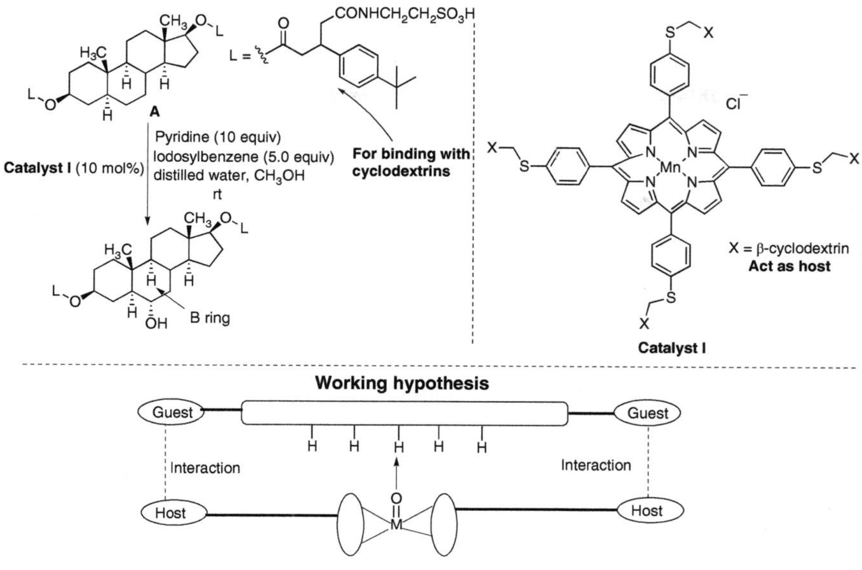

Scheme 1.46 Site-selective hydroxylation of steroids.

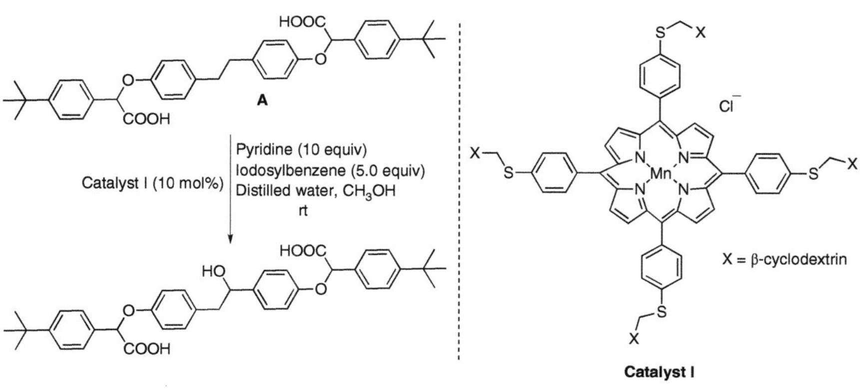

Scheme 1.47 An artificial cytochrome P450-catalyzed regio- and stereoselective hydroxylation of unactivated carbons.

In 2001, the same group (Breslow group) [76] developed a method for regioselective hydroxylation of unactivated C—H bonds of steroids by using new manganese-porphyrin-carrying metal coordinating groups. The auxiliary group of the substrate molecule coordinated with the metal ion of the porphyrin bipyridyl, which was preceded by a geometrically controlled oxidation. The metal chosen for coordination was Cu^{2+}. The androstane derivatives **A** underwent oxidation in the presence of 0.5 equiv of manganese porphyrin catalyst and 2 mmol of $Cu(OTf)_2$ and 10 equiv of iodosylbenzene (PhIO) as an oxidant. Under this condition, overall 20% of total conversion was found by the formation of a mixture of products **B**, **C**, and **D**, yielding 50%, 15%, and 35%, respectively (Scheme 1.48). They also explored the reaction with the epiandrosterone derivative.

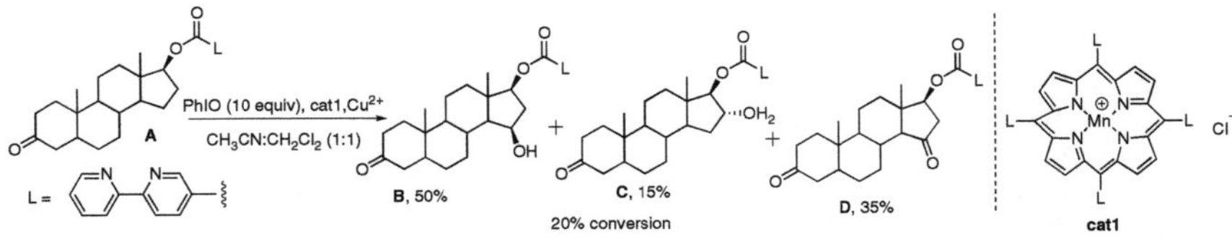

Scheme 1.48 Manganese-porphyrin-catalyzed regioselective hydroxylation.

Crabtree and co-workers [77] used molecular recognition through a hydrogen bonding strategy for regioselective C(sp^3)—H using —COOH group as a recognition group. In 2007, they reported the nonporphyrin-di-μ-oxo dimanganese-catalyzed highly regioselective oxygenation at saturated C—H bonds of ibuprofen and (4-methylcyclohexyl)acetic acid (both *cis* and *trans*) (Scheme 1.49). The turnover number of this reaction was 9100. Control experiments suggested that hydrogen bonding played a key role in orienting the substrate toward high selectivity.

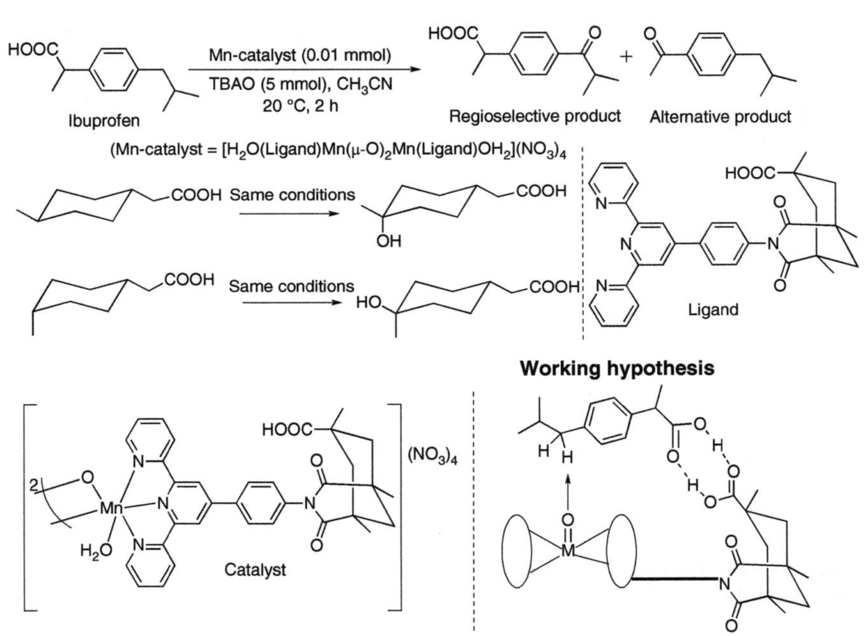

Scheme 1.49 Di-μ-oxo dimanganese-catalyzed regioselective oxygenation.

In 2007, Chen and White [78] developed an electrophilic iron catalyst [Fe(PDP)] having a bulky ligand framework for unactivated sp^3 C—H bond oxidation using hydrogen peroxide (H_2O_2) as an inexpensive, environmentally friendly oxidant. A broad variety of substrates got oxidized at the unactivated C(sp^3)—H bond under the optimized reaction conditions. The site of oxidation of the bonds depended upon the electronic and steric environment of the C—H bond. When carboxylic functionality was present in the substrate, it led to the formation of lactone through oxidation (Schemes 1.50 and 1.51).

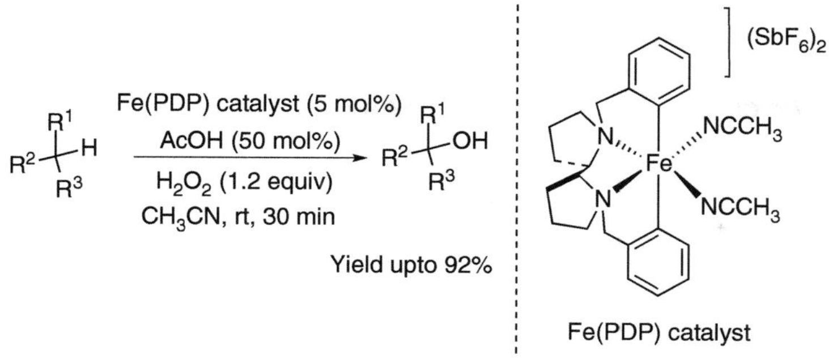

Scheme 1.50 Fe-catalyzed hydroxylation of unactivated sp^3 C—H bonds.

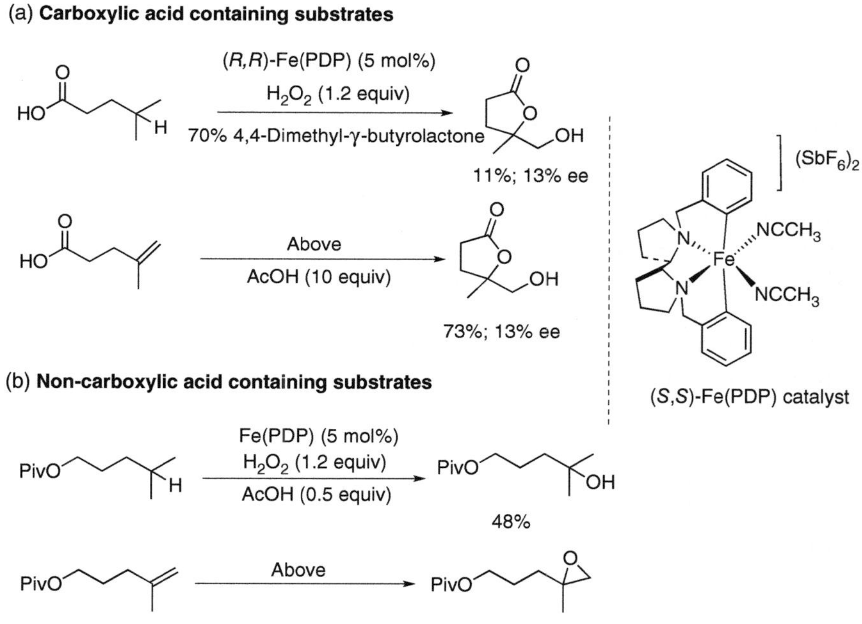

Scheme 1.51 Fe-catalyzed regio- and stereoselective hydroxylation of unactivated carbons.

In 2011, the same group (White group) [79] reported a nonheme iron hydroxylase complex, [Fe(PDP)]-catalyzed site-selective aliphatic C—H hydroxylation of carboxylic and non-carboxylic acid derivatives, using H_2O_2 as a terminal oxidant (Scheme 1.52). They performed a taxane-based radical trap experiment, and under the reaction condition, it rearranges to a nortaxane skeleton. They demonstrated that 13% ee was obtained when they used a chiral Fe(PDP) catalyst.

The following year (2012), the same group (White group) [80] described nonheme iron complex, Fe(PDP)-catalyzed C—H lactonization followed by hydroxylation of carboxylic acids derivatives. Iron complex Fe(PDP) catalyst was suitable for overcoming the unfavorable electronic, steric, and stereoelectronic biases of carboxylic

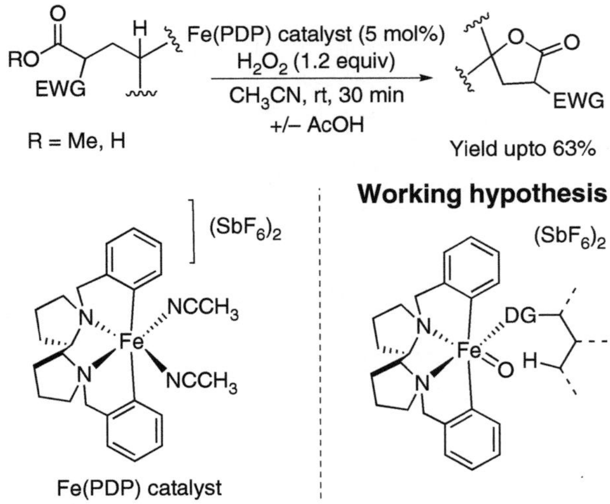

Scheme 1.52 Di-μ-oxo dimanganese-catalyzed regioselective C–H hydroxylation.

acid moieties for the C—H hydroxylations. In a demonstration of the power of this directing effect, C—H oxidation was diverted away from an electronically favored C-1 H abstraction/rearrangement pathway in the paclitaxel framework to enable installation of C-2 oxidation in the naturally occurring oxidation state and stereoconfiguration. A series of linear methyl esters having an electron-withdrawing acetoxy (-OAc) group were favorable for this reaction. Interestingly, they also succeeded in C-2 hydroxylation on a medicinally important class of molecules like taxane.

1.6 Conclusions

Transition-metal-catalyzed directed C—H bond activation for construction of C—C and C—heteroatom bond has widely expanded in the last few decades, and now, it has become a major research topic among synthetic organic chemists. The pioneering work in this field was reported by Lewis group [81] in 1985 for the Ru-catalyzed *ortho*-C—H deuteration of phenol using phosphite as a transient directing group. After that, a variety of functionalization like arylation, alkylation, alkynylation, olefination, carbonylation, amination, and halogenations have been developed by different research groups. In this chapter, we tried to give an overview on different strategies for directed selective and efficient C—H bond functionalization via C—H activation. We included early discoveries on stoichiometric metal-promoted proximal C—H bond activation, directing group (both built-in functional group and removal directing group)-assisted catalytic proximal C—H bond functionalization, and directing group-assisted distal (both *meta* & *para*) C—H bond activation. We believe that this chapter will motivate the researcher to further explore this field. Although this field has been explored extensively, according to green chemistry aspect, the development of user-friendly waste-free reaction conditions and modification of catalysts, which are not air or moisture sensitive, is highly desirable and is a challenging work in the near future. We envisage that numerous environmentally friendly C—H activation reactions with industrial potential will be explored.

Acknowledgments

A. Hajra acknowledges the financial support from CSIR and SERB, New Delhi.

References

1 (a) Sinha, S.K., Guin, S., Maiti, S. et al. (2022). Toolbox for distal C—H bond functionalizations in organic molecules. *Chem. Rev.* 122 (6): 5682–5841. (b) Davies, H.M.L., Du Bois, J., and Yu, J.-Q. (2011). C–H functionalization in organic synthesis. *Chem. Soc. Rev.* 40 (4): 1855–1856. (c) (2016). Recent advances in C–H functionalization. *J. Org. Chem.* 81 (2): 343–350. (d) Ghosh, A., Pyne, P., Ghosh, S. et al. (2022). Visible-light-induced metal-free coupling of $C(sp^3)$–H sources with heteroarenes. *Green Chem.* 24 (8): 3056–3080. (e) Ghosh, S., Ghosh, A., Pyne, P. et al. (2022). Asymmetric $C(sp^3)$–H borylation: an update. *Org. Biomol. Chem.* 20 (22): 4496–4511. (f) Guillemard, L., Kaplaneris, N., Ackermann, L. et al. (2021). Late-stage C–H functionalization offers new opportunities in drug discovery. *Nat. Rev. Chem.* 5 (8): 522–545. (g) Gutekunst, W.R. and Baran, P.S. (2011). C–H functionalization logic in total synthesis. *Chem. Soc. Rev.* 40 (4): 1976–1991.

2 (a) Hartwig, J.F. and Larsen, M.A. (2016). Undirected, homogeneous C—H bond functionalization: challenges and opportunities. *ACS Cent. Sci.* 2 (5): 281–292. (b) Periana, R.A., Bhalla, G., Tenn, W.J. et al. (2004). Perspectives on some challenges and approaches for developing the next generation of selective, low temperature, oxidation catalysts for alkane hydroxylation based on the CH activation reaction. *J. Mol. Catal. A: Chem.* 220 (1): 7–25.

3 Sambiagio, C., Schönbauer, D., Blieck, R. et al. (2018). A comprehensive overview of directing groups applied in metal-catalysed C–H functionalisation chemistry. *Chem. Soc. Rev.* 47 (17): 6603–6743.

4 (a) Xie, J. and Zhu, C. (2016). *Sustainable* $C(sp^3)$*—H Bond Functionalization.* Berlin, Heidelberg: Springer https://doi.org/10.1007/978-3-662-49496-7. (b) Rej, S., Das, A., and Chatani, N. (2021). Strategic evolution in transition metal-catalyzed directed C—H bond activation and future directions. *Coord. Chem. Rev.* 431: 213683. (c) Rani, G., Luxami, V., and Paul, K. (2020). Traceless directing groups: a novel strategy in regiodivergent C–H functionalization. *Chem. Commun.* 56 (83): 12479–12521.

5 Yadav, M.R., Rit, R.K., Shankar, M. et al. (2015). Reusable and removable directing groups for $C(sp^2)$—H bond functionalization of arenes. *Asian J. Org. Chem.* 4 (9): 846–864.

6 Lewis, L.N. and Smith, J.F. (1986). Catalytic carbon-carbon bond formation via *ortho*-metalated complexes. *JACS* 108 (10): 2728–2735.

7 (a) (2017). Introduction: CH activation. *Chem. Rev.* 117 (13): 8481–8482. (b) Dalton, T., Faber, T., and Glorius, F. (2021). C—H activation: toward sustainability and applications. *ACS Cent. Sci.* 7 (2): 245–261.

8 (a) Neogi, S., Bhunya, S., Kumar Ghosh, A. et al. (2024). Silylarylation of alkenes via meta-selective C—H activation of arenes under ruthenium/iron cooperative catalysis: mechanistic insights from combined experimental and computational studies. *ACS Catal.* 14 (7): 4510–4522. (b) Mondal, K., Ghosh, S., and Hajra, A. (2022). Transition-metal-catalyzed *ortho* C–H functionalization of 2-arylquinoxalines. *Org. Biomol. Chem.* 20 (37): 7361–7376. (c) Ghosh, S., Pyne, P., Ghosh, A. et al. (2022). *Ortho* C—H functionalizations of 2-aryl-2H-indazoles. *Chem. Rec.* 22 (11): e202200158. (d) Ghosh, S., Laru, S., and Hajra, A. (2022). *Ortho* C–H functionalization of 2-arylimidazo[1,2-a]pyridines. *Chem. Rec.* 22 (3): e202100240. (e) Das, K.K., Ghosh, A.K., and Hajra, A. (2022). Late-stage *ortho*-C–H alkenylation of 2-arylindazoles in aqueous medium by manganese(i)-catalysis. *RSC Adv.* 12 (30): 19412–19416. (f) Kumar Ghosh, A., Kanta Das, K., and Hajra, A. (2021). *Ortho*-allylation of 2-arylindazoles with vinyl cyclic carbonate and diallyl carbonate via manganese-catalyzed C—H bond activation. *Adv. Synth. Catal.* 363 (21): 4974–4981. (g) Laru, S., Bhattacharjee, S., Ghosh, S. et al. (2021). One-pot construction of Indolo[2,3-b]quinoxalines through ruthenium-catalyzed *ortho* C—H bond functionalization of 2-arylquinoxalines with sulfonyl azides. *Org. Lett.* 23 (19): 7624–7629.

9 Farkas, A. and Farkas, L. (1937). The catalytic interaction of heavy hydrogen and benzene on platinum. *Trans. Faraday Soc.* 33 (0): 827–837.

10 Garnett, J.L. (1972). π-Complex intermediates in homogeneous and heterogeneous catalytic exchange reactions of hydrocarbons and derivatives with metals. *Catal. Rev.* 5 (1): 229–267.

11 Parshall, G.W. (1975). Homogeneous catalytic activation of carbon-hydrogen bonds. *Acc. Chem. Res.* 8 (4): 113–117.

12 (a) Murahashi, S. (1955). Synthesis of phthalimidines from Schiff bases and carbon monoxide. *JACS* 77 (23): 6403–6404. (b) Murahashi, S. and Horiie, S. (1956). The reaction of azobenzene and carbon monoxide. *JACS* 78 (18): 4816–4817.

13 Rosenthal, A., Astbury, R.F., and Hubscher, A. (1958). Reaction of aromatic ketoximes with carbon monoxide and hydrogen1. *J. Org. Chem.* 23 (7): 1037–1039.

14 Bagga, M.M., Flannigan, W.T., Knox, G.R. et al. (1968). Iron carbonyl complexes from Schiff bases. *J. Chem. Soc. C* 0: 36–40.

15 Kleiman, J.P. and Dubeck, M. (1963). The preparation of cyclopentadienyl [o-(phenylazo)phenyl]nickel. *JACS* 85 (10): 1544–1545.

16 Bagga, M.M., Pauson, P.L., Preston, F.J. et al. (1965). Complexes of azobenzenes and Schiff bases. *Chem. Commun. (London)* 21: 543–544.

17 Cope, A.C. and Siekman, R.W. (1965). Formation of covalent bonds from platinum or palladium to carbon by direct substitution. *JACS* 87 (14): 3272–3273.

18 Fahey, R. and D. (1971). The coordination-catalyzed *ortho*-halogenation of azobenzene. *J. Organomet. Chem.* 27 (2): 283–292.

19 Komiya, S. and Yamamoto, A. (2006). A novel oxidative addition involving C—H bond cleavage adjacent to a C=C double bond. *Chem. Lett.* 4 (5): 475–478.

20 Nugent, W.A., Ovenall, D.W., and Holmes, S.J. (1983). Catalytic C—H activation in early transition-metal dialkylamides and alkoxides. *Organometallics* 2 (1): 161–162.

21 Murai, S., Kakiuchi, F., Sekine, S. et al. (1993). Efficient catalytic addition of aromatic carbon-hydrogen bonds to olefins. *Nature* 366 (6455): 529–531.

22 Daugulis, O. and Zaitsev, V.G. (2005). Anilide *ortho*-arylation by using C–H activation methodology. *Angew. Chem. Int. Ed.* 44 (26): 4046–4048.

23 Kuninobu, Y., Nishina, Y., Takeuchi, T. et al. (2007). Manganese-catalyzed insertion of aldehydes into a C—H bond. *Angew. Chem. Int. Ed.* 46 (34): 6518–6520.

24 Zhang, Y.-H. and Yu, J.-Q. (2009). Pd(II)-catalyzed hydroxylation of arenes with 1 atm of O_2 or air. *JACS* 131 (41): 14654–14655.

25 Yang, L., Correia, C.A., and Li, C.-J. (2011). Grignard-type arylation of aldehydes via a rhodium-catalyzed C—H activation under mild conditions. *Adv. Synth. Catal.* 353 (8): 1269–1273.

26 Li, Y., Zhang, X.-S., Chen, K. et al. (2012). N-directing group assisted rhodium-catalyzed aryl C—H addition to aryl aldehydes. *Org. Lett.* 14 (2): 636–639.

27 Thirunavukkarasu, V.S., Hubrich, J., and Ackermann, L. (2012). Ruthenium-catalyzed oxidative C(sp^2)—H bond hydroxylation: site-selective C—O bond formation on benzamides. *Org. Lett.* 14 (16): 4210–4213.

28 Zhang, X.-S., Zhu, Q.-L., Luo, F.-X. et al. (2013). Aromatic C—H addition to ketones: the effect of directing groups. *Eur. J. Org. Chem.* 2013 (29): 6530–6534.

29 Yan, Y., Feng, P., Zheng, Q.-Z. et al. (2013). PdCl2 and N-hydroxyphthalimide co-catalyzed C–H hydroxylation by dioxygen activation. *Angew. Chem. Int. Ed.* 52 (22): 5827–5831.

30 Chen, J., Chen, P., Song, C. et al. (2014). Rhodium(III)-catalyzed N-nitroso-directed C—H addition to ethyl 2-oxoacetate for cycloaddition/fragmentation synthesis of indazoles. *Chem. - Eur. J.* 20 (44): 14245–14249.

31 Qi, Z., Yu, S., and Li, X. (2015). Rh(III)-catalyzed oxidative annulation of 2-phenylimidazo[1,2-a]pyridines with alkynes: mono versus double C—H activation. *J. Org. Chem.* 80 (7): 3471–3479.

32 Ghosh, M., Naskar, A., Mishra, S. et al. (2015). Palladium-catalyzed dehydrogenative annulation of imidazo[1,2-a]pyridines with diarylalkynes. *Tetrahedron Lett.* 56 (27): 4101–4104.

33 Sawant, D., Singh, I., Tulsyan, G. et al. (2015). Ruthenium-catalyzed oxidative C—H bond alkenylation of 2-phenylimidazo[1,2-a]pyridine. *Synlett* 26 (12): 1671–1676.

34 Shakoor, S.M.A., Kumari, S., Khullar, S. et al. (2016). Ruthenium(II)-catalyzed regioselective *ortho* amidation of imidazo heterocycles with isocyanates. *J. Org. Chem.* 81 (24): 12340–12349.

35 Samanta, S., Mondal, S., Ghosh, D. et al. (2019). Rhodium-catalyzed directed C—H amidation of imidazoheterocycles with dioxazolones. *Org. Lett.* 21 (12): 4905–4909.

36 Ghosh, A.K., Ghosh, P., and Hajra, A. (2020). Rhodium-catalyzed directed C(sp^2)—H bond addition of 2-arylindazoles to N-sulfonylformaldimines and activated aldehydes. *J. Org. Chem.* 85 (23): 15752–15759.

37 Ghosh, P., Samanta, S., and Hajra, A. (2020). Rhodium(iii)-catalyzed *ortho*-C—H amidation of 2-arylindazoles with a dioxazolone as an amidating reagent. *Org. Biomol. Chem.* 18 (9): 1728–1732.

38 Breslow, R. (1972). Centenary lecture. Biomimetic chemistry. *Chem. Soc. Rev.* 1 (4): 553–580.

39 Desai, L.V., Hull, K.L., and Sanford, M.S. (2004). Palladium-catalyzed oxygenation of unactivated sp^3 C—H bonds. *JACS* 126 (31): 9542–9543.

40 Giri, R., Liang, J., Lei, J.-G. et al. (2005). Pd-catalyzed stereoselective oxidation of methyl groups by inexpensive oxidants under mild conditions: a dual role for carboxylic anhydrides in catalytic C—H bond oxidation. *Angew. Chem. Int. Ed.* 44 (45): 7420–7424.

41 Li, X., Liu, Y.-H., Gu, W.-J. et al. (2014). Copper-mediated hydroxylation of arenes and heteroarenes directed by a removable bidentate auxiliary. *Org. Lett.* 16 (15): 3904–3907.

42 Sun, S.-Z., Shang, M., Wang, H.-L. et al. (2015). Cu(II)-mediated C(sp^2)–H hydroxylation. *J. Org. Chem.* 80 (17): 8843–8848.

43 Liu, J., Xie, Y., Zeng, W. et al. (2015). Pd(II)-catalyzed pyridine N-oxides directed arylation of unactivated Csp^3—H bonds. *J. Org. Chem.* 80 (9): 4618–4626.

44 Singh, B.K. and Jana, R. (2016). Ligand-enabled, copper-promoted regio- and chemoselective hydroxylation of arenes, aryl halides, and aryl methyl ethers. *J. Org. Chem.* 81 (3): 831–841.

45 Zhu, C., Stangier, M., Oliveira, J.C.A. et al. (2019). Iron-electrocatalyzed C—H arylations: mechanistic insights into oxidation-induced reductive elimination for ferraelectrocatalysis. *Chem. - Eur. J.* 25 (71): 16382–16389.

46 Maehara, A., Tsurugi, H., Satoh, T. et al. (2008). Regioselective C—H functionalization directed by a removable carboxyl group: palladium-catalyzed vinylation at the unusual position of indole and related heteroaromatic rings. *Org. Lett.* 10 (6): 1159–1162.

47 (a) Pichette Drapeau, M. and Gooßen, L.J. (2016). Carboxylic acids as directing groups for C—H bond functionalization. *Chem. - Eur. J.* 22 (52): 18654–18677. (b) Font, M., Quibell, J.M., Perry, G.J.P. et al. (2017). The use of carboxylic acids as traceless directing groups for regioselective C—H bond functionalisation. *Chem. Commun.* 53 (41): 5584–5597.

48 Chen, S.-Q., Li, X.-R., Li, C.-J. et al. (2020). Aldehyde as a traceless directing group for regioselective C—H alkylation catalyzed by rhodium(III) in air. *Org. Lett.* 22 (4): 1259–1264.

49 Hu, X., Chen, X., Shao, Y. et al. (2018). Co(III)-catalyzed coupling-cyclization of aryl C—H bonds with α-diazoketones involving wolff rearrangement. *ACS Catal.* 8 (2): 1308–1312.

50 Huang, C., Chattopadhyay, B., and Gevorgyan, V. (2011). Silanol: a traceless directing group for Pd-catalyzed o-alkenylation of phenols. *JACS* 133 (32): 12406–12409.

51 (a) Gao, D.-W., Zheng, J., Ye, K.-Y. et al. (2015). Asymmetric functionalization of C—H bonds via a transient carbon–metal (C–M) species. In: *Asymmetric Functionalization of C—H Bonds* (ed. S.-L. You), 0. The Royal Society of Chemistry. (b) Rasheed, O.K. and Sun, B. (2018). Advances in development of C–H activation/functionalization using a catalytic directing group. *ChemistrySelect* 3 (21): 5689–5708. (c) Zhao, Q., Poisson, T., Pannecoucke, X. et al. (2017). The transient directing group strategy: a new trend in transition-metal-catalyzed C—H bond functionalization. *Synthesis* 49 (21): 4808–4826.

52 Parshall, G.W., Knoth, W.H., and Schunn, R.A. (1969). Ligand-metal hydrogen-transfer reactions in triphenyl phosphite and triphenylphosphine complexes. *JACS* 91 (18): 4990–4995.

53 Bedford, R.B., Coles, S.J., Hursthouse, M.B. et al. (2003). The catalytic intermolecular *ortho* arylation of phenols. *Angew. Chem. Int. Ed.* 42 (1): 112–114.

54 (a) Oi, S., Watanabe, S.-i., Fukita, S. et al. (2003). Rhodium-HMPT-catalyzed direct *ortho* arylation of phenols with aryl bromides. *Tetrahedron Lett.* 44 (48): 8665–8668. (b) Lewis, J.C., Wu, J., Bergman, R.G. et al. (2005). Preagostic Rh–H interactions and C—H bond functionalization: a combined experimental and theoretical investigation of rhodium(I) phosphinite complexes. *Organometallics* 24 (23): 5737–5746. (c) Carrión, M.C. and Cole-Hamilton, D.J. (2006). Halide-free ethylation of phenol by multifunctional catalysis using phosphinite ligands. *Chem. Commun.* 43: 4527–4529. (d) Bedford, R.B., Betham, M., Caffyn, A.J.M. et al. (2008). Simple rhodium–chlorophosphine pre-catalysts for the *ortho*-arylation of phenols. *Chem. Commun.* 8: 990–992. (e) Yang, J.-F., Wang, R.-H., Wang, Y.-X. et al. (2016). Ligand-accelerated direct C—H arylation of BINOL: a rapid one-step synthesis of racemic 3,3′-diaryl BINOLs. *Angew. Chem. Int. Ed.* 55 (45): 14116–14120.

55 Jun, C.-H., Lee, H., and Hong, J.-B. (1997). Chelation-assisted intermolecular hydroacylation: direct synthesis of ketone from aldehyde and 1-alkene. *J. Org. Chem.* 62 (5): 1200–1201.

56 Jun, C.-H., Hong, J.-B., Kim, Y.-H. et al. (2000). The catalytic alkylation of aromatic imines by Wilkinson's complex: the domino reaction of hydroacylation and *ortho*-alkylation. *Angew. Chem. Int. Ed.* 39 (19): 3440–3442.

57 St John-Campbell, S. and Bull, J.A. (2018). Transient imines as 'next generation' directing groups for the catalytic functionalisation of C—H bonds in a single operation. *Org. Biomol. Chem.* 16 (25): 4582–4595.

58 Wang, Z., Reinus, B.J., and Dong, G. (2012). Catalytic intermolecular C-alkylation of 1,2-diketones with simple olefins: a recyclable directing group strategy. *JACS* 134 (34): 13954–13957.

59 Mo, F. and Dong, G. (2014). Regioselective ketone α-alkylation with simple olefins via dual activation. *Science* 345 (6192): 68–72.

60 Wang, P., Li, G.-C., Jain, P. et al. (2016). Ligand-promoted meta-C—H amination and alkynylation. *JACS* 138 (42): 14092–14099.

61 Pati, T.K., Debnath, S., Kundu, M. et al. (2018). 3-Amino-1-methyl-1H-pyridin-2-one-directed PdII catalysis: C(sp^3)–H activated diverse arylation reaction. *Org. Lett.* 20 (13): 4062–4066.

62 Jin, Z., Chu, L., Chen, Y.-Q. et al. (2018). Pd-catalyzed remote meta-C–H functionalization of phenylacetic acids using a pyridine template. *Org. Lett.* 20 (2): 425–428.

63 Fang, S., Wang, X., Yin, F. et al. (2019). Palladium-catalyzed meta-C–H olefination of arene-tethered diols directed by well-designed pyrimidine-based group. *Org. Lett.* 21 (6): 1841–1844.

64 Porey, S., Zhang, X., Bhowmick, S. et al. (2020). Alkyne linchpin strategy for drug:pharmacophore conjugation: experimental and computational realization of a meta-selective inverse Sonogashira coupling. *JACS* 142 (8): 3762–3774.

65 Cong, X. and Zeng, X. (2014). Iron-catalyzed, chelation-induced remote C–H allylation of quinolines via 8-amido assistance. *Org. Lett.* 16 (14): 3716–3719.

66 Whiteoak, C.J., Planas, O., Company, A. et al. (2016). A first example of cobalt-catalyzed remote C—H functionalization of 8-aminoquinolines operating through a single electron transfer mechanism. *Adv. Synth. Catal.* 358 (10): 1679–1688.

67 Hoque, M.E., Bisht, R., Haldar, C. et al. (2017). Noncovalent interactions in Ir-catalyzed C—H activation: L-shaped ligand for *para*-selective borylation of aromatic esters. *JACS* 139 (23): 7745–7748.

68 (a) Pal, P.P., Ghosh, S., and Hajra, A. (2023). Recent advances in carbosilylation of alkenes and alkynes. *Org. Biomol. Chem.* 21 (11): 2272–2294. (b) Ghosh, S., Lai, D., and Hajra, A. (2021). Visible-light-induced silylation: an update. *Org. Biomol. Chem.* 19 (11): 2399–2415.

69 Maji, A., Guin, S., Feng, S. et al. (2017). Experimental and computational exploration of *para*-selective silylation with a hydrogen-bonded template. *Angew. Chem. Int. Ed.* 56 (47): 14903–14907.

70 Mondal, S. and Hajra, A. (2018). Ruthenium(ii)-catalyzed remote C–H addition of 8 aminoquinoline amide to activated aldehyde. *Org. Biomol. Chem.* 16 (16): 2846–2850.

71 Pimparkar, S., Bhattacharya, T., Maji, A. et al. (2020). *Para*-selective cyanation of arenes by H-bonded template. *Chem. - Eur. J.* 26 (50): 11558–11564.

72 Chen, X., Fan, S., Zhang, M. et al. (2021). Palladium-catalyzed remote *para*-C—H activation of arenes assisted by a recyclable pyridine-based template. *Chem. Sci.* 12 (11): 4126–4131.

73 Dutta, U., Prakash, G., Devi, K. et al. (2023). Directing group assisted *para*-selective C—H alkynylation of unbiased arenes enabled by rhodium catalysis. *Chem. Sci.* 14 (41): 11381–11388.

74 Breslow, R., Zhang, X., and Huang, Y. (1997). Selective catalytic hydroxylation of a steroid by an artificial cytochrome P-450 enzyme. *JACS* 119 (19): 4535–4536.

75 Breslow, R., Huang, Y., Zhang, X. et al. (1997). An artificial cytochrome P450 that hydroxylates unactivated carbons with regio- and stereoselectivity and useful catalytic turnovers. *Proc. Natl. Acad. Sci.* 94 (21): 11156–11158.

76 Belvedere, S. and Breslow, R. (2001). Regioselective oxidation of steroids by a manganese porphyrin carrying metal coordinating groups. *Bioorg. Chem.* 29 (5): 321–331.

77 Das, S., Incarvito, C.D., Crabtree, R.H. et al. (2006). Molecular recognition in the selective oxygenation of saturated C—H bonds by a dimanganese catalyst. *Science* 312 (5782): 1941–1943.

78 Chen, M.S. and White, M.C. (2007). A predictably selective aliphatic C—H oxidation reaction for complex molecule synthesis. *Science* 318 (5851): 783–787.

79 Bigi, M.A., Reed, S.A., and White, M.C. (2011). Diverting non-haem iron catalysed aliphatic C—H hydroxylations towards desaturations. *Nat. Chem.* 3 (3): 216–222.

80 Bigi, M.A., Reed, S.A., and White, M.C. (2012). Directed metal (oxo) aliphatic C—H hydroxylations: overriding substrate bias. *JACS* 134 (23): 9721–9726.

81 Lewis, L.N. (1985). Reexamination of the deuteration of phenol catalyzed by an orthometalated ruthenium complex. *Inorg. Chem.* 24 (25): 4433–4435.

2

Pd-catalyzed In-built Functional Group-directed C–H Functionalization

Ananya Dutta and Masilamani Jeganmohan

Department of Chemistry, Indian Institute of Technology Madras, Chennai 600036, Tamil Nadu, India

2.1 Introduction

Transition metal-catalyzed C—H bond functionalization reactions have witnessed immense growth in the field of synthetic organic chemistry due to their potential of constructing complex molecules from feedstocks in a laboratory-friendly condition [1–6]. The ubiquitous nature of C—H bonds in an organic molecule makes the regioselective functionalization of any particular C—H bond difficult. In order to combat the regioselective issue, directing group-assisted C—H bond activation strategies are adopted as they would lead to the selective activation of a particular type of C—H bond followed by functionalization. The directing group attached to a molecule consists of a Lewis basic motif that will coordinate with the metal center and orients the metal toward a particular C—H bond, thus selectively activating and functionalizing a particular C—H bond via the formation of metallacycle [7]. Directing groups present in any molecule can be selectively divided into two categories – (i) The first category could be presented as the in-built functional group where the directing ability of a functional group already present in the molecule could be utilized to coordinate with the metal center and perform the functionalization later; (ii) the second category could be referred to the removal directing group where a functional group could be pre-incorporated in the molecule or could be incorporated *in situ* in a molecule and then desired C—H bond functionalization could be performed (Figure 2.1) [8]. Both the categories of the directing group have been continuously explored to carry out various C—H functionalization reactions of C(sp^2)—H and C(sp^3)—H bonds at the proximal and distal positions of the molecule. It has been noted that the in-built functional groups coordinate weakly to the metal center, whereas the removable directing group exhibits optimum binding to the metal center, thus providing an easy pathway for various C—H functionalization transformations. In spite of the advantages of the removable directing group, some extra steps regarding the inclusion and exclusion of such functional moieties in the molecule are involved, which limits its use (Figure 2.2). So, the directing ability of the in-built functional motifs has been exploited for many years as they would

Directed C–H Bond Functionalization: Concepts and Applications, First Edition.
Edited by Debabrata Maiti and Supriya Rej.

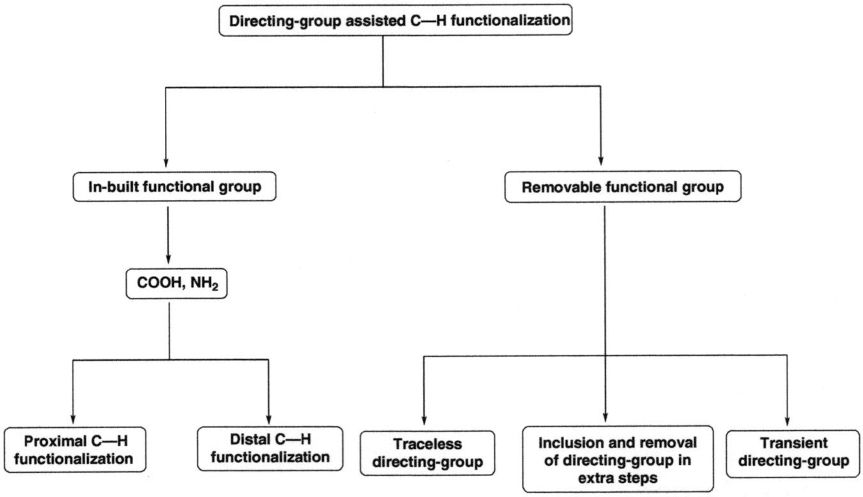

Figure 2.1 In-built vs removable functional group-assisted C–H functionalization.

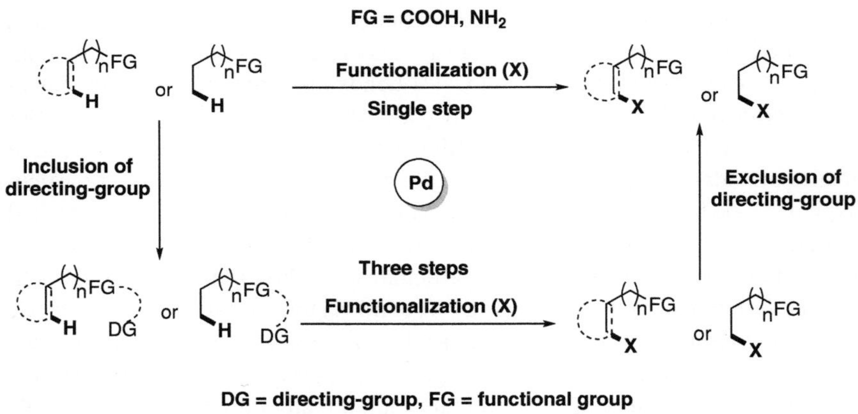

Figure 2.2 Schematic representation of directed C–H functionalization.

lead to desired C—H functionalization transformations in a single step and also in an atom-economical manner. Various functional groups in organic chemistry like carbonyl, amine, carboxylic acid, hydroxyl, and sulfonic acid derivatives have performed very well as in-built functional-group-directed C—H functionalization reactions. Apart from these, some heterocyclic scaffolds present in a molecule could also act as an efficient in-built functional group [9–15]. Among all the in-built functional groups, carboxylic acids and amines have played a vital role in these C—H functionalizations as they would bind strongly to the metal centers, thus leading to many novel transformations in the field of C(sp^2)—H and C(sp^3)—H functionalization reactions. Carboxylic acids and amines are one of the most preferred scaffolds in organic chemistry due to their vast utilities in pharmaceuticals and natural products. Among all the transition metals, palladium complexes have led to many novel transformations via carboxylic acids and amine-assisted C—H functionalizations that involve cleavage of C(sp^2)—H and C(sp^3)—H bonds at proximal and distal positions of the molecule. For functionalization at the β-position (proximal), a thermodynamically favored five-membered palladacycle was involved, whereas for functionalization at the γ- and δ-positions, larger palladacycles are involved. Hence, in this chapter, we would like to highlight how the directing abilities of the in-built carboxylic acid and amine functional groups of the molecules have been utilized to perform various Pd-catalyzed C—H functionalization reactions.

2.2 In-built Nitrogen Atom in a Heterocycle as the Efficient Directing Group

The initial reports on Pd-catalyzed in-built functional-group-directed C—H functionalization were reported by Sanford group in 2004, where they performed Pd-catalyzed regio- and chemoselective arene C(sp^2)—H functionalizations and benzylic C(sp^3)—H functionalizations aided by in-built sp^2 nitrogen atom of the heterocyclic ring (Scheme 2.1) [16]. The same group also carried out numerous Pd-catalyzed functionalizations using an in-built functional motif as the potential directing group [17–23].

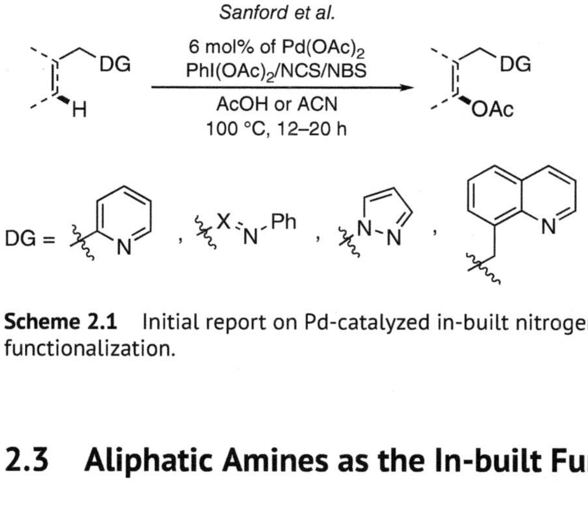

Scheme 2.1 Initial report on Pd-catalyzed in-built nitrogen atom-directed functionalization.

2.3 Aliphatic Amines as the In-built Functional Group

The amine functional group is an important moiety in drugs. Over the years, researchers have been working on different approaches to the synthesis of such derivatized amine moieties. One such approach is the amine-directed C—H functionalization where various C—C, C—N, and C—O bonds can be generated. The amine-directed C—H functionalization has been explored in three ways: (i) the innate nature of the amine has been utilized for directing ability; (ii) subsequently, installation and removable of any external functional group to the amine moiety; and (iii) transient directing group approach that includes *in situ* installation and removal of the directing group to the amine moiety [24]. In this chapter, we would only like to focus on the various Pd-catalyzed transformations where the inner directing ability of the amine motif has played the key role in coordinating to Pd(II)-salts and further directing it to selectively functionalize a particular proximal or distal C(sp^3)—H bond. The amine functional group coordinates strongly to the electrophilic Pd-center and facilitates the desired C—H activation via the formation of four/five-membered or larger palladacycles. The palladacycles could then lead to many functionalization reactions like arylation, acetoxylation, olefination, and also intramolecular C—N bond formation reactions in the presence of Pd(II)-salts and in some cases external ligands. The only disadvantage is that the amine group could also form a stable bis-diamine palladium complexes other than the required palladacycle, thus providing hindrance to the desired functionalization [25]. Overcoming all these drawbacks, research groups like Gaunt group and Yu group have developed many interesting strategies to exploit the directing ability of the in-built amine motifs in the field of proximal and distal C(sp^3)—H bond functionalization.

2.3.1 Pd-catalyzed Amine-directed Intramolecular C(sp^3)–H Amination

The first example of Pd-catalyzed C—H functionalization of unprotected aliphatic amines as the in-built directing group was reported by Gaunt group in 2014 (Scheme 2.2) [26]. Secondary amines containing fully substituted carbon centers at the α-position with respect to amines were utilized for the desired aziridine product. The sterically crowded cyclic amine underwent four-membered cyclopalladation via β-C(sp^3)—H activation followed by oxidation with PhI(OAc)$_2$ to generate the strained aziridine product. The aziridine product further provided access to various functionalized moieties by sequential ring opening carried out by the other nucleophiles.

Based on this pioneering report, the same group in 2017 carried out the enantioselective aziridine formation using a similar type of sterically hindered cyclic amines and a chiral BINOL-derived phosphoric acid ligand, (*R*)-TRIP [27]. This was the first enantioselective desymmetrization reaction of prochiral methyl groups using Pd(II)/Pd(IV) catalysis pathway. In the subsequent year, the same group utilized a similar type of strategy and performed Pd-catalyzed intramolecular γ-C(sp^3)—H amination to synthesize substituted azetidine ring [28].

2.3.2 Pd-catalyzed In-built Amine-directed C–H Arylation

Over many years, chemists have exploited the directing properties of many functional groups and ligands to carry out desired C—H bond functionalization. Hence, in 2015, Gaunt and co-workers adopted steric parameters to perform C—H functionalization of amino alcohols (Scheme 2.3) [29]. The amino alcohol was treated with cyclohexanone as a steric tether to form a hindered *N*,*O*-ketal moiety. The hindered ketal motif will bind with Pd(OAc)$_2$ to form a lower energy monoamine Pd(II) complex, where hydrogen bonding was involved between the palladium-bound amine and the acetate ligand. This intermediate further facilitated the desired

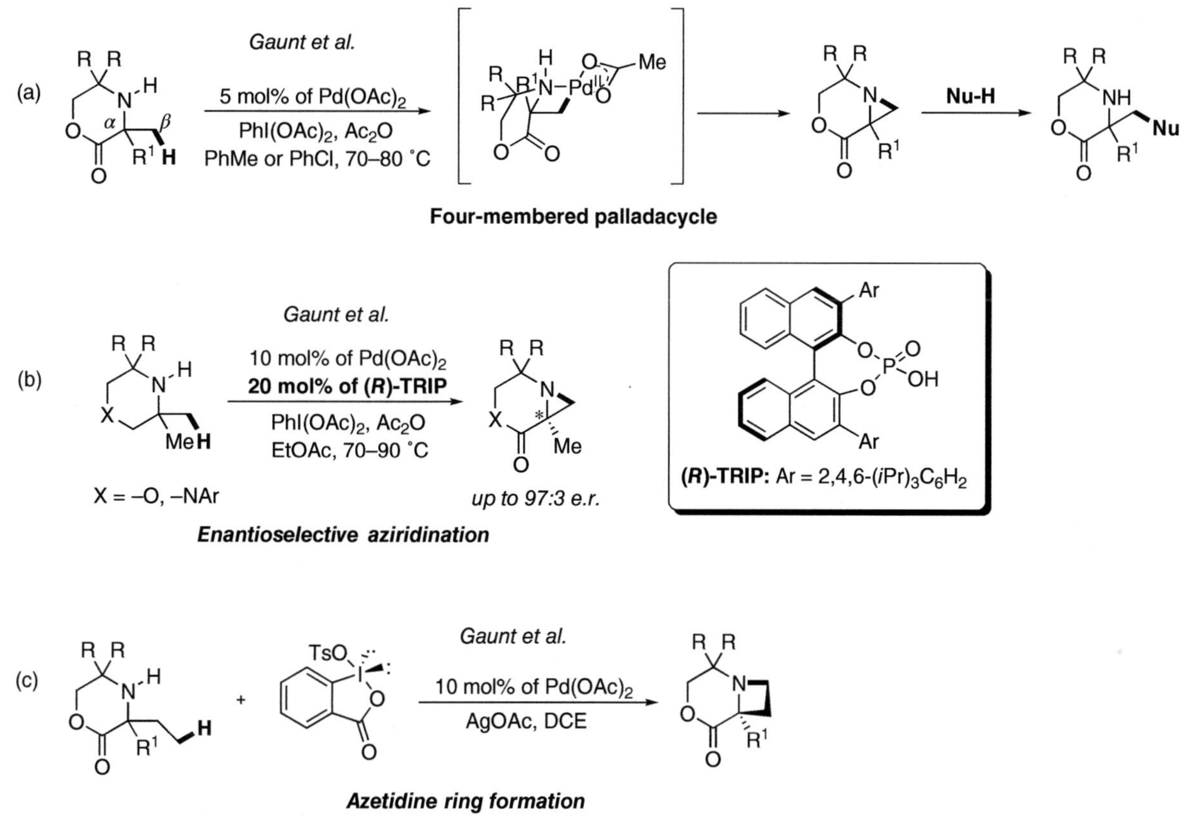

Scheme 2.2 Pd(II)-catalyzed intramolecular C(sp^3)–H amination.

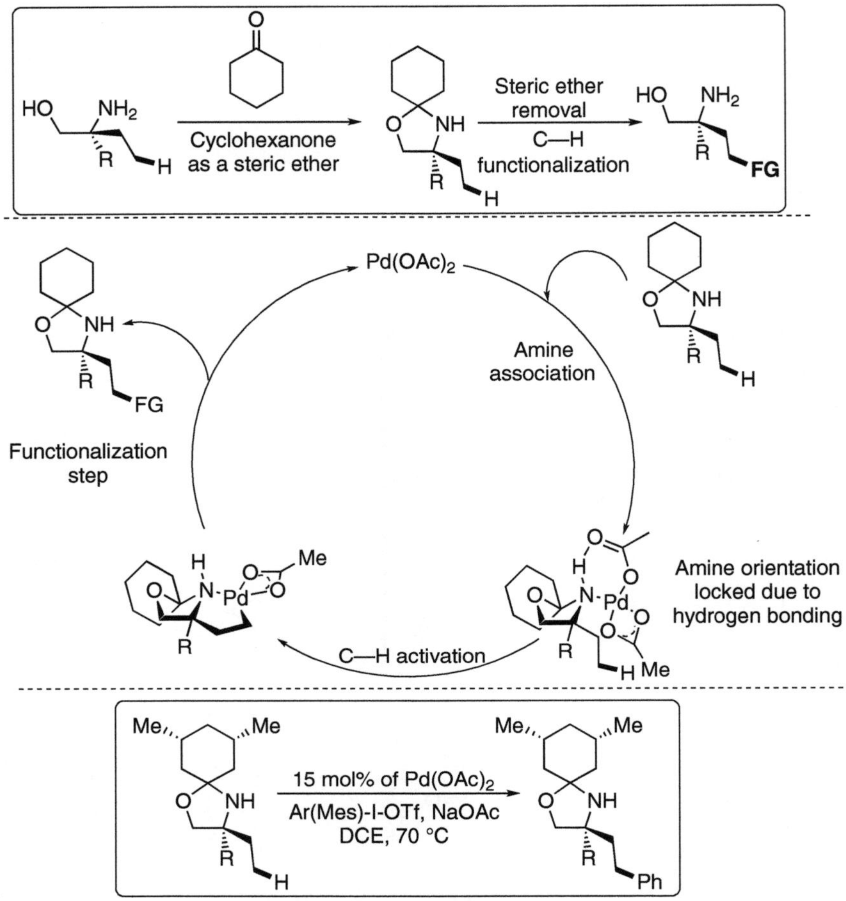

Scheme 2.3 Pd-catalyzed C–H bond arylation of primary amino alcohols using a steric tether.

C—H activation and led to the formation of palladacycle. The palladacycle upon treatment with the hypervalent iodine reagent diphenyliodonium triflate resulted in the γ-C(sp^3)—H arylation.

The same group further demonstrated the arylation of hindered piperidine type of substrates via β-C(sp^3)—H arylation aided by *N*-acetyl phenylalanine as a ligand (Scheme 2.4) [30]. In this reaction, arylboronic esters were utilized as the arylating agents in the presence of silver carbonate and benzoquinone. The newly generated arylamine product was further exploited, subsequent C—H functionalizations like carbonylation, alkenylation, and alkylation were performed with different coupling partners, and a range of architecturally complex scaffolds were synthesized.

In comparison with γ-C(sp^3)—H activation, δ-C(sp^3)—H activation is very much less explored due to the formation of kinetically less-favored six-membered palladacycle [31–34]. In order to combat this, Bannister group in 2019 reported the δ-C(sp^3)—H arylation of primary aliphatic amines with heteroaryl iodides as

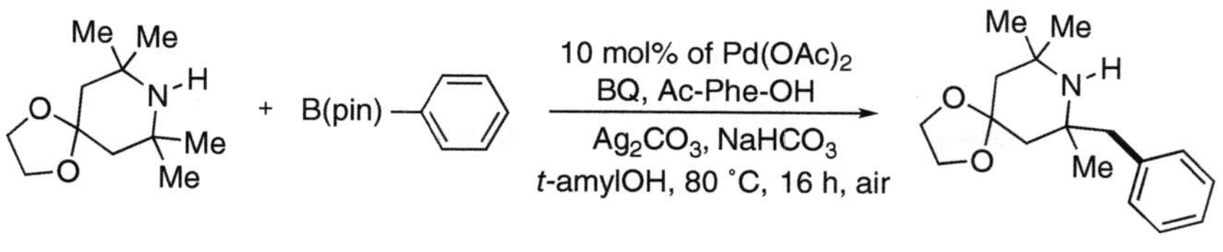

Scheme 2.4 Pd-catalyzed β-C(sp^3)—H bond arylation.

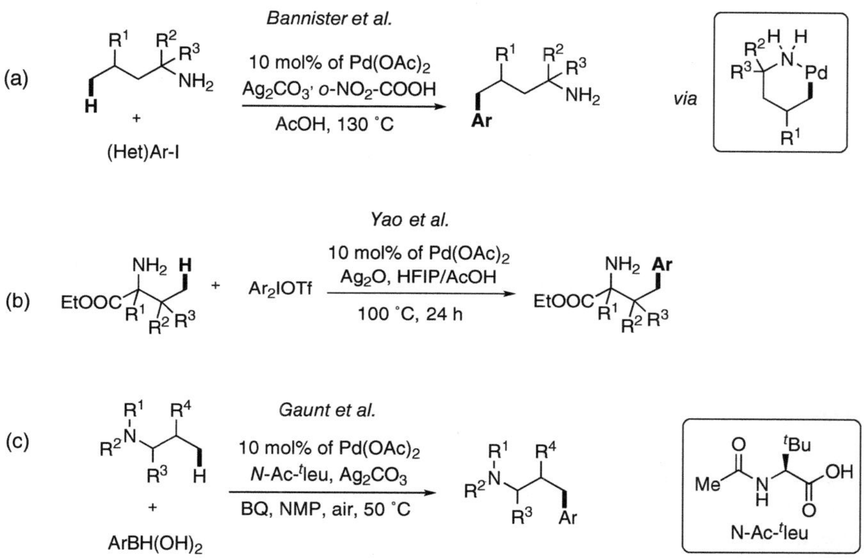

Scheme 2.5 Pd-catalyzed C(sp^3)—H bond arylation of free aliphatic amines.

the coupling partner (Scheme 2.5) [35]. Free aliphatic amines are typically not good substrates as they can form stable bis-amine palladium complexes, thus providing hindrance to the activation of desired C—H activation. So, weak acid additives were used in the reaction as they would aid in the dissociation of the stable bis-amine complex, enabling the formation of C—H activation reaction. In the same year, Yao group reported γ-C(sp^3)—H arylation of α-amino esters using diaryliodonium triflates, Ag_2O, and HFIP/AcOH cosolvent system [36]. This protocol needed the presence of tertiary carbon center at the α-site of amino esters. The reaction was also compatible with a wide variety of amino acids and various aryl partners. Later in 2020, for the first time, Gaunt and co-workers demonstrated the concept of γ-C(sp^3)—H arylation of tertiary amines utilizing arylboronic acids as the coupling partner [37]. This strategy involves the requirement of *N-Ac-^{t}*Leu ligand, which will distort the coplanarity of the Pd(II)-ligand complex, thereby eliminating the chances of β-hydride elimination and favoring the desired C(sp^3)—H arylation.

In 2020, Yu group reported an efficient protocol for enantioselective γ-C(sp^3)—H bond arylation of cyclopropylmethyl amines with the help of external bidentate

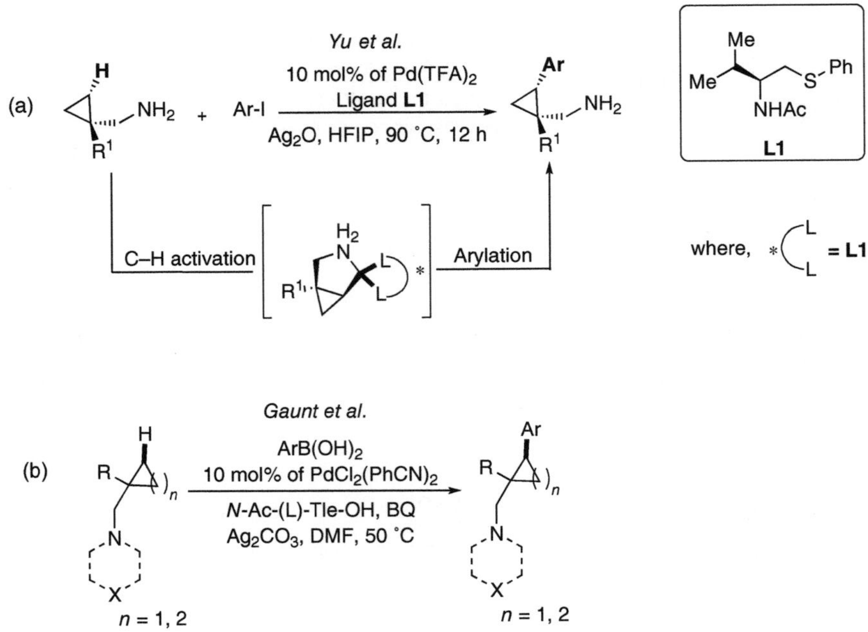

Scheme 2.6 Pd-catalyzed γ-C(sp^3)—H bond arylation of free cyclopropylmethyl amines.

thioether ligand **L1** (Scheme 2.6) [38]. It is proposed that the ligand **L1** promoted the formation of the mono(amine)–Pd(II) intermediate through its strong σ-donor properties. The protocol utilizes Pd(TFA)$_2$ as catalyst and Ag$_2$O, and HFIP were found to be crucial for Pd(II)/Pd(IV) catalytic cycle. This strategy was also found to be compatible with a wide range of aryl iodides giving the desired arylation in good yields and enantioselectivities >95% ee. Subsequently, in 2022, Gaunt and co-workers illustrated the enantioselective γ-C(sp^3)—H bond arylation of similar aminomethyl cyclopropane and cyclobutane scaffolds [39]. The ligand *N*-Ac-(L)-Tle-OH was crucial for facilitating γ-C—H activation in place of β-hydride elimination and also for inducing enantioselectivity in the reaction.

2.3.3 Pd-catalyzed Amine-directed C—H Acetoxylation

In 2015, Gaunt group performed acetoxylation of amino alcohols by attachment of steric tether to generate *N*,*O*-ketal motifs (Scheme 2.7) [29]. The reaction was hypothesized to proceed via a high valent Pd intermediate where subsequent reductive elimination resulted in the desired C(sp^3)—H acetoxylated product. The same group in 2019 performed a similar type of γ-C(sp^3)—H acetoxylation on morpholinone derivatives [40]. Density functional theory (DFT) and kinetic studies stated that the reaction proceeded through a five-membered palladacycle where C—H activation was found to be the rate-determining step. Subsequently, in 2017, Shi group demonstrated γ-C(sp^3)—H acetoxylation of free aliphatic amines under weak acid conditions like acetic acid [41]. The hypervalent iodine species led to the formation

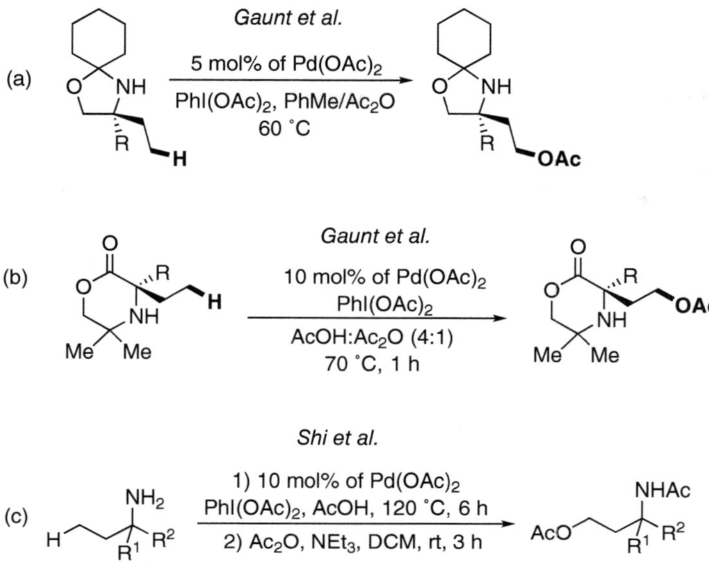

Scheme 2.7 Pd-catalyzed γ-C(sp^3)—H bond acetoxylation of aliphatic amines.

of a five-membered palladacycle intermediate (**C**) that further led to acetoxylation via reductive elimination pathway. This strategy even provided a lucid approach to synthesize γ-amino alcohols from primary amines. The proposed reaction pathway has been depicted here in Scheme 2.8.

2.3.4 Pd-catalyzed Amine-directed Alkenylation Reaction

The C—C bond formation combining C(sp^3)—H bonds with alkene has always been a challenging task for chemists. To overcome such challenges, Gaunt and his co-workers in 2015 utilized the steric tethering approach to carry out selective alkenylation of γ-C(sp^3)—H bond of amino alcohols [29]. Upon treatment of *N*,*O*-ketal moiety with acrylates in the presence of Pd(OAc)$_2$, AgOAc, and DCE solvent, the desired γ-C(sp^3)—H alkenylated product was obtained. This alkenylation was further accompanied by an intramolecular aza-Michael reaction that furnished the pyrrolidine compounds. Further, the *N*,*O*-ketal moiety was cleaved under mild conditions to regenerate the functionalized amino alcohols (Scheme 2.9). Subsequently, in 2017, the same group performed alkenylation reaction of morpholinone moieties bearing aliphatic amines [42]. They even used an external amino acid ligand in this strategy to understand the reversible C—H bond activation of morpholinone moieties to furnish various pyrrolidine scaffolds. A wide range of substituted morpholinone moieties and alkene molecules were found to be compatible with this protocol. They also suggested that the alkenylation proceeded through a five-membered palladacycle formation, followed by alkene insertion and aza-Michael cyclization to form a pyrrolidine ring. Later on, Yu group in 2020 successfully demonstrated the asymmetric olefination of free cyclopropylmethyl

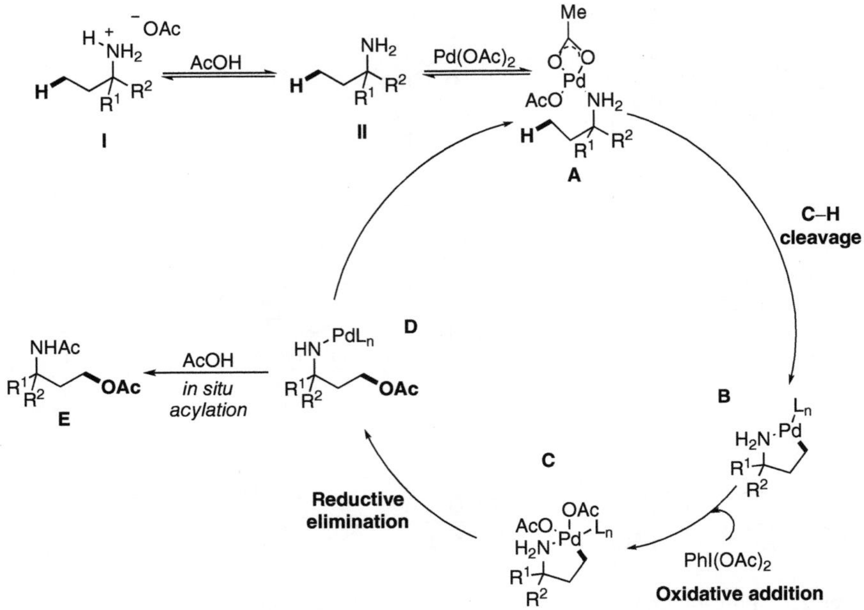

Scheme 2.8 Plausible mechanism of γ-C(sp^3)–H bond acetoxylation of free aliphatic amines.

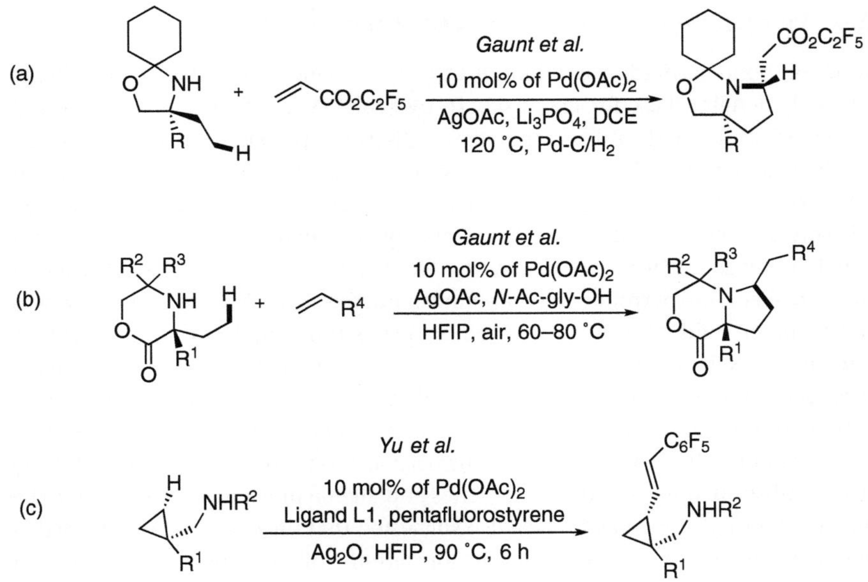

Scheme 2.9 Pd-catalyzed γ-C(sp^3)–H bond alkenylation of aliphatic amines.

amines involving pentafluorostyrene as the coupling partner and ligand **L1** as the external ligand [38]. Various cyclopropyl amines generated the olefinated products in moderate yields and excellent enantioselectivity.

2.3.5 Amine Group-directed Carbonylation Reactions

Gaunt and co-workers have devoted significant efforts to carry out β-C(sp^3)—H bond carbonylation of hindered piperidine substrates with the help of CO and $Cu(OAc)_2$ as a suitable terminal oxidant (Scheme 2.10) [26]. In a mechanistic point of view, the hindered cyclic amine-directed C—H activation led to the formation of a four-membered palladacycle, which was then accompanied by CO insertion and reductive elimination to generate the desired β-lactam. In the subsequent year, the same group experimented the strategy of β-C(sp^3)—H bond carbonylation with β-methylene bonds of secondary aliphatic amines. The use of bis(phosphine) ligand xantphos was found to be crucial in this strategy for obtaining good yields [43]. Notably, in 2017, Gaunt group demonstrated that under the catalytic Pd(II)/xantphos conditions, β-methylene-C(sp^3)—H bond is found to undergo carbonylation rather than β-methyl-C(sp^3)—H bond. This remarkable β-methylene selectivity is attributed to the hydrogen-bonding interaction between the carbamoyl carbonyl and a ligated amine as shown in Scheme 2.10 (intermediates **I** and **II**), which locks their relative conformations. Hence, this protocol generated a wide range of β-lactam products in good yields [44].

Gaunt and his co-workers further adopted the C(sp^3)—H bond carbonylation to transform acyclic secondary amines to trans-substituted γ-lactams in good yields and with high diastereoselectivity (Scheme 2.11). The protocol utilized $Pd(OPiv)_2$ as the catalyst and $Cu(OAc)_2$ as the oxidant. The use of sterically hindered 2,4,6-triisopropylbenzoic acid (TRIP-OH) in the presence of quinuclidine further enhanced the diastereoselectivity and yield of the desired γ-lactam scaffolds [45]. Prior to this study, the same group in 2015 carried out γ-C(sp^3)—H carbonylation of *N*,*O*-ketal motifs derived from amino alcohols via steric tethering approach [29]. They envisioned that such *N*,*O*-ketal motifs after γ-C(sp^3)—H activation would generate a five-membered palladacycle, which upon CO insertion could lead to one carbon homologation and further cyclization would result in the formation of pyrrolidinone scaffold. Therefore, upon the reaction of the *N*,*O*-ketal moiety with $Pd(OAc)_2$, AgOAc as the terminal oxidant in the presence of CO, the desired pyrrolidinone scaffolds were obtained in good yields. Notably, some small amount of the α, β-unsaturated pyrrolidinone product was observed in this reaction, which was further converted to the desired pyrrolidinone by direct hydrogenation of the reaction mixture.

Additionally, Yu group made a significant contribution to this field by demonstrating an efficient strategy of free aliphatic amine-directed carbonylation reaction in an asymmetric fashion. The strategy focused on the formation of bicyclic lactams via significant γ-C(sp^3)—H bond activation of cyclopropyl methyl amines, using $Pd(OAc)_2$ as the effective catalyst, $Mo(CO)_6$ as a CO source, and ligand **L1** as the suitable external amino acid ligand [38].

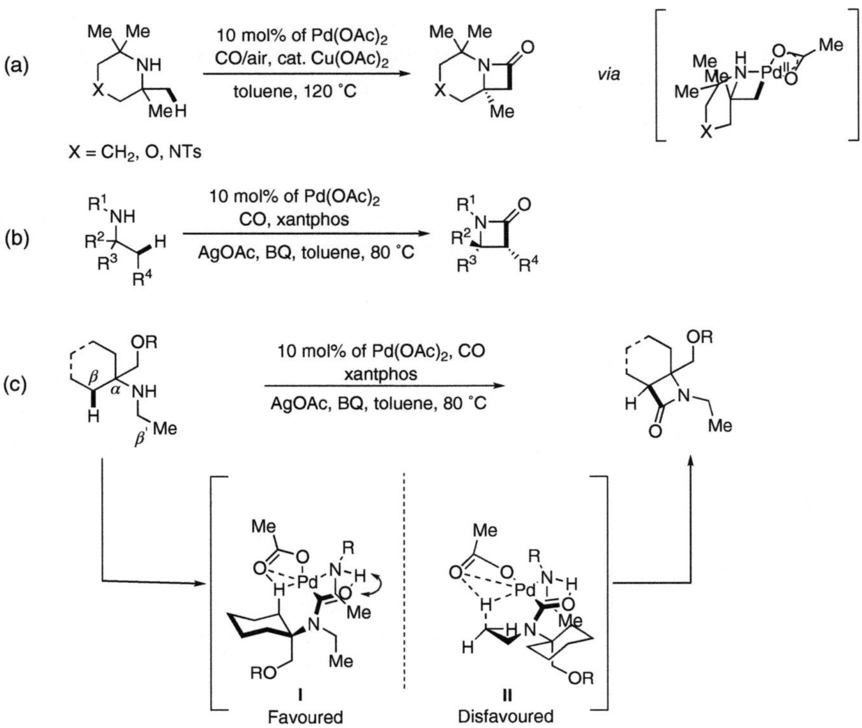

Scheme 2.10 Pd-catalyzed β-C(sp^3)—H bond carbonylation.

2.4 Carboxylic Acids as the In-built Functional Group in C—H Activation

Carboxylic acids moieties are the most abundant and versatile functional groups in organic chemistry and are found in abundance in pharmaceuticals, natural products, and many bioactive molecules. So, derivatization of free carboxylic acid-bearing molecules has attracted many researchers, and a lot of approaches have been developed over the years. One of the approaches was C—H activation methods where the directing ability of the carboxyl acid moiety could be utilized to bind with the Pd(II)-salts and further orient Pd(II) to selectively activate a particular C—H bond. The carboxylic acids could also act as traceless directing groups, but in this chapter, we would only highlight the in-built functional group characteristics of carboxylic acids. The acid moiety binds to the Pd(II) salts and could activate C(sp^2)—H and C(sp^3)—H bonds at the proximal position and the distal C(sp^3)—H bonds via the formation of five- and six-membered palladacycle, respectively. The carboxylate moiety can undergo two types of coordination modes in the presence of Pd-catalyst – one is κ^2 mode where both the oxygen atoms would bind with the Pd-metal, whereas another is κ^1 mode where only one oxygen atom would bind with the metal center [46].

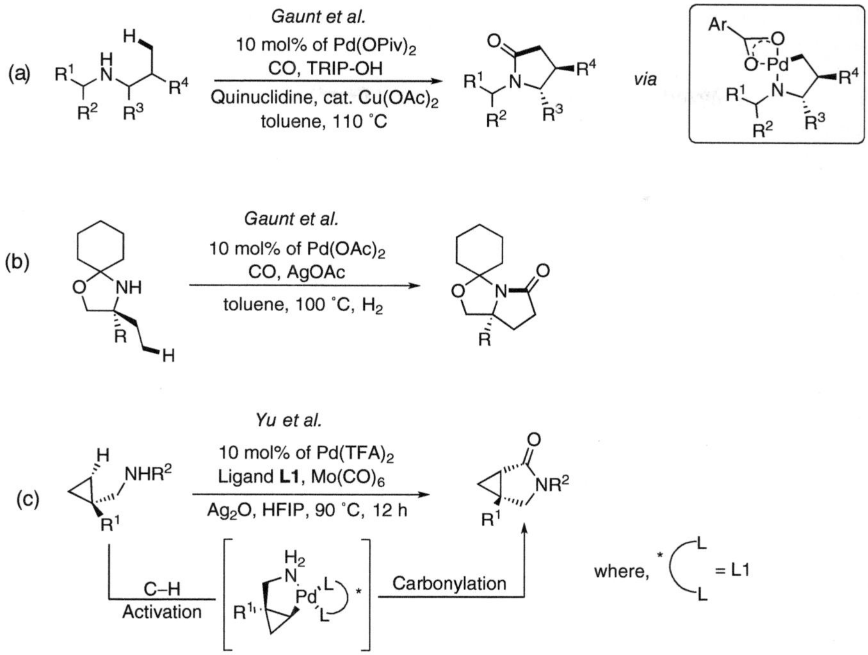

Scheme 2.11 Pd-catalyzed carbonylation of distal $C(sp^3)$–H bonds.

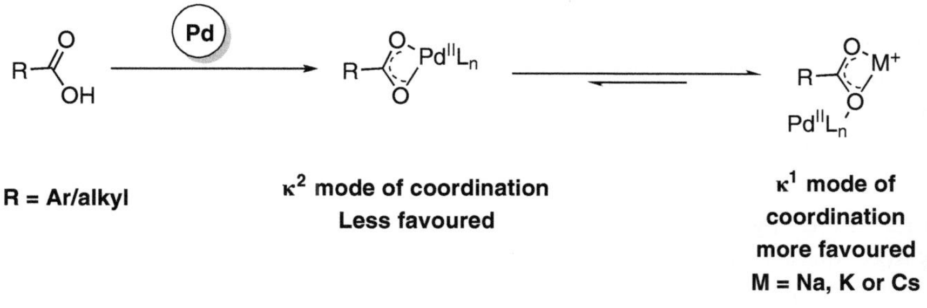

Figure 2.3 Binding modes of carboxylate to palladium.

The κ^1 mode of carboxylates predominates only in the presence of some external alkali metal ions (Na^+, Cs^+, K^+) as these metal ions would coordinate with both the oxygen atoms instead of Pd, thus orientating the Pd(II) toward the activation of the targeted $C(sp^2)$—H and $C(sp^3)$—H bonds. Hence, the carboxylic acid motif has been utilized widely to functionalize various proximal and distal C—H bonds via its innate directing ability. It has also been noted that for carboxylate-assisted C—H bond functionalization, an external ligand source and a base additive are always required for the predominance of the κ^1 mode in the reaction (Figure 2.3).

2.4.1 Pd-catalyzed C(sp^2)–H Bond Functionalization of Benzoic and Phenyl Acetic Acids

2.4.1.1 Pd-catalyzed Carboxylate Group-directed C(sp^2)–H Bond Arylation of Benzoic Acids

In 2007, Yu group first reported the Pd-catalyzed C(sp^2)—H arylation at the *ortho*-position of simple benzoic acids via carboxylate-directed moiety (Scheme 2.12). The strategy involved simple phenyl boronates as the coupling partners in the presence of silver carbonate and benzoquinone as the oxidant [47]. Various benzoic acids were found to be compatible with this protocol, which further led to the desired arylated and methylated products in good-to-excellent yields. In the same year, Daugulis and his co-workers made a significant contribution toward the direct *ortho*-C(sp^2)—H activation of benzoic acids [48]. The first method is applied for the arylation of electron-rich benzoic acids with the help of aryl iodides and in the presence of acetic acid as solvent and silver acetate additive for the removal of iodides. The second protocol bearing *n*-butyl-di-1-adamantylphosphine as a ligand, cesium carbonate as a base, and DMF as a solvent was found to be very much compatible for electron-rich as well as electron-deficient benzoic acids. In the subsequent years, Larrosa group adopted two different protocols for the synthesis of sterically hindered 2-aryl,6-substituted benzoic acids [49]. In one of the protocols, they used silver carbonate as an oxidant and potassium carbonate as a base additive, whereas the other protocol involved silver salts free route with the usage of inexpensive and organic analog, tetramethylammonium acetate (NMe_4OAc) for *in situ* catalyst generation. An important contribution was also made by Su group where they utilized an external amino acid ligand **L2** for the selective C(sp^2)—H arylation of electron-deficient benzoic acids. HFIP was used as the solvent in their protocol as it is a strong polar protic solvent that could bind with the metal and orient it to selectively activate one particular C(sp^2)—H bond [50]. Further, kinetic studies revealed the role of ligands in the C—H activation step for that protocol.

2.4.1.2 Pd-catalyzed Carboxylate Group-directed Benzolactone and Isocoumarin Formation Using Benzoic Acids

In 2007, along with the arylation report, Yu group reported methylation of simple benzoic acids using methyl boronic acids as the coupling partner to synthesize *o*-toluic acids through Pd(II)/Pd(0) pathway [47]. Later, in 2009, they developed another alkylation strategy by reacting benzoic acids with alkyl halides under silver-free conditions but in the presence of potassium biphosphate or potassium bicarbonate as a base [51a]. This protocol led to the formation of *ortho*-alkylation initially, which further underwent an S_N2 reaction to give the corresponding five- or six-membered benzolactones. The other competing side reaction for the protocol was the S_N2 reaction between the substituted acid and alkyl chloride/bromide, which was reduced by the type of bases used. In 2015, the same group demonstrated a new approach to C(sp^2)—H alkylation by employing an epoxide as the coupling partner and without the use of any silver salts [51b]. They hypothesized that the arylpalladium intermediate reacted with epoxide via

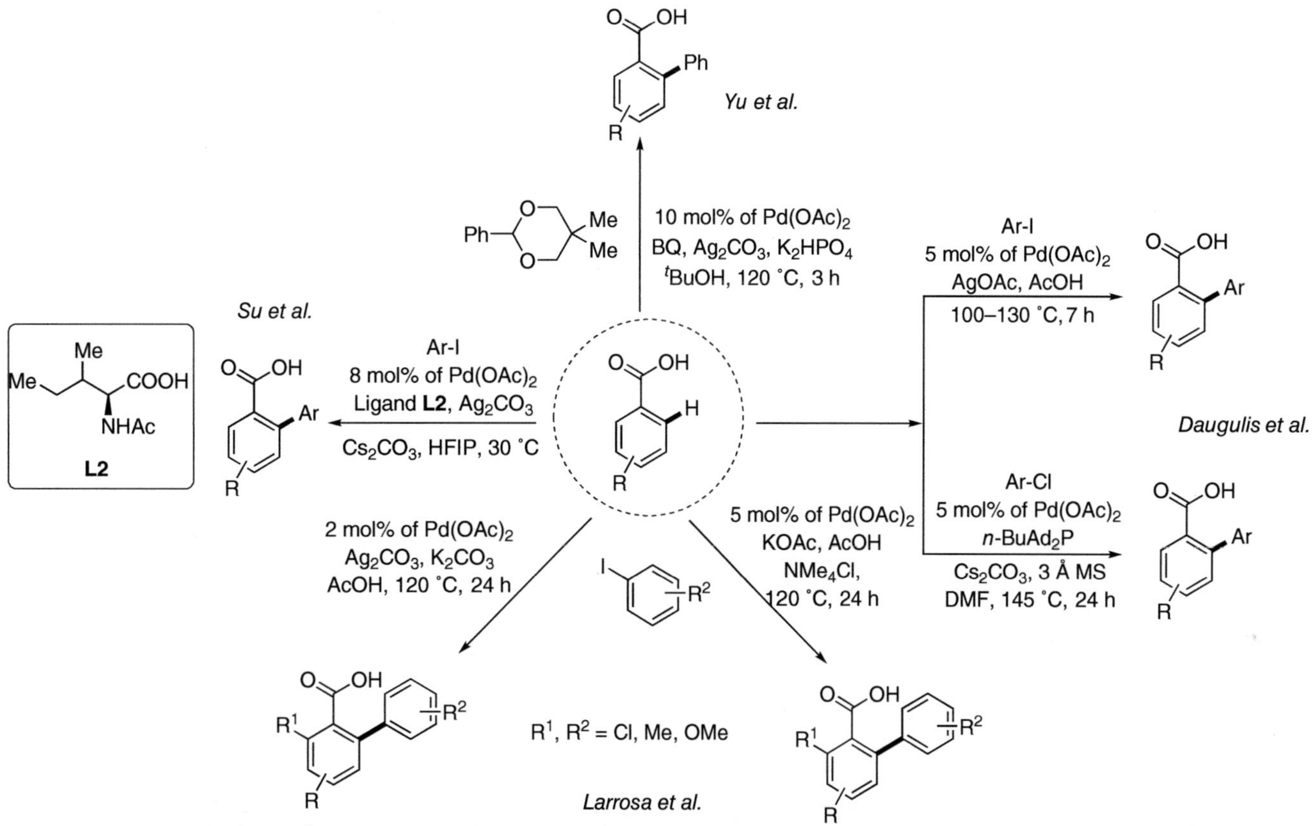

Scheme 2.12 Pd-catalyzed carboxylate-directed arylation of C(sp^2)–H bonds.

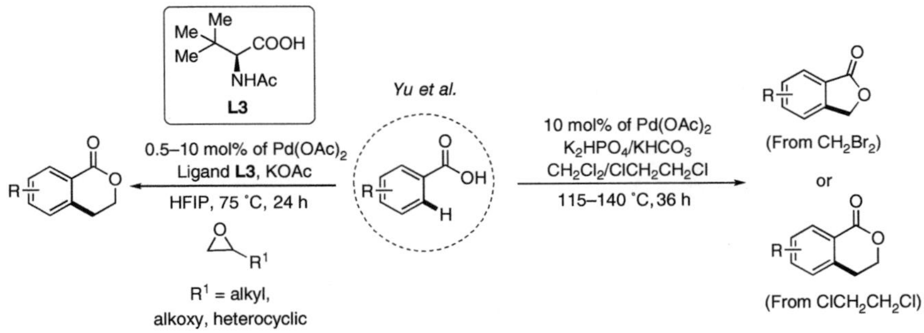

Scheme 2.13 Pd-catalyzed carboxylate-directed benzolactone formation.

a redox-neutral S_N2 nucleophilic ring-opening process, and then subsequent lactonization led to the benzolactone formation (Scheme 2.13).

In 1984, Larock group first reported C(sp^2)—H olefination of aryl carboxylic acids. Substituted benzoic acids were first thallated by thallium(III) trifluoroacetate and then reacted with palladium chloride and simple olefins to synthesize isocoumarins [52] (Scheme 2.14). In 2013, Lee group developed a protocol for the synthesis of isocoumarins and 3-benzylidenephthalides through Pd-catalyzed C—H olefination accompanied by oxidative coupling of benzoic acids and vinyl arenes [53]. Electron-rich benzoic acids resulted in the isocoumarin formation in high yields, whereas electron-deficient arenes were incompatible for the protocol. Upon reaction of *ortho*-substituted benzoic acids with arenes, the corresponding isomeric (*Z*)-3-benzylidenephthalides were isolated in good yields instead of isocoumarin. These results revealed the directing effect of the different benzene ring substituents on the regioselective nucleophilic attack of the carboxylate atom on the olefin that led to different product formation. Following this in 2019, Zeng and co-workers designed a carboxylate-moiety-directed C(sp^2)—H activation strategy to synthesize Pyrano[4,3-*b*]Indol-1(5*H*)-ones, a multicyclic indole scaffold. This strategy involved reaction of readily available indole-3-carboxylic acids and olefins via carboxylate-directed sequential *ortho*-C(sp^2)—H alkenylation followed by remote olefinic C(sp^2)—H activation [54].

Aided by mechanistic studies, a plausible mechanism was proposed by Zeng group. First, $Pd(OAc)_2$ reacted with a mono-protected amino acid ligand to generate the active catalyst (**I**). Then, through an irreversible *ortho*-aryl C—H activation with (**1a**), a five-membered palladacycle intermediate (**II**) was formed. Further, migratory insertion and β-hydride elimination of acrylate (**2a**) generated the key intermediate product (**III**). The intermediate (**III**) further led to the seven-membered palladacycle intermediate (**IV**) via the secondary remote olefinic C—H bond activation at the δ-position. The intermediate (**IV**) then finally generated the desired product (**3a**) via reductive elimination. The active palladium catalyst (**I**) was regenerated with the help of an oxidant (Scheme 2.15) [54].

After the numerous contributions on arylation and lactone formation of aryl carboxylates by various groups, Yu group in 2008 adopted C(sp^2)—H halogenation strategy by the usage of iodoacetic acid (IOAc) and *N*-bromosuccinimide (NBS) as

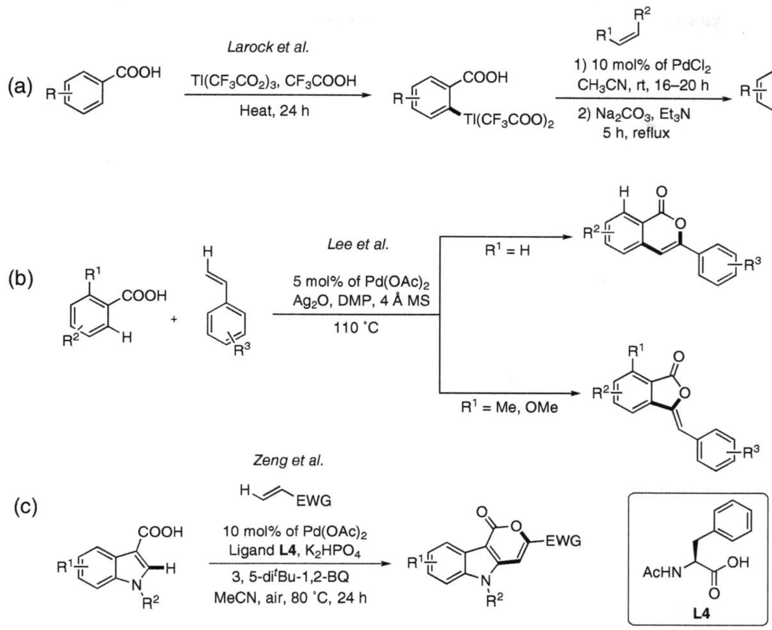

Scheme 2.14 Pd-catalyzed carboxylate-directed isocoumarin formation.

the halogenating source (Scheme 2.16) [55]. Yu and co-workers in 2012 synthesized anthranilic acids from lithium carboxylates using ethyl mesitylsulfonyloxycarbamates as the suitable amidating agents via C(sp^2)—H activation [56].

In 2008, Yu group developed a new strategy for the synthesis of 1,2- and 1,3-dicarboxylic acids via Pd(II)-catalyzed carboxylation of benzoic and phenyl acetic acids. The protocol was further extended to carboxylate vinyl C—H bonds. A large number of benzoic acids and phenyl acetic acids were found to be compatible with this strategy (Scheme 2.17a) [57]. Later on, in 2013, Ge group designed a protocol for acylation of substituted benzoic acids with the help of α-oxocarboxylic acids via C(sp^2)—H bond functionalization (Scheme 2.17b) [58].

A plausible reaction mechanism was proposed by Ge group for C(sp^2)—H bond acylation. The transformation started with the palladacycle formation of silver benzoate (**A**) that resulted in the intermediate (**B**). The intermediate (**B**) then underwent transmetalation with the acyl-silver species to form the intermediate (**D**). The intermediate (**D**) then generated the silver salt (**E**) and Pd(0) after reductive elimination. The Pd(0) was then reoxidized to Pd(II) by a silver-based oxidant (Scheme 2.18) [58].

2.4.1.3 Pd-catalyzed Carboxylate-assisted Halogenation, Amidation, Carboxylation, and Acylation Reaction of Benzoic Acids

Yu group first explored the hydroxylation of substituted benzoic acids in 2009 (Scheme 2.19) [59]. Various substituted benzoic acids were found to be compatible with the developed protocol. Isotope-labeling studies revealed the role of oxygen

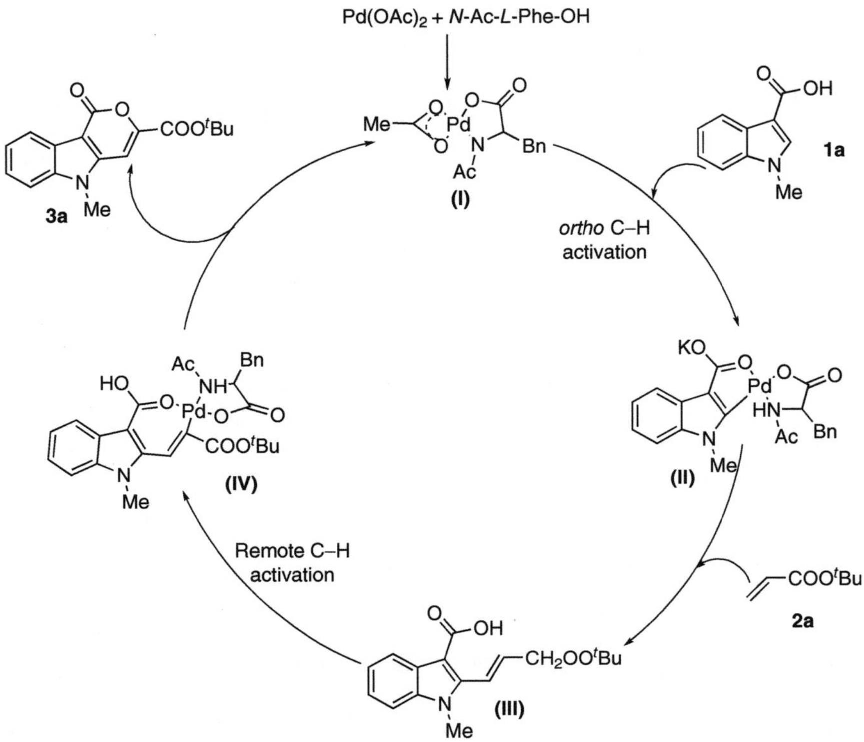

Scheme 2.15 Plausible mechanism of Pd-catalyzed carboxylate-directed isocoumarin formation by Zeng group.

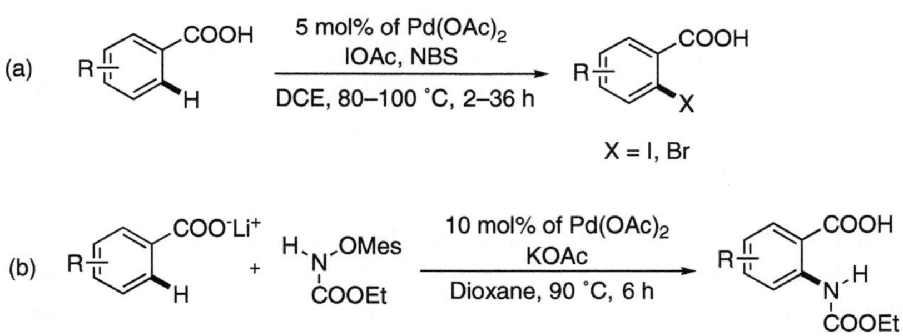

Scheme 2.16 Carboxylate-directed C—H halogenation and amidation.

in the hydroxylation reaction. In the recent year, the same group designed a bifunctional ligand **L5** containing pyridine and pyridone moieties to selectively hydroxylate $C(sp^2)$—H bond. The two heterocyclic moieties present in the ligand are known for their dual role where the pyridine part helps in oxygen activation, whereas the pyridone part accelerates the C–H activation [60].

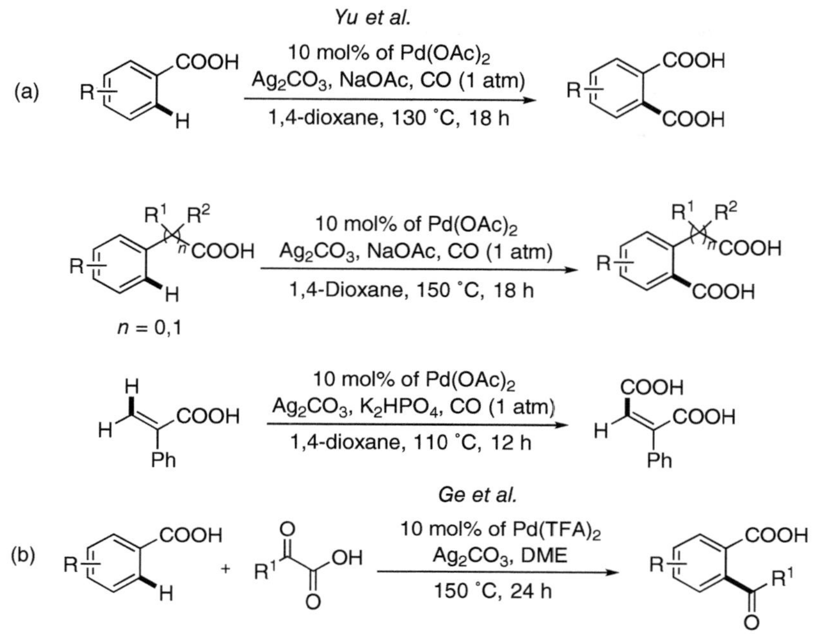

Scheme 2.17 Carboxylate-directed C(sp^2)–H bond carboxylation and acylation.

2.4.1.4 Pd-catalyzed Carboxylate-assisted C(sp^2)–H Bond Arylation of Phenyl Acetic Acids

In 2008, Yu group developed arylation protocol for substituted benzoic acids and phenyl acetic acids utilizing aryltrifluoroborates as the coupling partner (Scheme 2.20) [61]. Organic oxidants like benzoquinone were used along with oxygen as an oxidant. In the subsequent year, the same group extended the protocol for selective *ortho*-arylation of phenyl acetic acid derivatives using the external *N*-protected amino acid ligand **L2** and a silver-based oxidant [62]. In 2015, the same group performed arylation reactions using pivaloyl or acetyl-protected aryl acetic acids without the assistance of any external amino acid ligand. A large number of substituted aryl acetic acids were found to be compatible with this protocol [63]. In 2013, the same group designed the amino acid ligand **L6** for carrying out *ortho*-alkylation of benzoic as well as phenyl acetic acids in good-to-excellent yields [64]. In the reaction bearing alkyltrifluoroborates as the suitable alkylating coupling partners, lithium carbonate as base and silver carbonate as oxidant were found to be highly compatible with electron-donating and electron-withdrawing acids. In 2017, the same group made a tremendous contribution toward the ligand-controlled C(sp^2)—H bond arylation by performing *meta*-C(sp^2)—H activation. They used NBE-COOMe as a transient mediator and ligand **L7** as the external amino acid ligand in this protocol. This reaction was also further extended to arylation of mandelic acid and phenyl glycine at the *meta*-position [65].

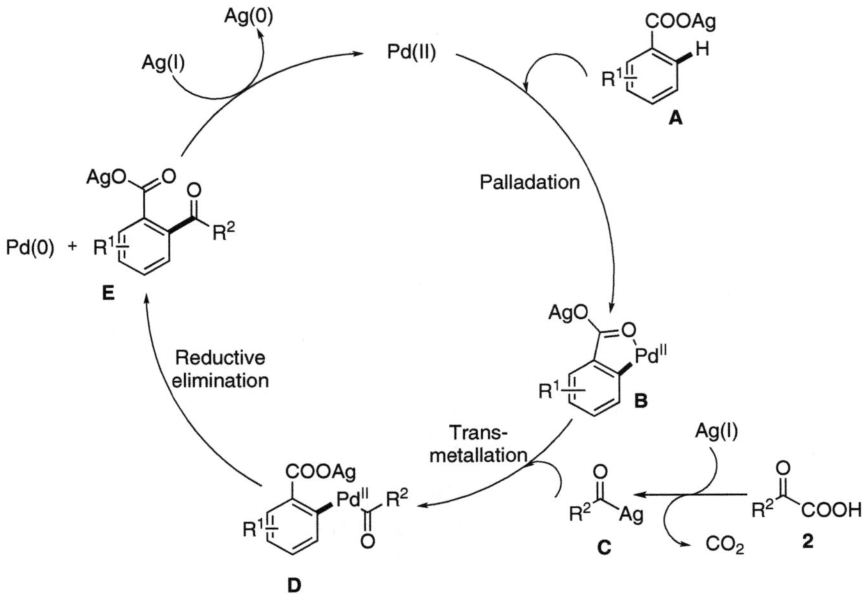

Scheme 2.18 Plausible mechanism of carboxylate-directed C(sp^2)—H bond acylation.

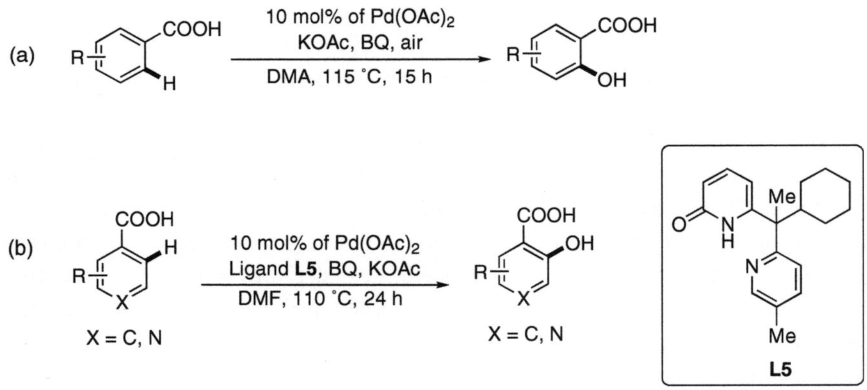

Scheme 2.19 Carboxylate-directed C(sp^2)—H bond hydroxylation.

Yu and his co-workers even proposed a plausible mechanism for their arylation reaction with aryltrifluoroboronates using ligand **L2**. They envisioned that the reaction has proceeded through Pd(II)/Pd(0) catalytic cycle, where the coordination of phenyl acetic acid along with C—H activation led to the formation of intermediate **B**. Subsequent transmetalation with aryltrifluoroboronate generated the intermediate **C**. Then, reductive elimination of **C** finally led to the desired arylation in good-to-excellent yields. Pd(0) was then reoxidized to Pd(II) either with Ag(I) salts or with oxygen and benzoquinone (Scheme 2.21) [62].

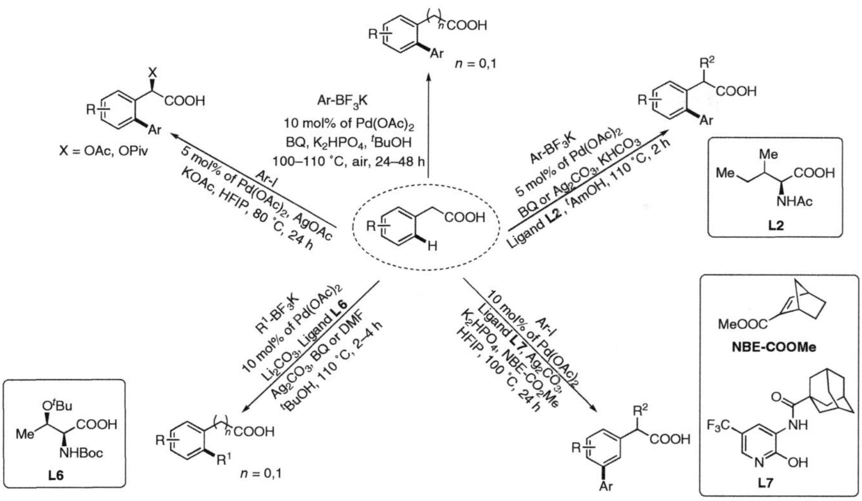

Scheme 2.20 Carboxylate-directed C(sp^2)–H bond arylation.

2.4.1.5 Ligand-assisted Pd-catalyzed Olefination of Substituted Phenyl Acetic Acids

In 2010, Yu group developed an external ligand-assisted olefination of substituted aryl acetic acids. They used benzoquinone as the external ligand source and oxygen as the suitable oxidant [66]. It was discovered later that the use of the external ligand **L2** along with benzoquinone further brought a lot of impact on the yield of the olefination reaction. Based on this observation, the C—H activation mechanism was changed from electrophilic cyclopalladation to concerted metalation deprotonation, which they highlighted in one more article in 2010 [67].

A similar concept has also been utilized in the alkenylation reaction of aryl acetic acids using external ligand **L8**. The utilization of ligand **L8** led to the diolefinated product in presence of oxygen, $Pd(OAc)_2$, and $KHCO_3$. Sequential *ortho*-C(sp^2)—H olefination has also been reported using the same set of reaction conditions (Scheme 2.22) [68].

External ligand **L2**-assisted concerted metalation deprotonation pathway of C—H activation was explained by Yu group in their work where they performed olefination of aryl acetic acids using ligand **L2** [67]. The transformation started with the coordination of substrate **1** to Pd(II), and then carboxylate-directed C—H cleavage generated the cyclopalladacycle intermediate **A**. The olefin insertion to **A** led to the formation of intermediate **C**, which underwent β-hydride elimination to form the olefinated product (**3**). The Pd(0) species was further oxidized to Pd(II) by using oxygen as the oxidant, which would initiate the next catalytic cycle (Scheme 2.23).

2.4.1.6 Pd-catalyzed *ortho*-C(sp^2)–H Functionalizations of Phenyl Acetic Acids

In 2010, Yu group reported mono-selective iodination of aryl acetic acids by utilizing $PhI(OAc)_2$ and I_2 as the suitable iodinating agents in the absence of light (Scheme 2.24) [69]. The *ortho*-C(sp^2)—H iodinated aryl acetic acids were further

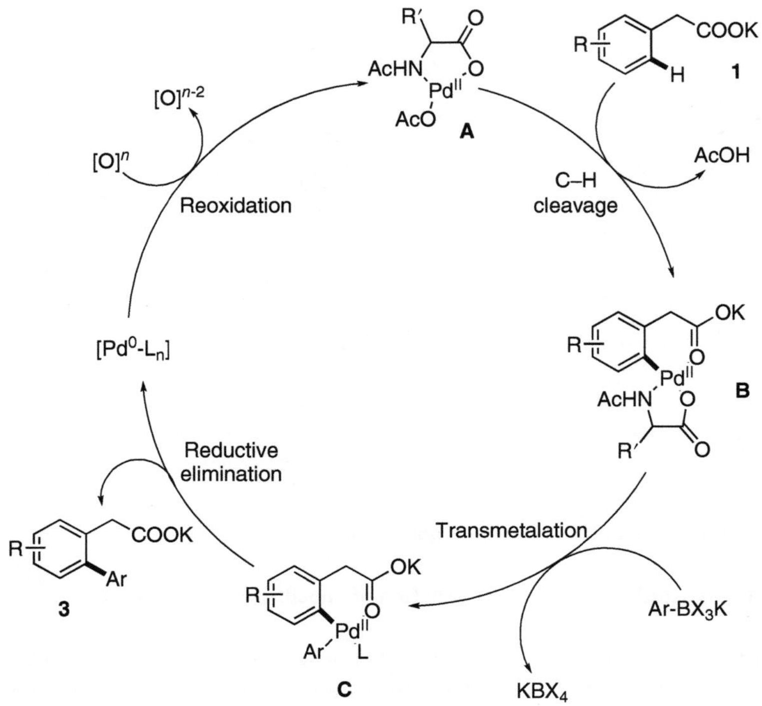

Scheme 2.21 Plausible mechanism of external ligand-assisted cross-coupling of C(sp^2)—H bond with arylboron reagents.

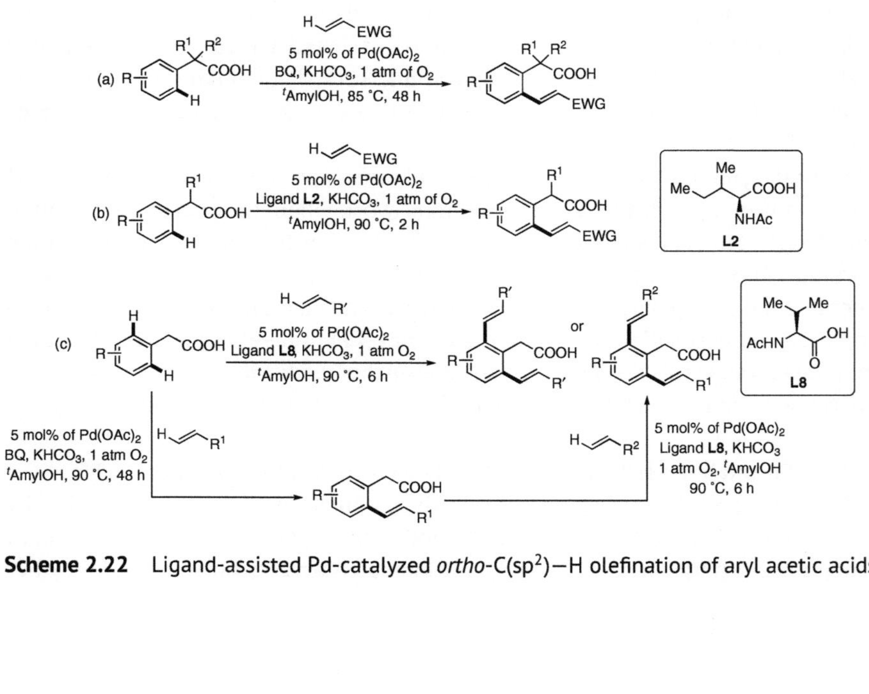

Scheme 2.22 Ligand-assisted Pd-catalyzed *ortho*-C(sp^2)—H olefination of aryl acetic acids.

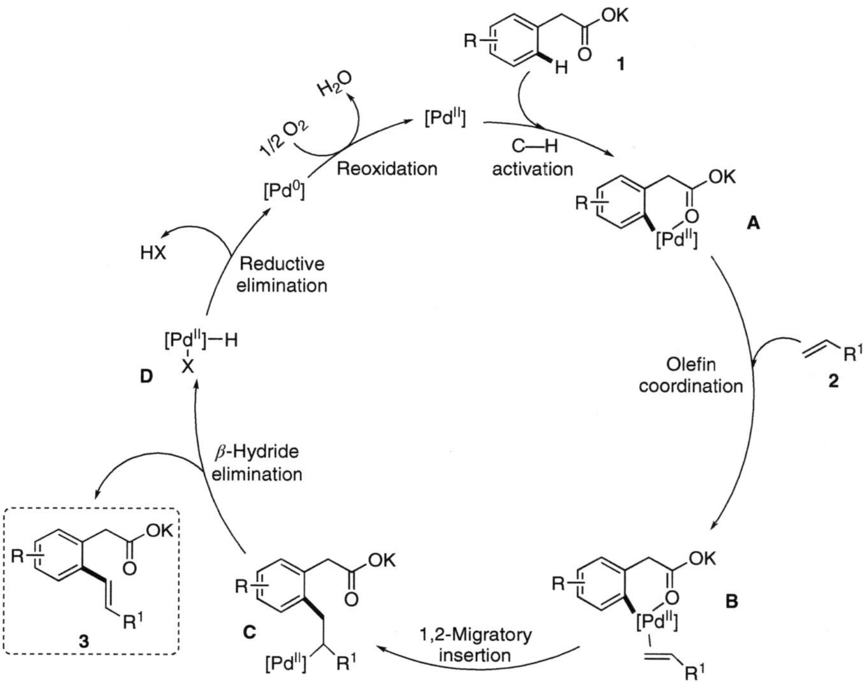

Scheme 2.23 Mechanism associated with ligand-assisted Pd-catalyzed *ortho*-C(sp^2)–H olefination of aryl acetic acids.

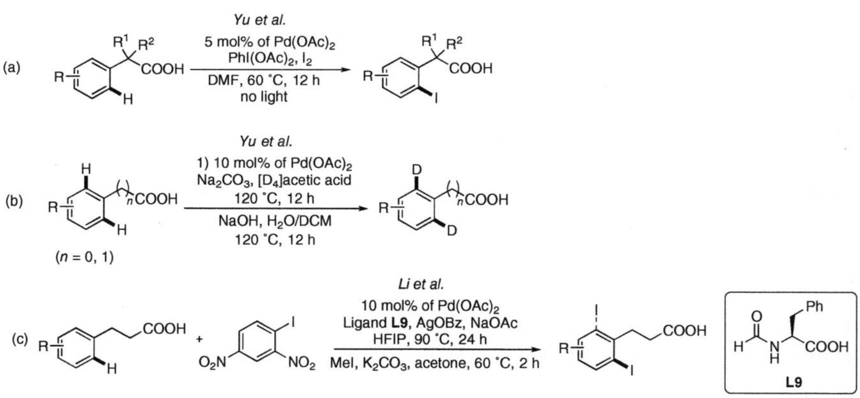

Scheme 2.24 Pd-catalyzed carboxylate assisted various *ortho*-C(sp^2)–H functionalization reactions.

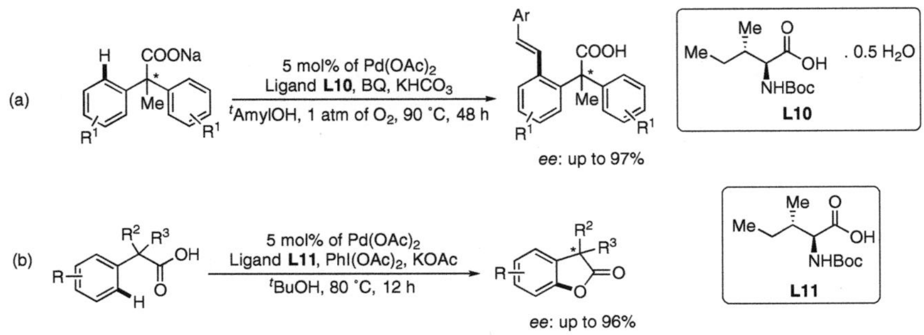

Scheme 2.25 Pd-catalyzed enantioselective *ortho*-C(sp^2)—H functionalization of diphenylacetic acids.

exposed to various transformations that led to the synthesis of many commercial drugs. In 2014, the same group reported *ortho*-C(sp^2)—H deuteration using [D_4] acetic acid, and numerous deuterated benzoic acid and phenyl acetic acids were synthesized in good-to-excellent yields [70]. Recently, Li group reported the *ortho*-C(sp^2)—H iodination of hydrocinnamic acids using ligand **L9** as the external ligand and aryl iodide as the iodinating agent [71].

The enantioselective version of carboxylate-assisted *ortho*-C(sp^2)—H functionalization was first introduced by Yu group. In 2010, they developed a strategy for enantioselective olefination of diphenyl acetic acids using ligand **L10** as the chiral ligand [72]. Similarly, in 2013, they adopted the same strategy and utilized ligand **L11** as the chiral ligand to synthesize chiral benzofuranone via intramolecular C—O bond formation (Scheme 2.25) [73].

2.4.2 Benzylic C(sp^3)—H Activation of Carboxylate Directing Group

2.4.2.1 External ligand-assisted Benzylic C(sp^3)—H Activation of Carboxylate Motifs

Toward carboxylate-assisted benzylic-C(sp^3)—H activation, Martin group in 2011 developed a novel strategy for the synthesis of biologically active benzolactone motifs in excellent yields via the assistance of an external amino acid ligand **L12** (Scheme 2.26) [74]. They even performed various mechanistic studies and revealed that C(sp^3)—O bond-forming reductive elimination step was the rate-determining step. They also demonstrated a mechanistic cycle where Pd(II)/Pd(0) catalytic cycle was involved. After that, recently in 2020, Pd-catalyzed benzolactone formation was reported by Yu and his co-workers with the help of ligand **L13**, which is a pyridone-based ligand [75]. In the same year, Yu group also carried out successful benzylic C(sp^3)—H olefination of *o*-toluic acids in order to synthesize δ-lactones in excellent yields by utilizing amino acid ligand **L4** as the external ligand source [76].

Based on the mechanistic studies, Martin group proposed a catalytic cycle where Pd(II) binds with silver benzoate to produce the intermediate **B**. Intermediate **B** formed the palladacycle intermediate **C**, which will lead to C(sp^3)—O bond reductive elimination to form the desired benzolactone (Scheme 2.27) [74].

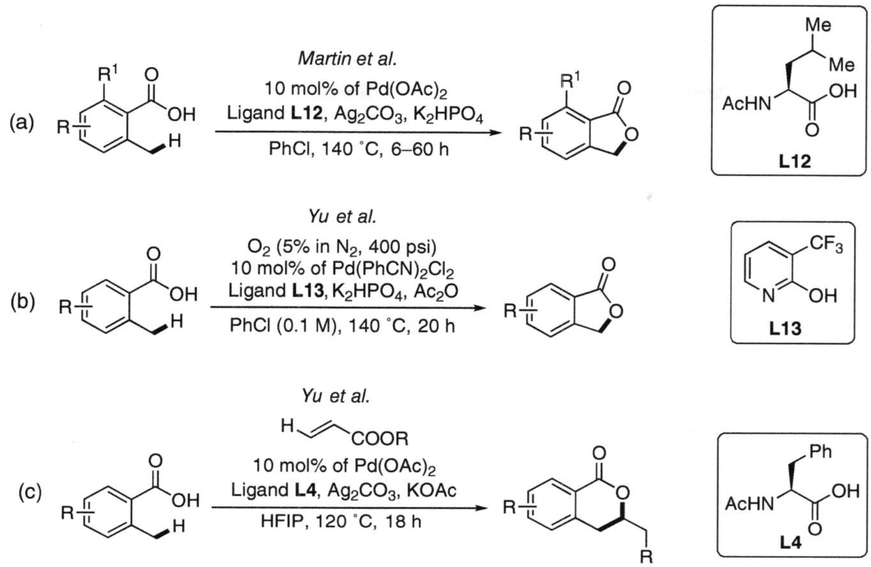

Scheme 2.26 Pd-catalyzed external ligand-assisted lactone-formation of benzylic-C(sp^3)–H bonds.

2.4.3 Pd-catalyzed C(sp^3)–H Activation of Aliphatic Acids Assisted by In-built Carboxylate Group

Aliphatic C(sp^3)—H bonds are comparatively stable and non-polar; also, the absence of any π-conjugation in the aliphatic system makes the cleavage of such C—H bonds difficult. So, to perform Pd-catalyzed functionalization of such C(sp^3)—H bonds more easily, directing group-assisted C—H activation has been exploited. Since carboxylate moiety coordinates weakly to Pd-salts, external ligands were utilized to increase the binding capability of carboxylates and further facilitate the cleavage of C(sp^3)—H bonds. Various research groups have contributed largely to the significant functionalization of proximal ($-\beta$) and distal ($-\gamma$ and -δ) C(sp^3)—H bonds. The initial report related to C(sp^3)—H bond functionalization was in 2006, where White and co-workers revealed an interesting strategy of C(sp^3)—H activation via Pd-catalyzed intramolecular lactonization of ω-alkenoic acids utilizing $Pd(OAc)_2$ with phenyl bis-sulfoxide as the catalyst (Scheme 2.28) [77]. In 2011, the group also further extended this protocol toward the synthesis of an erythromycin precursor [78].

2.4.3.1 Pd-catalyzed Carboxylate-assisted Arylation of Proximal Aliphatic C(sp^3)–H Bonds

The first report on aliphatic C(sp^3)—H bond functionalization was illustrated by Yu group in 2007 [47]. They carried out arylation reactions with the help of phenyl boronates in the presence of $Pd(OAc)_2$ as the catalyst and without any external ligand. They observed that this protocol was only applicable to α-bearing quaternary centers, and selectively only mono-arylation was observed under the given reaction

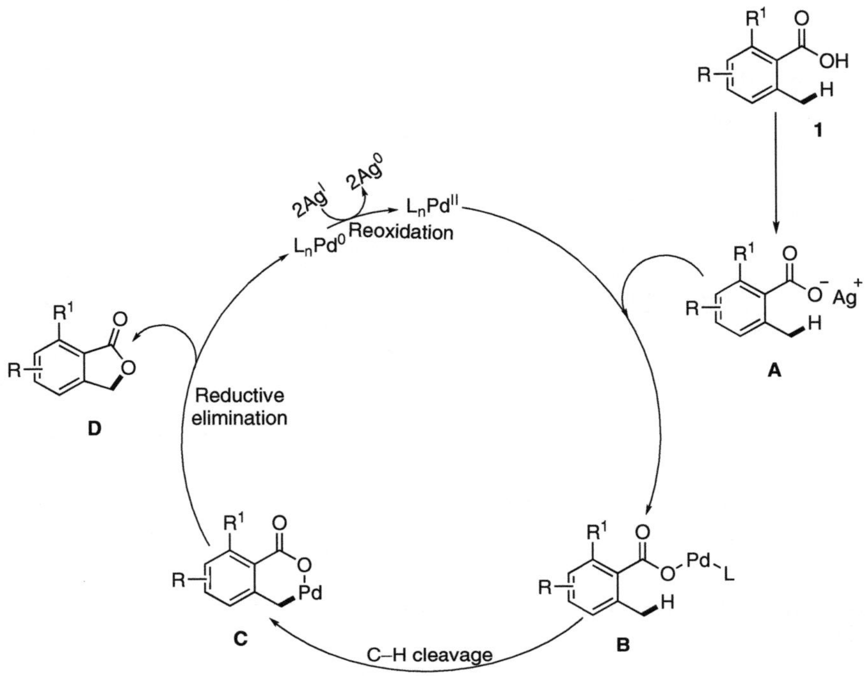

Scheme 2.27 Pd-catalyzed external ligand-assisted lactone formation of benzylic-C(sp^3)–H bonds.

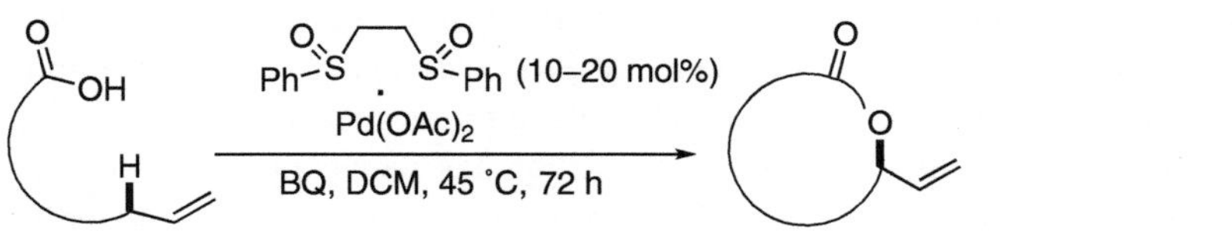

Scheme 2.28 Pd-catalyzed external ligand-assisted intramolecular lactonization.

parameters. Further, they also performed arylation under almost the same reaction parameters with the help of aryl iodides as the potential coupling partner, and mono- and di-arylation was observed in good yields. They also proposed that the reaction must have followed through Pd(II)/Pd(IV) catalytic cycle (Scheme 2.29).

After many years in 2017, Zhao group made an important contribution to the field of carboxylate-assisted β-C(sp^3)—H bond functionalization and reported β-C(sp^3)—H bond arylation with the help of acetyl-protected glycine ligand **L14** (Scheme 2.30) [79]. The arylated aliphatic acids were obtained in good-to-excellent yields, and also, the substrates bearing α-hydrogen were found to be very compatible with the reaction conditions. In the same year, Gemmeren group demonstrated β-C(sp^3)—H bond arylation with the assistance of protected β-alanine ligand **L15** as the exogenous ligand, and the desired arylated acids were observed in good-to-excellent yields under the reaction conditions [80]. Also, aliphatic acids bearing α-hydrogens were also found to be suitable for the reaction.

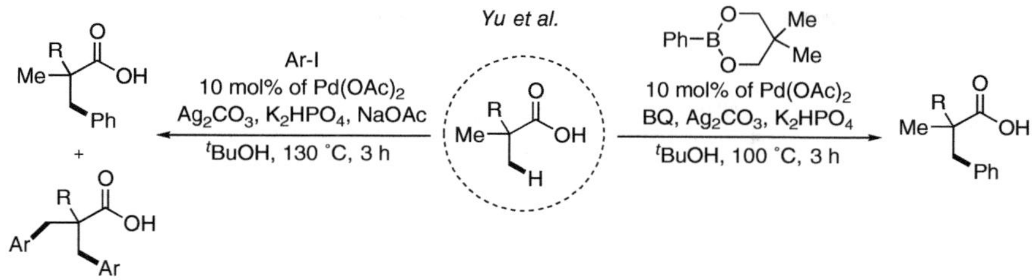

Scheme 2.29 Pd-catalyzed external ligand-free arylation of aliphatic carboxylic acids.

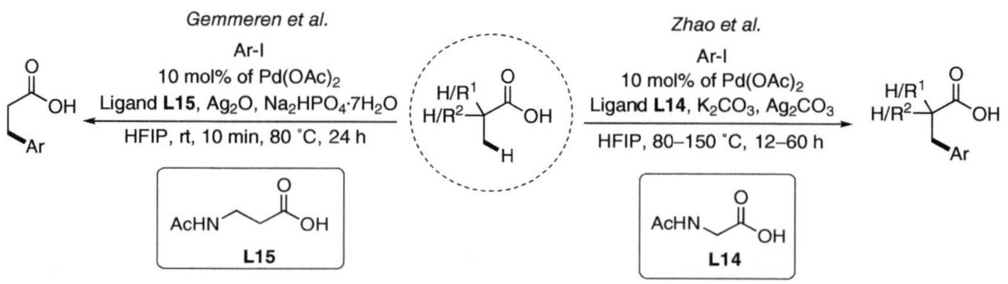

Scheme 2.30 Pd-catalyzed external ligand-assisted arylation of aliphatic carboxylic acids.

2.4.3.2 Pd-catalyzed External Ligand-assisted Lactonization of Proximal C(sp^3)–H Bonds

In 2018, Yu and co-workers designed a new thioether-based ligand **L16** to synthesize five-membered lactone scaffolds in good-to-excellent yields [81]. They envisioned that the reaction proceeded through Pd(II)/Pd(0) catalytic cycle, where initially, olefination occurred at the β-C(sp^3)—H bond of the aliphatic carboxylic acid followed by subsequent lactonization. Later on, in 2020, the same group adopted a new strategy for the intramolecular lactonization of substituted pivalic acids through Pd(II)/Pd(IV) catalytic cycle, where protected amino acid **L17** was used as the exogenous ligand [82]. In the reaction, TBHP was used for the oxidation of Pd(II) to Pd(IV) in the catalytic cycle. It was envisioned that the intermediate **IV** would have been formed in the reaction and would further undergo reductive elimination to generate the desired four-membered lactone. A large variety of substituted aliphatic acids were found to be compatible with this strategy that resulted in the diverse set of β-lactones in good-to-excellent yields (Scheme 2.31).

2.4.3.3 Pd-catalyzed Carboxylate-assisted β-C(sp^3)–H Acetoxylation

In 2019, Gemmeren group designed a novel strategy for β-C(sp^3)—H acetoxylation of aliphatic carboxylic acids without the help of any external ligand where PhI(OCOR)$_2$ was utilized as an acetoxylating agent and the sodium salt of HFIP was used as an additive (Scheme 2.32) [83]. The only drawback with this strategy is that this could only be applied to substrates bearing α-quaternary centers. So, in order to combat the limitations of this protocol, Yu group planned to use one external ligand **L18** in the β-C(sp^3)—H acetoxylation reaction [84]. In view of this, Yu group

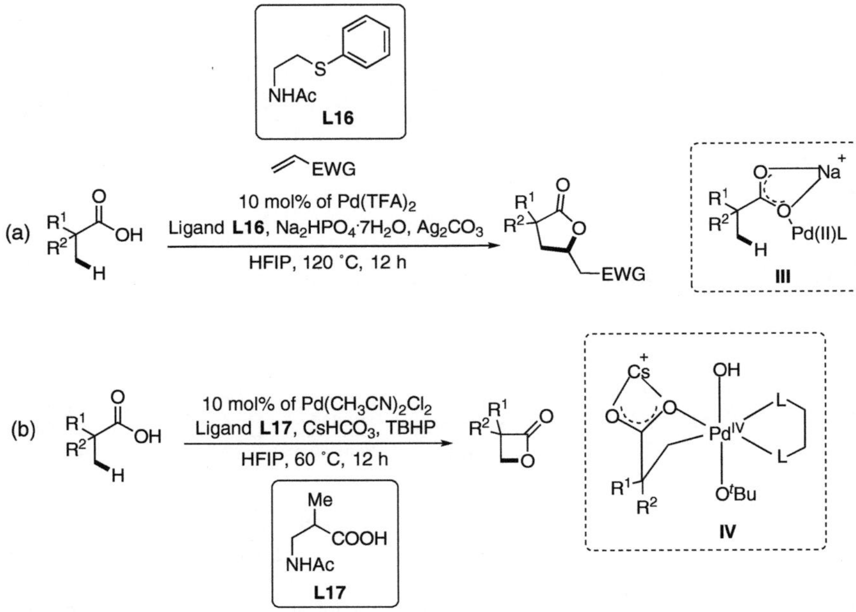

Scheme 2.31 Pd-catalyzed external ligand-assisted lactonization reactions.

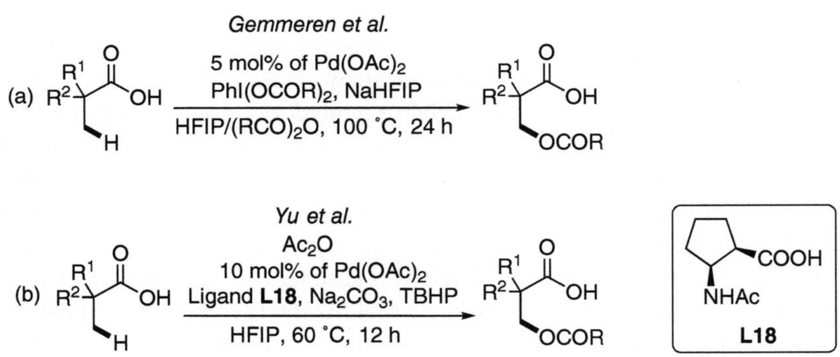

Scheme 2.32 Pd-catalyzed external ligand-assisted acetoxylation reactions.

in 2020 reported β-acetoxylation reaction where acetic anhydride was used as the potential acetoxylating agent and TBHP was used as an oxidizing agent. They even proposed a catalytic cycle for the successful acetoxylation of aliphatic acids, where they claimed that the reaction proceeded through Pd(II)/Pd(IV) catalytic cycle.

The catalytic cycle initiated with the ligand coordination with $Pd(OAc)_2$ that generated the active Pd(II) species. The sodium salt of the substituted aliphatic acid resulted in the formation of intermediate **A** after the β-C(sp^3)—H activation. Then, the oxidative addition of intermediate **A** via TBHP led to the formation of intermediate **B**. The intermediate **B** after ligand exchange with acetic anhydride further

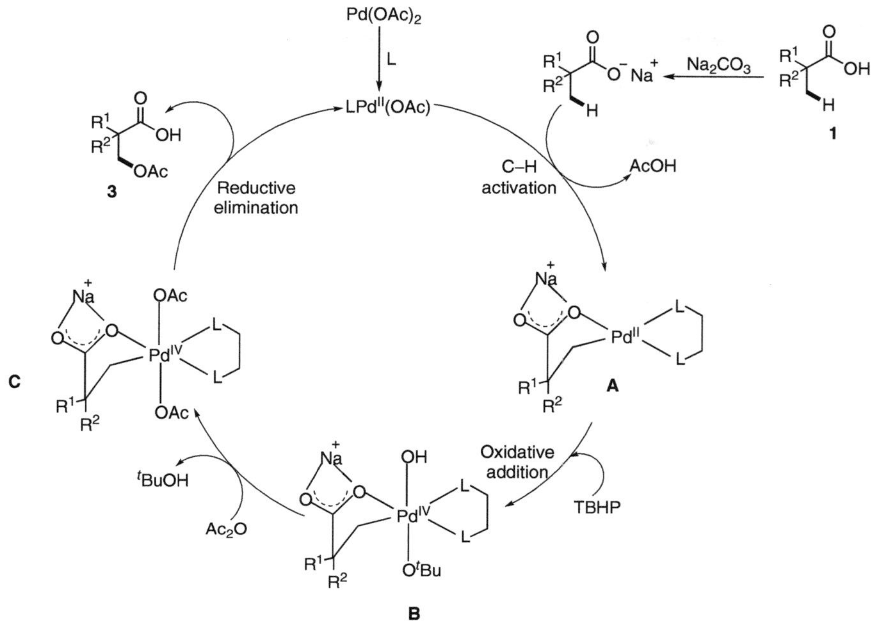

Scheme 2.33 Proposed mechanism for Pd-catalyzed β-C(sp^3)–H acetoxylation of aliphatic acids.

generated the intermediate **C**. Intermediate **C** would finally lead to the β-C(sp^3)—H acetoxylated product **3** after the reductive elimination of **C** (Scheme 2.33) [84].

Over many years, there has been a lot of contribution from Yu and Gemmeren groups on arylation, acetoxylation, and lactonization reactions of carboxylic acids that proceeded via β-C(sp^3)—H bond activation followed by desired functionalization reactions. So, toward carboxylate-directed β-C(sp^3)—H alkynylation reactions, Gemmeren group in 2020 designed an interesting protocol where they utilized protected ethylenediamine-based ligands **L19** and **L20** [85]. The external ligands aided in the formation of the desired alkynylation reaction at the β-position in good to excellent yields (Scheme 2.34). The carboxylic acids with or without α-quaternary centers were found to be highly compatible with the designed protocol. Subsequently, in 2021, the same group performed β-C(sp^3)—H deuteration of substituted pivalic acids using deuterated hexafluoroisopropanol as the deuterating source and ligand.

2.4.3.4 Pd-catalyzed β-C(sp^3)–H Alkynylation and Deuteration of Free Carboxylic Acids

L21 as the suitable exogenous ligand [86]. In comparison with the previous reports, for the first time, the β-C(sp^3)—H bond of unactivated and sterically hindered methylene moiety was also functionalized under the deuteration reaction conditions.

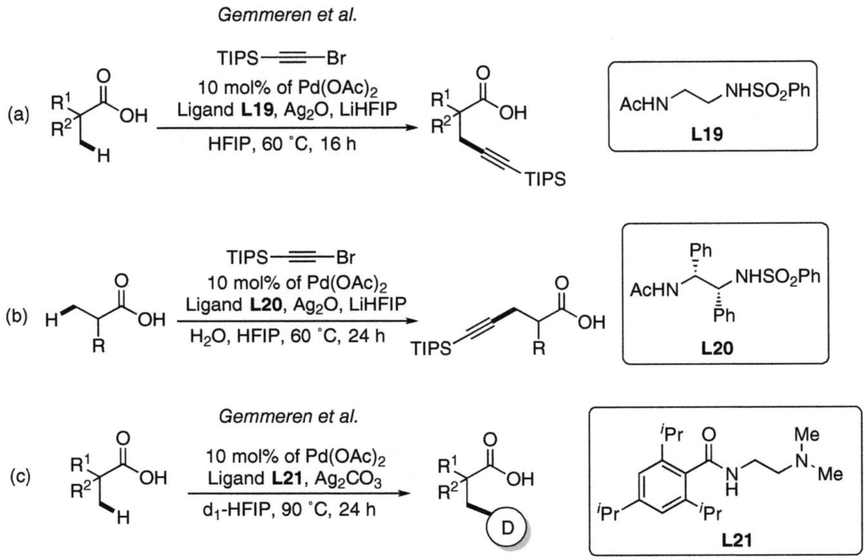

Scheme 2.34 Pd-catalyzed β-C(sp^3)—H functionalization of aliphatic acids.

In 2019, Maiti and co-workers reported distal γ-C(sp^3)—H bond functionalization with the aryl iodides in the presence of the ligand **L14** and a silver additive that resulted in monoarylation at the γ-position in good-to-excellent yields (Scheme 2.35) [87]. The monoarylated product was further treated with similar reaction parameters but at high temperature to synthesize γ-diarylated butyric acids in excellent yields. Subsequently, in 2020, Shi group reported γ-C(sp^3)—H arylation of substituted amino acids and peptide derivatives aided by the external ligand **L22** and silver phosphate as the additive that has been proposed to have played dual role [88].

2.4.3.5 Pd-catalyzed Ligand-assisted Distal C(sp^3)—H Bond Arylation

Maiti group even proposed a mechanism on how the γ-C(sp^3)—H bond of substituted butyric acids was selectively activated on the basis of the exogenous ligand **L14**. They claimed that the mechanism proceeded through Pd(II)/Pd(IV) catalytic cycle, where initial C—H activation of **1** led to the formation of six-membered palladacycle **A**. The palladacycle **A** gave rise to intermediate **B** after oxidative addition of aryl iodides. The intermediate **B** after reductive elimination generated the final γ-arylated product **2**, and the Pd(II) further initiated the next catalytic cycle (Scheme 2.36) [87].

2.4.3.6 Pd-catalyzed Ligand-assisted Distal C(sp^3)—H Bond Lactonization

In 2020, Gemmeren group, Yu group, and Maiti group demonstrated γ-C(sp^3)—H bond olefination followed by lactonization of substituted butyric acids that led to the formation of δ-lactones in good-to-excellent yields (Scheme 2.37). The very first contribution toward this field was by Gemmeren group who designed the ligand **L15** and revealed that the reaction followed the Pd(II)/Pd(0) catalytic cycle for

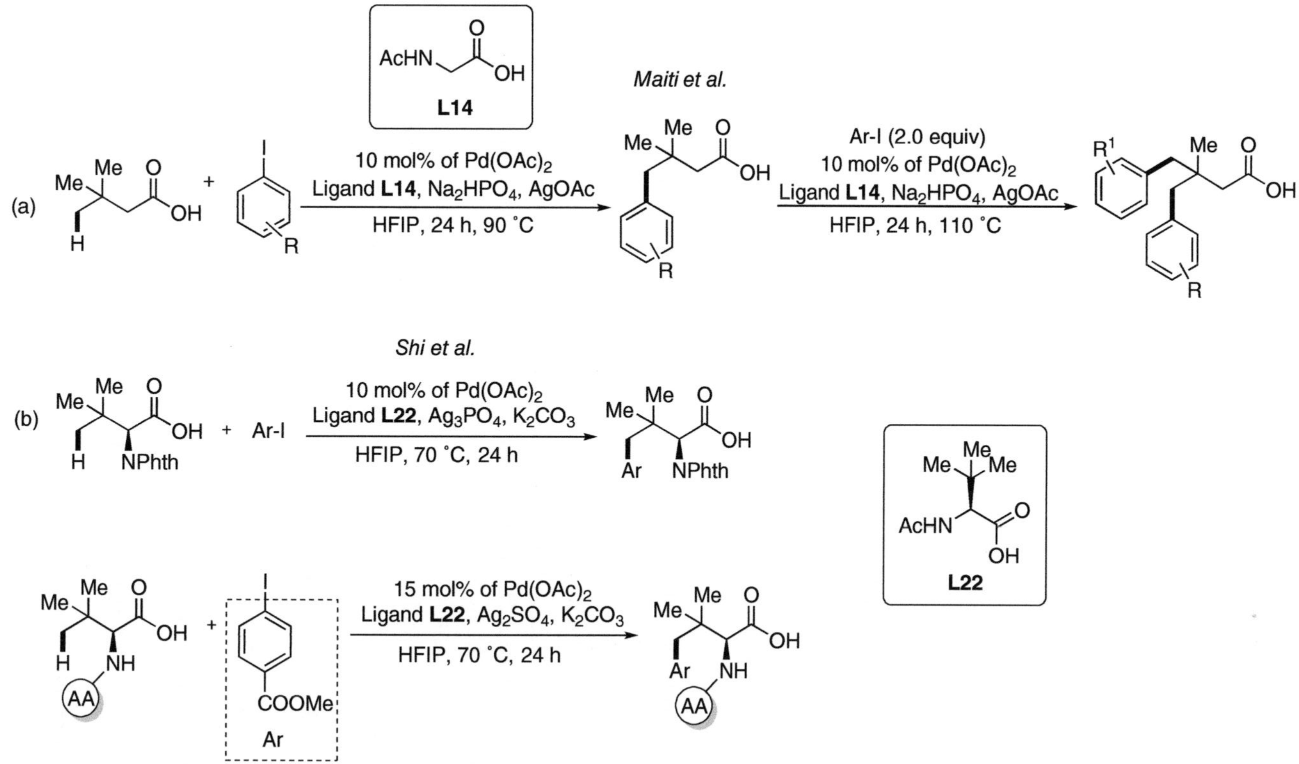

Scheme 2.35 Pd-catalyzed γ-C(sp^3)−H bond arylation of aliphatic acids.

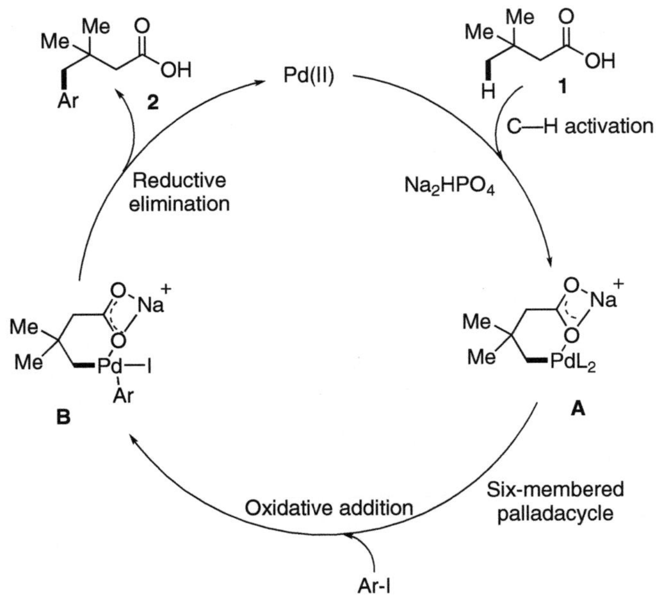

Scheme 2.36 Plausible mechanism of Pd-catalyzed γ-C(sp^3)–H arylation of aliphatic acids.

γ-C(sp^3)—H olefination reaction [89]. In the same year, Maiti group and Yu group also reported γ-C(sp^3)—H bond olefination followed by intramolecular Michael addition by utilizing ligand **L8** and ligand **L4** [76, 90]. Maiti group also utilized maleimides along with the acrylates as the coupling partners for the following lactonization process. Based on the reaction conditions, Maiti group also demonstrated a mechanism where Pd(II)/Pd(0) catalytic cycle was followed (Scheme 2.38) [90]. Pd(II) initially coordinated with **1** to generate the intermediate **A**, which would then undergo C—H activation to form the intermediate **B**. Then, olefin coordination to intermediate **B** would lead to intermediate **C** formation, which would then undergo migratory insertion to form the cyclopalladacycle **D**. The palladacycle **D** would then undergo β-hydride elimination to form the olefinated product **E**, followed by intramolecular Michael addition to form the desired lactonized moiety.

In 2020, Gemmeren group reported γ-C(sp^3)—H alkynylation of butyric acid derivatives with the help of protected ethylene diamine-based ligand **L20** as the suitable ligand and LiHFIP as the additive. Various butyric acid derivatives were found to be compatible with the reaction parameters, and the desired alkynylation was also observed in good-to-excellent yields (Scheme 2.39) [85].

2.4.3.7 Pd-catalyzed Enantioselective Carboxylate-directed C(sp^3)–H Activation

The first enantioselective C(sp^3)—H functionalization was performed by Yu group in 2018. They particularly designed the chiral ligand **L23** to carry out enantioselective arylation with the aryl iodides at the β-C(sp^3)—H bond of cyclopropane

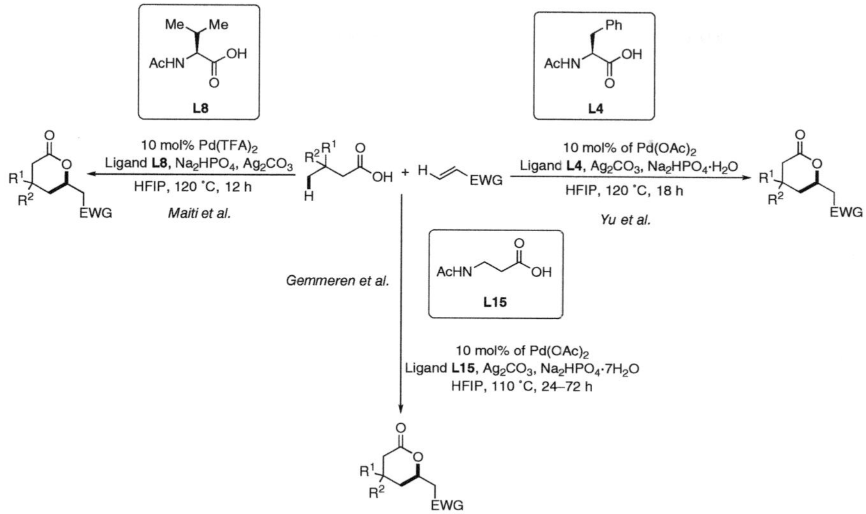

Scheme 2.37 Pd-catalyzed external ligand-assisted γ-C(sp^3)—H olefination of aliphatic acids.

carboxylic acids (Scheme 2.40) [91]. Later on, in 2019, a similar methodology was also applied to cyclopropane and cyclobutane carboxylic acids where the chiral ligands **L24**, **L25**, and **L26** helped in the formation of enantioselective β-C(sp^3)—H monoarylation and monovinylation in good-to-excellent yields [92].

2.5 Aldehyde as the In-built Functional Group in C—H Activation

Aromatic aldehydes are ubiquitous structural motifs that are found in biologically active compounds and also serve as intermediates for the synthesis of important compounds. The inherent directing ability of the aldehyde functional group has been very well utilized by various researchers toward transition metal-catalyzed C—H functionalization reactions. In this chapter, we have selectively discussed the Pd-catalyzed C—H functionalization of aldehydes where the in-built nature of the functional group has been utilized [93].

The C—H activation process of aldehydes has been broadly classified into three types. The first strategy consists of masked aldehyde derivatives where the aldehyde moiety is initially converted to aldoxime or aldimine, which subsequently directs C—H activation. The aldehyde functional group could be further obtained by hydrolysis. The only drawback relies on the extra synthetic steps for the preparation of imines or oximes. The second strategy involves the usage of a transient mediator that consists of a catalytic directing group (transient directing group, TDG) that can reversibly bind to the aldehyde *in situ* and perform desired C—H activation. Various aniline derivatives and amino acids have been studied as transient directing groups for aldehydes. The third strategy involves the usage of directing ability of free

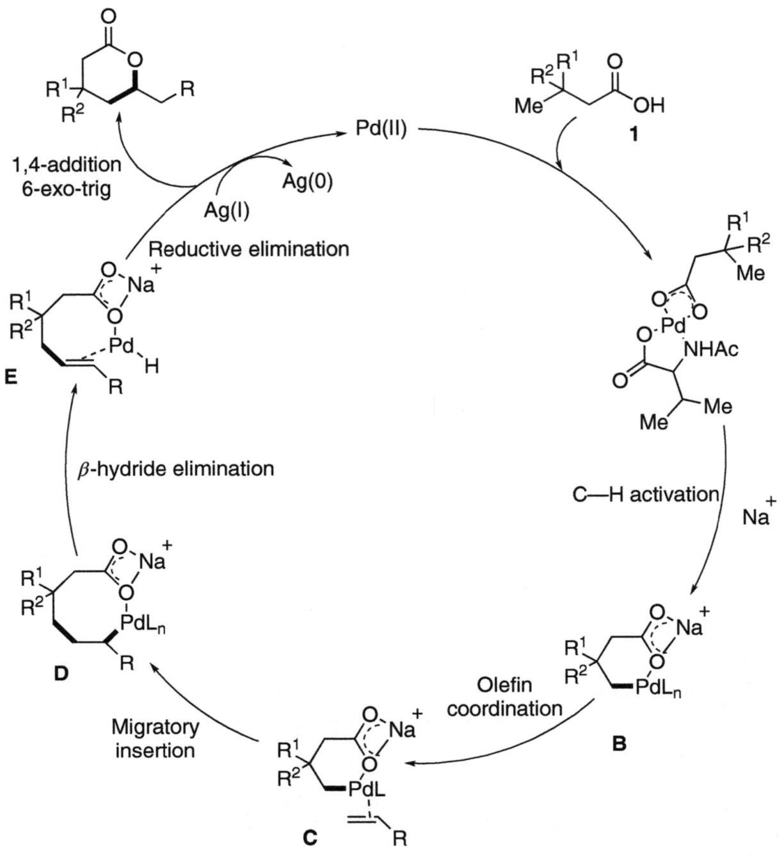

Scheme 2.38 Pd-catalyzed external ligand-assisted γ-C(sp^3)—H lactonization of aliphatic acids.

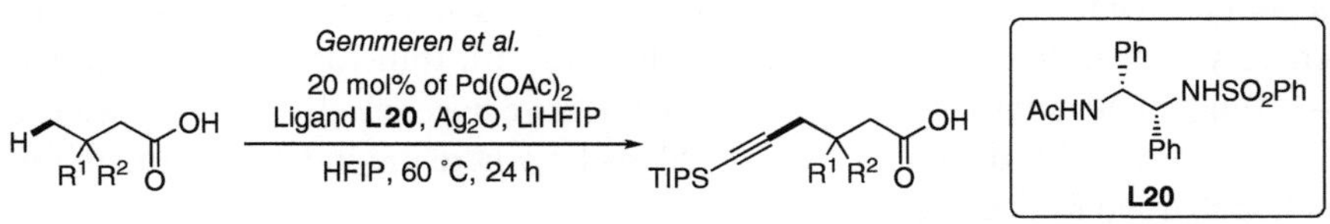

Scheme 2.39 Pd-catalyzed external ligand-assisted γ-C(sp^3)—H alkynylation.

aldehydes without the involvement of any mediator. This type of functionalization is still challenging as free aldehyde moiety is prone to overoxidation and decarbonylation under transition metal-catalyzed reactions. Hence, in this chapter, C—H functionalization involving the inherent directing ability of aldehydes has been discussed.

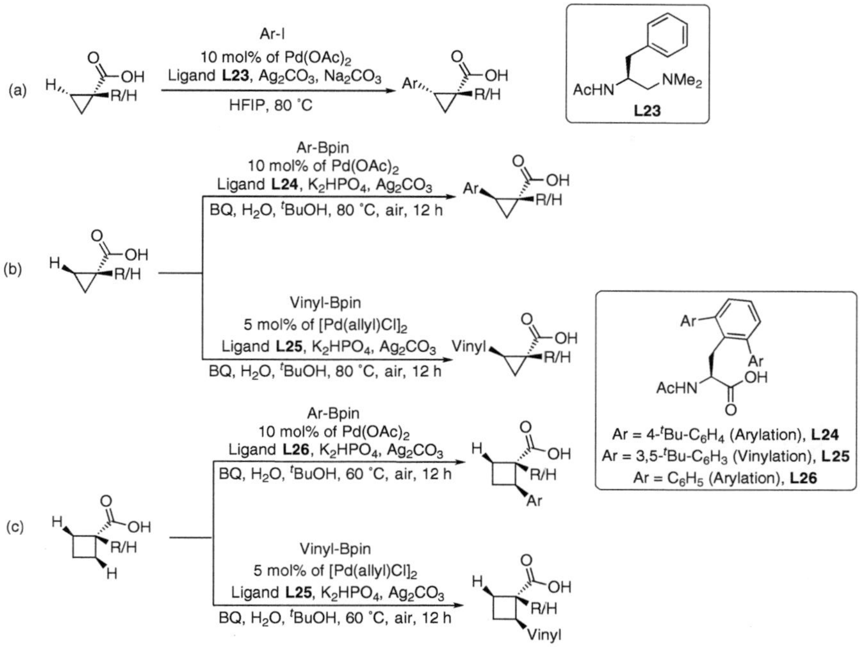

Scheme 2.40 Pd-catalyzed enantioselective arylation and vinylation of C(sp^3)—H bonds of aliphatic acids.

2.5.1 Pd-catalyzed C(sp^2)—H Functionalization of Free Aldehydes

The methods for Pd-catalyzed C—H activation of aromatic aldehydes utilizing the inherent nature have been rarely reported. In 2005, Cetinkaya and co-workers performed an efficient *N*-heterocyclic-carbene-based Pd-catalyzed selective mono- and di- *ortho*-arylation of benzaldehydes [94–96] (Scheme 2.41). In 2023, Saracoglu and co-workers developed a methodology for C4-arylation of free indole moiety with the assistance of aryl iodides via free aldehyde-directed C(sp^2)—H functionalization [97].

2.6 Sulfonic Acid as the In-built Functional Group in C—H Activation

2.6.1 Pd-catalyzed C(sp^2)—H Functionalization of Free Sulfonic Acids

In this chapter, we would also like to highlight the in-built nature of sulfonic acids. *p*-Toluenesulfonic acid has been found to increase the electrophilicity of Pd-center and has been used very often in Pd-catalyzed reactions, but the reports on the inherent directing nature of sulfonic acids are limited. In this regard, Guanwu and co-workers in 2012 revealed Pd-catalyzed synthesis of aromatic sultones via utilization of the inner directing nature of sulfonic acids [98, 99] (Scheme 2.42).

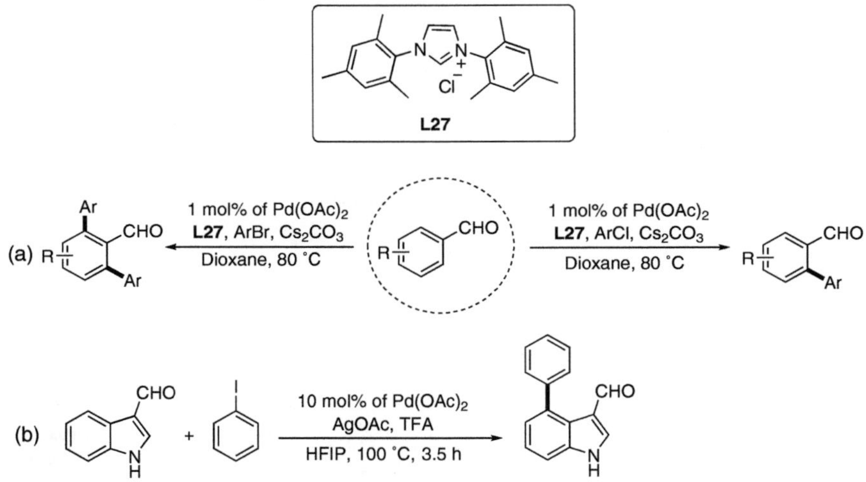

Scheme 2.41 Pd-catalyzed arylation of C(sp^2)–H bonds of aromatic aldehyde.

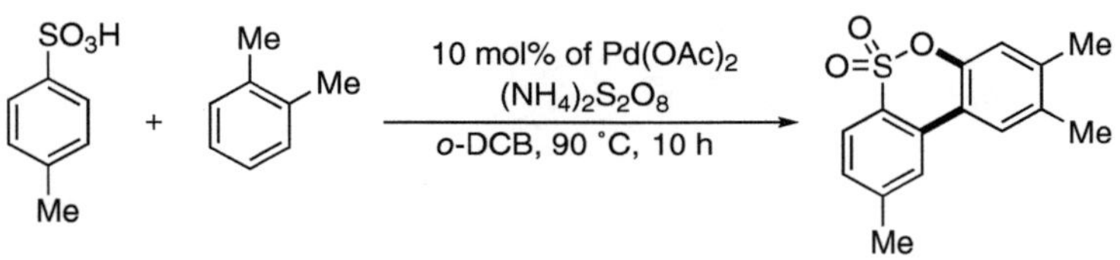

Scheme 2.42 Pd-catalyzed synthesis of aromatic sultones from TsOH.

2.7 Alcohols as the In-built Functional Group in C–H Activation

Alcohols are the most important functional groups found in a wide range of molecules, and they can also be easily transformed into many functional groups. In this chapter, we intend to focus mainly on the Pd-catalyzed inherent directing ability of alcohols toward selective C—H functionalization [100].

2.7.1 Phenethyl Alcohol as the In-built Functional Group

In 2010, Yu and co-workers began their exploration for the native directing nature of alcohols with the report of Pd-catalyzed C—H olefination reaction of phenethyl alcohols [101]. The overall yield of the olefinated product was very less, but this drawback was overcome by the usage of monoprotected amino acids ligand (+)-Menthyl(O_2C)-Leu-OH, which successfully facilitated the product yield. In the same year, Yu group reported phenethyl-alcohol-directed Pd-catalyzed synthesis of dihydrobenzofuran and its spirocyclic analogs with the help of $PhI(OAc)_2$, Li_2CO_3, and hexafluorobenzene solvent [102]. Further studies revealed that tertiary alcohols generated the desired products in high yields compared with the secondary alcohols. In the subsequent year, the same group reported phenethyl-alcohol-directed

C—H carbonylation in order to generate 1-isochromanones and its derivatives via the assistance of the external mono-*N*-protected amino acid ligands under CO (Scheme 2.43) [103].

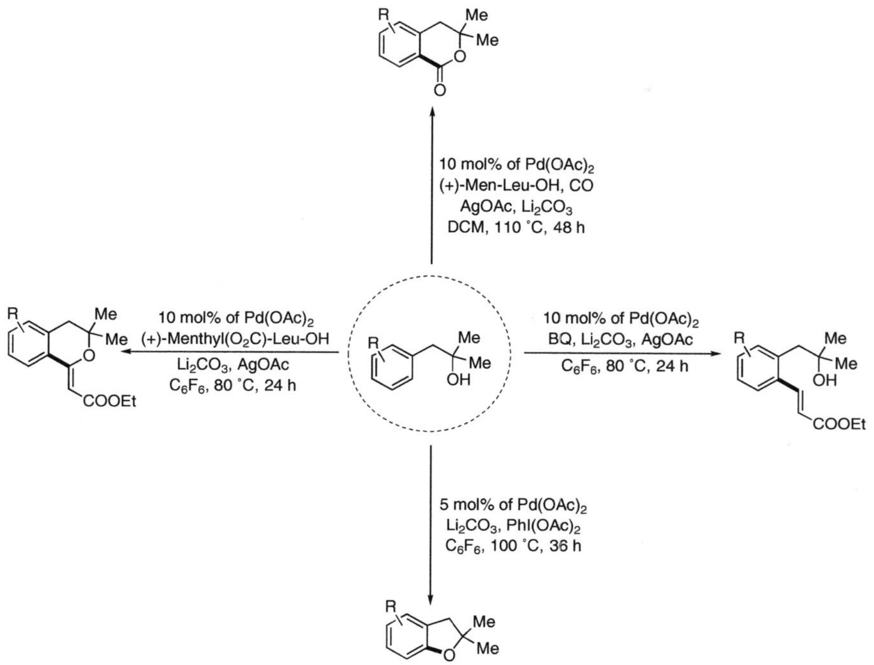

Scheme 2.43 Pd-catalyzed C(sp^2)—H bond functionalization assisted by phenethyl alcohol directing group.

2.7.2 Phenol as the In-built Functional Group

In 1997, Miura and co-workers performed Pd-catalyzed arylation of 2-aryl phenols using the phenolic —OH as the directing group with Cs_2CO_3 as the base [104]. In the same year, the same group reported phenol-directed cyclization with the help of $Pd(OAc)_2$ as the catalyst and $Cu(OAc)_2.H_2O$ as the oxidant in DMF solvent [105]. In 2014, Sun and co-workers reported a similar kind of phenol-directed C—H olefination under different set of reaction conditions. Olefination was observed instead of cyclization for 2-phenylphenols [106]. In 2011, Liu and co-workers reported Pd-catalyzed C—H activation followed by intramolecular C—O cyclization of 2-arylphenols to synthesize dibenzofurans [107]. In the same year, Yoshikai demonstrated a similar protocol for phenol-directed dibenzofuran formation under a different set of reaction conditions [108]. Later, in 2013, Shi group reported Pd-catalyzed carbonylative C—H cyclization to generate dibenzopyranones in good yields (Scheme 2.44) [109].

Zhu and co-workers in 2019 demonstrated Pd-catalyzed *ortho*-C—H alkenylation of phenols via the assistance of phenolic hydroxyl as the directing group [110]. Previously, in 2013, Iwasawa group disclosed phenolic-hydroxyl-directed carboxylation of alkenyl C(sp^2)—H bond with CO_2 to effectively generate coumarins

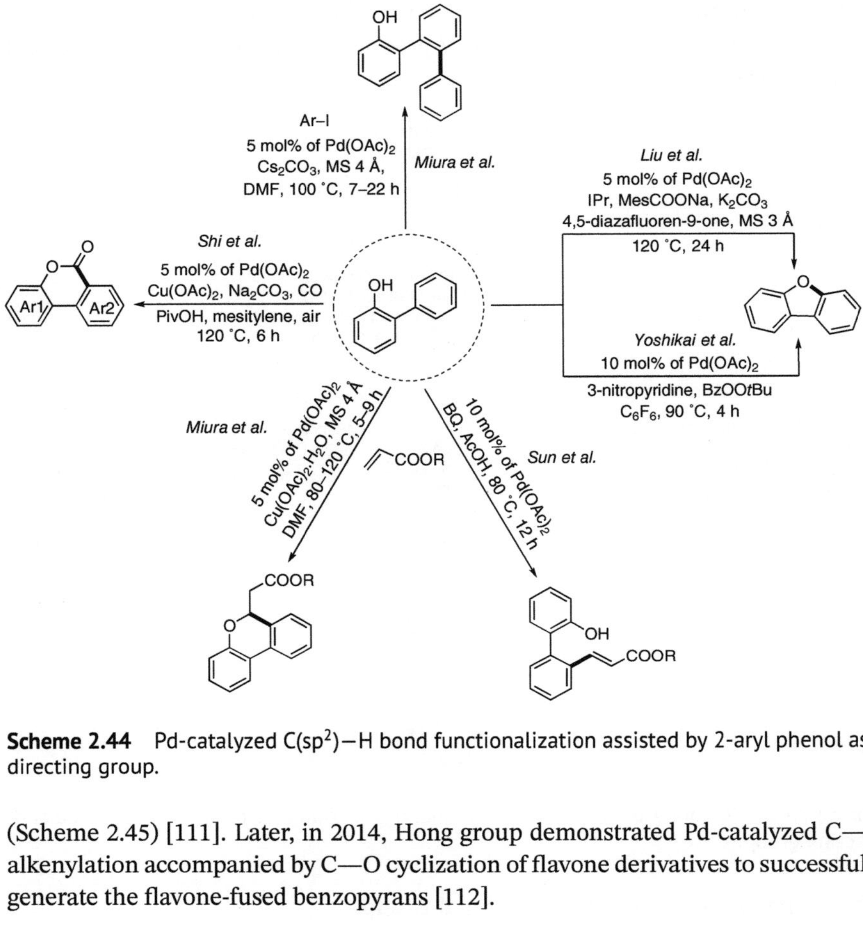

Scheme 2.44 Pd-catalyzed C(sp^2)–H bond functionalization assisted by 2-aryl phenol as a directing group.

(Scheme 2.45) [111]. Later, in 2014, Hong group demonstrated Pd-catalyzed C—H alkenylation accompanied by C—O cyclization of flavone derivatives to successfully generate the flavone-fused benzopyrans [112].

2.7.3 Hydroxyl Moiety of Salicylaldehyde as the In-built Functional Group

In 1996, Miura and co-workers demonstrated a Pd-catalyzed cross-coupling reaction of 2-hydroxybenzaldehydes with aryl iodides where the free hydroxyl assists the C—H activation of aldehyde [113]. After many years in 2009, Peng group demonstrated a methodology for the synthesis of xanthones from salicylaldehydes via Pd-catalyzed reactions (Scheme 2.46) [114].

2.7.4 Miscellaneous Examples of Free Alcohol as the In-built Functional Group

In 2013, Lam group demonstrated a different methodology for the synthesis of benzopyrans from 2-aryl-3-hydroxy-2-cyclo-hexenones. Ru was also found to be very effective as a catalyst for this transformation (Scheme 2.47) [115].

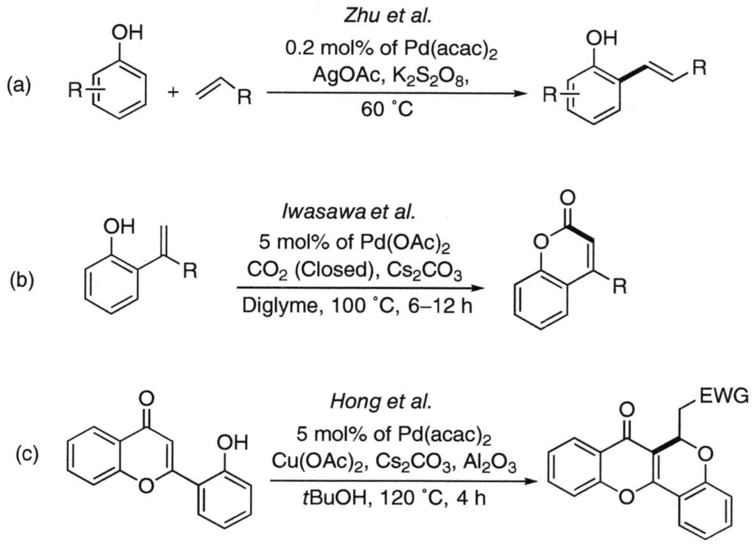

Scheme 2.45 Pd-catalyzed C(sp^2)—H bond functionalization assisted by phenolic-OH as a directing group.

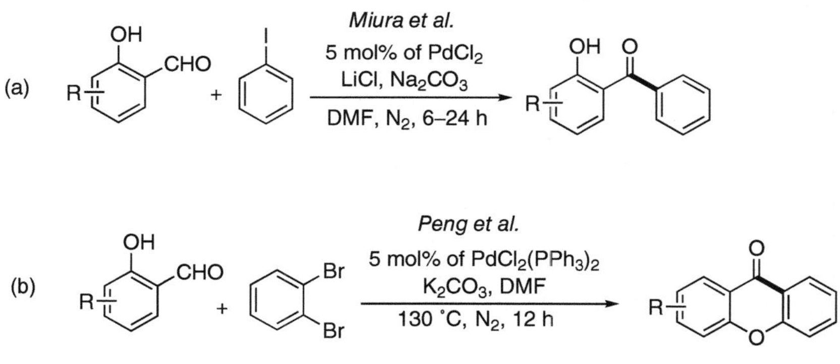

Scheme 2.46 Pd-catalyzed C(sp^2)—H bond functionalization of salicylaldehyde.

2.8 Conclusion

The acid and the amine functional moieties present in the molecules have served as the significant directing group in the Pd(II)-catalyzed chelation-assisted C—H bond functionalization chemistry. These groups bind to the Pd(II)-center and further assist in the selective desired functionalization of C(sp^2)—H and C(sp^3)—H bonds. Functionalization reactions such as arylation, alkylation, acetoxylation, and olefination reactions have been lucidly achieved via Pd(II)-catalyzed in-built directing

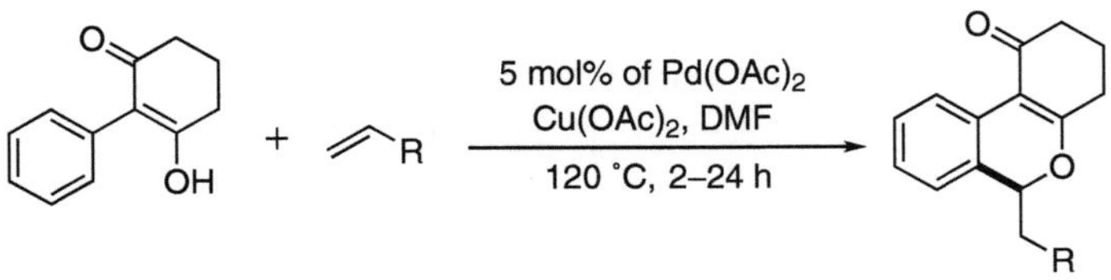

Scheme 2.47 Pd-catalyzed C–H alkenylation.

group chemistry. Apart from the acid- and amine-assisted C—H bond functionalization, this chapter has also highlighted the significant reports of Pd-catalyzed C—H bond functionalization reactions assisted by aldehydes, alcohols, sulfonic acids, and a few nitrogen-based heterocycles as the suitable in-built directing group. Hence, in this chapter, different strategies developed by various researchers in the Pd(II)-catalyzed in-built directing group-assisted chemistry have been summarized.

References

1 Jia, C., Kitamura, T., and Fujiwara, Y. (2001). Catalytic functionalization of arenes and alkanes via C—H bond activation. *Acc. Chem. Res.* 34 (8): 633–639.

2 Liu, C., Yuan, J., Lei, A. et al. (2015). Oxidative coupling between two hydrocarbons: an update of recent C–H Functionalizations. *Chem. Rev.* 115 (22): 12138–12204.

3 Le Bras, J. and Muzart, J. (2011). Intermolecular dehydrogenative heck reactions. *Chem. Rev.* 111 (3): 1170–1214.

4 Yu, J.Q. and Shi, Z.J. (2010). *C-H Activation*. New York: Springer.

5 Gensch, T., Hopkinson, M.N., Glorius, F. et al. (2016). Mild metal-catalyzed C–H activation: examples and concepts. *Chem. Soc. Rev.* 45 (10): 2900–2936.

6 Gandeepan, P., Müller, T., Zell, D. et al. (2019). 3d transition metals for C–H activation. *Chem. Rev.* 119 (4): 2192–2452.

7 Rej, S., Ano, Y., and Chatani, N. (2020). Bidentate directing groups: an efficient tool in C—H bond functionalization chemistry for the expedient construction of C—C bonds. *Chem. Rev.* 120: 1788–1887.

8 Rej, S., Das, A., and Chatani, N. (2021). Strategic evolution in transition metal-catalyzed directed C—H bond activation and future directions. *Coord. Chem. Rev.* 431: 213683.

9 Li, H., Lia, B.J., and Shi, Z.J. (2011). Challenge and progress: palladium-catalyzed sp^3 C-H activation. *Catal. Sci. Technol.* 1: 191–206.

10 Huang, Z. and Dong, G. (2014). Catalytic C—C bond forming transformations via direct β-C-H functionalization of carbonyl compounds. *Tetrahedron Lett.* 55: 5869–5889.

11 Liu, B., Engle, K.M., Shi, B.F. et al. (2021). Transition-metal-catalyzed, coordination-assisted functionalization of nonactivated $C(sp^3)$—H bonds. *Chem. Rev.* 121: 14957–15074.

12 Zhang, Q. and Shi, B.F. (2021). Site-selective functionalization of remote aliphatic C—H bonds via C–H metalation. *Chem. Sci.* 12: 841.
13 Li, Y., Zhang, Q., and Shi, B.F. (2021). Directing group assisted distal C(sp^3)–H functionalization of aliphatic substrates. In: *Remote C—H Bond Functionalizations: Methods and Strategies in Organic Synthesis* (ed. D. Maiti and S. Guin), 279–314. John Wiley & Sons, Ltd.
14 Ni, S.F., Li, M., Dang, L. et al. (2022). Recent advances in γ-C(sp^3)—H bond activation of amides, aliphatic amines, sulfanilamides and amino acids. *Coord. Chem. Rev.* 455: 214255.
15 He, J., Wasa, M., Yu, J.Q. et al. (2017). Palladium-catalyzed transformations of alkyl C—H bonds. *Chem. Rev.* 117: 8754–8786.
16 Dick, A.R., Hull, K.L., and Sanford, M.S. (2004). A highly selective catalytic method for the oxidative functionalization of C—H bonds. *JACS* 126 (8): 2300–2301.
17 Kalyani, D., Deprez, N.R., Sanford, M.S. et al. (2005). Oxidative C–H activation/C—C bond forming reactions: synthetic scope and mechanistic insights. *JACS* 127 (20): 7330–7331.
18 Kalyani, D. and Sanford, M.S. (2005). Regioselectivity in palladium-catalyzed C–H activation/oxygenation reactions. *Org. Lett.* 7 (19): 4149–4152.
19 Dick, A.R., Kampf, J.W., and Sanford, M.S. (2005). Platinum model studies for palladium-catalyzed oxidative functionalization of C—H bonds. *Organometallics* 24 (4): 482–485.
20 Hull, K.L., Anani, W.Q., and Sanford, M.S. (2006). Palladium-catalyzed fluorination of carbon–hydrogen bonds. *JACS* 128 (22): 7134–7135.
21 Hull, K.L. and Sanford, M.S. (2007). Catalytic and highly Regioselective cross-coupling of aromatic C–H substrates. *JACS* 129 (39): 11904–11905.
22 Desai, L.V., Stowers, K.J., and Sanford, M.S. (2008). Insights into directing group ability in palladium-catalyzed C—H bond functionalization. *JACS* 130 (40): 13285–13293.
23 Deprez, N.R. and Sanford, M.S. (2009). Synthetic and mechanistic studies of Pd-catalyzed C–H arylation with diaryliodonium salts: evidence for a bimetallic high oxidation state Pd intermediate. *JACS* 131 (31): 11234–11241.
24 Saha, S., Alsulami, Q.A., Maiti, D. et al. (2023). Native functional group directed distal C(sp^3)-H activation of aliphatic systems. *Catal. Sci. Technol.* 13: 11–27.
25 He, C., Whitehurst, W.G., and Gaunt, M.J. (2019). Palladium-catalyzed C(sp^3)—H bond functionalization of aliphatic amines. *Chem* 5: 1031–1058.
26 McNally, A., Haffemayer, B., Gaunt, M.J. et al. (2014). Palladium-catalysed C–H activation of aliphatic amines to give strained nitrogen heterocycles. *Nature* 510: 129–133.
27 Smalley, A.P., Cuthbertson, J.D., and Gaunt, M.J. (2017). Palladium-catalyzed enantioselective C–H activation of aliphatic amines using chiral anionic BINOL-phosphoric acid ligands. *JACS* 139 (4): 1412–1415.
28 Nappi, M., He, C., Gaunt, M.J. et al. (2018). Selective reductive elimination at alkyl palladium(IV) by dissociative ligand ionization: catalytic C(sp^3)–H amination to azetidines. *Angew. Chem. Int. Ed.* 130 (12): 3232–3236.

29 Calleja, J., Pla, D., Gaunt, M.J. et al. (2015). A steric tethering approach enables palladium-catalysed C–H activation of primary amino alcohols. *Nat. Chem.* 7: 1009–1016.

30 He, C. and Gaunt, M.J. (2015). Ligand-enabled catalytic C–H arylation of aliphatic amines by a four-membered-ring cyclopalladation pathway. *Angew. Chem. Int. Ed.* 54 (52): 15840–15844.

31 He, G., Zhao, Y., Chen, G. et al. (2012). Highly efficient syntheses of azetidines, Pyrrolidines, and Indolines via palladium catalyzed intramolecular amination of $C(sp^3)$—H and $C(sp^2)$—H bonds at γ and δ positions. *JACS* 134 (1): 3–6.

32 Nadres, E.T. and Daugulis, O. (2012). Heterocycle synthesis via direct C–H/N–H coupling. *JACS* 134 (1): 7–10.

33 Xu, J.W., Zhang, Z.Z., Shi, B.F. et al. (2016). Site-selective alkenylation of δ-$C(sp^3)$—H bonds with alkynes via a six-membered palladacycle. *JACS* 138 (34): 10750–10753.

34 Zhan, B.B., Li, Y., Shi, B.F. et al. (2018). Site-selective δ-$C(sp^3)$–H alkylation of amino acids and peptides with maleimides via a six-membered palladacycle. *Angew. Chem. Int. Ed.* 57 (20): 5858–5862.

35 Lin, H., Pan, X., Bannister, T.D. et al. (2019). Native directed site-selective δ-$C(sp^3)$–H and δ-$C(sp^2)$–H arylation of primary amines. *ACS Catal.* 9 (6): 4887–4891.

36 Pramanick, P.K., Zhou, Z., Yao, B. et al. (2019). Free amino group-directed γ-$C(sp^3)$–H Arylation of α-amino esters with diaryliodonium Triflates by palladium catalysis. *J. Org. Chem.* 84 (9): 5684–5694.

37 Rodrigalvarez, J., Nappi, M., Gaunt, M.J. et al. (2020). Catalytic $C(sp^3)$—H bond activation in tertiary alkylamines. *Nat. Chem.* 12: 76–81.

38 Zhuang, Z. and Yu, J.Q. (2020). Pd(II)-catalyzed enantioselective γ-$C(sp^3)$–H functionalizations of free cyclopropylmethylamines. *JACS* 142 (28): 12015–12019.

39 Rodrigalvarez, J., Reeve, L.A., Gaunt, M.J. et al. (2022). Pd(II)-catalyzed enantioselective $C(sp^3)$–H arylation of cyclopropanes and cyclobutanes guided by tertiary alkylamines. *JACS* 144 (9): 3939–3948.

40 Buettner, C.S., Willcox, D., Gaunt, M.J. et al. (2019). Mechanistic investigation into the $C(sp^3)$–H acetoxylation of morpholinone. *Chem. Sci.* 10: 83–89.

41 Chen, K., Wang, D., Shi, Z.J. et al. (2017). Palladium catalyzed $C(sp^3)$–H acetoxylation of aliphatic primary amines to γ-amino alcohol derivatives. *Org. Chem. Front.* 4: 2097–2101.

42 He, C. and Gaunt, M.J. (2017). Ligand-assisted palladium-catalyzed C–H alkenylation of aliphatic amines for the synthesis of functionalized pyrrolidines. *Chem. Sci.* 8: 3586–3592.

43 Cabrera-Pardo, J.R., Trowbridge, A., Gaunt, M.J. et al. (2017). Selective palladium(II)-catalyzed carbonylation of methylene β-C—H bonds in aliphatic amines. *Angew. Chem. Int. Ed.* 129 (39): 12120–12124.

44 Hogg, K.F., Trowbridge, A., Gaunt, M.J. et al. (2017). The α-tertiary amine motif drives remarkable selectivity for Pd-catalyzed carbonylation of β-methylene C—H bonds. *Chem. Sci.* 8: 8198–8203.

45 Png, Z.M., Cabrera-Pardo, J.R., Gaunt, M.J. et al. (2018). Diastereoselective C–H carbonylative annulation of aliphatic amines: a rapid route to functionalized γ-lactams. *Chem. Sci.* 9: 7628–7633.

46 Dutta, S., Maiti, D., Werz, D.B. et al. (2022). Pd-catalysed C–H functionalisation of free carboxylic acids. *Chem. Sci.* 13: 2551.

47 Giri, R., Maugel, N., Yu, J.Q. et al. (2007). Palladium-catalyzed methylation and arylation of sp^2 and sp^3 C—H bonds in simple carboxylic acids. *JACS* 129 (12): 3510–3511.

48 Chiong, H.A., Pham, Q.N., and Daugulis, O. (2007). Two methods for direct *ortho*-arylation of benzoic acids. *JACS* 129 (32): 9879–9884.

49 (a) Arroniz, C., Ironmonger, A., Larrosa, I. et al. (2013). Direct *ortho*-arylation of *ortho*-substituted benzoic acids: overriding Pd-catalyzed protodecarboxylation. *Org. Lett.* 15 (4): 910–913. (b) Arroniz, C., Rassias, G., Larrosa, I. et al. (2014). An organic cation as a silver(I) analogue for the arylation of sp^2 and sp^3 C—H bonds with iodoarenes. *Chem. Sci.* 5: 3509–3514.

50 Zhu, C., Zhao, H., Su, W. et al. (2015). Ambient-temperature *ortho* C–H arylation of benzoic acids with aryl iodides with ligand-supported palladium catalyst. *Org. Lett.* 17 (14): 3418–3421.

51 (a) Zhang, Y.H., Shi, B.F., and Yu, J.Q. (2009). Pd(II)-catalyzed *ortho*-alkylation of benzoic acids with alkyl halides. *Angew. Chem. Int. Ed.* 48 (33): 6097–6100. (b) Cheng, G., Li, T.J., and Yu, J.Q. (2015). Practical Pd(II)-catalyzed C–H alkylation with epoxides: one-step syntheses of 3,4-dihydroisocoumarins. *JACS* 137 (34): 10950–10953.

52 Larock, R.C., Varaprath, S., Fellows, C.A. et al. (1984). Synthesis of isocoumarins via thallation-olefination of benzoic acids. *JACS* 106: 5274–5284.

53 Nandi, D., Ghosh, D., Lee, H.M. et al. (2013). One-step synthesis of isocoumarins and 3-benzylidenephthalides via ligandless Pd-catalyzed oxidative coupling of benzoic acids and vinylarenes. *J. Org. Chem.* 78 (7): 3445–3451.

54 Wang, H., Liu, Y.J., Zeng, M.H. et al. (2019). Carboxylate-assisted Pd(II)-catalyzed *ortho*-C–H and remote C–H activation: economical synthesis of pyrano[4,3-b]Indol-1(5H)-ones. *Org. Lett.* 21 (8): 2847–2850.

55 Mei, T.S., Giri, R., Yu, J.Q. et al. (2008). Pd^{II}-catalyzed monoselective *ortho* halogenation of C—H bonds assisted by counter cations: a complementary method to directed *ortho* lithiation. *Angew. Chem. Int. Ed.* 47 (28): 5215–5219.

56 Ng, K.H., Ng, F.N., and Yu, W.Y. (2012). A convenient synthesis of anthranilic acids by Pd-catalyzed direct intermolecular *ortho*-C–H amidation of benzoic acids. *Chem. Commun.* 48: 11680–11682.

57 Giri, R. and Yu, J.Q. (2008). Synthesis of 1,2- and 1,3-dicarboxylic acids via Pd(II)-catalyzed carboxylation of aryl and vinyl C—H bonds. *JACS* 130 (43): 14082–14083.

58 Miao, J. and Ge, H. (2013). Palladium-catalyzed chemoselective decarboxylative *ortho* acylation of benzoic acids with α-oxocarboxylic acids. *Org. Lett.* 15 (12): 2930–2933.

59 Zhang, Y.H. and Yu, J.Q. (2009). Pd(II)-catalyzed hydroxylation of arenes with 1 atm of O_2 or air. *JACS* 131 (41): 14654–14655.

60 Li, Z., Wang, Z., Yu, J.Q. et al. (2021). A tautomeric ligand enables directed C–H hydroxylation with molecular oxygen. *Science* 372: 1452.

61 Wang, D.H., Mei, T.S., and Yu, J.Q. (2008). Versatile Pd(II)-catalyzed C–H activation/aryl–aryl coupling of benzoic and phenyl acetic acids. *JACS* 130 (52): 17676–17677.

62 Engle, K.M., Thuy-Boun, P.S., Yu, J.Q. et al. (2011). Ligand-accelerated cross-coupling of C(sp^2)—H bonds with arylboron reagents. *JACS* 133 (45): 18183–18193.

63 Dastbaravardeh, N., Toba, T., Yu, J.Q. et al. (2015). Monoselective o-C–H functionalizations of mandelic acid and α-phenylglycine. *JACS* 137 (13): 9877–9884.

64 Thuy-Boun, P.S., Villa, G., Yu, J.Q. et al. (2013). Ligand-accelerated *ortho*-C–H alkylation of arylcarboxylic acids using alkyl boron reagents. *JACS* 135 (46): 17508–17513.

65 Li, G.C., Wang, P., Yu, J.Q. et al. (2017). Ligand-enabled auxiliary-free *meta*-C–H arylation of phenylacetic acids. *Angew. Chem. Int. Ed.* 129 (24): 6978–6981.

66 Wang, D.H., Engle, K.M., Yu, J.Q. et al. (2010). Ligand-enabled reactivity and selectivity in a synthetically versatile aryl C–H olefination. *Science* 327 (5963): 315–319.

67 Engle, K.M., Wang, D.H., and Yu, J.Q. (2010). Ligand-accelerated C–H activation reactions: evidence for a switch of mechanism. *JACS* 132 (40): 14137–14151.

68 Engle, K.M., Wang, D.H., and Yu, J.Q. (2010). Constructing multiply substituted arenes using sequential Pd(II)-catalyzed C–H olefination. *Angew. Chem. Int. Ed.* 49 (35): 6169–6173.

69 Mei, T.S., Wang, D.H., and Yu, J.Q. (2010). Expedient drug synthesis and diversification via *ortho*-C–H iodination using recyclable PdI_2 as the precatalyst. *Org. Lett.* 12 (14): 3140–3143.

70 Ma, S., Villa, G., and Yu, J.Q. (2014). Palladium-catalyzed *ortho*-selective C-H deuteration of arenes: evidence for superior reactivity of weakly coordinated palladacycles. *Angew. Chem. Int. Ed.* 53 (3): 734–737.

71 Li, S., Wang, X., Li, G. et al. (2021). Arene C–H iodination using aryl iodides. *CCS Chem.* 4 (6): 1889–1900.

72 Shi, B.F., Zhang, Y.H., Yu, J.Q. et al. (2010). Pd(II)-catalyzed enantioselective C–H olefination of diphenylacetic acids. *JACS* 132 (2): 460–461.

73 Cheng, X.F., Wang, X.S., Yu, J.Q. et al. (2013). Pd(II)-catalyzed enantioselective C–H activation/C—O bond formation: synthesis of chiral benzofuranones. *JACS* 135 (4): 1236–1239.

74 Novak, P., Correa, A., Martin, R. et al. (2011). Synergistic palladium-catalyzed C(sp^3)-H activation/C(sp^3)—O bond formation: a direct, step-economical route to benzolactones. *Angew. Chem. Int. Ed.* 123 (51): 12444–12447.

75 Qian, S., Chen, J.S., Yu, J.Q. et al. (2020). Ligand-enabled Pd(II)-catalyzed C(sp^3)–H lactonization using molecular oxygen as oxidant. *Org. Lett.* 22 (10): 3960–3963.

76 Park, H.S., Fan, Z., Yu, J.Q. et al. (2020). Distal γ-C(sp^3)–H olefination of ketone derivatives and free carboxylic acids. *Angew. Chem. Int. Ed.* 132 (31): 12953–12959.

77 Fraunhoffer, K.J., Sirois, L.E., White, M.C. et al. (2006). Macrolactonization via hydrocarbon oxidation. *JACS* 128 (28): 9032–9033.

78 Stang, E.M. and White, M.C. (2011). On the macrocyclization of the erythromycin core: preorganization is not required. *Angew. Chem. Int. Ed.* 50 (9): 2094–2097.

79 Zhu, Y., Chen, X., Zhao, Y. et al. (2017). Pd-catalysed ligand-enabled carboxylate-directed highly regioselective arylation of aliphatic acids. *Nat. Commun.* 8: 14904.

80 Ghosh, K.K. and van Gemmeren, M. (2017). Pd-catalyzed β-C(sp^3)–H arylation of propionic acid and related aliphatic acids. *Chem. Eur. J.* 23 (70): 17697–17700.

81 Zhuang, Z., Yu, C.B., Yu, J.Q. et al. (2018). Ligand-enabled β-C(sp^3)–H olefination of free carboxylic acids. *JACS* 140 (32): 10363–10367.

82 Zhuang, Z. and Yu, J.Q. (2020). Lactonization as a general route to β-C(sp^3)–H functionalization. *Nature* 577: 656–659.

83 Ghosh, K.K., Uttry, A., van Gemmeren, M. et al. (2019). Direct β-C(sp^3)–H acetoxylation of aliphatic carboxylic acids. *Org. Lett.* 21 (17): 7154–7157.

84 Zhuang, Z., Herron, A.N., Yu, J.Q. et al. (2020). Ligand-enabled monoselective β-C(sp^3)–H acyloxylation of free carboxylic acids using a practical oxidant. *JACS* 142 (14): 6769–6776.

85 Ghiringhelli, F., Uttry, A., van Gemmeren, M. et al. (2020). Direct β- and γ-C(sp^3)–H alkynylation of free carboxylic acids. *Angew. Chem. Int. Ed.* 59 (51): 23127–23131.

86 Uttry, A., Mal, S., and van Gemmeren, M. (2021). Late-stage β-C(sp^3)–H deuteration of carboxylic acids. *JACS* 143 (29): 10895–10901.

87 Dolui, P., Das, J., Maiti, D. et al. (2019). Ligand-enabled Pd^{II}-catalyzed iterative γ-C(sp^3)-H arylation of free aliphatic acid. *Angew. Chem. Int. Ed.* 58 (39): 13773–13777.

88 Liu, L., Liu, Y.H., and Shi, B.F. (2020). Synthesis of amino acids and peptides with bulky side chains via ligand-enabled carboxylate-directed γ-C(sp^3)–H arylation. *Chem. Sci.* 11: 290–294.

89 Ghosh, K.K., Uttry, A., van Gemmeren, M. et al. (2020). Ligand-enabled γ-C(sp^3)–H olefination of free carboxylic acids. *Angew. Chem. Int. Ed.* 59 (31): 12848–12852.

90 Das, J., Dolui, P., Maiti, D. et al. (2020). A direct route to six and seven membered lactones via γ-C(sp^3)–H activation: a simple protocol to build molecular complexity. *Chem. Sci.* 11: 9697–9702.

91 Shen, P.X., Hu, L., Yu, J.Q. et al. (2018). Pd(II)-catalyzed enantioselective C(sp^3)–H arylation of free carboxylic acids. *JACS* 140 (21): 6545–6549.

92 Hu, L., Qiao, J.X., Yu, J.Q. et al. (2019). Pd^{II}-catalyzed enantioselective C(sp^3)–H activation/cross-coupling reactions of free carboxylic acids. *Angew. Chem. Int. Ed.* 58 (7): 2134–2138.

93 Luo, J. and Fu, Q. (2021). Aldehyde-directed C(sp^2)-H functionalization under transition-metal catalysis. *Adv. Synth. Catal.* 363: 3868–3878.

94 Gurbuz, N., Ozdemira, I., and Cetinkaya, B. (2005). Selective palladium-catalyzed arylation(s) of benzaldehyde derivatives by *N*-heterocarbene ligands. *Tetrahedron Lett.* 46: 2273.

95 Ozdemir, I., Demira, S., and Cetinkaya, B. (2005). Use of tetrahydropyrimidinium salts for highly efficient palladium-catalyzed cross-coupling reactions of aryl bromides and chlorides. *Tetrahedron* 61 (41): 9791–9798.

96 Dogan, O., Gurbuz, N., Cetinkaya, B. et al. (2008). Palladium *N*-heterocyclic-carbene-catalyzed *ortho*-arylation of benzaldehyde derivatives. *Heteroat. Chem* 19 (6): 549–648.

97 Taskesenligil, Y., Aslan, M., Saracoglu, N. et al. (2023). Directed C-H functionalization of C3-aldehyde, ketone, and acid/ester-substituted free (NH) indoles with iodoarenes *via* a palladium catalyst system. *J. Org. Chem.* 88: 1299–1318.

98 Fei, L., Taoshan, J., Guanwu, W. et al. (2012). Palladium-catalyzed synthesis of aromatic sultones via sulfonic acid group-directed C-H activation. *Chin. J. Chem.* 30: 2041–2046.

99 Li, F., Liu, T.X., and Guanwu, W. (2012). Synthesis of [60] fullerene-fused sultones via sulfonic acid group-directed C—H bond activation. *Org. Lett.* 14 (8): 2176–2179.

100 Keshri, R., Rana, D., Maiti, D. et al. (2023). Free amine and alcohol as the director for regioselective C(sp^2)—H bond functionalization. *ACS Catal.* 13: 4500–4516.

101 Lu, Y., Engle, K.M., Yu, J.Q. et al. (2010). Pd(II)-catalyzed hydroxyl-directed C–H olefination enabled by monoprotected amino acid ligands. *JACS* 132 (16): 5916–5921.

102 Wang, X., Dai, H., Yu, J.Q. et al. (2010). Pd(II)-catalyzed hydroxyl-directed C–H activation/C–O cyclization: expedient construction of dihydrobenzofurans. *JACS* 132 (35): 12203–12205.

103 Lu, Y., Engle, K.M., Yu, J.Q. et al. (2011). Hydroxyl-directed C–H carbonylation enabled by mono N-protected amino acid ligands: an expedient route to 1-Isochromanones. *Chem. Sci.* 2: 967–971.

104 Satoh, T., Miura, M., Nomura, M. et al. (1997). Palladium-catalyzed regioselective mono- and diarylation reactions of 2-phenylphenols and naphthols with aryl halides. *Angew. Chem. Int. Ed.* 36 (16): 1740–1742.

105 Miura, M., Tsuda, T., Nomura, M. et al. (1997). Palladium-catalyzed oxidative cross-coupling of 2-phenylphenols with alkenes. *Chem. Lett.* 26 (11): 1103–1104.

106 Zhang, C., Ji, J., and Sun, P. (2014). Palladium-catalyzed alkenylation via sp^2 C—H bond activation using phenolic hydroxyl as the directing group. *J. Org. Chem.* 79 (7): 3200–3205.

107 Xiao, B., Xu, J., Liu, L. et al. (2011). Synthesis of dibenzofurans via palladium-catalyzed phenol-directed C–H activation/C–O cyclization. *JACS* 133 (24): 9250–9253.

108 Wei, Y. and Yoshikai, N. (2011). Oxidative cyclization of 2-arylphenols to dibenzofurans under Pd(II)/peroxybenzoate catalysis. *Org. Lett.* 13 (20): 5504–5507.

109 Luo, S., Zhang, X.S., Shi, Z.J. et al. (2013). Synthesis of dibenzopyranones through palladium-catalyzed directed C–H activation/carbonylation of 2-arylphenols. *Angew. Chem. Int. Ed.* 52 (40): 10598–10601.

110 Dou, Y., Liu, J., Zhu, Q. et al. (2019). Late-stage direct *o*-alkenylation of phenols by Pd^{II}-catalyzed C–H functionalization. *Chem. Eur. J.* 25 (28): 6896–6901.

111 Sasano, K., Takaya, J., and Iwasawa, N. (2013). Palladium(II)-catalyzed direct carboxylation of alkenyl C—H bonds with CO_2. *JACS* 135 (30): 10954–10957.

112 Kim, Y., Moon, Y., Hong, S. et al. (2014). Synthesis of heterocyclic-fused benzopyrans via the Pd(II)-catalyzed C–H alkenylation/C–O cyclization of flavones and coumarins. *Org. Biomol. Chem.* 12: 3413–3422.

113 Satoh, T., Miura, M., Nomura, M. et al. (1996). Palladium-catalyzed coupling reaction of salicylaldehydes with aryl iodides *via* cleavage of the aldehyde C—H bond. *Chem. Lett.* 25 (9): 823–824.

114 Wang, S., Guo, C.C., Peng, Z. et al. (2009). One-step preparation of xanthones *via* Pd-catalyzed annulation of 1,2-dibromoarenes and salicylaldehydes. *Chem. Commun.* 6469–6471.

115 Chidipudi, S.R., Khan, I., Lam, H.W. et al. (2013). Synthesis of benzopyrans by Pd(II) or Ru(II)–catalyzed enolate–directed C–H alkenylation. *Org. Lett.* 15 (3): 570–573.

3

Traceless Directing Group in C—H Bond Functionalization

Shuvojit Haldar and Debasis Banerjee

Department of Chemistry, Laboratory of Catalysis and Organic Synthesis, Indian Institute of Technology Roorkee, Roorkee 247667, Uttarakhand, India

3.1 Introduction

Functionalization of C—H bond to form a new C—C or C—X (X = heteroatom) bond often needs a pre-activated group and is widely employed in the synthesis of complex molecular architecture [1–6]. However, functionalization of chemically inactive C—H bonds has proven to be an effective and environmentally benign alternative, which improves the atom economy of a reaction as well as the overall efficiency of the organic transformation [7–10]. In this direction, in the case of arenes or aliphatic compounds, functionalization of a specific C—H bond is quite challenging even though nucleophilicity and/or acidity can influence the transformations. However, in the case of heteroarenes, the presence of heteroatoms influences the reactivity based on their electron density toward the specific bonds. This heteroatom or the neighboring functionality present in the molecule can bring the metal center close to the reactive C—H bond and facilitate high regioselective outcomes. Thus, the presence of heteroatom or the functional group can influence the functionalization of a specific C—H bond [11–16].

In the 1990s, an initial report by Murai on Ru(II)-catalyzed *ortho*-alkylation of arenes with alkenes directed by ketones attracted significant attention [17]. Thereafter, several groups demonstrated the potential of this area, and directing group-assisted C—H bond functionalization was widely employed for organic transformations [18]. Directing groups can be grouped into several categories, such as non-removable directing groups, which remain intact or are used to form the cyclized product or heterocycles (Scheme 3.1). Thus, it is not possible to remove them from molecular scaffolds. So, this strategy is not viable for the process, where the directing group has to be functionalized or to be removed from the reactive moiety [15, 19–23]. Another kind of directing group is known as removable directing group. This methodology is not atom economic for industrial applications,

Directed C–H Bond Functionalization: Concepts and Applications, First Edition.
Edited by Debabrata Maiti and Supriya Rej.

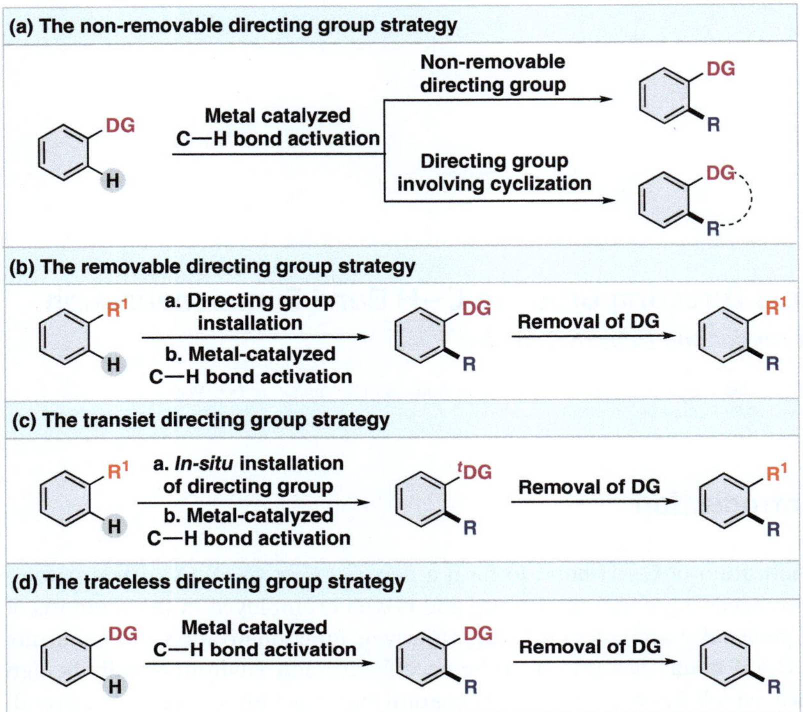

Scheme 3.1 Transition metal-catalyzed C–H bond functionalization using directing group strategy.

as the directing group must be installed and removed after the reaction. [18, 24] The third category of the directing group is called transient directing group (TGD), which is temporarily attached to the pre-existed group; thereafter, it directs the incoming group for the installation and then itself gets relieved from its position. However, some of the functional groups tethered themselves permanently, which limits their application. Moreover, in the case of TGD, a weakly coordinating group is transformed into a better σ-donor motif and assists the desired functionalization [25–32]. Pioneering work by Parshall in 1969 explored the application of tri-phenylphosphite complexes of ruthenium, rhodium, and cobalt for *ortho*-C—H bond activation, a five- or six-membered metallacycle is proposed, and such chelation assistance (thermodynamically and kinetically favorable) can promote the C—H bond functionalization [33]. However, the formation of TDG (i) should be reversible, (ii) should have high chemoselectivity, and (iii) should not react within the reaction system to behave as a TGD. Another class of functional group is pre-existing in the molecule or the directing group is introduced *in situ* followed by C—H bond activation, and insertion of the functional group is followed by the removal of the directing functional group. This type of directing group is called a traceless functional group. The reaction is performed under stoichiometric conditions, making the process a better atom and step economic one [34–37].

3.2 Classification of the Traceless Groups

It was observed that, for a highly chemoselective C—H bond activation, the donating group through its σ-coordination to the metal center facilitates the specific bond, where C—H bond activation needs to be done [38–40]. Several functional groups have been found to have a high potential for bond activation. However, the choice of such functional groups is crucial as well as its presence in the initial substrate. Thus, to participate as an effective traceless directing character, it should meet some of the following criteria: (i) the directing group must be stable and should not be cleaved without directing the metal center toward the reactive center, and (ii) after the reaction, the directing group should be extruded from the substrate without any side reaction. Thus, the functional groups containing nitrogen, oxygen, sulfur, phosphorus, silicon, π-chelation elements, etc. have been utilized for the traceless C—H bond activation. Some of the directing groups being utilized for the bond activation are likely carboxylic acids, carbon dioxide, aldehydes, ester, amide-based groups, silicon tethered groups, *N*-oxides, nitroso, and α-imino oxy acids [36, 37].

3.3 Carbonyl Group as a Traceless Directing Group

3.3.1 Carboxylic Acid as a Traceless Directing Group for Various Organic Transformations

3.3.1.1 Carboxylic Acid as a Traceless Directing Group Toward Biaryl Synthesis

Bi-aryl moiety is found in various drug compounds, polymeric materials, and ligands Figure 3.1 [41]. They are commonly synthesized through cross-coupling reactions, where excess amounts of organometallic reagents are used and make the synthesis process less atom economic. However, for the non-activated aryl system, synthesis of biaryls through C—H activation is very challenging and the process is less regioselective [42, 43]. Thus, a traceless directing group is a better pathway for the introduction of the aryl group in a specific position. Carboxylic acid functionality is one of the abundant and easily available moieties, which can be utilized for the introduction of the aryl functionalities.

Glorius in 2009 reported the tandem decarboxylation/C—H bond activation for the synthesis of dibenzofuran. 2-Phenoxybenzoic acid with electron-donating as well as electron-withdrawing substituents can sustain the catalytic transformations including halogen and acids having internal double bonds [44]. Larossa in 2011 explored the application of *ortho*-hydroxy-substituted benzoic acid for the generation of *meta*-equivalent anions to allow *meta*-aryl phenol compounds [45, 46]. Directing group relay strategy was utilized for the double arylation by Zhang and access to a series of *ortho*-di(2-pyridyl)arenes (Scheme 3.2). The carboxyl group directs the C—H bond activation and results in *ortho*-pyridination, which accelerates the decarboxylation and *ortho*-incorporation of the other pyridyl group [47]. Decarboxylative C—H/C—H cross-coupling reactions are rare with the use of carboxylic acids as a traceless directing group. Computational studies showed that *ortho*-substituted acids are reactive toward decarboxylative *ortho*-thiopenation [48].

Traditionally, the synthesis of dibenzofuran required pre-activated halogenation or the presence of a good leaving group for S_N2 displacement strategy. However, tandem decarboxylation/C—H bond activation using —COOH as a better leaving group provides highly regioselective dibenzofuran synthesis in one-pot operation as presented in Scheme 3.3 [44]. Although *meta*-functionalization can be difficult due to the close proximity of the carboxyl group toward *ortho*-position, pyridination at *ortho*-position promotes decarboxylation and brings the metal to close proximity of *meta*-position, resulting in subsequent *meta*-alkylation [49].

Etherification is normally performed by Williamson synthesis or coupling reaction between alkyl halide with alcohols or phenol. Gooßen reported the decarboxylative alkoxylation of carboxylates using Cu salts. Initially formed Cu-benzoate promotes the alkoxylation using trimethylborate, which further allows the one-pot synthesis of borate ester and subsequent alkoxylation, resulting in late-stage functionalization to various important derivatives (Scheme 3.4) [50]. Following a similar strategy, Zhang reported on decarboxylative *ortho*-selective acetylation using keto acid and subsequently transformed into decarboxylation

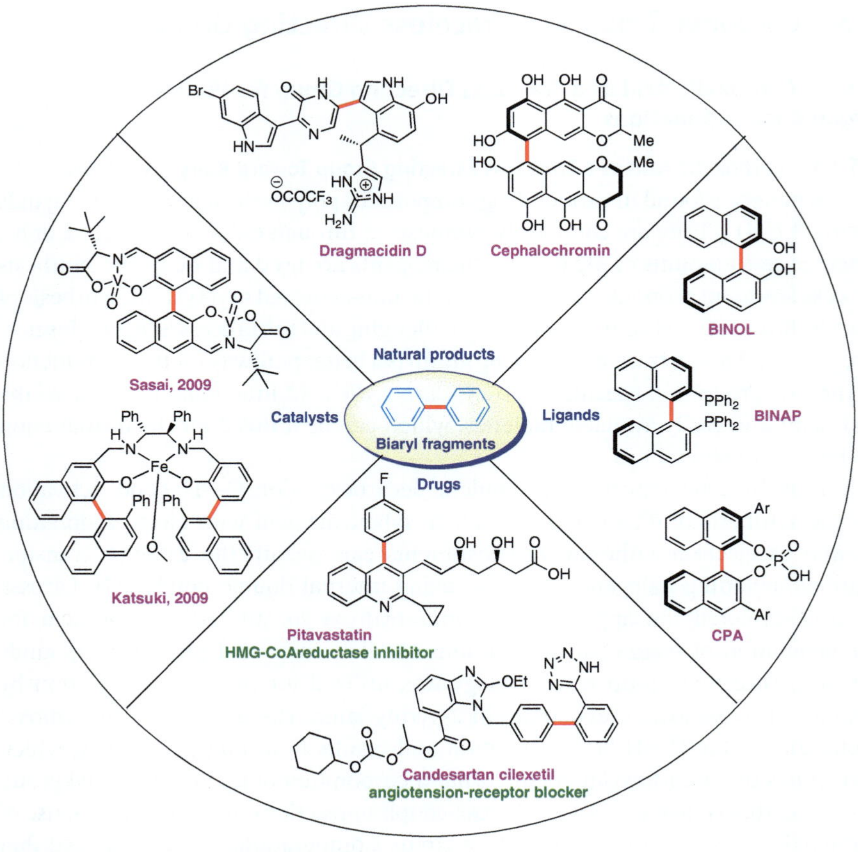

Figure 3.1 Applications of bi-aryl fragments.

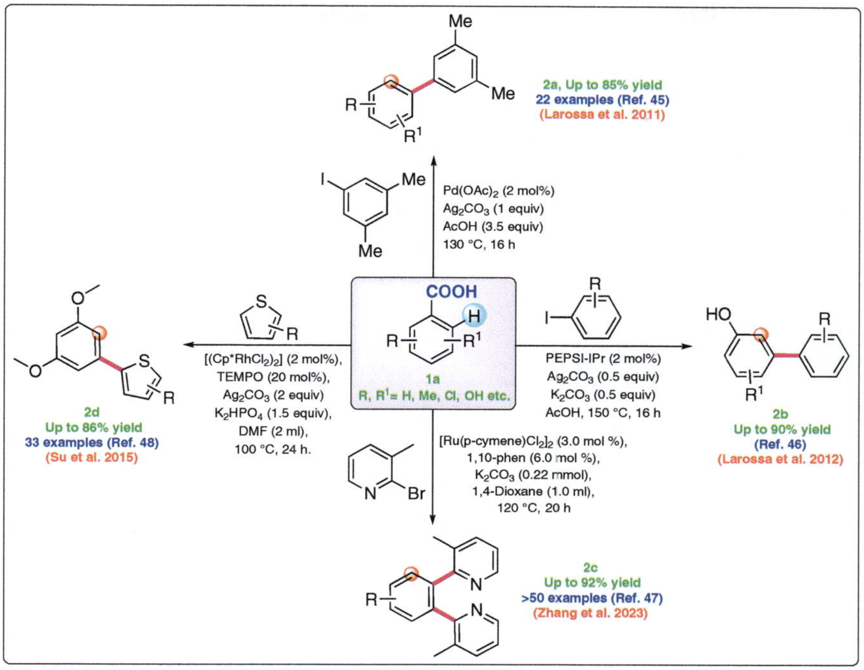

Scheme 3.2 Pd-catalyzed decarboxylative arylation: carboxylic acid as a traceless directing group.

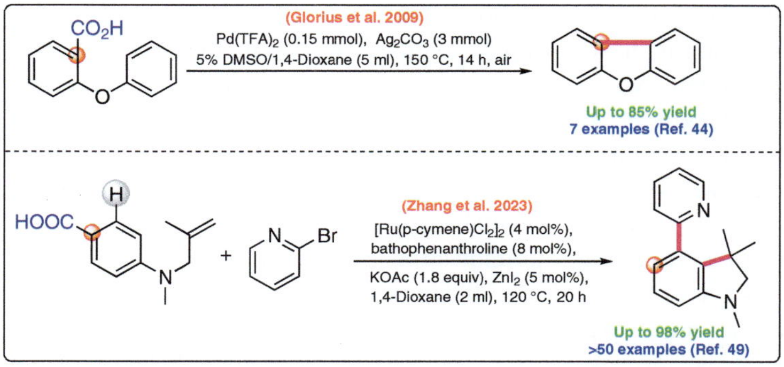

Scheme 3.3 Pd-catalyzed decarboxylative cyclization to heteroaryls.

para-substituted arenes [51]. Secondary amide was also achieved using phenyl isocyanate by Li and co-workers [52].

Chang in 2015 explored the application of two-metal two-step processes, where one metal leads to the C—H bond activation amination, followed by decarboxylation by another metal to the desired amine [54]. Cai group reported the halogenation of *o*-nitro benzoic acid, directed by the carboxyl group. The presence of bismuth salts increases the electrophilicity of the Pd catalyst, whereas Cu catalyst acts as a promoter for the decarboxylation process [53].

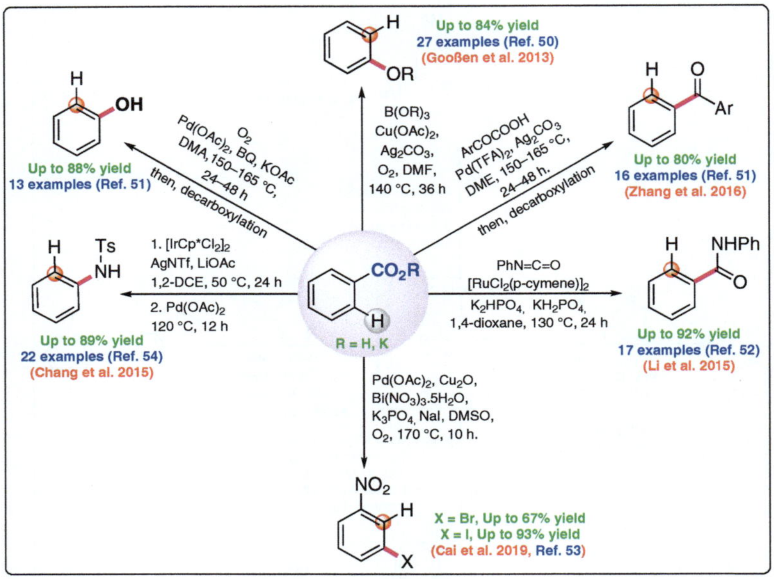

Scheme 3.4 *Ortho*-functionalization using carboxylic acid as a traceless directing group.

3.3.1.2 Carboxylic Acid as a Traceless Directing Group: Alkylation of Indole

Sequential functionalization decarboxylation to important *N*-heteroarenes may occur in several ways. One of them involves a multi-metallic system, which induces the alkylation process followed by decarboxylation in a single step or in a consecutive two-step process, resulting in the desired products. Another methodology involves a single metallic system, which promotes both steps in a single-pot operation. In this direction, Dauglis in 2007 first reported the Pd-catalyzed *ortho*-arylation of benzoic acid [55]. Later on, Miura and coworkers in 2008 reported C—H bond functionalization followed by decarboxylation to 2-alkenylated indoles.

This method involves a two-step bimetallic system, where Pd serves as a catalyst for the C—H bond activation and promotes the decarboxylation process. On the other hand, copper acts as an oxidant. It was proposed that, initially, $Pd(OAc)_2$ coordinates the carbonyl oxygen to palladium(II) carboxylate, which is then directed to form a palladacycle intermediate and facilitate the alkylation at the C-2 position. Thereafter, subsequent alkene insertion and β-hydride elimination resulted in hydrido-palladium carboxylate, which upon decarboxylation and reductive elimination forms 2-vinylindole (Scheme 3.5). It is to be noted that Cu(II) salt regenerates $Pd(OAc)_2$ by oxidation of Pd(0), whereas acetate ions from LiOAc prevent the deactivation of Pd(0) to metallic species [56]. Decarboxylative C—C coupling was developed by Kumar using $Pd(OAc)_2$ as catalysts in water. Stable and highly electrophilic diaryl-iodonium salt is utilized as the alkylating agent. Arylating salts having electron-rich, electron-poor, or heterocyclic groups can sustain the catalytic reaction. It was proposed that the catalytic system with Pd catalyst is involved through Pd^{II}/Pd^{IV} pathway (Scheme 3.5) [57]. Liu also reported that the synthesis

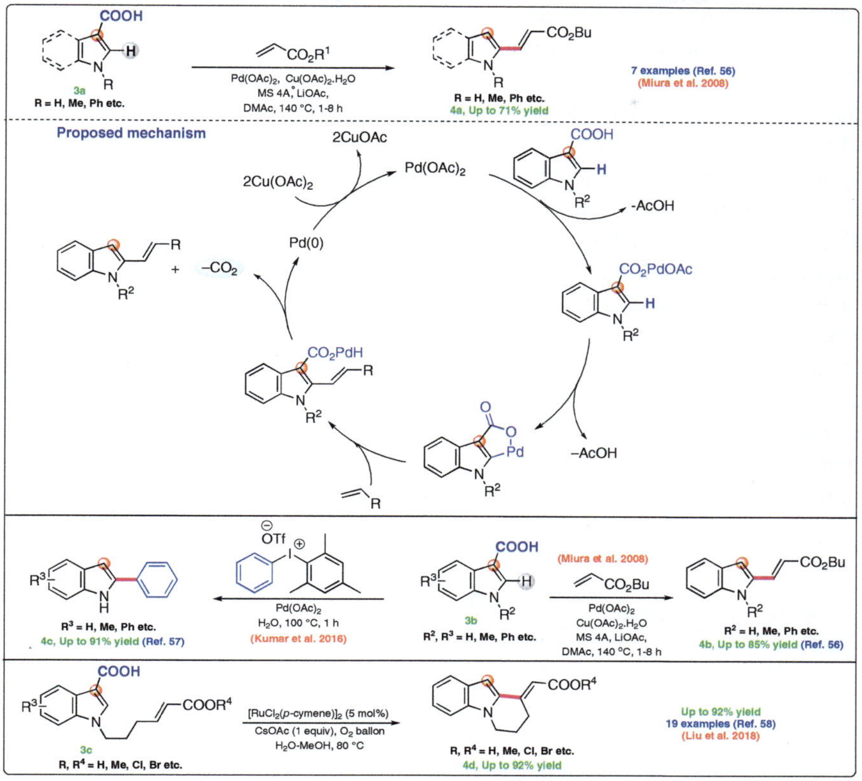

Scheme 3.5 Transition metal-catalyzed C2-alkylation of indoles.

of indole-fused cyclohexanyl core is useful for the synthesis of several drug-based molecules and natural products. It was observed that —COOH facilitates the C—H bond activation at the C-2 position. *In situ* generated metallacycle intermediate facilitates the cyclization and forms a six-membered ring (Scheme 3.5) [58].

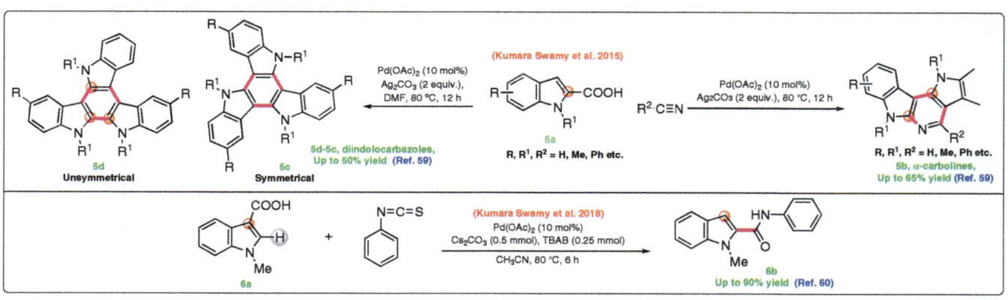

Scheme 3.6 Pd-catalyzed C—H bond functionalization of Indoles.

Kumara Swamy reported the synthesis of indolocarbolines and tri-indoles through decarboxylative C—H bond activation followed by nitrile insertion. Indole-2-carboxylic acid undergoes decarboxylation by Ag salt and then forms bi-indolyl species, which undergoes nitrile insertion. Next, intramolecular C—H bond activation forms α-carboline [59]. Later, they reported that the insertion of C—H bond into isothiocyanates leads to amide synthesis, although isothiocyanate tends to be less sensitive to moisture Scheme 3.6 [60].

Huang utilized electricity as a traceless oxidant and carboxylic group as the traceless directing group for the C-alkenylation of indoles in water. Application of facile undivided cell with graphite felt (GF) as an anode and platinum (Pt) as a cathode, and they prepared a diverse range of C-2 vinylated indoles. Initially, active rhodium(III) species undergo C—H bond activation to form a five-membered rhoda(III) cycle as presented in Scheme 3.7. Thereafter, the formation of a seven-membered rhoda(III) cycle, which undergoes β-H elimination and subsequent reductive elimination, leads to C-2 alkenylated indole-3-carboxylic acid. Lastly, decarboxylation leads to 2-alkenylated-*N*-methyl-1H-indole, while electrooxidation of the Cp*Rh(I) intermediate takes place at the anode and allows regeneration of the active Cp*Rh(III) catalyst. The reaction was facilitated by carboxylate base, and other inorganic bases did not result in any product [61].

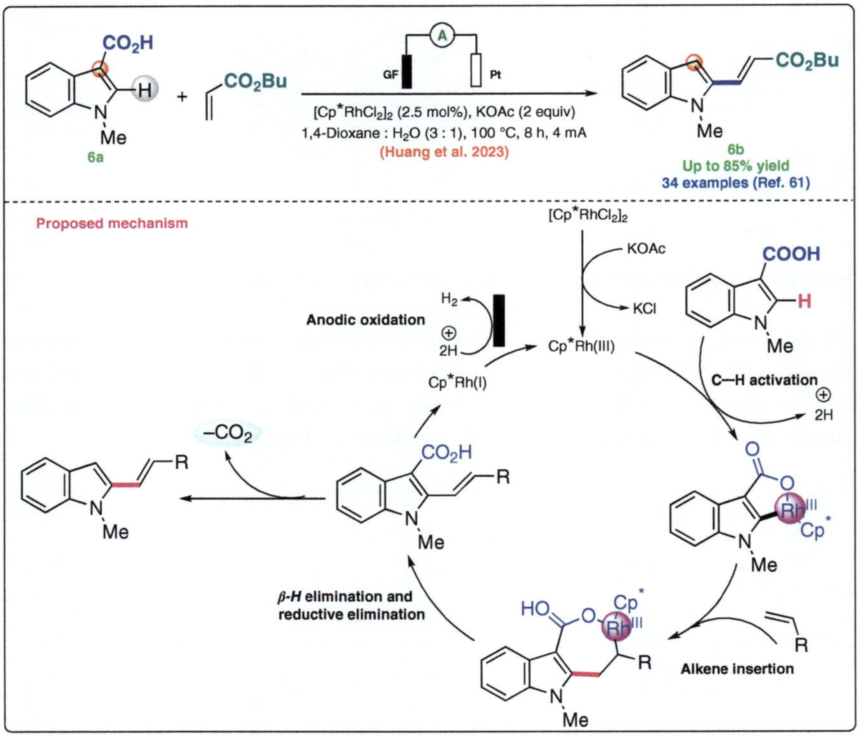

Scheme 3.7 Electrocatalytic decarboxylative alkenylation of indoles.

Functionalization of isooxazole moiety using carboxylic acid as a traceless directing group was reported by Nakamura. Synthesis of pyranoisoxazolones was obtained as a single regioisomer. They reported the decarboxylative alkenylation of isoxazolyl-4-carboxylic acid with ethyl acrylate in the presence of $[Cp^*RhCl_2]_2$

catalyst, and the products were obtained in high *trans*-selectivity. Further, carboxylate-directed acyl methylation was also observed in the case of phenyl isoxazole (Scheme 3.8) [62, 63].

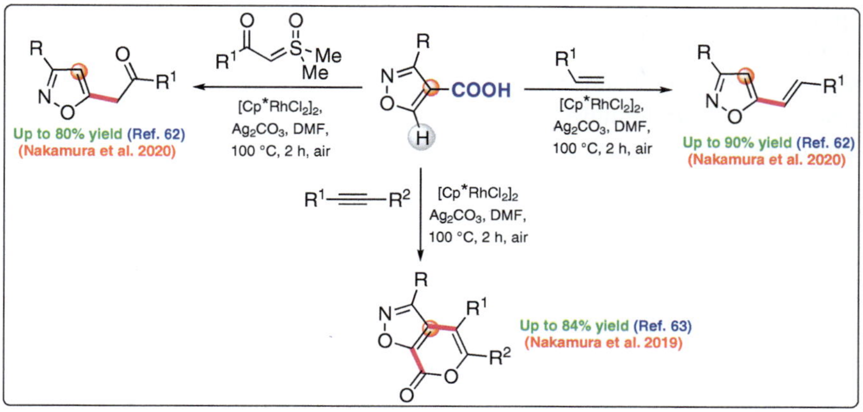

Scheme 3.8 Diverse functionalization of isooxazole using carboxylic acid as a traceless directing group.

3.3.1.3 Carboxylic Acid as a Traceless Directing Group in Alkylation/Alkenylation

Polyaromatic hydrocarbons are naturally obtained from coal, crude oil, and gasoline, and they have broad applications from medicine to plasticizers. Thus, their synthesis can be achieved through a cross-coupling reaction. Another approach to reach this goal is through selective C—H bond functionalization of suitable aromatic compounds. In this direction, carboxyl acid-directed C—H bond activation for the synthesis of linear aromatics compounds attracted significant attention. Glorius reported the formal [4 + 2] cyclization of 2-phenylbenzoic acids with alkynes for the synthesis of phenanthrene [64]. Pd-catalyzed decarboxylative alkylation followed by cyclization resulted in the desired products in good yield (Scheme 3.9).

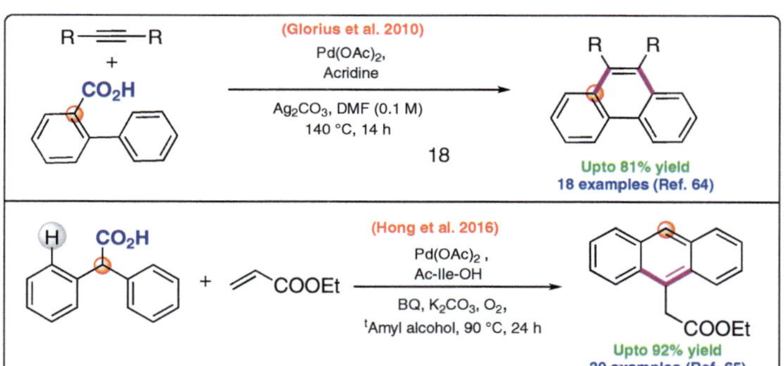

Scheme 3.9 Transition metal-catalyzed C—H functionalization using directing group strategy.

Another report by Hong on the sequential alkenylation-cyclization/decarboxylation to achieve desired acene derivatives is noteworthy [65]. Later on, Ackermann explored the application of Ru(II)-bis(carboxylate)catalyst for the decarboxylative alkenylation, and the process does not need any additives. Further, they extended the protocol to the synthesis of isocoumarin derivatives [66].

Further, synthesis of 3-aryl succinimides, commonly found in drugs and bioactive natural products, was obtained using carboxylic acid as a traceless directing group. A series of 3-aryl succinimides were obtained using a combination of benzoic acid and maleimide as presented in Scheme 3.10 [67–69].

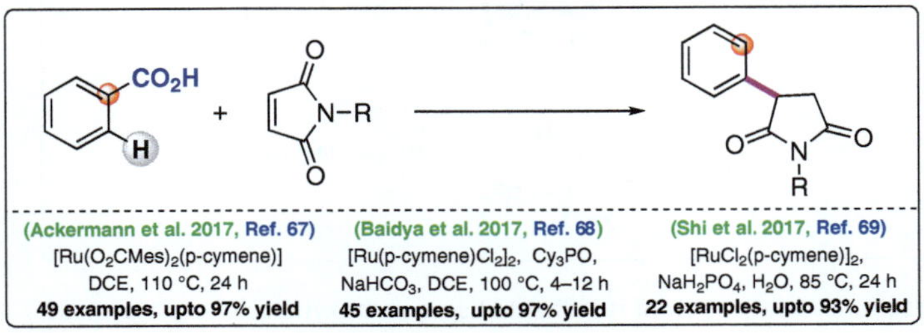

Scheme 3.10 Transition metal-catalyzed C–H bond functionalization to 3-aryl succinimides.

Stilbene and di-styryl-benzene are omnipresent in various natural products and have potential applications in organic materials. However, often, their synthesis is based on alkene metathesis [70], Mizoraki–Heck coupling [71], or Fujiwara–Moritani reaction [72]. These protocols involved a stoichiometric amount of waste generation and direct use of alkene raised serious concerns on product selectivity. Therefore, the application of a traceless directing group reduced the amount of waste and executed the product formation regioselectively. In this regard, Miura and co-workers [73], Maiti and co-workers [74], and Snurch and co-workers [75] have explored the application of —COOH as a traceless directing

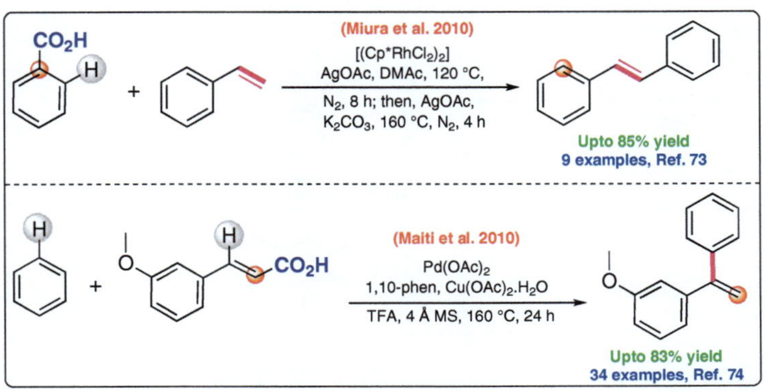

Scheme 3.11 Transition metal-catalyzed C–H bond functionalization to stilbene and di-styryl-benzene.

group for the synthesis of stilbene, di-styryl, and 1,1-disubstituted benzene using alkene as the substrate (Scheme 3.11).

3.3.2 Aldehyde and Ketone as a Traceless Directing Group

The carboxylic group has been employed extensively as the traceless directing group due to easy removal. Aldehyde was also prevalent in nature and served as a versatile motif for valuable organic transformations. In this direction, You explored the application of the —CHO group as a traceless directing group for the synthesis of indolo[1,2-*a*]quinolines (Scheme 3.12). It was observed that the aldehyde group promotes Rh(III)-catalyzed C—H activation to form a five-membered rhodacycle intermediate followed by cyclization with alkynes leads to indolo[1,2-*a*]quinolines. Also, they accomplished the concise synthesis of indolo[1,2-*a*]quinoline-based sensitizers used for organic optoelectronic materials [76]. Later on, Kathiravan and Nicholls extended the process for the di-sulfenation of indole at the C-2 and C-4 positions and the use of pentamethylcyclopentadienyl iridium dichloride dimer ($[Cp^*IrCl_2]_2$) catalyst and di-sulfides facilitated the catalytic transformations. It was proposed that the mechanism proceeds through a tandem catalytic pathway, and initial coordination of the catalyst with either the C-4 or C-2 C—H bond is reported. Next, electron-rich ketone directs the formation of a favorable six-membered metallacycle to the desired product. Further, oxidative sulfenation leads to product formation (Scheme 3.12) [77].

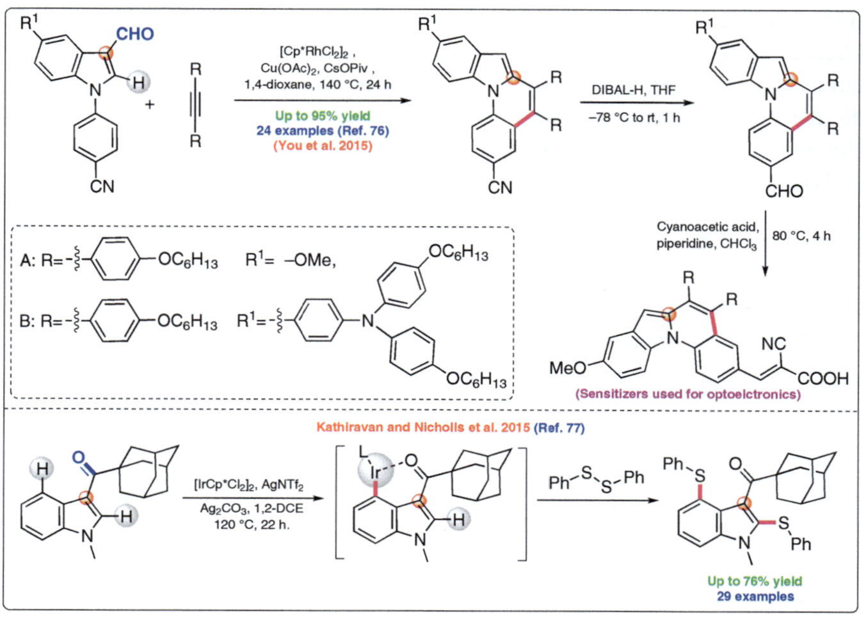

Scheme 3.12 Ru-catalyzed C—H bond functionalization using aldehyde as a traceless directing group.

3.3.3 Ester as a Traceless Directing Group

Glorius executed the use of tertiary propargylic carbonate as the alkyne substrate with a traceless donating group for the synthesis of isoquinolines. They extended the protocol for the synthesis of pharmaceutical anti-tumor derivative PK11209 and moxaverine (Scheme 3.13) [78].

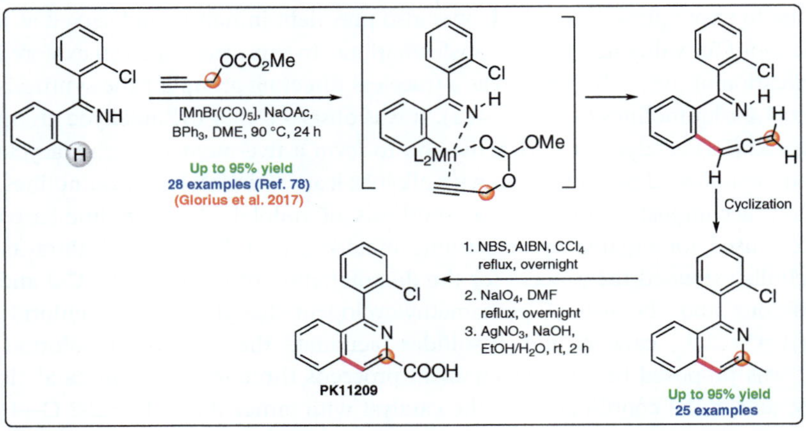

Scheme 3.13 Ester as traceless directing group in C—H bond functionalization.

3.3.4 Amide as a Traceless Directing Group

Amide bond is the strongest bond among the carboxylic acid derivatives. Cobalt, nickel, or copper complexes of aminoquinoline-derived amides have previously been used for the C(sp^2)—H bond activation of amides for the synthesis of phthalic acid derivative. These phthalic derivatives can be converted to *N*-substituted indolinones with a stoichiometric amount of waste generation. To overcome this issue, Zhong reported that picolinamide-derived TDG leads to C—H bond

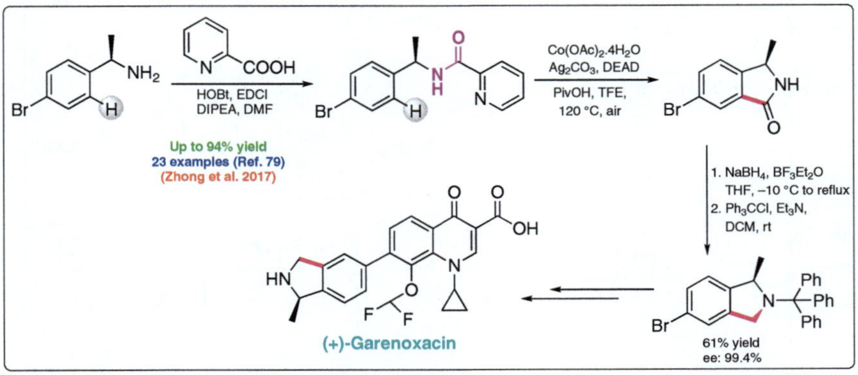

Scheme 3.14 Amide as a traceless directing group in the synthesis of *N*-substituted indolinones.

activation and carbonylation followed by reductive elimination and hydrolysis leads to the desired product. They performed a gram-scale reaction for the synthesis of a key intermediate of (+)-Garenoxacin (Scheme 3.14) [79].

3.3.5 CO_2 as a Traceless Directing Group in C—H Bond Activation

CO_2 can be introduced transiently to control the selectivity of an aromatic system. This allows the metal catalyst to come closer to the *ortho*-position, which governs the C—H activation process. In this direction, Larossa achieved *meta*-selective arylation of phenol and fluoroarenes. For the phenol functionalization, iodoarene derivatives with several functionalities even with F-, Cl-, and Br-containing iodoarene can sustain the reaction condition. The protocol works smoothly with *meta*-arylation. Initially, CO_2 was introduced to the *ortho*- to -OH/–F group, followed by carboxyl-group-directed C—H bond activation, and arylation leads to a *meta*-selective product (Scheme 3.15) [80–82].

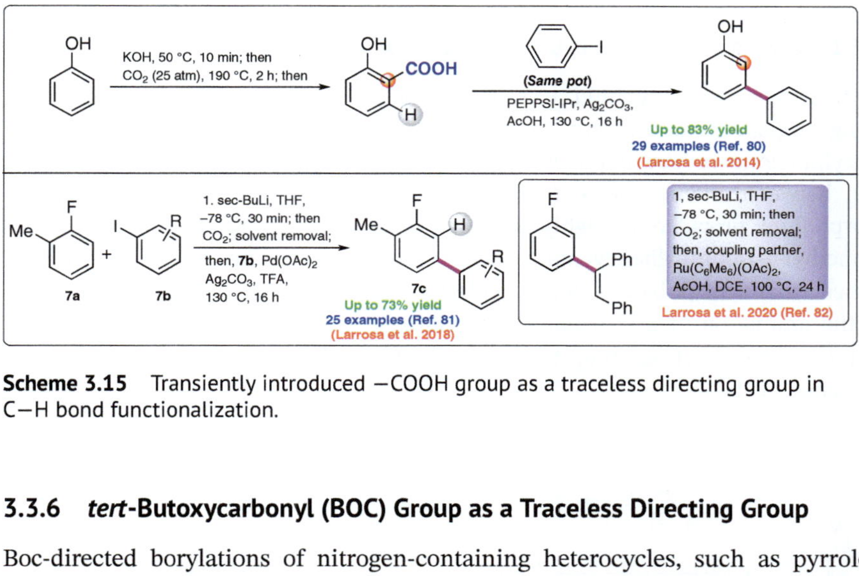

Scheme 3.15 Transiently introduced —COOH group as a traceless directing group in C—H bond functionalization.

3.3.6 *tert*-Butoxycarbonyl (BOC) Group as a Traceless Directing Group

Boc-directed borylations of nitrogen-containing heterocycles, such as pyrroles, indoles, azaindoles, and pyrazoles, are well studied. However, all these processes required another extra synthetic step to install and remove the Boc group. In this direction, Glorius reported redox neutral Boc as a traceless directing group for the C—H bond functionalization using highly active Lewis acidic [Cp*Co(CO)I_2] as a catalyst. The active cobalt species cause C—H metalation to form the cobaltacycle, which is stabilized by the Boc group. Following this strategy, alkyne insertion took place and a seven-membered intermediate was generated. Thereafter, the seven-membered ring rearranges into a six-membered ring intermediate through *N*-ligand shifting, and reductive elimination followed by an oxidative addition and protonation resulted in the desired product (Scheme 3.16) [83].

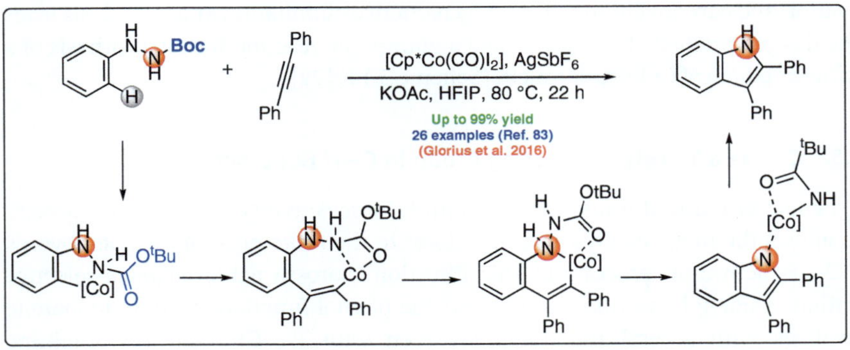

Scheme 3.16 *tert*-Butoxycarbonyl-directed strategy for the synthesis of indole.

3.4 Nitrogen-containing Functional Groups as a Traceless Directing Group

3.4.1 Amine as a Traceless Directing Group

Hydroarylation strategy could be used to synthesize ketone from an alkyne. Recently, Dong proposed amine-controlled selective arylation of alkyne, which resulted in α-arylphenone or α,α-diarylketone. Palladium-catalyzed three-component reaction between alkyne and aryl triflates in the presence of a secondary amine morpholine provided a highly selective α-arylphenone product. By contrast, *N*-methyl-α-isopropylbenzylamine favors α,α-diary acetone formation. It was noticed that a weak base suppresses the formation of Buchwald–Hartwig product of the amination of aryl triflates. Therefore, it was observed that amine works as a traceless directing group to control the regioselectivity of product formation (Scheme 3.17) [84].

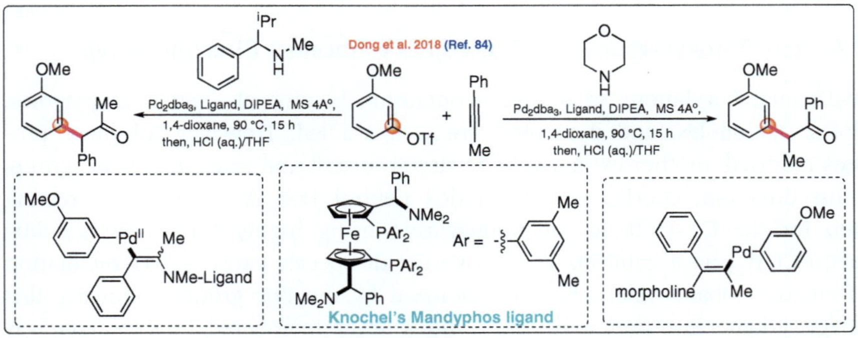

Scheme 3.17 Transition metal-catalyzed C–H bond functionalization using amine as a traceless directing group.

3.4.2 Hydrazone as a Traceless Directing Group

To prevent the application of any external oxidant and reduce the waste generation in indole synthesis, Glorius reported the use of a directing and oxidizing group (DG^{ox}) as an internal oxidant in C—H bond functionalization. Application of 2-acetyl-1-arylhydrazines in combination with electron-rich, electron-poor, or sterically demanding functional group and symmetrical or unsymmetrical alkyne can provide a diverse range of unprotected indoles. NHAc group work as a directing group and get released after breaking of the N—N bond. Isotope labeling experiments showed that the nitrogen comes from the hydrazine nitrogen directly attached to the aromatic ring (Scheme 3.18) [85]. By contrast, Zhu implemented the use of a bidentate directing group (2-hydrazinylpyridine), which prevents the need for the use of any external oxidant and a directing group for C—H bond functionalization. *In situ* generated (*E*)-2-(2-(1-phenylethylidene)hydrazinyl) pyridine reacts with an alkyne derivative in the presence of a cobalt catalyst, leading to isoquinoline synthesis. Although the reaction protocol sustains electron-withdrawing and -donating group, *ortho*-substitution resulted in lowering yield due to steric interaction. The formation of 2-aminopyridine as a side product proves the internal oxidation mechanism and a traceless directing group behavior. Moreover, if the traceless directing group is gaseous, the reaction is much cleaner [86].

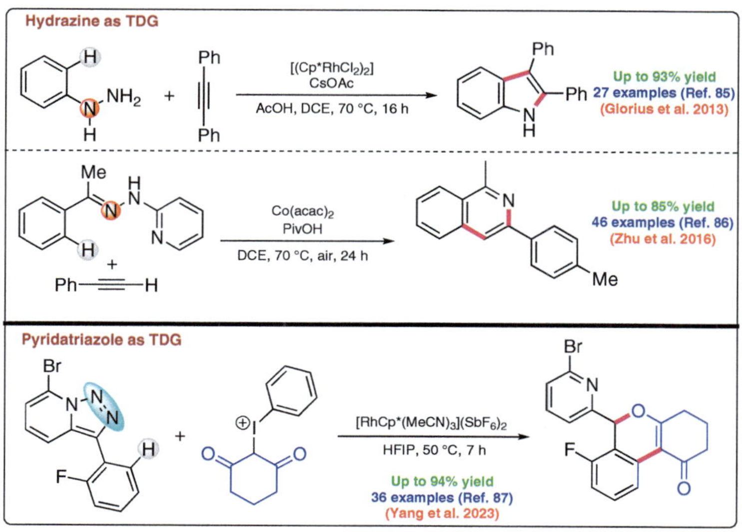

Scheme 3.18 N–N group as an internal oxidant and traceless directing group.

Pyridatriazole can also act as a directing group for the C—H bond activation using $[Cp^*Rh(MeCN)_3](SbF_6)_2$ catalyst and coupling with iodonium ylides.

The reaction was performed in HFIP solvent without any additives, but substitution at the C7 position of pyridatriazole was required for the activation of the ring. Metal promotes the decomposition of triazole to release nitrogen through diazo pathway (Scheme 3.18) [87].

3.4.3 N—O Group as a Traceless Directing Group

Being a better oxygen atom transfer group, *N*-oxide can act as an internal oxidant as well as a directing group for C—H bond functionalization. You reported *ortho*-alkenylation of a tertiary amine; however, due to their electronic and steric effect, often, the *para*-selective alkylation occurs. It was proposed that *N*-oxide reacts with Rh catalyst to form a cyclorhodium intermediate through electrophilic C—H bond activation and favoring *ortho*-alkenylation (Scheme 3.19) [88]. Later, the scope was broadened for the synthesis of *N*-protected indole by Wang using alkyne as the coupling partner, and they employed nitrone being the imine *N*-oxide as the

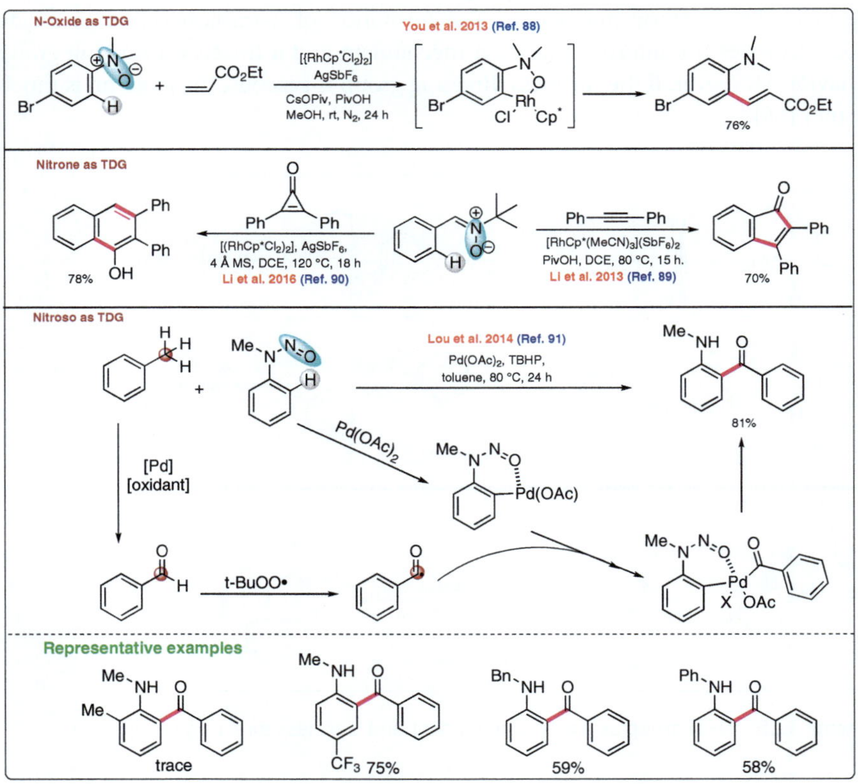

Scheme 3.19 N—O groups as traceless directing groups in C—H bond functionalization.

N—O bond is electrophilic in nature. Next, Li selectively synthesized indenone through C—H bond functionalization with alkyne as a coupling partner [89]. Thus, just by controlling the nature of the substrate, the properties of the traceless directing group of nitrone were tuned. Initial mechanistic studies suggested that the reaction goes through cyclometallation and coordination of the diphenylcyclopropenone, followed by migratory insertion and β-carbon elimination, leading to 1-naphthol [90].

N-nitroso group possesses exceptional properties as a nitrogen-based functional group and has potential application as a directing group. In this direction, Zhu reported the application of the *N*-nitroso group on *ortho*-alkenylation using Rh(III) catalyst. Normally, it appears as the *N*-nitroso of aniline, where it has an excellent character to direct the metal for C—H bond activation as well as act as an internal oxidant in catalyst recycling. Thus, in several cases, the nitroso group has been employed as a traceless directing for the synthesis of heterocyclic scaffolds like indoles and quinoline (Scheme 3.18). This scaffold can be synthesized from 2-amino benzophenone following Friedel–Crafts acylation of anthranilic acid; however, the process is associated with the generation of a stoichiometric amount of waste. To overcome such limitations, Luo in 2014 reported the utilization of the *N*-nitroso group as the traceless directing group for the synthesis of the *N*-alkyl-2-aminobenzophenones and employed toluene derivatives as acyl sources. Various functional groups at different positions of toluene can sustain the reaction conditions. They proposed that the reaction goes through oxidation of toluene to aldehyde in the presence of Pd(II) catalyst at higher temperature, whereas *t*-BuOO• radical generates acyl radical from aldehyde. In the next step, the nitroso group directs the process for the electrophilic palladation, and a palladacycle is formed. Thereafter, migratory insertion of acyl radical followed by reductive elimination resulted in the *N*-alkyl-2-aminobenzophenones and indicated the involvement of a Pd(IV) or a dimeric Pd(III) pathway as presented in Scheme 3.19 [91].

3.4.4 Alkene-tethered Aldoxime as a Traceless Directing Group

Nitrile synthesis from aldehyde is one of the important pathways to generate new C—N bonds in organic molecules, which can be further functionalized. Several catalytic processes have been employed to synthesize nitrile from aldoxime. In this direction, Kawatsura reported that alkene can trigger transition-metal-catalyzed nitrile synthesis from alkene-tethered oxime ethers, prepared from aldehyde with *O*-allylhydroxylamine hydrochloride, as presented in Scheme 3.20. Studies based on various precious metal catalysts, like $[Cp^*Ru(CH_3CN)_3]PF_6$, $[RhCl(cod)]_2$, and $[IrCl(cod)]_2$, did not improve the yield, whereas $Ni(cod)_2$/DPEphos resulted in 94% yield to the desired product. Preliminary mechanistic studies established that the reaction goes through oxidative addition of an oxime C—H bond to Ni(0) followed by β-hydride elimination or β-oxygen elimination/reductive elimination to yield the desired nitrile product [92].

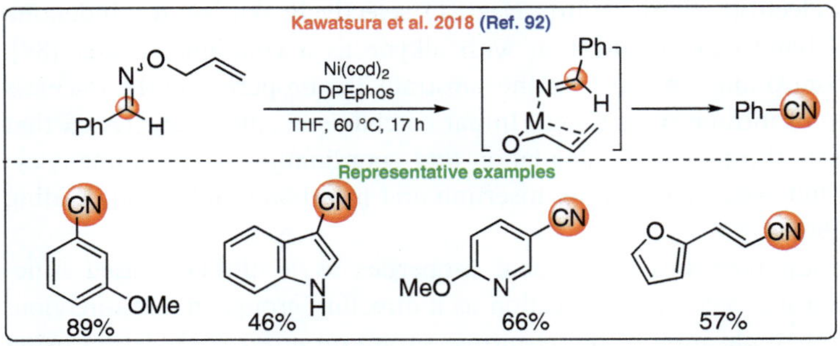

Scheme 3.20 Alkene-tethered aldoxime as a traceless directing group for C–H bond functionalization.

3.5 Miscellaneous Groups as a Traceless Directing Group

3.5.1 ((Pinacolato)boron (Bpin)) Group as a Traceless Directing Group

Borylation of —NH or —OH bond is favorable due to acidic in nature of N—H/O—H bond and facilitate a variety of C-2 alkylation to *N*-heterocycles directed by these -NH_2/-OH group. Smith III explored such a concept and reported HBpin-directed borylation of indole at the C-3 position. Triethylamine base was used to increase the nucleophilicity of HBpin and make B—H bond more hydridic for *N*-heterocycles with less acidic NH groups. Borenium cations and electrophilic ruthenium catalysts borylates the C-3 position, whereas for the aniline, less polar solvents and more electron-rich ligands do affect the *ortho*-selectivity. Finally, methanol treatment breaks down B—N bond and the yielded borylated product as presented in Scheme 3.21 [93].

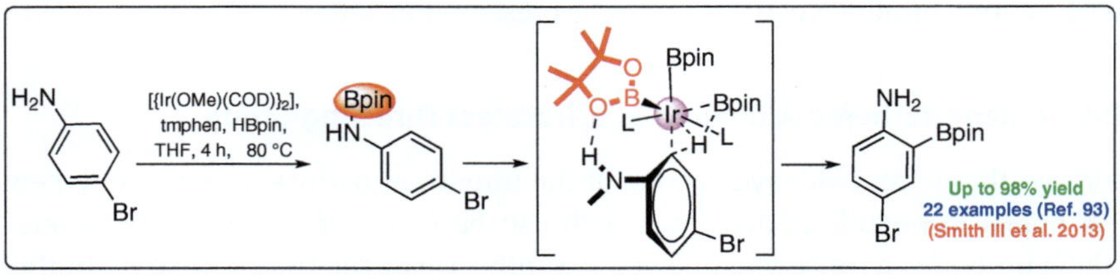

Scheme 3.21 Transition metal-catalyzed C–H bond functionalization using directing group strategy.

3.5.2 Acetal as a Traceless Directing Group

Generally, *O*-silylation of phenol requires a sub-stoichiometric amount of base. However, for the synthesis of structurally diverse molecules, directing group-mediated C—H functionalization found a potential application to install a silyl group. However, such processes required extra synthesis steps for the installation and as well as the removal of the directing groups. Jeon established sequential single-pot two-metal mediated conversion for the silylation process. Iridium catalyst is involved in the hydrosilylation of the phenyl acetate, whereas Rh-catalyzed C—H silylation provides benzo-di-oxasilines behaving as a halosilane; subsequently, the addition of a nucleophile removes the acetal group and facilitates the functionalization to silyl moiety (Scheme 3.22) [94].

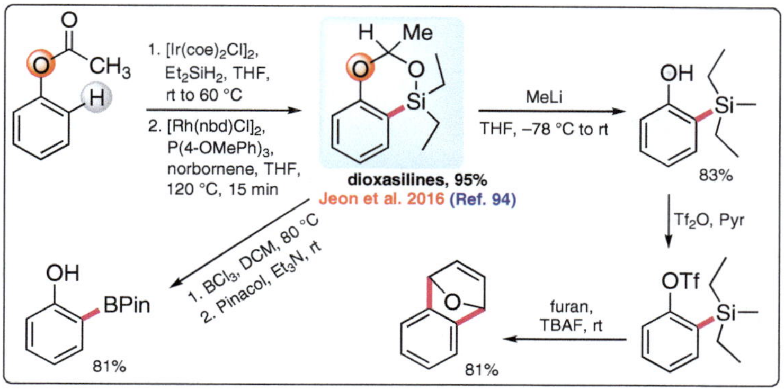

Scheme 3.22 Acetal group in C—H bond functionalization and silylation reaction.

3.5.3 Sulfur-based Group as a Traceless Directing Group

Application of sulfur-containing functional group is not so prominent in organic synthesis although the metal-coordinating capability of sulfoxides or sulfones is well established in literature [95]. Recently, Sahoo explored the reactivity and studied the property of methylphenyl sulfoximine (MPS) as a traceless directing group in a sequential C—C and C—N bond formation strategy through C—H bond activation. They established the synthesis of pyrido-infused iso-quinolinone motif from MPS-derived acrylamides with alkyne. They validated the protocol again with benzopyran derivatives, which are present in various natural products of biological significance. They obtained highly regioselective annulated products with methylphenyl acetylene as a coupling partner (Scheme 3.23) [96]. Recently, Prabhu and co-workers [97] and Li and co-workers [98] independently reported the synthesis of 1-naphthol derivatives from alkynes and sulfoxonium yields as a traceless directing group through C—H activation. The redox neutral annulation using a Rh(III) catalyst with di-arylacetylenes bearing both electron-releasing and electron-withdrawing groups proceeded smoothly. The reaction goes through the coordination of sulfoxonium oxygen to Rh(III) followed by C—H bond activation to form a rhodacycle followed by insertion of alkyne and isomerization to a seven-membered rhodacycle. The internal oxidation eliminates DMSO and later proto-demetalation leads to the formation of the product and facilitates the generation of the active catalyst as presented in Scheme 3.23.

3.5.4 Silicon Group as a Traceless Directing Group

Si atom can also serve as a traceless directing group and have potential application in organic synthesis. They are easily removable by fluoride source under milder condition. In this direction, Gevorgyan reported the use of silanol as a traceless

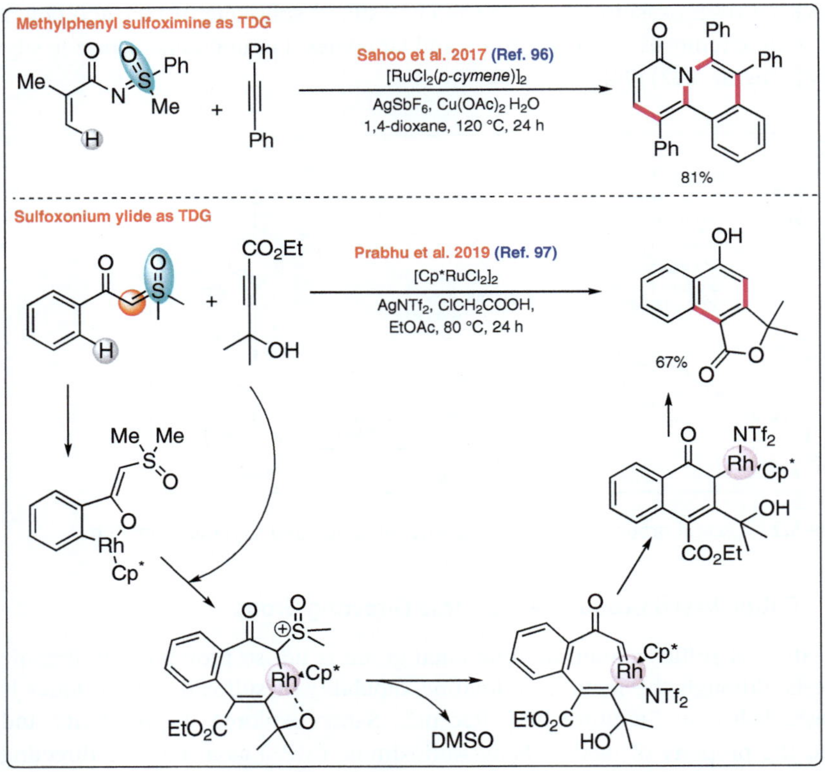

Scheme 3.23 Sulfur-containing functional group as a traceless directing group in C–H bond functionalization.

directing group for the *ortho*-alkenylation of phenol using butyl acrylate. They extended the application of Pd(II)-catalyzed alkenylation methodology for the synthesis of benzofuranone using silanol as a traceless directing group [99]. Further, extensions on the use of silanol for the carboxylation using carbon mono-oxide as a carbonyl source were performed and resulted in salicylic acid. The process is highly selective for the *ortho*-carboxylation with excellent functional group tolerance. The traceless directing group should undergo removal in a single-pot reaction; however, in the case of the silanol group, an additive is added at the end of the reaction and may be noted as a semi-one-pot reaction (Scheme 3.24) [100].

3.5.5 Halides as a Traceless Directing Group

Cyclic hypervalent iodine compounds are commonly utilized for the functionalization of alkenes. In 2018, Gulder employed fluorobenziodoxole for chemodivergent cascade reactions of *N*-substituted styrene. The reaction goes through a spiro-cyclic *F*-cyclopropane. The chemoselectivity for the C—H fluorination and C—H amination is being determined by the *N*-substitution and its control over cationic charge delocalization. Functional groups, such as Me, TBDMS, and others, determine the carbo-cationic character of the nitrogen atom

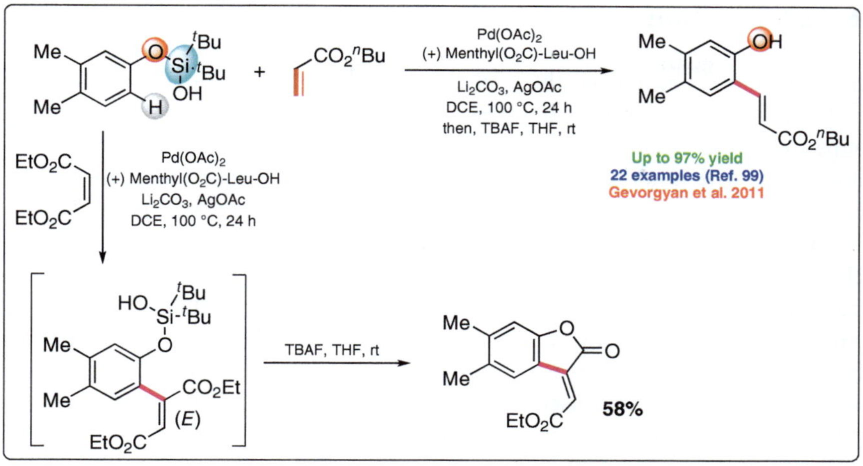

Scheme 3.24 Silanol group as a traceless directing group in *ortho*-alkenylation of phenol.

and allow regiodivergent synthetic routes. Thus, by selective trapping of *F*-based carbocations, a diverse range of indoles and tryptophan could be achieved as presented in Scheme 3.25 [101].

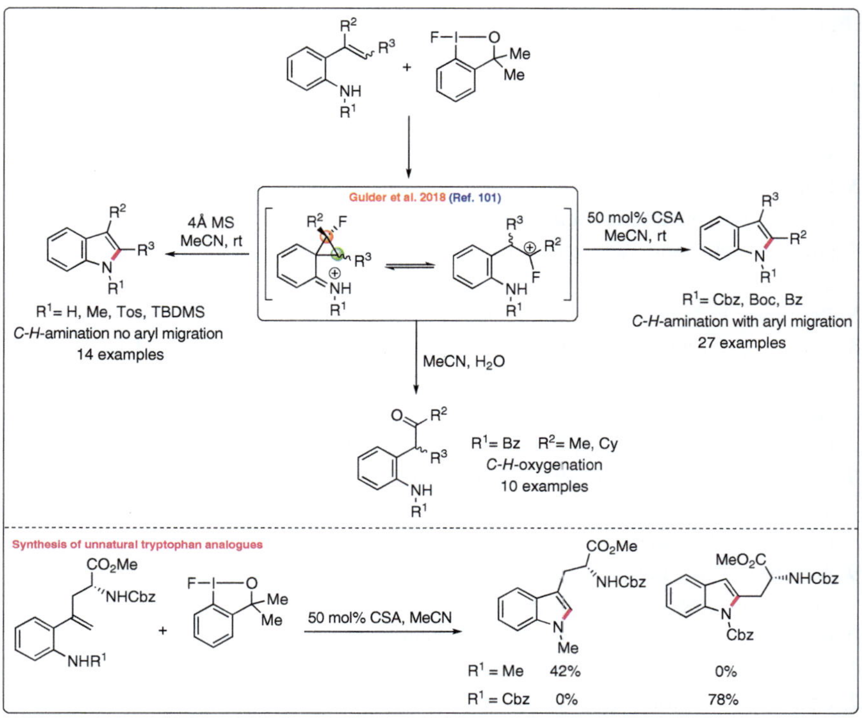

Scheme 3.25 Halides as a traceless directing group in C–H bond functionalization.

3.6 Conclusions

Herein, we have summarized the application of site-selective functionalization of C—H bonds using a series of traceless directing groups. Such traceless directing groups are easy to install and could be removed and do not participate in any chemical reactions with other functionalities present in the molecules. Often, this C—H bond functionalization is considered to be step and atom economic. This chapter summarized the importance and applications of a few traceless directing groups, such as carboxylic acids, aldehydes, esters, amides, and CO_2 including BOC. Further, extension toward the synthetic potential of nitrogen-containing functional groups, such as amines, hydrazine, and N–O group including alkene-tethered aldoxime and siloxane has been discussed. Preliminary mechanistic studies have been considered and discussed as and when necessary including potential applications toward organic transformations.

Acknowledgments

The authors thank SERB India (CRG/2021/002686) and MoE-STARS/STARS-2/2023-1003 for financial support. IIT Roorkee is gratefully acknowledged for the infrastructure facilities. S. Haldar thanks CSIR New Delhi for a fellowship.

References

1 Labinger, J.A. and Bercaw, J.E. (2002). *Nature* 417: 507–514.

2 Olah, G.A. and Molnar, A. (2003). *Hydrocarbon Chemistry*. Hoboken, NJ: Wiley-Interscience.

3 Hopf, H. (2000). *Classics in Hydrocarbon Chemistry: Syntheses, Concepts, Perspectives*. Weinheim: Wiley-VCH.

4 Gradassi, M.J. and Wayne Green, N. (1995). *Fuel Process. Technol.* 42 (2–3): 65–83.

5 Bergman, R.G. (2007). *Nature* 446: 391–393.

6 Shilov, A.E. (2006). *Activation and Catalytic Reactions of Saturated Hydrocarbons in the Presence of Metal Complexes*. Springer Science Business Media B.V. 2000.

7 Bromberg, S.E., Yang, H., Asplund, M.C. et al. (1997). *Science* 278: 260–263.

8 Goldman, A.S. and Goldberg, K.I. *Organometallic C—H Bond Activation: An Introduction, ACS Symposium Series*, vol. 885, ISBN13:, 1–43. ISBN: 9780841238497.

9 Arndtsen, B.A., Bergman, R.G., Mobley, T.A., and Peterson, T.H. (1995). *Acc. Chem. Res.* 28 (3): 154–162.

10 Busca, G. (2007). *Chem. Rev.* 107 (11): 5366–5410.

11 Jia, C., Kitamura, T., and Fujiwara, Y. (2001). *Acc. Chem. Res.* 34 (8): 633–639.

12 Conley, B.L., Tenn, W.J., Young, K.J.H. et al. (2006). *J. Mol. Catal. A Chem.* 251 (1–2): 8–23.

13 Seregin, I.V. and Gevorgyan, V. (2007). *Chem. Soc. Rev.* 36: 1173–1193.

14 Daugulis, O., Do, H., and Shabashov, D. (2009). *Acc. Chem. Res.* 42 (8): 1074–1086.

15 Chen, X., Engle, K.M., Wang, D., and Yu, J. (2009). *Angew. Chem. Int. Ed.* 48: 5094–5115.

16 Giri, R., Shi, B.F., Engle, K.M. et al. (2009). *Chem. Soc. Rev.* 38 (11): 3242–3272.

17 Murai, S., Kakiuchi, F., Sekine, S. et al. (1993). *Nature* 366 (6455): 529–531.

18 Chen, Z., Wang, B., Zhang, J. et al. (2015). *Org. Chem. Front.* 2 (9): 1107–1295.

19 Colby, D.A., Bergman, R.G., and Ellman, J.A. (2010). *Chem. Rev.* 110 (2): 624–655.

20 Lyons, T.W. and Sanford, M.S. (2010). *Chem. Rev.* 110 (2): 1147–1169.

21 Wencel-Delord, J., Droge, T., Liu, F., and Glorius, F. (2011). *Chem. Soc. Rev.* 40: 4740–4761.

22 Arockiam, P.B., Bruneau, C., and Dixneuf, P.H. (2012). *Chem. Rev.* 112 (11): 5879–5918.

23 Ackermann, L. (2014). *Acc. Chem. Res.* 47 (2): 281–295.

24 Rousseau, G. and Breit, B. (2011). *Angew. Chem. Int. Ed.* 50 (11): 2450–2494.

25 Zhang, F.-L., Hong, K., Li, T.-J. et al. (2016). *Science* 351: 252–256.

26 Wu, Y., Chen, Y.-Q., Liu, T. et al. (2016). *J. Am. Chem. Soc.* 138 (44): 14554–14557.

27 Wang, X., Song, S., and Jiao, N. (2018). *Chin. J. Chem.* 36: 213–216.

28 Gandeepan, P. and Ackermann, L. (2018). *Chem* 4: 199–222.

29 Goswami, N., Bhattacharya, T., and Maiti, D. (2016). *Nat. Rev. Chem.* 5: 646–659.

30 Kuang, G., Liu, D., Chen, X. et al. (2021). *Org. Lett.* 23 (21): 8402–8406.

31 Zhao, Q., Poisson, T., Pannecoucke, X., and Besset, T. (2017). *Synthesis (Germany)* 49 (21): 4808–4826.

32 Lapuh, M.I., Mazeh, S., and Besset, T. (2020). *ACS Catal.* 10 (21): 12898–12919.

33 Parshall, G.W., Knoth, W.H., and Schunn, R.A. (1969). *J. Am. Chem. Soc.* 91 (18): 4990–4499.

34 Zhang, F. and Spring, D.R. (2014). *Chem. Soc. Rev.* 43 (20): 6906–6919.

35 Font, M., Quibell, J.M., Perry, G.J.P., and Larrosa, I. (2017). *Chem. Commun.* 53 (41): 5584–5597.

36 Rani, G., Luxami, V., and Paul, K. (2020). *Chem. Commun.* 56 (83): 12479–12521.

37 Zarkadoulas, A., Zgouleta, I., Tzouras, N.V., and Vougioukalakis, G.C. (2021). *Catalysts* 11 (5): 554–591.

38 Altus, K.M. and Love, J.A. (2021). *Commun. Chem.* 4 (1): 1–11.

39 Dalton, T., Faber, T., and Glorius, F. (2021). *ACS Cent. Sci.* 7 (2): 245–261.

40 Rogge, T. et al. (2021). *Nat. Rev. Methods Primers* 1: 43.

41 Shuo, Y., Chang, J., and Yu, B. (2020). *Top Curr. Chem. (Z)* 378: 23–93.

42 Bringmann, G., Walter, R., and Weirich, R. (1990). *Angew. Chem. Int. Ed.* 29 (9): 977–991.

43 Zhang, Y.-F. and Shi, Z.-J. (2019). *Acc. Chem. Res.* 52 (1): 161–169.

44 Wang, C., Piel, I., and Glorius, F. (2009). *J. Am. Chem. Soc.* 131 (12): 4194–4195.

45 Cornella, J., Righi, M., and Larrosa, I. (2011). *Angew. Chem. Int. Ed.* 50 (40): 9429–9432.

46 Luo, J., Preciado, S., and Larrosa, I. (2015). *Chem. Commun.* 51 (15): 3127–3130.

47 Chen, M., Li, M., and Zhang, J. *ACS Catal.* 13 (22): 14907–14913.

48 Zhang, Y., Zhao, H., Zhang, M., and Su, W. (2015). *Angew. Chem. Int. Ed.* 54 (12): 3817–3821.

49 Li, X., Wang, X., and Zhang, J. (2023). *Chem. Sci.* 14 (20): 5470–5476.

50 Bhadra, S., Dzik, W.I., and Gooßen, L.J. (2013). *Angew. Chem. Int. Ed.* 52 (10): 2959–2962.

51 Pan, S., Zhou, B., Zhang, Y. et al. (2016). *Synlett* 27 (2): 277–281.

52 Shi, X.-Y., Dong, X.-F., Fan, J. et al. (2015). *Sci. China Chem.* 58 (8): 1286–1291.

53 Fu, Z., Jiang, Y., Wang, S. et al. (2009). *Org. Lett.* 21 (9): 3003–3007.

54 Lee, D. and Chang, S. (2015). *Chem. Eur. J.* 21 (14): 5364–5368.

55 Chiong, H.A., Pham, Q.N., and Daugulis, O. (2007). *J. Am. Chem. Soc.* 129 (32): 9879–9884.

56 Maehara, A., Tsurugi, H., Satoh, T., and Miura, M. (2008). *Org. Lett.* 10 (6): 1159–1162.

57 Arun, V., Pilania, M., and Kumar, D. (2016). *Chem. Asian J.* 11 (23): 3345–3349.

58 Jin, X.-Y., Xie, L.-J., Cheng, H.-P. et al. (2018). *J. Org. Chem.* 83 (14): 7514–7522.

59 Tulichala, R.N.P. and Kumara Swamy, K.C. (2015). *Chem. Commun.* 51 (60): 12008–12011.

60 Tulichala, R.N.P., Shankar, M., and Kumara Swamy, K.C. (2018). *J. Org. Chem.* 83 (8): 4375–4383.

61 Zhang, J., Xu, W., Zhuang, W. et al. (2023). *J. Org. Chem.* 88 (21): 15198–15208.

62 Yugandar, S., Morita, T., and Nakamura, H. (2020). *Org. Biomol. Chem.* 18 (42): 8625–8628.

63 Yugandar, S. and Nakamura, H. (2019). *Chem. Commun.* 55 (58): 8382–8385.

64 Wang, C., Rakshit, S., and Glorius, F. (2010). *J. Am. Chem. Soc.* 132 (40): 14006–14008.

65 Kim, K., Vasu, D., Im, H., and Hong, S. (2016). *Angew. Chem. Int. Ed.* 55 (30): 8652–8655.

66 Kumar, N.Y.P., Bechtoldt, A., Raghuvanshi, K., and Ackermann, L. (2016). *Angew. Chem. Int. Ed.* 55 (24): 6929–6932.

67 Kumar, N.Y.P., Rogge, T., Reddy Yetra, S. et al. (2017). *Chem. Eur. J.* 23 (69): 17449–17453.

68 Mandal, A., Sahoo, H., Dana, S., and Baidya, M. (2017). *Org. Lett.* 19 (15): 4138–4141.

69 Pu, F., Liu, Z.-W., Zhang, L.-Y. et al. (2019). *ChemCatChem* 11 (16): 4116–4122.

70 Schrock, R.R. (1986). *Acc. Chem. Res.* 19 (11): 342–348.

71 Heck, R.F. (1968). *J. Am. Chem. Soc.* 90 (20): 5518–5526.

72 Moritani, I. and Fujiwara, Y. (1967). *Tetrahedron Lett.* 8 (12): 1119–1122.

73 Mochida, S., Hirano, K., Satoh, T., and Miura, M. (2010). *Org. Lett.* 12 (24): 5776–5779.

74 Agasti, S., Dey, A., and Maiti, D. (2016). *Chem. Commun.* 52 (82): 12191–12194.

75 Pourkaveh, R., Podewitz, M., and Schnürch, M. (2023). *Eur. J. Org. Chem.* 26 (8): 1–6.

76 Liu, X., Liu, X., Li, X. et al. (2015). *Org. Lett.* 17 (12): 2936–2939.

77 Kathiravan, S., Anaspure, A., Zhang, T., and Nicholls, I.A. (2021). *Org. Lett.* 23 (9): 3331–3336.

78 Lu, Q., Greßies, S., Cembellin, S. et al. (2017). *Angew. Chem. Int. Ed.* 56: 12778–12782.

79 Ling, F., Ai, C., Lv, Y., and Zhonga, W. (2017). *Adv. Synth. Catal.* 2017: 3707–3712.

80 Luo, J., Preciado, S., and Larrosa, I. (2014). *J. Am. Chem. Soc.* 136 (11): 4109–4112.

81 Font, M., Spencer, A.R.A., and Larrosa, I. (2018). *Chem. Sci.* 9 (35): 7133–7137.

82 Spencer, A.R.A., Korde, R., Font, M., and Larrosa, I. (2020). *Chem. Sci.* 11 (16): 4204–4208.

83 Lerchen, A., Vasquez-Cespedes, S., and Glorius, F. (2016). *Angew. Chem. Int. Ed.* 55: 3208–3211.

84 Park, J.-W., Kang, B., and Dong, V.Y. (2018). *Angew. Chem. Int. Ed.* 130 (41): 13786–13790.

85 Zhao, D., Shi, Z., and Glorius, F. (2013). *Angew. Chem. Int. Ed.* 52: 12426–12429.

86 Zhou, S., Wang, M., Wang, L. et al. (2016). *Org. Lett.* 18 (21): 5632–5635.

87 Lv, G., Zhang, Q., Zhang, C. et al. (2023). *Org. Lett.* 25 (22): 4022–4027.

88 Huang, X., Huang, J., Du, C. et al. (2013). *Angew. Chem. Int. Ed.* 52: 12970–12974.

89 Qi, Z., Wang, M., and Li, X. (2013). *Org. Lett.* 15 (21): 5440–5443.

90 Xie, F., Yu, S., Qi, Z., and Li, X. (2016). *Angew. Chem. Int. Ed.* 55: 15351–15355.

91 Wu, Y., Feng, L.-J., Lu, X. et al. (2014). *Chem. Commun.* 50: 15352–15354.

92 Takahashi, Y., Tsuji, H., and Kawatsura, M. (2020). *J. Org. Chem.* 85 (4): 2654–2665.

93 Preshlock, S.M., Plattner, D.L., Maligres, P.E. et al. (2013). *Angew. Chem. Int. Ed.* 52 (49): 12915–12919.

94 Hua, Y., Asgari, P., Avullala, T., and Jeon, J. (2016). *J. Am. Chem. Soc.* 138 (25): 7982–7991.

95 Gómez-Arrayas, R. and Rodríguez, N. (2017). *Sci. Synth.* 4: 449–478.

96 Shankar, M., Guntreddi, T., Ramesh, E., and Sahoo, A.K. (2017). *Org. Lett.* 19 (20): 5665–5668.

97 Hanchate, V., Kumar, A., and Prabhu, K.R. (2019). *Org. Lett.* 21 (20): 8424–8428.

98 Xu, Y., Yang, X., Zhou, X. et al. (2017). *Org. Lett.* 19 (16): 4307–4310.

99 Huang, C., Chattopadhyay, B., and Gevorgyan, V. (2011). *J. Am. Chem. Soc.* 133 (32): 12406–12409.

100 Wang, Y. and Gevorgyan, V. (2015). *Angew. Chem. Int. Ed.* 54 (7): 2255–2259.

101 Andries-Ulmer, A., Brunner, C., Rehbein, J., and Gulder, T. (2018). *J. Am. Chem. Soc.* 140 (40): 13034–13041.

4

Removable Directing Group in Proximal C—H Functionalization

Vikash Kumar, Malati Das, Sivakumar Sudharsan, and Parthasarathy Gandeepan

Department of Chemistry, Indian Institute of Technology Tirupati, Yerpedu-Venkatagiri Road, Yerpedu Post, Tirupati 517619, Andhra Pradesh, India

4.1 Introduction

Transition-metal (TM) catalysis has become a crucial technique in modern chemical synthesis, significantly improving production efficiency and reducing costs, ultimately benefiting human life [1]. Over the past several decades, TM-catalyzed C—H activation and functionalization have emerged as powerful methods in organic synthesis, allowing for the preparation of complex organic molecules from simple and less functionalized starting materials [2]. Despite the advantages of C—H activation, achieving positional selectivity remains a major challenge due to the presence of multiple C—H bonds with similar reactivity within a molecule.

Chelation-controlled regio- and stereoselective transformations are well known in traditional organic synthesis [3]. For TM-catalyzed reactions, reversible chelation of the catalyst to a functional group (FG) is crucial for efficient catalyst turnover. The use of native FGs like aldehyde, ketone [4], carboxylic acid [5], amide [6], amine [7], hydroxyl group [8], alkene [9], cyano [10], and others [11] as directing groups (DGs) in C—H activation has been extensively studied. However, these groups often exhibit either weak or excessively strong interactions with the catalysts or form energetically unfavorable metallacycles. These issues can be addressed by introducing a specific DG to the substrate or modifying an existing native FG with the DG (Figure 4.1) [12]. Generally, an ideal DG for C—H activation should have the following properties: (i) reversible coordination with the catalyst, (ii) efficient regiocontrol, and (iii) ease of installation and removal. Over the past three decades, numerous TM-catalyzed regioselective C—H activation and functionalization reactions using removable directing groups (RDGs) have been developed [13]. This chapter focuses on important *ortho*-selective C—H activation reactions facilitated by RDGs. Reactions involving native FGs, non-removable DGs, transient DGs, and template-directed remote functionalization are beyond the scope of this chapter. Furthermore, only selected examples are discussed, rather than providing

Directed C–H Bond Functionalization: Concepts and Applications, First Edition.
Edited by Debabrata Maiti and Supriya Rej.

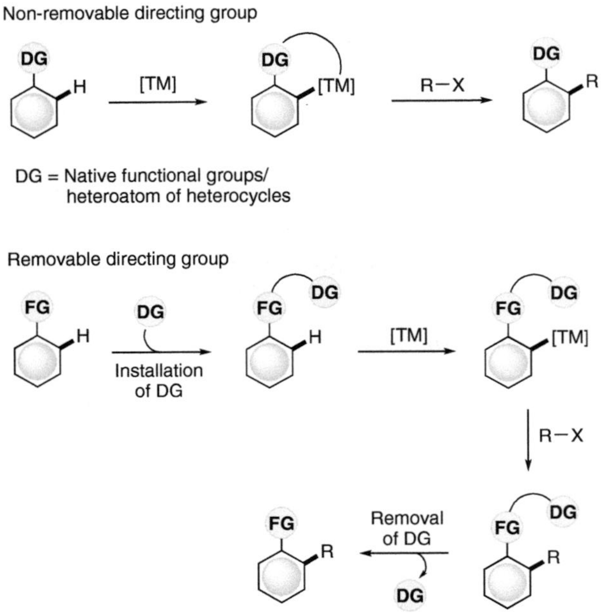

Figure 4.1 Non-removable and removable directing group strategy in C—H activation. DG, directing group; FG, functional group; TM, transition metal.

a comprehensive summary of all reactions for each DG. For further exploration, readers are directed to existing literature reviews.

4.2 Removable Directing Groups

4.2.1 C—H Functionalization of Amino Compounds

The amine (-NH_2) group is the most common FG in organic compounds. Regioselective functionalization of C—H bonds in both aromatic and aliphatic amines is well studied. A wide range of RDGs has been developed and employed to control the positional selectivity of the C—H bond to be functionalized (Figure 4.2). In an early example, Leeuwen achieved the *ortho*-selective alkenylation of acyl-protected aniline derivatives **1** in the presence of a palladium(II) catalyst (Scheme 4.1) [14]. Various anilides, such as acetanilide, formanilide, and benzanilide, were found to be suitable RDGs. In subsequent years, numerous oxidative C—H alkenylations were realized using anilide DGs [15]. The anilide DGs were not limited to alkenylation reactions. A wide range of *ortho*-selective C—H functionalization, such as arylation [16], acylation [17], amination [18], amidation [19], ethoxycarboxylation [20], carboxylation [21], alkoxylation [22], alkylation [23], halogenation [24], and annulation reactions [25], were also achieved (Figure 4.3). The acyl protection of anilines can be easily removed by acid or base treatment to obtain the free-NH_2 group.

Figure 4.2 Most commonly used RDGs for *ortho*-C—H functionalization of aniline derivatives.

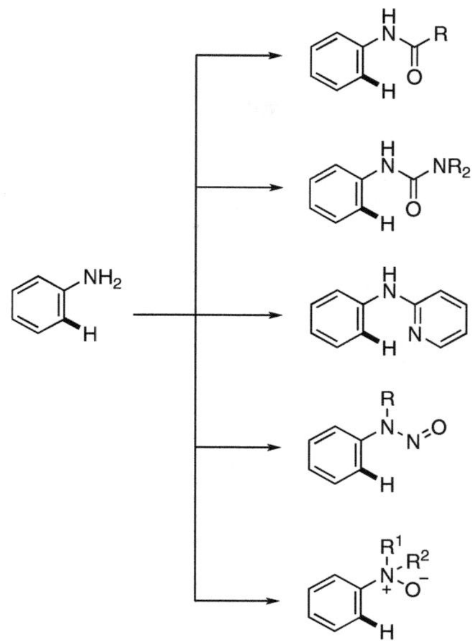

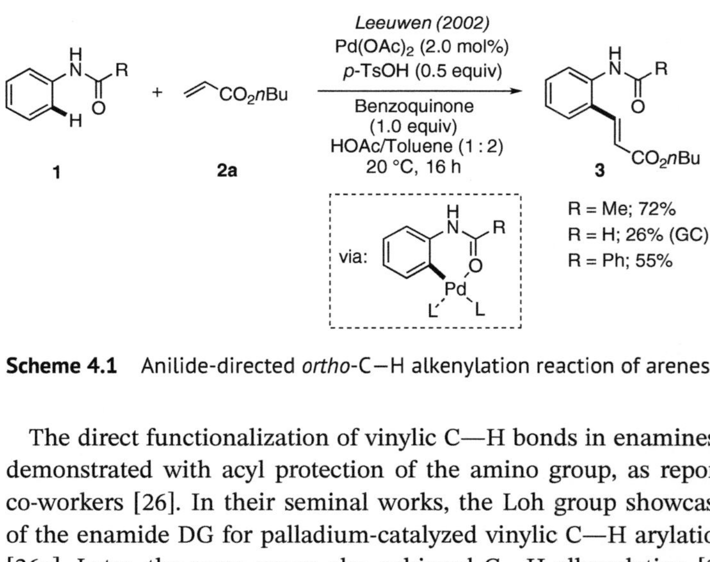

Scheme 4.1 Anilide-directed *ortho*-C—H alkenylation reaction of arenes.

The direct functionalization of vinylic C—H bonds in enamines **4** has also been demonstrated with acyl protection of the amino group, as reported by Loh and co-workers [26]. In their seminal works, the Loh group showcased the potential of the enamide DG for palladium-catalyzed vinylic C—H arylations (Scheme 4.2) [26g]. Later, the same group also achieved C—H alkenylation [26e] and alkynylation reactions [26a, b]. It was evidenced that the acetyl group of the enamide is involved in the cyclometallation step. Subsequently, a wide range of vinylic C—H functionalizations were realized using the enamide DG [27].

The amino group can be readily converted into urea derivatives and vice versa. Hence, urea derivatives of anilines are well studied as DGs in C—H activation reactions [28]. For example, Lipshutz demonstrated the C—H arylation of aryl urea **8** in water as the reaction medium (Scheme 4.3a) [28b]. In subsequent efforts, the *N*-nitroso derivative of anilines **11** has also proven to be a suitable RDG for

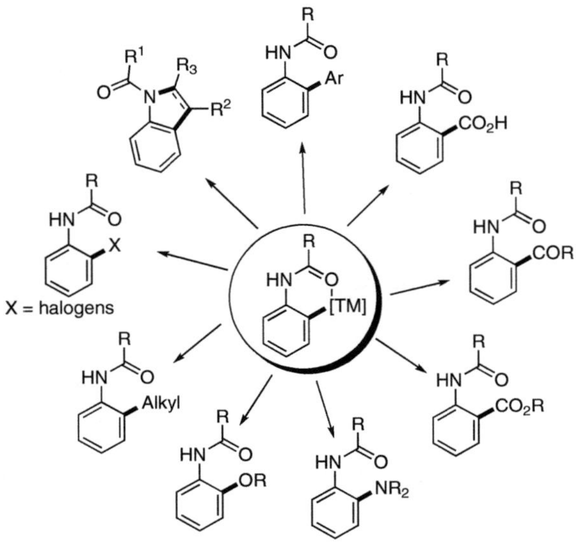

Figure 4.3 *Ortho*-selective C–H functionalization of anilides.

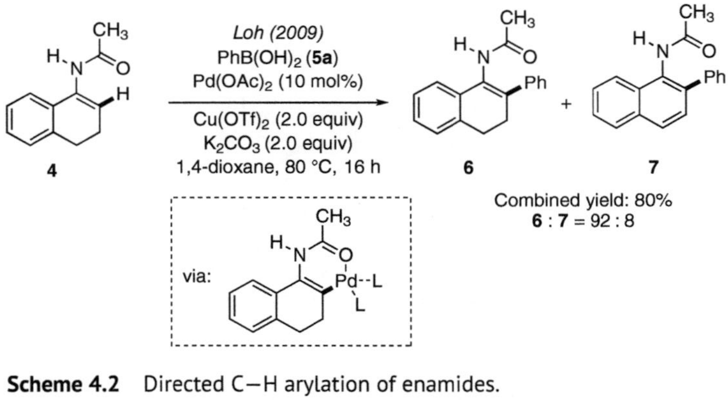

Scheme 4.2 Directed C–H arylation of enamides.

TM-catalyzed C—H activation reactions [29]. For instance, Zhu showcased the directing ability of the *N*-nitroso group for rhodium(III)-catalyzed C—H olefination and alkyne annulation reactions (Scheme 4.3b) [29c]. The authors have also demonstrated the denitrosation of **12** under mild reaction conditions [29c]. In another study, the You group reported a C—H olefination reaction using *N*-oxide as the RDG for the synthesis of *ortho*-alkenylated tertiary anilines **14** (Scheme 4.3c) [30].

The Carretero group developed *ortho*-functionalization of anilines **15** via *N*-(2-pyridyl)sulfonyl-group-directed C—H metalation [31]. This strategy was successfully applied to alkenylation and halogenation reactions (Scheme 4.4). This *N*-(2-pyridyl)sulfonyl DG could be easily removed to obtain a free amine under reductive conditions using an excess of zinc.

N-Pyridyl and *N*-pyrimidyl are the most studied RDGs for the *ortho*-selective C—H functionalization of anilines [32–36]. These DGs are well suited for 3d- and 4d-TM-catalyzed C—H activation reactions. A wide range of reactions,

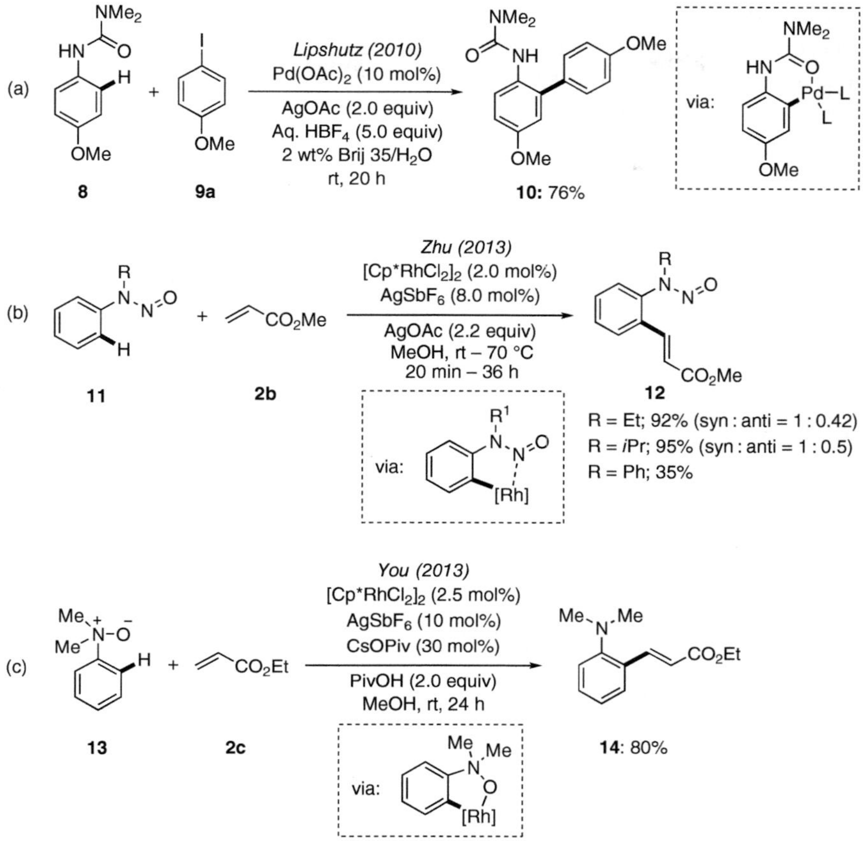

Scheme 4.3 (a) Urea-directed C—H activation. (b) *N*-Nitroso-directed C—H activation. (c) *N*-Oxide-directed C—H activation.

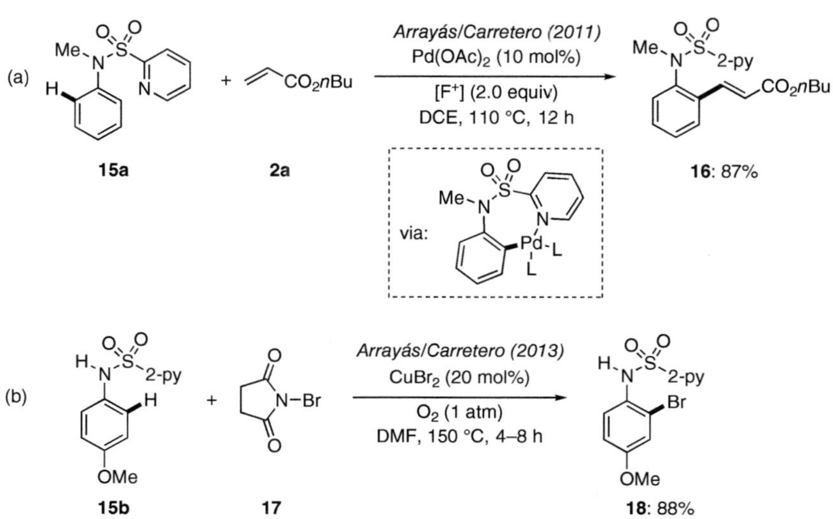

Scheme 4.4 *N*-(2-Pyridyl)sulfonyl-group-directed C—H activation reactions. (a) Alkenylation; (b) halogenation. [F$^+$], *N*-Fluoro-2,4,6-trimethylpyridinium triflate.

including amination [32], annulation [33], alkylation [34], alkynylation [35], and chalcogenation [36], have been explored (Scheme 4.5). Removal of *N*-pyrimidyl DG can be achieved in a traceless fashion under acidic microwave reaction conditions (Scheme 4.6) [32, 34, 36b].

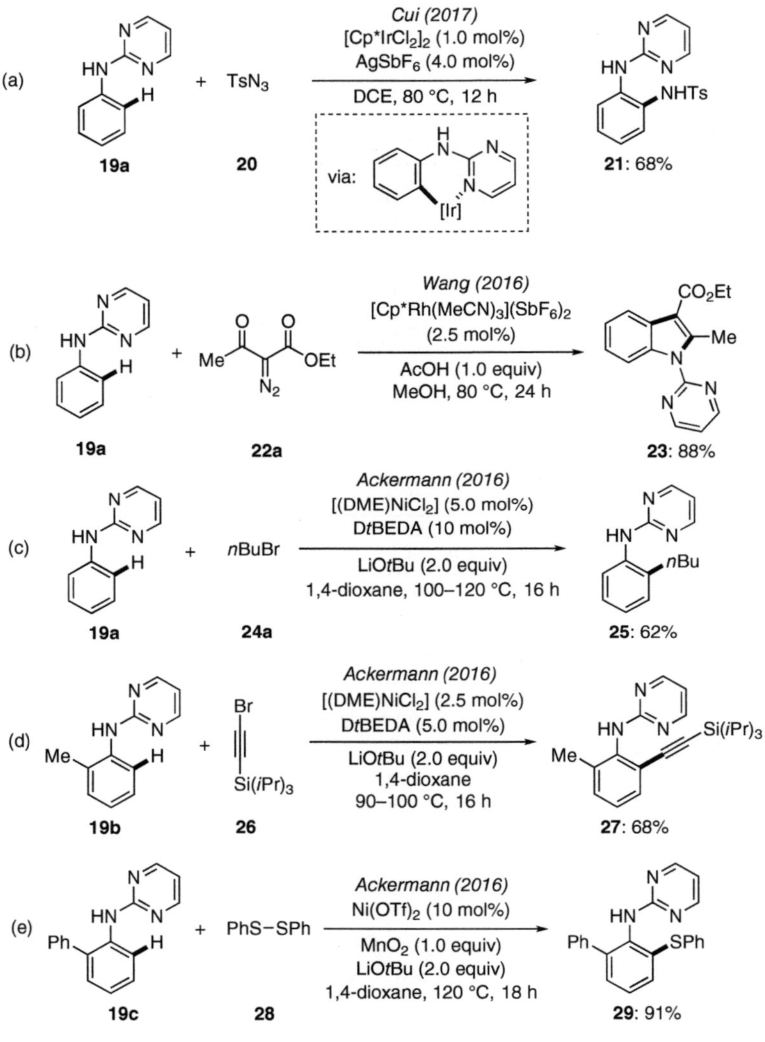

Scheme 4.5 *N*-(2-Pyrimidyl)-group-directed *ortho*-C–H activation of anilines. (a) Amidation; (b) annulation; (c) alkylation; (d) alkynylation; (e) chalcogenation. D*t*BEDA, *N*,*N'*-Di-*tert*-butylethylenediamine.

4.2.2 C–H Functionalization of Hydroxyl Compounds

Organic compounds containing hydroxyl (–OH) FG are essential in various fields due to their high reactivity and hydrogen bonding capabilities. Continuous efforts

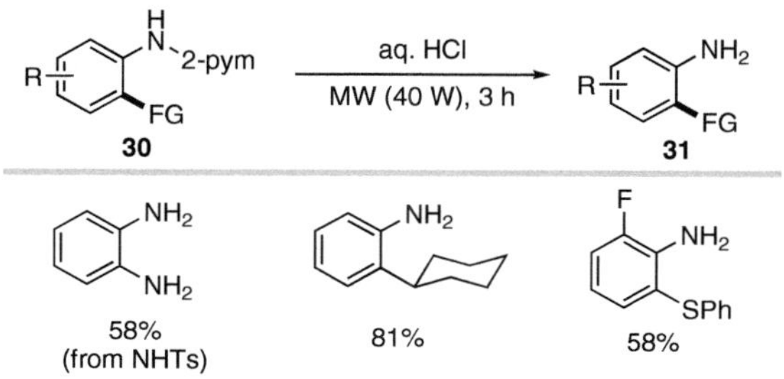

Scheme 4.6 Removal of *N*-pyrimidyl DG in a traceless fashion.

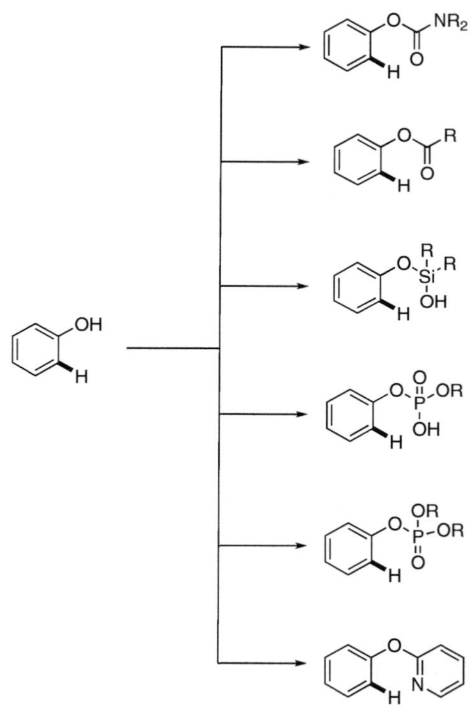

Figure 4.4 Most commonly used RDGs for *ortho*-C—H functionalization of phenol derivatives.

are ongoing to develop novel and efficient methods to access highly functionalized phenolic compounds [37]. In this context, several C—H activation methods have also been developed to access useful compounds [8c, d, 38]. However, these known methods often allow access to a limited range of functionalities and can be challenging to achieve high efficiencies in directed C—H activation. Hence, a wide range of RDGs have been installed by protecting the -OH FG to perform *ortho*-selective C—H bond activation of phenolic compounds (Figure 4.4).

Phenols can be readily converted into *O*-phenyl carbamates by treating them with carbamoyl chlorides in the presence of a base. Similarly, the reversal of *O*-phenyl carbamates into phenols can be readily achieved by reacting them with NaOH/EtOH. Hence, the carbamate DG is widely used as an RDG for the *ortho*-functionalization of phenols [8b, 39]. A wide range of transformations, including arylation [40], alkenylation [41], halogenation [42], and others [43], have been established in the presence of TM catalysts (Scheme 4.7). Furthermore, phenol *ortho*-C—H functionalizations have also been achieved via phenol ester form **37** (Scheme 4.8) [44]. Here, the acyloxy group acts as the RDG.

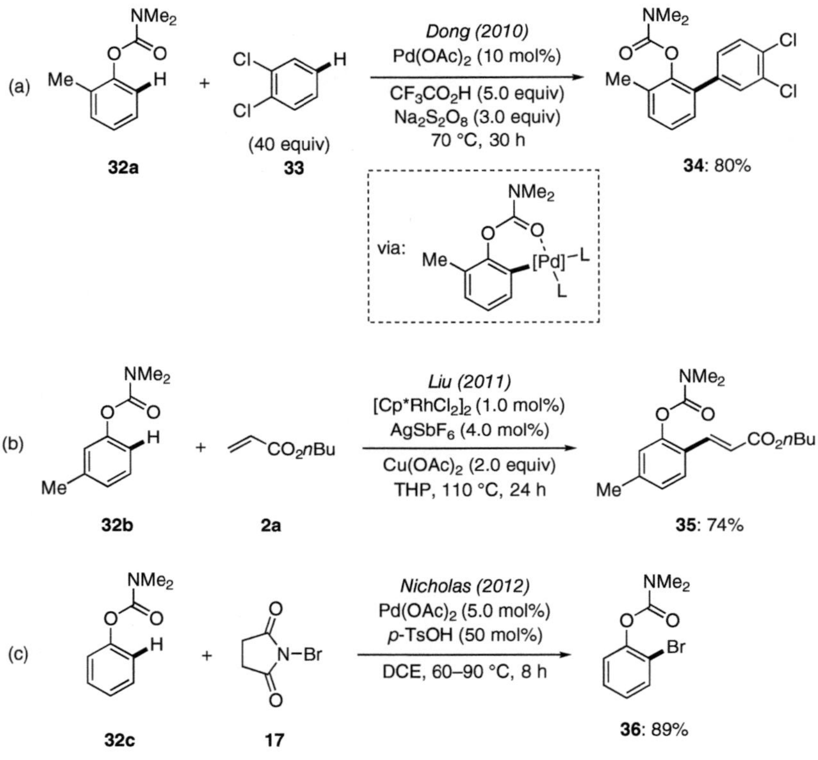

Scheme 4.7 Carbamate-directed *ortho*-C—H functionalization of phenol derivatives. (a) Arylation; (b) alkenylation; (c) bromination. THP, Tetrahydropyran.

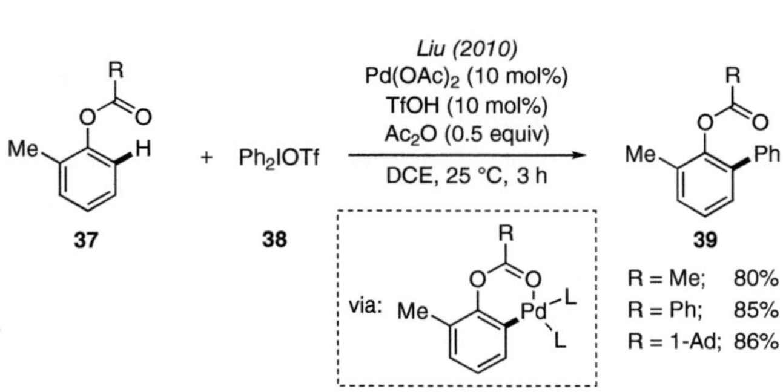

Scheme 4.8 Acyloxy-group-directed *ortho*-C—H arylation of phenol esters.

In further developments, Gevorgyan demonstrated that silanol **40** could act as an RDG for *ortho*-selective C—H activation (Scheme 4.9a) [45]. The authors also demonstrated the deprotection of silanol DG at room temperature using tetrabutylammonium fluoride (TBAF). Similarly, phosphates **43** [46] and mono-phosphoric acid groups [47] were employed as RDGs in the *ortho*-selective C—H functionalization of phenols (Scheme 4.9b).

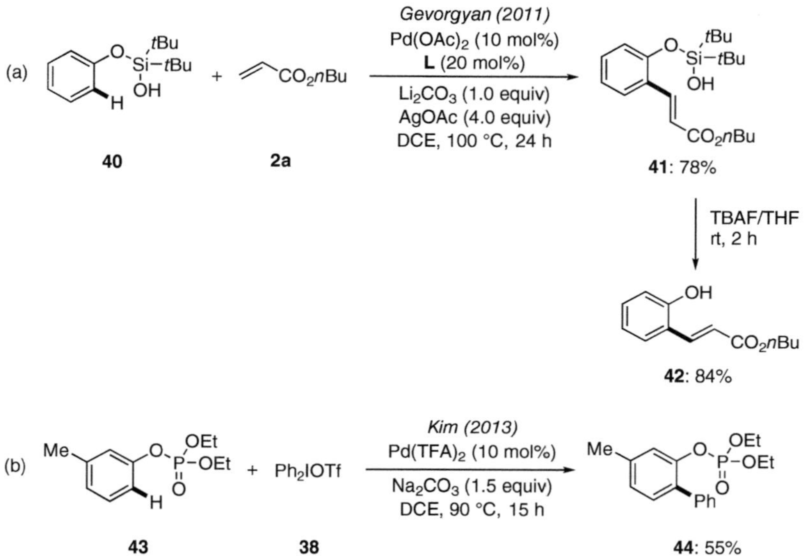

Scheme 4.9 *Ortho*-selective C—H functionalization of phenols using (a) silanol DG and (b) phosphate DG. **L**, (+) Menthyl(O_2C)-Leu-OH.

Regioselective C—H functionalization of phenols has also been realized using a pyridine DG. For instance, 2-phenoxypyridines **45** were conveniently functionalized via pyridine-directed C—H metalation. This strategy enabled a wide range of transformations, including arylation [48], alkenylation [49], acylation [50], nitration [51], and many others [52] (Scheme 4.10). The depyridinylation of functionalized 2-phenoxypyridines **52** was achieved by treating them with methyl trifluoromethanesulfonate (MeOTf) in toluene followed by Na/MeOH solution (Scheme 4.11) [48c, 49a, 51b].

4.2.3 C—H Functionalization of Aldehyde and Ketone Compounds

Aldehydes and ketones are important carbonyl functional groups widely present in diverse bioactive compounds and functional materials [53]. They are also key starting materials in a variety of organic transformations. Consequently, there is a high demand for widely decorated aromatic aldehydes and ketones. In this context, carbonyl-group-directed C—H activation strategies have been well exploited for the functionalization of these compounds [4, 54]. However, the weak coordination

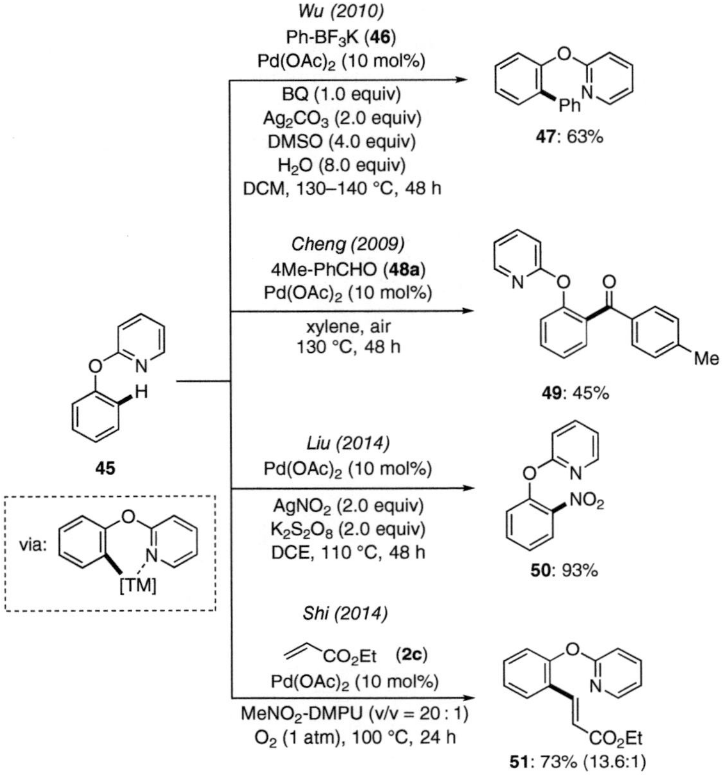

Scheme 4.10 Pyridine-directed C–H functionalization of phenols. BQ, Benzoquinone. DMPU, *N,N′*-Dimethylpropyleneurea.

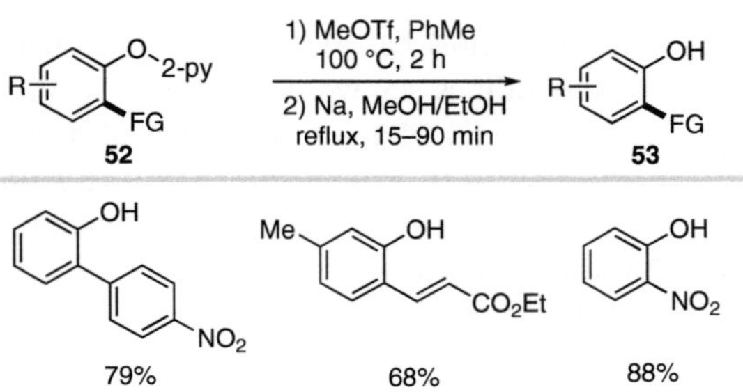

Scheme 4.11 Traceless removal of *O*-pyridyl DG.

ability of the carbonyl group (C=O) often makes these reactions less efficient, requiring harsh reaction conditions and late TM catalysts [55]. To overcome these limitations, a wide range of easily accessible RDGs have been derived from carbonyl compounds to utilize directed C—H activation effectively (Figure 4.5).

In an early work, Satoh and Miura observed that *NH*-imine could be the DG in rhodium(I)-catalyzed C—H arylation reactions (Scheme 4.12) [56]. Subsequently,

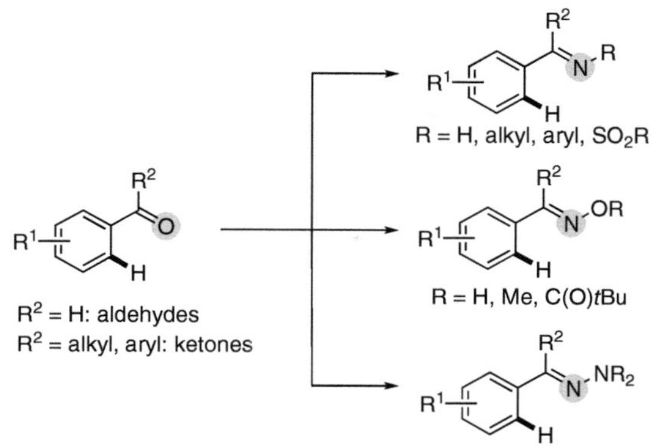

Figure 4.5 RDGs for *ortho*-C—H functionalization of aldehyde and ketone derivatives.

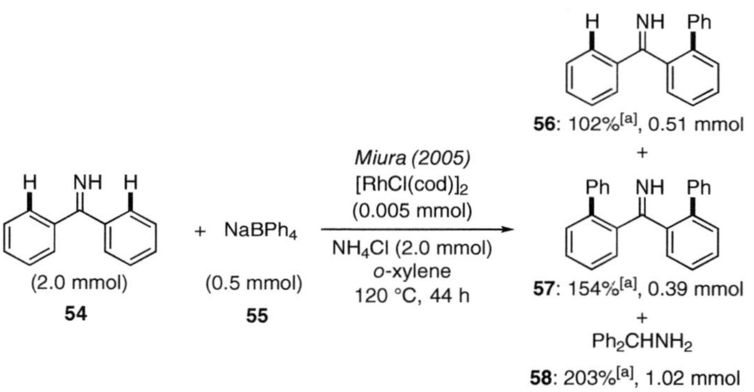

Scheme 4.12 *NH*-Imine-directed C—H arylation with aryl boron compound. [a] GC yield based on the amount of aryl boron reagent used.

several annulation reactions were reported between aromatic *NH*-ketimines **54** and various π-components, such as alkynes and allenes [57]. Further improvements were achieved in the *ortho*-C—H functionalization of carbonyl compounds using easily accessible *N*-alkyl or *N*-aryl imines. For instance, the *ortho*-selective C—H alkylation of ketimines **59** with olefins **60** was realized using Wilkinson's complex $[Rh(PPh_3)_3Cl]$ (Scheme 4.13a) [58]. Related transformations were also accomplished in an intramolecular fashion [59]. Later, Yoshikai achieved C—H alkylation and alkenylation of aromatic imines **62** with alkyl halides **24** through cobalt catalysis (Scheme 4.13b) [60]. Furthermore, a wide range of isoquinoline derivatives were synthesized by the reaction of aromatic ketimines with alkynes [61]. In addition to imines derived from alkyl or aryl amines, *N*-sulfonyl imines were also proved to be an efficient directing for TM-catalyzed C—H activation reactions [62]. For example, the synthesis of *ortho*-olefinated benzaldehydes was achieved via rhodium(III)-catalyzed olefination of *N*-sulfonyl imines **64** with activated alkenes **2c** (Scheme 4.14) [62b]. All the above-mentioned imine derivatives can readily revert to the original aldehyde or ketone FGs through simple aqueous acid hydrolysis.

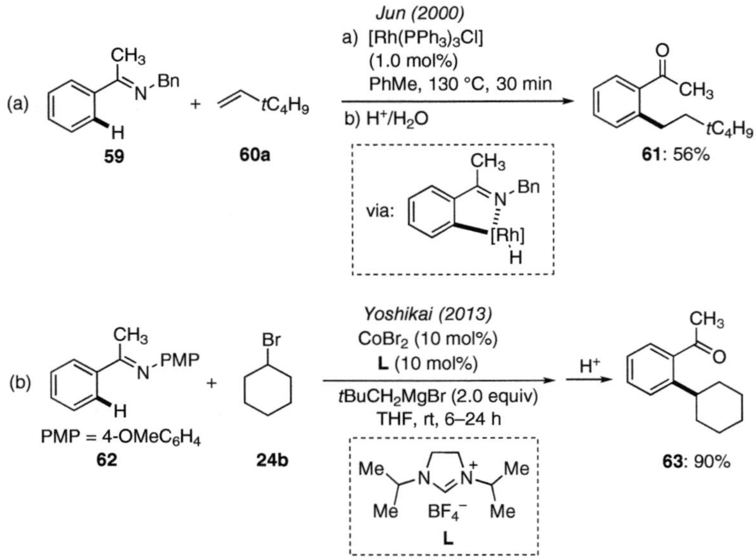

Scheme 4.13 *N*-Alkyl or *N*-aryl-imine-directed C–H alkylation. (a) Using alkenes. (b) Using alkyl bromide.

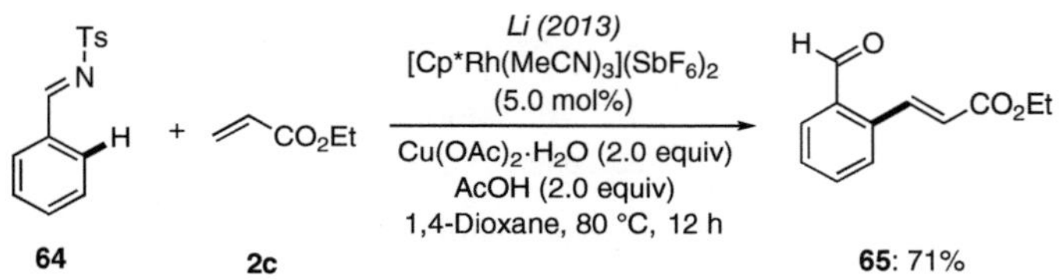

Scheme 4.14 *N*-Sulfonyl-imine-directed C–H alkenylation.

Oximes are another class of protecting groups for carbonyl compounds, easily synthesized from aldehydes or ketones and hydroxylamine or its derivatives. They are more stable and can be synthesized under mild reaction conditions. Oxime derivatives **66** of carbonyl compounds have been widely used as DGs for TM-catalyzed C—H activation reactions. In an early study, Cheng synthesized substituted pyridines from α,β-unsaturated ketoximes and alkynes under rhodium catalysis [63], where oxime-directed C—H regioselective metalation was proposed as a key intermediate. Subsequent efforts led to the achievement of several related annulations [64]. Following these annulation transformations [65], a wide range of C—C and C-heteroatom bond formation reactions were achieved using oxime ether derivatives of aromatic aldehydes and ketones (Scheme 4.15) [66]. In addition to imines and oximes, hydrazones are another common DGs derived from aldehydes, and ketones were also well exploited in directed C—H activation reactions [67].

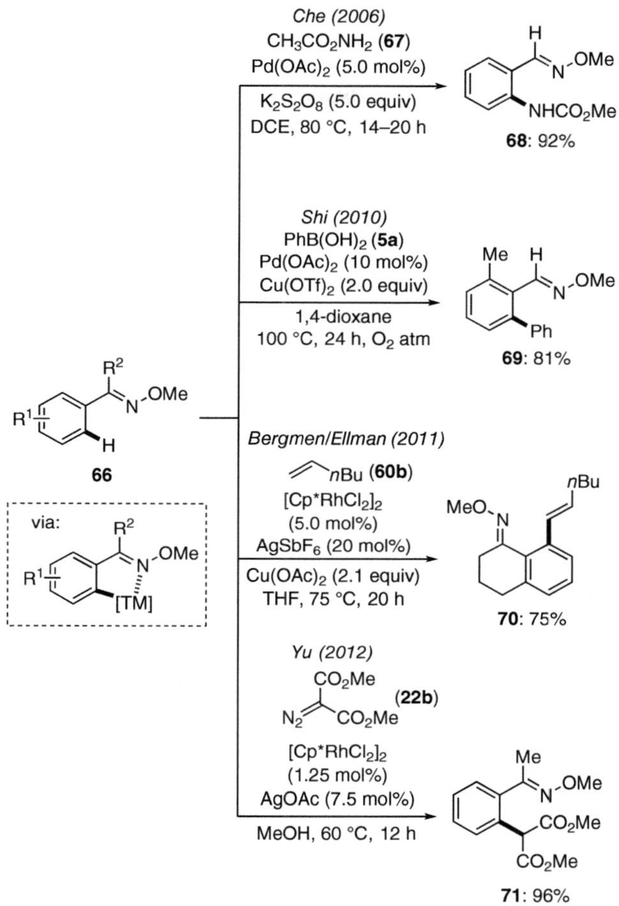

Scheme 4.15 Selected examples of oxime-ether-directed C—H activation reactions.

4.2.4 C—H Functionalization of Carboxylic Acids

Carboxylic acids are among the most common and important building blocks in organic chemistry for the synthesis of useful compounds [68]. Consequently, continuous efforts have been made to develop efficient methods to synthesize diversely substituted benzoic acids. Notably, several C—H activation transformations have been reported using carboxylic acid as a native DG [5b, 68b, 69]. However, the use of weakly coordinating carboxylic acid DG often requires 4d- and 5d-TM catalysts, harsh reaction conditions, and offers limited scope. Therefore, converting the weakly coordinating carboxylic acid FG into a strongly coordinating amide group is a widely used strategy in directed C—H activation reactions. This amide DG can act as a monodentate or bidentate chelating group, as illustrated in Figure 4.6.

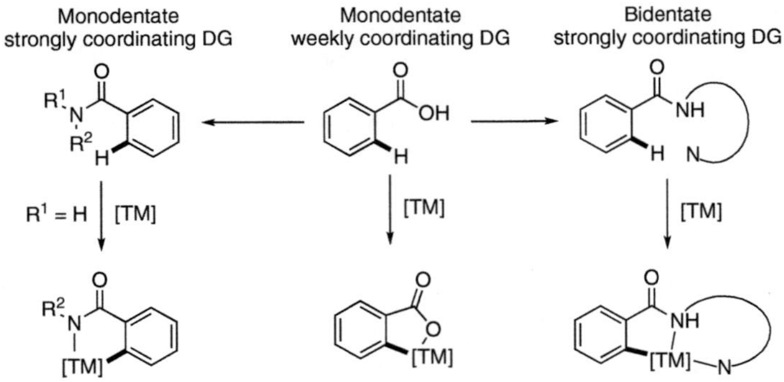

Figure 4.6 RDG strategies for carboxylic acid C–H functionalization.

Given the large number of C—H activation reactions reported using amide DGs, this section will discuss only a few representative examples.

So far, a wide range of monodentate amide DGs have been developed for chelation-assisted C—H activation reactions (Figure 4.7) [6a, b, 70]. For example, primary amide **72** (—$CONH_2$) has been used as the removable DG for palladium-catalyzed *ortho*-C—H arylation reactions (Scheme 4.16) [71]. While primary amides have served as DGs, secondary or tertiary amides are more commonly employed in directed C—H activation. For instance, a wide range of *N*-alkyl or *N*-aryl secondary benzamides **74** have been efficiently functionalized at the *ortho*

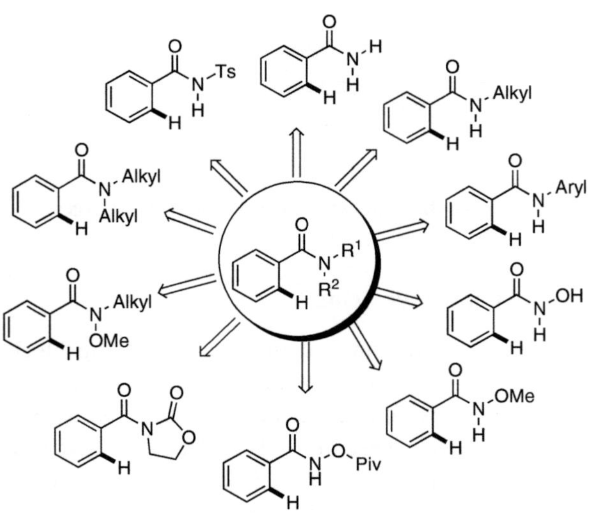

Figure 4.7 Commonly used monodentate amide DGs for transition-metal-catalyzed C–H activation.

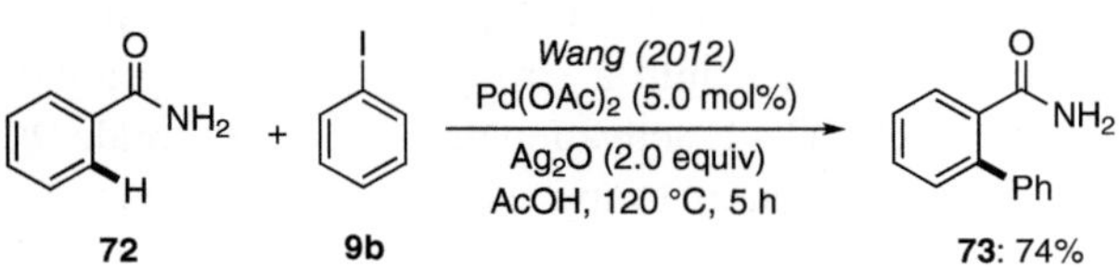

Scheme 4.16 Primary amide-directed C–H arylation reaction.

position through different types of reactions, including arylation [72], alkylation [73], alkenylation [74], amination [75], halogenation [76], borylation [77], and others [78] (Scheme 4.17). In addition to *N*-alkyl or *N*-aryl secondary amides, *N*-alkoxy benzamides have also been well established as removable DGs for the directed C—H activation reactions [79]. Similarly, *N*-Ts-substituted benzamides have proven to be viable FGs for the chelation-assisted C—H activation reactions [80]. Moreover, catalytic C—H activation assisted by *N,N*-di-alkyl tertiary amides has been widely utilized for various transformations, including arylation, acylation, allylation, alkynylation, halogenation, and hydroxylation [81]. Furthermore, Weinreb amides, another class of tertiary amides, have also been demonstrated as convenient substrates for directed C—H activation reactions [6d, 82].

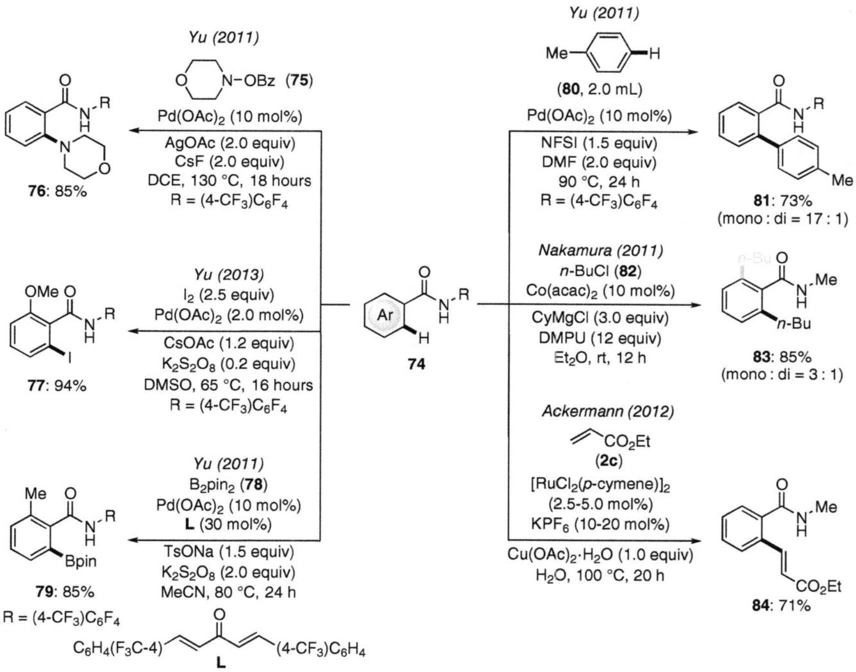

Scheme 4.17 Selected examples of secondary amide-directed C—H activation reactions. NFSI, *N*-Fluorobenzenesulfonimide.

Despite significant advances in monodentate amide DG-assisted C—H activation reactions, most of the methods still rely on late TM catalysts. To improve the chelation efficiency of amide DGs suitable to a wide range of reactions and to enable the use of 3d-TM catalysts, many amide-based bidentate DGs have been developed (Figure 4.8) [83]. The amide group derived from 8-aminoquinoline (8-AQ) auxiliary has been used as monoanionic *N,N*-coordinating DG for close-proximity C—H bond activation using both early and late-TM catalysts [83f]. The earliest example of 8-AQ-directed C—H activation was reported by Daugulis, where arylation of aliphatic **85** and aromatic amides **87** with aryl iodides **9** was achieved through

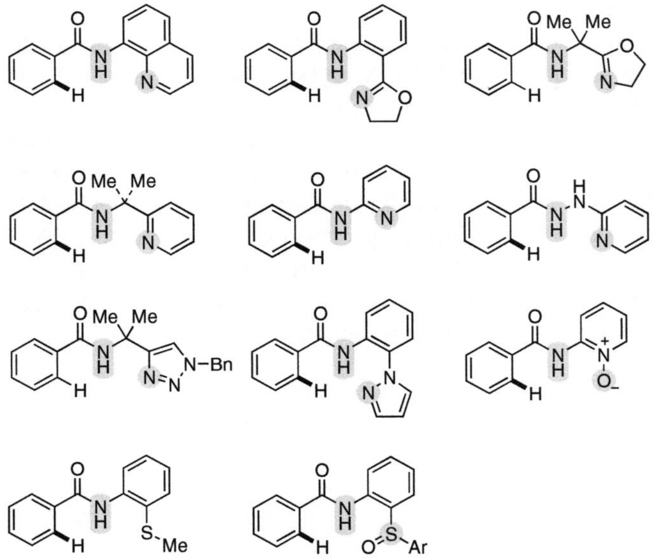

Figure 4.8 Most common bidentate DGs for aromatic C—H activation.

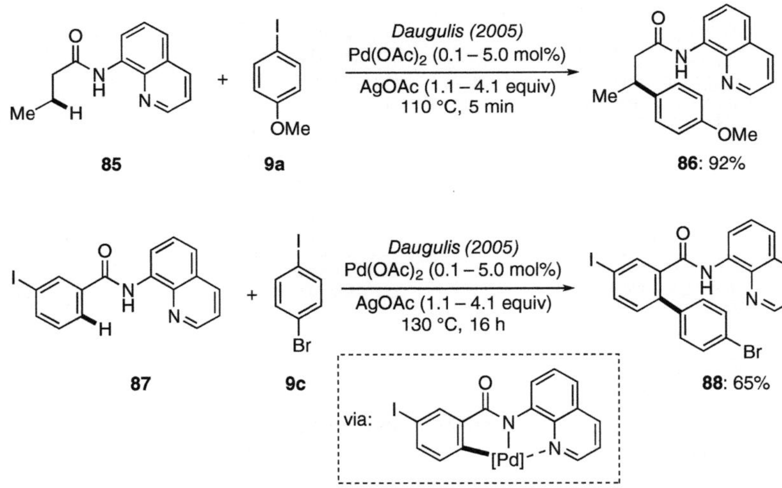

Scheme 4.18 Pioneering examples of 8-aminoquinoline auxiliary-directed C—H activation of aliphatic and aromatic amides.

palladium catalysis (Scheme 4.18) [84]. Following this breakthrough, several C—H activation reactions assisted by 8-AQ auxiliary were developed, which are nicely summarized in the review article by Chatani [83c]. The 8-AQ DG can be readily removed by the base hydrolysis [13a]. For example, the treatment of compound **89** with NaOH/EtOH afforded the corresponding carboxylic acid **90** in 91% yield (Scheme 4.19) [85].

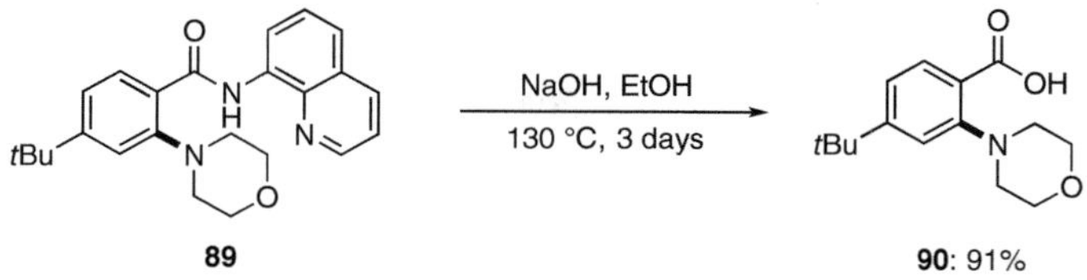

Scheme 4.19 Removal of 8-AQ DG.

In 2008, Yu introduced a 2-aminophenyloxazoline auxiliary for directed C—H activation reactions (Scheme 4.20) [86]. This *N,N*-bidentate DG forms a 5,6-membered metallacycle intermediate via C—H metalation with TM catalysts. Inspired by this seminal work, several C—C and C–heteroatom bond-forming reactions have since been reported using a 2-aminophenyloxazoline as a DG [87e]. Removal of 2-aminophenyloxazoline DG was conveniently achieved by base hydrolysis (Scheme 4.20c) [87c, h]. In 2009, Chatani reported the first 2-pyridylmethylamine (PM)-directed C—H activation of aromatic amides **98** in the presence of a ruthenium catalyst (Scheme 4.21a) [88]. In subsequent studies, 2-pyridylisopropylamine (PIP) [83a] derived *N,N*-bidentate DGs have proven suitable for various TM-catalyzed C—H activation reactions (Scheme 4.21) [89]. Like that of 8-AQ or 2-aminophenyloxazoline DGs, the PIP DG is also readily cleaved by acid or base hydrolysis to afford the respective carboxylic acid compounds [89n, q]. Later, the 2-hydrazinylpyridine moiety was also shown to function as an *N,N*-bidentate DG for C—H activation reactions [90]. In this context, an early example of 2-hydrazinylpyridine auxiliary was reported by Zhai in 2018 (Scheme 4.22) [90b].

In 2014, Ackermann demonstrated C–H activation of aromatic **114** and aliphatic **118** amides using triazolyldimethylmethyl (TAM) amine auxiliary as an *N,N*-bidentate DG (Scheme 4.23) [91]. Following this, the potential of TAM-based DGs was extensively explored in various TM-catalyzed reactions [92]. Notably, this triazole-based DG has been well-studied in iron-catalyzed C—H activation reactions. The TAM DG can be easily removed in a traceless fashion through aqueous acid hydrolysis [91]. Continued research into the development of removable bidentate DG for C—H activation reactions led to the discovery of many other suitable auxiliaries, including pyrazole-based DGs [93], 2-methoxyiminoacetyl [94], 2-amino-3-fluoro-*N,N*-dimethylbenzamide [95], 2-dimethylaminoethylamine [96], isoleucine-NH_2 [97], and *N*-(2-aminophenyl)acetamide [98]. Additionally, several other DGs have been identified as suitable removable options for amide C—H activation [99].

In addition to *N,N*-bidentate DGs, *N,O*-bidentate auxiliaries have also been widely studied for regioselective C—H activation reactions [83c]. Among the various *N,O*-bidentate ligands developed, 2-aminopyridine-*N*-oxide **120** has been well explored for positional-selective C–H activation reactions with different metal catalysts (Scheme 4.24) [100]. Similarly, *N,S*-bidentate systems, such as 2-thiomethylaniline-based [101] and 2-sulfinylaniline-based auxiliaries [102], have proven to be effective for chelation-assisted C—H activation reactions

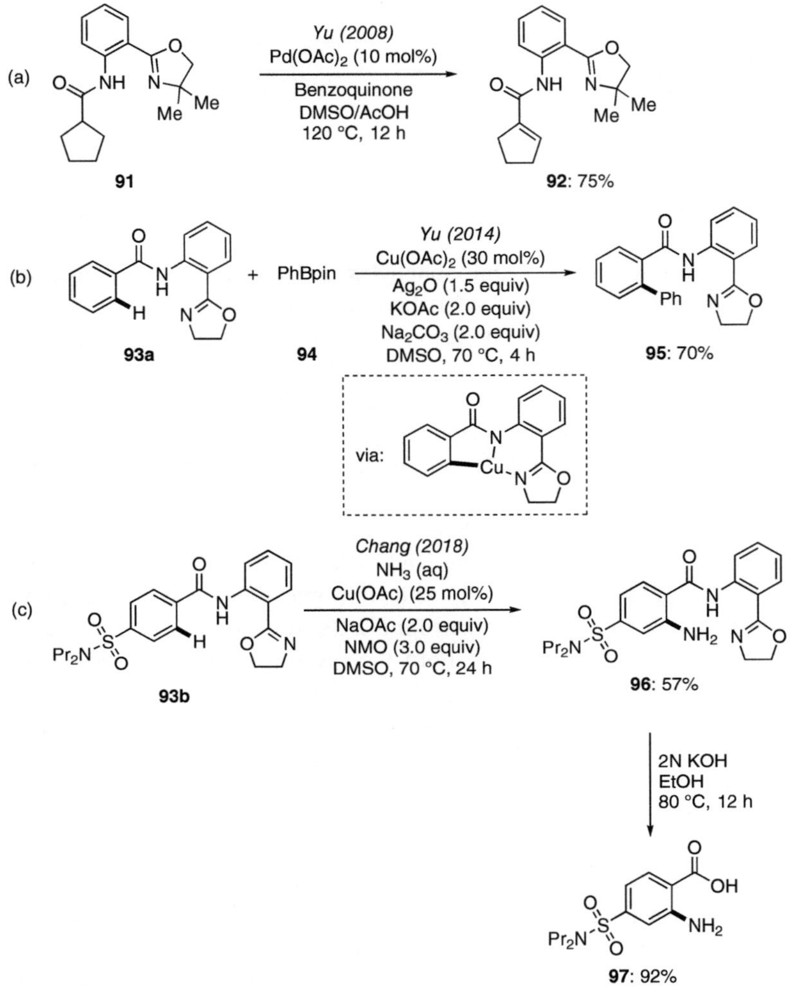

Scheme 4.20 Examples of 2-aminophenyloxazoline DG-assisted C—H activation. (a) Palladium(II)-catalyzed dehydrogenation reaction. (b) Copper(II)-catalyzed arylation reaction. (c) Copper-catalyzed amination using ammonia. NMO, *N*-Methylmorpholine *N*-oxide.

(Scheme 4.25). The removal of *N,O*- and *N,S*-bidentate DGs can be achieved by basic hydrolysis using NaOH/EtOH [100m].

4.2.5 C—H Functionalization of Sulfonic Acid

Sulfonic acid and its derivatives are important FGs in organic synthesis. Consequently, direct functionalization of these compounds has been achieved [103]. For example, Yu has reported a palladium-catalyzed C—H activation method using sulfonamide DG [103c] (Scheme 4.26). The sulfonamide DG is effective for alkylation,

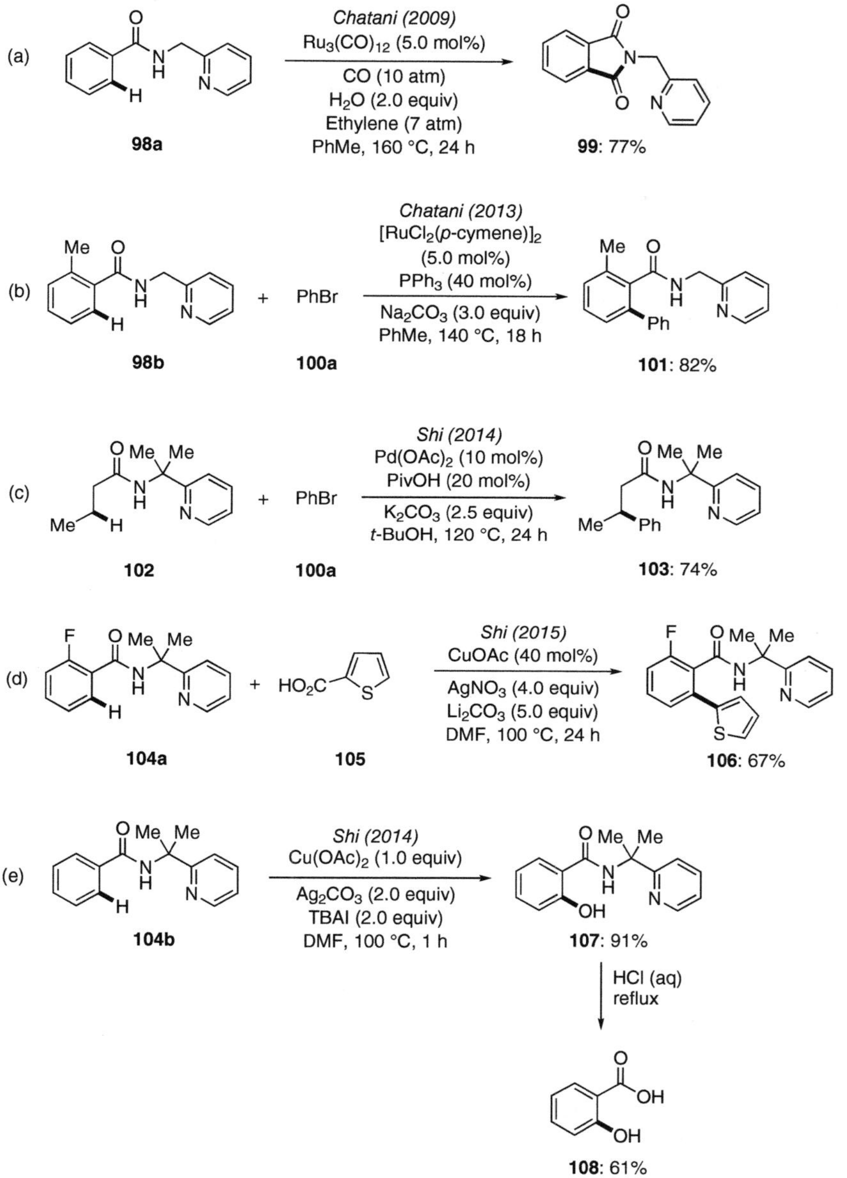

Scheme 4.21 Pyridine-based *N,N*-bidentate removable DGs for C—H activation of amides. (a) Ruthenium-catalyzed carbonylation. (b) Ruthenium-catalyzed arylation. (c) Palladium-catalyzed C(sp^3)—H arylation. (d) Copper-catalyzed decarboxylative arylation. (e) Copper-catalyzed hydroxylation. TBAI, Tetrabutylammonium iodide.

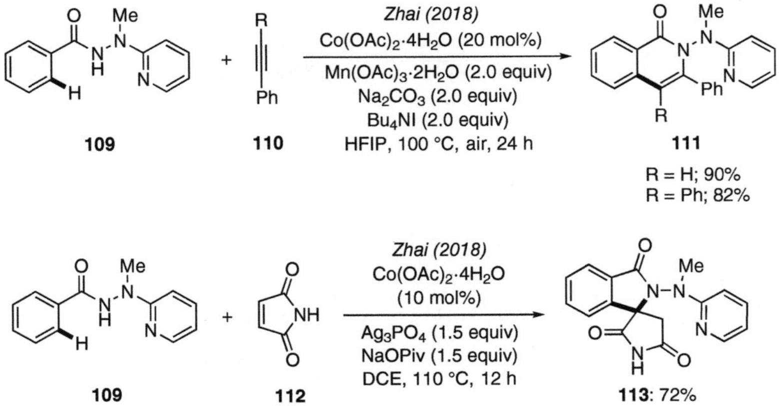

Scheme 4.22 2-Hydrazinylpyridine auxiliary chelation-assisted C—H activation reactions.

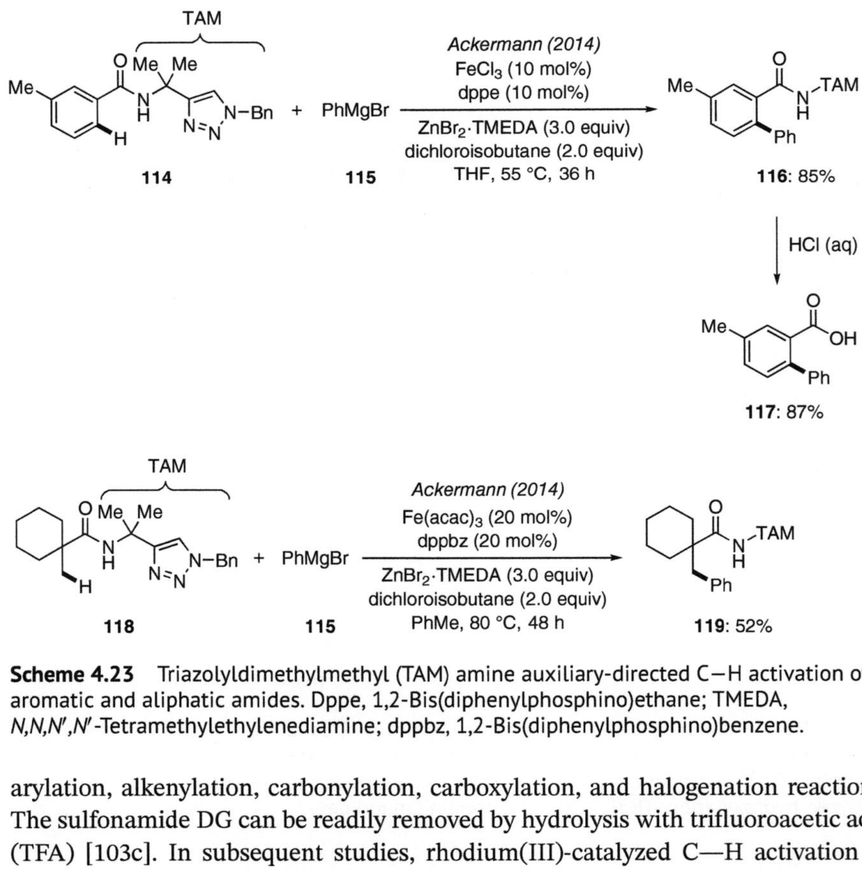

Scheme 4.23 Triazolyldimethylmethyl (TAM) amine auxiliary-directed C—H activation of aromatic and aliphatic amides. Dppe, 1,2-Bis(diphenylphosphino)ethane; TMEDA, *N,N,N′,N′*-Tetramethylethylenediamine; dppbz, 1,2-Bis(diphenylphosphino)benzene.

arylation, alkenylation, carbonylation, carboxylation, and halogenation reactions. The sulfonamide DG can be readily removed by hydrolysis with trifluoroacetic acid (TFA) [103c]. In subsequent studies, rhodium(III)-catalyzed C—H activation of sulfonamide with internal alkynes, leading to the formation of benzosultams, has also been reported [103b, 104]. Furthermore, the groups of Wang/Li and co-workers [105], Xu and co-workers [106], Chatani and co-workers [107], Jeganmohan and

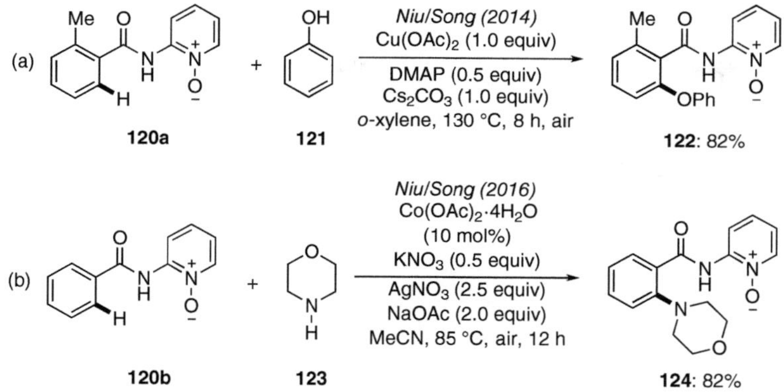

Scheme 4.24 Selected examples of 2-aminopyridine-*N*-oxide auxiliary-assisted C—H activation reactions. (a) Copper-catalyzed C—O bond formation. (b) Cobalt-catalyzed amination.

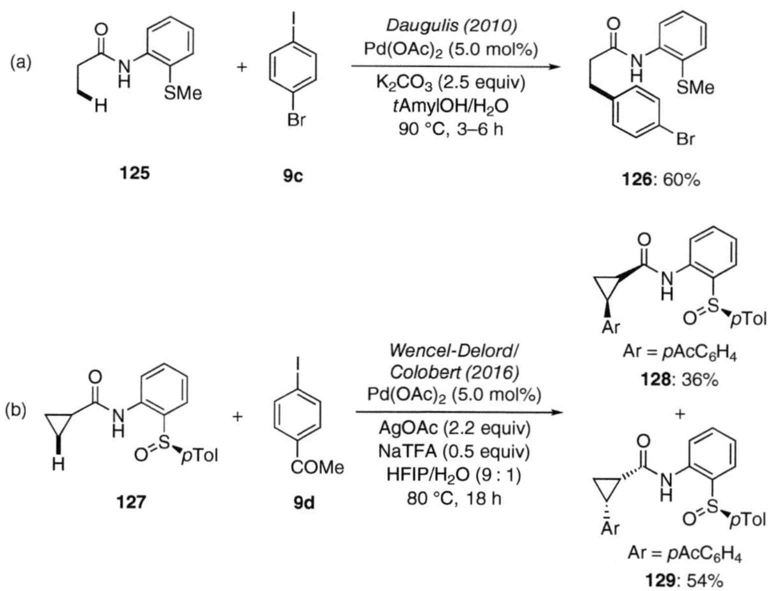

Scheme 4.25 Examples of *N,S*-bidentate directing groups for C—H activation reactions. (a) Thioether assisted reaction. (b) Sulfoxide assisted reaction.

co-workers [108], and others [109] have expanded the utility of the sulfonamide DG for a wide range of C—H functionalization reactions (Scheme 4.27).

4.2.6 C—H Functionalization of Heterocycles

Regioselective functionalization of heterocycles has been well established through TM-catalyzed chelation-assisted C—H activation reactions [110]. Specifically, functionalization of particular C—H bonds of indoles, indolines, pyridines, quinolines,

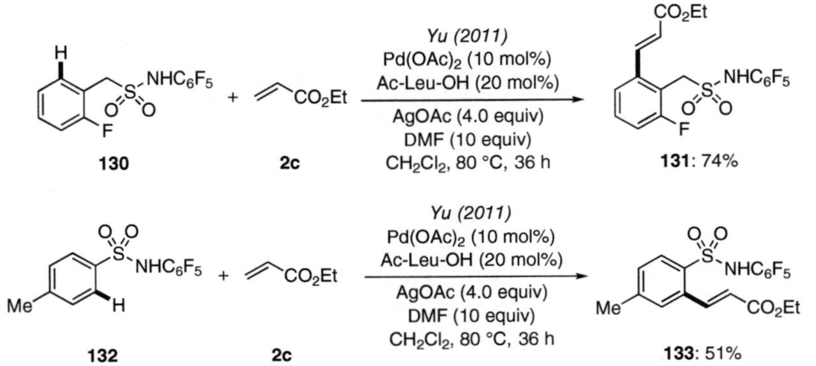

Scheme 4.26 Selected examples of palladium-catalyzed sulfonamide-directed C–H activation reactions.

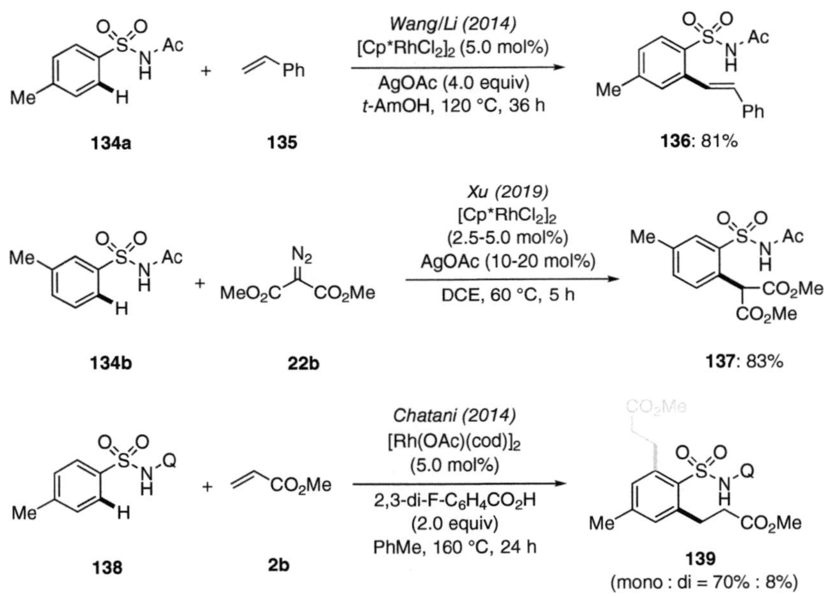

Scheme 4.27 Selected examples of rhodium-catalyzed sulfonamide-directed C–H activation reactions.

and carbazoles has been achieved with the assistance of removable DGs. In this section, we will highlight the commonly used removable DGs for the heterocycle C—H activation reactions.

The diversification of indoles through direct C—H activation has been extensively studied in the literature [111]. Installing a removable DG at the *N*-atom of indole enables regiocontrolled C—H activation at the C2 or C7 positions. As shown in Figure 4.9, a wide range of functional motifs serve as the DGs for C2–H or C7–H functionalization, as summarized in recent reviews [112]. Among various FGs, *N*-pyridyl or *N*-pyrimidyl, *N*-acyl, and *N*-carboxamides have been found to be the

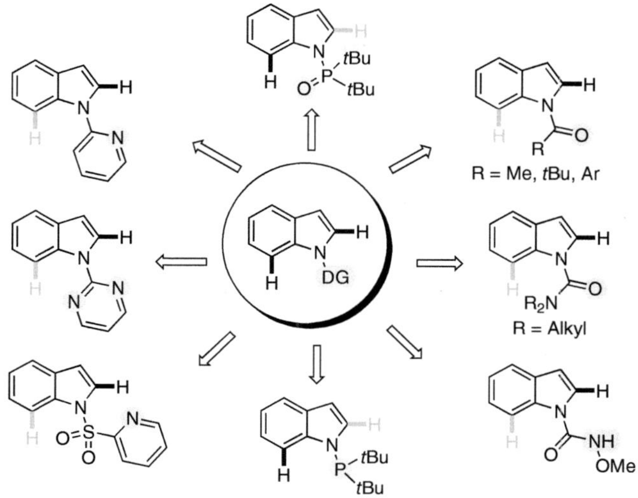

Figure 4.9 Removable DGs for regioselective C–H activation indoles.

most common removable DGs for C2 selective C—H functionalization of indoles (Scheme 4.28) [112e, 113]. On the other hand, sterically hindered di-*tert*-butyl phosphine (*N*-P*t*Bu_2) or di-*tert*-butyl phosphinoyl (*N*-P[O]*t*Bu_2) moieties are more suitable for regioselective C7–H activation reactions (Scheme 4.29) [114].

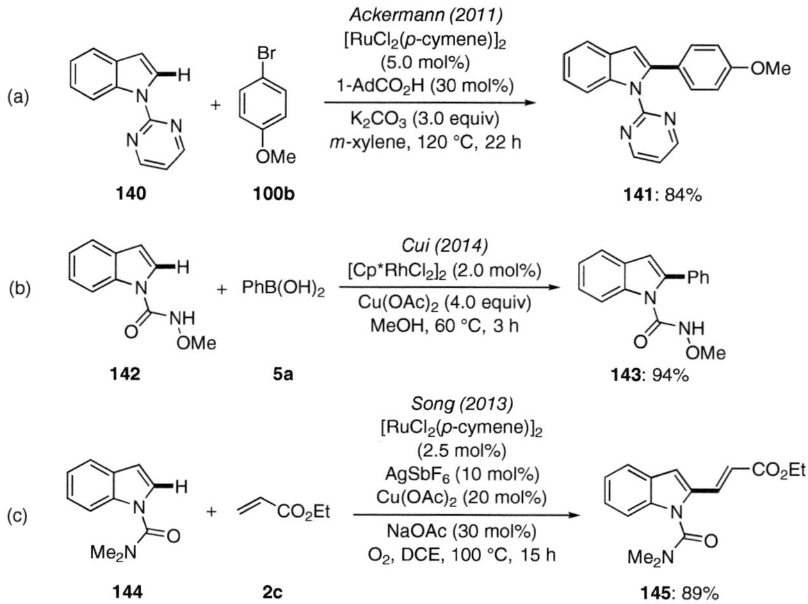

Scheme 4.28 Selected examples of indole C2–H activation reactions using RDGs. (a) *N*-pyrimidyl DG assisted reaction. (b) *N*-Carboxamide DG assisted reaction. (c) *N*-Dimethylcarbamoyl DG assisted reaction.

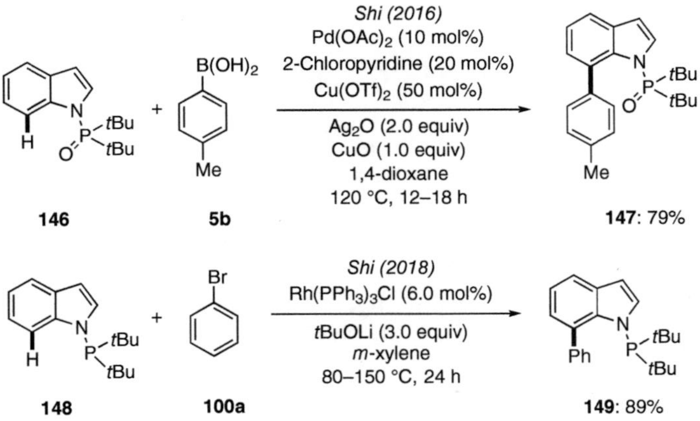

Scheme 4.29 P(V)- and P(III)-directed C7–H activation of indoles.

The DGs used for C—H activations in indolines are similar to those in indoles [112f]. The *N*-pyridyl/*N*-pyrimidyl group is the most common removable DG utilized for C7-H functionalization (Scheme 4.30) [115]. This DG is suitable for a wide range of C—H functionalization reactions using 3d- and 4d-TM catalysts. In addition to powerful *N*-pyridyl/*N*-pyrimidyl DGs, acyl [116], carbamoyl [116b, 117], and other [118] DGs have also been found suitable, although primarily for late-TM catalysts.

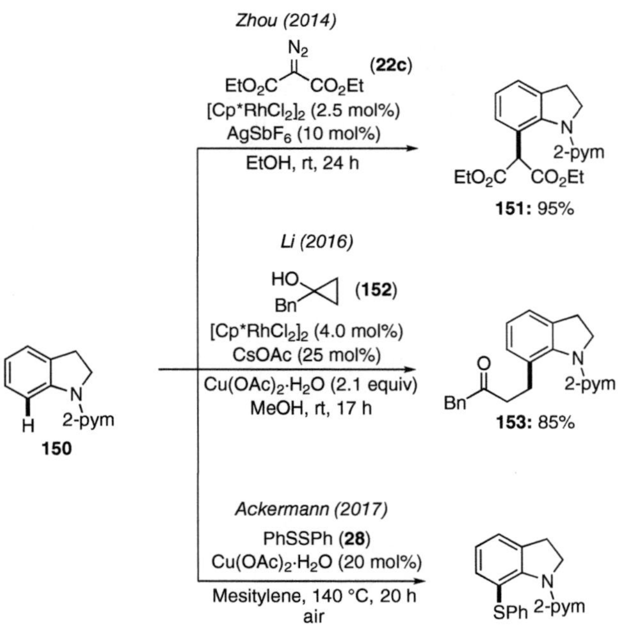

Scheme 4.30 Selected examples of chelation-assisted C7–H activation of indolines.

The 2-pyridyl or 2-pyrimidyl DG has also been utilized for the regioselective C—H functionalization of heterocycles such as carbazoles **155** and pyridones **159**. For instance, the incorporation of aryl [119], acyl [120], alkynyl [121], thioaryl [122], acyloxy [123], and amino [124] FGs has been achieved with the assistance of 2-pyridyl DG (Scheme 4.31). Similarly, the diversification of 2-pyridones **159** with a wide range of FGs has also been accomplished using 2-pyridyl DG (Scheme 4.32) [125].

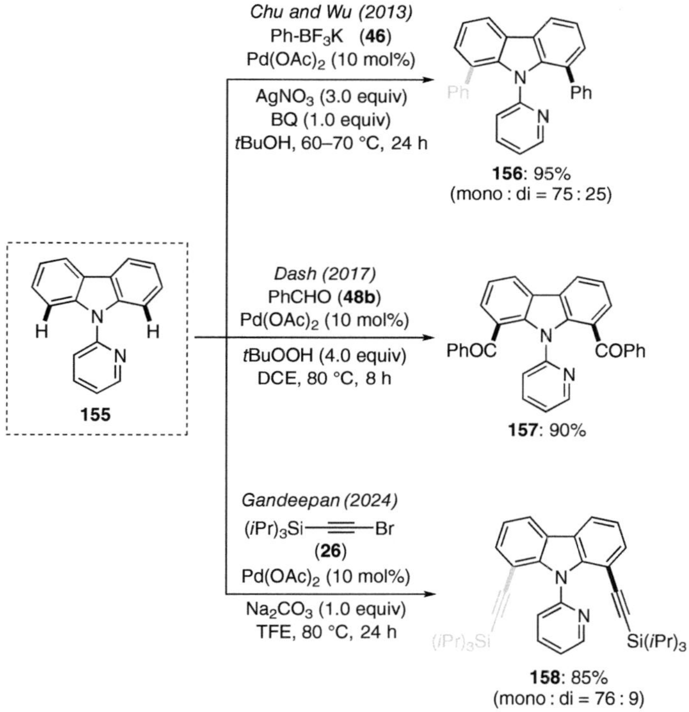

Scheme 4.31 Selected examples of *N*-pyridyl DG-assisted C—H activation of carbazoles.

Regioselective direct functionalization of electron-deficient heteroarenes, such as pyridines and quinolines, has been achieved using *N*-oxide DGs with TM catalysis. These heterocyclic *N*-oxides are conveniently prepared by treating the heteroarenes with peroxide reagents. In 2005, Fagnou demonstrated the C2-selective arylation of pyridines **165** with aryl bromides **100** in the presence of a palladium catalyst (Scheme 4.33) [126]. Following this breakthrough, several C—H activation reactions for the functionalization of pyridines, quinolines, and other heterocycles have been developed, many of which are already summarized in the literature [127].

4.2.7 Silicon Tethers for C—H Functionalization

The Gevorgyan group has introduced silicon-tethered DGs for the *ortho*-selective C—H functionalization of arenes [128]. As shown in Scheme 4.34, the DG can

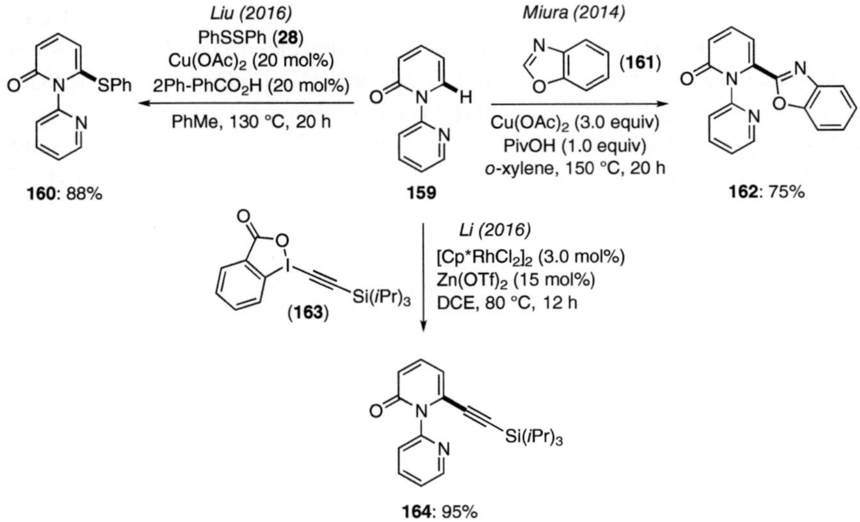

Scheme 4.32 Selected examples of pyridine-directed C—H activation of pyridones.

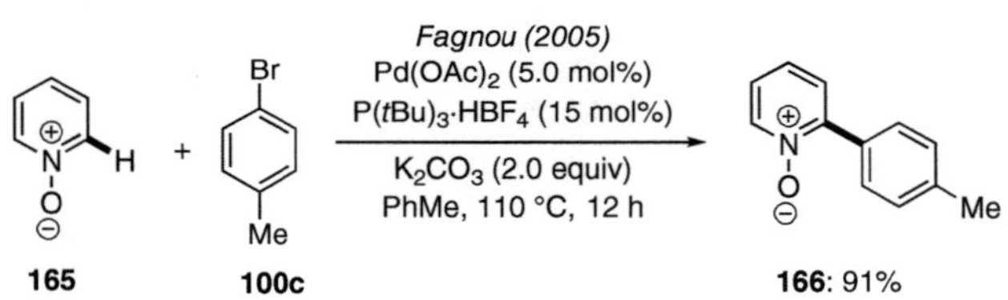

Scheme 4.33 An early example of *N*-oxide-directed C—H activation of pyridines.

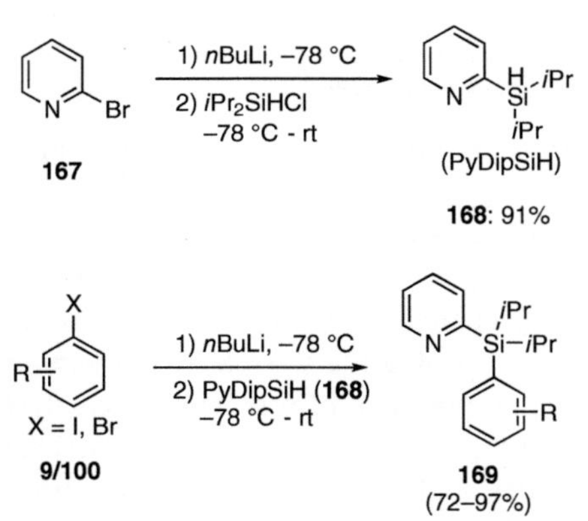

Scheme 4.34 Synthesis and installation of silyl-tethered directing group.

be readily installed by functionalizing aryl **9/100** and heteroaryl **167** halides (Scheme 4.34) [128e]. Using this RDG, several C—H transformation reactions including acyloxylation [128e], halogenation [128d], alkylation [128c], and alkoxycarbonylation [128a] have been reported (Scheme 4.35).

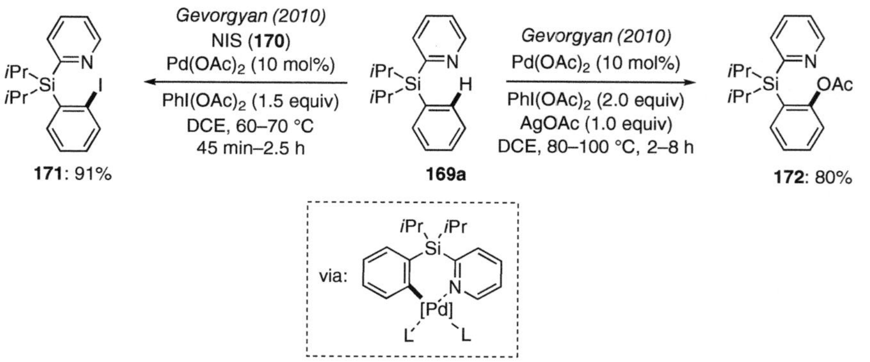

Scheme 4.35 Silyl-tethered pyridine DG-assisted C—H functionalization of arenes.

4.3 Summary and Conclusions

This chapter highlights the significant role of RDGs in achieving regioselective C—H activation reactions via TM catalysis. It addresses the core challenge of positional selectivity, where multiple C—H bonds within a molecule exhibit similar reactivity, making selective functionalization difficult. The chapter thoroughly discusses the application of RDGs across various FGs, including amines, hydroxyls, aldehydes, ketones, and carboxylic acids, showcasing how traditional FGs can be converted into highly effective DGs. Using RDGs offers advantages like reversible coordination with catalysts, efficient regio-control, and ease of removal, making them versatile tools for organic synthesis.

A key focus is the *ortho*-selective C—H activation of aromatic compounds, where RDGs have enabled a wide range of transformations, such as arylation, alkylation, amination, and more. It also covers the development of different bidentate DGs and their ability to enhance regioselectivity and increase the efficiency of reactions involving 3d-TM catalysts.

In conclusion, RDGs play a transformative role in organic synthesis by overcoming challenges in regioselective C—H activation. Their ability to facilitate selective transformations across a broad spectrum of FGs significantly enhances the scope of TM catalysis. These advancements offer valuable methodologies for constructing complex molecular architectures, contributing to developing novel pharmaceuticals, materials, and other organic compounds.

References

1 (a) Crawley, M.L. and Trost, B.M. (2012). *Applications of Transition Metal Catalysis in Drug Discovery and Development*. New York: Wiley. (b) Boström, J., Brown, D.G., Young, R.J. et al. (2018). Expanding the medicinal chemistry synthetic toolbox. *Nat. Rev. Drug Discov.* 17 (10): 709–727. (c) Brown, D.G. and Boström, J. (2015). Analysis of past and present synthetic methodologies on medicinal chemistry: where have all the new reactions gone? *J. Med. Chem.* 59 (10): 4443–4458. (d) Roughley, S.D. and Jordan, A.M. (2011). The medicinal chemist's toolbox: an analysis of reactions used in the pursuit of drug candidates. *J. Med. Chem.* 54 (10): 3451–3479.

2 (a) Sinha, S.K., Ghosh, P., Jain, S. et al. (2023). Transition-metal catalyzed C–H activation as a means of synthesizing complex natural products. *Chem. Soc. Rev.* 52 (21): 7461–7503. (b) Rogge, T., Kaplaneris, N., Chatani, N. et al. (2021). C–H activation. *Nat. Rev. Methods Primers* 1 (1): 43. (c) Abrams, D.J., Provencher, P.A., and Sorensen, E.J. (2018). Recent applications of C–H functionalization in complex natural product synthesis. *Chem. Soc. Rev.* 47 (23): 8925–8967. (d) Cernak, T., Dykstra, K.D., Tyagarajan, S. et al. (2016). The medicinal chemist's toolbox for late stage functionalization of drug-like molecules. *Chem. Soc. Rev.* 45 (3): 546–576. (e) Wencel-Delord, J. and Glorius, F. (2013). C–H bond activation enables the rapid construction and late-stage diversification of functional molecules. *Nat. Chem.* 5 (5): 369–375. (f) Yamaguchi, J., Yamaguchi, A.D., and Itami, K. (2012). C—H bond functionalization: emerging synthetic tools for natural products and pharmaceuticals. *Angew. Chem. Int. Ed.* 51 (36): 8960–9009. (g) McMurray, L., O'Hara, F., and Gaunt, M.J. (2011). Recent developments in natural product synthesis using metal-catalysed C—H bond functionalisation. *Chem. Soc. Rev.* 40 (4): 1885–1898.

3 Hoveyda, A.H., Evans, D.A., and Fu, G.C. (2002). Substrate-directable chemical reactions. *Chem. Rev.* 93 (4): 1307–1370.

4 (a) Zheng, Q.-Z. and Jiao, N. (2014). Transition-metal-catalyzed ketone-directed *ortho*-C—H functionalization reactions. *Tetrahedron Lett.* 55 (6): 1121–1126. (b) Huang, Z., Lim, H.N., Mo, F. et al. (2015). Transition metal-catalyzed ketone-directed or mediated C–H functionalization. *Chem. Soc. Rev.* 44 (21): 7764–7786.

5 (a) Shi, G. and Zhang, Y. (2014). Carboxylat-directed C-H functionalization. *Adv. Synth. Catal.* 356 (7): 1419–1442. (b) Pichette Drapeau, M. and Gooßen, L.J. (2016). Carboxylic acids as directing groups for C–H bond functionalization. *Chem. Eur. J.* 22 (52): 18654–18677.

6 (a) Zhu, C., Wang, R., and Falck, J.R. (2012). Amide-directed tandem C-C/C-N bond formation through C-H activation. *Chem. Asian J.* 7 (7): 1502–1514. (b) Das, R., Kumar, G.S., and Kapur, M. (2017). Amides as weak coordinating groups in proximal C—H bond activation. *Eur. J. Org. Chem.* 2017 (37): 5439–5459. (c) Subhedar, D.D., Mishra, A.A., and Bhanage, B.M. (2019). *N*-methoxybenzamide: a versatile directing group for palladium-, rhodium- and ruthenium-catalyzed C—H bond activations. *Adv. Synth. Catal.* 361 (18):

4149–4195. (d) Kalepu, J. and Pilarski, L. (2019). Weinreb amides as directing groups for transition metal-catalyzed C-H functionalizations. *Molecules* 24 (5): 830.

7 (a) Tischler, M.O., Tóth, M.B., and Novák, Z. (2016). Mild palladium catalyzed ortho C—H bond Functionalizations of aniline derivatives. *Chem. Rec.* 17 (2): 184–199. (b) He, C., Whitehurst, W.G., and Gaunt, M.J. (2019). Palladium-catalyzed C(sp^3)—H bond functionalization of aliphatic amines. *Chem* 5 (5): 1031–1058.

8 (a) Mo, F., Tabor, J.R., and Dong, G. (2014). Alcohols or masked alcohols as directing groups for C—H bond functionalization. *Chem. Lett.* 43 (3): 264–271. (b) Xu, X. and Luo, J. (2019). Transition metal-catalyzed directing-group-assisted C–H activation of phenols. *ChemSusChem* 12 (20): 4601–4616. (c) Al Mamari, H.H., Štefane, B., and Žugelj, H.B. (2020). Metal-catalyzed C—H bond functionalization of phenol derivatives. *Tetrahedron* 76 (9): 130925. (d) Huang, Z. and Lumb, J.-P. (2018). Phenol-directed C–H functionalization. *ACS Catal.* 9 (1): 521–555.

9 Gandeepan, P. and Cheng, C.-H. (2015). Transition-metal-catalyzed π-bond-assisted C—H bond functionalization: an emerging trend in organic synthesis. *Chem. Asian J.* 10 (4): 824–838.

10 Ping, Y., Wang, L., Ding, Q. et al. (2017). Nitrile as a versatile directing group for C(sp^2)-H functionalizations. *Adv. Synth. Catal.* 359 (19): 3274–3291.

11 (a) Swamy, T., Subba Reddy, B.V., Grée, R. et al. (2018). Substrate-directed C-H functionalization of 2-aryl pyridines by transition metal complexes. *ChemistrySelect* 3 (1): 47–70. (b) Nguyen, T.H.L., Gigant, N., and Joseph, D. (2018). Advances in direct metal-catalyzed functionalization of azobenzenes. *ACS Catal.* 8 (2): 1546–1579. (c) Pulis, A.P. and Procter, D.J. (2016). C–H coupling reactions directed by sulfoxides: teaching an old functional group new tricks. *Angew. Chem. Int. Ed.* 55 (34): 9842–9860. (d) Mondal, K., Ghosh, S., and Hajra, A. (2022). Transition-metal-catalyzed ortho C–H functionalization of 2-arylquinoxalines. *Org. Biomol. Chem.* 20 (37): 7361–7376. (e) Tang, K.X., Wang, C.M., Gao, T.H. et al. (2018). Transition metal-catalyzed C—H bond functionalizations by use of sulfur-containing directing groups. *Adv. Synth. Catal.* 361 (1): 26–38. (f) Manikandan, R. and Jeganmohan, M. (2017). Recent advances in the ruthenium(II)-catalyzed chelation-assisted C–H olefination of substituted aromatics, alkenes and heteroaromatics with alkenes via the deprotonation pathway. *Chem. Commun.* 53 (64): 8931–8947.

12 (a) Lapuh, M.I., Mazeh, S., and Besset, T. (2020). Chiral transient directing groups in transition-metal-catalyzed enantioselective C—H bond functionalization. *ACS Catal.* 10 (21): 12898–12919. (b) Sambiagio, C., Schönbauer, D., Blieck, R. et al. (2018). A comprehensive overview of directing groups applied in metal-catalysed C–H functionalisation chemistry. *Chem. Soc. Rev.* 47 (17): 6603–6743. (c) Gandeepan, P. and Ackermann, L. (2018). Transient directing groups for transformative C–H activation by synergistic metal catalysis. *Chem* 4 (2): 199–222. (d) Kim, D.-S., Park, W.-J., and Jun, C.-H. (2017). Metal-organic cooperative catalysis in C-H and C-C bond activation. *Chem. Rev.* 117 (13): 8977–9015. (e) Besset, T., Zhao, Q., Poisson, T. et al. (2017). The transient directing group strategy: a new trend in transition-metal-catalyzed C—H bond functionalization. *Synthesis* 49 (21): 4808–4826. (f) Rit, R.K., Yadav, M.R., Ghosh, K. et al. (2015). Reusable directing groups [8-aminoquinoline, picolinamide, sulfoximine] in C(sp^3)—H bond activation: present and future. *Tetrahedron* 71 (26–27): 4450–4459. (g) Huang, Y. and Sun, H. (2015). Recent progress in the development of multitasking directing groups for carbon–hydrogen activation reactions. *Synlett* 26 (20): 2751–2762. (h) Chen, Z., Wang, B., Zhang, J. et al. (2015). Transition metal-catalyzed C—H bond functionalizations by the use of diverse directing groups. *Org. Chem. Front.* 2 (9): 1107–1295. (i) Zhang, F. and Spring, D.R. (2014). Arene C–H functionalisation using a removable/modifiable or a traceless directing group strategy. *Chem. Soc. Rev.* 43 (20): 6906–6919.

13 (a) Fitzgerald, L.S. and O'Duill, M.L. (2021). A guide to directing group removal: 8-aminoquinoline. *Chem. Eur. J.* 27 (33): 8411–8436. (b) Carvalho, R.L., Almeida, R.G., Murali, K. et al. (2021). Removal and modification of directing groups used in metal-catalyzed C–H functionalization: the magical step of conversion into 'conventional' functional groups. *Org. Biomol. Chem.* 19 (3): 525–547. (c) Ma, W., Gandeepan, P., Li, J. et al. (2017). Recent advances in positional-selective alkenylations: removable guidance for twofold C–H activation. *Org. Chem. Front.* 4 (7): 1435–1467. (d) Rousseau, G. and Breit, B. (2011). Removable directing groups in organic synthesis and catalysis. *Angew. Chem. Int. Ed.* 50 (11): 2450–2494.

14 Boele, M.D.K., van Strijdonck, G.P.F., de Vries, A.H.M. et al. (2002). Selective Pd-catalyzed oxidative coupling of anilides with olefins through C–H bond activation at room temperature. *J. Am. Chem. Soc.* 124 (8): 1586–1587.

15 (a) Gao, Y., Huang, Y., Wu, W. et al. (2014). Pd-catalyzed C–H activation/oxidative cyclization of acetanilide with norbornene: concise access to functionalized indolines. *Chem. Commun.* 50 (61): 8370–8373. (b) Schmidt, B. and Elizarov, N. (2012). Selective arene functionalization through sequential oxidative and non-oxidative heck reactions. *Chem. Commun.* 48 (36): 4350–4352. (c) Liu, X. and Hii, K.K. (2011). Alternative to benzoquinone for room-temperature Fujiwara–Moritani reactions. *J. Org. Chem.* 76 (19): 8022–8026. (d) Chng, L.L., Zhang, J., Yang, J. et al. (2011). C-C bond formation via C-H activation and C—N bond formation via oxidative amination catalyzed by palladium-polyoxometalate nanomaterials using dioxygen as the terminal oxidant. *Adv. Synth. Catal.* 353 (16): 2988–2998. (e) Rauf, W., Thompson, A.L., and Brown, J.M. (2010). Anilide activation of adjacent C—H bonds in the palladium-catalysed Fujiwara–Moritani reaction. *Dalton Trans.* 39 (43): 10414–10421. (f) Patureau, F.W. and Glorius, F. (2010). Rh catalyzed olefination and vinylation of unactivated acetanilides. *J. Am. Chem. Soc.* 132 (29): 9982–9983. (g) Kim, B.S., Jang, C., Lee, D.J. et al. (2010). Highly effective Pd-catalyzed orthoOlefination of acetanilides: broad substrate scope and high tolerability. *Chem. Asian J.* 5 (11): 2336–2340. (h) Wang, J.-R., Yang, C.-T., Liu, L. et al. (2007). Pd-catalyzed aerobic oxidative coupling of anilides with olefins through regioselective C—H bond activation. *Tetrahedron Lett.* 48 (31): 5449–5453. (i) Amatore, C., Cammoun, C., and Jutand, A. (2007). Electrochemical recycling of benzoquinone in the Pd/benzoquinone-catalyzed heck-type reactions from arenes. *Adv. Synth. Catal.* 349 (3): 292–296.

16 (a) Li, D., Xu, N., Zhang, Y. et al. (2014). A highly efficient Pd-catalyzed decarboxylative ortho-arylation of amides with aryl acylperoxides. *Chem. Commun.* 50 (94): 14862–14865. (b) Yang, F., Song, F., Li, W. et al. (2013). Palladium-catalyzed C–H activation of anilides at room temperature: ortho-arylation and acetoxylation. *RSC Adv.* 3 (25): 9649–9652. (c) Yeung, C.S., Zhao, X., Borduas, N. et al. (2010). Pd-catalyzed ortho-arylation of phenylacetamides, benzamides, and anilides with simple arenes using sodium persulfate. *Chem. Sci.* 1 (3): 331–336. (d) Li, B.J., Tian, S.L., Fang, Z. et al. (2008). Multiple C-H activations to construct biologically active molecules in a process completely free of organohalogen and organometallic components. *Angew. Chem. Int. Ed.* 47 (6): 1115–1118. (e) Brasche, G., García-Fortanet, J., and Buchwald, S.L. (2008). Twofold C–H functionalization: palladium-catalyzed ortho arylation of anilides. *Org. Lett.* 10 (11): 2207–2210. (f) Yang, S., Li, B., Wan, X. et al. (2007). Ortho arylation of acetanilides via Pd(II)-catalyzed C–H functionalization. *J. Am. Chem. Soc.* 129 (19): 6066–6067. (g) Shi, Z., Li, B., Wan, X. et al. (2007). Suzuki–Miyaura coupling reaction by PdII-catalyzed aromatic C—H bond activation directed by an N-alkyl acetamino group. *Angew. Chem. Int. Ed.* 46 (29): 5554–5558. (h) Daugulis, O. and Zaitsev, V.G. (2005). Anilide ortho-arylation by using C-H activation methodology. *Angew. Chem. Int. Ed.* 44 (26): 4046–4048.

17 (a) Kim, M., Kumar Mishra, N., Park, J. et al. (2014). Decarboxylative acylation of indolines with α-keto acids under palladium catalysis: a facile strategy for the synthesis of 7-substituted indoles. *Chem. Commun.* 50 (91): 14249–14252. (b) Wu, Y., Choy, P.Y., Mao, F. et al. (2013). Toluene derivatives as simple coupling precursors for cascade palladium-catalyzed oxidative C—H bond acylation of acetanilides. *Chem. Commun.* 49 (7): 689–691. (c) Wu, Y., Li, B., Mao, F. et al. (2011). Palladium-catalyzed oxidative C—H bond coupling of steered acetanilides and aldehydes: a facile access to ortho-acylacetanilides. *Org. Lett.* 13 (12): 3258–3261. (d) Li, C., Wang, L., Li, P. et al. (2011). Palladium-catalyzed ortho-acylation of acetanilides with aldehydes through direct C—H bond activation. *Chem. Eur. J.* 17 (37): 10208–10212. (e) Chan, C.W., Zhou, Z., and Yu, W.Y. (2011). Palladium(II)-catalyzed direct ortho-C-H acylation of anilides by oxidative cross-coupling with aldehydes using tert-butyl hydroperoxide as oxidant. *Adv. Synth. Catal.* 353 (16): 2999–3006. (f) Fang, P., Li, M., and Ge, H. (2010). Room temperature palladium-catalyzed decarboxylative ortho-acylation of acetanilides with α-oxocarboxylic acids. *J. Am. Chem. Soc.* 132 (34): 11898–11899.

18 (a) Sun, K., Li, Y., Xiong, T. et al. (2011). Palladium-catalyzed C–H aminations of anilides with N-fluorobenzenesulfonimide. *J. Am. Chem. Soc.* 133 (6): 1694–1697. (b) Ng, K.-H., Chan, A.S.C., and Yu, W.-Y. (2010). Pd-catalyzed intermolecular ortho-C–H amidation of anilides by N-nosyloxycarbamate. *J. Am. Chem. Soc.* 132 (37): 12862–12864.

19 Hesp, K.D., Bergman, R.G., and Ellman, J.A. (2011). Expedient synthesis of N-acyl anthranilamides and β-enamine amides by the Rh(III)-catalyzed amidation of aryl and vinyl C—H bonds with isocyanates. *J. Am. Chem. Soc.* 133 (30): 11430–11433.

20 (a) Huang, Y., Li, G., Huang, J. et al. (2014). Palladium-catalyzed direct ortho-C–H ethoxycarboxylation of anilides at room temperature. *Org. Chem. Front.* 1 (4): 347–350. (b) Wang, S., Yang, Z., Liu, J. et al. (2012). Efficient synthesis of anthranilic esters via Pd-catalyzed dehydrogenative/decarbonylative coupling of anilides and glyoxylates. *Chem. Commun.* 48 (79): 9924–9926.

21 Giri, R., Lam, J.K., and Yu, J.-Q. (2009). Synthetic applications of Pd(II)-catalyzed C–H carboxylation and mechanistic insights: expedient routes to anthranilic acids, oxazolinones, and quinazolinones. *J. Am. Chem. Soc.* 132 (2): 686–693.

22 Jiang, T.-S. and Wang, G.-W. (2012). Palladium-catalyzed ortho-alkoxylation of anilides via C–H activation. *J. Org. Chem.* 77 (21): 9504–9509.

23 (a) Gandeepan, P., Rajamalli, P., and Cheng, C.-H. (2016). Palladium-catalyzed C–H activation and cyclization of anilides with 2-iodoacetates and 2-iodobenzoates: an efficient method toward oxindoles and phenanthridones. *Synthesis* 48 (12): 1872–1879. (b) Zhang, L.-S., Chen, K., Chen, G. et al. (2012). Palladium-catalyzed trifluoromethylation of aromatic C—H bond directed by an acetamino group. *Org. Lett.* 15 (1): 10–13.

24 (a) Bedford, R.B., Haddow, M.F., Mitchell, C.J. et al. (2011). Mild C-H halogenation of anilides and the isolation of an unusual palladium(I)–palladium(II) species. *Angew. Chem. Int. Ed.* 50 (24): 5524–5527. (b) Wan, X., Ma, Z., Li, B. et al. (2006). Highly selective C–H functionalization/halogenation of acetanilide. *J. Am. Chem. Soc.* 128 (23): 7416–7417.

25 (a) Manikandan, R. and Jeganmohan, M. (2014). Ruthenium-catalyzed cyclization of anilides with substituted propiolates or acrylates: an efficient route to 2-quinolinones. *Org. Lett.* 16 (13): 3568–3571. (b) Stuart, D.R., Alsabeh, P., Kuhn, M. et al. (2010). Rhodium(III)-catalyzed arene and alkene C—H bond functionalization leading to indoles and pyrroles. *J. Am. Chem. Soc.* 132 (51): 18326–18339. (c) Stuart, D.R., Bertrand-Laperle, M., Burgess, K.M.N. et al. (2008). Indole synthesis via rhodium catalyzed oxidative coupling of acetanilides and internal alkynes. *J. Am. Chem. Soc.* 130 (49): 16474–16475.

26 (a) Xu, Y.H., Zhang, Q.C., He, T. et al. (2014). Palladium-catalyzed direct alkynylation of N-vinylacetamides. *Adv. Synth. Catal.* 356 (7): 1539–1543. (b) Feng, C., Feng, D., and Loh, T.-P. (2014). Rhodium(III)-catalyzed olefinic C–H alkynylation of enamides at room temperature. *Chem. Commun.* 50 (69): 9865–9868. (c) Pankajakshan, S., Xu, Y.H., Cheng, J.K. et al. (2012). Palladium-catalyzed direct C—H arylation of enamides with simple arenes. *Angew. Chem. Int. Ed.* 51 (23): 5701–5705. (d) Feng, C. and Loh, T.-P. (2012). Copper-catalyzed olefinic trifluoromethylation of enamides at room temperature. *Chem. Sci.* 3 (12): 3458–3462. (e) Xu, Y.-H., Chok, Y.K., and Loh, T.-P.

(2011). Synthesis and characterization of a cyclic vinylpalladium(II) complex: vinylpalladium species as the possible intermediate in the catalytic direct olefination reaction of enamide. *Chem. Sci.* 2 (9): 1822–1825. (f) Zhou, H., Xu, Y.H., Chung, W.J. et al. (2009). Palladium-catalyzed direct arylation of cyclic enamides with aryl silanes by sp^2 C-H activation. *Angew. Chem. Int. Ed.* 48 (29): 5355–5357. (g) Zhou, H., Chung, W.-J., Xu, Y.-H. et al. (2009). Direct arylation of cyclic enamides via Pd(II)-catalyzed C–H activation. *Chem. Commun.* 23: 3472–3474.

27 (a) Zhao, M.-N., Ren, Z.-H., Wang, Y.-Y. et al. (2014). Pd-catalyzed oxidative coupling of enamides and alkynes for synthesis of substituted pyrroles. *Org. Lett.* 16 (2): 608–611. (b) Yu, W., Chen, J., Gao, K. et al. (2014). Amide-assisted acetoxylation of vinyl $C(sp^2)$—H bonds by rhodium catalysis. *Org. Lett.* 16 (18): 4870–4873. (c) Murugan, K. and Liu, S.-T. (2013). Preparation of substituted pyrroles via Ru(II)-catalyzed coupling of alkynes with enamides. *Tetrahedron Lett.* 54 (21): 2608–2611. (d) Wang, H., Guo, L.-N., and Duan, X.-H. (2012). Decarboxylative acylation of cyclic enamides with α-oxocarboxylic acids by palladium-catalyzed C–H activation at room temperature. *Org. Lett.* 14 (17): 4358–4361. (e) Li, B., Wang, N., Liang, Y. et al. (2012). Ruthenium-catalyzed pyrrole synthesis via oxidative annulation of enamides and alkynes. *Org. Lett.* 15 (1): 136–139.

28 (a) Jiang, Z., Zhang, L., Dong, C. et al. (2013). Direct synthesis of 8-aryl tetrahydroquinolines via Pd-catalyzed ortho-arylation of arylureas in water. *RSC Adv.* 3 (4): 1025–1028. (b) Nishikata, T., Abela, A.R., and Lipshutz, B.H. (2010). Room temperature C-H activation and cross-coupling of aryl ureas in water. *Angew. Chem. Int. Ed.* 49 (4): 781–784. (c) Nishikata, T., Abela, A.R., Huang, S. et al. (2010). Cationic palladium(II) catalysis: C–H activation/Suzuki–Miyaura couplings at room temperature. *J. Am. Chem. Soc.* 132 (14): 4978–4979.

29 (a) Wang, C. and Huang, Y. (2013). Traceless directing strategy: efficient synthesis of N-alkyl indoles via redox-neutral C–H activation. *Org. Lett.* 15 (20): 5294–5297. (b) Liu, B., Song, C., Sun, C. et al. (2013). Rhodium(III)-catalyzed indole synthesis using N—N bond as an internal oxidant. *J. Am. Chem. Soc.* 135 (44): 16625–16631. (c) Liu, B., Fan, Y., Gao, Y. et al. (2012). Rhodium(III)-catalyzed N-nitroso-directed C–H olefination of arenes. High-yield, versatile coupling under mild conditions. *J. Am. Chem. Soc.* 135 (1): 468–473.

30 Huang, X., Huang, J., Du, C. et al. (2013). N-oxide as a traceless oxidizing directing group: mild rhodium(III)-catalyzed C-H olefination for the synthesis of ortho-alkenylated tertiary anilines. *Angew. Chem. Int. Ed.* 52 (49): 12970–12974.

31 (a) Hernando, E., Castillo, R.R., Rodríguez, N. et al. (2014). Copper-catalyzed mild nitration of protected anilines. *Chem. Eur. J.* 20 (43): 13854–13859. (b) Urones, B., Martínez, Á.M., Rodríguez, N. et al. (2013). Copper-catalyzed ortho-halogenation of protected anilines. *Chem. Commun.* 49 (94): 11044–11046. (c) García-Rubia, A., Urones, B., Gómez Arrayás, R. et al. (2011). PdII-catalyzed

C-H olefination of N-(2-pyridyl)sulfonyl anilines and arylalkylamines. *Angew. Chem. Int. Ed.* 50 (46): 10927–10931.

32 Wang, L., Yang, Z., Yang, M. et al. (2017). Iridium-catalyzed direct C–H amidation of anilines with sulfonyl azides: easy access to 1,2-diaminobenzenes. *Org. Biomol. Chem.* 15 (39): 8302–8307.

33 (a) Xia, J., Yang, X., Li, Y. et al. (2017). Iridium(III)-catalyzed synthesis of benzimidazoles via C–H activation and amidation of aniline derivatives. *Org. Lett.* 19 (12): 3243–3246. (b) Wu, Z.-J., Huang, K.L., and Huang, Z.-Z. (2017). A cascade C–H functionalization/cyclization reaction of N-arylpyridin-2-amines with α,β-unsaturated aldehydes for the synthesis of dihydroquinolinone derivatives under rhodium catalysis. *Org. Biomol. Chem.* 15 (23): 4978–4983. (c) Hu, X., Chen, X., Zhu, Y. et al. (2017). Rh(III)-catalyzed carboamination of propargyl cycloalkanols with arylamines via Csp^2–H/Csp^3–Csp^3 activation. *Org. Lett.* 19 (13): 3474–3477. (d) Bai, S., Chen, X., Hu, X. et al. (2017). An Ir(III)-catalyzed aryl C—H bond carbenoid functionalization cascade: access to 1,3-dihydroindol-2-ones. *Org. Biomol. Chem.* 15 (17): 3638–3647. (e) Yu, K., Liang, Y., Li, B. et al. (2016). Rhodium(III)-catalyzed synthesis of indole derivatives from pyrimidyl-substituted anilines and diazo compounds. *Adv. Synth. Catal.* 358 (4): 661–666. (f) Tang, G.-D., Pan, C.-L., and Li, X. (2016). Iridium(III)- and rhodium(III)-catalyzed coupling of anilines with α-diazoesters via chelation-assisted C–H activation. *Org. Chem. Front.* 3 (1): 87–90. (g) Jiang, H., Gao, S., Xu, J. et al. (2016). Multiple roles of the Pyrimidyl Group in the Rhodium-Catalyzed Regioselective Synthesis and Functionalization of Indole-3-carboxylic acid esters. *Adv. Synth. Catal.* 358 (2): 188–194. (h) Reddy, P.V., Annapurna, M., Srinivas, P. et al. (2015). Nanocrystalline magnesium oxide-stabilized palladium(0): an efficient and reusable catalyst for synthesis of N-(2-pyridyl)indoles. *New J. Chem.* 39 (5): 3399–3404.

34 Ruan, Z., Lackner, S., and Ackermann, L. (2016). A general strategy for the nickel-catalyzed C-H alkylation of anilines. *Angew. Chem. Int. Ed.* 55 (9): 3153–3157.

35 Ruan, Z., Lackner, S., and Ackermann, L. (2016). Nickel-catalyzed C–H alkynylation of anilines: expedient access to functionalized indoles and purine nucleobases. *ACS Catal.* 6 (7): 4690–4693.

36 (a) Yang, S., Feng, B., and Yang, Y. (2017). Rh(III)-catalyzed direct ortho-chalcogenation of phenols and anilines. *J. Org. Chem.* 82 (23): 12430–12438. (b) Müller, T. and Ackermann, L. (2016). Nickel-catalyzed C–H chalcogenation of anilines. *Chem. Eur. J.* 22 (40): 14151–14154.

37 Floris, B., Galloni, P., Conte, V. et al. (2021). Tailored functionalization of natural phenols to improve biological activity. *Biomolecules* 11 (9): 1325.

38 (a) Brufani, G., Di Erasmo, B., Li, C.-J. et al. (2024). Csp2–H functionalization of phenols: an effective access route to valuable materials via Csp2–C bond formation. *Chem. Sci.* 15 (11): 3831–3871. (b) Li, Y., Huang, Y., Li, Z. et al. (2023). Recent advances in regioselective C—H bond functionalization of free phenols. *Molecules* 28 (8): 3397.

39 (a) Yang, X., Sun, Y., Chen, Z. et al. (2014). A general approach towards catechol and pyrogallol through ruthenium- and palladium-catalyzed C-H hydroxylation by weak coordination. *Adv. Synth. Catal.* 356 (7): 1625–1630. (b) Jansen-van Vuuren, R.D., Liu, S., Miah, M.A.J. et al. (2024). The versatile and strategic O-carbamate directed metalation group in the synthesis of aromatic molecules: an update. *Chem. Rev.* 124 (12): 7731–7828.

40 (a) Bedford, R.B., Webster, R.L., and Mitchell, C.J. (2009). Palladium-catalysed ortho-arylation of carbamate-protected phenols. *Org. Biomol. Chem.* 7 (23): 4853–4857. (b) Zhao, X., Yeung, C.S., and Dong, V.M. (2010). Palladium-catalyzed ortho-arylation of O-phenylcarbamates with simple arenes and sodium persulfate. *J. Am. Chem. Soc.* 132 (16): 5837–5844.

41 (a) Reddy, M.C. and Jeganmohan, M. (2013). Ruthenium-catalyzed highly regio- and stereoselective hydroarylation of aryl carbamates with alkynes via C—H bond activation. *Chem. Commun.* 49 (5): 481–483. (b) Reddy, M.C. and Jeganmohan, M. (2013). Ruthenium-catalyzed selective aerobic oxidative ortho-alkenylation of substituted phenols with alkenes through C—H bond activation. *Eur. J. Org. Chem.* 2013 (6): 1150–1157. (c) Li, B., Ma, J., Liang, Y. et al. (2013). Regio- and stereoselective olefination of phenol carbamates through C—H bond functionalization. *Eur. J. Org. Chem.* 2013 (10): 1950–1962. (d) Li, J., Kornhaaß, C., and Ackermann, L. (2012). Ruthenium-catalyzed oxidative C–H alkenylation of aryl carbamates. *Chem. Commun.* 48 (92): 11343–11345. (e) Gong, T.-J., Xiao, B., Liu, Z.-J. et al. (2011). Rhodium-catalyzed selective C–H activation/olefination of phenol carbamates. *Org. Lett.* 13 (12): 3235–3237. (f) Feng, C. and Loh, T.-P. (2011). Rhodium-catalyzed direct ortho C–H olefination of phenol derivatives. *Chem. Commun.* 47 (37): 10458–10460.

42 (a) Sun, X., Sun, Y., Zhang, C. et al. (2014). Room-temperature Pd-catalyzed C–H chlorination by weak coordination: one-pot synthesis of 2-chlorophenols with excellent regioselectivity. *Chem. Commun.* 50 (10): 1262–1264. (b) John, A. and Nicholas, K.M. (2012). Palladium catalyzed C–H functionalization of O-arylcarbamates: selective ortho-bromination using NBS. *J. Org. Chem.* 77 (13): 5600–5605.

43 (a) Sharma, S., Kim, A., Park, E. et al. (2013). Palladium-catalyzed decarboxylative acylation of O-phenyl carbamates with α-oxocarboxylic acids at room temperature. *Adv. Synth. Catal.* 355 (4): 667–672. (b) Liu, W. and Ackermann, L. (2013). Ortho- and para-selective ruthenium-catalyzed C(sp^2)–H oxygenations of phenol derivatives. *Org. Lett.* 15 (13): 3484–3486. (c) Yamazaki, K., Kawamorita, S., Ohmiya, H. et al. (2010). Directed ortho borylation of phenol derivatives catalyzed by a silica-supported iridium complex. *Org. Lett.* 12 (18): 3978–3981.

44 Xiao, B., Fu, Y., Xu, J. et al. (2009). Pd(II)-catalyzed C–H activation/aryl–aryl coupling of phenol esters. *J. Am. Chem. Soc.* 132 (2): 468–469.

45 (a) Huang, C., Chattopadhyay, B., and Gevorgyan, V. (2011). Silanol: a traceless directing group for Pd-catalyzed o-alkenylation of phenols. *J. Am. Chem. Soc.* 133 (32): 12406–12409. (b) Huang, C., Ghavtadze, N., Chattopadhyay, B. et al. (2011). Synthesis of catechols from phenols via Pd-catalyzed silanol-directed C–H oxygenation. *J. Am. Chem. Soc.* 133 (44): 17630–17633.

46 (a) Chan, L.Y., Cheong, L., and Kim, S. (2013). Pd(II)-catalyzed ortho-arylation of aryl phosphates and aryl hydrogen phosphates with diaryliodonium triflates. *Org. Lett.* 15 (9): 2186–2189. (b) Jeon, W.H., Lee, T.S., Kim, E.J. et al. (2013). Palladium(II)-catalyzed ortho-arylation via phosphate-group-directed C–H activation. *Tetrahedron* 69 (25): 5152–5159.

47 Chan, L.Y., Kim, S., Ryu, T. et al. (2013). Palladium-catalyzed ortho-alkenylation of aryl hydrogen phosphates using a new mono-phosphoric acid directing group. *Chem. Commun.* 49 (41): 4682–4684.

48 (a) Zhang, L., Zhu, L., Zhang, Y. et al. (2018). Experimental and theoretical studies on Ru(II)-catalyzed oxidative C–H/C–H coupling of phenols with aromatic amides using air as oxidant: scope, synthetic applications, and mechanistic insights. *ACS Catal.* 8 (9): 8324–8335. (b) Wu, Y., Li, W., Jiang, L. et al. (2018). Rhodium-catalyzed ortho-heteroarylation of phenols: directing group-enabled switching of the electronic bias for heteroaromatic coupling partner. *Chem. Sci.* 9 (33): 6878–6882. (c) Chu, J.-H., Lin, P.-S., and Wu, M.-J. (2010). Palladium(II)-catalyzed Ortho Arylation of 2-Phenoxypyridines with potassium aryltrifluoroborates via C–H functionalization. *Organometallics* 29 (18): 4058–4065.

49 (a) Liu, B., Jiang, H.-Z., and Shi, B.-F. (2014). Palladium-catalyzed oxidative olefination of phenols bearing removable directing groups under molecular oxygen. *J. Org. Chem.* 79 (3): 1521–1526. (b) Cong, X., You, J., Gao, G. et al. (2013). 2-Pyridylmethyl ether: a readily removable and efficient directing group for amino acid ligand accelerated ortho-C–H olefination of phenols. *Chem. Commun.* 49 (7): 662–664.

50 (a) Chu, J.-H., Chen, S.-T., Chiang, M.-F. et al. (2015). Palladium-catalyzed direct ortho aroylation of 2-phenoxypyridines with aldehydes and catalytic mechanism investigation. *Organometallics* 34 (5): 953–966. (b) Jia, X., Zhang, S., Wang, W. et al. (2009). Palladium-catalyzed acylation of sp^2 C—H bond: direct access to ketones from aldehydes. *Org. Lett.* 11 (14): 3120–3123.

51 (a) Liang, Y.-F., Li, X., Wang, X. et al. (2015). Aerobic oxidation of PdII to Pd^{IV} by active radical reactants: direct C–H nitration and acylation of arenes via oxygenation process with molecular oxygen. *ACS Catal.* 5 (3): 1956–1963. (b) Zhang, W., Zhang, J., Ren, S. et al. (2014). Palladium-catalyzed aromatic C—H bond nitration using removable directing groups: regiospecific synthesis of substituted o-nitrophenols from related phenols. *J. Org. Chem.* 79 (23): 11508–11516.

52 (a) Lou, S.-J., Chen, Q., Wang, Y.-F. et al. (2015). Selective C—H bond fluorination of phenols with a removable directing group: late-stage fluorination of 2-phenoxyl nicotinate derivatives. *ACS Catal.* 5 (5): 2846–2849. (b) Zhang, C. and Sun, P. (2014). Palladium-catalyzed direct $C(sp^2)$–H Alkoxylation of 2-Aryloxypyridines using 2-Pyridyloxyl as the directing group. *J. Org. Chem.* 79 (17): 8457–8461. (c) Xu, Y., Liu, P., Li, S.-L. et al. (2014). Palladium-catalyzed ortho-sulfonylation of 2-aryloxypyridines and subsequent formation of ortho-sulfonylated phenols. *J. Org. Chem.* 80 (2): 1269–1274.

53 (a) Sk, B., Ravindran, E., Deori, U. et al. (2022). A deep blue thermally activated delayed fluorescence emitter: balance between charge transfer and color purity. *J. Mater. Chem. C* 10 (12): 4886–4893. (b) Sk, B., Thangaraji, V., Yadav, N. et al. (2021). High performance non-doped green organic light emitting diodes via delayed fluorescence. *J. Mater. Chem. C* 9 (43): 15583–15590. (c) Mancha, S.R., Regnery, C.M., Dahlke, J.R. et al. (2013). Antiviral activity of (+)-sattabacin against varicella zoster. *Bioorg. Med. Chem. Lett.* 23 (2): 562–564. (d) Misawa, N., Nakamura, R., Kagiyama, Y. et al. (2005). Synthesis of vicinal diols from various arenes with a heterocyclic, amino or carboxyl group by using recombinant *Escherichia coli* cells expressing evolved biphenyl dioxygenase and dihydrodiol dehydrogenase genes. *Tetrahedron* 61 (1): 195–204. (e) Melvin, L.S., Johnson, M.R., Harbert, C.A. et al. (2002). A cannabinoid derived prototypical analgesic. *J. Med. Chem.* 27 (1): 67–71. (f) Roth, G.N., Chandra, A., and Nair, M.G. (1998). Novel bioactivities of *Curcuma longa* constituents. *J. Nat. Prod.* 61 (4): 542–545.

54 (a) Amistadi-Revol, H., Liu, S., and Prévost, S. (2023). C–H functionalization of aldehydes and ketones with transient directing groups: recent developments. *Eur. J. Org. Chem.* 26 (31): e202300582. (b) Taskesenligil, Y., Aslan, M., Cogurcu, T. et al. (2022). Directed C–H functionalization of C3-aldehyde, ketone, and acid/ester-substituted free (NH) indoles with iodoarenes via a palladium catalyst system. *J. Org. Chem.* 88 (3): 1299–1318.

55 (a) Thirunavukkarasu, V.S. and Ackermann, L. (2012). Ruthenium-catalyzed C—H bond oxygenations with weakly coordinating ketones. *Org. Lett.* 14 (24): 6206–6209. (b) Li, H., Zhu, R.-Y., Shi, W.-J. et al. (2012). Synthesis of fluorenone derivatives through Pd-catalyzed dehydrogenative cyclization. *Org. Lett.* 14 (18): 4850–4853. (c) Gandeepan, P., Hung, C.-H., and Cheng, C.-H. (2012). Pd-catalyzed double C—H bond activation of diaryl ketones for the synthesis of fluorenones. *Chem. Commun.* 48 (75): 9379–9381. (d) Patureau, F.W., Besset, T., and Glorius, F. (2010). Rhodium-catalyzed oxidative olefination of C—H bonds in acetophenones and benzamides. *Angew. Chem. Int. Ed.* 50 (5): 1064–1067. (e) Kakiuchi, F., Kochi, T., Mizushima, E. et al. (2010). Room-temperature regioselective C–H/olefin coupling of aromatic ketones using an activated ruthenium catalyst with a carbonyl ligand and structural elucidation of key intermediates. *J. Am. Chem. Soc.* 132 (50): 17741–17750. (f) Gandeepan, P., Parthasarathy, K., and Cheng, C.-H. (2010). Synthesis of phenanthrone derivatives from sec-alkyl aryl ketones and aryl halides via a palladium-catalyzed dual C—H bond activation and enolate cyclization. *J. Am. Chem. Soc.* 132 (25): 8569–8571. (g) Kakiuchi, F., Matsuura, Y., Kan, S. et al. (2005). A $RuH_2(CO)(PPh_3)_3$-catalyzed regioselective arylation of aromatic ketones with arylboronates via carbon–hydrogen bond cleavage. *J. Am. Chem. Soc.* 127 (16): 5936–5945. (h) Murai, S., Kakiuchi, F., Sekine, S. et al. (1993). Efficient catalytic addition of aromatic carbon-hydrogen bonds to olefins. *Nature* 366 (6455): 529–531.

56 Ueura, K., Satoh, T., and Miura, M. (2005). Rhodium-catalyzed arylation using arylboron compounds: efficient coupling with aryl halides and unexpected multiple arylation of benzonitrile. *Org. Lett.* 7 (11): 2229–2231.

57 (a) He, R., Huang, Z.T., Zheng, Q.Y. et al. (2014). Manganese-catalyzed dehydrogenative [4+2] annulation of N-H imines and alkynes by C-H/N-H activation. *Angew. Chem. Int. Ed.* 53 (19): 4950–4953. (b) Zhang, J., Ugrinov, A., and Zhao, P. (2013). Ruthenium(II)/N-heterocyclic carbene catalyzed [3+2] carbocyclization with aromatic N-H ketimines and internal alkynes. *Angew. Chem. Int. Ed.* 52 (26): 6681–6684. (c) Tran, D.N. and Cramer, N. (2013). Rhodium-catalyzed dynamic kinetic asymmetric transformations of racemic allenes by the [3+2] annulation of aryl Ketimines. *Angew. Chem. Int. Ed.* 52 (40): 10630–10634. (d) Tran, D.N. and Cramer, N. (2011). Enantioselective rhodium(I)-catalyzed [3+2] annulations of aromatic ketimines induced by directed C-H activations. *Angew. Chem. Int. Ed.* 50 (47): 11098–11102. (e) Tran, D.N. and Cramer, N. (2010). Syn-selective rhodium(I)-catalyzed allylations of ketimines proceeding through a directed C-H activation/Allene addition sequence. *Angew. Chem. Int. Ed.* 49 (44): 8181–8184. (f) Sun, Z.M., Chen, S.P., and Zhao, P. (2010). Tertiary carbinamine synthesis by rhodium-catalyzed [3+2] annulation of N-unsubstituted aromatic ketimines and alkynes. *Chem. Eur. J.* 16 (8): 2619–2627.

58 Jun, C.-H., Hong, J.-B., Kim, Y.-H. et al. (2000). The catalytic alkylation of aromatic imines by Wilkinson's complex: the domino reaction of hydroacylation and ortho-alkylation. *Angew. Chem. Int. Ed.* 39 (19): 3440–3442.

59 (a) Wilson, R.M., Thalji, R.K., Bergman, R.G. et al. (2006). Enantioselective synthesis of a PKC inhibitor via catalytic C—H bond activation. *Org. Lett.* 8 (8): 1745–1747. (b) O'Malley, S.J., Tan, K.L., Watzke, A. et al. (2005). Total synthesis of (+)-lithospermic acid by asymmetric intramolecular alkylation via catalytic C—H bond activation. *J. Am. Chem. Soc.* 127 (39): 13496–13497. (c) Thalji, R.K., Ellman, J.A., and Bergman, R.G. (2004). Highly efficient and enantioselective cyclization of aromatic imines via directed C—H bond activation. *J. Am. Chem. Soc.* 126 (23): 7192–7193. (d) Ahrendt, K.A., Bergman, R.G., and Ellman, J.A. (2003). Synthesis of a tricyclic mescaline analogue by catalytic C—H bond activation. *Org. Lett.* 5 (8): 1301–1303. (e) Thalji, R.K., Ahrendt, K.A., Bergman, R.G. et al. (2001). Annulation of aromatic imines via directed C–H activation with Wilkinson's catalyst. *J. Am. Chem. Soc.* 123 (39): 9692–9693.

60 (a) Gao, K. and Yoshikai, N. (2013). Cobalt-catalyzed ortho alkylation of aromatic imines with primary and secondary alkyl halides. *J. Am. Chem. Soc.* 135 (25): 9279–9282. (b) Lee, P.-S., Fujita, T., and Yoshikai, N. (2011). Cobalt-catalyzed, room-temperature addition of aromatic imines to alkynes via directed C—H bond activation. *J. Am. Chem. Soc.* 133 (43): 17283–17295.

61 (a) Senthilkumar, N., Gandeepan, P., Jayakumar, J. et al. (2014). Rh(III)-catalyzed synthesis of 1-substituted isoquinolinium salts via a C—H bond activation reaction of ketimines with alkynes. *Chem. Commun.* 50 (23): 3106–3108. (b) Jayakumar, J., Parthasarathy, K., and Cheng, C.H. (2011). One-pot synthesis of isoquinolinium salts by rhodium-catalyzed C—H bond activation: application to the total synthesis of oxychelerythrine. *Angew. Chem. Int. Ed.* 51 (1): 197–200. (c) Lim, S.-G., Lee, J.H., Moon, C.W. et al. (2003). Rh(I)-catalyzed direct ortho-alkenylation of aromatic ketimines with alkynes and its application to the synthesis of isoquinoline derivatives. *Org. Lett.* 5 (15): 2759–2761. (d) Fukutani, T., Umeda, N., Hirano, K. et al. (2009). Rhodium-catalyzed oxidative coupling of aromatic imines with internal alkynes via regioselective C—H bond cleavage. *Chem. Commun.* 34: 5141–5143. (e) Guimond, N. and Fagnou, K. (2009). Isoquinoline synthesis via rhodium-catalyzed oxidative cross-coupling/cyclization of aryl aldimines and alkynes. *J. Am. Chem. Soc.* 131 (34): 12050–12051.

62 (a) Hung, C.H., Gandeepan, P., and Cheng, C.H. (2014). Ruthenium(II)-catalyzed C—H bond activation: an efficient route toward indenamines. *ChemCatChem* 6 (9): 2692–2697. (b) Zhang, T., Wu, L., and Li, X. (2013). Rh(III)-catalyzed olefination of N-sulfonyl imines: synthesis of ortho-olefinated benzaldehydes. *Org. Lett.* 15 (24): 6294–6297. (c) Zhao, P., Wang, F., Han, K. et al. (2012). Ruthenium- and sulfonamide-catalyzed cyclization between N-sulfonyl imines and alkynes. *Org. Lett.* 14 (21): 5506–5509.

63 Parthasarathy, K., Jeganmohan, M., and Cheng, C.-H. (2007). Rhodium-catalyzed one-pot synthesis of substituted pyridine derivatives from α,β-unsaturated ketoximes and alkynes. *Org. Lett.* 10 (2): 325–328.

64 (a) Shi, Z., Koester, D.C., Boultadakis-Arapinis, M. et al. (2013). Rh(III)-catalyzed synthesis of multisubstituted Isoquinoline and pyridine N-oxides from oximes and diazo compounds. *J. Am. Chem. Soc.* 135 (33): 12204–12207. (b) Kornhaaß, C., Li, J., and Ackermann, L. (2012). Cationic ruthenium catalysts for alkyne annulations with oximes by C–H/N–O functionalizations. *J. Org. Chem.* 77 (20): 9190–9198. (c) Zhang, X., Chen, D., Zhao, M. et al. (2011). Synthesis of isoquinolines via rhodium(III)-catalyzed dehydrative C-C and C-N coupling between oximines and alkynes. *Adv. Synth. Catal.* 353 (5): 719–723. (d) Hyster, T.K. and Rovis, T. (2011). Pyridine synthesis from oximes and alkynes via rhodium(III) catalysis: Cp* and Cpt provide complementary selectivity. *Chem. Commun.* 47 (43): 11846–11848.

65 (a) Lian, Y., Huber, T., Hesp, K.D. et al. (2012). Rhodium(III)-catalyzed alkenyl C—H bond functionalization: convergent synthesis of furans and pyrroles. *Angew. Chem. Int. Ed.* 52 (2): 629–633. (b) Lian, Y., Bergman, R.G., and Ellman, J.A. (2012). Rhodium(III)-catalyzed synthesis of phthalides by cascade addition and cyclization of benzimidates with aldehydes. *Chem. Sci.* 3 (10): 3088–3092. (c) Thirunavukkarasu, V.S. and Cheng, C.H. (2011). Pd-catalyzed multiple C-H functionalization to construct biologically active compounds from aryl aldoxime ethers with arenes. *Chem. Eur. J.* 17 (52): 14723–14726. (d) Thirunavukkarasu, V.S., Parthasarathy, K., and Cheng, C.H. (2008). Synthesis of fluorenones from aromatic aldoxime ethers and aryl halides by palladium-catalyzed dual C–H activation and heck cyclization. *Angew. Chem. Int. Ed.* 47 (49): 9462–9465.

66 (a) Yu, S., Wan, B., and Li, X. (2014). Rh(III)-catalyzed selenylation of arenes with selenenyl chlorides/diselenides via C–H activation. *Org. Lett.* 17 (1): 58–61. (b) Lou, S.J., Xu, D.Q., and Xu, Z.Y. (2014). Mild and versatile nitrate-promoted C—H bond fluorination. *Angew. Chem. Int. Ed.* 53 (39): 10330–10335. (c) Zhang, W., Lou, S., Liu, Y. et al. (2013). Palladium-catalyzed chelation-assisted aromatic C–H nitration: regiospecific synthesis of nitroarenes free from the effect of the orientation rules. *J. Org. Chem.* 78 (12): 5932–5948. (d) Ryu, J., Shin, K., Park, S.H. et al. (2012). Rhodium-catalyzed direct C-H amination of benzamides with aryl azides: a synthetic route to diarylamines. *Angew. Chem. Int. Ed.* 51 (39): 9904–9908. (e) Chan, W.-W., Lo, S.-F., Zhou, Z. et al. (2012). Rh-catalyzed intermolecular carbenoid functionalization of aromatic C—H bonds by α-diazomalonates. *J. Am. Chem. Soc.* 134 (33): 13565–13568. (f) Tsai, A.S., Brasse, M., Bergman, R.G. et al. (2010). Rh(III)-catalyzed oxidative coupling of unactivated alkenes via C–H activation. *Org. Lett.* 13 (3): 540–542. (g) Chan, C.-W., Zhou, Z., Chan, A.S.C. et al. (2010). Pd-catalyzed ortho-C–H acylation/cross coupling of aryl ketone O-methyl oximes with aldehydes using tert-butyl Hydroperoxide as oxidant. *Org. Lett.* 12 (17): 3926–3929. (h) Sun, C.-L., Liu, N., Li, B.-J. et al. (2009). Pd-catalyzed C–H functionalizations of O-methyl oximes with arylboronic acids. *Org. Lett.* 12 (1): 184–187. (i) Thu, H.-Y., Yu, W.-Y., and Che, C.-M. (2006). Intermolecular amidation of unactivated sp2 and sp3 C—H bonds via palladium-catalyzed cascade C–H activation/nitrene insertion. *J. Am. Chem. Soc.* 128 (28): 9048–9049.

67 (a) Pawar, A.B., Agarwal, D., and Lade, D.M. (2016). Cp*Co(III)-catalyzed C–H/N–N functionalization of arylhydrazones for the synthesis of isoquinolines. *J. Org. Chem.* 81 (22): 11409–11415. (b) Huang, X.-C., Yang, X.-H., Song, R.-J. et al. (2014). Rhodium-catalyzed synthesis of isoquinolines and indenes from benzylidenehydrazones and internal alkynes. *J. Org. Chem.* 79 (3): 1025–1031. (c) Han, W., Zhang, G., Li, G. et al. (2014). Rh-catalyzed sequential oxidative C—H and N—N bond activation: conversion of azines into isoquinolines with air at room temperature. *Org. Lett.* 16 (13): 3532–3535. (d) Chuang, S.-C., Gandeepan, P., and Cheng, C.-H. (2013). Synthesis of isoquinolines via Rh(III)-catalyzed C–H activation using hydrazone as a new oxidizing directing group. *Org. Lett.* 15 (22): 5750–5753.

68 (a) Cai, H., Gan, X., Jin, Z. et al. (2023). Carboxylic acid derivatives in herbicide development. *J. Agric. Food Chem.* 71 (26): 9973–9993. (b) Chen, N., Ye, Z., and Zhang, F. (2021). Recent progress on electrochemical synthesis involving carboxylic acids. *Org. Biomol. Chem.* 19 (25): 5501–5520. (c) Varenikov, A., Shapiro, E., and Gandelman, M. (2020). Decarboxylative halogenation of organic compounds. *Chem. Rev.* 121 (1): 412–484.

69 (a) Gooßen, L. and Biafora, A. (2017). Thieme chemistry journals awardees – where are they now? New reaction mode in carboxylate-directed C–H functionalizations: carboxylates as deciduous directing groups. *Synlett* 28 (15): 1885–1890. (b) Shi, G. and Zhang, Y. (2014). Carboxylate-directed C-H functionalization. *Adv. Synth. Catal.* 356 (7): 1419–1442.

70 Zhu, R.Y., Farmer, M.E., Chen, Y.Q. et al. (2016). A simple and versatile amide directing group for C–H functionalizations. *Angew. Chem. Int. Ed.* 55 (36): 10578–10599.

71 (a) Kondapalli, V., Yu, X., Yamamoto, Y. et al. (2017). Synthesis of 5H-dibenzo[*c,e*]azepine-5,7(6*H*)-diones from benzamides via palladium-catalyzed double C—H bond activation. *J. Org. Chem.* 82 (4): 2288–2293. (b) Jaiswal, Y., Kumar, Y., Thakur, R. et al. (2016). Primary amide directed regioselective ortho-C–H-arylation of (aryl)acetamides. *J. Org. Chem.* 81 (24): 12499–12505. (c) Li, D.-D., Yuan, T.-T., and Wang, G.-W. (2012). Palladium-catalyzed ortho-arylation of benzamides via direct sp^2 C—H bond activation. *J. Org. Chem.* 77 (7): 3341–3347.

72 (a) Daniels, M.H., Armand, J.R., and Tan, K.L. (2016). Sequential regioselective C–H functionalization of thiophenes. *Org. Lett.* 18 (14): 3310–3313. (b) Yang, W., Ye, S., Fanning, D. et al. (2015). Orchestrated triple C-H activation reactions using two directing groups: rapid assembly of complex pyrazoles. *Angew. Chem. Int. Ed.* 54 (8): 2501–2504. (c) Chinnagolla, R.K. and Jeganmohan, M. (2012). Regioselective ortho-arylation and alkenylation of N-alkyl benzamides with boronic acids via ruthenium-catalyzed C—H bond activation: an easy route to fluorenones synthesis. *Org. Lett.* 14 (20): 5246–5249. (d) Wang, X., Leow, D., and Yu, J.-Q. (2011). Pd(II)-catalyzed para-selective C–H arylation of monosubstituted arenes. *J. Am. Chem. Soc.* 133 (35): 13864–13867.

73 (a) Li, J. and Ackermann, L. (2015). Carboxylate-assisted ruthenium(II)-catalyzed C–H activations of monodentate amides with conjugated alkenes. *Org. Chem. Front.* 2 (9): 1035–1039. (b) Xia, Y., Liu, Z., Feng, S. et al. (2014). Ir(III)-catalyzed aromatic C—H bond functionalization via metal carbene migratory insertion. *J. Org. Chem.* 80 (1): 223–236. (c) Chen, Q., Ilies, L., Yoshikai, N. et al. (2011). Cobalt-catalyzed coupling of alkyl grignard reagent with benzamide and 2-phenylpyridine derivatives through directed C—H bond activation under air. *Org. Lett.* 13 (12): 3232–3234. (d) Chen, Q., Ilies, L., and Nakamura, E. (2010). Cobalt-catalyzed ortho-alkylation of secondary benzamide with alkyl chloride through directed C—H bond activation. *J. Am. Chem. Soc.* 133 (3): 428–429.

74 (a) Wang, D., Yu, X., Xu, X. et al. (2016). Non-coordinating-anion-directed reversal of activation site: selective C—H bond activation of N-aryl rings. *Chem. Eur. J.* 22 (25): 8663–8668. (b) Ackermann, L., Wang, L., Wolfram, R. et al. (2012). Ruthenium-catalyzed oxidative C–H alkenylations of anilides and benzamides in water. *Org. Lett.* 14 (3): 728–731.

75 (a) Yoo, E.J., Ma, S., Mei, T.-S. et al. (2011). Pd-catalyzed intermolecular C–H amination with alkylamines. *J. Am. Chem. Soc.* 133 (20): 7652–7655. (b) Ryu, J., Kwak, J., Shin, K. et al. (2013). Ir(III)-catalyzed mild C–H amidation of arenes and alkenes: an efficient usage of acyl azides as the nitrogen source. *J. Am. Chem. Soc.* 135 (34): 12861–12868. (c) Shang, M., Zeng, S.-H., Sun, S.-Z. et al. (2013). Ru(II)-catalyzed ortho-C–H amination of arenes and heteroarenes at room temperature. *Org. Lett.* 15 (20): 5286–5289. (d) Kim, H., Shin, K., and Chang, S. (2014). Iridium-catalyzed C–H amination with anilines at room temperature: compatibility of iridacycles with external oxidants. *J. Am. Chem. Soc.* 136 (16): 5904–5907.

76 Wang, X.-C., Hu, Y., Bonacorsi, S. et al. (2013). Pd(II)-catalyzed C–H iodination using molecular I2 as the sole oxidant. *J. Am. Chem. Soc.* 135 (28): 10326–10329.

77 Dai, H.-X. and Yu, J.-Q. (2011). Pd-catalyzed oxidative ortho-C–H borylation of arenes. *J. Am. Chem. Soc.* 134 (1): 134–137.

78 Zhang, X.-G., Dai, H.-X., Wasa, M. et al. (2012). Pd(II)-catalyzed ortho trifluoromethylation of arenes and insights into the coordination mode of acidic amide directing groups. *J. Am. Chem. Soc.* 134 (29): 11948–11951.

79 (a) Senthilkumar, N., Parthasarathy, K., Gandeepan, P. et al. (2013). Synthesis of phenanthridinones from N-methoxybenzamides and aryltriethoxysilanes through Rh^{III}-catalyzed C—H and N—H bond activation. *Chem. Asian J.* 8 (9): 2175–2181. (b) Zeng, R., Wu, S., Fu, C. et al. (2013). Room-temperature synthesis of trisubstituted allenylsilanes via regioselective C–H functionalization. *J. Am. Chem. Soc.* 135 (49): 18284–18287. (c) Fabry, D.C., Zoller, J., Raja, S. et al. (2014). Combining rhodium and photoredox catalysis for C-H functionalizations of arenes: oxidative heck reactions with visible light. *Angew. Chem. Int. Ed.* 53 (38): 10228–10231. (d) Boobalan, R., Gandeepan, P., and Cheng, C.-H. (2016). Ruthenium-catalyzed C–H alkynylation of aromatic amides with hypervalent iodine–alkyne reagents. *Org. Lett.* 18 (14): 3314–3317. (e) Chen, Z., Hu, L.A., Zeng, F. et al. (2017). Selective mono-alkylation of N-methoxybenzamides. *Chem. Commun.* 53 (30): 4258–4261.

80 (a) Péron, F., Fossey, C., Sopkova-de Oliveira Santos, J. et al. (2014). Room-temperature ortho-alkoxylation and -halogenation of N-tosylbenzamides by using palladium(II)-catalyzed C-H activation. *Chem. Eur. J.* 20 (24): 7507–7513. (b) Feng, C. and Loh, T.P. (2013). Directing-group-assisted copper-catalyzed olefinic trifluoromethylation of electron-deficient alkenes. *Angew. Chem. Int. Ed.* 52 (47): 12414–12417. (c) Zhu, C. and Falck, J.R. (2011). N-acylsulfonamide assisted tandem C–H olefination/annulation: synthesis of isoindolinones. *Org. Lett.* 13 (5): 1214–1217.

81 (a) Collins, K.D., Lied, F., and Glorius, F. (2014). Preparation of conjugated 1,3-enynes by Rh(III)-catalysed alkynylation of alkenes via C–H activation. *Chem. Commun.* 50 (34): 4459–4461. (b) Wang, H., Schröder, N., and Glorius, F. (2013). Mild rhodium(III)-catalyzed direct C-H allylation of arenes with allyl carbonates. *Angew. Chem. Int. Ed.* 52 (20): 5386–5389. (c) Park, J., Kim, M., Sharma, S. et al. (2013). Pd(II)-catalyzed decarboxylative acylation of phenylacetamides with α-oxocarboxylic acids via C—H bond activation. *Chem. Commun.* 49 (16): 1654–1656. (d) Wencel-Delord, J., Nimphius, C., Wang, H. et al. (2012). Rhodium(III) and hexabromobenzene—a catalyst system for the cross-dehydrogenative coupling of simple arenes and heterocycles with arenes bearing directing groups. *Angew. Chem. Int. Ed.* 51 (52): 13001–13005. (e) Wencel-Delord, J., Nimphius, C., Patureau, F.W. et al. (2012). [Rh^{III}Cp*]-catalyzed dehydrogenative aryl-aryl bond formation. *Angew. Chem. Int. Ed.* 51 (9): 2247–2251. (f) Wencel-Delord, J., Nimphius, C., Patureau, F.W. et al. (2012). Undirected arene and chelate-assisted olefin C—H bond activation: [Rh^{III}Cp*]-catalyzed dehydrogenative alkene–arene coupling as a new pathway for the selective synthesis of highly substituted Z olefins. *Chem. Asian J.* 7 (6): 1208–1212. (g) Thirunavukkarasu, V.S., Hubrich, J., and Ackermann, L. (2012). Ruthenium-catalyzed oxidative C(sp^2)—H bond hydroxylation: site-selective C—O bond formation on benzamides. *Org. Lett.* 14 (16): 4210–4213. (h) Schröder, N., Wencel-Delord, J., and Glorius, F. (2012).

High-yielding, versatile, and practical [Rh(III)Cp*]-catalyzed ortho bromination and iodination of arenes. *J. Am. Chem. Soc.* 134 (20): 8298–8301.

82 (a) Kawai, K., Bunno, Y., Yoshino, T. et al. (2018). Weinreb amide directed versatile C—H bond functionalization under (η^5-pentamethylcyclopentadienyl) cobalt(III) catalysis. *Chem. Eur. J.* 24 (40): 10231–10237. (b) Das, R. and Kapur, M. (2017). Palladium-catalyzed, ortho-selective C–H halogenation of benzyl nitriles, aryl weinreb amides, and anilides. *J. Org. Chem.* 82 (2): 1114–1126. (c) Das, R. and Kapur, M. (2016). Product control using substrate design: ruthenium-catalysed oxidative C–H olefinations of cyclic Weinreb amides. *Chem. Eur. J.* 22 (47): 16986–16990. (d) Wang, Y., Zhou, K., Lan, Q. et al. (2015). Pd(II)-catalyzed C–H arylation of aryl and benzyl Weinreb amides. *Org. Biomol. Chem.* 13 (2): 353–356. (e) Das, R. and Kapur, M. (2015). Fujiwara–Moritani reaction of Weinreb amides using a ruthenium-catalyzed C–H functionalization reaction. *Chem. Asian J.* 10 (7): 1505–1512. (f) Yang, F. and Ackermann, L. (2013). Ruthenium-catalyzed C–H oxygenation on aryl Weinreb amides. *Org. Lett.* 15 (3): 718–720. (g) Wang, Y., Li, C., Li, Y. et al. (2013). Rhodium-catalyzed C-H olefination of aryl Weinreb amides. *Adv. Synth. Catal.* 355 (9): 1724–1728.

83 (a) Zhang, Q. and Shi, B.-F. (2021). 2-(pyridin-2-yl)isopropyl (PIP) amine: an enabling directing group for divergent and asymmetric functionalization of unactivated methylene C(sp3)—H bonds. *Acc. Chem. Res.* 54 (12): 2750–2763. (b) Jeon, J., Lee, C., Park, I. et al. (2021). Regio- and stereoselective functionalization enabled by bidentate directing groups. *Chem. Rec.* 21 (12): 3613–3627. (c) Rej, S., Ano, Y., and Chatani, N. (2020). Bidentate directing groups: an efficient tool in C—H bond functionalization chemistry for the expedient construction of C—C bonds. *Chem. Rev.* 120 (3): 1788–1887. (d) Zhang, Q. and Shi, B.F. (2019). From reactivity and regioselectivity to stereoselectivity: an odyssey of designing PIP amine and related directing groups for C–H activation. *Chin. J. Chem* 37 (7): 647–656. (e) Daugulis, O., Roane, J., and Tran, L.D. (2015). Bidentate, monoanionic auxiliary-directed functionalization of carbon–hydrogen bonds. *Acc. Chem. Res.* 48 (4): 1053–1064. (f) Corbet, M. and De Campo, F. (2013). 8-Aminoquinoline: a powerful directing group in metal-catalyzed direct functionalization of C—H bonds. *Angew. Chem. Int. Ed.* 52 (38): 9896–9898.

84 Zaitsev, V.G., Shabashov, D., and Daugulis, O. (2005). Highly regioselective arylation of sp^3 C—H bonds catalyzed by palladium acetate. *J. Am. Chem. Soc.* 127 (38): 13154–13155.

85 Tran, L.D., Roane, J., and Daugulis, O. (2013). Directed amination of non-acidic arene C—H bonds by a copper-silver catalytic system. *Angew. Chem. Int. Ed.* 52 (23): 6043–6046.

86 Giri, R., Maugel, N., Foxman, B.M. et al. (2008). Dehydrogenation of inert alkyl groups via remote C–H activation: converting a propyl group into a π-allylic complex. *Organometallics* 27 (8): 1667–1670.

87 (a) Xu, L.-L., Wang, X., Ma, B. et al. (2018). Copper mediated C–H amination with oximes: en route to primary anilines. *Chem. Sci.* 9 (23): 5160–5164.

(b) Kong, W.-J., Shao, Q., Li, M.-H. et al. (2018). Copper-mediated diastereoselective C–H thiolation of ferrocenes. *Organometallics* 37 (17): 2832–2836. (c) Kim, H., Heo, J., Kim, J. et al. (2018). Copper-mediated amination of aryl C—H bonds with the direct use of aqueous ammonia via a disproportionation pathway. *J. Am. Chem. Soc.* 140 (43): 14350–14356. (d) Shang, M., Wang, M.M., Saint-Denis, T.G. et al. (2017). Copper-mediated late-stage functionalization of heterocycle-containing molecules. *Angew. Chem. Int. Ed.* 56 (19): 5317–5321. (e) Maity, S., Dolui, P., Kancherla, R. et al. (2017). Introducing unactivated acyclic internal aliphatic olefins into a cobalt catalyzed allylic selective dehydrogenative heck reaction. *Chem. Sci.* 8 (7): 5181–5185. (f) Liang, S., Liu, N.W., and Manolikakes, G. (2015). Copper-mediated sulfonylation of aryl C(sp^2)—H bonds with sodium and lithium sulfinates. *Adv. Synth. Catal.* 358 (1): 159–163. (g) Shang, M., Wang, H.-L., Sun, S.-Z. et al. (2014). Cu(II)-mediated ortho C–H alkynylation of (hetero)arenes with terminal alkynes. *J. Am. Chem. Soc.* 136 (33): 11590–11593. (h) Shang, M., Sun, S.-Z., Dai, H.-X. et al. (2014). $Cu(OAc)_2$-catalyzed coupling of aromatic C—H bonds with arylboron reagents. *Org. Lett.* 16 (21): 5666–5669. (i) Shang, M., Sun, S.-Z., Dai, H.-X. et al. (2014). Cu(II)-mediated C–H amidation and amination of arenes: exceptional compatibility with heterocycles. *J. Am. Chem. Soc.* 136 (9): 3354–3357.

88 Inoue, S., Shiota, H., Fukumoto, Y. et al. (2009). Ruthenium-catalyzed carbonylation at ortho C—H bonds in aromatic amides leading to phthalimides: C—H bond activation utilizing a bidentate system. *J. Am. Chem. Soc.* 131 (20): 6898–6899.

89 (a) Liu, S.-L., Li, X.-H., Zhang, S.-S. et al. (2017). Nickel-catalyzed sulfonylation of C(sp^2)—H bonds with sodium sulfinates. *Adv. Synth. Catal.* 359 (13): 2241–2246. (b) Li, Y., Liu, Y.-J., and Shi, B.-F. (2017). Copper-mediated thiolation of unactivated heteroaryl C—H bonds with disulfides under ligand- and metal-oxidant-free conditions. *Adv. Synth. Catal.* 359 (23): 4117–4121. (c) Li, Q., Hu, W., Hu, R. et al. (2017). Cobalt-catalyzed cross-dehydrogenative coupling reaction between unactivated C(sp^2)—H and C(sp^3)—H bonds. *Org. Lett.* 19 (17): 4676–4679. (d) Zhao, S., Liu, B., Zhan, B.-B. et al. (2016). Nickel-catalyzed *ortho*-arylation of unactivated (hetero)aryl C—H bonds with arylsilanes using a removable auxiliary. *Org. Lett.* 18 (18): 4586–4589. (e) Zhan, B.-B., Liu, Y.-H., Hu, F. et al. (2016). Nickel-catalyzed *ortho*-halogenation of unactivated (hetero)aryl C—H bonds with lithium halides using a removable auxiliary. *Chem. Commun.* 52 (27): 4934–4937. (f) Matsubara, T., Ilies, L., and Nakamura, E. (2016). Oxidative C–H activation approach to pyridone and isoquinolone through an iron-catalyzed coupling of amides with alkynes. *Chem. Asian J.* 11 (3): 380–384. (g) Li, Q., Li, Y., Hu, W. et al. (2016). Cobalt-catalyzed C(sp^2)–H methylation by using dicumyl peroxide as both the methylating reagent and hydrogen acceptor. *Chem. Eur. J.* 22 (35): 12286–12289. (h) Zhang, Q., Yin, X.-S., Chen, K. et al. (2015). Stereoselective synthesis of chiral β-fluoro α-amino acids via Pd(II)-catalyzed fluorination of unactivated methylene C(sp^3)—H bonds: scope and mechanistic studies. *J. Am. Chem. Soc.* 137 (25): 8219–8226. (i) Yang, K., Wang, Y., Chen, X. et al. (2015).

Nickel-catalyzed and benzoic acid-promoted direct sulfenylation of unactivated arenes. *Chem. Commun.* 51 (17): 3582–3585. (j) Miao, J., Yang, K., Kurek, M. et al. (2015). Palladium-catalyzed site-selective fluorination of unactivated C(sp^3)—H bonds. *Org. Lett.* 17 (15): 3738–3741. (k) Liu, Y.-J., Liu, Y.-H., Yan, S.-Y. et al. (2015). A sustainable and simple catalytic system for direct alkynylation of C(sp^2)—H bonds with low nickel loadings. *Chem. Commun.* 51 (29): 6388–6391. (l) Liu, Y.-H., Liu, Y.-J., Yan, S.-Y. et al. (2015). Ni(II)-catalyzed dehydrative alkynylation of unactivated (hetero)aryl C—H bonds using oxygen: a user-friendly approach. *Chem. Commun.* 51 (58): 11650–11653. (m) Liu, Y.J., Liu, Y.H., Yin, X.S. et al. (2014). Copper/silver-mediated direct ortho-ethynylation of unactivated (hetero)aryl C—H bonds with terminal alkyne. *Chem. Eur. J.* 21 (1): 205–209. (n) Li, X., Liu, Y.-H., Gu, W.-J. et al. (2014). Copper-mediated hydroxylation of arenes and heteroarenes directed by a removable bidentate auxiliary. *Org. Lett.* 16 (15): 3904–3907. (o) Chen, F.-J., Liao, G., Li, X. et al. (2014). Cu(II)-mediated C—S/N—S bond formation *via* C–H activation: access to benzoisothiazolones using elemental sulfur. *Org. Lett.* 16 (21): 5644–5647. (p) Shiota, H., Ano, Y., Aihara, Y. et al. (2011). Nickel-catalyzed chelation-assisted transformations involving ortho C—H bond activation: regioselective oxidative cycloaddition of aromatic amides to alkynes. *J. Am. Chem. Soc.* 133 (38): 14952–14955. (q) Zhao, S., Liu, Y.-J., Yan, S.-Y. et al. (2015). Copper-/silver-mediated arylation of C(sp^2)—H bonds with 2-thiophenecarboxylic acids. *Org. Lett.* 17 (13): 3338–3341. (r) Aihara, Y. and Chatani, N. (2013). Ruthenium-catalyzed direct arylation of C—H bonds in aromatic amides containing a bidentate directing group: significant electronic effects on arylation. *Chem. Sci.* 4 (2): 664–670. (s) Zhang, Q., Yin, X.-S., Zhao, S. et al. (2014). Pd(II)-catalyzed arylation of unactivated methylene C(sp^3)—H bonds with aryl halides using a removable auxiliary. *Chem. Commun.* 50 (61): 8353–8355.

90 (a) Zhao, H., Shao, X., Qing, Z. et al. (2019). Cobalt-catalyzed 2-(1-methylhydrazinyl)pyridine-assisted direct C–H/N–H functionalization of benzoyl hydrazide with isocyanide: efficient synthesis of iminoisoindolinone derivatives. *Adv. Synth. Catal.* 361 (7): 1678–1682. (b) Zhai, S., Qiu, S., Chen, X. et al. (2018). 2-(1-methylhydrazinyl)pyridine as a reductively removable directing group in a cobalt-catalyzed C(sp^2)—H bond alkenylation/annulation cascade. *Chem. Commun.* 54 (1): 98–101. (c) Zhai, S., Qiu, S., Chen, X. et al. (2018). Trifunctionalization of allenes via cobalt-catalyzed MHP-assisted C—H bond functionalization and molecular oxygen activation. *ACS Catal.* 8 (7): 6645–6649. (d) Qiu, S., Zhai, S., Wang, H. et al. (2018). Efficient synthesis of phthalimides via cobalt-catalyzed C(sp2)–H carbonylation of benzoyl hydrazides with carbon monoxide. *Adv. Synth. Catal.* 360 (17): 3271–3276. (e) Meyer, T.H., Oliveira, J.C.A., Sau, S.C. et al. (2018). Electrooxidative allene annulations by mild cobalt-catalyzed C–H activation. *ACS Catal.* 8 (10): 9140–9147. (f) Mei, R., Sauermann, N., Oliveira, J.C.A. et al. (2018). Electroremovable traceless hydrazides for cobalt-catalyzed electro-oxidative C–H/N–H activation with internal alkynes. *J. Am. Chem. Soc.*

140 (25): 7913–7921. (g) Zhao, H., Shao, X., Wang, T. et al. (2018). A 2-(1-methylhydrazinyl)pyridine-directed C–H functionalization/spirocyclization cascade: facile access to spirosuccinimide derivatives. *Chem. Commun.* 54 (39): 4927–4930.

91 Gu, Q., Al Mamari, H.H., Graczyk, K. et al. (2014). Iron-catalyzed C(sp^2)-H and C(sp^3)-H arylation by triazole assistance. *Angew. Chem. Int. Ed.* 53 (15): 3868–3871.

92 (a) Mo, J., Müller, T., Oliveira, J.C.A. et al. (2018). 1,4-iron migration for expedient allene annulations through iron-catalyzed C–H/N–H/C–O/C–H functionalizations. *Angew. Chem. Int. Ed.* 57 (26): 7719–7723. (b) Bauer, M., Wang, W., Lorion, M.M. et al. (2018). Internal peptide late-stage diversification: peptide-isosteric triazoles for primary and secondary C(sp^3)–H activation. *Angew. Chem. Int. Ed.* 57 (1): 203–207. (c) Shen, Z., Cera, G., Haven, T. et al. (2017). Tri-substituted triazole-enabled C–H activation of benzyl and aryl amines by iron catalysis. *Org. Lett.* 19 (14): 3795–3798. (d) Liu, W., Cera, G., Oliveira, J.C.A. et al. (2017). $MnCl_2$-catalyzed C–H alkylations with alkyl halides. *Chem. Eur. J.* 23 (48): 11524–11528. (e) Cera, G., Haven, T., and Ackermann, L. (2017). Iron-catalyzed C–H alkynylation through triazole assistance: expedient access to bioactive heterocycles. *Chem. Eur. J.* 23 (15): 3577–3582. (f) Cera, G., Haven, T., and Ackermann, L. (2017). Iron-catalyzed C-H/N-H activation by triazole guidance: versatile alkyne annulation. *Chem. Commun.* 53 (48): 6460–6463. (g) Ye, X., Zhang, Y., He, Y. et al. (2016). 1,2,3-triazole amine as directing group in promoting catalytic oxidative C–H olefination under aerobic conditions. *Tetrahedron* 72 (22): 2756–2762. (h) Ye, X., Xu, C., Wojtas, L. et al. (2016). Silver-free palladium-catalyzed sp^3 and sp^2 C–H alkynylation promoted by a 1,2,3-triazole amine directing group. *Org. Lett.* 18 (12): 2970–2973. (i) Santrač, D., Cella, S., Wang, W. et al. (2016). Palladium-catalyzed C–H arylation of amides by triazole assistance. *Eur. J. Org. Chem.* 2016 (32): 5429–5436. (j) Cera, G., Haven, T., and Ackermann, L. (2016). Expedient iron-catalyzed C–H allylation/alkylation by triazole assistance with ample scope. *Angew. Chem. Int. Ed.* 55 (4): 1484–1488. (k) Zhang, G., Xie, X., Zhu, J. et al. (2015). Pd(II)-catalyzed C(sp^3)–H arylation of amino acid derivatives with click-Triazoles as removable directing groups. *Org. Biomol. Chem.* 13 (19): 5444–5449. (l) Al Mamari, H.H., Diers, E., and Ackermann, L. (2014). Triazole-assisted ruthenium-catalyzed C—H arylation of aromatic amides. *Chem. Eur. J.* 20 (31): 9739–9743. (m) Ye, X., He, Z., Ahmed, T. et al. (2013). 1,2,3-triazoles as versatile directing group for selective sp^2 and sp^3 C–H activation: cyclization vs substitution. *Chem. Sci.* 4 (9): 3712–3716.

93 (a) Mandal, A., Selvakumar, J., Dana, S. et al. (2018). A cross-dehydrogenative annulation strategy towards synthesis of polyfluorinated phenanthridinones with copper. *Chem. Eur. J.* 24 (14): 3448–3454. (b) Lee, W.-C.C., Wang, W., and Li, J.J. (2018). Copper(II)-mediated ortho-selective C(sp^2)–H tandem alkynylation/annulation and ortho-hydroxylation of anilides with 2-aminophenyl-1*H*-pyrazole as a directing group. *J. Org. Chem.* 83 (4): 2382–2388.

(c) Ji, Y.-F., Hu, Y.-H., Xu, Z. et al. (2018). Palladium-catalyzed arylation of aromatic amides directed by a [4-chloro-2-(1H-pyrazol-1-yl)phenyl]amine auxiliary. *Synlett* 29 (14): 1875–1880. (d) Grandhi, G.S., Selvakumar, J., Dana, S. et al. (2018). Directed C—H bond functionalization: a unified approach to formal syntheses of amorfrutin A, cajaninstilbene acid, hydrangenol, and macrophyllol. *J. Org. Chem.* 83 (19): 12327–12333. (e) Shen, Y., Cindy Lee, W.-C., Gutierrez, D.A. et al. (2017). Palladium-catalyzed direct C(sp2)–H ortho-arylation of anilides using 2-aminophenylpyrazole as the directing group. *J. Org. Chem.* 82 (21): 11620–11625. (f) Li, J., Lee, W.-C., and Tehrani, A. (2017). Copper-mediated sp^2 C–H chlorination with trichloroacetamide using a removable directing group. *Synthesis* 49 (13): 2865–2872. (g) Selvakumar, J., Grandhi, G.S., Sahoo, H. et al. (2016). Copper-mediated etherification of arenes with alkoxysilanes directed by an (2-aminophenyl)pyrazole group. *RSC Adv.* 6 (83): 79361–79365. (h) Lee, W.-C.C., Shen, Y., Gutierrez, D.A. et al. (2016). 2-aminophenyl-1H-pyrazole as a removable directing group for copper-mediated C–H amidation and sulfonamidation. *Org. Lett.* 18 (11): 2660–2663.

94 Fan, M. and Ma, D. (2013). Palladium-catalyzed direct functionalization of 2-aminobutanoic acid derivatives: application of a convenient and versatile auxiliary. *Angew. Chem. Int. Ed.* 52 (46): 12152–12155.

95 Cai, Z., Li, S., Gao, Y. et al. (2018). Weak, bidentate chelating group assisted cross-coupling of $C(sp^3)$–H bonds in aliphatic acid derivatives with aryltrifluoroborates. *Chem. Commun.* 54 (90): 12766–12769.

96 Lou, J., Wang, Q., He, Y. et al. (2018). A simple aliphatic diamine auxiliary for palladium-catalyzed arylation of unactivated β-$C(sp^3)$—H bonds. *Adv. Synth. Catal.* 360 (23): 4571–4584.

97 Kim, J., Sim, M., Kim, N. et al. (2015). Asymmetric C–H functionalization of cyclopropanes using an isoleucine-NH_2 bidentate directing group. *Chem. Sci.* 6 (6): 3611–3616.

98 Reddy, M.D., Blanton, A.N., and Watkins, E.B. (2017). Palladium-catalyzed, N-(2-aminophenyl)acetamide-assisted ortho-arylation of substituted benzamides: application to the synthesis of urolithins B, M6, and M7. *J. Org. Chem.* 82 (10): 5080–5095.

99 (a) Yu, J., Lv, W., and Cheng, G. (2018). Palladium-catalyzed site-selective C–H arylation of 2,2′-bipyridine-6-carboxamides via a rollover cyclometalation pathway. *Org. Lett.* 20 (16): 4732–4735. (b) Usui, K., Haines, B.E., Musaev, D.G. et al. (2018). Understanding regiodivergence in a Pd(II)-mediated site-selective C–H alkynylation. *ACS Catal.* 8 (5): 4516–4527. (c) Bai, Z., Cai, C., Yu, Z. et al. (2018). Backbone-enabled directional peptide macrocyclization through late-stage palladium-catalyzed δ-$C(sp^2)$–H olefination. *Angew. Chem. Int. Ed.* 57 (42): 13912–13916. (d) Viart, H.M.F., Bachmann, A., Kayitare, W. et al. (2017). β-carboline amides as intrinsic directing groups for $C(sp^2)$–H functionalization. *J. Am. Chem. Soc.* 139 (3): 1325–1329. (e) Reddy, C., Bisht, N., Parella, R. et al. (2016). 4-amino-2,1,3-benzothiadiazole as a removable bidentate directing group for the Pd(II)-catalyzed arylation/oxygenation of sp^2/sp^3 β-C—H bonds of carboxamides. *J. Org. Chem.* 81 (24): 12143–12168.

(f) Pasunooti, K.K., Banerjee, B., Yap, T. et al. (2015). Auxiliary-directed Pd-catalyzed γ-C(sp^3)—H bond activation of α-aminobutanoic acid derivatives. *Org. Lett.* 17 (24): 6094–6097.

100 (a) Sauermann, N., Mei, R., and Ackermann, L. (2018). Electrochemical C–H amination by cobalt catalysis in a renewable solvent. *Angew. Chem. Int. Ed.* 57 (18): 5090–5094. (b) Li, M. and Wang, J.J. (2018). Cobalt-catalyzed direct C–H thiolation of aromatic amides with disulfides: application to the synthesis of quetiapine. *Org. Lett.* 20 (20): 6490–6493. (c) Tan, G., Zhang, L., Liao, X. et al. (2017). Copper- or nickel-enabled oxidative cross-coupling of unreactive C(sp^3)—H bonds with azole C(sp^2)—H bonds: rapid access to β-azolyl propanoic acid derivatives. *Org. Lett.* 19 (18): 4830–4833. (d) Sauermann, N., Meyer, T.H., Tian, C. et al. (2017). Electrochemical cobalt-catalyzed C–H oxygenation at room temperature. *J. Am. Chem. Soc.* 139 (51): 18452–18455. (e) Zheng, X.-X., Du, C., Zhao, X.-M. et al. (2016). Ni(II)-catalyzed C(sp^2)–H alkynylation/annulation with terminal alkynes under an oxygen atmosphere: a one-pot approach to 3-methyleneisoindolin-1-one. *J. Org. Chem.* 81 (10): 4002–4011. (f) Zhang, L.-B., Zhang, S.-K., Wei, D. et al. (2016). Cobalt(II)-catalyzed C–H amination of arenes with simple alkylamines. *Org. Lett.* 18 (6): 1318–1321. (g) Hao, X.-Q., Du, C., Zhu, X. et al. (2016). Cobalt(II)-catalyzed decarboxylative C–H activation/annulation cascades: regioselective access to isoquinolones and isoindolinones. *Org. Lett.* 18 (15): 3610–3613. (h) Zhang, S.-K., Yang, X.-Y., Zhao, X.-M. et al. (2015). Palladium-catalyzed selective β-arylation of aliphatic amides using a removable N, O-bidentate auxiliary. *Organometallics* 34 (17): 4331–4339. (i) Zhang, L.B., Hao, X.Q., Liu, Z.J. et al. (2015). Cobalt(II)-catalyzed C-H alkynylation/annulation with terminal alkynes: selective access to 3-methyleneisoindolin-1-one. *Angew. Chem. Int. Ed.* 54 (34): 10012–10015. (j) Zhang, L.-B., Hao, X.-Q., Zhang, S.-K. et al. (2015). Cobalt-catalyzed C(sp^2)-H alkoxylation of aromatic and olefinic carboxamides. *Angew. Chem. Int. Ed.* 54 (1): 272–275. (k) Guo, X.-K., Zhang, L.-B., Wei, D. et al. (2015). Mechanistic insights into cobalt(II/III)-catalyzed C–H oxidation: a combined theoretical and experimental study. *Chem. Sci.* 6 (12): 7059–7071. (l) Zhang, L.-B., Hao, X.-Q., Zhang, S.-K. et al. (2014). Copper-mediated direct alkoxylation of arenes using an N, O-bidentate directing system. *J. Org. Chem.* 79 (21): 10399–10409. (m) Hao, X.-Q., Chen, L.-J., Ren, B. et al. (2014). Copper-mediated direct aryloxylation of benzamides assisted by an N, O-bidentate directing group. *Org. Lett.* 16 (4): 1104–1107.

101 (a) Gopalakrishnan, B., Mohan, S., Parella, R. et al. (2016). Diastereoselective Pd(II)-catalyzed sp^3 C–H arylation followed by ring opening of cyclopropanecarboxamides: construction of anti β-acyloxy carboxamide derivatives. *J. Org. Chem.* 81 (19): 8988–9005. (b) Padmavathi, R., Sankar, R., Gopalakrishnan, B. et al. (2015). $Pd(OAc)_2$/AgOAc catalytic system based bidentate ligand directed regiocontrolled C–H arylation and alkylation of the C-3 position of thiophene- and furan-2-carboxamides. *Eur. J. Org. Chem.* 2015 (17): 3727–3742. (c) Babu, S. and Parella, R. (2014). Palladium-catalyzed double activation and arylation of 2° and 3° C(sp^3)—H bonds of the norbornane system: formation of a

C—C bond at the bridgehead carbon and bridgehead quaternary stereocenter. *Synlett* 25 (10): 1395–1402. (d) Parella, R., Gopalakrishnan, B., and Babu, S.A. (2013). Auxiliary-enabled Pd-catalyzed direct arylation of methylene $C(sp^3)$—H bond of cyclopropanes: highly diastereoselective assembling of di- and trisubstituted cyclopropanecarboxamides. *Org. Lett.* 15 (13): 3238–3241. (e) Parella, R., Gopalakrishnan, B., and Babu, S.A. (2013). Direct bis-arylation of cyclobutanecarboxamide via double C–H activation: an auxiliary-aided diastereoselective Pd-catalyzed access to trisubstituted cyclobutane scaffolds having three contiguous stereocenters and an all-cis stereochemistry. *J. Org. Chem.* 78 (23): 11911–11934. (f) Tran, L.D. and Daugulis, O. (2012). Nonnatural amino acid synthesis by using carbon–hydrogen bond functionalization methodology. *Angew. Chem. Int. Ed.* 51 (21): 5188–5191. (g) Shabashov, D. and Daugulis, O. (2010). Auxiliary-assisted palladium-catalyzed arylation and alkylation of sp^2 and sp^3 carbon–hydrogen bonds. *J. Am. Chem. Soc.* 132 (11): 3965–3972.

102 (a) Wang, R., Ding, Y., and Li, G. (2017). Sulfinyl isobutyramide as an auxiliary for palladium(II)-catalyzed C–H arylation and iodination of benzylamine derivatives. *Org. Biomol. Chem.* 15 (23): 4966–4970. (b) Jerhaoui, S., Chahdoura, F., Rose, C. et al. (2016). Enantiopure sulfinyl aniline as a removable and recyclable chiral auxiliary for asymmetric $C(sp^3)$—H bond activation. *Chem. Eur. J.* 22 (48): 17397–17406.

103 (a) Qi, Z., Wang, M., and Li, X. (2014). Rh(III)-catalyzed synthesis of sultones through C–H activation directed by a sulfonic acid group. *Chem. Commun.* 50 (68): 9776–9778. (b) Pham, M.V., Ye, B., and Cramer, N. (2012). Access to sultams by rhodium(III)-catalyzed directed C-H activation. *Angew. Chem. Int. Ed.* 51 (42): 10610–10614. (c) Dai, H.-X., Stepan, A.F., Plummer, M.S. et al. (2011). Divergent C–H functionalizations directed by sulfonamide pharmacophores: late-stage diversification as a tool for drug discovery. *J. Am. Chem. Soc.* 133 (18): 7222–7228.

104 (a) Thrimurtulu, N., Nallagonda, R., and Volla, C.M.R. (2017). Cobalt-catalyzed aryl C–H activation and highly regioselective intermolecular annulation of sulfonamides with allenes. *Chem. Commun.* 53 (11): 1872–1875. (b) Lan, T., Wang, L., and Rao, Y. (2017). Regioselective annulation of aryl sulfonamides with allenes through cobalt-promoted C–H functionalization. *Org. Lett.* 19 (5): 972–975. (c) Ran, Y., Yang, Y., and Zhang, L. (2016). Sodium chlorate as a viable substoichiometric oxidant for cobalt-catalyzed oxidative annulation of aryl sulfonamides with alkynes. *Tetrahedron Lett.* 57 (30): 3322–3325. (d) Planas, O., Whiteoak, C.J., Company, A. et al. (2015). Regioselective access to sultam motifs through cobalt-catalyzed annulation of aryl sulfonamides and alkynes using an 8-aminoquinoline directing group. *Adv. Synth. Catal.* 357 (18): 4003–4012. (e) Kalsi, D. and Sundararaju, B. (2015). Cobalt catalyzed C—H and N—H bond annulation of sulfonamide with terminal and internal alkynes. *Org. Lett.* 17 (24): 6118–6121.

105 Xie, W., Yang, J., Wang, B. et al. (2014). Regioselective ortho olefination of aryl sulfonamide via rhodium-catalyzed direct C—H bond activation. *J. Org. Chem.* 79 (17): 8278–8287.

106 (a) Dong, Y., Zhang, X., Chen, J. et al. (2019). Switching the site-selectivity of C–H activation in aryl sulfonamides containing strongly coordinating N-heterocycles. *Chem. Sci.* 10 (38): 8744–8751. (b) Dong, Y., Chen, J., and Xu, H. (2019). Rhodium(III)-catalyzed sulfonamide directed ortho C–H carbenoid functionalization via metal carbene migratory insertion. *Chem. Commun.* 55 (14): 2027–2030. (c) Dong, Y., Chen, J., Cui, Y. et al. (2019). $Cp^{*}Rh^{III}$-catalyzed sulfonamide-directed ortho arene C–H carbenoid functionalization with pyridotriazoles. *Org. Lett.* 22 (3): 772–775.

107 (a) Yamazaki, K., Rej, S., Ano, Y. et al. (2021). An unusual perpendicular metallacycle intermediate is the origin of branch selectivity in the Rh(II)-catalyzed C–H alkylation of aryl sulfonamides with vinylsilanes. *Organometallics* 40 (23): 3935–3942. (b) Rej, S. and Chatani, N. (2020). Rh(II)-catalyzed branch-selective C–H alkylation of aryl sulfonamides with vinylsilanes. *Chem. Sci.* 11 (2): 389–395. (c) Rej, S. and Chatani, N. (2019). Rhodium(I)-catalyzed mono-selective C–H alkylation of benzenesulfonamides with terminal alkenes. *Chem. Commun.* 55 (71): 10503–10506.

108 Chakraborty, T., Naskar, G., and Jeganmohan, M. (2024). Palladium-catalyzed selective benzylic C–H alkylation of aromatic sulfonamides with maleimides. *J. Org. Chem.* 89 (15): 10624–10638.

109 (a) Li, X., Dong, Y., Qu, F. et al. (2014). Synthesis of benzofused five-ring sultams via Rh-catalyzed C–H olefination directed by an N-Ac-substituted sulfonamide group. *J. Org. Chem.* 80 (2): 790–798. (b) Ding, Q., Liu, T., Zheng, Q. et al. (2014). Rh(III)-catalyzed tandem oxidative olefination-cyclization of aryl sulfonamides. *RSC Adv.* 4 (93): 51309–51314.

110 (a) Liu, L., Durai, M., and Doucet, H. (2022). Transition metal-catalyzed regiodivergent C−H arylations of aryl-substituted azoles. *Eur. J. Org. Chem.* 2022 (12). (b) Sarkar, T., Shah, T.A., Maharana, P.K. et al. (2021). Transition-metal-catalyzed directing group assisted (hetero)aryl C−H functionalization: construction of C−C/C-heteroatom bonds. *Chem. Rec.* 21 (12): 3758–3778. (c) Akulov, A.A., Varaksin, M.V., Mampuys, P. et al. (2021). C(sp2)–H functionalization in non-aromatic azomethine-based heterocycles. *Org. Biomol. Chem.* 19 (2): 297–312. (d) Morgan, D., Yarwood, S.J., and Barker, G. (2020). Recent developments in C−H functionalisation of benzofurans and benzothiophenes. *Eur. J. Org. Chem.* 2021 (7): 1072–1102. (e) Mao, S., Li, H., Shi, X. et al. (2018). Environmentally benign arylations of 5-membered ring heteroarenes by Pd-catalyzed C—H bonds activations. *ChemCatChem* 11 (1): 269–286. (f) Das, R. and Kapur, M. (2018). Transition-metal-catalyzed C−H functionalization reactions of π-deficient heterocycles. *Asian J. Org. Chem.* 7 (7): 1217–1235. (g) Yamaguchi, J. and Itami, K. (2017). Toward an ideal synthesis of (bio)molecules through direct arene assembling reactions. *Bull. Chem. Soc. Jpn.* 90 (4): 367–383. (h) Gayakhe, V., Sanghvi, Y.S., Fairlamb, I.J.S. et al. (2015). Catalytic C—H bond functionalisation of purine and pyrimidine nucleosides: a synthetic and mechanistic perspective. *Chem. Commun.* 51 (60): 11944–11960. (i) Bellina, F., Rossi, R., and Lessi, M. (2010). Alkenylation reactions of heteroarenes by transition-metal catalysts. *Synthesis* 2010 (24): 4131–4153.

111 (a) Cacchi, S. and Fabrizi, G. (2005). Synthesis and functionalization of indoles through palladium-catalyzed reactions. *Chem. Rev.* 105 (7): 2873–2920. (b) Bandini, M. and Eichholzer, A. (2009). Catalytic functionalization of indoles in a new dimension. *Angew. Chem. Int. Ed.* 48 (51): 9608–9644.

112 (a) Awasthi, S.K., Mishra, A., and Gope, B. (2024). Rhodium(III)-catalyzed C–H activation in indole: a comprehensive report (2017–2022). *Synthesis* 56 (12): 1815–1842. (b) Kumar, V., Maayuri, R., Mantry, L. et al. (t). *Transition-Metal-Catalyzed C-H Functionalization of Heterocycles* (ed. T. Punniyamurthy and A. Kumar), 155–191. NJ: Wiley. (c) Wen, J. and Shi, Z. (2021). From C4 to C7: innovative strategies for site-selective functionalization of indole C—H bonds. *Acc. Chem. Res.* 54 (7): 1723–1736. (d) Urbina, K., Tresp, D., Sipps, K. et al. (2021). Recent advances in metal-catalyzed functionalization of indoles. *Adv. Synth. Catal.* 363 (11): 2723–2739. (e) Kumar, P., Nagtilak, P.J., and Kapur, M. (2021). Transition metal-catalyzed C–H functionalizations of indoles. *New J. Chem.* 45 (31): 13692–13746. (f) Pradhan, S., De, P.B., Shah, T.A. et al. (2020). Recent advances in metal-catalyzed alkylation, alkenylation and alkynylation of indole/indoline benzenoid nucleus. *Chem. Asian J.* 15 (24): 4184–4198. (g) Jagtap, R.A. and Punji, B. (2020). C–H functionalization of indoles by 3d transition-metal catalysis. *Asian J. Org. Chem.* 9 (3): 326–342. (h) Shah, T.A., De, P.B., Pradhan, S. et al. (2019). Transition-metal-catalyzed site-selective C7-functionalization of indoles: advancement and future prospects. *Chem. Commun.* 55 (5): 572–587. (i) Yang, Y. and Shi, Z. (2018). Regioselective direct arylation of indoles on the benzenoid moiety. *Chem. Commun.* 54 (14): 1676–1685. (j) Kalepu, J., Gandeepan, P., Ackermann, L. et al. (2018). C4–H indole functionalisation: precedent and prospects. *Chem. Sci.* 9 (18): 4203–4216. (k) Petrini, M. (2017). Regioselective direct C-alkenylation of indoles. *Chem. Eur. J.* 23 (64): 16115–16151. (l) Osipov, S. and Vorobyeva, D. (2017). Selective synthesis of 2- and 7-substituted indole derivatives via chelation-assisted metallocarbenoid C—H bond functionalization. *Synthesis* 50 (02): 227–240. (m) Leitch, J.A., Bhonoah, Y., and Frost, C.G. (2017). Beyond C2 and C3: transition-metal-catalyzed C–H functionalization of indole. *ACS Catal.* 7 (9): 5618–5627. (n) Sandtorv, A.H. (2015). Transition metal-catalyzed C-H activation of indoles. *Adv. Synth. Catal.* 357 (11): 2403–2435.

113 (a) Zheng, J., Zhang, Y., and Cui, S. (2014). Rh(III)-catalyzed selective coupling of N-methoxy-1H-indole-1-carboxamides and aryl boronic acids. *Org. Lett.* 16 (13): 3560–3563. (b) Zhang, L.-Q., Yang, S., Huang, X. et al. (2013). Aerobic Ru-catalyzed direct C2-olefination of N-heteroarenes with alkenes directed by a removable N-dimethylcarbamoyl group. *Chem. Commun.* 49 (78): 8830–8832. (c) Ackermann, L. and Lygin, A.V. (2011). Ruthenium-catalyzed direct C—H bond arylations of heteroarenes. *Org. Lett.* 13 (13): 3332–3335.

114 (a) Dong, B., Qian, J., Li, M. et al. (2020). External oxidant-compatible phosphorus(III)-directed site-selective C–H carbonylation. *Sci. Adv.* 6 (51): eaau6468. (b) Yang, Y., Qiu, X., Zhao, Y. et al. (2016). Palladium-catalyzed C–H arylation of indoles at the C7 position. *J. Am. Chem. Soc.* 138 (2): 495–498.

115 (a) Jagtap, R.A., Samal, P.P., Vinod, C.P. et al. (2020). Iron-catalyzed C(sp2)–H alkylation of indolines and benzo[h]quinoline with unactivated alkyl chlorides through chelation assistance. *ACS Catal.* 10 (13): 7312–7321. (b) De, P.B., Atta, S., Pradhan, S. et al. (2020). Cp*Co(III)-catalyzed C-7 C–C coupling of indolines with aziridines: merging C–H activation and ring opening. *J. Org. Chem.* 85 (7): 4785–4794. (c) Banerjee, S., Vivek Kumar, S., and Punniyamurthy, T. (2020). Site-selective Rh-catalyzed C-7 and C-6 dual C–H functionalization of indolines: synthesis of functionalized pyrrolocarbazoles. *J. Org. Chem.* 85 (4): 2793–2805. (d) Wang, Q., Zhi, C.L., Lu, P.P. et al. (2019). Rhodium(III)-catalyzed direct C7 allylation of indolines via sequential C–H and C–C activation. *Adv. Synth. Catal.* 361 (6): 1253–1258. (e) Wang, L., Xiong, D., Jie, L. et al. (2018). Rhodium-catalyzed oxidative homologation of N-pyrimidyl indolines with alkynes via dual C-H activation: facile synthesis of benzo[g]indolines. *Chin. Chem. Lett.* 29 (6): 907–910. (f) Gandeepan, P., Koeller, J., and Ackermann, L. (2017). Expedient C–H chalcogenation of indolines and indoles by positional-selective copper catalysis. *ACS Catal.* 7 (2): 1030–1034. (g) Zhou, X., Yu, S., Qi, Z. et al. (2016). Rhodium(III)-catalyzed mild alkylation of (hetero)arenes with cyclopropanols via C–H activation and ring opening. *J. Org. Chem.* 81 (11): 4869–4875. (h) Premi, C., Dixit, A., and Jain, N. (2015). Palladium-catalyzed regioselective decarboxylative alkylation of arenes and heteroarenes with aliphatic carboxylic acids. *Org. Lett.* 17 (11): 2598–2601. (i) Ai, W., Yang, X., Wu, Y. et al. (2014). Rhodium(III)- and iridium(III)-catalyzed C7 alkylation of indolines with diazo compounds. *Chem. Eur. J.* 20 (52): 17653–17657. (j) Yang, X.-F., Hu, X.-H., Feng, C. et al. (2015). Rhodium(III)-catalyzed C7-position C–H alkenylation and alkynylation of indolines. *Chem. Commun.* 51 (13): 2532–2535.

116 (a) Pan, S., Wakaki, T., Ryu, N. et al. (2014). Ir(III)-catalyzed C7-position-selective oxidative C-H alkenylation of indolines with alkenes in air. *Chem. Asian J.* 9 (5): 1257–1260. (b) Jiao, L.-Y. and Oestreich, M. (2013). Oxidative palladium(II)-catalyzed C-7 alkenylation of indolines. *Org. Lett.* 15 (20): 5374–5377. (c) Pan, S., Ryu, N., and Shibata, T. (2014). Iridium(I)-catalyzed direct C-H bond alkylation of the C-7 position of indolines with alkenes. *Adv. Synth. Catal.* 356 (5): 929–933. (d) Zhang, L., Chen, C., Han, J. et al. (2016). Ru-catalyzed selective C–H oxidative olefination with N-heteroarenes directed by pivaloyl amide. *Org. Chem. Front.* 3 (10): 1271–1275. (e) Zhou, T., Wang, Y., Li, B. et al. (2016). Rh(III)-catalyzed carbocyclization of 3-(indolin-1-yl)-3-oxopropanenitriles with alkynes and alkenes through C–H activation. *Org. Lett.* 18 (19): 5066–5069. (f) Banjare, S.K., Chebolu, R., and Ravikumar, P.C. (2019). Cobalt catalyzed hydroarylation of Michael acceptors with indolines directed by a weakly coordinating functional group. *Org. Lett.* 21 (11): 4049–4053.

117 (a) Song, Z., Samanta, R., and Antonchick, A.P. (2013). Rhodium(III)-catalyzed direct regioselective synthesis of 7-substituted indoles. *Org. Lett.* 15 (22): 5662–5665. (b) Manoharan, R. and Jeganmohan, M. (2015).

Ruthenium-catalyzed cyclization of N-carbamoyl indolines with alkynes: an efficient route to pyrroloquinolinones. *Org. Biomol. Chem.* 13 (35): 9276–9284.

118 (a) Park, J., Mishra, N.K., Sharma, S. et al. (2015). Mild Rh(III)-catalyzed C7-allylation of indolines with allylic carbonates. *J. Org. Chem.* 80 (3): 1818–1827. (b) Wang, X., Tang, H., Feng, H. et al. (2015). Access to six- and seven-membered 1,7-fused indolines via Rh(III)-catalyzed redox-neutral C7-selective C–H functionalization of indolines with alkynes and alkenes. *J. Org. Chem.* 80 (12): 6238–6249. (c) Yang, D., Mao, S., Gao, Y.-R. et al. (2015). Palladium-catalyzed C-7 alkenylation of indolines using molecular oxygen as the sole oxidant. *RSC Adv.* 5 (30): 23727–23736.

119 (a) Shahid, M., Muthuraja, P., and Gopinath, P. (2024). Substituent-controlled regioselective arylation of carbazoles using dual catalysis. *Org. Biomol. Chem.* 22 (4): 753–758. (b) Harsha Vardhan Reddy, K., Kumar, R.U., Reddy, V.P. et al. (2016). Ru-catalyzed highly site-selective C—H bond arylation of 9-(pyrimidin-2-yl)-9*H*-carbazole. *RSC Adv.* 6 (59): 54431–54434. (c) Reddy, V.P., Qiu, R., Iwasaki, T. et al. (2013). Rhodium-catalyzed intermolecular oxidative cross-coupling of (hetero)arenes with chalcogenophenes. *Org. Lett.* 15 (6): 1290–1293. (d) Chu, J.-H., Wu, C.-C., Chang, D.-H. et al. (2012). Direct ortho arylation of 9-(pyridin-2-yl)-9H-carbazoles bearing a removable directing group via palladium(II)-catalyzed C—H bond activation. *Organometallics* 32 (1): 272–282.

120 (a) Shahid, M., Punnya, A.J., Babu, S.S. et al. (2023). Dual palladium-photoredox-mediated regioselective acylation of carbazoles and indolines. *J. Org. Chem.* 88 (19): 13686–13698. (b) Maiti, S., Mandal, T., Dash, B.P. et al. (2020). Site-selective aerobic C–H monoacylation of carbazoles using palladium catalysis. *J. Org. Chem.* 86 (2): 1396–1407. (c) Maiti, S., Burgula, L., Chakraborti, G. et al. (2017). Palladium-catalyzed pyridine-directed regioselective oxidative C–H acylation of carbazoles by using aldehydes as the acyl source. *Eur. J. Org. Chem.* 2017 (2): 332–340.

121 (a) Dharaniyedath, J., Kumar, V., and Gandeepan, P. (2024). Palladium-catalyzed direct alkynylation of carbazoles with alkynyl bromides. *Eur. J. Org. Chem.* 27 (40): e202400649. (b) Li, T., Wang, Z., Qin, W.B. et al. (2016). Rhodium-catalyzed/copper-mediated selective C2 alkynylation of indoles and C1 alkynylation of carbazoles with γ-substituted tert-propargyl alcohols. *ChemCatChem* 8 (12): 2146–2154.

122 Qiu, R., Reddy, V.P., Iwasaki, T. et al. (2014). The palladium-catalyzed intermolecular C–H chalcogenation of arenes. *J. Org. Chem.* 80 (1): 367–374.

123 Okada, T., Nobushige, K., Satoh, T. et al. (2016). Ruthenium-catalyzed regioselective C—H bond acetoxylation on carbazole and indole frameworks. *Org. Lett.* 18 (5): 1150–1153.

124 Banerjee, S., De, P.B., Pradhan, S. et al. (2019). Ru^{II}-catalysed regioselective C—N bond formation of indolines and carbazole with acyl azides. *Eur. J. Org. Chem.* 2019 (7): 1677–1684.

125 (a) Hirano, K. and Miura, M. (2018). A lesson for site-selective C–H functionalization on 2-pyridones: radical, organometallic, directing group and steric controls. *Chem. Sci.* 9 (1): 22–32. (b) Ni, J., Zhao, H., and Zhang, A. (2017). Manganese(I)-catalyzed C–H 3,3-difluoroallylation of pyridones and indoles. *Org. Lett.* 19 (12): 3159–3162. (c) Miura, W., Hirano, K., and Miura, M. (2017). Nickel-catalyzed directed C6-selective C–H alkylation of 2-pyridones with dienes and activated alkenes. *J. Org. Chem.* 82 (10): 5337–5344. (d) Biswas, A., Giri, D., Das, D. et al. (2017). A mild rhodium catalyzed direct synthesis of quinolones from pyridones: application in the detection of nitroaromatics. *J. Org. Chem.* 82 (20): 10989–10996. (e) Anil Kumar, K., Kannaboina, P., and Das, P. (2017). Ruthenium-catalyzed site-selective C–H arylation of 2-pyridones and 1-isoquinolinones. *Org. Biomol. Chem.* 15 (26): 5457–5461. (f) Peng, P., Wang, J., Li, C. et al. (2016). Cu(II)-catalyzed C6-selective C–H thiolation of 2-pyridones using air as the oxidant. *RSC Adv.* 6 (62): 57441–57445. (g) Peng, P., Wang, J., Jiang, H. et al. (2016). Rhodium(III)-catalyzed site-selective C–H alkylation and arylation of pyridones using organoboron reagents. *Org. Lett.* 18 (20): 5376–5379. (h) Miura, W., Hirano, K., and Miura, M. (2016). Rhodium-catalyzed C6-selective C–H borylation of 2-pyridones. *Org. Lett.* 18 (15): 3742–3745. (i) Li, Y., Xie, F., and Li, X. (2016). Formal gold- and rhodium-catalyzed regiodivergent C–H alkynylation of 2-pyridones. *J. Org. Chem.* 81 (2): 715–722. (j) Odani, R., Hirano, K., Satoh, T. et al. (2014). Copper-mediated C6-selective dehydrogenative heteroarylation of 2-pyridones with 1,3-azoles. *Angew. Chem. Int. Ed.* 53 (40): 10784–10788.

126 Campeau, L.-C., Rousseaux, S., and Fagnou, K. (2005). A solution to the 2-pyridyl organometallic cross-coupling problem: regioselective catalytic direct arylation of pyridine *N*-oxides. *J. Am. Chem. Soc.* 127 (51): 18020–18021.

127 (a) Corio, A., Gravier-Pelletier, C., and Busca, P. (2021). Regioselective functionalization of quinolines through C-H activation: a comprehensive review. *Molecules* 26 (18). (b) da Silva Júnior, E.N., Jardim, G.A.M., Gomes, R.S. et al. (2018). Weakly-coordinating N-oxide and carbonyl groups for metal-catalyzed C–H activation: the case of A-ring functionalization. *Chem. Commun.* 54 (54): 7398–7411. (c) Yan, G., Borah, A.J., and Yang, M. (2014). Recent advances in catalytic functionalization of N-oxide compounds via C—H bond activation. *Adv. Synth. Catal.* 356 (11–12): 2375–2394.

128 (a) Wang, Y. and Gevorgyan, V. (2017). Synthesis of active hexafluoroisopropyl benzoates through a hydrogen-bond-enabled palladium(II)-catalyzed C–H alkoxycarbonylation reaction. *Angew. Chem. Int. Ed.* 56 (12): 3191–3195. (b) Parasram, M. and Gevorgyan, V. (2017). Silicon-tethered strategies for C–H functionalization reactions. *Acc. Chem. Res.* 50 (8): 2038–2053. (c) Sarkar, D. and Gevorgyan, V. (2016). Pd-catalyzed C–H alkylation of arenes using PyrDipSi, a transformable and removable silicon-tethered directing group. *Chem. Eur. J.* 22 (32): 11201–11204. (d) Dudnik, A.S., Chernyak, N., Huang, C. et al. (2010). A general strategy toward aromatic 1,2-ambiphilic synthons: palladium-catalyzed ortho-halogenation of PyDipSi-arenes. *Angew. Chem. Int. Ed.* 49 (46): 8729–8732. (e) Chernyak, N., Dudnik, A.S., Huang, C. et al. (2010). PyDipSi: a general and easily modifiable/traceless Si-tethered directing group for C–H acyloxylation of arenes. *J. Am. Chem. Soc.* 132 (24): 8270–8272.

5

Removable Template-assisted Transition Metal-catalyzed Distal C–H Functionalization

Ke Yang[1], *Dan Yuan*[1], *Faith Herington*[2], *and Haibo Ge*[2]

[1]*Jiangsu Key Laboratory of Advanced Catalytic Materials & Technology, School of Petrochemical Engineering, Changzhou University, 1 Gehu Road, Changzhou, Jiangsu 213164, China*
[2]*Department of Chemistry and Biochemistry, Texas Tech University, Lubbock, TX 79409, USA*

5.1 Introduction

Over the past two decades, transition-metal-catalyzed C—H bond functionalization has gained recognition as one of the most effective methods for site-selective C—C and C—X bond formation [1–6]. Despite significant developments in this field on both C(sp^2)—H and C(sp^3)—H bonds using substrates containing various templates/directing groups, the majority of studies have primarily focused on activating neighboring C—H bonds including *ortho*-C(sp^2)—H bonds of arenes, β-C(sp^3)—H bonds of aliphatic acids, and γ-C(sp^3)—H bonds of aliphatic amines, among others [7–13]. Typically, these C—H bonds are activated through the formation of a five-membered metallacycle intermediate. Nonetheless, advancements in this subject for a distal C—H bond had been considerably impeded due to the difficulties of generating the necessary six- or larger-membered metallacycle intermediate, which is a kinetically unfavorable process.

Distal C—H functionalization was initially introduced by the Breslow group and later advanced by the Carbtree and White groups, all utilizing well-designed transition-metal catalysts. The Breslow group designed a β-cyclodextrin-containing Mn-porphyrin catalyst to enable the hydroxylation of distal C(sp^3)—H bonds [14–16]. Likewise, Carbtree and co-workers reported a Mn-catalyzed distal C(sp^3)—H oxidation using hydrogen bonding interactions [17]. Subsequently, the White group successfully developed an Fe-catalyzed distal C(sp^3)—H oxidation protocol in the presence of hydrogen peroxide [18–20]. Following these advancements, numerous chemists collaborated to further develop a diverse range of innovative removable substrate templates to facilitate distal C—H bond functionalization via six- or larger-membered C—H cyclometallation [21–28]. This chapter highlights the recent progress in removable template-assisted

Directed C–H Bond Functionalization: Concepts and Applications, First Edition.
Edited by Debabrata Maiti and Supriya Rej.

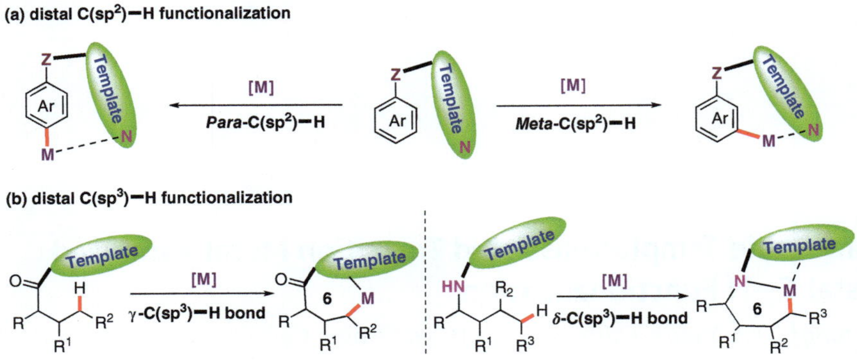

Figure 5.1 Removable template-assisted transition-metal-catalyzed distal C–H functionalization. (a) Distal C(sp^2)–H functionalization; (b) distal C(sp^2)–H functionalization.

transition-metal-catalyzed *meta-* and *para*-C(sp^2)—H bond functionalization of arenes, γ-C(sp^3)—H bond functionalization of aliphatic acids and ketones, and δ-C(sp^3)—H bond functionalization of aliphatic amines (Figure 5.1).

5.2 Distal C(sp^2)–H Bond Functionalization

5.2.1 Distal C(sp^2)–H Functionalization of Arylalkyl and Aryl Acid Derivatives

Arylalkyl acids can be identified as arylacetic acids, arylpropionic acids, and arylbutyric acid, among other derivatives. These compounds form the fundamental structural frameworks found in non-steroidal anti-inflammatory drugs (NSAIDs) such as indomethacin, sulindac, and diclofenac sodium, which are known to exhibit potent anti-inflammatory, analgesic, and antipyretic properties [29].

In 2012, the Yu group achieved groundbreaking advancements in directing group-assisted C—H functionalization, successfully demonstrating the Pd(II)-catalyzed *meta*-C(sp^2)—H bond olefination of arylpropionic acids using easily removable nitrile templates [30]. Notably, the *N*-acetyl glycine ligand was found to facilitate C(sp^2)—H olefination in this process, with the reaction occurring at the remote aryl ring of the biphenyl substrate. The weakly coordinated linear nitrile template is believed to facilitate the proximity of the catalyst to the distal C(sp^2)—H bond, alleviate ring tension in the macrocyclic metal transition state, and enhance *meta*-C(sp^2)—H bond activation. Furthermore, these nitrile templates can be easily removed via hydrolysis with LiOH as the base. In 2013, the Yu group employed the same nitrile templates to achieve the *meta*-C(sp^2)—H bond arylation and methylation of 3-phenylpropanoic acid derivatives [31]. The utilization of a loosely binding U-shaped nitrile template in conjunction with a mono-protected amino acid (Ac-Gly-OH) ligand played a pivotal role in facilitating the cross-coupling of organoboron compounds with distal $C(sp^2)$—H bonds (Scheme 5.1).

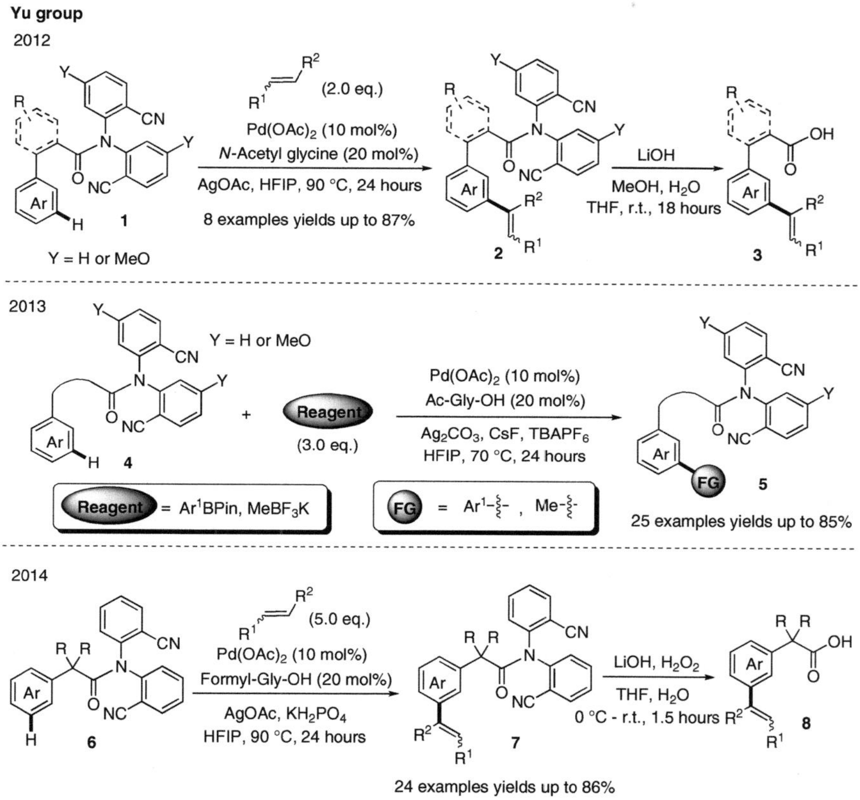

Scheme 5.1 Pd(II)-catalyzed *meta*-$C(sp^2)$–H bond olefination of arylalkyl acids utilizing easily removable nitrile templates.

In 2014, the Yu group additionally reported Pd(II)-catalyzed *meta*-$C(sp^2)$—H bond olefination of phenylacetic acid derivatives using the same nitrile-containing template [32]. This method highlighted the crucial role of the *N*-formyl-protected glycine ligand (Formyl-Gly-OH) along with the KH_2PO_4 additive. Notably, the removal of the nitrile-containing template from *meta*-olefinated phenylacetic acids only occurs under Evans' hydrolytic conditions ($LiOH/H_2O_2$). The previously reported conditions utilizing LiOH in a solvent mixture of MeOH, THF, and H_2O resulted in the formation of an unidentified compound mixture (Scheme 5.1).

The proposed mechanism initially begins with the coordination of the nitrile template with the Pd(II) catalyst, followed by subsequent *meta*-C—H bond activation to generate an 11-membered palladacycle intermediate **1A**. Due to the bis-dentate coordination mode with a monoprotected amino acid ligand, the olefin likely displaces the nitrile group before migratory insertion occurs. Consequently, the palladacycle intermediate **1A** involves an initial olefin-binding step, followed by a subsequent 1,2-migratory insertion, resulting in the formation of the palladacycle intermediate **1C**. Next, intermediate **1C** undergoes *β*-hydride elimination to produce the desired product **7** and [Pd(II)—H] intermediate. Finally, the Pd(II) catalyst is regenerated through reductive elimination and oxidation steps (Scheme 5.2).

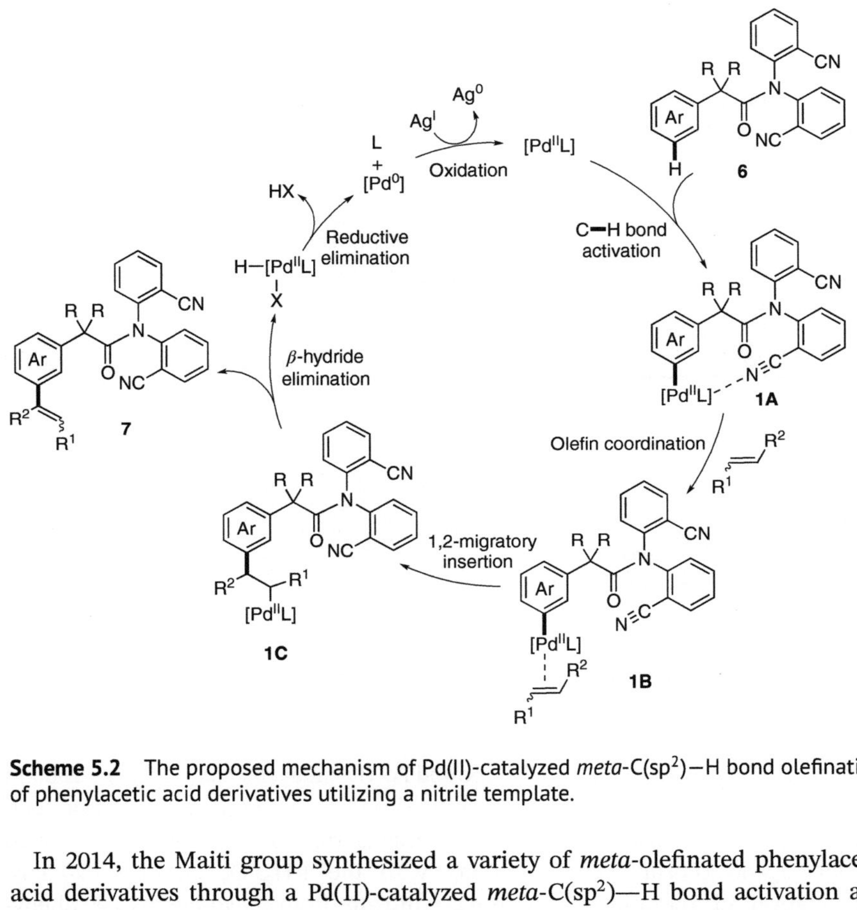

Scheme 5.2 The proposed mechanism of Pd(II)-catalyzed *meta*-C(sp^2)–H bond olefination of phenylacetic acid derivatives utilizing a nitrile template.

In 2014, the Maiti group synthesized a variety of *meta*-olefinated phenylacetic acid derivatives through a Pd(II)-catalyzed *meta*-C(sp^2)—H bond activation and cascade transesterification reaction facilitated by a 2-hydroxybenzonitrile template [33]. Notably, the HFIP solvent was incorporated into the product, forming a hexafluoroisopropyl ester group, which could further be readily hydrolyzed at room temperature using LiOH, resulting in the production of *meta*-substituted phenylacetic acid with high yield (Scheme 5.3a). Following this, the Yu group developed a Pd(II)-catalyzed distal *meta*-C(sp^2)—H bond functionalization of phenylacetic acid derivatives using a specifically designed U-shaped pyridine template [34]. This innovative pyridine template not only enabled *meta*-C—H olefination but also facilitated *meta*-C—H arylation and iodination reactions by utilizing $ArBF_3K$ and 1,3-diiodo-5,5-dimethylhydantoin (DIH) as coupling partners. Additionally, it can be efficiently removed under the HCl-EtOH system with high yield (Scheme 5.3b).

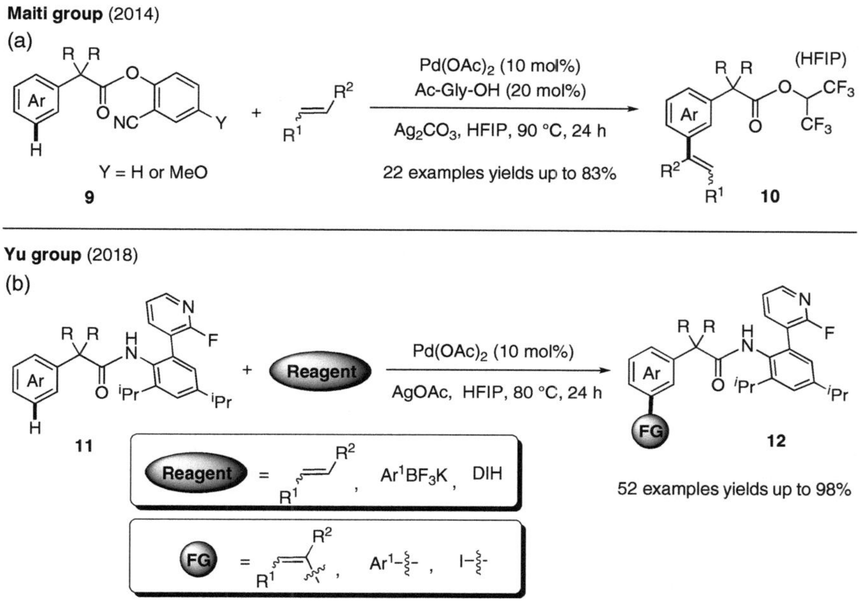

Scheme 5.3 (a) Pd(II)-catalyzed *meta*-C(sp^2)–H bond olefination of phenylacetic acid derivatives using a nitrile template; (b) Pd(II)-catalyzed *meta*-C(sp^2)–H bond functionalization of phenylacetic acid derivatives using a U-shaped pyridine template.

In 2017, the Yu and Houk group introduced a conformationally flexible nitrile template for the Pd(II)-catalyzed *meta*-C(sp^2)—H bond olefination of benzoic acids [35]. The use of Ac-Gly-OH ligand provided the desired products with good-to-excellent yields and high *meta*-selectivity. Furthermore, density functional theory (DFT) calculations provided insights into the roles of the nitrile template, Ac-Gly-OH ligand, and silver salt in the catalytic transformation (Scheme 5.4a). In 2019, the Yu and Dai group developed the first example of Pd(II)-catalyzed acetoxylation of electron-deficient benzoic acids at the *para*-C(sp^2)—H bond using a nitrile template [36]. This innovative strategy successfully overcame steric and electron interference, resulting in the production of various *para*-selective acetoxylated products with good-to-excellent yields. Moreover, *para*-hydroxylated benzoic acids can be easily obtained under basic hydrolysis conditions (Scheme 5.4b).

In 2015, Maiti and co-workers reported the first Pd(II)-catalyzed *meta*-C(sp^2)—H bond olefination of benzyl sulfonic acid derivatives using the 2-hydroxybenzonitrile template [37]. The researchers observed that both mono- and di-olefinated benzyl sulfonic acid derivatives were produced in good yields with high *meta*-selectivity (*meta*/others > 20 : 1). Additionally, the 2-hydroxybenzonitrile template was found to be readily eliminated via a hydrolysis process by exerting KOH in methanol.

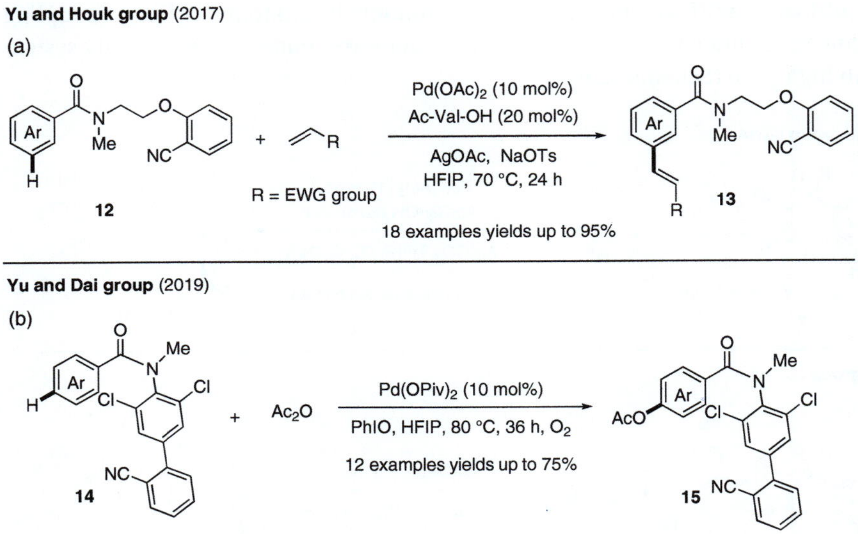

Scheme 5.4 (a) Pd(II)-catalyzed *meta*-C(sp^2)–H bond olefination of benzoic acids using a nitrile template; (b) Pd(II)-catalyzed *para*-C(sp^2)–H bond acetoxylation of benzoic acids using a nitrile template.

In 2016, they also used the same substrate to achieve the earliest cases of Pd(II)-catalyzed *meta*-C(sp^2)—H hydroxylation and acetoxylation reactions [38]. Mechanistic studies revealed the crucial role of the HFIP solvent in the catalytic cycle (Scheme 5.5).

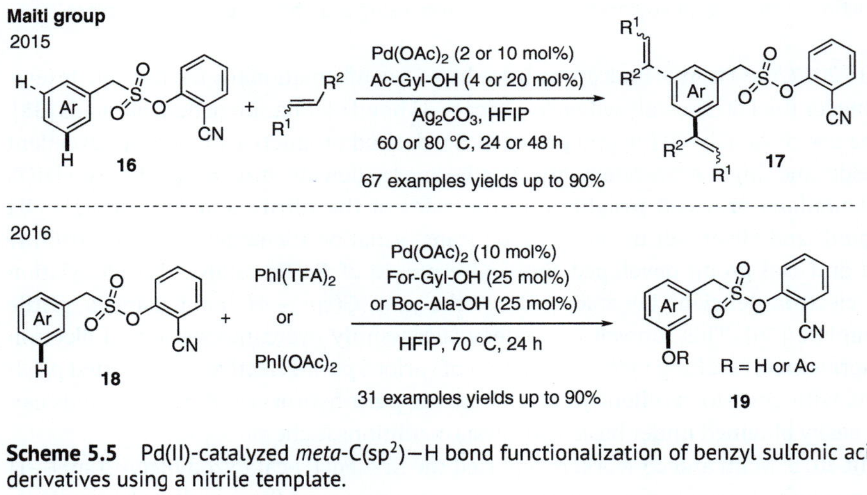

Scheme 5.5 Pd(II)-catalyzed *meta*-C(sp^2)–H bond functionalization of benzyl sulfonic acid derivatives using a nitrile template.

The following year, the same group discovered that an 8-nitroquinoline-based template could be employed as a strong σ-donating directing group to accomplish Pd(II)-catalyzed *meta*-C(sp^2)—H bond olefination and acetoxylation of benzyl sulfonic acid derivatives [39]. The authors proposed the reaction mechanism based on kinetic experiments, ESI-MS, NMR, and DFT studies (Scheme 5.6). Kinetic studies

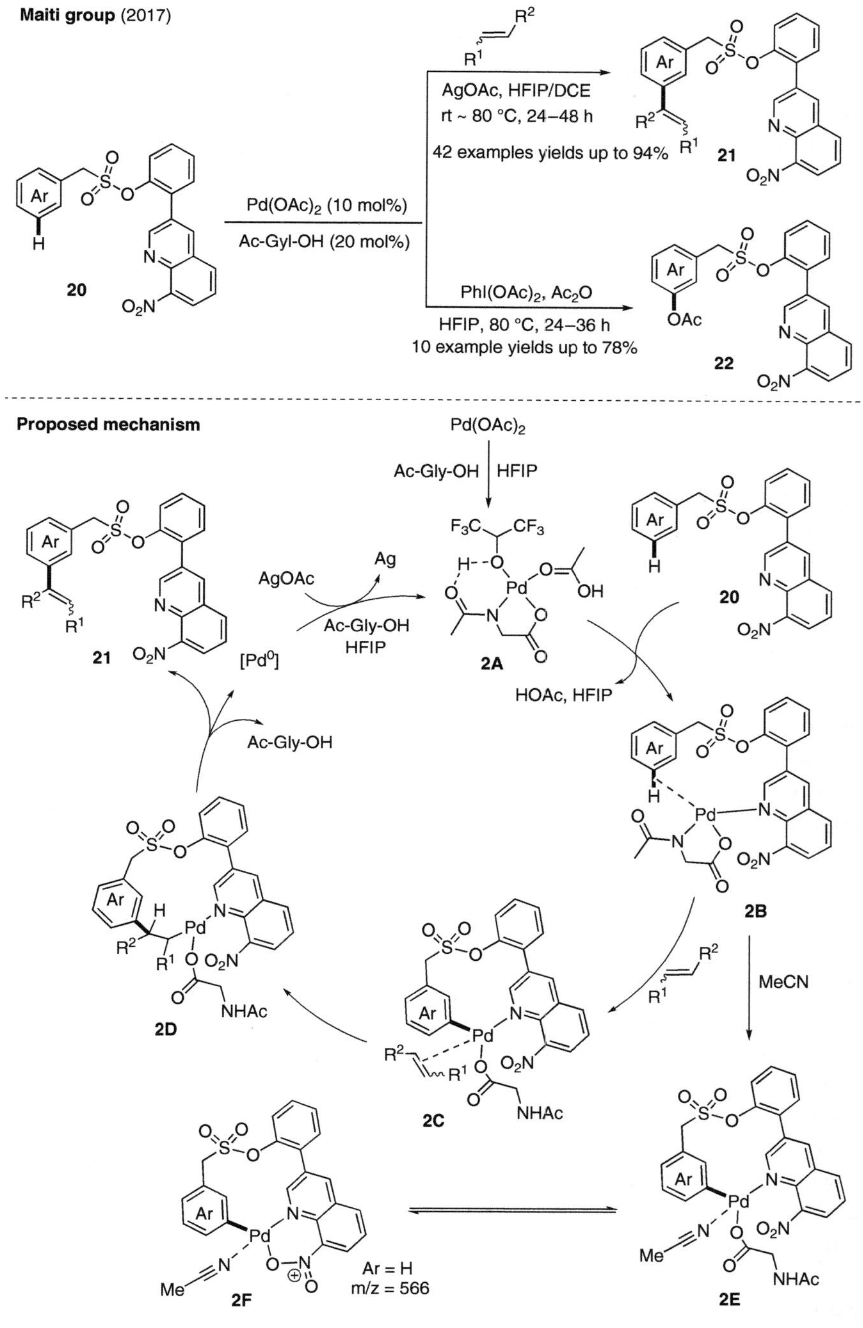

Scheme 5.6 Pd(II)-catalyzed *meta*-C(sp^2)–H bond functionalization of benzyl sulfonic acid derivatives using an 8-nitroquinoline-based template.

indicate that the first-order rate depends on the substrate **20**. NMR studies show that the coordination between the Pd catalyst and quinoline group is formed. Additionally, DFT studies provide a reasonable catalytic cycle. First, Pd(II) catalyst **2A** is formed via the coordination of Ac-Gly-OH and HFIP to $Pd(OAc)_2$. Subsequently, Pd(II) catalyst **2A** coordinates with the 8-nitroquinoline on substrate **20** to generate intermediate **2B**. Next, intermediate **2B** undergoes *meta*-C—H bond activation, followed by olefin coordination to generate intermediate **2C**. The subsequent 1,2-migratory insertion of intermediate **2C** produces intermediate **2D**, which is further converted into the final product **21** through *β*-hydride elimination. The *in situ* formed Pd(0) intermediate is then oxidized to the Pd(II) catalyst by a silver salt. Notably, intermediate **2B** can coordinate with MeCN to form intermediate **2E** and its tautomer **2F**. The presence of the tautomer **2F** can be detected using ESI-MS.

The Maiti group further expanded their utilization of the same 2-hydroxybenzonitrile template from previous years to accomplish *meta*-C(sp^2)—H bond olefination of 2-phenethylsulfonic acid and 3-phenylpropanoic acid derivatives [40]. This method enabled the synthesis of exclusive mono-olefinated products with moderate-to-good yields, while also enabling the preparation of various di-olefinated products through a cascade olefination reaction. Mechanistic studies suggest that this process may involve a 12-membered palladacycle intermediate.

Furthermore, they have also designed a pyrimidine-based template to attain *meta*-C(sp^2)—H bond alkylation and alkenylation [41]. Using allyl alcohols as partners, a variety of arylalkyl sulfonic acid, carboxylic acid, and phosphonic acid derivatives were converted to the corresponding *β*-aryl aldehydes and ketone derivatives with good yields and site-selectivity through *meta*-C(sp^2)—H bond alkylation. It was noted that the use of benzyl-protected allyl alcohols could afford the corresponding *α*,*β*-unsaturated aldehyde derivatives via a *meta*-C(sp^2)—H bond alkenylation (Scheme 5.7).

In addition to palladium catalysts, rhodium catalysts have proven effective in *meta*-C(sp^2)—H bond functionalization reactions. In 2017, the Maiti group reported the Rh-catalyzed *meta*-C(sp^2)—H bond olefination of benzylsulfonic acid and phenylacetic acid derivatives employing a 2-hydroxy-4-methoxy benzonitrile template [42]. In this reaction, XPhos plays a vital role as a ligand by substantially enhancing the reaction yield. Unlike previous Pd catalysis systems, Rh catalysis often employs V_2O_5 and Cu salts as fundamental oxidants to generate the active Rh(III) species from the Rh(I) catalyst (Scheme 5.8a).

Meanwhile, Yu and co-workers have developed a meticulously designed nitrile template for the Rh-catalyzed *meta*-C(sp^2)—H bond olefination of phenylpropionic acid derivatives, emphasizing the essentiality of Cu salt as an oxidizing agent in the process [43]. The researchers subsequently reported an Rh-catalyzed *meta*-C(sp^2)—H bond olefination reaction with di-substituted alkynes directed by the same nitrile template [44]. This novel approach permits both aryl and alkyl-disubstituted alkynes to efficiently construct tri-substituted olefins with

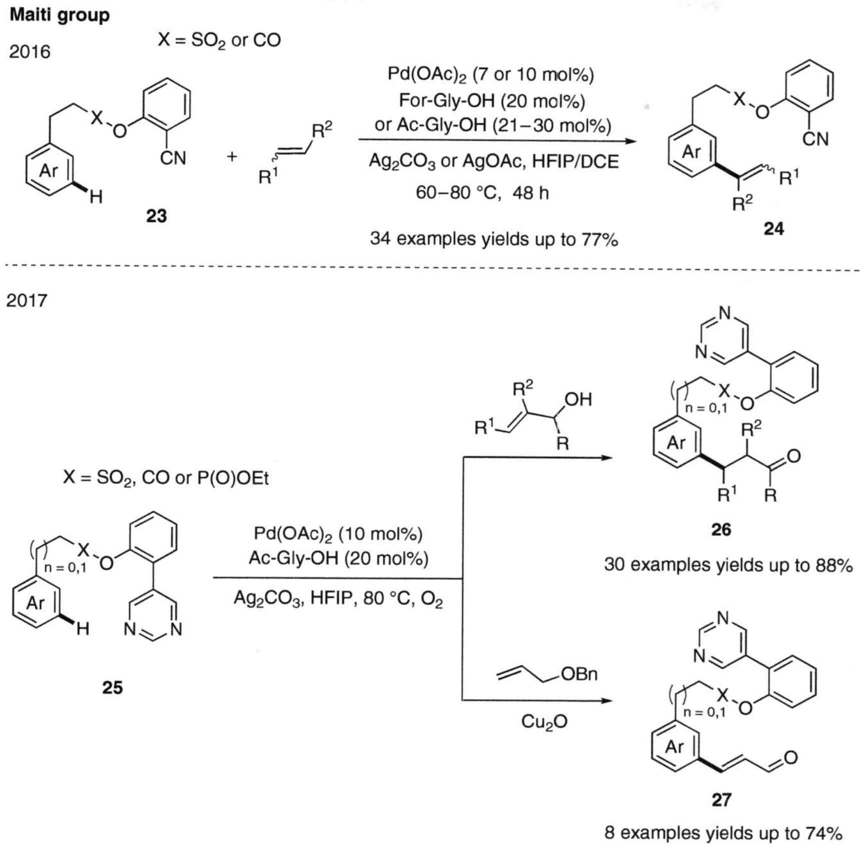

Scheme 5.7 Pd(II)-catalyzed *meta*-C(sp^2)–H bond functionalization of 2-phenethylsulfonic acid and 3-phenylpropanoic acid derivatives using a nitrile and pyrimidine template.

high yields and good selectivity. Moreover, the protocol is compatible with propargyl alcohols, leading to the formation of corresponding allylic alcohols (Scheme 5.8b).

The proposed mechanism suggests that the coupling of $[RhCp^*Cl_2]_2$ with $AgSbF_6$ generates the active catalytic species $[Cp^*Rh](SbF_6)_2$. Subsequently, the nitrile template coordinates to $[Cp^*Rh](SbF_6)_2$ before undergoing *meta*-C—H bond activation, generating 12-membered Rh intermediate **3A**. Lastly, intermediate **3A** experiences alkyne insertion followed by protonation to yield the desired product and regenerate the active Rh(III) species (Scheme 5.9).

5.2.2 Distal C(sp^2)–H Functionalization of Arylalkyl and Aryl Amines

Amine derivatives are commonly found in pharmaceutical and biological compounds, exhibiting diverse biological activities [45, 46]. Due to their widespread

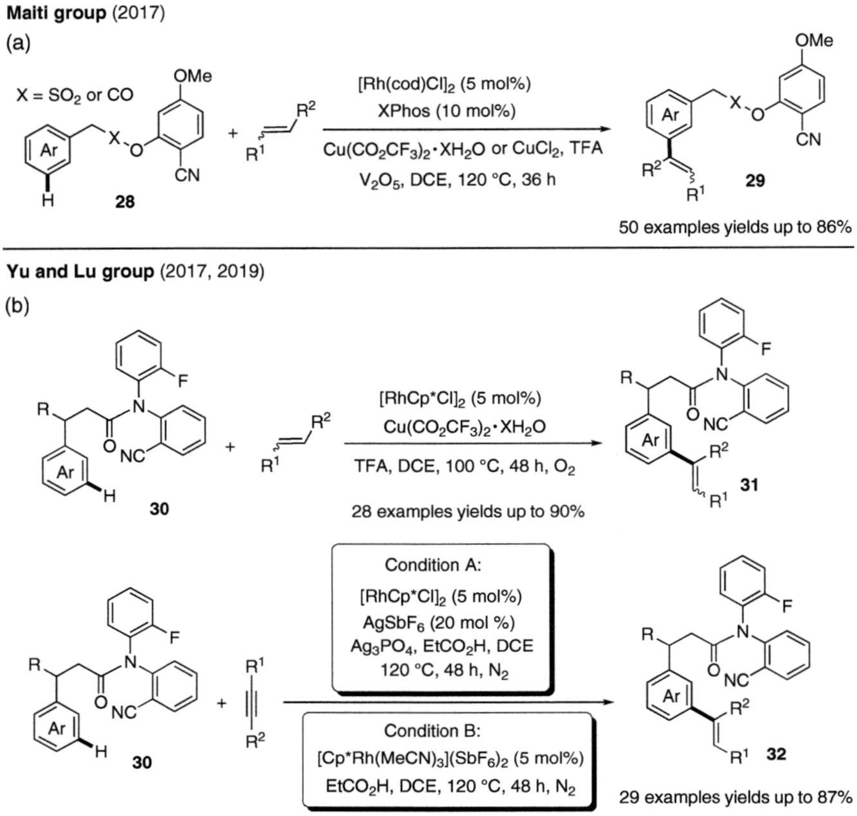

Scheme 5.8 (a) Rh-catalyzed *meta*-C(sp^2)–H bond olefination of benzylsulfonic acid and phenylacetic acid derivatives with olefins using a nitrile template; (b) Rh-catalyzed *meta*-C(sp^2)–H bond olefination of phenylpropionic acid derivatives using a nitrile template.

usage and significance, chemists and pharmacists consistently prioritize the development of efficient and straightforward strategies for synthesizing amine derivatives.

In 2014, the Yu group first developed the Pd-catalyzed *meta*-C(sp^2)—H bond olefination and acetoxylation of aniline derivatives using a nitrile U-shaped template and Ac-Gly-OH ligand [47]. They also found that benzylamine derivatives were also suitable for the *meta*-C(sp^2)—H acetoxylation process, providing the desired products with moderate yields and good selectivity. Concurrently, the Yu group also reported Pd(II)-catalyzed *meta*-C(sp^2)—H olefination, arylation, and acetoxylation of indolines using an analogous nitrile template attached to the indolinyl nitrogen via a sulfonamide linkage [48]. The use of the Ac-Gly-OH ligand with the nitrile directing group proved to be imperative for achieving *meta*-selective C(sp^2)—H functionalization of electron-rich indolines. In 2018, they further employed the same nitrile template to achieve Pd-catalyzed *meta*-C(sp^2)—H bond olefination of

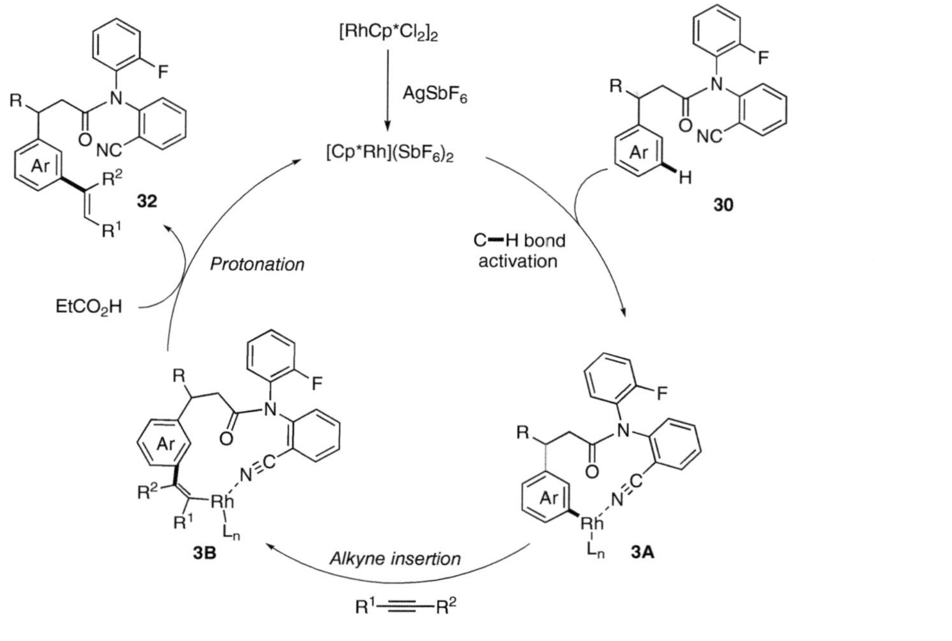

Scheme 5.9 The proposed mechanism of Rh-catalyzed *meta*-C(sp^2)–H bond olefination of phenylpropionic acid derivatives with di-substituted alkynes using a nitrile template.

different *N*-heterocycles including tetrahydroisoquinoline, isoindoline, indoline, indole, 3,4-dihydro-2*H*-benzo[*b*][1,4]oxazine, and 2-phenylpyrrolidine derivatives [49] (Scheme 5.10).

In 2017, the Li group pioneered the use of carbon dioxide (CO_2) as a building block for synthesizing a novel carbamate nitrile template. This innovative method allowed them to achieve palladium-catalyzed olefination and acetoxylation of *meta*-C(sp^2)—H bonds of aniline derivatives [50]. They additionally employed the same carbamate nitrile template to develop Pd-catalyzed bromination and chlorination of *meta*-C(sp^2)—H bonds of aniline derivatives using commercially available *N*-bromophthalimide (NBP) and 1,3-dichloro-5,5-dimethylhydantoin (DCH) [51]. They found that both the nosyl (Ns) group and carbamate nitrile template could be easily removed under mild conditions (Scheme 5.11).

The proposed mechanism of Pd-catalyzed *meta*-C(sp^2)—H bond bromination of aniline derivatives commences with the coordination of the aniline substrate **42** to the Pd(II) catalyst, resulting in the formation of intermediate **4A**. Subsequently, cleavage of the *meta*-C(sp^2)—H bond in intermediate **4A** leads to the formation of a 12-membered Pd intermediate **4B**. Simultaneously, NBP is activated by H^+ to generate intermediate **4C**. Following this, oxidative addition of **4B** with **4C** gives rise to Pd(IV) intermediate **4D**, which then undergoes reductive elimination to afford the desired product **44a** and regenerate the Pd(II) catalyst (Scheme 5.12).

In 2019, Jin and colleagues demonstrated the olefination reaction of aryl tertiary amines via Pd-catalyzed *meta*-C(sp^2)—H activation, directed by a

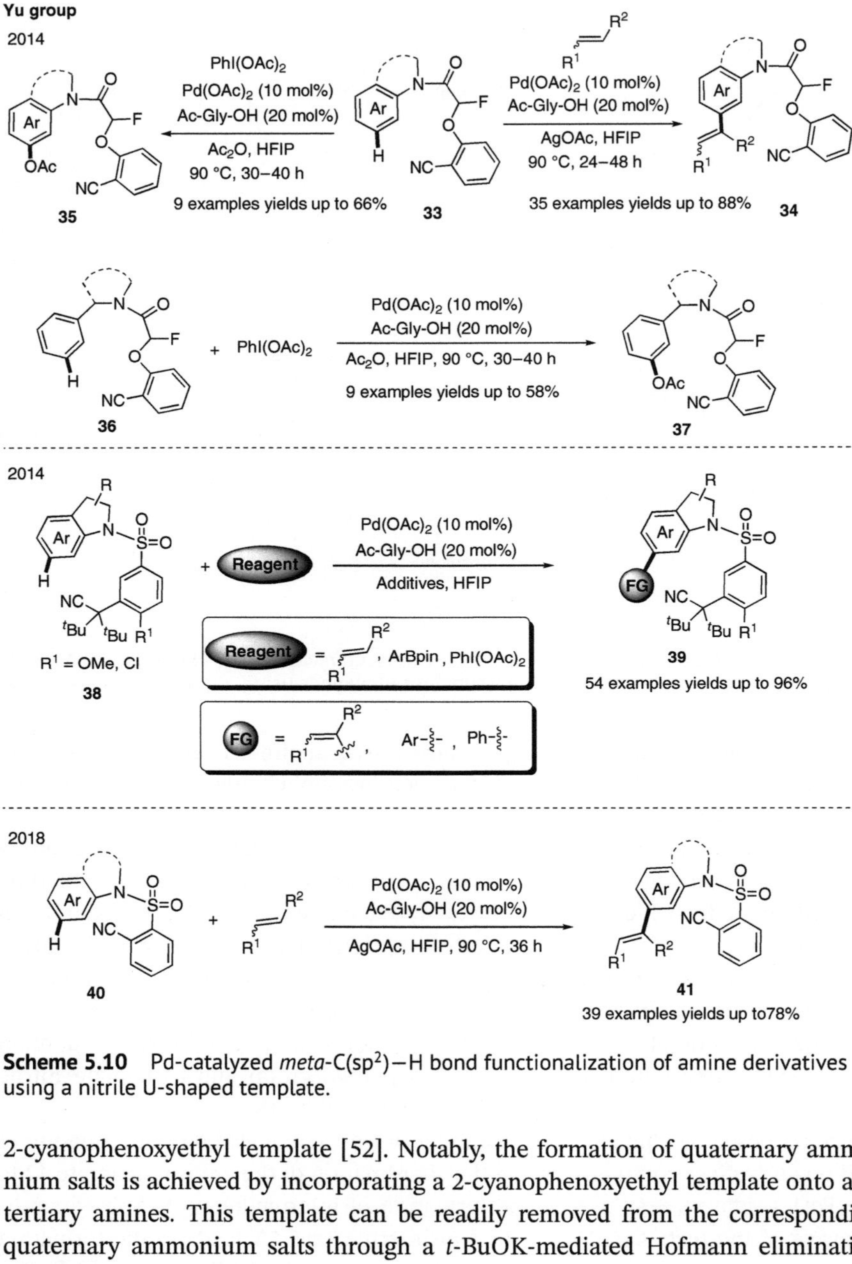

Scheme 5.10 Pd-catalyzed *meta*-C(sp^2)–H bond functionalization of amine derivatives using a nitrile U-shaped template.

2-cyanophenoxyethyl template [52]. Notably, the formation of quaternary ammonium salts is achieved by incorporating a 2-cyanophenoxyethyl template onto aryl tertiary amines. This template can be readily removed from the corresponding quaternary ammonium salts through a *t*-BuOK-mediated Hofmann elimination process (Scheme 5.13).

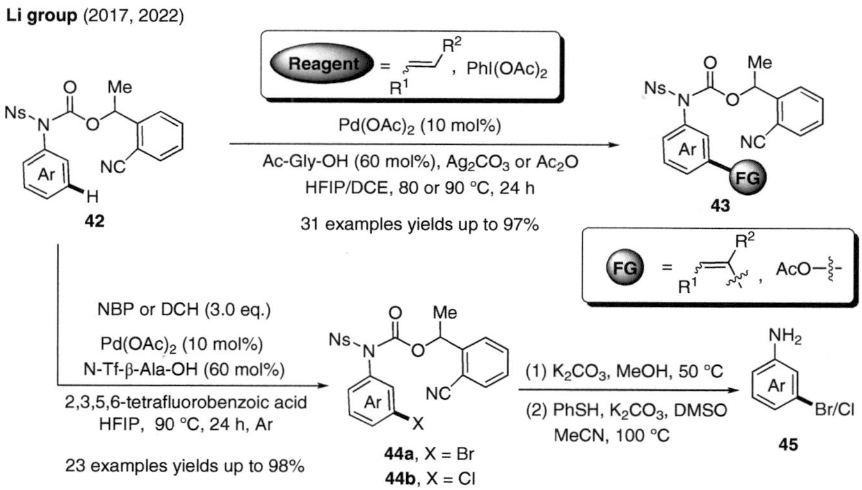

Scheme 5.11 Pd-catalyzed *meta*-C(sp^2)–H bond functionalization of aniline derivatives using a novel carbamate nitrile template.

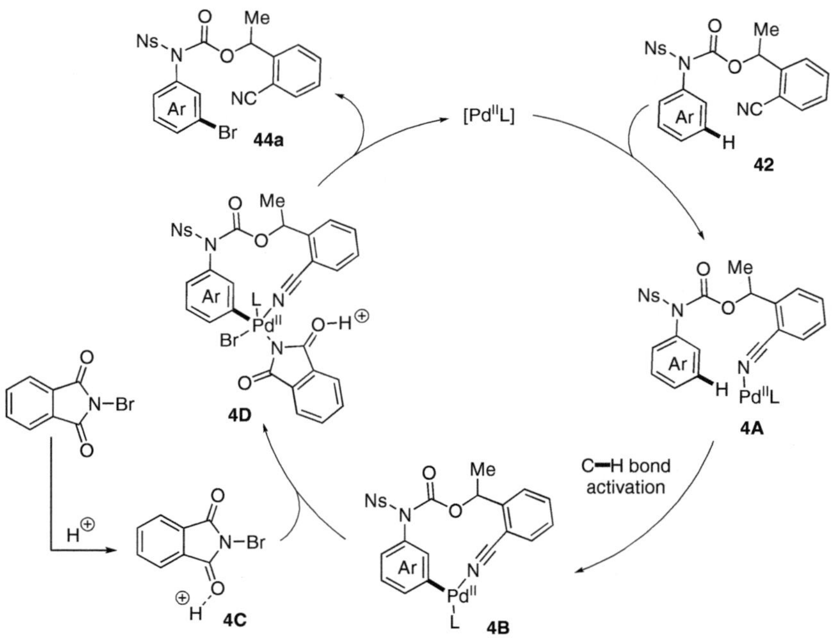

Scheme 5.12 The proposed mechanism of Pd-catalyzed *meta*-C(sp^2)–H bond bromination of aniline derivatives using a novel carbamate nitrile template.

Jin group (2019)

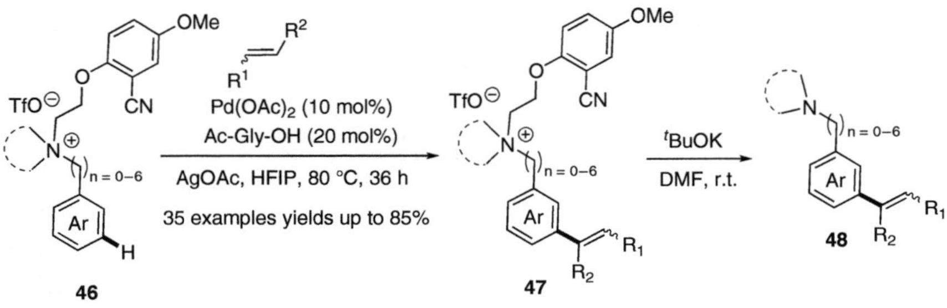

Scheme 5.13 Pd-catalyzed *meta*-C(sp^2)–H bond olefination of aryl tertiary amines using a 2-cyanophenoxyethyl template.

5.2.3 Distal C(sp^2)–H Functionalization of Arylalkyl Alcohols and Phenols

In 2013, Tan and co-workers disclosed the first example of Pd-catalyzed *meta*-C(sp^2)—H bond olefination of aryl methanol derivatives using a nitrile template [53]. The novel nitrile template incorporates a silicon atom to facilitate the attachment of aryl methanol compounds. Notably, this template can be effectively cleaved by tetrabutylammonium fluoride (TBAF) at room temperature within one hour. Additionally, this strategy exhibits good functional group compatibility, generating the desired products in good yields along with high *meta*-selectivity (Scheme 5.14).

In 2015, the Yu group designed a pyridine template to carry out *meta*-C(sp^2)—H bond olefination and iodination of benzyl and phenyl ethyl alcohol derivatives via Pd catalysis [54]. Unlike nitrile templates, this template utilizes a stronger σ-coordinating pyridine for these reactions. Results indicate that in the presence of KOH, *meta*-iodinated free alcohols can be obtained in good yields via hydrolysis (Scheme 5.14).

In 2017, Jin and co-workers developed the Pd-catalyzed *meta*-C(sp^2)—H bond olefination of aryl alkyl alcohol derivatives using a salicylonitrile template [55]. Various aryl alkyl alcohol derivatives including 2-phenylethyl and 3-phenylpropyl alcohols, as well as their long-chain homologues, were investigated. DFT computational results highlighted the necessity of AgOAc in assisting the Pd catalyst in C—H activation. Furthermore, ceric ammonium nitrate (CAN) can effectively remove this template at room temperature (Scheme 5.14).

In 2019, Kong and colleagues reported Pd-catalyzed *meta*-C(sp^2)—H bond olefination of aryl alkyl diol derivatives using a pyridine template [56] (Scheme 5.15). This protocol not only yielded the corresponding *meta*-alkenylated products but also afforded free diol compounds. Moreover, the pyridine template was found to be removable when TsOH was combined in a solvent mixture of H_2O and MeOH at 80 °C. The mechanism was proposed based on experimental and computational studies. First, the coordination of Ac-Gly-OH and HFIP on $Pd(OAc)_2$ generates the Pd(II) catalyst **5A**. Next, the pyridine template coordinates with Pd(II)

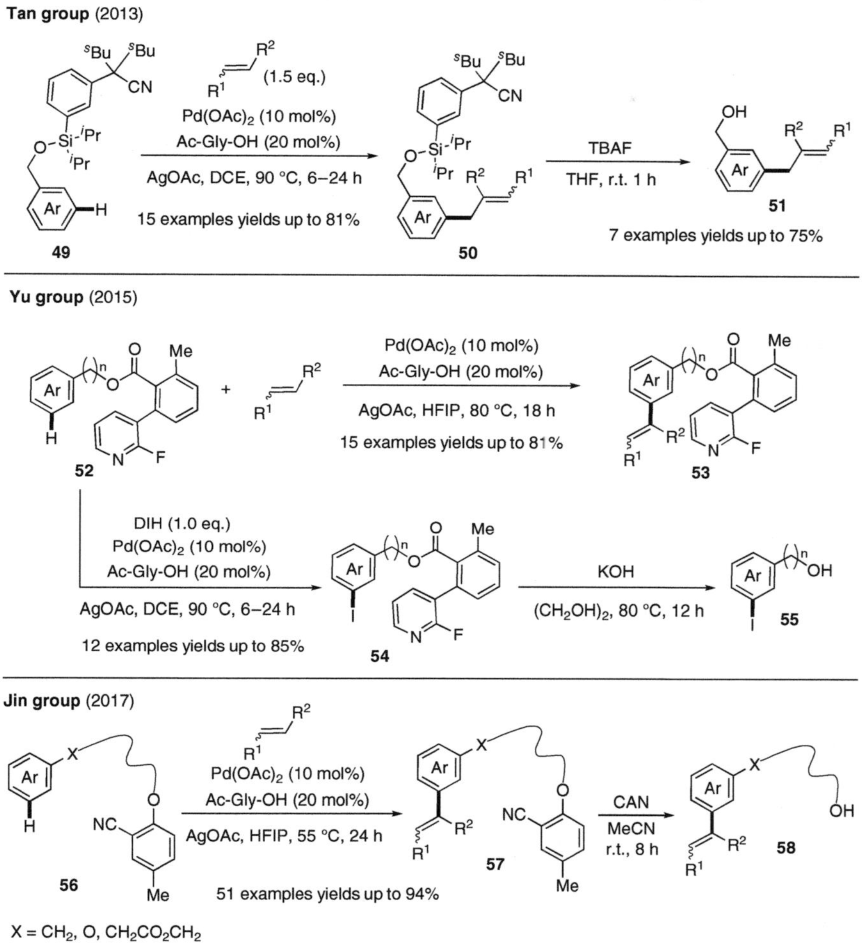

Scheme 5.14 Pd-catalyzed *meta*-C(sp^2)–H bond olefination of aryl alkyl alcohol derivatives using a nitrile template.

catalyst **5A**, followed by *meta*-C—H bond activation to form 13-membered palladacycle intermediate **5B**. The olefin then coordinates and undergoes subsequent 1,2-migratory insertion to produce intermediate **5D**. The *β*-hydride elimination of intermediate **5D** yields the final product and [Ac-Gly-O-Pd(II)-H] intermediate. Finally, reductive elimination of the [Ac-Gly-O-Pd(II)-H] intermediate generates the Pd(0) intermediate, which is re-oxidized by a silver salt to form the Pd(II) catalyst. It is noteworthy that the hydrolysis process may concurrently generate the hydroxylated product.

Phenol derivatives have attracted substantial attention due to their extensive synthetic versatility [57]. Previous studies have primarily focused on the functionalization of *ortho*-C(sp^2)—H bonds, while the activation of *meta*-C(sp^2)—H bonds poses a significant challenge due to the strong electron-donating capability exhibited by the phenoxyl functional group.

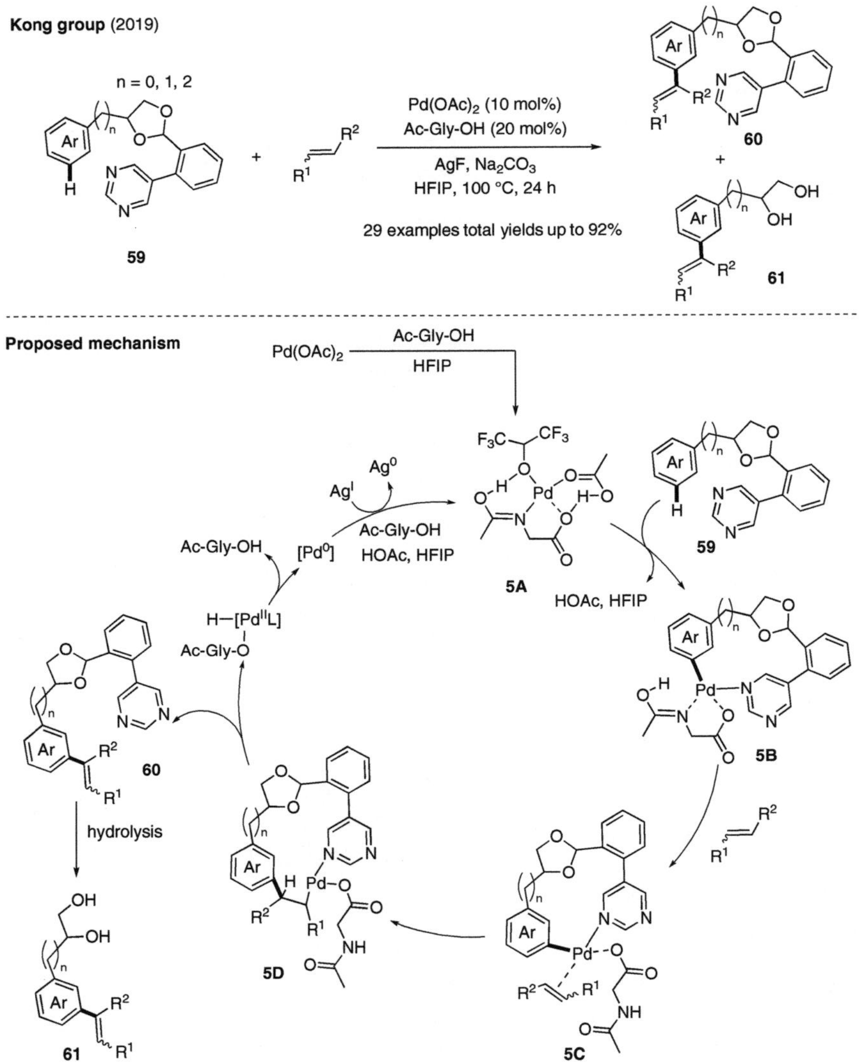

Scheme 5.15 Pd(II)-catalyzed *meta*-C(sp^2)—H bond olefination of phenylacetic acid derivatives utilizing a nitrile template.

In 2013, the Yu group reported the first example of Pd(II)-catalyzed *meta*-C(sp^2)—H bond olefination of phenol derivatives utilizing a nitrile template [58]. The nitrile template is connected to the phenoxyl group by an acetic acid moiety. In the presence of DPPA, this template can be removed to provide the *meta*-alkylated phenols in good yields.

In 2016, the Maiti group demonstrated the Pd-catalyzed *meta*-C(sp^2)—H bond olefination of biphenyl phenols with the assistance of a 2-cyano benzoic acid directing template [59]. Different electron-withdrawing olefins were well tolerated in this process, providing the desired products in moderate to good yields.

In 2019, the Yu group used a 1,3,5-triazine-derived biaryl nitrile template to execute the *meta*-C(sp²)—H bond olefination of phenol derivatives [60]. This well-designed biaryl nitrile template was prepared using commercially available 2,4-dichloro-6-methoxy-1,3,5-triazine with 2-cyanophenylboronic acid. Furthermore, the nickel-catalyzed *ipso*-C—O arylation of *meta*-olefinated phenol derivatives with arylboronic acids provided a variety of important substituted arenes with good yields (Scheme 5.16).

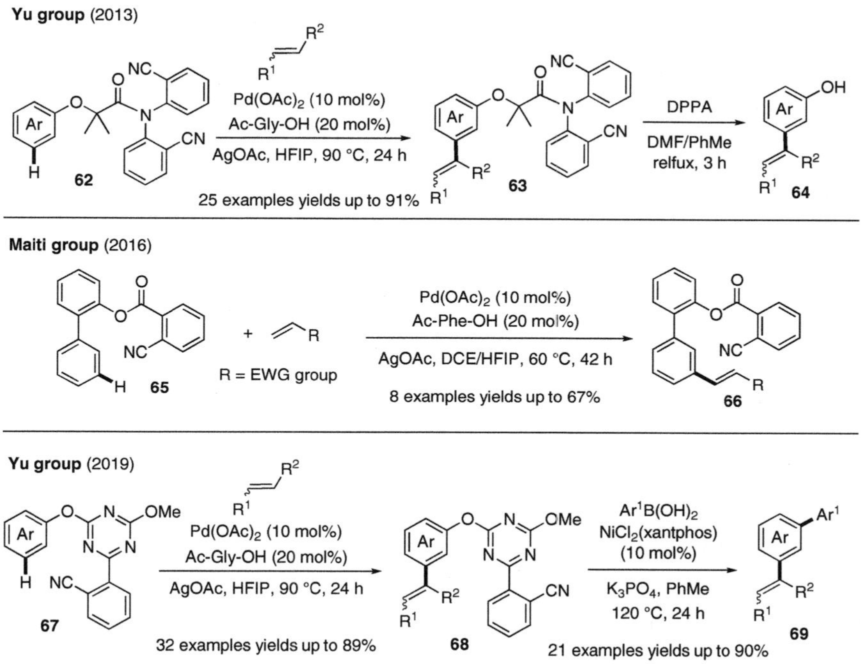

Scheme 5.16 Pd(II)-catalyzed *meta*-C(sp²)−H bond olefination of phenol derivatives utilizing a nitrile template.

Compared with *meta*-C(sp²)—H bond functionalization, there have been only a few reports on *para*-C(sp²)—H bond functionalization. In 2016, the Maiti group first demonstrated the Pd-catalyzed *para*-C(sp²)—H bond olefination of phenol derivatives using a silicon-containing biphenyl nitrile template [61]. They provided a total of 41 examples to depict the applicability of this approach, including various natural products. Furthermore, the silyl-biphenyl nitrile template could be readily removed by implementing TBAF in THF (Scheme 5.17).

5.2.4 Distal C(sp²)−H Functionalization of Arylalkyl Silanes

In 2012, the Yu group first developed a nitrile template featuring an ether linkage for attaching aryl alkane derivatives, which thus achieved Pd(II)-catalyzed *meta*-C(sp²)—H bond olefination [30]. Various aryl alkane derivatives substituted

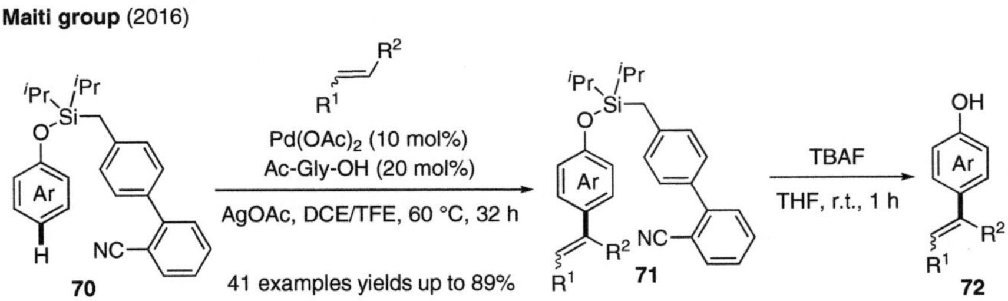

Scheme 5.17 Pd(II)-catalyzed *para*-C(sp^2)–H bond olefination of phenol derivatives utilizing a silicon-containing biphenyl nitrile template.

with different functional groups were successfully converted into the desired products with high yields and *meta*-selectivity. Additionally, *ortho*, *ortho*-disubstituted substrates were compatible under the same conditions. The removal of this template could be easily conquered through Pd/C-mediated hydrogenolysis, resulting in excellent yields of meta-alkylated arylmethanes (Scheme 5.18).

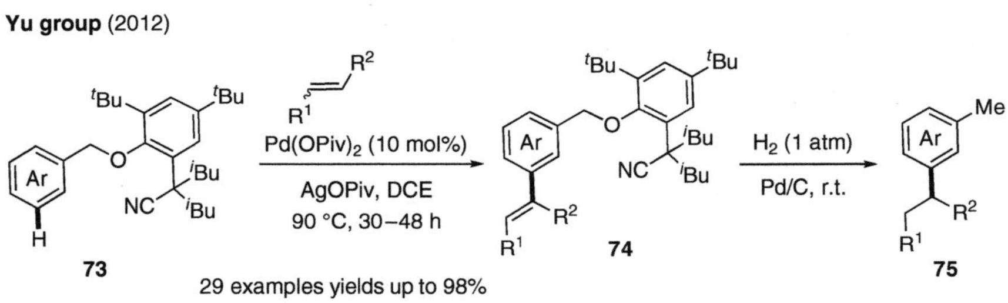

Scheme 5.18 Pd(II)-catalyzed *meta*-C(sp^2)–H bond olefination of aryl alkane derivatives utilizing a nitrile template.

In 2016, the Maiti group used the 4-methoxy-2-cyanophenol template to perform Pd(II)-catalyzed *meta*-C(sp^2)—H bond olefination of aryl alkane derivatives by attaching the substrates to the nitrile template via a Si—O bond [62]. After treating product **78** with TBAF, *meta*-olefinated toluene was isolated in a 94% yield. Meanwhile, *meta*-olefinated benzyl silanol was produced in 76% yield after employing a catalytic amount of TsOH. Subsequently, they reported Pd(II)-catalyzed *meta*-C(sp^2)—H bond cyanation of aryl alkane derivatives via a Si—O linkage from a biphenyl pyrimidine template along with a CuCN cyanating agent [63] (Scheme 5.19).

In addition, the authors provided a plausible catalytic cycle, in which the initial coordination is followed by successive *meta*-C—H bond activation to form 12-membered palladacycle intermediate **6A**. Next, ligand exchange between palladacycle intermediate **6A** and CuCN leads to the formation of palladacycle intermediate **6B**. The reductive elimination of **6B** then generates the desired product **82** and Pd(0) catalyst. Finally, a silver salt re-oxidizes the Pd(II) catalyst, completing the catalytic cycle (Scheme 5.20).

Maiti group (2016 and 2017)

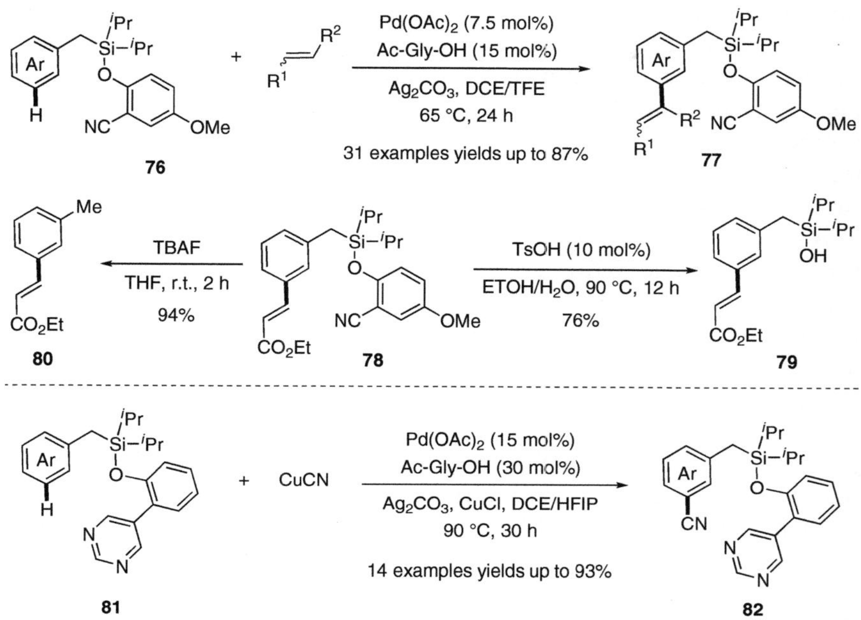

Scheme 5.19 Pd(II)-catalyzed *meta*-C(sp^2)–H bond functionalization of aryl alkane derivatives using a nitrile or pyrimidine template.

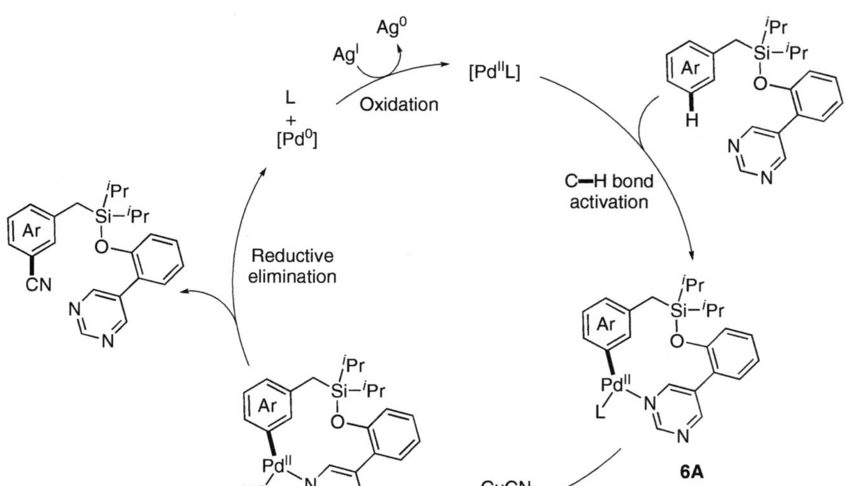

Scheme 5.20 The proposed mechanism of Pd(II)-catalyzed *meta*-C(sp^2)–H bond cyanation of aryl alkane derivatives using a biphenyl pyrimidine template.

In 2015, the Maiti group reported the Pd(II)-catalyzed *para*-C(sp^2)—H bond olefination and acetoxylation of arylmethane derivatives via Si—O bond linkage to a D-shaped biphenyl nitrile template [64] (Scheme 5.20). Unlike previous U-shaped templates, D-shaped templates can facilitate *para*-C(sp^2)—H bond functionalization by overcoming electronic and steric biases. Furthermore, the removal of the biphenyl nitrile template can be accomplished by using TBAF in THF. Later, the same group used a similar biphenyl nitrile template to carry out *para*-C(sp^2)—H bond ketonization of arylmethane derivatives using vinyl ethers as carbonyl sources [65]. Additionally, they disclosed the Pd(II)-catalyzed *para*-C(sp^2)—H bond cyanation of arylmethane derivatives using CuCN along with a biphenyl nitrile directing group [66] (Scheme 5.21).

In 2019, the Maiti group additionally introduced the first example of Rh(III)-catalyzed *para*-C(sp^2)—H bond olefination of arylmethane derivatives using the same biphenyl nitrile template [67]. Results suggest that electron-withdrawing olefins can be effectively coupled with various arylmethane derivatives, resulting in high *para*-selectivity and good yields for the desired products. Moreover, mechanistic studies suggest that activation of the C—H bond may serve as the rate-determining step (Scheme 5.22).

5.3 Distal C(sp^3)—H Bond Functionalization

5.3.1 γ-C(sp^3)—H Bond Functionalization of Carboxylic Acids

Compared with the functionalization of distal C(sp^2)—H bonds, unactivated distal C(sp^3)—H bonds often require bidentate directing groups. In 2011, the Chatani group initially explored the 8-aminoquinoline bidentate template (8-AQ) in perpetrating Pd(II)-catalyzed alkynylation of γ-C(sp^3)—H bond of 3,3-dimethylbutanoic acid with (bromoethynyl)triisopropylsilane. Unfortunately, only 33% of the alkynylated product was isolated. Moreover, the researchers found the removal of the 8-aminoquinoline template to be feasible [68].

In 2013, Chen and co-workers disclosed Pd(II)-catalyzed intramolecular γ-C(sp^3)—H bond amination of aliphatic carboxylic acid derivatives using the same 8-aminoquinoline template. They additionally found the 2-pyridylmethyl amine (PM) template to be successful, while also obtaining a variety of pyrrolidones in high yields through the formation of a [6,5]-bicyclic palladium intermediate [69].

In 2017, the Maiti group reported Pd(II)-catalyzed intermolecular γ-C(sp^3)—H bond olefination of aliphatic carboxylic acid derivatives with acrylates and vinyl iodides employing an 8-aminoquinoline template as the directing group [70]. The crucial ligand in the olefination reaction of aliphatic carboxylic acid derivatives with acrylates was determined to be 4,4′-di-*tert*-butyl-2,2′-bipyridine (DTBD). The utilization of vinyl iodides as olefination sources obviated the need for any additional ligands.

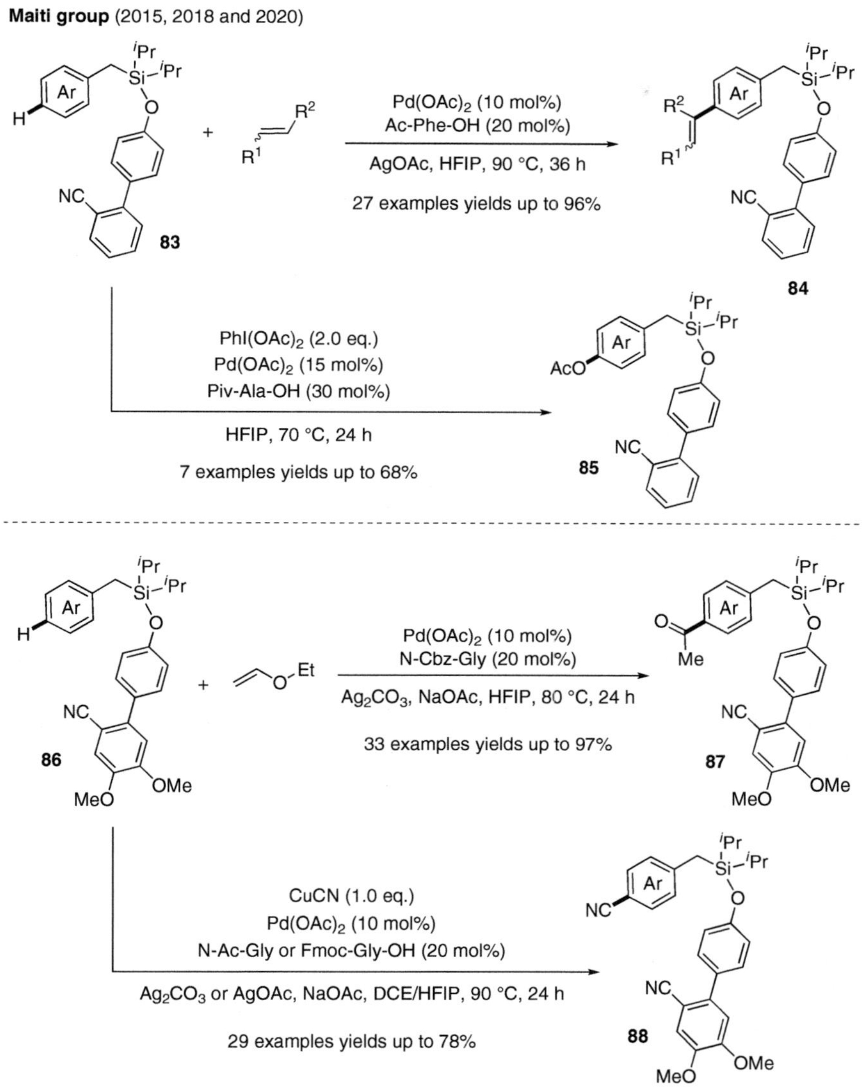

Scheme 5.21 Pd(II)-catalyzed *para*-C(sp^2)–H bond functionalization of aryl alkane derivatives using a nitrile template.

The researchers additionally applied a similar palladium catalysis strategy to perform γ-C(sp^3)—H bond silylation and germanylation of aliphatic carboxylic acid derivatives, identifying 2-chloroquinoline as the optimal palladium ligand in the process [71].

In 2020, the Shi and Yu group illustrated an 8-aminoquinoline-directed Pd-catalyzed γ-C(sp^3)—H alkylation of aliphatic carboxylic acid with strained

Maiti group (2019)

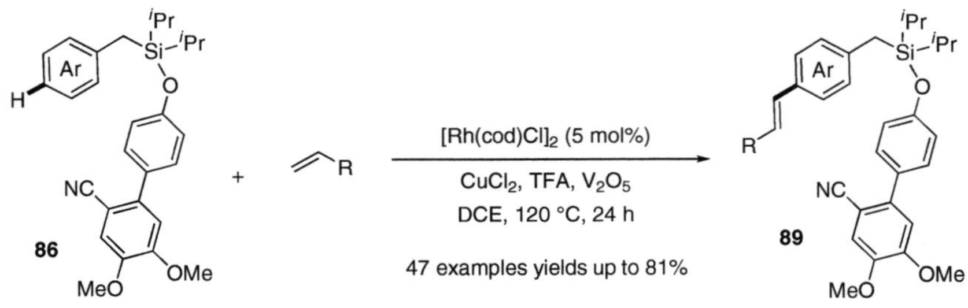

Scheme 5.22 Rh(III)-catalyzed *para*-C(sp^2)–H bond olefination of aryl alkane derivatives using a nitrile template.

bicyclic alkenes [72]. The inclusion of pyridine-2-one as an external palladium ligand was crucial to the success of this protocol, as its absence resulted in only a trace amount of product being detected (Scheme 5.23).

In addition, the 8-aminoquinoline template can be extended to functionalize the benzylic γ-C(sp^3)—H bond (Scheme 5.23). For example, in 2017, the Badu group reported Pd(II)-catalyzed arylation and cascade arylation/intramolecular amidation of benzylic γ-C(sp^3)—H bonds with aryl iodides assisted by the 8-aminoquinoline template [73]. Direct arylation of benzylic γ-C(sp^3)—H provided arylheteroarylmethane derivatives within a reaction time ranging from 4 to 48 hours. Extending the reaction period to 48–70 hours resulted in cascade reactions, facilitating the formation of both C—C and C—N bonds while also yielding pyrrolidone-ring annulated thiophene/furan-based heterocyclic scaffolds.

In 2019, the Chatani group also employed the 8-aminoquinoline template to achieve Pd(II)-catalyzed benzylic γ-C(sp^3)—H alkylation of *ortho*-methyl-substituted aromatic acids with maleimides [74]. Additionally, DFT calculations support the energetic preference for the insertion of maleimide into the six-membered palladacycle. The reaction mechanism hypothesizes that the reaction is initiated by the coordination of **102** to the Pd(II) catalyst, followed by cleavage of a benzylic γ-C(sp^3)—H bond to generate intermediate **7B**, a six-membered palladacycle. The insertion of maleimide into intermediate **7B** produces intermediate **7C**, which then undergoes protonolysis to yield the desired product **104** (Scheme 5.24).

In contrast to bidentate directing groups, the combination of monodentate directing groups with external ligands can be administered for distal C(sp^3)—H bond activation. In 2014, the Yu group reported the first instance of Pd(II)-catalyzed γ-C(sp^3)—H olefination of aliphatic carboxylic acid derivatives using a 2,3,5,6-tetrafluoro-4-(trifluoromethyl)aniline (Ar_F) monodentate directing group [75]. They found the utilization of a quinoline-derived ligand, TDMPQ, to be indispensable for conducting this transformation. Furthermore, they broadened the same strategy to accomplish γ-C(sp^3)—H carbonylation using CO as the carbonyl source as well as Pd(II)-catalyzed γ-C(sp^3)—H arylation of aliphatic carboxylic acid derivatives [76]. Most notably, two quinoline-derived ligands, TDMPQ and DMPQ, were essential in this approach (Scheme 5.25).

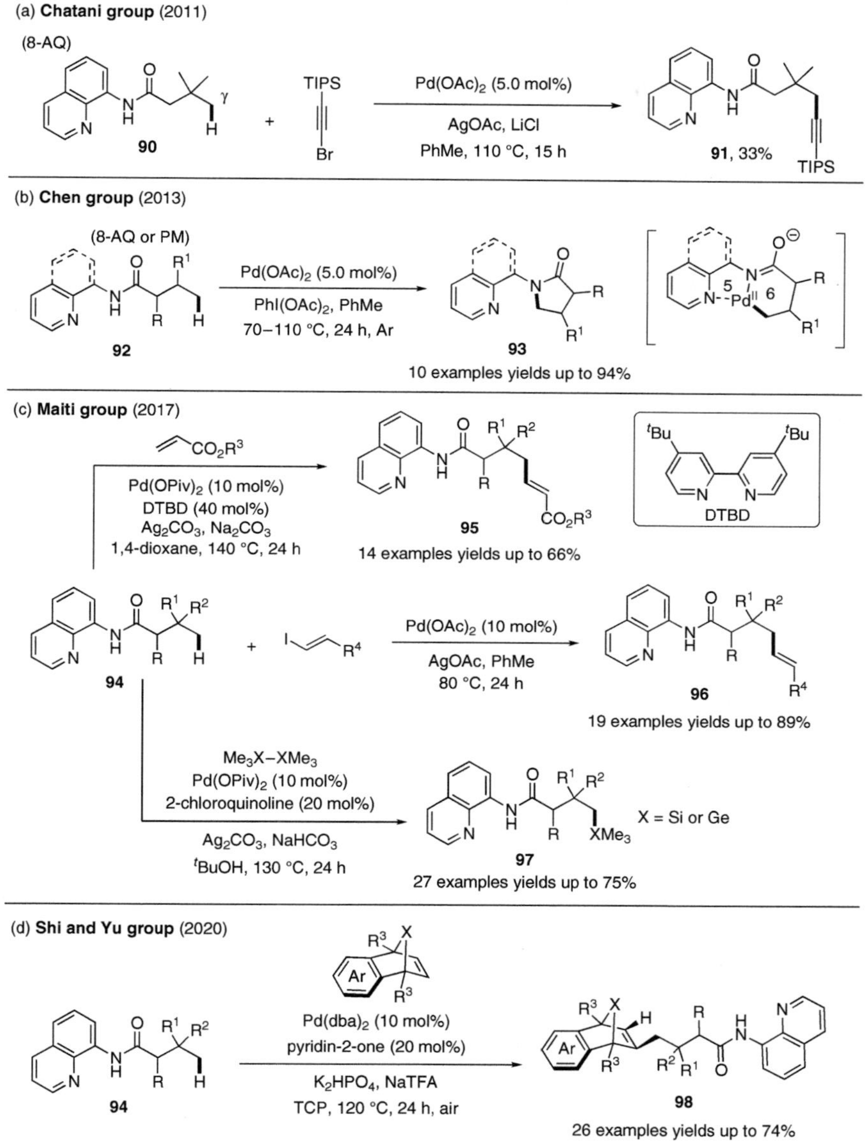

Scheme 5.23 Pd-catalyzed γ-C(sp^3)–H bond functionalization of aliphatic carboxylic acid derivatives by using an 8-aminoquinoline or 2-pyridylmethyl amine templates. (a) γ-C(sp^3)–H alkynylation; (b) intramolecular γ-C(sp^3)–H amination; (c) intermolecular γ-C(sp^3)–H olefination, silylation and germanylation; (d) γ-C(sp^3)–H alkylation.

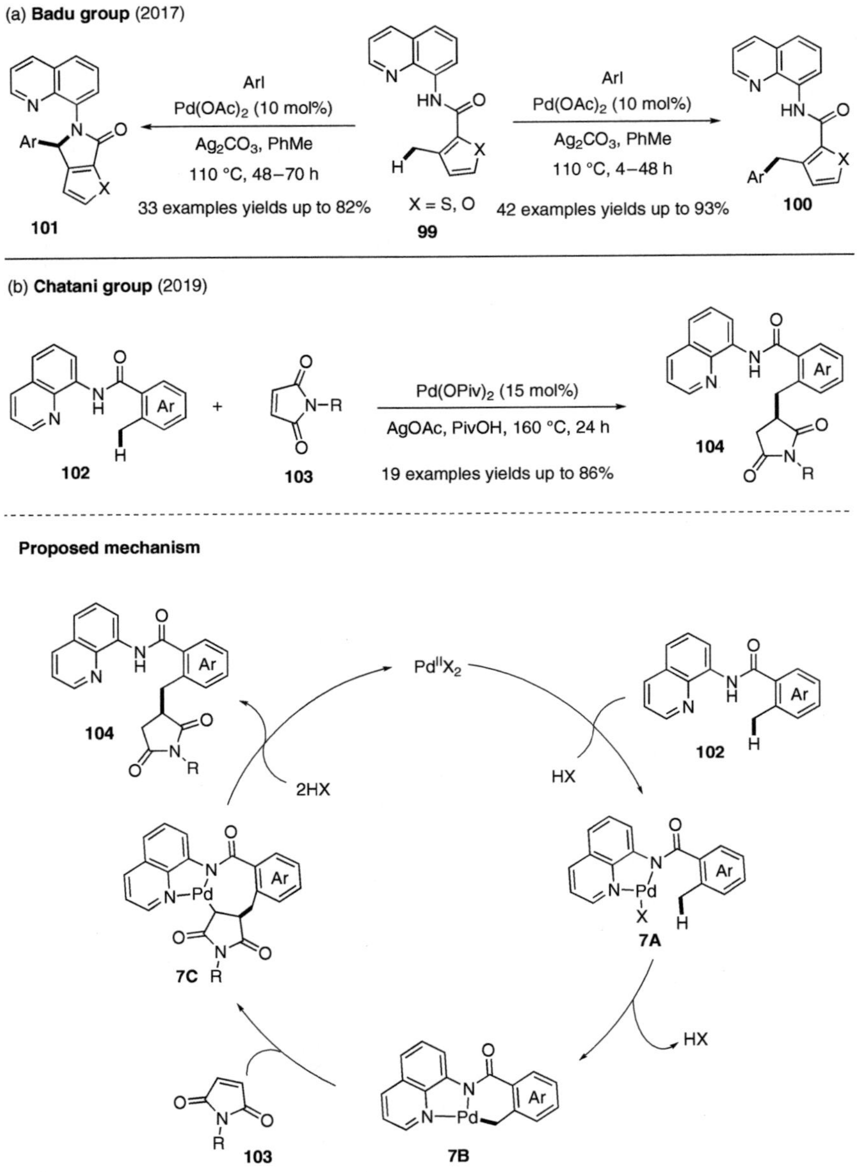

Scheme 5.24 8-Aminoquinoline-directed Pd(II)-catalyzed benzylic γ-C(sp^3)–H functionalization of *ortho*-methyl-substituted aromatic acids. (a) Benzylic γ-C(sp^3)–H arylation and cascade arylation/intramolecular amidation; (b) benzylic γ-C(sp^3)–H alkylation.

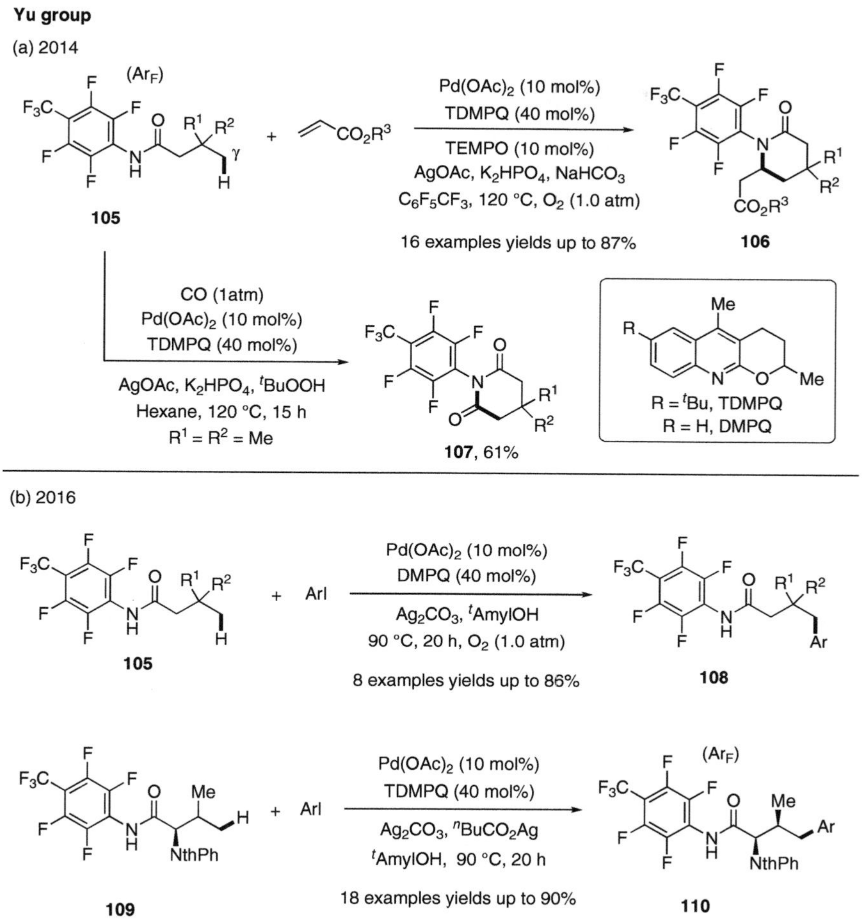

Scheme 5.25 Pd(II)-catalyzed γ-C(sp^3)−H functionalization of aliphatic carboxylic acid derivatives using 2,3,5,6-tetrafluoro-4-(trifluoromethyl)aniline (Ar$_F$) template and quinoline-derived ligands. (a) γ-C(sp^3)−H olefination and carbonylation; (b) γ-C(sp^3)−H arylation.

In 2021, the Yu group disclosed Pd(II)-catalyzed intramolecular γ-C(sp^3)—H amination of carboxylic acid derivatives enabled by 2-pyridone ligands and amino acids templates [77]. The amino acids can serve as partners for cyclization, while *tert*-butyl hydrogen peroxide (TBHP) serves as an environmentally friendly oxidant. The crucial 6-methyl-4-trifluoromethyl-2-pyridone not only promotes C—H cleavage but also facilitates intramolecular reductive elimination. Moreover, various types of *N*-containing heterocyclic compounds, including γ-lactams and isoindolinones, were obtained in moderate-to-good yields (Scheme 5.26). The postulated catalytic cycle suggests that the 2-pyridone ligand initially coordinates with Pd(OAc)$_2$ to form Pd(II) catalyst **8A**. In the presence of $CsHCO_3$, the substrate coordinates with Pd(II) catalyst **8A** to form Pd(II) intermediate **8B**, which is

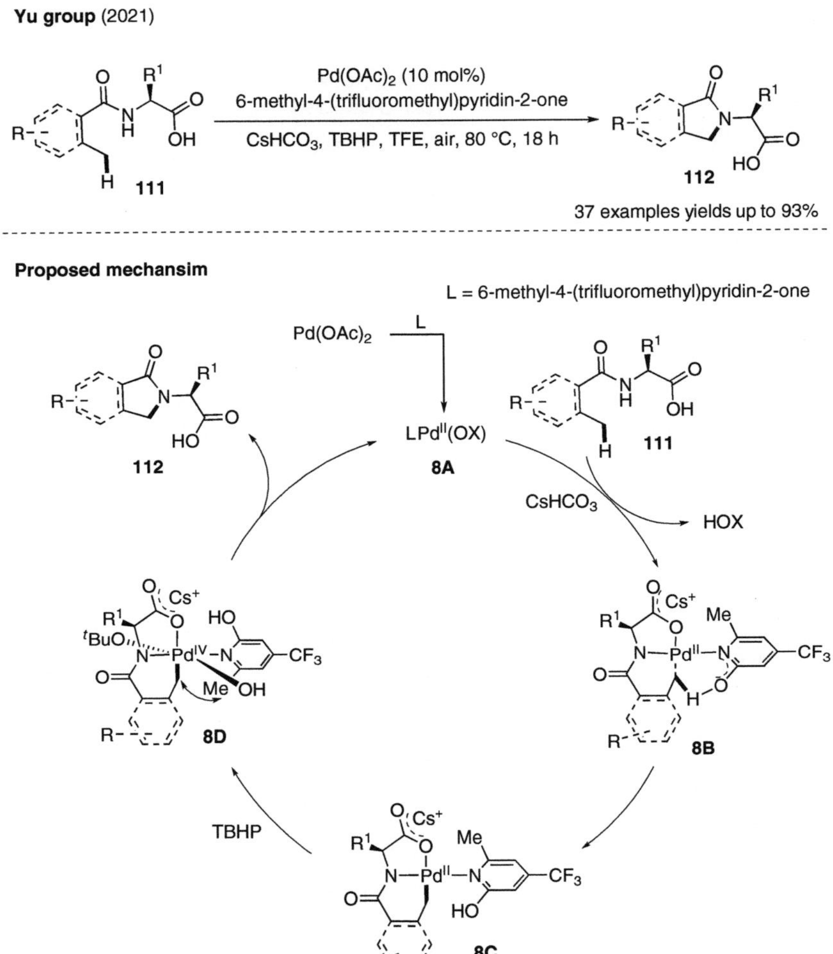

Scheme 5.26 Pd(II)-catalyzed and 2-pyridone ligand-enabled intramolecular γ-C(sp^3)–H amination of carboxylic acid derivatives using amino acid templates.

then converted to Pd(II) intermediate **8C** through γ-C(sp^3)—H bond activation assisted by a 2-pyridone ligand. Subsequently, oxidation of Pd(II) intermediate **7C** with TBHP generates Pd(IV) intermediate **8D**. Finally, intramolecular reductive elimination of Pd(IV) intermediate **8D** yields the desired product along with the regenerated Pd(II) catalyst (Scheme 5.26).

5.3.2 γ-C(sp^3)–H Bond Functionalization of Aliphatic Ketones

In 2018, the Yu group reported Pd(II)-catalyzed and 5-nitro-2-pyridone ligand-enabled γ-C(sp^3)—H (hetero)arylation of aliphatic ketone derivatives via a 2,2-dimethyl aminooxyacetic acid template [78]. They were the first group to prove that the electron-deficient 5-nitro-2-pyridone is capable of inducing γ-C(sp^3)—H activation. The 2,2-dimethyl aminooxyacetic acid template is readily removable by employing $Mn(OAc)_2$ as the catalyst via radical decarboxylation. Furthermore, this strategy could also achieve the γ-C(sp^3)—H olefination of ketone with 3-iodocyclopent-2-enone. Subsequently, the same group used a similar template 2,2-dimethyl aminooxyacetic amide to perform a Pd(II)-catalyzed γ-C(sp^3)—H olefination of aliphatic ketone derivatives [79]. The combination of a 5-(trifluoromethyl)-3-nitro-2-pyridone ligand with a commercial polypeptide, Ac-Ala-OH, effectively facilitated the γ-C(sp^3)—H activation in the process (Scheme 5.27).

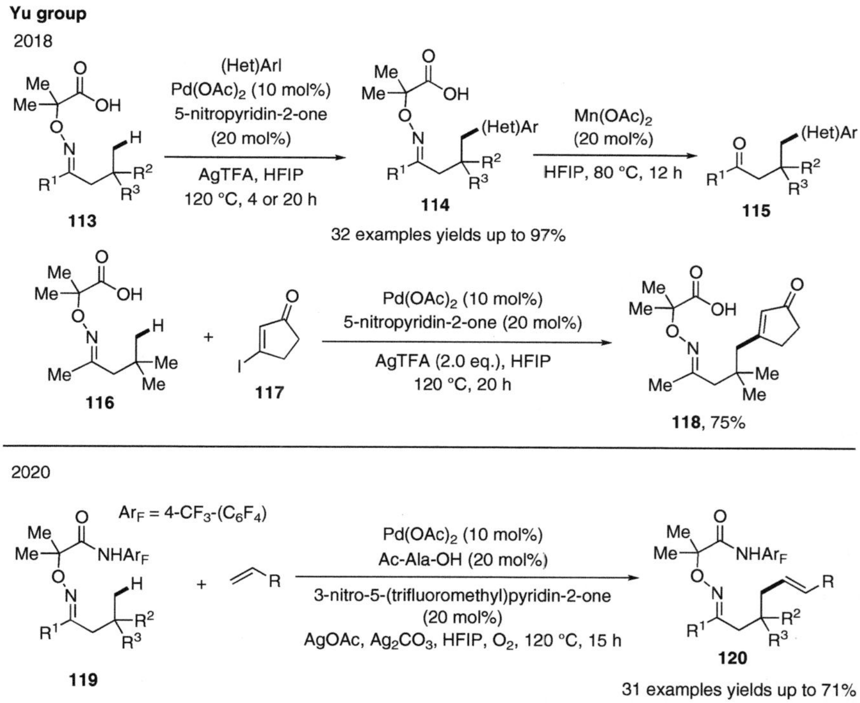

Scheme 5.27 Pd(II)-catalyzed and 2-pyridone ligand-enabled γ-C(sp^3)–H functionalization of aliphatic ketone derivatives using 2,2-dimethyl aminooxyacetic acid and amide templates.

The proposed mechanism of Pd-catalyzed γ-C(sp^3)—H (hetero)arylation of aliphatic ketone derivatives begins with the interrelation of the 5-nitro-2-pyridone ligand with both the substrate and $Pd(OAc)_2$ to form the Pd(II) intermediate **9A**. Subsequently, the same ligand facilitates γ-C(sp^3)—H cleavage of this intermediate to generate Pd(II) intermediate **9B**. Lastly, oxidation addition of Pd(II) intermediate **9B** followed by reductive elimination gives rise to the desired product (Scheme 5.28).

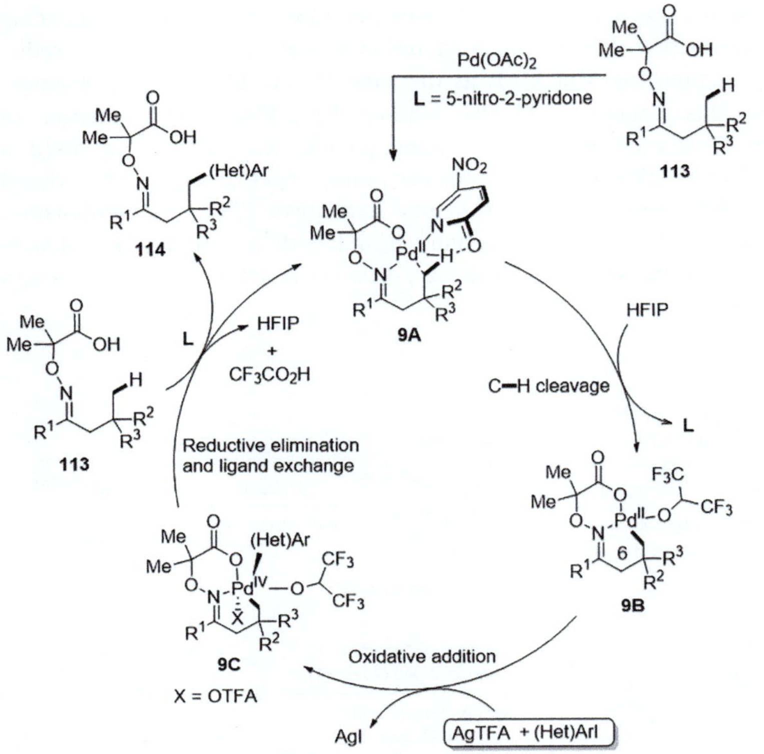

Scheme 5.28 Proposed mechanism of Pd-catalyzed and 5-nitro-2-pyridone ligand-enabled γ-C(sp^3)—H (hetero)arylation of aliphatic ketone derivatives.

5.3.3 δ-C(sp³)—H Bond Functionalization of Aliphatic Amines

Aliphatic amines are widely present in numerous biologically active molecules. However, their synthesis and modification through transition metal catalysis pose significant challenges due to the strong coordination ability of these compounds. The γ-C(sp^3)—H functionalization of aliphatic amines is often preferred, utilizing their amine-based directing groups to form five-membered metallocycle intermediates. In 2012, the Daugulis and Chen groups independently reported the first example of Pd(II)-catalyzed intramolecular δ-C(sp^3)—H amination of aliphatic amine derivatives using a 2-picolinic acid template [80, 81]. The Chen group found that adding AcOH could promote this process. The six-membered palladacycle intermediate is formed with the assistance of 2-picolinic acid, which facilitates δ-C(sp^3)—H activation (Scheme 5.29).

In 2014, Wang and co-workers utilized the 2-picolinic acid template to develop Pd(II)-catalyzed δ-C(sp^3)—H arylation of 3-pinanamine derivatives [82]. In this reaction, various aryl iodides and bromides were well tolerated, resulting in the formation of δ-arylated products with moderate-to-good yields. In 2019, the Maiti and Paton group reported the 2-picolinic-acid-directed and Pd(II)-catalyzed

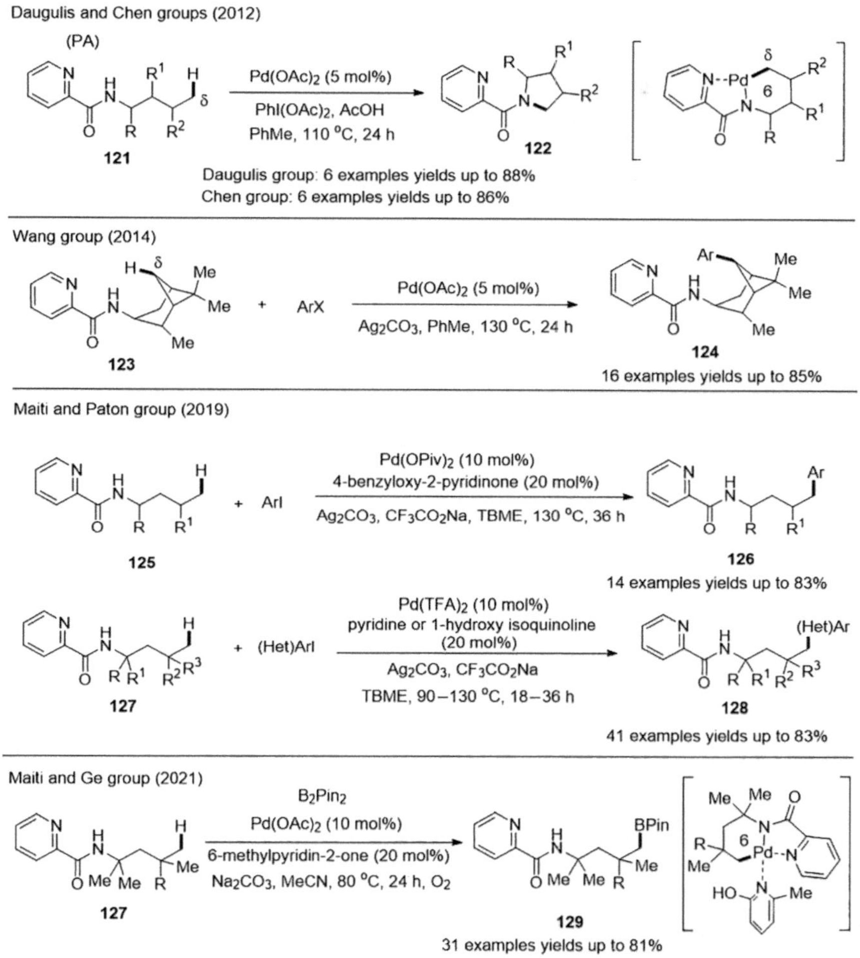

Scheme 5.29 Pd(II)-catalyzed δ-C(sp^3)–H functionalization of aliphatic amine derivatives using a 2-picolinic acid template.

δ-C(sp^3)—H arylation of aliphatic amine derivatives with (hetero)aryl iodides [83]. Different pyridine- and pyridone-based ligands played a crucial role in activating the δ-C(sp^3)—H bond. Specifically, 4-benzyloxy-2-pyridone was suitable for arylation reactions involving alicyclic amines, amino acids, and structurally analogous amine derivatives. Pyridine was found to be optimal for arylation, while 1-hydroxy isoquinoline was preferred for heteroarylation of the 2,4,4-trimethylpentan-2-amine substrate (Scheme 5.29).

In 2021, the Maiti and Ge groups also reported a Pd(II)-catalyzed δ-C(sp^3)—H borylation of aliphatic amine derivatives using 2-picolinic acid as the optimal directing template [84]. The effective ligand identified for accessing a variety of δ-boronated aliphatic amines was 6-methyl-2-pyridone, which forms a six-membered palladacycle intermediate (Scheme 5.29). The proposed mechanism commences with the correlation of $Pd(OAc)_2$ with 6-methyl-2-pyridone to generate Pd(II) catalyst **10A**. The amine substrate then coordinates with Pd(II) catalyst **10A**, followed by activation of the δ-C(sp^3)—H bond to form six-membered palladacycle intermediate **10B**. Subsequently, borylation of intermediate **10B** generates intermediate **10C**. Finally, reductive elimination leads to the desired δ-borylated product formation. The presence of oxygen facilitates the regeneration of the active catalyst and enables the initiation of a new cycle (Scheme 5.30).

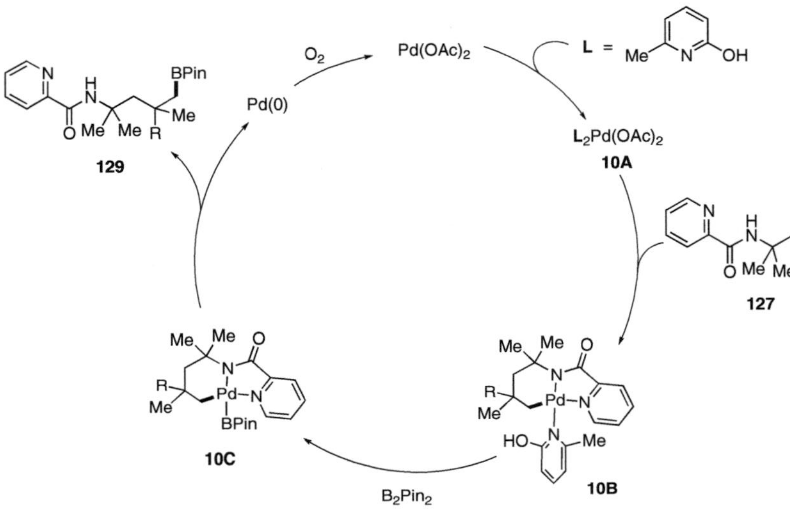

Scheme 5.30 Proposed mechanism of Pd(II)-catalyzed δ-C(sp^3)–H borylation of aliphatic amine derivatives using a 2-picolinic acid template.

In 2016, the Shi group first exhibited a Pd(II)-catalyzed δ-C(sp^3)—H alkenylation of amino acid derivatives with alkynes employing the 2-picolinic acid template [85]. 2,6-Dimethylbenzoquinone (2,6-DMBQ) was observed to enhance the reaction, potentially acting as both a ligand and a co-oxidant. Following the same strategy, site-selective δ-C(sp^3)—H alkylation was also accomplished by employing maleimides and *p*-benzoquinone (BQ) [86] (Scheme 5.31).

In addition to the 2-picolinic acid template, other templates such as 2-(diisopropylamino)-2-oxoacetic acid (OA) and 4,4-dimethyl-4,5-dihydrooxazole-2-carboxylic acid (Oxa) can serve as directing groups for achieving δ-C(sp^3)—H activation (Scheme 5.32). In 2014, the Zhao group reported the OA-directed Pd(II)-catalyzed intramolecular δ-C(sp^3)—H amination of aliphatic amine derivatives. A variety of pyrrolidine derivatives were isolated in good yields, and the OA template was willingly removed under basic conditions [87]. Furthermore, in 2015, Shi and colleagues described the use of the Oxa template in the Pd(II)-catalyzed δ-C(sp^3)—H arylation reaction. However, only one illustrative example was provided with a modest isolated yield of 30% [88]. The removal of the Oxa template proved to be feasible under acidic conditions (Scheme 5.32).

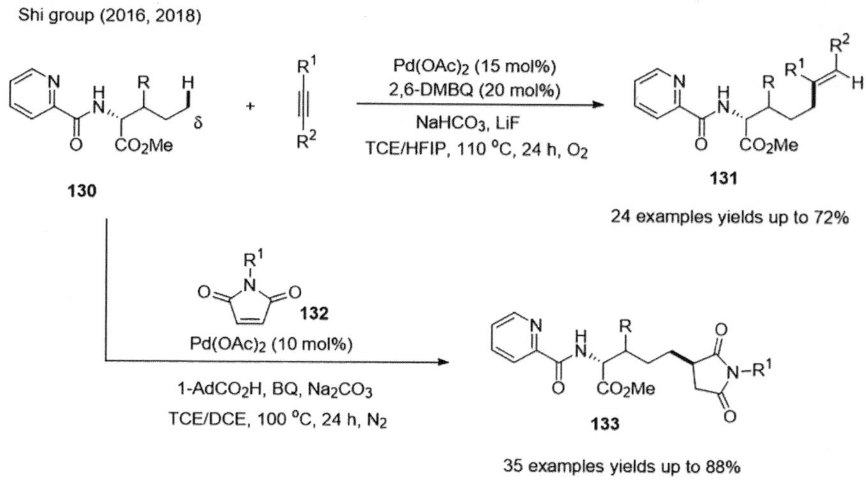

Scheme 5.31 Pd(II)-catalyzed δ-C(sp^3)—H alkenylation and alkylation of amino acid derivatives with alkynes employing a 2-picolinic acid template.

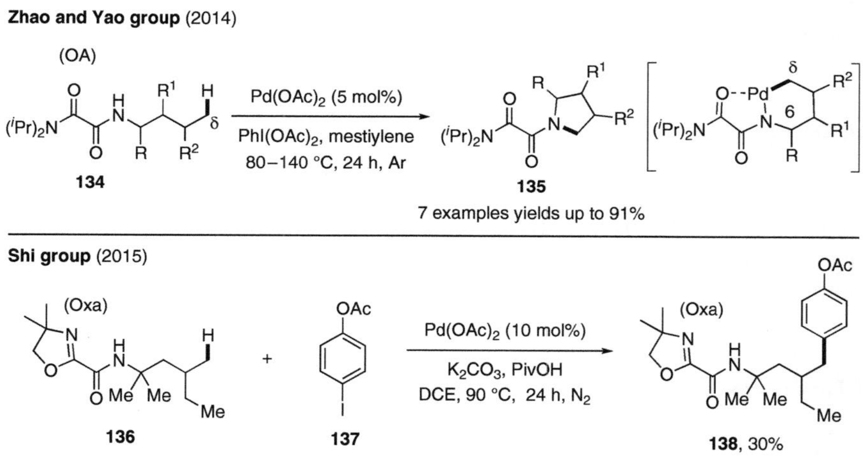

Scheme 5.32 Pd(II)-catalyzed δ-C(sp^3)—H functionalization of amine derivatives employing 2-(diisopropylamino)-2-oxoacetic acid (OA) and 4,4-dimethyl-4,5-dihydrooxazole-2-carboxylic acid (Oxa) templates.

5.4 Conclusions

In this chapter, we have summarized and discussed recent progress in the field of removable template-assisted transition-metal-catalyzed distal C—H functionalization reactions. The first section of this chapter mainly focused on the Pd or Rh-catalyzed distal C(sp^2)—H bond functionalization of arenes via larger-membered C—H cyclometallation. As demonstrated, a wide range of acids, amines, alcohols, and alkanes containing an aromatic ring have successfully achieved *meta-* or *para-*C(sp^2)—H bond functionalization by utilizing a nitrile, pyridine, or pyrimidine template. The latter half of this chapter focused on Pd-catalyzed distal $C(sp^3)$—H bond functionalization of aliphatic acids, ketones, and amines via six-membered C—H cyclometallation. Evidence revealed that γ-$C(sp^3)$—H bond functionalization of carboxylic acids mainly uses the 8-aminoquinoline (8-AQ) or 2,3,5,6-tetrafluoro-4-(trifluoromethyl)aniline (ArF) template, while the use of 2-dimethylaminooxyacetic acid or amide templates facilitates γ-$C(sp^3)$—H bond functionalization of ketones. Furthermore, δ-$C(sp^3)$—H bond functionalization of aliphatic amines can be achieved by using 2-picolinic acid (PA), 2-(diisopropylamino)-2-oxoacetic acid (OA), or 4,4-dimethyl-4,5-dihydrooxazole-2-carboxylic acid (Oxa). Despite the notable achievements made in this field, the practicality of this powerful approach still requires further improvement. The presence of a template is often necessary for optimal performance but results typically decrease overall process efficiency due to the requirement of additional synthetic steps for its installation and removal. Therefore, the transient directing group strategy can serve as an optimal approach to enhance the efficacy of this synthetic methodology. The studies included in this chapter are expected to serve as a source of inspiration for future researchers, encouraging them to explore and develop more innovative and versatile transition-metal-catalyzed distal C—H functionalization reactions.

Funding

Haibo Ge acknowledges NSF (CHE-2029932), the Robert A. Welch Foundation (D-2034-20230405), and Texas Tech University for financial support. Ke Yang is grateful for financial support from the "Advanced Catalysis and Green Manufacturing Collaborative Innovation Center, Changzhou University."

References

1 Godula, K. and Sames, D. (2006). C—H bond functionalization in complex organic synthesis. *Science* 312 (5770): 67–72.
2 Abrams, D.J., Provencher, P.A., and Sorensen, E.J. (2018). Recent application of C—H functionalization in complex natural product synthesis. *Chem. Soc. Rev.* 47 (23): 8925–8967.
3 Chen, Z., Rong, M.Y., Nie, J. et al. (2019). Catalytic alkylation of unactivated $C(sp^3)$—H bonds for $C(sp^3)$—$C(sp^3)$ bond formation. *Chem. Soc. Rev.* 48 (18): 4921–4942.
4 He, J., Wasa, M., Chan, K.S.L. et al. (2017). Palladium-catalyzed transformations of alkyl C—H bonds. *Chem. Rev.* 117 (13): 8754–8786.

5 Yang, K., Song, M.J., Ma, Z.Y. et al. (2019). The decarboxylative C—H heteroarylation of azoles catalysed by nickel catalysts to access unsymmetrical biheteroaryls. *Org. Chem. Front.* 6 (24): 3996–3999.

6 Khake, S.M. and Chatani, N. (2020). Nickel-catalyzed C—H functionalization using a non-directed strategy. *Chem* 6 (5): 1056–1081.

7 Chen, Z.K., Wang, B.J., Zhang, J.T. et al. (2015). Transition metal-catalyzed C—H bond functionalizations by the use of diverse directing groups. *Org. Chem. Front.* 2 (9): 1107–1295.

8 Yang, K., Song, M.J., Liu, H., and Ge, H.B. (2020). Palladium-catalyzed direct asymmetric C—H bond functionalization enabled by the directing group strategy. *Chem. Sci.* 11 (47): 12616–12632.

9 Niu, B., Yang, K., Lawrence, B., and Ge, H.B. (2019). Transient ligand-enabled transition metal-catalyzed C—H functionalization. *ChemSusChem* 12 (13): 2955–2969.

10 Yang, K., Li, Z., Liu, C. et al. (2022). Ligand-promoted palladium-catalyzed β-methylene C—H arylation of primary aldehydes. *Chem. Sci.* 13 (20): 5938–5943.

11 Li, B.J., Ali, A.I.M., and Ge, H.B. (2020). Recent advances in using transition-metal-catalyzed C—H functionalization to build fluorescent materials. *Chem* 6 (10): 2591–2657.

12 Zhang, M., Zhang, Y.F., Jie, X.M. et al. (2014). Recent advances in directed C—H functionalizations using monodentate nitrogen-based directing groups. *Org. Chem. Front.* 1 (7): 843–895.

13 Rouquet, G. and Chatani, N. (2013). Catalytic functionalization of C(sp^2)—H and C(sp^3)—H bonds by using bidentate directing groups. *Angew. Chem. Int. Ed.* 52 (45): 11726–11743.

14 Breslow, R., Zhang, X., and Huang, Y. (1997). Selective catalytic hydroxylation of a steroid by an artificial cytochrome P-450 enzyme. *J. Am. Chem. Soc.* 119 (19): 4535–4536.

15 Breslow, R., Huang, Y., Zhang, X., and Yang, J. (1997). An artificial cytochrome P450 that hydroxylates unactivated carbons with regio- and stereoselectivity and useful catalytic turnover. *Proc. Natl. Acad. Sci.* 94 (21): 11156–11158.

16 Belvedere, S. and Breslow, R. (2001). Regioselective oxidation of steroids by a manganese porphyrin carrying metal coordinating groups. *Bioorg. Chem.* 29 (5): 321–331.

17 Das, S., Incarvito, C.D., Carbtree, R.H., and Brudvig, G.W. (2006). Molecular recognition in the selective oxygenation of saturated C—H bonds by a dimanganese catalyst. *Science* 312 (5782): 2004–2007.

18 Bigi, M.A., Reed, S.A., and White, M.C. (2012). Directed metal (oxo) aliphatic C—H hydroxylations: overriding substrate bias. *J. Am. Chem. Soc.* 134 (23): 9721–9726.

19 Bigi, M.A., Reed, S.A., and White, M.C. (2011). Diverting non-haem iron catalysed aliphatic C—H hydroxylations towards desaturations. *Nat. Chem.* 3 (3): 216–222.

20 Chen, M.S. and White, M.C. (2007). A predictably selective aliphatic C—H oxidation reaction for complex molecule synthesis. *Science* 318 (5851): 783–788.

21 Dey, A., Sinha, S.K., Achar, T.K., and Maiti, D. (2019). Accessing remote *meta*-and *para*-C(sp^2)—H bonds with covalently attached directing groups. *Angew. Chem. Int. Ed.* 58 (32): 10820–10843.

22 Wu, M., Huang, X.P., Zhang, H.B., and Li, P.F. (2019). Advances on directing-group assisted *meta*-C—H functionalization catalyzed by transition metal. *Chin. J. Org. Chem.* 39 (11): 3114–3131.

23 Das, J., Guin, S., and Maiti, D. (2020). Diverse strategies for transition metal catalyzed distal C(sp^3)—H functionalizations. *Chem. Sci.* 11 (40): 10887–10909.

24 Zhang, Q. and Shi, B.F. (2020). Site-selective functionalization of remote aliphatic C—H bonds via C—H metallation. *Chem. Sci.* 12 (3): 841–852.

25 Mingo, M.M., Rodríguez, N., Arrayás, R.G., and Carretero, J.C. (2021). Remote C(sp^3)—H functionalization via catalytic cyclometallation: beyond five-membered ring metallacycle intermediates. *Org. Chem. Front.* 8 (17): 4914–4946.

26 Li, B.J., Elsaid, M., and Ge, H.B. (2022). Transition-metal-catalyzed site-selective γ- and δ-C(sp^3)—H functionalization reactions. *Chem* 8 (5): 1254–1360.

27 Dutta, U., Maiti, S., Bhattacharya, T., and Maiti, D. (2021). Arene diversification through distal C(sp^2)—H functionalization. *Science* 372 (6543): eabd5992.

28 Feng, X.Q., Wang, H.C., Li, Z. et al. (2022). Transition-metal-catalyzed remote C—H functionalization of thioethers. *RSC Adv.* 12 (17): 10835–10845.

29 Sehajpal, S., Prasad, D.N., and Singh, R.K. (2018). Prodrugs of non-steroidal anti-inflammatory drugs (NSAIDs): a long march towards synthesis of safer NSAIDs. *Mini-Rev. Med. Chem.* 18 (14): 1199–1219.

30 Leow, D., Li, G., Mei, T.S., and Yu, J.Q. (2012). Activation of remote *meta*-C—H bonds assisted by an end-on template. *Nature* 486: 518–522.

31 Li, W., Dastbaravardeh, N., Li, G., and Yu, J.Q. (2013). Cross-coupling of remote *meta*-C—H bonds directed by a u-shaped template. *J. Am. Chem. Soc.* 135 (48): 18056–18059.

32 Deng, Y.Q. and Yu, J.Q. (2014). Remote *meta*-C—H olefination of phenylacetic acids directed by a versatile U-shaped template. *Angew. Chem. Int. Ed.* 54 (3): 888–891.

33 Bera, M., Modak, A., Patra, T. et al. (2014). Meta-selective arene C—H bond olefination of arylacetic acid using a nitrile-based directing group. *Org. Lett.* 16 (21): 5760–5763.

34 Jin, Z., Chu, L., Chen, Y.Q., and Yu, J.Q. (2018). Pd-catalyzed remote *meta*-C—H functionalization of phenylacetic acids using a pyridine template. *Org. Lett.* 20 (2): 425–428.

35 Fang, L.Z., Saint-Denis, T.G., Taylor, B.L.H. et al. (2017). Experimental and computational development of a conformationally flexible template for the *meta*-C—H functionalization of benzoic acids. *J. Am. Chem. Soc.* 139 (31): 10702–10714.

36 Li, M.H., Shang, M., Xu, H. et al. (2019). Remote *para*-C—H acetoxylation of electron-deficient arenes. *Org. Lett.* 21 (2): 540–544.

37 Bera, M., Maji, A., Sahoo, S.K., and Maiti, D. (2015). Palladium(II)-catalyzed *meta*-C—H olefination: constructing multisubstituted arenes through homo-diolefination and sequential hetero-diolefination. *Angew. Chem. Int. Ed.* 54 (29): 8515–8519.

38 Maji, A., Bhaskararao, B., Singha, S. et al. (2016). Directing group assisted *meta*-hydroxylation by C—H activation. *Chem. Sci.* 7 (5): 3147–3153.

39 Dutta, U., Modak, A., Bhaskararao, B. et al. (2017). Catalytic arene *meta*-C—H functionalization exploiting a quinoline-based template. *ACS Catal.* 7 (5): 3162–3168.

40 Modak, A., Mondal, A., Watile, R. et al. (2016). Remote *meta* C—H bond functionalization of 2-phenethylsulphonic acid and 3-phenylpropanoic acid derivatives. *Chem. Commun.* 52 (96): 13916–13919.

41 Bag, S., Jayarajan, R., Mondal, R., and Maiti, D. (2017). Template-assisted *meta*-C—H alkylation and alkenylation of arenes. *Angew. Chem. Int. Ed.* 56 (12): 3182–3186.

42 Bera, M., Agasti, S., Chowdhury, R. et al. (2017). Rhodium-catalyzed *meta*-C—H functionalization of arenes. *Angew. Chem. Int. Ed.* 56 (19): 5272–5276.

43 Xu, H.J., Lu, Y., Farmer, M.E. et al. (2017). Rh(III)-catalyzed *meta*-C—H olefination directed by a nitrile template. *J. Am. Chem. Soc.* 139 (6): 2200–2203.

44 Xu, H.J., Kang, Y.S., Shi, H. et al. (2019). Rh(III)-catalyzed *meta*-C—H alkenylation with alkynes. *J. Am. Chem. Soc.* 141 (1): 76–79.

45 Muthukrishnan, I., Sridharan, V., and Menéndez, J.C. (2019). Progress in the chemistry of tetrahydroquinolines. *Chem. Rev.* 119 (8): 5057–5191.

46 Aleku, G.A., Tichiner, G.R., Roberts, G.W. et al. (2022). Enzymatic *N*-allylation of primary and secondary amines using renewable cinnamic acids enabled by bacterial reductive aminases. *ACS Sustain. Chem. Eng.* 10 (20): 6794–6806.

47 Tang, R.Y., Li, G., and Yu, J.Q. (2014). Conformation-induced remote *meta*-C—H activation of amines. *Nature* 507: 215–220.

48 Yang, G.Q., Lindovska, P., Zhu, D.J. et al. (2014). Pd(II)-catalyzed *meta*-C—H olefination, arylation, and acetoxylation of indolines using a U-shaped template. *J. Am. Chem. Soc.* 136 (30): 10807–10813.

49 Yang, G.P., Zhu, D.J., Wang, P. et al. (2018). Remote C—H activation of various *N*-heterocycles using a single template. *Chem. Eur. J.* 24 (14): 3434–3438.

50 Yang, L., Fu, L., and Li, G. (2017). Incorporation of carbon dioxide into carbamate directing groups: palladium-catalyzed *meta*-C—H olefination and acetoxylation of aniline derivatives. *Adv. Synth. Catal.* 359 (13): 2235–2240.

51 Wang, H., Fu, L., Zhou, C.L., and Li, G. (2022). Pd(II)-catalyzed *meta*-C—H bromination and chlorination of aniline and benzoic acid derivatives. *Chem. Sci.* 13 (29): 8686–8692.

52 Wang, B., Zhou, Y., Xu, N.N. et al. (2019). Palladium(II)-catalyzed remote *meta*-C—H functionalization of aromatic tertiary amines. *Org. Lett.* 21 (6): 1885–1889.

53 Lee, S.G., Lee, H.L., and Tan, K.L. (2013). *Meta*-selective C—H functionalization using a nitrile-based directing group and cleavable Si-tether. *J. Am. Chem. Soc.* 135 (50): 18778–18781.

54 Chu, L., Shang, M., Tanaka, K. et al. (2015). Remote *meta*-C—H activation using a pyridine-based template: achieving site-selectivity via the recognition of distance and geometry. *ACS Cent. Sci.* 1 (7): 394–399.

55 Zhang, L.L., Zhao, C.Y., Liu, Y. et al. (2017). Activation of remote *meta*-C—H bonds in arenes with tethered alcohols: a salicylonitrile template. *Angew. Chem. Int. Ed.* 56 (40): 12245–12249.

56 Fang, S.Q., Wang, X.B., Yin, F.C. et al. (2019). Palladium-catalyzed *meta*-C—H olefination of arene-tethered diols directed by well-designed pyrimidine-based group. *Org. Lett.* 21 (6): 1841–1844.

57 Decostanzi, M., Auvergne, R., Boutevina, B., and Caillol, S. (2019). Biobased phenol and furan derivative coupling for the synthesis of functional monomers. *Green Chem.* 21 (4): 724–747.

58 D, H.X., Li, G., Zhang, X.G., Stepan, A.F., Yu, J.Q. (2013). Pd(II)-catalyzed *ortho*- or *meta*-C—H olefination of phenol derivatives. *J. Am. Chem. Soc.* 135 (20): 7567–7571.

59 Maity, S., Hoque, E., Dhawa, U., and Mati, D. (2016). Palladium catalyzed selective distal C—H olefination of biaryl systems. *Chem. Commun.* 52 (97): 14003–14006.

60 Xu, J.C., Chen, J.J., Gao, F. et al. (2019). Sequential functionalization of *meta*-C—H and *ipso*-C—O bonds of phenols. *J. Am. Chem. Soc.* 141 (5): 1903–1907.

61 Patra, T., Bag, S., Kancherla, R. et al. (2016). Palladium-catalyzed directed *para* C—H functionalization of phenols. *Angew. Chem. Int. Ed.* 55 (27): 7751–7755.

62 Patra, T., Watile, R., Agasti, S. et al. (2016). Sequential *meta*-C—H olefination of synthetically versatile benzyl silanes: effective synthesis of *meta*-olefinated toluene, benzaldehyde and benzyl alcohols. *Chem. Commun.* 52 (10): 2027–2030.

63 Bag, S., Jayarajan, R., Dutta, U. et al. (2017). Remote *meta*-C—H cyanation of arenes enabled by a pyrimidine-based auxiliary. *Angew. Chem. Int. Ed.* 56 (41): 12538–12542.

64 Bag, S., Patra, T., Modak, A. et al. (2015). Remote *para*-C—H functionalization of arenes by a D-shaped biphenyl template-based assembly. *J. Am. Chem. Soc.* 137 (37): 11888–11891.

65 Maji, A., Dahiya, A., Lu, G. et al. (2018). H-bonded reusable template assisted *para*-selective ketonisation using soft electrophilic vinyl ethers. *Nat. Commun.* 9: 3582.

66 Pimparkar, S., Bhattacharya, T., Maji, A. et al. (2020). *Para*-selective cyanation of arenes by H-bonded template. *Chem. Eur. J.* 26 (50): 11558–11564.

67 Dutta, U., Maiti, S., Pimparkar, S. et al. (2019). Rhodium catalyzed template-assisted distal *para*-C—H olefination. *Chem. Sci.* 10 (31): 7426–7432.

68 Ano, Y., Tobisu, M., and Chatani, N. (2011). Palladium-catalyzed direct ethynylation of C(sp^3)—H bonds in aliphatic carboxylic acid derivatives. *J. Am. Chem. Soc.* 133 (33): 12984–12986.

69 He, G., Zhang, S.Y., Nack, W.A. et al. (2013). Use of a readily removable auxiliary group for the synthesis of pyrrolidones by the palladium-catalyzed intramolecular amination of unactivated γ-C(sp^3)—H bonds. *Angew. Chem. Int. Ed.* 52 (42): 11124–11128.

70 Thrimurtulu, N., Khan, S., Maity, S. et al. (2017). Palladium catalyzed direct aliphatic γ-C(sp^3)—H alkenylation with alkenes and alkenyl iodides. *Chem. Commun.* 53 (92): 12457–12460.

71 Deb, A., Singh, S., Seth, K. et al. (2017). Experimental and computational studies on remote γ-C(sp^3)—H silylation and germanylation of aliphatic carboxamides. *ACS Catal.* 7 (12): 8171–8175.

72 Li, Y., Zhang, P., Liu, Y.J. et al. (2020). Remote γ-C(sp^3)—H alkylation of aliphatic carboxamides via an unexpected regiodetermining Pd migration process: reaction development and mechanistic study. *ACS Catal.* 10 (15): 8212–8222.

73 Parella, R. and Babu, S.A. (2017). Pd(II)-catalyzed arylation and intramolecular amidation of γ-C(sp^3)—H bonds: en route to arylheteroarylmethane and pyrrolidone ring annulated furan/thiophene scaffolds. *J. Org. Chem.* 82 (14): 7123–7150.

74 He, Q.Y., Ano, Y., and Chatani, N. (2019). The Pd-catalyzed C—H alkylation of *ortho*-methylsubstituted aromatic amides with maleimide occurs preferentially at the *ortho*-methyl C—H bond over the *ortho*-C—H bond. *Chem. Commun.* 55 (67): 9983–9986.

75 Li, S.H., Chen, G., Feng, C.G. et al. (2014). Ligand enabled γ-C—H olefination and carbonylation: construction of γ-quaternary carbon centers. *J. Am. Chem. Soc.* 136 (14): 5267–5270.

76 Li, S.H., Zhu, R.Y., Xiao, K.J., and Yu, J.Q. (2016). Ligand-enabled arylation of γ-C—H bonds. *Angew. Chem. Int. Ed.* 55 (13): 4317–4321.

77 Liu, S., Zhuang, Z., Qiao, J.X. et al. (2021). Ligand enabled Pd(II)-catalyzed γ-C(sp^3)—H lactamization of native amides. *J. Am. Chem. Soc.* 143 (51): 21657–21666.

78 Zhu, R.Y., Li, Z.Q., Park, H.S. et al. (2018). Ligand-enabled γ-C(sp^3)—H activation of ketones. *J. Am. Chem. Soc.* 140 (10): 3564–3568.

79 Park, H.S., Fan, Z.L., Zhu, R.Y., and Yu, J.Q. (2020). Distal γ-C(sp^3)—H olefination of ketone derivatives and free carboxylic acids. *Angew. Chem. Int. Ed.* 59 (31): 12853–12859.

80 Nadres, E.T. and Daugulis, O. (2012). Heterocycle synthesis via direct C—H/N—H coupling. *J. Am. Chem. Soc.* 134 (1): 7–10.

81 He, G., Zhao, Y.S., Zhang, S.Y. et al. (2012). Highly efficient syntheses of azetidines, pyrrolidines, and indolines via palladium catalyzed intramolecular amination of C(sp^3)—H and C(sp^2)—H bonds at γ and δ positions. *J. Am. Chem. Soc.* 134 (1): 3–6.

82 Cui, W., Chen, S.W., Wu, J.Q. et al. (2014). Palladium-catalyzed remote C(sp^3)—H arylation of 3-pinanamine. *Org. Lett.* 16 (16): 4288–4291.

83 Guin, S., Dolui, P., Zhang, X.L. et al. (2019). Iterative arylation of amino acids and aliphatic amines via δ-C(sp^3)—H activation: experimental and computational exploration. *Angew. Chem. Int. Ed.* 58 (17): 5633–5638.

84 Chandrashekar, H.B., Dolui, P., Li, B.J. et al. (2021). Ligand-enabled δ-C(sp^3)—H borylation of aliphatic amines. *Angew. Chem. Int. Ed.* 60 (33): 18194–18200.

85 Xu, J.W., Zhang, Z.Z., Rao, W.H., and Shi, B.F. (2016). Site-selective alkenylation of δ-C(sp^3)—H bonds with alkynes via a six-membered palladacycle. *J. Am. Chem. Soc.* 138 (34): 10750–10753.

86 Zhan, B.B., Li, Y., Xu, J.W. et al. (2018). Site-selective δ-C(sp^3)—H alkylation of amino acids and peptides with maleimides via a six-membered palladacycle. *Angew. Chem. Int. Ed.* 57 (20): 5858–5862.

87 Wang, C., Chen, C.P., Zhang, J.Y. et al. (2014). Easily accessible auxiliary for palladium-catalyzed intramolecular amination of C(sp^2)—H and C(sp^3)—H bonds at δ- and ε-positions. *Angew. Chem. Int. Ed.* 53 (37): 9884–9888.

88 Ling, P.X., Fang, S.L., Yin, X.S. et al. (2015). Palladium catalyzed arylation of unactivated γ-methylene C(sp^3)—H and δ-C—H bonds with an oxazoline-carboxylate auxiliary. *Chem. Eur. J.* 21 (48): 17503–17507.

6

Non-covalent Template-assisted C—H Bond Functionalization

Yoichiro Kuninobu

Institute for Materials Chemistry and Engineering, Kyushu University, 6-1 Kasugakoen, Kasuga-shi, Fukuoka 816-8580, Japan

6.1 Introduction

The control of site selectivity is important in synthetic organic reactions. This is because if the reaction does not proceed site selectively, the product is obtained as a mixture of multiple compounds. The separation of only the desired compound from such a mixture is laborious and not always possible. In many excellent synthetic organic reactions that have been developed through the efforts of predecessors, the control of site selectivity has not been a problem because the reaction is carried out on the basis of reaction sites, such as carbon–halogen bonds and carbonyl groups, which are more reactive than other bonds or functional groups (of course, there are several reactions, such as the Diels–Alder reaction, in which mixtures of stereoisomers and/or regioisomers are obtained depending on the direction of reaction between the substrate and the reactant).

Recently, carbon–hydrogen (C—H) bond transformations have attracted much attention as a more efficient synthetic reaction than many conventional synthetic organic reactions, and active research is being conducted worldwide. C—H transformations do not require the prior conversion of a poorly reactive C—H bond into a highly reactive bond or functional group such as a carbon–halogen bond or carbonyl group but are a highly efficient method of directly converting a C—H bond into another bond. C—H transformation is a very efficient method and has the potential to replace some of the conventional synthetic organic reactions. However, direct conversion of C—H bonds inevitably faces the problem of site selectivity. For example, in the Friedel–Crafts reaction, which is not included in the category of C—H transformations discussed here but is always mentioned in organic chemistry textbooks, the reaction proceeds at the *ortho*- and *para*-positions of electron-rich aromatic rings and at the *meta*-position of electron-poor aromatic rings. However, in reality, these products are the main products, and in many cases,

Directed C–H Bond Functionalization: Concepts and Applications, First Edition.
Edited by Debabrata Maiti and Supriya Rej.

the reaction proceeds at *ortho*-, *meta*-, and *para*-positions. The situation is even more difficult in C—H transformations. In many organic compounds, there are many similar but less reactive C—H bonds in the same molecules. Therefore, even if C—H transformations can proceed, the reactions will proceed at various reaction sites to give a mixture of regioisomers.

Under such circumstances, the "directing group method" was very successful [1]. The directing group method is a method that facilitates the C—H transformations by coordinating a functional group in the substrates to the metal center of the catalysts, bringing the metal center closer to the C—H bond, the reaction site, and stabilizing the cyclic intermediate generated after the C—H bond activation. A variety of transformations have been realized, including arylation, alkenylation, alkylation, acylation, alkoxycarbonylation, aminocarbonylation, borylation, silylation, amination, alkoxylation, and halogenation. The directing group method is an excellent method that allows easy design of reaction systems and efficient progression of various C—H transformations. However, as with all things, there are several problems with directing groups. For example, (i) the directing group must be introduced to the substrate before the reaction, and if the structure of the directing group is not required for the product, the directing group must be removed from the product after the reaction; (ii) many directing groups are difficult to remove; (iii) basically, the C—H transformations are performed in the vicinity of the directing group (*ortho*-position for aromatic compounds, β-position of aliphatic compounds, etc.); and (iv) the site selectivity is controlled by the substrates. Currently, the directing group method is the most widely used and reliable method. In the C(sp^3)—H transformations, a number of intramolecular 1,5- or 1,6-hydrogen atom transfer (HAT) reactions have been reported in recent years [2]. In addition to the directing group and HAT methods, methods using steric repulsion between the catalytically active species and the substrate [3] and methods using the electronic properties of the substrate [4] have been reported as site-selectivity control methods. However, all of these methods control the site selectivity of the reactions by the substituents or electronic properties of the substrates (substrate control), and the range of substrate scope is inevitably limited. Therefore, it is desirable to control the site selectivity of the reactions not by the substrates but by the catalysts (catalyst control) (Figure 6.1).

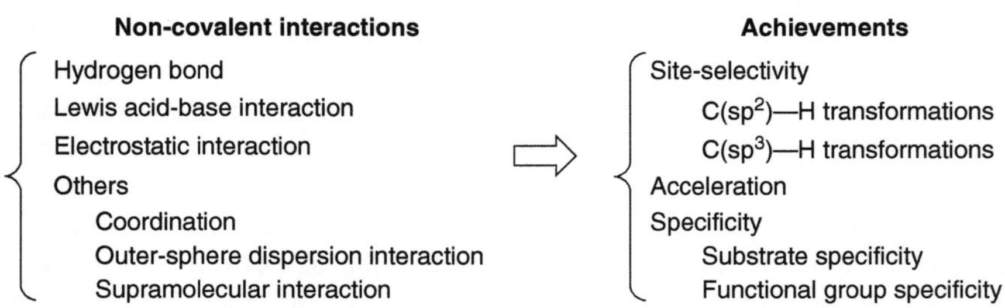

Figure 6.1 Several types of non-covalent interactions.

6.2 Control of Site Selectivity

The "non-covalent method" has recently been developed as a method to catalytically control site selectivity in C—H transformations and has attracted much attention, with many examples of researches reported by research groups around the world. In this method, the active site of the catalyst selectively approaches the desired reaction site of the substrates through non-covalent interaction between the catalyst and substrate, and the C—H transformations proceed in a site-selective manner. When the reaction is over, the product dissociates from the catalyst, and the unreacted substrate is incorporated into the catalyst, allowing the reaction to proceed catalytically. In this chapter, site-selective C–H transformations by non-covalent method reported so far are classified according to the type of non-covalent interaction between catalyst and substrate, in the order of C(sp^2)—H and C(sp^3)—H transformations.

6.2.1 C(sp^2)—H Transformations

6.2.1.1 Controlled by Hydrogen Bond

Among the above problems in the directing group method, one of the causes of the problem of attaching and detaching directing groups is that the directing groups and substrates are linked by covalent bonds. Kuninobu thought that this point could be solved by utilizing the non-covalent interaction between the catalyst and the substrate, as described below. The problem of the reaction sites being limited to the vicinity of the directing groups is expected to be solved by designing directing groups that can reach farther. In fact, Yu et al. realized a *meta*-selective C(sp^2)—H alkenylation using an elaborately designed and synthesized directing group (Scheme 6.1) [5]. However, because this method is the directing group method, it does not solve the problems of the directing group method described above.

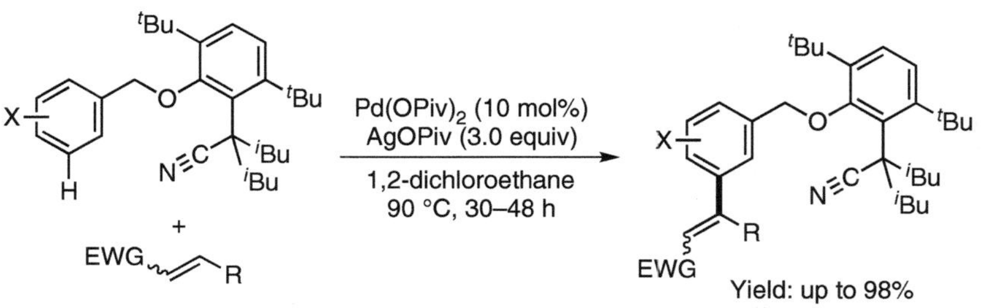

Scheme 6.1 Directing group-assisted palladium-catalyzed *meta*-selective oxidative C(sp^2)—H alkenylation of aromatic compounds.

Therefore, Kuninobu considered exploiting the non-covalent interactions between the catalyst and the substrate (Figure 6.2). The design of this catalytic system is as follows: (i) introduction of a substrate recognition site via an appropriate linker to the ligand in the metal catalyst (catalytically active site); (ii) the substrate recognition site recognizes and captures the substrate through non-covalent interactions such as hydrogen bonds, Lewis acid–base interactions, electrostatic interactions, and coordination bonds; (iii) the substrate is captured by the catalyst; (iv) if the catalytic design is appropriate, the catalytically active site will selectively approach the C—H bond to be reacted; (v) site-selective C—H transformations can proceed under the action of appropriate reagents; and (vi) non-covalent interactions can be used to capture the substrate. Since the non-covalent interaction does not form a strong bond, the reaction proceeds catalytically, with the product dissociating from the catalyst and the substrate being incorporated into the catalyst after the reaction. Such a catalyst can recognize and capture the substrate through non-covalent interactions and bring the desired reaction site close to the catalytically active site, thus eliminating the need to attach or detach directing groups as in the directing group method.

Based on the catalytic design shown in Figure 6.2, Kuninobu considered developing a remote *meta*-selective C(sp^2)—H borylation of aromatic compounds. Iridium-bipyridyl complex, which is commonly used in C(sp^2)—H borylation, was selected as catalytically active sites, and urea derivatives, which are known to hydrogen bond with carbonyl groups, were selected as substrate recognition sites, since the structure of the linker is important in order to bring the catalytically active site close to the *meta*-C(sp^2)—H bond.

When the borylation reagent B_2pin_2 was applied to aromatic amides in the presence of the bipyridine ligand dtbpy, the reaction proceeded at both *meta*- and *para*-positions as is well known, and the *para*-form was obtained as the main product (Scheme 6.2). On the other hand, when the reaction was carried out in

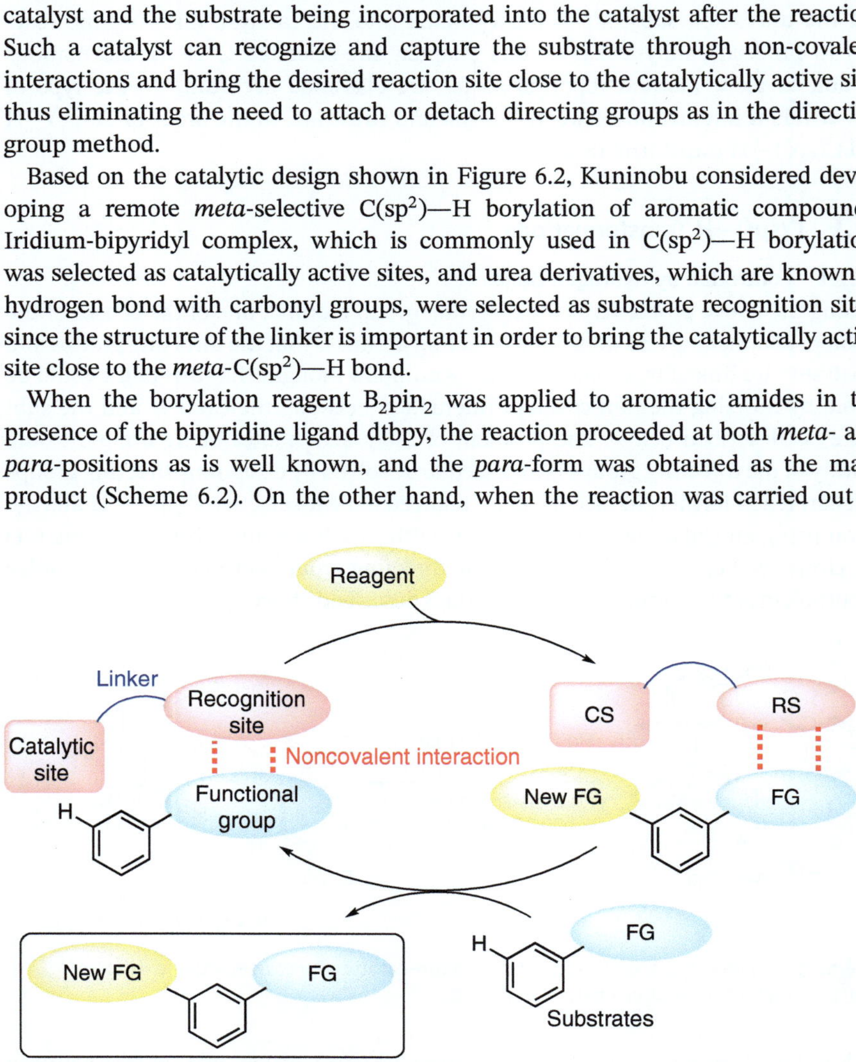

Figure 6.2 Design of a catalytic system that enables site-selective C−H transformations using non-covalent interaction between a catalyst and a substrate.

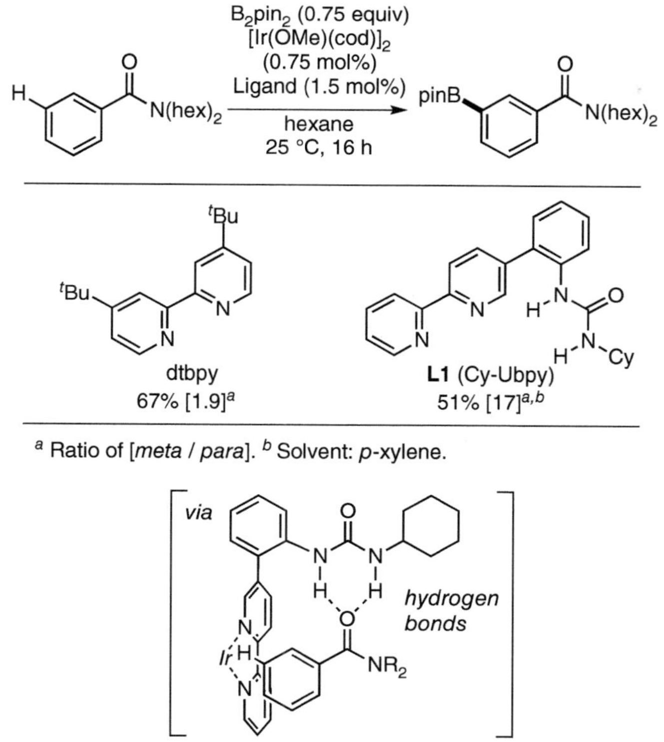

Scheme 6.2 *meta*-Selective C(sp^2)—H borylation of aromatic compounds using hydrogen bond between the catalyst and the substrate.

the presence of ligand **L1** (Cy-Ubpy), the site selectivity of the reaction changed dramatically, proceeding preferentially to the *meta*-position (Scheme 6.2) [6].

The catalyst system was particularly effective for aromatic amides and phosphorus compounds, giving C(sp^2)—H borylation products with high *meta*-selectivity (Scheme 6.3). Site-selective C(sp^2)—H borylation also proceeded in six- and five-membered heteroaromatic rings with amide groups (Scheme 6.3).

To confirm the importance of hydrogen bond in this catalytic reaction, several experiments were performed and the following results were obtained: (i) site selectivity is greatly affected by solvent polarity (the reaction proceeds *with* high meta-selectivity in nonpolar solvents, whereas the site selectivity is greatly reduced in highly polar solvents), (ii) the site selectivity is greatly affected by the reaction temperature (the selectivity decreases as the reaction temperature is increased), (iii) ^1H NMR shows that the NH proton signals of the urea moiety shift to lower field when the substrate (aromatic amide) is added dropwise to a benzene-d_6 solution of the ligand Cy-Ubpy, and (iv) no *meta*-selectivity is observed when the reaction is carried out with a ligand in which the two NH groups of the urea moiety are protected by methyl groups (Figure 6.3a). From the above experimental results, it is clear that hydrogen bond plays an important role in the site selectivity of this

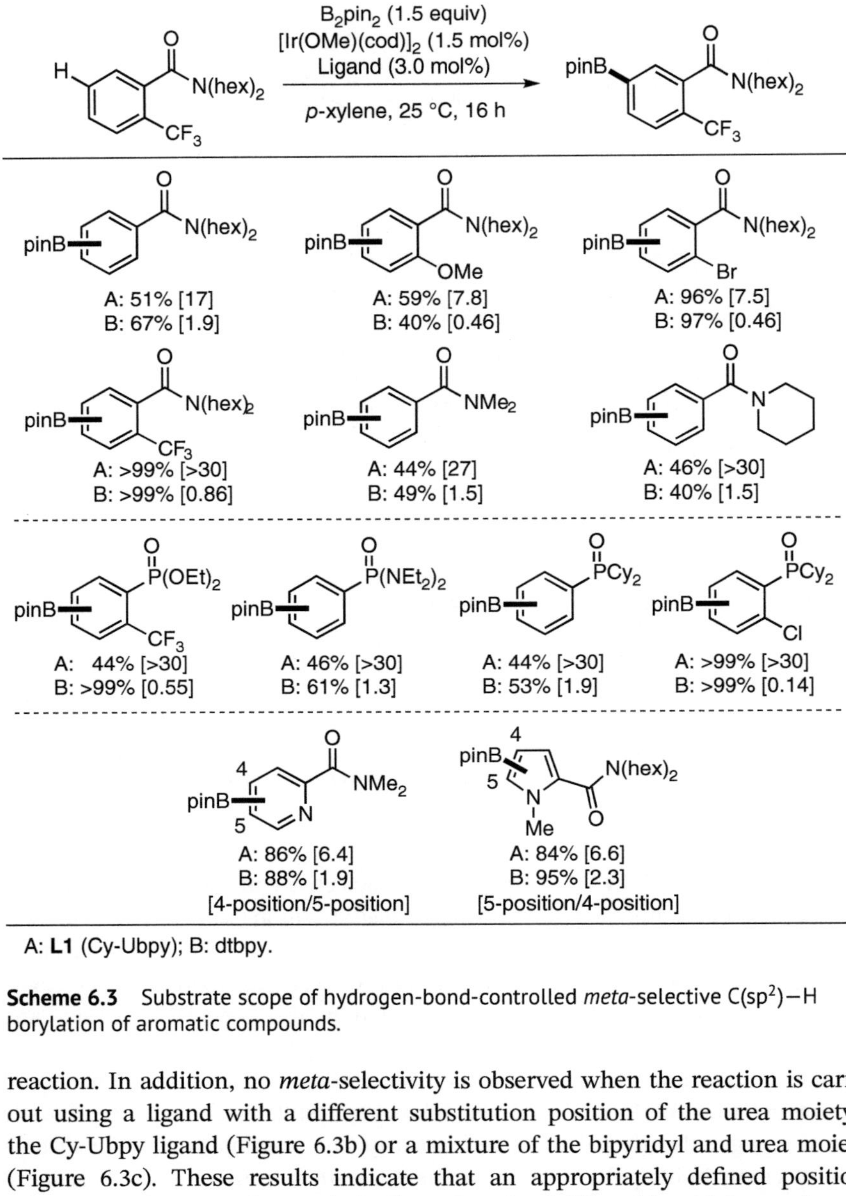

Scheme 6.3 Substrate scope of hydrogen-bond-controlled *meta*-selective $C(sp^2)$–H borylation of aromatic compounds.

reaction. In addition, no *meta*-selectivity is observed when the reaction is carried out using a ligand with a different substitution position of the urea moiety of the Cy-Ubpy ligand (Figure 6.3b) or a mixture of the bipyridyl and urea moieties (Figure 6.3c). These results indicate that an appropriately defined positional relationship between the catalytically active site (iridium-bipyridyl complex site) and the substrate recognition site (urea moiety) is important for the expression of *meta*-selectivity. Sunoj and Liang independently reported the DFT calculations of the reaction [7].

Phipps and co-workers reported *meta*-selective $C(sp^2)$—H borylation of aromatic compounds with an amide group as a hydrogen bond donor (Scheme 6.4) [8]. The reaction proceeded using amides derived from benzylamines, phenethylamines, and phenylpropylamines. The *meta*-selectivity is controlled by the hydrogen bond

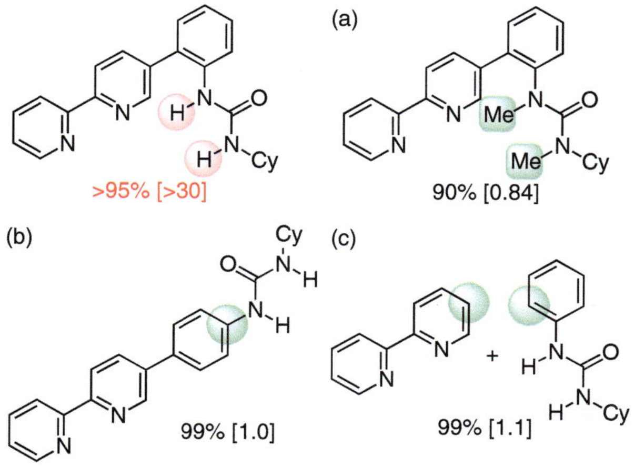

Figure 6.3 Several experiments to clarify the importance of hydrogen bond and defined positional relationship to control the site selectivity.

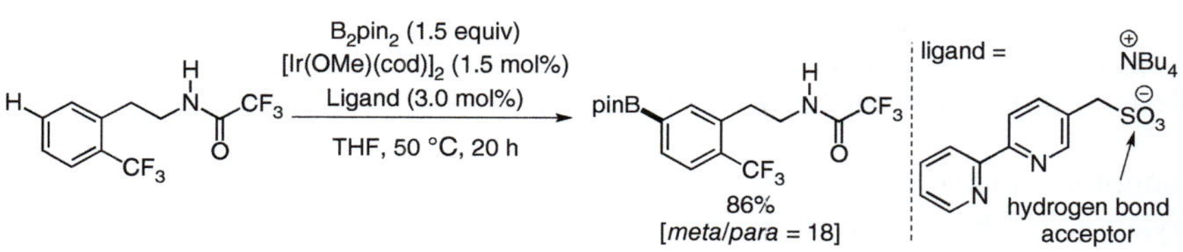

Scheme 6.4 Hydrogen-bond-controlled iridium-catalyzed *meta*-selective $C(sp^2)-H$ borylation of aromatic compounds with an amide group.

between the amide group and the sulfonate group of the bipyridine-type ligand, which was reported by the same group (see Scheme 6.17) [9].

Different from directing group-assisted $C(sp^2)$—H transformations, it was difficult to promote *ortho*-selective $C(sp^2)$—H borylation and the reactions generally proceed at both the *meta*- and *para*-positions, due to the steric repulsion between the catalytically active iridium-trisboryl species and a substituent(s) of substrates [3a, b]. Reek and co-workers developed *ortho*-selective $C(sp^2)$—H borylation of aromatic and 2-thiophenyl amides utilizing hydrogen bond between the indole amide moiety of the ligand and the carbonyl group of substrates (Scheme 6.5) [10]. Li and Lei reported the DFT calculations of the similar reaction system [11].

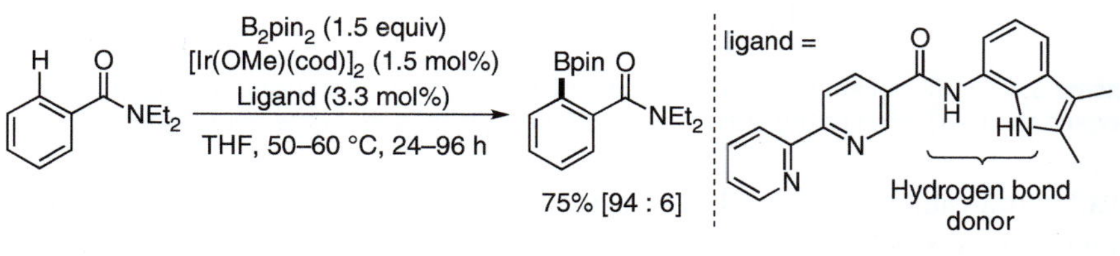

Scheme 6.5 Hydrogen-bond-controlled iridium-catalyzed *ortho*-selective $C(sp^2)-H$ borylation of aromatic amides.

Smith and co-workers succeeded in iridium-catalyzed *ortho*-selective C(sp^2)—H borylation of *N*-Boc Anilines (Scheme 6.6) [12]. The *ortho*-selectivity is controlled by an outer sphere mechanism by hydrogen bond between the N–H proton of the substrate and the oxygen atom of the boryl ligand.

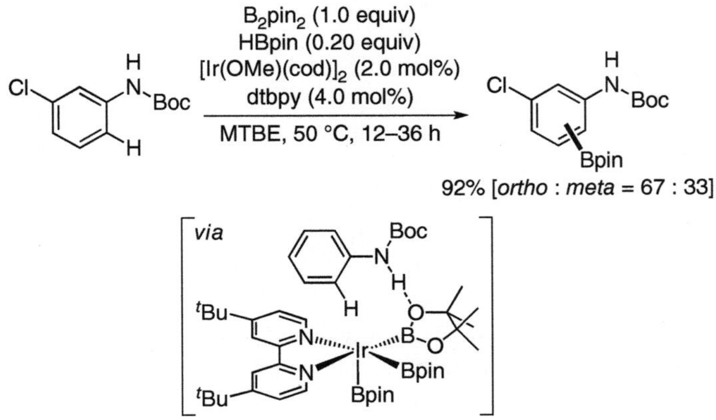

Scheme 6.6 Iridium-catalyzed *ortho*-selective C(sp^2)–H borylation by an outer sphere mechanism by hydrogen bond between the substrate and the boryl ligand.

Kuninobu and co-workers reported iridium-catalyzed *ortho*-selective C(sp^2)—H borylation of thioanisole derivatives (Scheme 6.7) [13]. The DFT calculations revealed that weak interactions, such as hydrogen bond between the C—H bond of the SCH_3 group and the oxygen atom of the boryl ligand, control the *ortho*-selectivity.

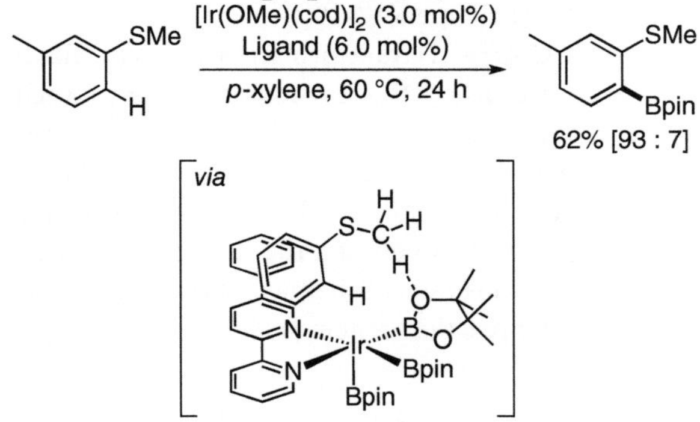

Scheme 6.7 Iridium-catalyzed *ortho*-selective C(sp^2)–H borylation using hydrogen bond between the C–H bond of the SCH_3 group and the oxygen atom of the boryl ligand.

Palladium-catalyzed *meta*-selective oxidative C(sp^2)—H alkenylation of aromatic compounds was achieved by directing template strategies [5, 14]. However, the strategies have drawbacks, such as the stoichiometric installation of directing groups into substrates and the removal of directing groups from products.

Jin, Xu, and co-workers developed a palladium-catalyzed *meta*-selective oxidative C(sp^2)—H alkenylation of aromatic compounds by utilizing hydrogen bond between the urea moiety of the ligand and the carbonyl group of substrates (Scheme 6.8) [15].

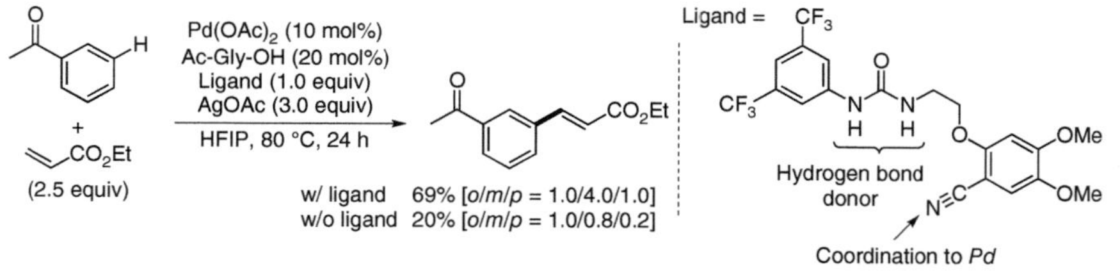

Scheme 6.8 Hydrogen-bond-controlled palladium-catalyzed *meta*-selective oxidative C(sp^2)−H alkenylation of aromatic amides.

Liang and co-workers designed ligands computationally and succeeded in hydrogen-bond-controlled *meta*-selective C(sp^2)—H borylation of aryl sulfones and aryl sulfonamides (Scheme 6.9) [16]. By changing the structure of the ligand, *para*-selective C(sp^2)—H borylation of aryl sulfones and aryl sulfonamides was also achieved (Scheme 6.9) [16]. The ligand has a urea moiety at the backbone of the phenanthloline-type ligand that enables C(sp^2)—H borylation at the remote position, such as *para*-position.

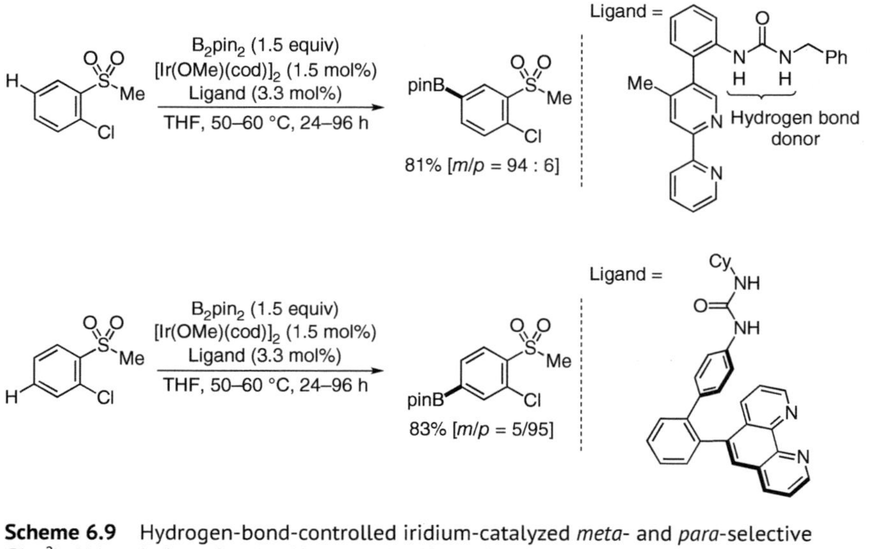

Scheme 6.9 Hydrogen-bond-controlled iridium-catalyzed *meta*- and *para*-selective C(sp^2)−H borylation of aryl sulfones and sulfonamides.

Liang and co-workers also designed two sets of ligands to achieve *meta*- and *para*-selective C(sp^2)—H borylation of aromatic compounds by DFT calculations (Scheme 6.10) [17]. In these reactions, *meta*- and *para*-selective C(sp^2)—H

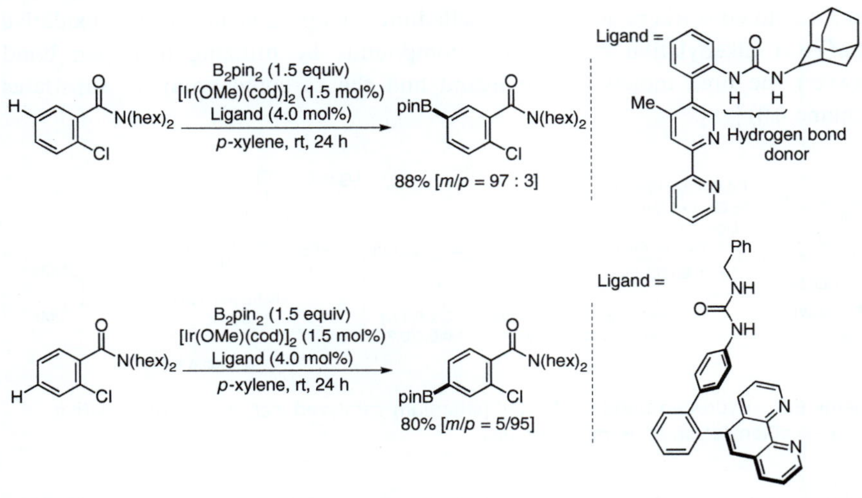

Scheme 6.10 Hydrogen-bond-controlled iridium-catalyzed *meta*- and *para*-selective C(sp^2)–H borylation of aryl and benzyl amides and aryl and benzyl phosphonates as substrates.

borylated products of aryl and benzyl amides and aryl and benzyl phosphonates were obtained using the similar ligand designs in Scheme 6.9.

Maiti and co-workers reported a hydrogen-bond-controlled palladium-catalyzed *meta*-selective C(sp^2)—H alkenylation of aromatic amines (Scheme 6.11) [18]. Experimental and computational studies suggest that the key to success in this reaction is the use of a strong hydrogen bond between the N–H bond of a substrate and the anionic functional group of a ligand.

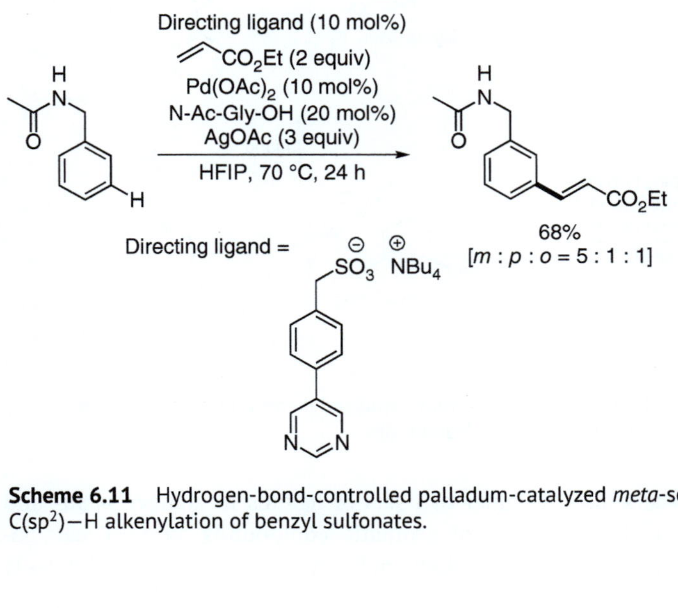

Scheme 6.11 Hydrogen-bond-controlled palladum-catalyzed *meta*-selective oxidative C(sp^2)–H alkenylation of benzyl sulfonates.

6.2.1.2 Controlled by Lewis Acid–Base Interaction

Although directing group-assisted C(sp^2)—H transformations generally occur at the *ortho*-position of aromatic substrates, iridium-catalyzed C(sp^2)—H borylation usually proceeds at the *meta*- and *para*-positions to avoid the steric repulsion between the bulky iridium catalytic species and substituent(s) of aromatic substrates. Sawamura et al. achieved *ortho*-selective C(sp^2)—H borylation using an iridium catalyst with a compact monodentate phosphine ligand supported on silica gel [19]. In contrast, we have found that *ortho*-selective C(sp^2)—H borylation of aromatic compounds proceeds in good yield under mild conditions at room temperature by utilizing the Lewis acid–base interaction between the aromatic substrate and the borylation reagent [20].

Another way to realize *ortho*-selective C(sp^2)—H borylation is to utilize non-covalent interaction between catalyst and substrate. In order for the *ortho*-C(sp^2)—H borylation to proceed, the non-covalent interaction must overcome the steric repulsion between the catalytically active species and the functional group of the substrate, which is stronger than the hydrogen bonding interaction. Therefore, Lewis acid–base interaction was used in the reaction and a new ligand was designed (Scheme 6.12) [21]. In this reaction, *ortho*-selective C(sp^2)—H borylation

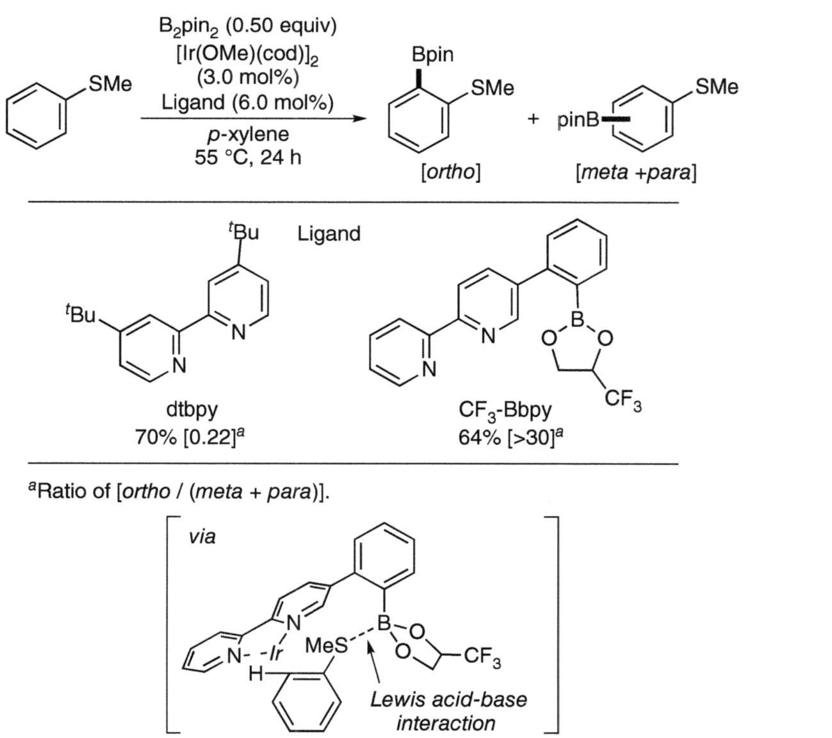

Scheme 6.12 *ortho*-Selective C(sp^2)–H borylation using Lewis acid–base interaction between catalyst and substrate.

of thioanisole derivatives proceeds using Lewis acid–base interaction between the boryl group of the ligand and the sulfur atom of the substrate.

Interestingly, the reaction sites can be switched by changing the ligands (Scheme 6.13) [21]. C(sp^2)—H borylation proceeds at the *meta*- and *para*-positions of aromatic ring without methylthio group of biphenyl methyl sulfide using dtbpy as a ligand. In contrast, the reaction occurred only at the *ortho*-position of the methylthio group of the substrate using the CF_3-Bbpy ligand.

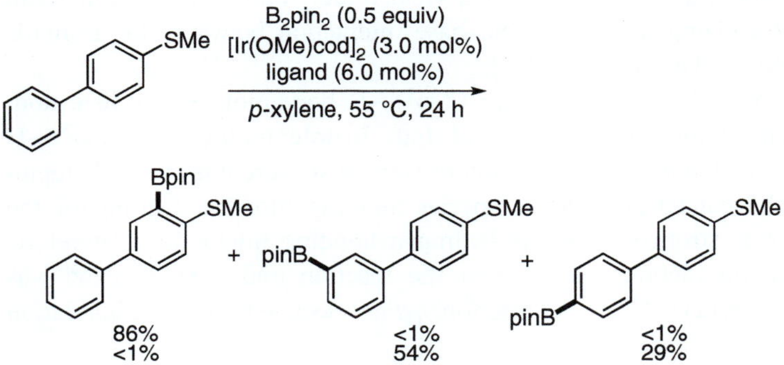

Scheme 6.13 Switching of reaction sites by changing ligands in iridium-catalyzed C(sp^2)—H borylation.

As mentioned earlier, C(sp^2)—H borylation at the *ortho*-position is usually difficult due to steric repulsion between the ligand of the catalyst and the substrate. However, in this catalytic system, the Lewis acid–base interaction overcomes the steric hindrance problem. The fact that the *ortho*-selectivity is not observed by a ligand with the boryl group at a different position on the ligand (Figure 6.4a) or when the reaction was carried out with a mixture of the bipyridine and borylbenzene (Figure 6.4b) indicates that the positional relationship between the bipyridyl moiety (iridium-bipyridyl complex moiety) and the boryl group should be defined appropriately to determine the reaction site.

Nakao and co-workers developed *meta*-selective C(sp^2)—H borylation of benzamides using Lewis acid–base interaction between the aluminum moiety of the

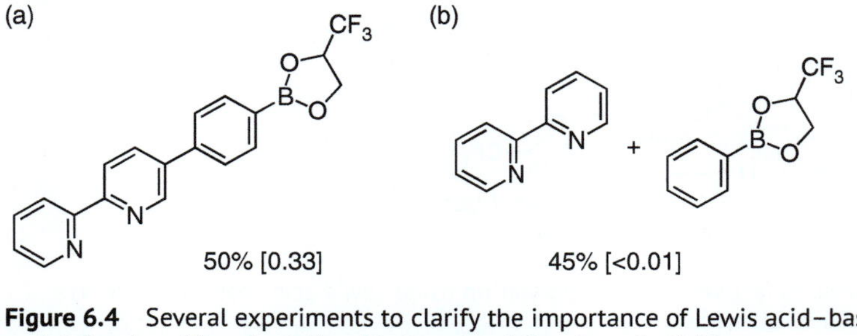

Figure 6.4 Several experiments to clarify the importance of Lewis acid–base interaction and defined positional relationship to control the site selectivity.

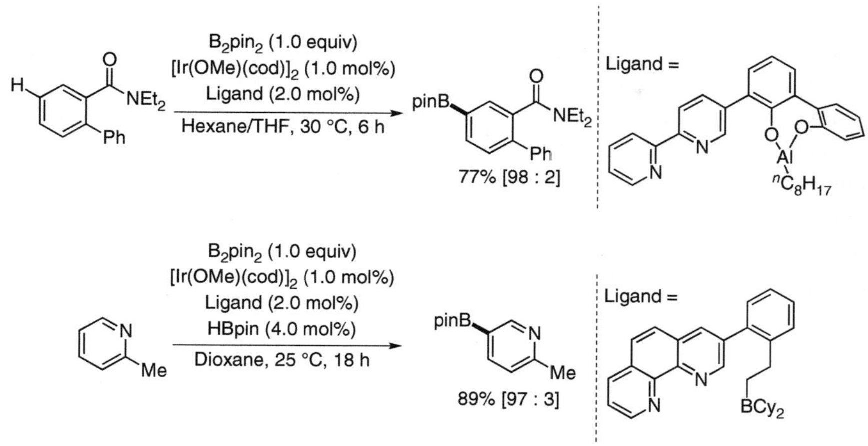

Scheme 6.14 Iridium-catalyzed *meta*-selective C(sp^2)—H borylation using Lewis acid–base interaction between catalyst and substrate.

ligand and the carbonyl group of substrates (Scheme 6.14) [22]. 3-Position-selective C(sp^2)—H borylation of pyridines was also achieved using phenanthroline-type ligand with a boryl group as a Lewis acid moiety (Scheme 6.14) [22].

Yu and co-workers succeeded in the development of 3-position-selective C(sp^2)—H alkenylation of pyridines with internal alkynes (Scheme 6.15) [23]. The site selectivity is controlled by the Lewis acid–base interaction between the Lewis acidic aluminum moiety of a nickel catalyst and Lewis basic pyridines. This reaction is effective for late-stage 3-position-selective C(sp^2)—H alkenylation of diverse complex pyridine motifs in bioactive molecules.

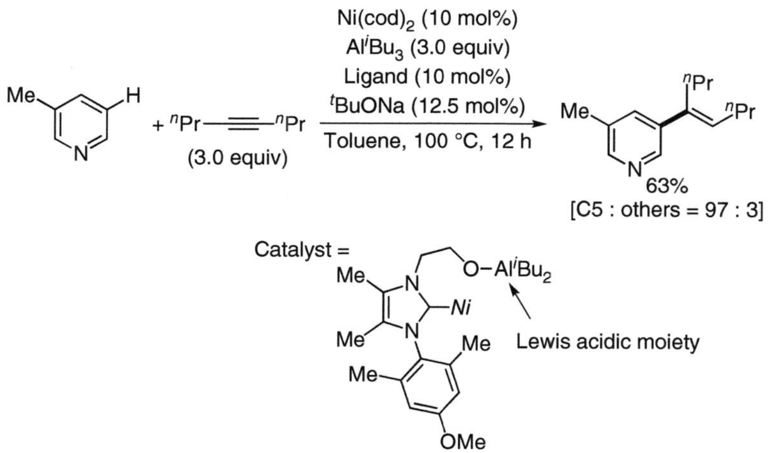

Scheme 6.15 Nickel-catalyzed 3-position-selective C(sp^2)—H alkenylation using Lewis acid–base interaction between catalyst and substrate.

Shi and co-workers reported nickel-catalyzed 2-position-selective C(sp^2)—H alkylation of pyridines with terminal alkenes (Scheme 6.16) [24]. The site selectivity is controlled by the Lewis acid–base interaction between the Lewis acidic aluminum moiety and Lewis basic pyridine. The site-selective C(sp^2)—H alkylation also occurred using azines such as pyrazines and pyrimidines. In this reaction, the reductive elimination step might be the rate-determining step.

Gramage–Doria and Poater developed iridium-catalyzed site-selective C(sp^2)—H borylation of pyridines and imidazoles (Scheme 6.17) [25]. The site selectivity is controlled by the Lewis acid–base interaction between the Lewis acidic

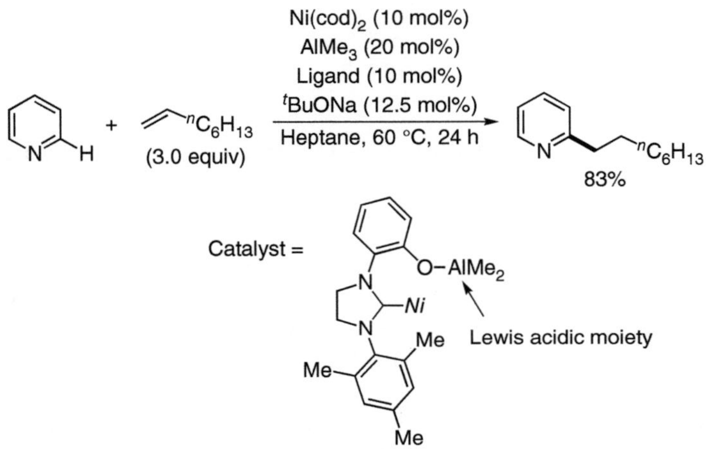

Scheme 6.16 Nickel-catalyzed 2-position-selective C(sp^2)—H alkylation using Lewis acid–base interaction between catalyst and substrate.

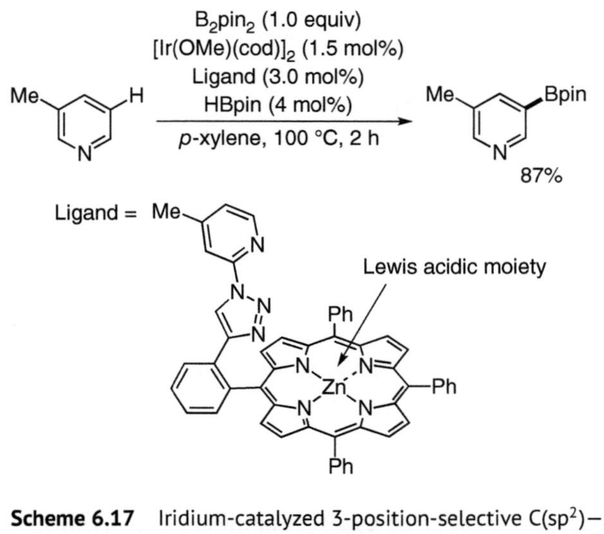

Scheme 6.17 Iridium-catalyzed 3-position-selective C(sp^2)—H borylation of pyridines using Lewis acid–base interaction between catalyst and substrate.

zinc-porphyrin moiety of the ligand and the Lewis basic pyridine of the imidazole substrate. The reaction mechanism was investigated by DFT calculations.

Nakao and Hiyama reported 2-position-selective C(sp^2)—H alkenylation of pyridine derivatives by nickel/Lewis acid catalysis (Scheme 6.18) [26]. In the case of using diorganozinc compounds as Lewis acid, monoalkenylated products are obtained. On the other hand, dienylated products are produced using $AlMe_3$ as a Lewis acid.

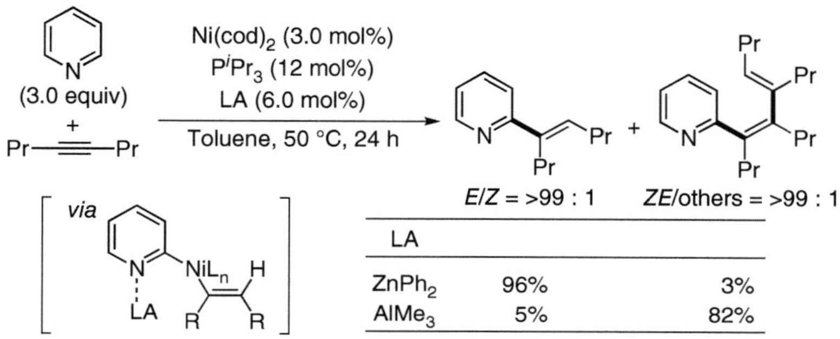

Scheme 6.18 Nickel-catalyzed 2-position-selective C(sp^2)—H alkenylation and dienylation of pyridines by Lewis acid–base interaction between the Lewis acidic additive and pyridine substrate.

Later, Nakao also achieved cooperative nickel/*N*-heterocyclic carbene (NHC) and aluminum-catalyzed *meta*- and *para*-selective C(sp^2)—H linear alkylation of anilides with alkents [27]. The alkylation proceeds mainly at the *meta*-position using a less bulky NHC ligand. In contrast, a bulky NHC ligand gives *para*-alkylated products.

Nakao reported 2-position-selective C(sp^2)—H alkylation of pyridine derivatives using a rhodium catalyst having a pincer ligand with an X-type aluminyl moiety bearing a tridentate NNN backbone (Scheme 6.19) [28]. IR spectrum and X-ray diffraction of the corresponding carbonyl complex indicate the considerable

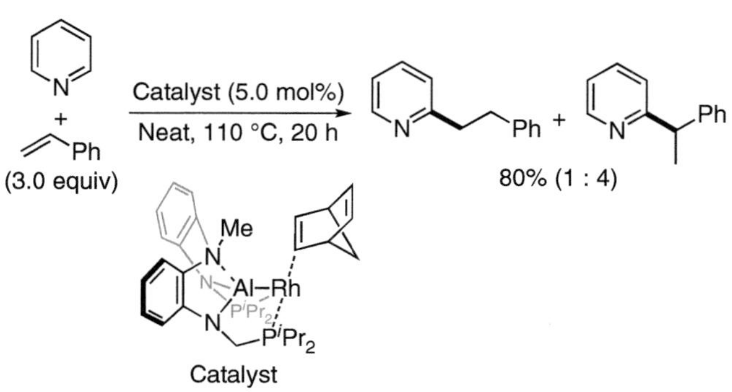

Scheme 6.19 Rhodium-catalyzed 2-position-selective C(sp^2)—H alkylation of pyridines by Lewis acid–base interaction between the aluminum center of the ligand and pyridine substrate.

σ-donating ability of the aluminyl ligand. The reaction proceeds via the coordination of pyridine to the aluminum center.

6.2.1.3 Controlled by Electrostatic Interaction

Phipps and co-workers succeeded in the development of iridium-catalyzed *meta*-selective C(sp^2)—H borylation of benzylammonium salts (Scheme 6.20) [9]. The site selectivity is controlled by electrostatic interaction between the anionic sulfonate group of the ligand and the cationic ammonium group of the substrate. Several control experiments revealed that the electrostatic interaction is essential to control the *meta*-selectivity.

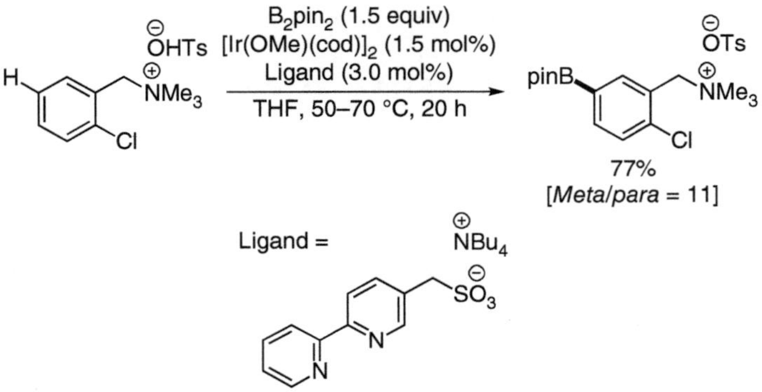

Scheme 6.20 Iridium-catalyzed *meta*-selective C(sp^2)—H borylation using electrostatic interaction between catalyst and substrate.

They also developed iridium-catalyzed *meta*-selective C(sp^2)—H borylation of flexible ammonium substrates (Scheme 6.21) [29]. The reaction proceeds using even flexible substrates with several freely rotable bonds even though the electrostatic interaction has low directionality. Datta and co-workers investigated the mechanism of the *meta*-selective C(sp^2)—H borylation of flexible ammonium substrates by DFT calculations [30].

They also reported iridium-catalyzed *meta*-selective C(sp^2)—H borylation of aromatic phosphonium salts (Scheme 6.22) [31]. The phosphonium salts show a higher *meta*-selectivity than the corresponding ammonium salts. They proposed that the high *meta*-selectivity is due to an attractive electrostatic interaction between the ligand and the substrate in the transition state of C(sp^2)—H borylation.

They achieved *para*-selective C(sp^2)—H borylation of aromatic compounds by utilizing the electrostatic interaction between the anionic substrate and bulky tetrabutylammonium cation (Scheme 6.23) [32]. In this reaction, the bulky cationic additive inhibits *meta*-C(sp^2)—H borylation by steric hindrance. It is necessary to introduce a substituent at the *ortho*-position of substrates to protect another *meta*-C(sp^2)—H bond and to increase the efficiency of the steric protection at the *meta*-C(sp^2)—H bond by the bulky cationic additive. At the same time, Maleczka Jr. and Smith also developed the similar *para*-selective C(sp^2)—H borylation of aromatic compounds using the same strategy [33].

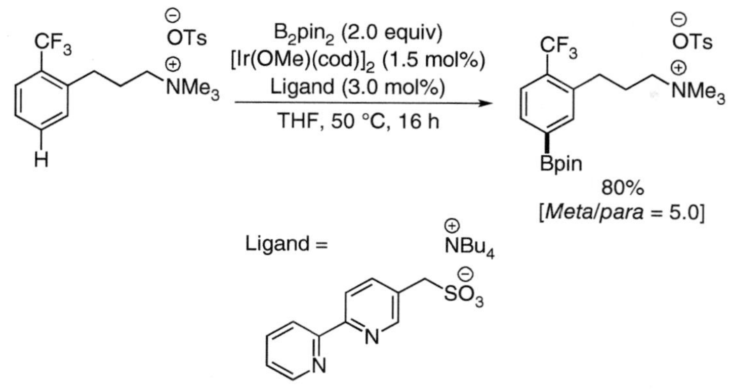

Scheme 6.21 Iridium-catalyzed *meta*-selective C(sp^2)–H borylation of flexible ammonium substrates using electrostatic interaction between catalyst and substrate.

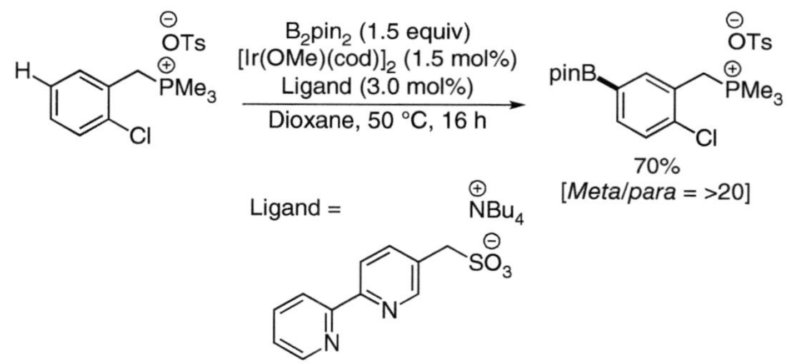

Scheme 6.22 Iridium-catalyzed *meta*-selective C(sp^2)–H borylation of aromatic phosphonium salts using electrostatic interaction between catalyst and substrate.

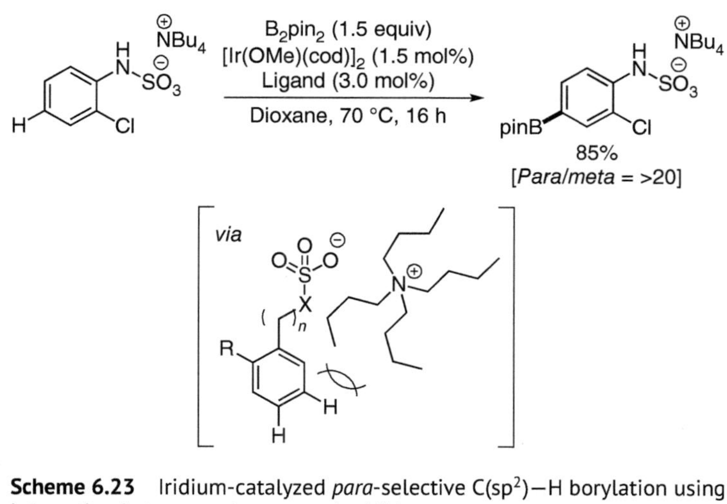

Scheme 6.23 Iridium-catalyzed *para*-selective C(sp^2)–H borylation using electrostatic interaction between substrate and additive.

Liang and co-workers developed *para*-selective C(sp^2)—H borylation of aromatic compounds using electrostatic interaction between anionic sulfamate ligand and cationic substrate with a quaternary ammonium or phosphonium group (Scheme 6.24) [34]. The key to success to exhibit *para*-selectivity is the use of a ligand with a biphenyl-phenanthroline skeleton. The origin of the *para*-selectivity was elucidated by mechanistic studies and computational analysis.

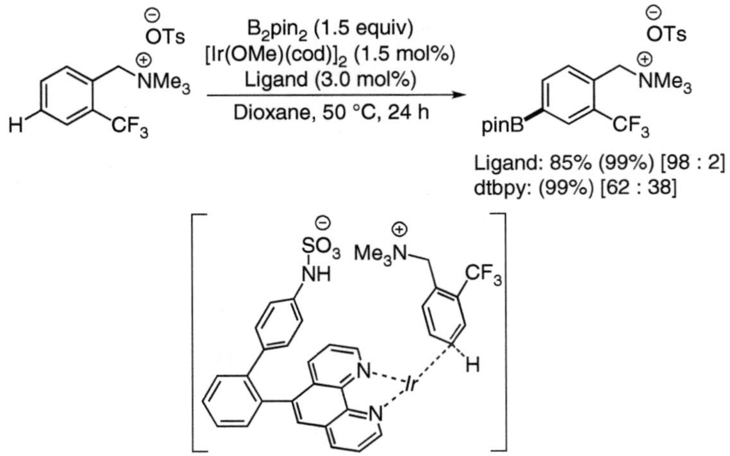

Scheme 6.24 Iridium-catalyzed *para*-selective C(sp^2)−H borylation using electrostatic interaction between ligand and substrate.

Phipps and co-worker investigated *para*-selective C(sp^2)—H borylation of benzylic ammonium salts using electrostatic interaction between the anionic ligand and the cationic substrate (Scheme 6.25) [35]. Although the reaction proceeds *para*-selectively, the site selectivity is not sufficient.

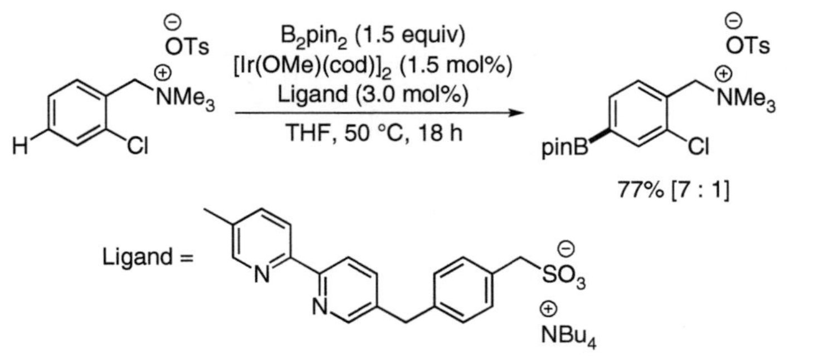

Scheme 6.25 Iridium-catalyzed *para*-selective C(sp^2)−H borylation using electrostatic interaction between ligand and substrate.

They applied electrostatic-interaction-controlled site-selective C(sp^2)—H borylation to enantioselective reactions. They achieved enantioselective *meta*-C(sp^2)—H borylation by constructing a chiral environment using electrostatic interaction between the anionic ligand and cationic chiral additive (Scheme 6.26) [36].

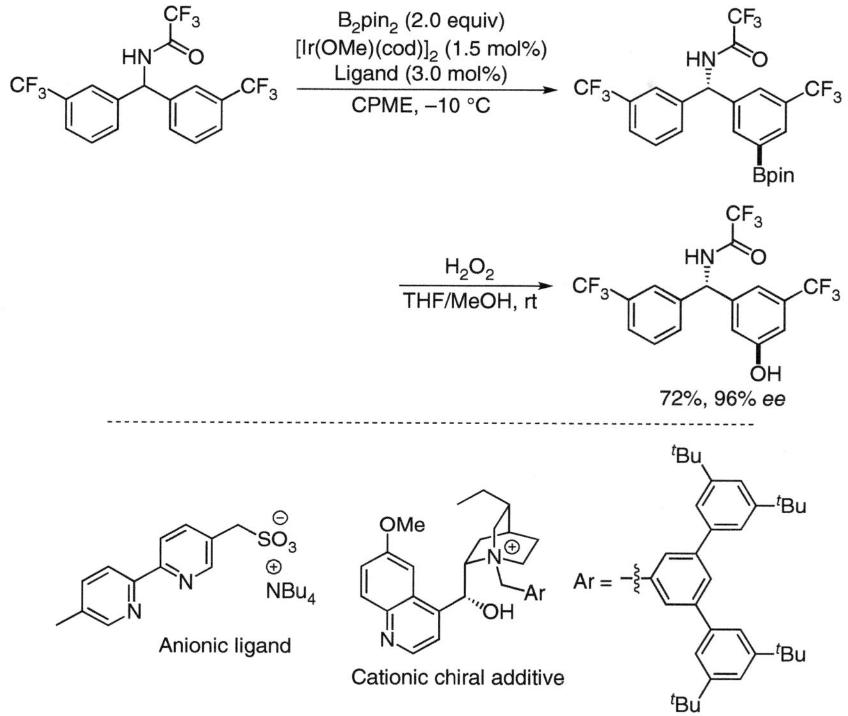

Scheme 6.26 Iridium-catalyzed enantioselective *meta*-selective C(sp^2)−H borylation using electrostatic interaction between anionic ligand and cationic chiral additive.

They also developed rhodium-catalyzed enantioselective benzylic C(sp^3)—H amination via the formation of rhodium-nitrene intermediate (Scheme 6.27) [37]. The enantioselectivity was achieved by the construction of a chiral environment by electrostatic interaction between the anionic rhodium catalyst and cationic chiral additive derived from quaternized chinchona alkaloids.

Chattopadhyay and co-workers reported iridium-catalyzed *meta*-selective C(sp^2)—H borylation of aromatic compounds (Scheme 6.28) [38]. The *meta*-selectivity is controlled by electrostatic interaction between the partially positive-charged phenanthroline ligand and the partially negative-charged carbonyl group of substrate

Maseras and van Gemmeren developed palladium-catalyzed *meta*-selective oxidative C(sp^2)—H alkenylation of benzylammonium salts using electrostatic interaction between the cationic ammonium group of substrate and the anionic carboxylate of the amino acid ligand (Scheme 6.29) [39]. The mechanistic studies proved the electrostatic interaction is responsible for the selectivity.

6.2.1.4 Controlled by Other Non-covalent Interactions

Yu and co-workers succeeded in palladium-catalyzed remote position-selective oxidative C(sp^2)—H alkenylation of 3-phenylpyridines using the coordination

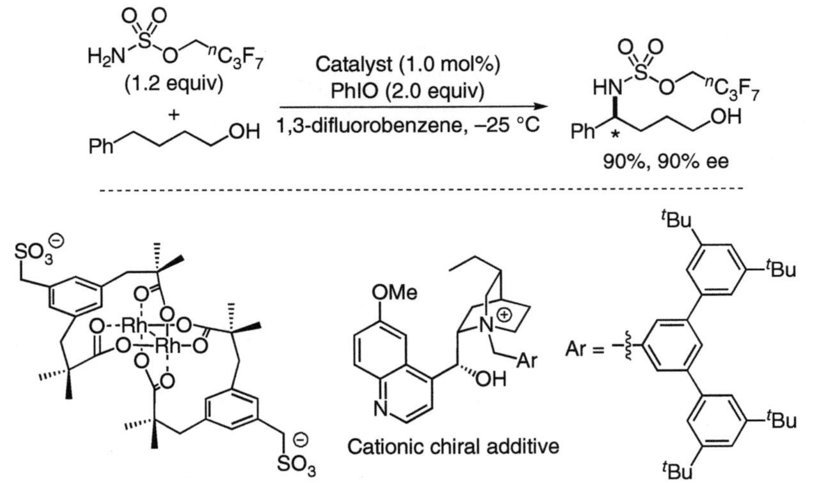

Scheme 6.27 Rhodium-catalyzed enantioselective benzylic C(sp^3)–H amination using electrostatic interaction between anionic catalyst and cationic chiral additive.

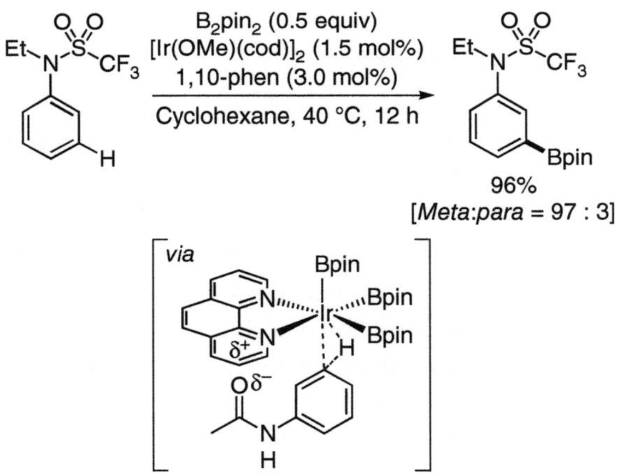

Scheme 6.28 Iridium-catalyzed *meta*-selective C(sp^2)–H borylation using electrostatic interaction between the ligand and the carbonyl group of substrate.

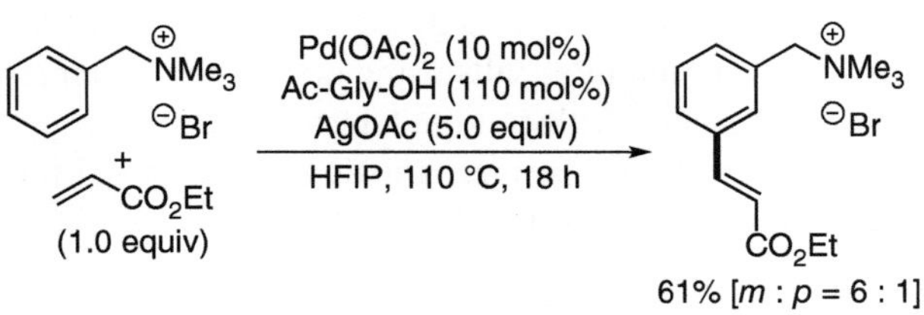

Scheme 6.29 Palladium-catalyzed *meta*-selective oxidative C(sp^2)–H alkenylation using electrostatic interaction between the cationic substrate and the anionic ligand.

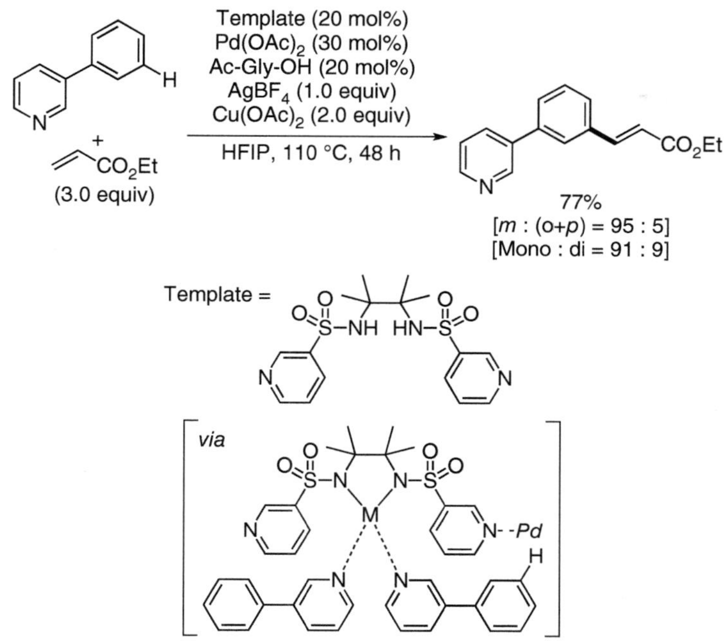

Scheme 6.30 Palladium-catalyzed *meta*-selective oxidative C(sp^2)—H alkenylation using bifunctional template.

of the pyridyl group of substrate and the metal center of the bifunctional template (Scheme 6.30) [40]. The two metal centers attached to the template work differently: one metal center anchors substrates near the catalyst and the other metal center cleaves the C(sp^2)—H bond and promotes further transformation. Maiti also developed palladium-catalyzed remote position-selective oxidative C(sp^2)—H alkenylation and alkylation of aromatic and heteroaromatic compounds [41].

Chattopadhyay and co-workers developed iridium-catalyzed *para*-selective C(sp^2)—H borylation of aromatic esters using L-shaped ligand (Scheme 6.31) [42]. The control experiments using a methoxy-protected ligand or a mixture of the L-shaped ligand and 18-crown-6 dramatically decreased the *para*-selectivity and revealed that the *para*-selectivity is controlled by the non-covalent interaction (C=O···K–O).

They also reported iridium-catalyzed *meta*-selective C(sp^2)—H borylation of aromatic amides using the same L-shaped ligand in Scheme 6.31 (Scheme 6.32) [43]. It is interesting that aromatic amides give *meta*-borylated products, whereas aromatic esters provide *para*-borylated products as shown in Scheme 6.31 using the same ligand. The *meta*-selectivity is achieved by non-covalent interaction (C=O···K–O) with the moderately distorted amide carbonyl group.

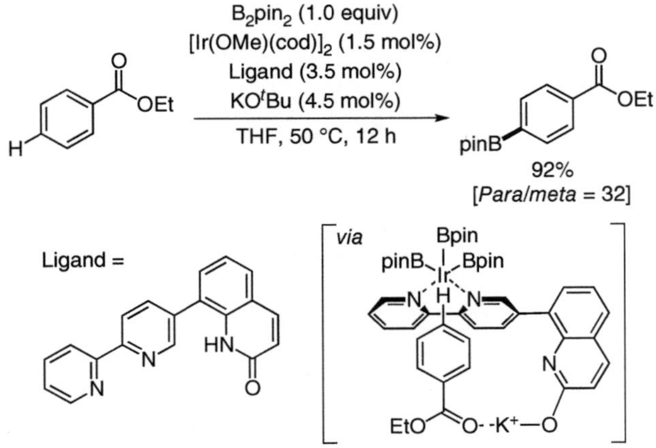

Scheme 6.31 Iridium-catalyzed *para*-selective $C(sp^2)$–H borylation using non-covalent interaction between the L-shaped ligand and the alkoxycarbonyl group of substrate.

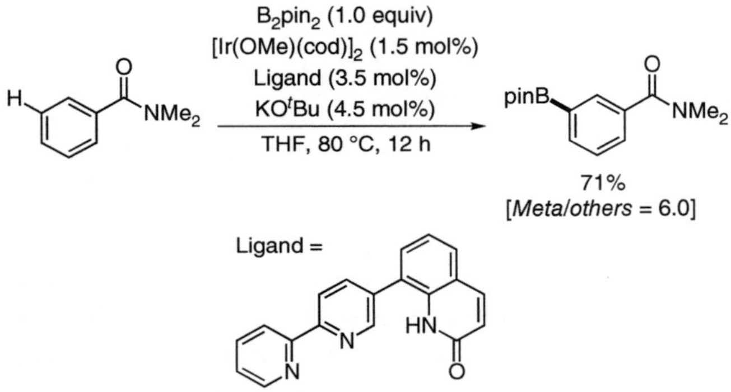

Scheme 6.32 Iridium-catalyzed *para*-selective $C(sp^2)$–H borylation using non-covalent interaction between the L-shaped ligand and the alkoxycarbonyl group of substrate.

Gramage-Doria and co-workers succeeded in iridium-catalyzed 3-position-selective $C(sp^2)$—H borylation of pyridines using coordination bond between pyridine substrate and the metal center of zinc porphyrin (Scheme 6.33) [44].

Houk, Yu, and co-workers succeeded in palladium-catalyzed C6-selective oxidative $C(sp^2)$—H alkenylation of quinolines (Scheme 6.34) [45]. The site selectivity is controlled by the coordination bond between quinoline and the palladium center of the template. C6-selective alkynylation also proceeded.

Jiménex-Osés and Mascareñas reported iridium-catalyzed *ortho*-selective $C(sp^2)$—H borylation of aromatic amides (Scheme 6.35) [46]. Computational studies suggest that the site selectivity is controlled by outer-sphere interaction between the carbonyl group of the substrate and the CF_3-substituted aryl ring of the bipyridine ligand.

Recently, Ilies, Asako, and co-workers reported iridium-catalyzed site-selective $C(sp^2)$—H borylation of 1,2-, 1,3-, and 1,4-disubstituted aromatic compounds

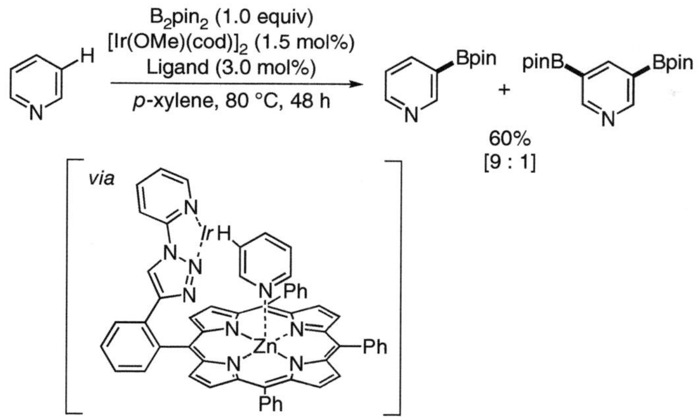

Scheme 6.33 Iridium-catalyzed 3-position-selective C(sp^2)–H borylation of pyridines using coordination bond between pyridine and the metal center of zinc porphirin.

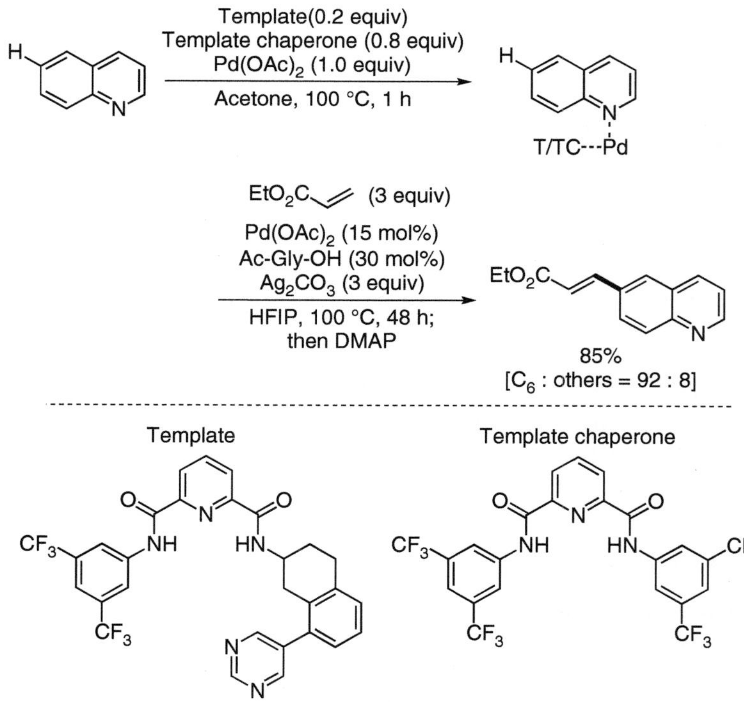

Scheme 6.34 Palladium-catalyzed C6-selective oxidative C(sp^2)–H alkenylation of quinolines using coordination bond between quinoline and the palladium center of template.

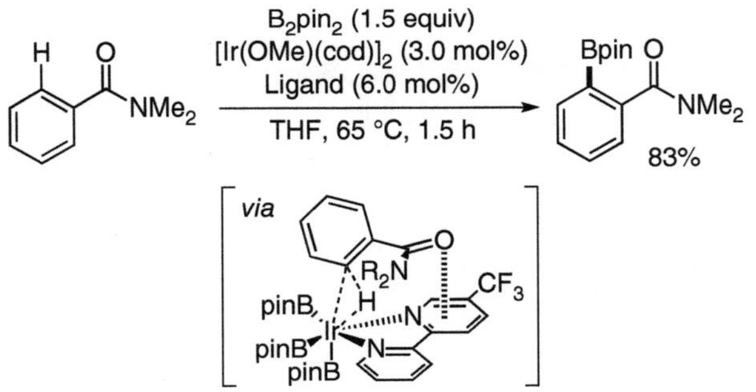

Scheme 6.35 Iridium-catalyzed *ortho*-selective C(sp^2)–H borylation of aromatic amides using outer-sphere dispersion interaction between the carbonyl group of the substrate and the CF_3-bipyridine ligand.

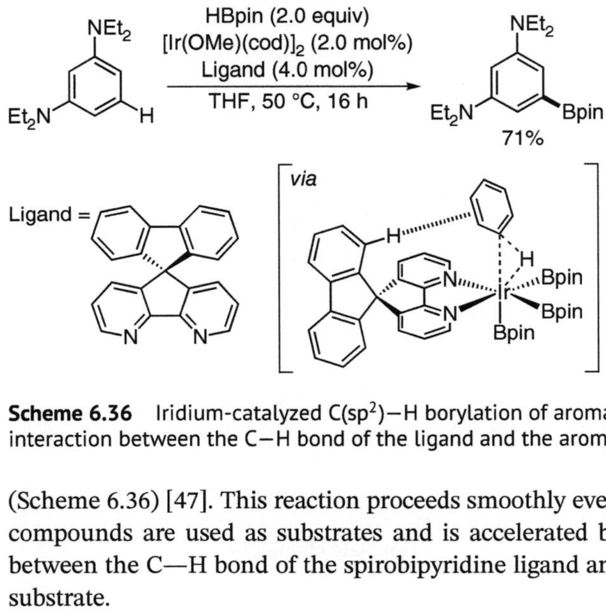

Scheme 6.36 Iridium-catalyzed C(sp^2)–H borylation of aromatic compounds by C–H/π interaction between the C–H bond of the ligand and the aromatic ring of the substrate.

(Scheme 6.36) [47]. This reaction proceeds smoothly even if electron-rich aromatic compounds are used as substrates and is accelerated by the C—H/π interaction between the C—H bond of the spirobipyridine ligand and the aromatic ring of the substrate.

6.2.2 C(sp^3)–H Transformations

6.2.2.1 Controlled by Hydrogen Bond

Although the benzylic C(sp^3)—H bond of ibuprofen is oxidized by dimanganese catalyst **Y** (favored product), decarboxylation and successive oxidation also proceeded to give disfavored product (Scheme 6.37). Brudvig, Crabtree, and co-workers succeeded in suppressing the formation of the disfavored product and producing the favored product selectively (Scheme 6.37) [48].

Costas, Stefano, and Olivo developed remote position-selective C(sp^3)—H oxidation of linear alkyl ammonium salts (Scheme 6.38) [49]. The site selectivity is

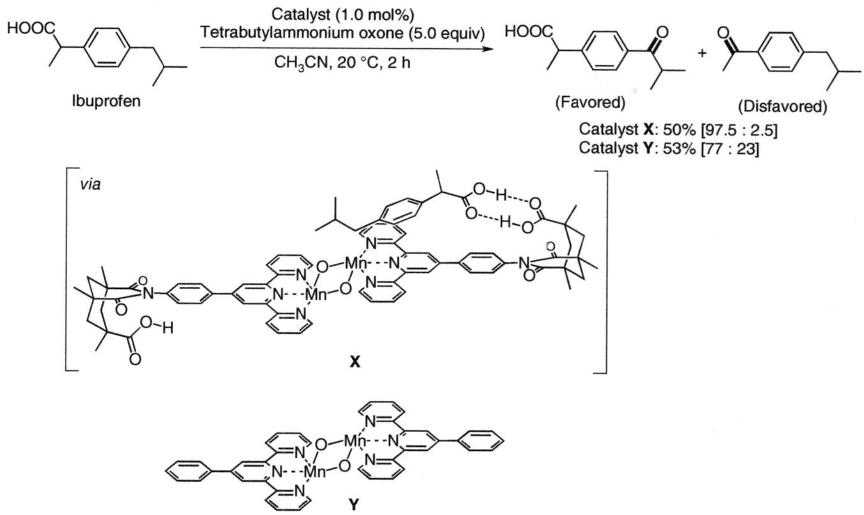

Scheme 6.37 Manganese-catalyzed site-selective C(sp^3)−H oxidation of carboxylic acids using hydrogen bond between the hydroxycarbonyl group of the substrate and the hydroxycarbonyl group of the ligand.

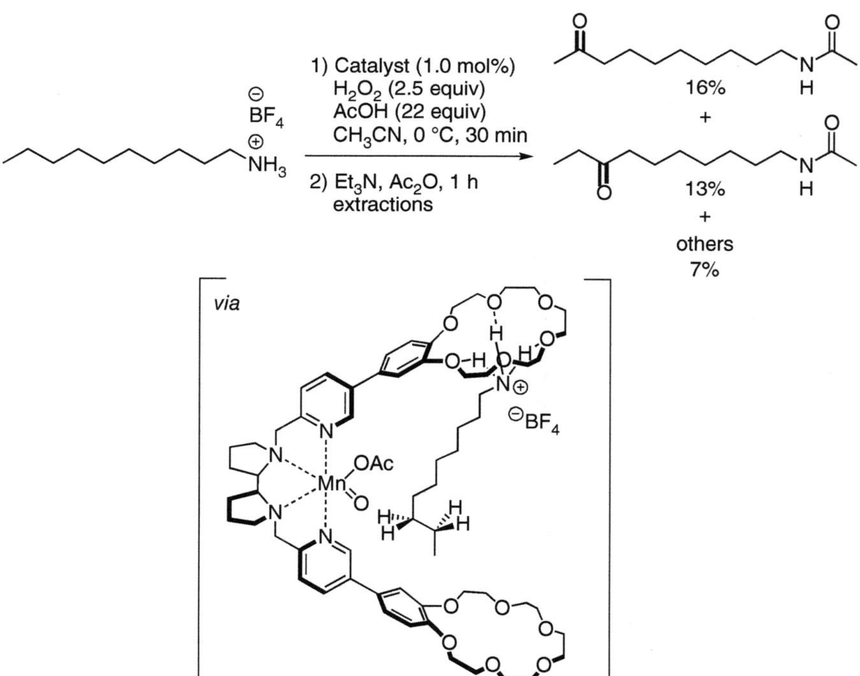

Scheme 6.38 Manganese-catalyzed remote position-selective C(sp^3)−H oxidation of linear alkylammonium salts using hydrogen bond between the ammonium group of the substrate and the crown ether moiety of the ligand.

controlled by hydrogen bond between the ammonium group of the substrate and the crown ether moiety of the ligand. Although the reaction proceeded remote position-selectively, the yield and site selectivity are not satisfactory.

They also reported site-selective C(sp^3)—H oxidation of steroids using the same catalyst in Scheme 6.38 (Scheme 6.39) [50]. The manganese catalyst gave the major product in 25% yield with 57% selectivity. On the other hand, a catalyst without having the crown ether moieties produced 10 products in 11% total yield.

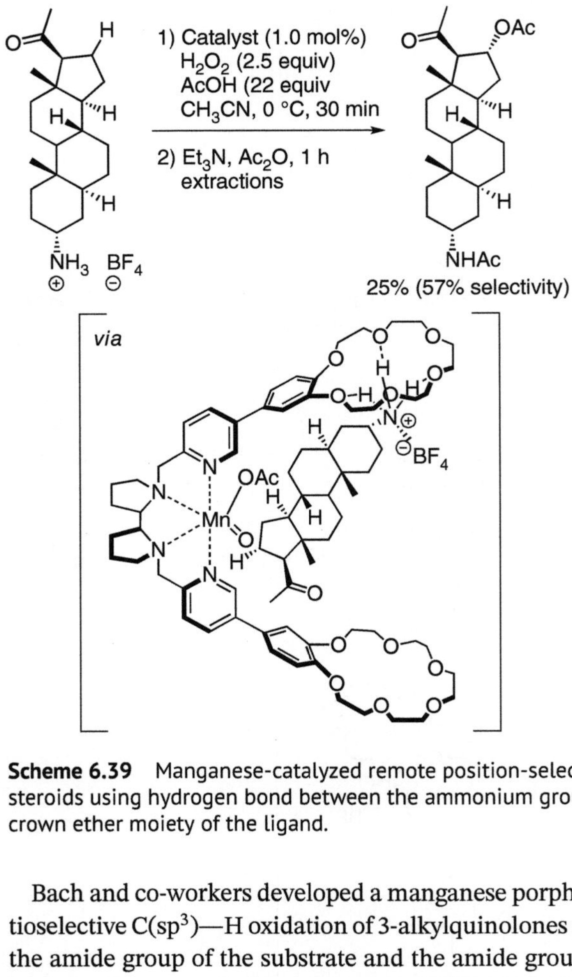

Scheme 6.39 Manganese-catalyzed remote position-selective C(sp^3)–H oxidation of steroids using hydrogen bond between the ammonium group of the substrate and the crown ether moiety of the ligand.

Bach and co-workers developed a manganese porphyrin-catalyzed site- and enantioselective C(sp^3)—H oxidation of 3-alkylquinolones using hydrogen bond between the amide group of the substrate and the amide group of the ligand (Scheme 6.40) [51]. In the case of using manganese(III)-5,10,15,20-tetrakis(pentafluorophenyl) porphyrin chloride (Mn[TPFPP]Cl), racemic C(sp^3)—H hydroxylation occurs at the benzylic position of ethyl group at the 7-position.

6.2.2.2 Controlled by Electrostatic Interaction

Kuninobu, Torigoe, and co-worker succeeded in *ortho*-selective benzylic C(sp^3)—H alkylation of anilinium salts (Scheme 6.41) [52]. The site selectivity is controlled by

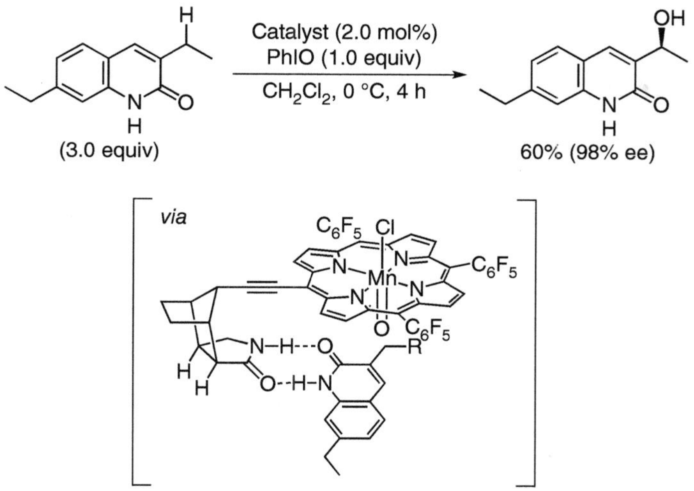

Scheme 6.40 Manganese-catalyzed site- and enantioselective C(sp^3)–H oxidation of 3-alkylquinolones using hydrogen bond between the amide group of the substrate and the amide group of the ligand.

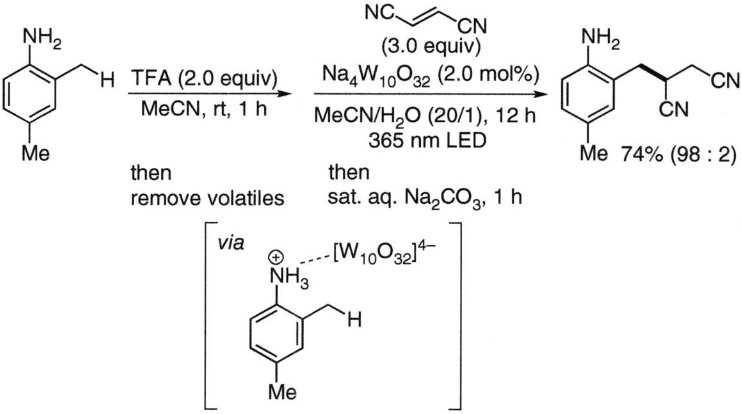

Scheme 6.41 Decatungstate-catalyzed site-selective C(sp^3)–H alkylation of anilinium salts using electrostatic interaction between the anionic catalyst and cationic substrate.

the electrostatic interaction between the anionic decatungstate photocatalyst and the cationic ammonium group of the substrate. In this reaction, the site selectivity can be controlled by utilizing the negative charge possessed by the catalyst, without the need to introduce a substrate recognition site on the catalyst.

The modification of amino acids and peptides is often carried out by the conversion of functional groups at the side-chain, [53] and there are few examples of amino acid or peptide modification by C—H transformations [54]. Kuninobu, Torigoe, and co-worker applied the reaction in Scheme 6.41 to the site-selective C(sp^3)—H alkylation of amino acids and peptides (Scheme 6.42) [55].

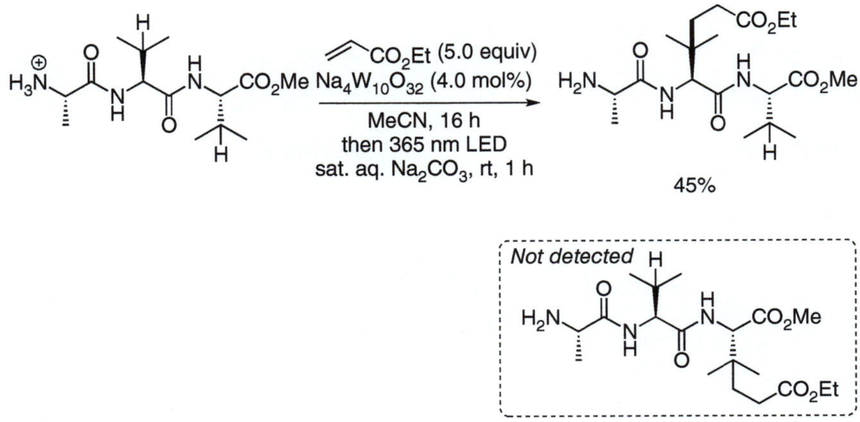

Scheme 6.42 Decatungstate-catalyzed site-selective C(sp^3)–H alkylation of peptides using electrostatic interaction between the anionic catalyst and cationic substrate.

6.2.2.3 Controlled by Other Non-covalent Interactions

Breslow and co-workers succeeded in a manganese-porphyrin-catalyzed site-selective C(sp^3)—H hydroxylation of a steroid derivative (Scheme 6.43) [56]. The site selectivity is controlled by the supramolecular interaction between the two β-cyclodextrin moieties of the catalyst and two 4-*tert*-butylphenyl groups of the substrate. In this reaction, the special substituents are required to be introduced into the steroid before the reaction, and the substituents are removed after the reaction.

6.3 Acceleration of Reactions and Substrate and Functional Group Specificities

Enzymes skillfully utilize non-covalent interactions such as hydrogen bond to recognize and trap specific substrates and bring them close to the reaction site in an appropriate orientation to rapidly carry out the desired reaction. Kuninobu and co-workers hypothesized that acceleration of reaction and substrate specificity should be observed in the same way as with enzymes if the target C—H transformation proceeds by recognizing and capturing the substrate through non-covalent interactions such as hydrogen bonding.

In the reaction shown in Schemes 6.2 and 6.3, it was assumed that the catalyst recognizes and traps the substrate by hydrogen bonding, and therefore, it was thought that there must be an acceleration effect of the reaction as in the case of enzymes. Taking ligand **L2** as the standard, Kuninobu and co-workers compared the initial reaction rates in the reactions with each ligand (Scheme 6.44) [57]. It was found that the reaction was accelerated about threefold using ligand **L1** (Cy-Ubpy)

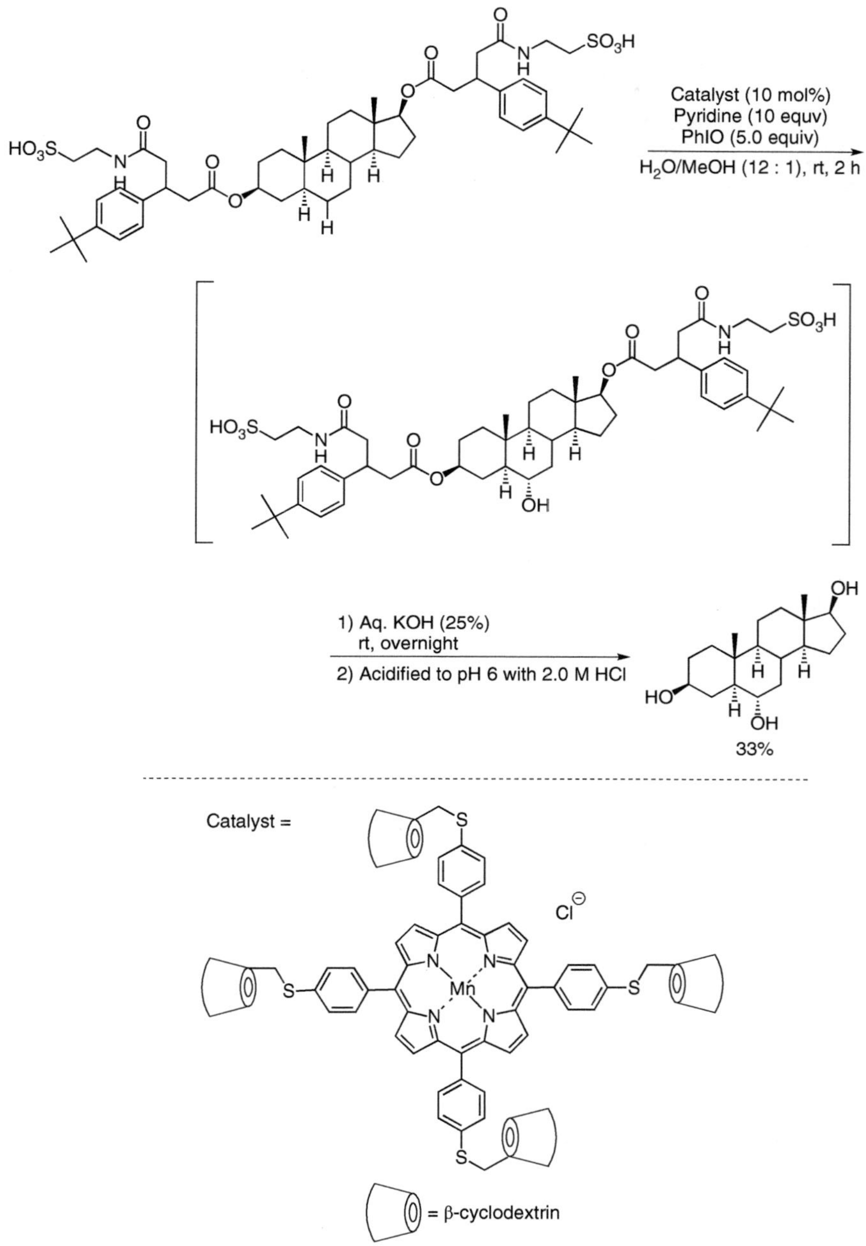

Scheme 6.43 Manganese Porphyrin-catalyzed site-selective C(sp^3)–H hydroxylation of a steroid derivative using supramolecular interaction between the two β-cyclodextrin moieties of the catalyst and the substrate.

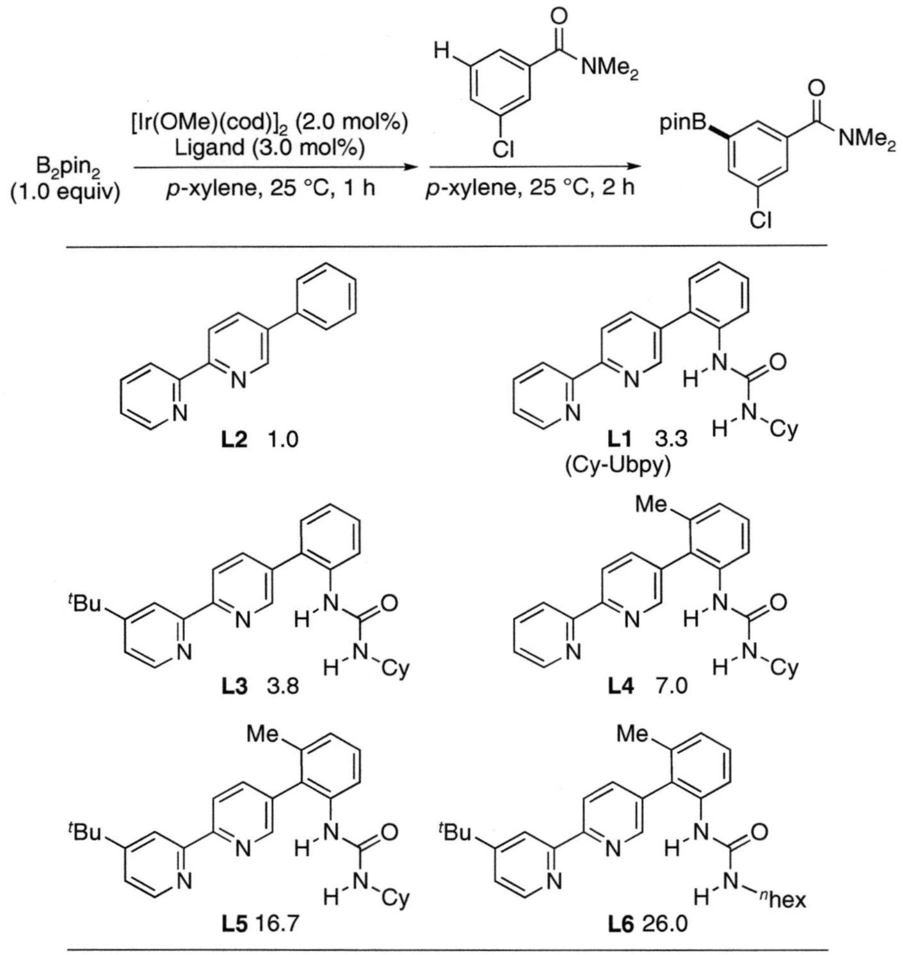

Scheme 6.44 Acceleration of reaction.

used in Scheme 6.3. Since it is known that the introduction of an electron-donating group to a bipyridyl ligand accelerates the C(sp^2)—H borylation, Kuninobu and co-workers synthesized ligand **L3** with a ^{t}Bu group on the pyridine ring and performed the reaction, and the reaction was slightly accelerated compared with ligand **L1**. As shown in Scheme 6.3, from the molecular model, it was assumed that the bipyridyl moiety is twisted with respect to the urea moiety, and the C(sp^2)—H borylation proceeds in a twisted state. In order to stabilize the twisted structure, ligand **L4** with a methyl group on the benzene ring was synthesized, and the reaction was accelerated compared with that of ligand **L1**. Ligand **L5**, which incorporates the above elements of ligands **L3** and **L4**, was found to further accelerate the reaction. The cyclohexyl group in the urea moiety of ligand **L5** is sterically bulky, making it difficult to form hydrogen bonds with the substrate, which may work against the acceleration of the reaction. Therefore, Kuninobu and co-workers developed ligand **L6** with a sterically smaller hexyl group in place of the cyclohexyl group. As expected, the reaction was accelerated compared with that of ligand **L5**, and the C(sp^2)—H borylation proceeded 26 times faster than with the reference ligand **L2**.

In the C(sp^2)—H borylation reaction of Scheme 6.3, amides with two substituents on the nitrogen atom gave *meta*-selective borylation products in good yields, whereas amides with only one substituent on the nitrogen atom hardly proceeded with the reaction. By using the newly developed ligand **L6**, the *meta*-borylation product was obtained in good yield even with amides having only one substituent on the nitrogen atom (Scheme 6.45) [57].

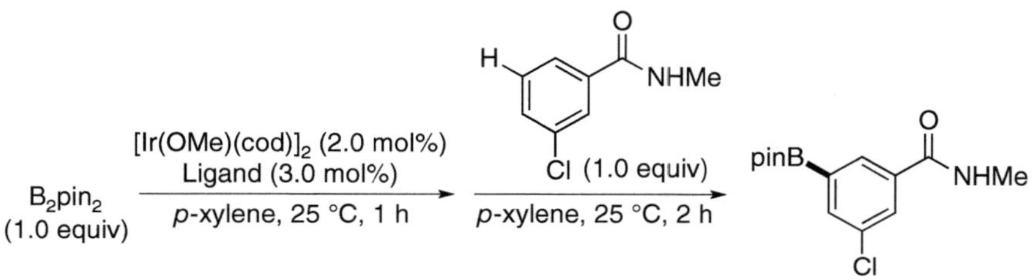

Ligand	Yield
L2	8%
L1	20%
L5	55%
L6	85%

Scheme 6.45 Improvement of yield.

To see if this catalytic system would allow for functional group specificity, C(sp^2)—H borylation was performed with 1 : 1 mixture of aromatic amide **1** and aromatic ester **2** (Scheme 6.46) [57]. For ligand **L2** without a urea moiety, the reaction proceeded with both substrates **1** and **2** to the same extent, and the total yield of borylated products was low. In contrast, when the reaction was carried out using ligand **L6** with a urea moiety, it was found that the reaction preferentially occurred with aromatic amide **1**. These results suggest that the urea moiety in the ligand selectively recognizes and traps the amide group rather than the ester group, and the reaction proceeds selectively at aromatic amide **2**.

The functional group specificity observed in Scheme 6.46 was more pronounced when C(sp^2)—H borylation was performed on aromatic substrates with both amide and ester groups in the molecule (Scheme 6.47) [57]. When Kuninobu and co-workers examined which *meta*-position was preferentially used for the reaction in terms of the amide and ester groups, they found that the reaction with ligand **L2**, which does not have a urea moiety, did not show selectivity, whereas the reaction with ligand **L6**, which has a urea moiety, proceeded almost exclusively at the *meta*-position of the amide group. This result, as in Scheme 6.46, can also be interpreted as the urea moiety selectively recognizing the amide group, and the C(sp^2)—H borylation reaction proceeded selectively at the *meta*-position of the amide.

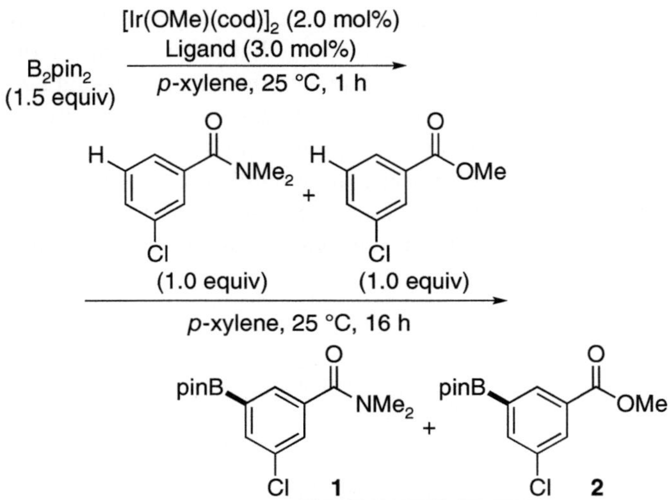

Ligand	1	2	1/2
L2	19%	17%	1.1
L1	69%	16%	4.3
L5	80%	12%	6.7
L6	94%	9%	10.4

Scheme 6.46 Expression of functional group specificity-1.

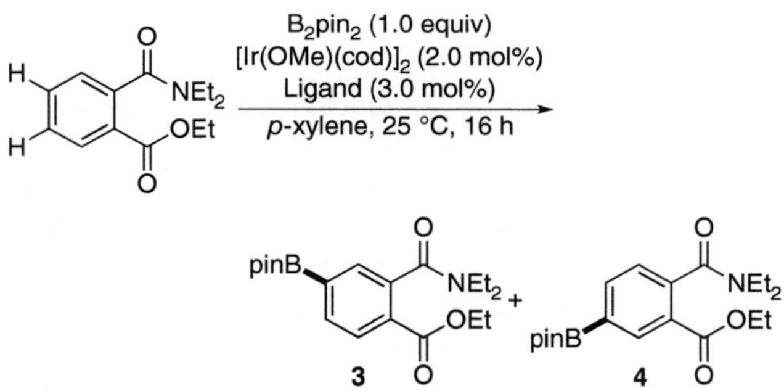

Ligand	3	4	3/4
L2	18%	8%	2.3
L1	68%	10%	6.8
L5	64%	3%	21
L6	66%	1%	66

Scheme 6.47 Expression of functional group specificity-2.

Next, to confirm whether substrate specificity was observed, two aromatic amides with different substituents on the nitrogen were mixed in a 1 : 1 ratio and subjected to C(sp^2)—H borylation (Scheme 6.48) [57]. With urea-free ligand **L2**, the two substrates could not be distinguished, and the C(sp^2)—H borylation proceeded in similar yields for both substrates. On the other hand, when reactions were carried

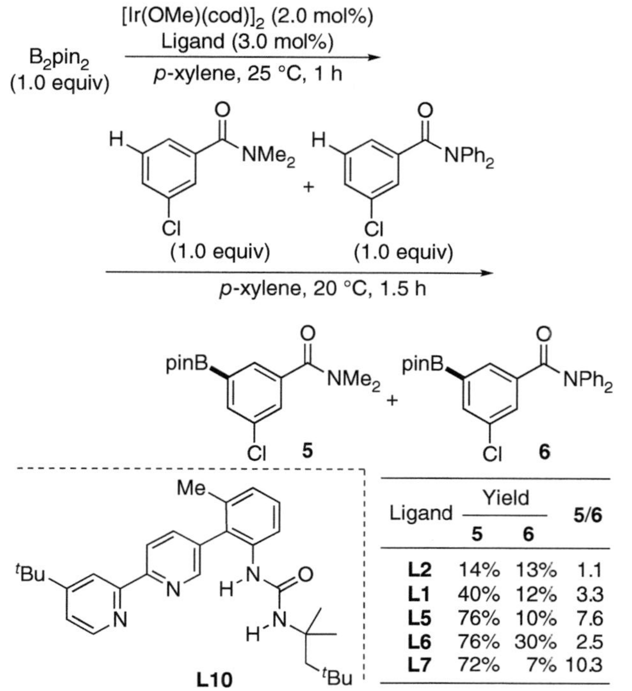

Ligand	Yield 5	Yield 6	5/6
L2	14%	13%	1.1
L1	40%	12%	3.3
L5	76%	10%	7.6
L6	76%	30%	2.5
L7	72%	7%	10.3

Scheme 6.48 Expression of substrate specificity.

out with ligands **L1**, **L5**, and **L6** having urea moieties, the reactions proceeded selectively with aromatic amides having two methyl groups on the nitrogen, but the selectivity was not satisfactory. In order to distinguish the two methyl and phenyl groups on the nitrogen atom of the aromatic amide not only electronically but also stereologically, Kuninobu and co-workers synthesized ligand **L7** in which the cyclohexyl group of the urea moiety was replaced with a bulkier substituent and performed the reaction. The reaction proceeded selectively with aromatic amides having two methyl groups on the nitrogen atom.

These investigations have revealed that our small molecular weight catalyst can accelerate reactions and exhibit substrate specificity like enzymes.

6.4 Summary and Conclusions

In this chapter, site-selective C—H transformations by utilizing non-covalent interactions such as hydrogen bond, Lewis acid–base interaction, and electrostatic interaction between catalyst and substrate as a method to control site selectivity in the C—H transformations (non-covalent method), which is different from the conventional directing group method. It is also demonstrated that the use of a catalyst with a urea moiety accelerates the reaction, improves the yield, and develops functional group and substrate specificity. It is expected that the above-mentioned research results must provide a solution to some of the problems with the directing group method and propose a guideline for future catalyst design in the research field of C—H transformations. It is hoped that further progress will be made in related research, that the non-covalent bond method will become a site-selective method that can play a twin role with the directing group method, and that the C—H transformations will become a truly practical synthetic organic reaction.

For more information on site-selective C—H bond conversion reactions using non-covalent interactions, please refer to the review articles [58].

References

1 (a) Rouquet, G. and Chatani, N. (2013). *Angew. Chem. Int. Ed.* 52: 11726–11743. (b) Rej, S., Ano, Y., and Chatani, N. (2020). *Chem. Rev.* 120: 1788–1887. (c) Rej, S., Das, A., and Chatani, N. (2021). *Coord. Chem. Rev.* 431: 213683. (d) Liu, B., Romine, A.M., Rubel, C.Z. et al. (2021). *Chem. Rev.* 121: 14957–15074.

2 (a) Hu, X.-Q., Chen, J.-R., and Xiao, W.-J. (2017). *Angew. Chem. Int. Ed.* 56: 1960–1962. (b) Goswami, N. and Maiti, D. (2020). *Isr. J. Chem.* 60: 303–312. (c) Kumar, G., Pradhan, S., and Chatterjee, I. (2020). *Chen. Asian J.* 15: 651–672. (d) Chen, H. and Yu, S. (2020). *Org. Biomol. Chem.* 18: 4519–4532. (e) Sarkar, S., Cheung, K.P.S., and Gevorgyan, V. (2020). *Chem. Sci.* 11: 12974–12993. (f) Zhang, J. and Rueping, M. (2023). *Chem. Soc. Rev.* 52: 4099–4120.

3 (a) Cho, J.-Y., Tse, M.K., Holmes, D. et al. (2002). *Science* 295: 305–308. (b) Ishiyama, T., Takagi, J., Ishida, K. et al. (2002). *J. Am. Chem. Soc.* 124: 390–391. (c) Saito, Y., Segawa, Y., and Itami, K. (2015). *J. Am. Chem. Soc.* 137: 5193–5198. (d) Yang, L., Semba, K., and Nakao, Y. (2017). *Angew. Chem. Int. Ed.* 56: 4853–4857. (e) Ramadoss, B., Jin, Y., Asako, S., and Ilies, L. (2022). *Science* 375: 658–663.

4 (a) Phipps, R.J. and Gaunt, M.J. (2009). *Science* 323: 1593–1597. (b) Saidi, O., Marafie, J., Ledger, A.E.W. et al. (2011). *J. Am. Chem. Soc.* 133: 19298–19301. (c) Hofmann, N. and Ackermann, L. (2013). *J. Am. Chem. Soc.* 135: 5877–5884. (d) Chen, M.S. and White, M.C. (2007). *Science* 318: 783–787.

5 (a) Leow, D., Li, G., Mei, T.-S., and Yu, J.-Q. (2012). *Nature* 486: 518–522. (b) Tang, R.-Y., Li, G., and Yu, J.-Q. (2014). *Nature* 507: 215–220.

6 Kuninobu, Y., Ida, H., Nishi, M., and Kanai, M. (2015). *Nat. Chem.* 7: 712–717.

7 (a) Unnikrishnan, A. and Sunoj, R.B. (2019). *Chem. Sci.* 10: 3826–3835. (b) Chang, W., Wang, Y., Chen, Y. et al. (2023). *Chin. Chem. Lett.* 34: 107879.

8 Davis, H.J., Genov, G.R., and Phipps, R.J. (2017). *Angew. Chem. Int. Ed.* 56: 13351–13355.

9 Davis, H.J., Mihai, M.T., and Phipps, R.J. (2016). *J. Am. Chem. Soc.* 138: 12759–12762.

10 Bai, S.-T., Bheeter, C.B., and Reek, J.N.H. (2019). *Angew. Chem. Int. Ed.* 58: 13039–13043.

11 Liu, C., Zhang, L., Li, L., and Lei, M. (2021). *J. Org. Chem.* 86: 16858–16866.

12 (a) Roosen, P.C., Kallepalli, V.A., Chattopadhyay, B. et al. (2012). *J. Am. Chem. Soc.* 134: 11350–11353. (b) Preshlock, S.M., Plattner, D.L., Maligres, P.E. et al. (2013). *Angew. Chem. Int. Ed.* 52: 12915–12919. (c) Smith, M.R. III, Bisht, R., Haldar, C. et al. (2018). *ACS Catal.* 8: 6126–6223.

13 Zeng, J., Naito, M., Torigoe, T. et al. (2020). *Org. Lett.* 22: 3485–3489.

14 (a) Bera, M., Modak, A., Patra, T. et al. (2014). *Org. Lett.* 16: 5760–5763. (b) Bera, M., Maji, A., Sahoo, S.K., and Maiti, D. (2015). *Angew. Chem. Int. Ed.* 54: 8515–8519.

15 Li, G., Yan, Y., Zhang, P. et al. (2021). *ACS Catal.* 11: 10460–10466.

16 Wang, Y., Chang, W., Qin, S. et al. (2022). *Angew. Chem. Int. Ed.* 61: e202206797.

17 Chang, W., Chen, Y., Lu, S. et al. (2022). *Chem* 8: 1775–1788.

18 Goswami, N., Sinha, S.K., Mondal, P. et al. (2023). *Chem* 9: 989–1003.

19 Kawamorita, S., Ohmiya, H., Hara, K. et al. (2009). *J. Am. Chem. Soc.* 131: 5058–5059.

20 (a) Kuninobu, Y., Iwanaga, T., Omura, T., and Takai, K. (2013). *Angew. Chem. Int. Ed.* 52: 4431–4434. (b) Yoshigoe, Y. and Kuninobu, Y. (2017). *Org. Lett.* 19: 3450–3453.

21 Li, H.-L., Kuninobu, Y., and Kanai, M. (2017). *Angew. Chem. Int. Ed.* 56: 1495–1499.

22 Yang, L., Uemura, N., and Nakao, Y. (2019). *J. Am. Chem. Soc.* 141: 7972–7979.

23 Zhang, T., Luan, Y.-X., Lam, N.Y.S. et al. (2021). *Nat. Chem.* 13: 1207–1213.

24 Jiang, B., Liu, J.-M., and Shi, S.-L. (2023). *ACS Catal.* 13: 6068–6075.

25 Trouvé, J., Rajeshwaran, P., Tomasini, M. et al. (2023). *ACS Catal.* 13: 7715–7729.

26 Nakao, Y., Kanyiva, K.S., and Hiyama, T. (2008). *J. Am. Chem. Soc.* 130: 2448–2449.

27 Okumura, S., Komine, T., Shigeki, E. et al. (2018). *Angew. Chem. Int. Ed.* 57: 929–932.

28 Hara, N., Saito, T., Semba, K. et al. (2018). *J. Am. Chem. Soc.* 140: 7070–7073.

29 Mihai, M.T., Davis, H.J., Genov, G.R., and Phipps, R.J. (2018). *ACS Catal.* 8: 3764–3769.

30 Mondal, P., Pal, R., Pal, K.A. et al. (2022). *J. Org. Chem.* 87: 9222–9231.

31 Lee, B., Mihai, M.T., Stojalnikova, V., and Phipps, R.J. (2019). *J. Org. Chem.* 84: 13124–13134.

32 Mihai, M.T., Williams, B.D., and Phipps, R.J. (2019). *J. Am. Chem. Soc.* 141: 15477–15482.

33 Bastidas, J.R.M., Oleskey, T.J., Miller, S.L. et al. (2019). *J. Am. Chem. Soc.* 141: 15483–15487.

34 Lu, S., Zheng, T., Ma, J. et al. (2022). *Angew. Chem. Int. Ed.* 61: e202201285.

35 Douthwaite, J.L. and Phipps, R.J. (2022). *Tetrahedron* 117–118: 132831.

36 Genov, G.R., Douthwaite, J.L., Lahdenpera, A.S.K. et al. (2020). *Science* 367: 1246–1251.

37 Fanourakis, A., Williams, B.D., Paterson, K.J. et al. (2021). *J. Am. Chem. Soc.* 143: 10070–10076.

38 (a) Chaturvedi, J., Haldar, C., Bisht, R. et al. (2021). *J. Am. Chem. Soc.* 143: 7604–7611. (b) Chaturvedi, J., Haldar, C., and Chattopadhyay, B. (2022). *Synlett* 33: 1108–1116.

39 Mondal, A., Díaz-Ruiz, M., Deufel, F. et al. (2023). *Chem* 9: 1004–1016.

40 Zhang, Z., Tanaka, K., and Yu, J.-Q. (2017). *Nature* 543: 538–542.

41 (a) Achar, T.K., Ramakrishna, K., Pal, T. et al. (2018). *Chem. Eur. J.* 24: 17906–17910. (b) Achar, T.K., Biswas, J.P., Porey, S. et al. (2019). *J. Org. Chem.* 84: 8315–8321. (c) Ramakrishna, K., Biswas, J.P., Jana, S. et al. (2019). *Angew. Chem. Int. Ed.* 58: 13808–13812.

42 Hoque, M.E., Bisht, R., Haldar, C., and Chattopadhyay, B. (2017). *J. Am. Chem. Soc.* 139: 7745–7748.

43 Bisht, R., Hoque, M.E., and Chattopadhyay, B. (2018). *Angew. Chem. Int. Ed.* 57: 15762–15766.

44 Trouvé, J., Zardi, P., Al-Shehimy, S. et al. (2021). *Angew. Chem. Int. Ed.* 60: 18006–18013.

45 Fan, Z., Chen, X., Tanaka, K. et al. (2022). *Nature* 610: 87–93.

46 Marcos-Atanes, D., Vidal, C., Navo, C.D. et al. (2023). *Angew. Chem. Int. Ed.* 62: e202214510.

47 Jin, Y., Ramadoss, B., Asako, S., and Ilies, L. (2024). *Nat. Commun.* 15: 2886.

48 Das, S., Incarvito, C.D., Crabtree, R.H., and Brudvig, G.W. (2006). *Science* 312: 1941–1943.

49 Olivo, G., Farinelli, G., Barbieri, A. et al. (2017). *Angew. Chem. Int. Ed.* 56: 16347–16351.

50 Olivo, G., Capocasa, G., Ticconi, B. et al. (2020). *Angew. Chem. Int. Ed.* 59: 12703–12708.

51 Burg, F., Breitenlechner, S., Jandl, C., and Bach, T. (2020). *Chem. Sci.* 11: 2121–2129.

52 Zeng, J., Torigoe, T., and Kuninobu, Y. (2022). *ACS Catal.* 12: 3058–3062.

53 deGruyter, J.N., Malins, L.R., and Baran, P. (2017). *Biochemistry* 56: 3863–3873.

54 (a) Gong, W., Zhang, G., Liu, T. et al. (2014). *J. Am. Chem. Soc.* 136: 16940–16946. (b) Osberger, T.J., Rogness, D.C., Kohrt, J.T. et al. (2016). *Nature* 537: 214–219.

55 Song, J., Torigoe, T., and Kuninobu, Y. (2023). *Org. Lett.* 25: 3708–3712.

56 (a) Breslow, R., Huang, Y., Zhang, X., and Yang, J. (1997). *Proc. Natl. Acad. Sci. USA* 94: 11156–11158. (b) Breslow, R., Zhang, X., and Huang, Y. (1997). *J. Am. Chem. Soc.* 119: 4535–4536.

57 Lu, X., Yoshigoe, Y., Ida, H. et al. (2019). *ACS Catal.* 9: 1705–1709.

58 (a) Kuninobu, Y. (2016). *J. Synth. Org. Chem. Jpn.* 74: 1058–1068. (b) Davis, H.J. and Phipps, R.J. (2017). *Chem. Sci.* 8: 864–877. (c) Kuninobu, Y. (2018). *Synlett* 29: 2093–2107. (d) Mihai, M.T., Genov, G.R., and Phipps, R.J. (2018). *Chem. Soc. Rev.* 47: 149–171. (e) Ghosh, M. and Sarkar, S.D. (2018). *Asian J. Org. Chem.* 7: 1236–1255. (f) Haldar, C., Hoque, M.E., Bisht, R., and Chattopadhyay, B. (2018). *Tetrahedron Lett.* 59: 1269–1277. (g) Kuroda, Y. and Nakao, Y. (2019). *Chem. Lett.* 48: 1092–1100. (h) Kuninobu, Y. and Torigoe, T. (2020). *Org. Biomol. Chem.* 18: 4126–4134. (i) Meng, G., Lam, N.Y.S., Lucas, E.L. et al. (2020). *J. Am. Chem. Soc.*

142: 10571–10591. (j) Fanourakis, A., Docherty, P.J., Chuentragool, P., and Phipps, R.J. (2020). *ACS Catal.* 10: 10672–10714. (k) Olivo, G., Capocasa, G., Giudice, D.D. et al. (2021). *Chem. Soc. Rev.* 50: 7681–7724. (l) Dutta, U., Maiti, S., Bhattacharya, T., and Maiti, D. (2021). *Science* 372: eabd5992. (m) Ali, R. and Siddiqui, R. (2021). *Adv. Synth. Catal.* 363: 1290–1316. (n) Haldar, C., Hoque, M.E., Chaturvedi, J. et al. (2021). *Chem. Commun.* 57: 13059–13074. (o) Pachisia, S. and Gupta, R. (2021). *Dalton Trans.* 50: 14951–14966. (p) Kuninobu, Y. (2022). *J. Synth. Org. Chem. Jpn.* 80: 421–430. (q) Bisht, R., Haldar, C., Hassan, M.M.M. et al. (2022). *Chem. Soc. Rev.* 51: 5042–5100. (r) Sinha, S.K., Guin, S., Maiti, S. et al. (2022). *Chem. Rev.* 122: 5682–5841. (s) Fernández-Figueiras, A., Ravutsov, M.A., and Simeonov, S.P. (2022). *ACS Omega* 7: 6439–6448. (t) Jiao, Y., Chen, X.-Y., and Stoddart, J.F. (2022). *Chem* 8: 414–438. (u) Reek, J.N.H., de Bruin, B., Pullen, S. et al. (2022). *Chem. Rev.* 122: 12308–12369. (v) Gillespie, J.E., Fanourakis, A., and Phipps, R.J. (2022). *J. Am. Chem. Soc.* 144: 18195–18211. (w) Kuninobu, Y. (2023). *Chem. Rec.* 23: e202300149. (x) Hassan, M.M.M., Guria, S., Dey, S. et al. (2023). *Sci. Adv.* 9: eadg3311.

7

Pd/Norbornene (NBE) Cooperative Catalysis in C—H Bond Activation

Zhibo Yan[1,2] and Zhe Dong[1,2]

[1] *Southern University of Science and Technology, Shenzhen Key Laboratory of Cross-Coupling Reactions, Xueyuan Road No. 1088, 518055, Shenzhen, China*
[2] *Southern University of Science and Technology, Department of Chemistry, Xueyuan Road No. 1088, 518055, Shenzhen, China*

7.1 Introduction

Directing group-assisted arene C—H bond activation proved to be the major breakthrough in the last two decades in the field of organometallic chemistry [1]. The directing group offered unique solutions to both the C—H bond's reactivity and selectivity issues. Thus, the directing group strategy has achieved tremendous success in developing the chemo/site/enantioselective C—H bond activation reaction [2]. However, the major concern of the directing group strategy was the step economy: additional chemical steps usually were required to install and remove the directing group [3]. This drawback has motivated synthetic chemists to develop the "*in situ* directing group" strategy and "traceless directing group" strategy (Scheme 7.1) [4]. The former one usually takes advantage of the dynamic-covalent bonding such as imine/enamine chemistry to install/remove the directing group. Among all the dynamic-covalent bonding systems, palladium/norbornene catalysis was the most unique one due to the reversible migratory insertion/β-carbon elimination step [5]. Palladium/norbornene cooperative catalysis offered a distinct way to install/remove directing group in one catalytic cycle and enable us to regioselectively install multiple functional groups onto arenes at the same time [5f].

The palladium/norbornene catalysis was originally discovered by Prof. Marta Catellani from Parma University Italy in 1997. Due to its rapid development in the last two decades, this type of transformation was systematically reviewed by Prof. Mark Lautens in 2011 and Prof. Guangbin Dong in 2019 [5g]. In this chapter, we will focus more on highlighting palladium/norbornene catalysis's mechanism insight and its application in the *meta*-selective arene activation.

Directed C–H Bond Functionalization: Concepts and Applications, First Edition.
Edited by Debabrata Maiti and Supriya Rej.

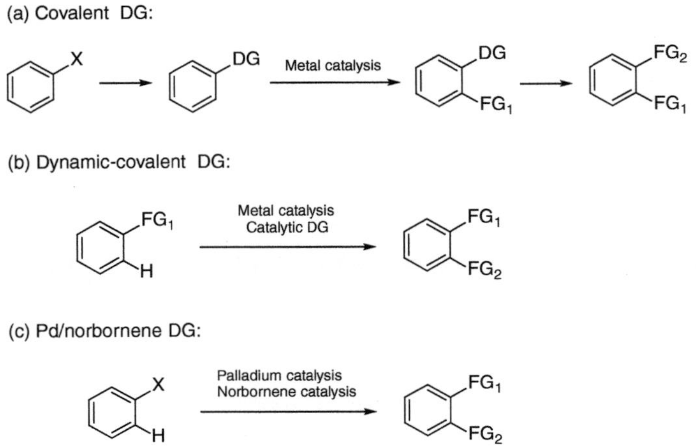

Scheme 7.1 Difference between the covalent directing group, dynamic-covalent directing group, and palladium/norbornene cooperative catalysis. (a) Covalent directing group strategy for the transition-metal catalyzed arene *ortho* C–H functionalization. (b) Dynamic-covalent group strategy for the transition-metal catalyzed arene *ortho* C–H functionalization. (c) Unique palladium/norbornene cooperative catalysis.

7.2 The Early Organometallic Study and Reaction Discovery

7.2.1 The Stoichiometric Organometallic Study

In 1974, the Inoue group reported an interesting reaction regarding the migratory insertion step: palladium phenyl chloride **2** could react with norbornene at room temperature to form sp^3—Pd bond species **3** (Scheme 7.2) [6]. They observed phenyl–norbornene–palladium dimer product (PNP dimer) instead of the common Heck Reaction product. They proposed that Pd-phenyl bond undergoes a stereoselective *syn-migratory insertion* with norbornene to form the PNP dimer. Pd(II) center in product **3** that does not have β-hydrogen is syn to the Pd center so the β-hydride elimination could not take place at this stage. This work built the cornerstone for all the Pd/NBE catalyses.

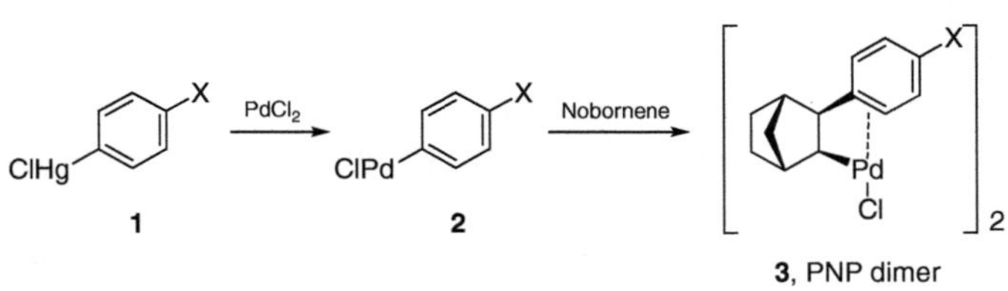

Scheme 7.2 Norbornene prevents β-hydride elimination from occurring on Pd(II) center.

In 1982, Catellani and co-workers reported the first catalytic C—H activation of aromatic C—H bond involving the palladium and norbornene (Scheme 7.3) [7].

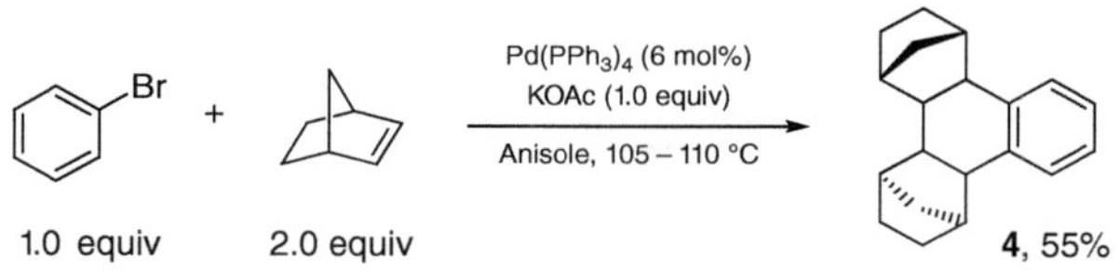

Scheme 7.3 Evidence that norbornene can serve as an intramolecular directing group.

When bromobenzene was treated with norbornene, potassium acetate, and 6 mol% $Pd(PPh_3)_4$ catalyst, a double-insertion product **4** involving *ortho*-C—H bond activation was isolated in 55% yield. This proved that norbornene can work as an intramolecular directing group for the aromatic *ortho*-C—H activation. The observed product diastereoselectivity was further approved for exclusive *syn-migratory insertion.*

In 1984, the Catellani group reported other interesting result where, under similar conditions, 4-nitrobromobenzene gave a cyclobutene-containing product **5** together with cyclohexane product **6** (Scheme 7.4) [8]. The formation of cyclobutene product strongly shows that C—H activation involves a five-membered metallocycle.

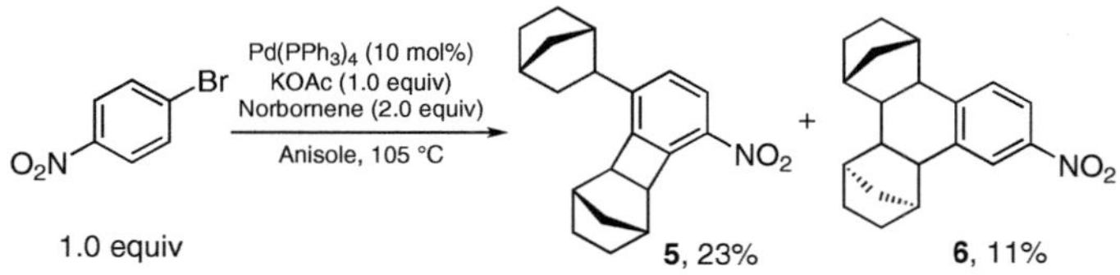

Scheme 7.4 First cyclobutane formation involving Pd/NBE catalysis.

In 1985, the Catellani group reported a complex tandem C—C bond formation involving the Pd/NBE-catalyzed C—H activation. When bromobenzene was treated with potassium *tert*-butlyoxide, a biaryl-containing cyclohexane product was isolated [9]. Interestingly, when *para*-substituted bromide was used, the product was a mixture of two regio-isomers **7** and **8** (Scheme 7.5).

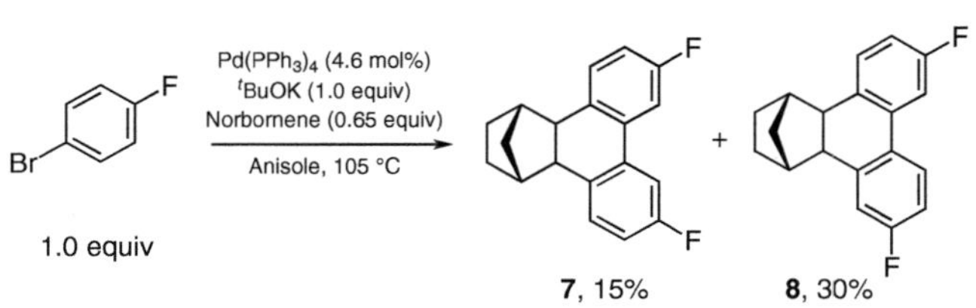

Scheme 7.5 Pd/Norbornene-mediated unselective aryl–aryl bond formation.

The proposed mechanism was shown in Scheme 7.6: Pd(0) oxidative addition into the aryl bromide bond and then *syn-migratory insertion* into the norbornene. Intramolecular C—H activation forms a key five-membered metallocycle. This aryl-norbornyl palladium(II) species (ANP) could further react with aryl bromide to form C—C bond at both aryl and norbornyl sites. From today's

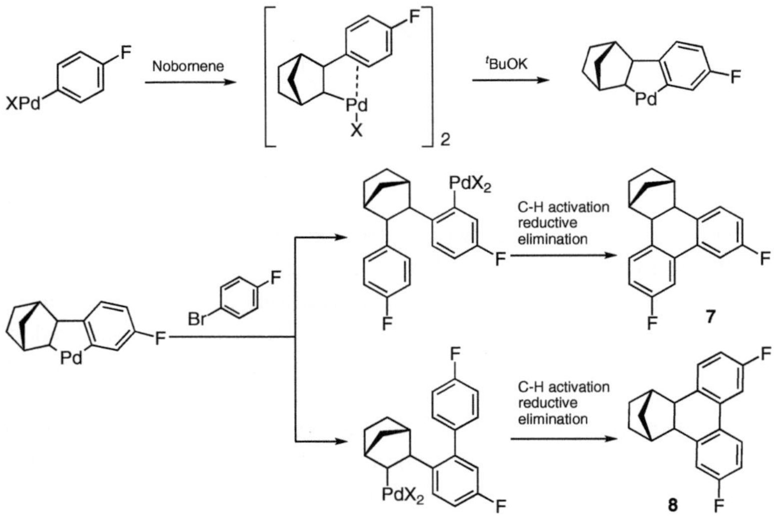

Scheme 7.6 Proposed mechanism for Pd/norbornene-mediated aryl–aryl bond formation.

mechanism view, this C—C bond formation step was very likely involving either further oxidative addition into Pd(IV) intermediates and then regio-unselective reductive elimination, or two aryl Pd(II) species underwent unselective transmetallation. The resulting Pd(II) species could undergo further intramolecular C—H activation and then reductive elimination to form another C—C bond to give the final six-membered ring product.

One important breakthrough is that Prof. Catellani group isolated the first intramolecular C—H activation intermediate with Pd/NBE in 1988 (Scheme 7.7) [10]. They found that when treating the PNP dimer with potassium phenolate and 1,10- phenanthroline, a five-membered cyclometalled compound 9 was isolated. The ANP has one sp^2 carbon and one sp^3 carbon as the X-type ligand, which did not undergo reductive elimination or β-hydride elimination at room temperature. This C—H activation occurred on mild basic conditions at room temperature, which was believed to undergo an electrophilic C—H bond palladation pathway.

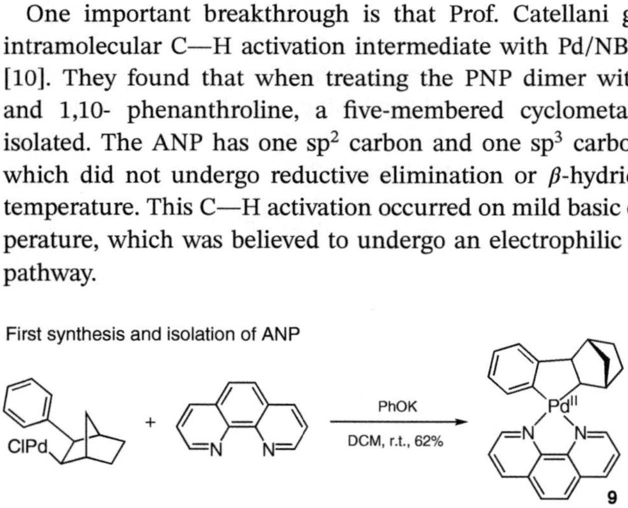

Scheme 7.7 Intramolecular Pd/NBE-mediated C—H activation.

Further studies by the Catellani group demonstrated that cyclometalled intermediates could undergo further oxidative addition with 4-nitro benzyl bromide at

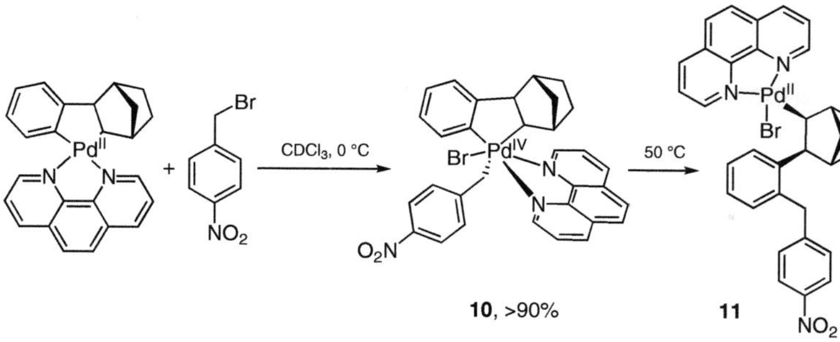

10, >90% **11**

Scheme 7.8 Palladium(IV) intermediate and its reductive elimination.

0 °C to form stable palladium(IV) intermediate **10** even at zero degree (Scheme 7.8) [11]. Interestingly, when warmed up to 50 °C, the palladium(IV) metal center undergoes selective sp^2–sp^3 reductive elimination to form benzylated arene product **11** [11b]. This provides us several key information: (i) ANP intermediate was highly electron-rich, and it can be oxidized to Pd(IV) by a weak oxidant such as benzyl halide at 0 °C. (ii) Reductive elimination from the palladium center was highly selective, predominantly taking place on the sp^2 carbon site. (iii) Aryl–Pd bond in the ANP was nucleophilic instead of common Ary-Pd-X (X≠C), which was electrophilic in most cross-coupling reactions.

All previous results always had a norbornene attached to the product, which makes the whole transformation much less attractive to the synthetic community. In 1994, Catellani and co-workers reported an elegant work to demonstrate that the norbornene can be removed from the arene product in the presence of palladium aryl species (Scheme 7.9) [12].

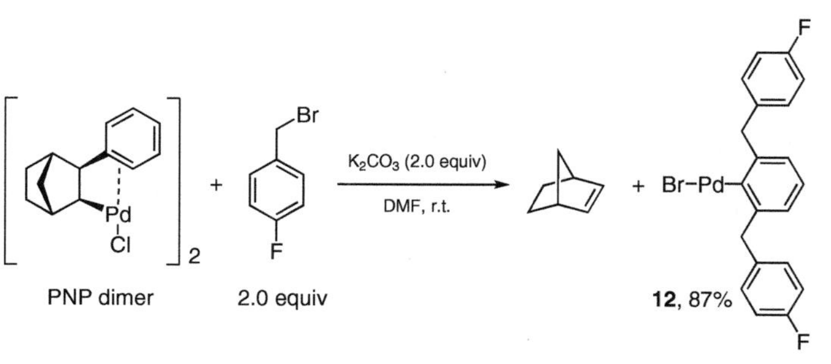

PNP dimer 2.0 equiv **12**, 87%

Scheme 7.9 First norbornene exclusion through *β*-carbon elimination.

The proposed mechanism is shown in Scheme 7.10. The PNP dimer could undergo C—H activation under basic conditions, and then the newly formed metallocycle would react with the benzyl bromide to form the Pd(IV) intermediate **13**. Selective sp^3–sp^2 carbon–carbon bond reductive elimination would give *ortho*-benzylated arene **14**. The same sequence could happen again to give 2,6-disubstituted arene **12** with norbornene attached to the middle carbon. Interestingly, due to the steric

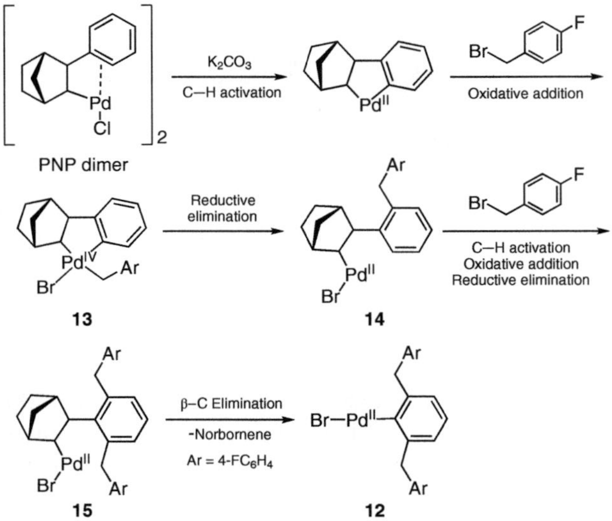

Scheme 7.10 First norbornene exclusion through β-carbon elimination.

repulsion between the 2,6-disubstituted arene and palladium(II) center, a β-carbon elimination would occur to release the norbornene, and palladium(II) would go back to the middle of the arene. These systematic studies already prove that it is possible to use catalytic amounts of palladium and norbornene to functionalize the arene's both *ipso*- and *ortho*-positions to introduce two different functional groups.

7.2.2 The Initial Reaction Discovery by Catellani

In 1997, the Catellani group reported an elegant work of palladium catalizing the *ortho*-alkylation/*ipso*-alkenylation of aryl iodide [13]. (Scheme 7.11) They found that the alkyl chain can be introduced at the *ortho*-position of aryl iodide, and Heck-type alkenylation could occur at the *ipso*-position. More interestingly, when aryl iodide without *ortho*-substitution was used, the double alkylation could happen at both *ortho*-positions. Although they used 4.0 equiv of norbornene in the system, they were able to recycle majority of norbornene after the reaction. This implies that norbornene actually could be a catalyst for this transformation.

Based on a previous study, the Catellani group proposed the mechanism shown in Scheme 7.12. A reaction was initiated with palladium(0) catalyst oxidative addition to the aryl–iodide bond to form aryl–Pd(II) species, which was followed by *syn*-migratory insertion of norbornene into the aryl–Pd bond to generate sp3-Pd(II) species. The migration was highly stereoselective, exclusively happening on the exo face of norbornene due to the exo face being less hindered than the endo face. Then, intramolecular cyclopalladation would happen to activate the *ortho*-C—H bond of the aromatic ring to form the key ANP intermediate. Then, an external electrophile such as alkyl iodide could oxidize the ANP to the Pd(IV). Then, selective reductive elimination occurs on the sp^2 carbon site to introduce

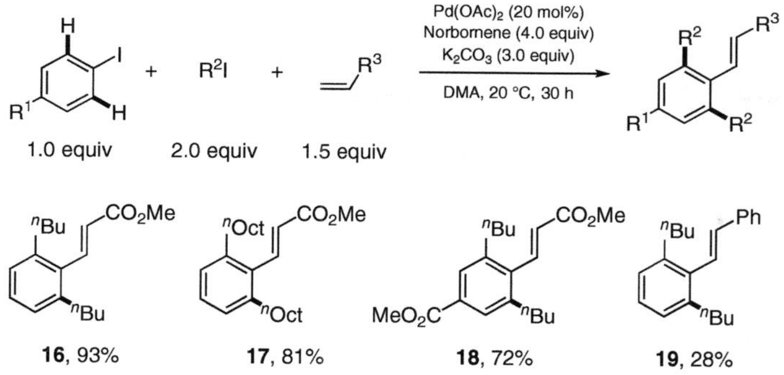

Scheme 7.11 First palladium-catalyzed *ortho-/ipso*-difunctionalization of aryl iodide.

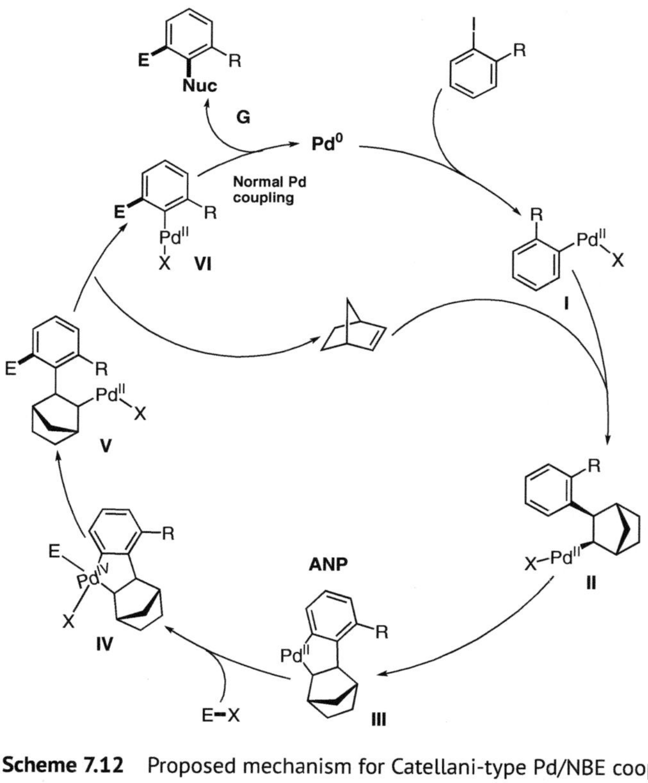

Scheme 7.12 Proposed mechanism for Catellani-type Pd/NBE cooperative catalysis.

the electrophile at the *ortho*-functional group. When arene's *ortho*-position was occupied, the sp^3-Pd(II) center preferred β-carbon elimination to form aryl–Pd(II). Newly formed aryl–Pd(II) could undergo normal cross-coupling to introduce the nucleophile at the *ipso*-position of the arene and regenerate Pd(0) catalyst.

This complex catalytic cycle contains almost all the elemental steps in the organometallic chemistry, and palladium catalyst's oxidative state changes among

Pd(0)/Pd(II)/Pd(IV) cycle. This catalytic cycle was further examined by Lautens [14] and other groups [15] both experimentally and computationally. Up to now, all studies have been consistent with Catellani group's original proposed mechanism. This correct mechanism provides us with plenty of information to further explore this wonderful and unique catalytic system.

Overall, palladium/norbornene cooperative catalysis could introduce the electrophile at the *ortho*-position and the nucleophile at the *ipso*-position. Compared with other arene C—H activation methods, it has several advantages: (i) Substrates do not have additional steps to install and remove the directing group, and simple aryl halides are widely available. (ii) One step could introduce two different functional groups on the vicinal positions with 100% regioselectivity. (iii) Reaction usually takes place under mild basic conditions and with a mild oxidant; thus, excellent functional group tolerance and board scope can be expected. (iv) The C—H activation only occurs at the *ortho*-position of aryl halide; this enables people to override the substrate's electronic and steric bias.

Based on the proposed mechanism, only a catalytic amount of norbornene is required for this transformation [16], although a stoichiometric amount was often used to achieve a satisfactory yield [5d, f]. Norbornene is not only the directing group for the *ortho*-position but also the protecting group for the *ipso*-position. Norbornene insertion only needs an activation energy of 17 kcal/mmol [15], which is significantly lower than the palladium-catalyzed common cross-coupling reaction. This provides the high regioselectivity for the difunctionalization. More importantly, the rigid backbond of norbornene makes Pd(II) center quite close to the arene's *ortho*-C—H bond, which significantly decreases the activation energy of key C—H activation [17]. After the migration insertion and C—H activation, norbornene actually becomes a catalytic directing chiral auxiliary. It can be expected that enantioselective catalytic C—H functionalization can be achieved using chiral norbornene or desymmetrization of simple norbornene.

7.3 Pd(0)/Pd(II)/Pd(IV) Catalytic Cycle: A Series of Chemoselectivity Puzzle

After Catellani group's initial discovery, accessing poly-substituted arenes from corresponding simple aryl iodide soon became a fast growing area. The discovery and application of this palladium/norbornene cooperative catalysis was majorly carried out by Prof. Marta Catellani and Prof. Mark Lautens. After Lautens group's discovery that tri(2-furyl)phosphine would significantly improve the reaction's catalytic performance, people soon realized that the nucleophile can be introduced at the *ipso*-position of aryl iodide which is actually extremely broad; almost all the palladium-catalyzed C—H bond could be merged together with Pd/norbornene-mediated C—H activation steps [18]. Based on the present understanding of the mechanism, the electron-deficient tri(2-furyl)phosphine ligand would significantly stabilize the palladium(0) catalyst to improve reaction scope (Scheme 7.13).

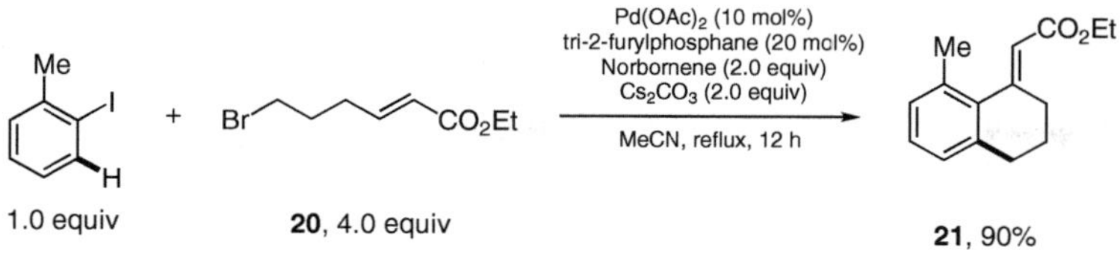

Scheme 7.13 Palladium-catalyzed synthesis of various carbocycles.

In sharp contrast, the electrophile, introduced at the *ortho*-position of aryl iodide, was quite limited due to the complex catalytic cycle. The unprecedented Pd(0)/Pd(II)/Pd(IV) cycle brought several selectivity issues and chemical compatibility issues. To further understand the key feature of palladium/norbornene catalysis, we will next focus on the selectivity issues in the catalytic cycle (Scheme 7.14). For redox chemistry, there were two different oxidants: aryl halides and external electrophile, and two different palladium species needed to be oxidized: Pd(0) and ANP intermediate. Aryl halide must oxidize the Pd(0) faster than the external electrophile, and external electrophile must oxidize the ANP to the Pd(IV) faster than the aryl halide. This is the most important rule for the success of the Pd/NBE cooperative catalysis. However, strong oxidants can oxidize Pd(II) to Pd(IV) and typically can also oxidize Pd(0) to Pd(II) very fast. This results in the very limited electrophile scope, the so-called **electrophile constraint**. From 1997 to 2013, the electrophile scope was limited to the alkyl halide and electron-deficient aryl bromide for more than one decade.

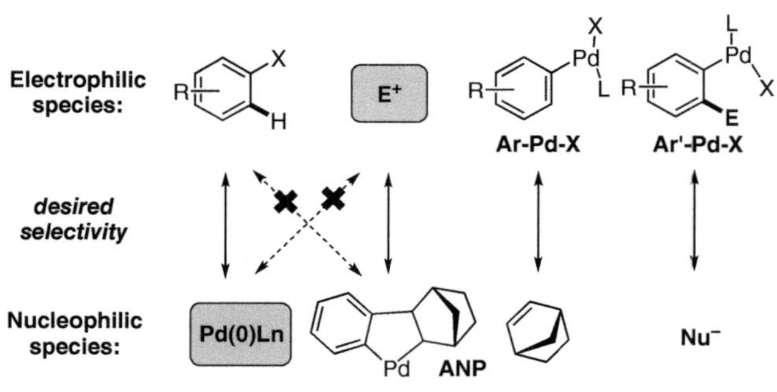

Scheme 7.14 Selectivity issues for Pd/NBE cooperative catalysis.

On the other hand, to ensure rapid oxidation of Pd(0) to aryl–Pd(II), a strong aryl halide was required. The majority of Pd/NBE cooperative catalyses have to use aryl iodide as the aryl source. There is only one elegant work from Lautens and co-workers that describes aryl triflate as the aryl source for the Pd/NBE cooperative catalysis. This is considered the **aryl iodide constraint**.

For non-redox chemistry, there were two aryl–Pd(II) species, one is before ortho-activation **Ar-Pd-X**, and another is after *ortho*-C—H activation **Ar'-Pd-X**. In the ideal catalytic system, aryl–Pd(II) before the C—H activation should chemoselectively react with norbornene to introduce the *ortho*-functional group.

Aryl–Pd(II), after the C—H activation, should only react with the nucleophile to regenerate the Pd(0) catalyst. In reality, the nucleophile will not be introduced until both *ortho*-positions are blocked. This was due to the fact that β-carbon elimination was favored only with 2,6-disubstituted arene. For catalysis, it means that aryl iodide has to have an *ortho*-substitution; otherwise, the C—H activation would happen twice. It was considered as the **ortho-constraint**, which greatly limited the scope and application of Pd/NBE cooperative catalysis.

7.3.1 The S_N-2-type Oxidative Addition vs Concerted Oxidative Addition: Electrophile Scope

In 2004, Catellani and co-workers reported an elegant work demonstrating that *ortho*-arylation of aryl iodide can be achieved using aryl bromide (Scheme 7.15) [19]. It is not surprising that aryl iodide reacts with Pd(0) faster than the aryl bromide, but it was really unique that aryl bromide reaction with ANP is faster than the aryl iodide. The authors found the main successful response on the aryl bromide's *ortho*-substitution [20]. A weak coordinating group such as the ester or hydroxyl group could significantly improve the efficiency and selectivity for the aryl–aryl coupling reaction. It is proposed that weak coordinating groups have a much better binding affinity with a palladium intermediate (ANP) than simple Pd(0). The coordination lowers the activation energy for the aryl bromide to oxidize the ANP to the Pd(IV).

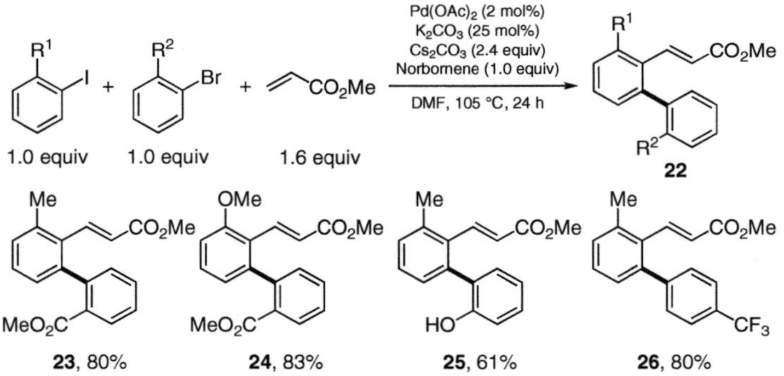

Scheme 7.15 *Ortho*-arylation of aryl iodide using aryl bromide.

In 2013, Dong and co-workers reported a major breakthrough in terms of overcoming electrophile constraints. *O*-benzoyl hydroxylamine proved to be an excellent external oxidation mechanism to introduce amine moiety at the *ortho*-position of aryl iodide (Scheme 7.16) [16]. This reaction provided an orthogonal regioselectivity compared with Buchwald–Hartwig coupling. They demonstrated a stepwise *meta*-selective C—H bond via the merging of unselective electrophilic arene C—H bond iodination with regio-convergent palladium/norbornene catalysis. Formally, it took steps to convert simple arene **27** to *meta*-aminated product **29** in good yield. Beyond the excellent scope of this transformation, due to the perfectly

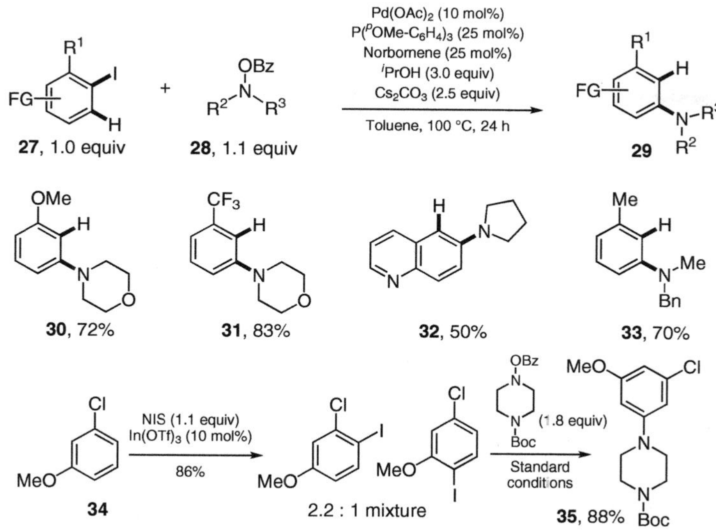

Scheme 7.16 *Ortho*-amination of aryl iodide using *O*-benzoyl hydroxylamine.

matched catalytic cycle, aryl iodide/external electrophile and nucleophile can be employed as equal stoichiometry, and the norbornene loading can be reduced to 25 mol%. This showed that high chemoselectivity had attracted physical organic chemists' interest to understand the mechanism insight behind this transformation.

Soon after this *ortho*-amination reaction, the same product also developed *ortho*-acylation and *ortho*-alkoxycarbonylation using similar reagent design strategies [21]. However, why could those oxidants selectively oxidize the ANP to introduce the external electrophile on the *ortho*-position of the arene? In 2020, Liu and Dong published a systematic mechanism study to understand this phenomenon (Scheme 7.17) [22]. They found that the external electrophile selectively could react with ANP intermediate with two distinct oxidative addition mechanisms. sp^3-type electrophilics such as alkyl halide and *O*-benzoyl hydroxylamine underwent an S_N-2-type transition state during the oxidative addition process of ANP to Pd(IV) (C-1 pathway in Scheme 7.17). This S_N-2-type oxidative addition mechanism was further proved by the stereochemistry outcome observed when stoichiometric ANP intermediate was reacted with primary chiral alkyl iodides [23]. On the other hand, the sp^2-type electrophile still participated in the more traditional concerted oxidative addition (C-2 and C-3 pathways in Scheme 7.17). In these cases, the electrophile's pre-coordination with ANP's palladium metal center played a critical role in lowering transition state energy to access Pd(IV), which ultimately controls the reaction's chemoselectivity.

Another key piece of mechanism insight was the active species of ANP intermediates. Although the isolated ANP species was usually stabilized by the addition of phosphine ligand, the experimental and computational results have demonstrated that iodide anion had played a critical role in both oxidative addition and reductive elimination steps. The ate complex formed between the iodide anion and ANP Pd(II)

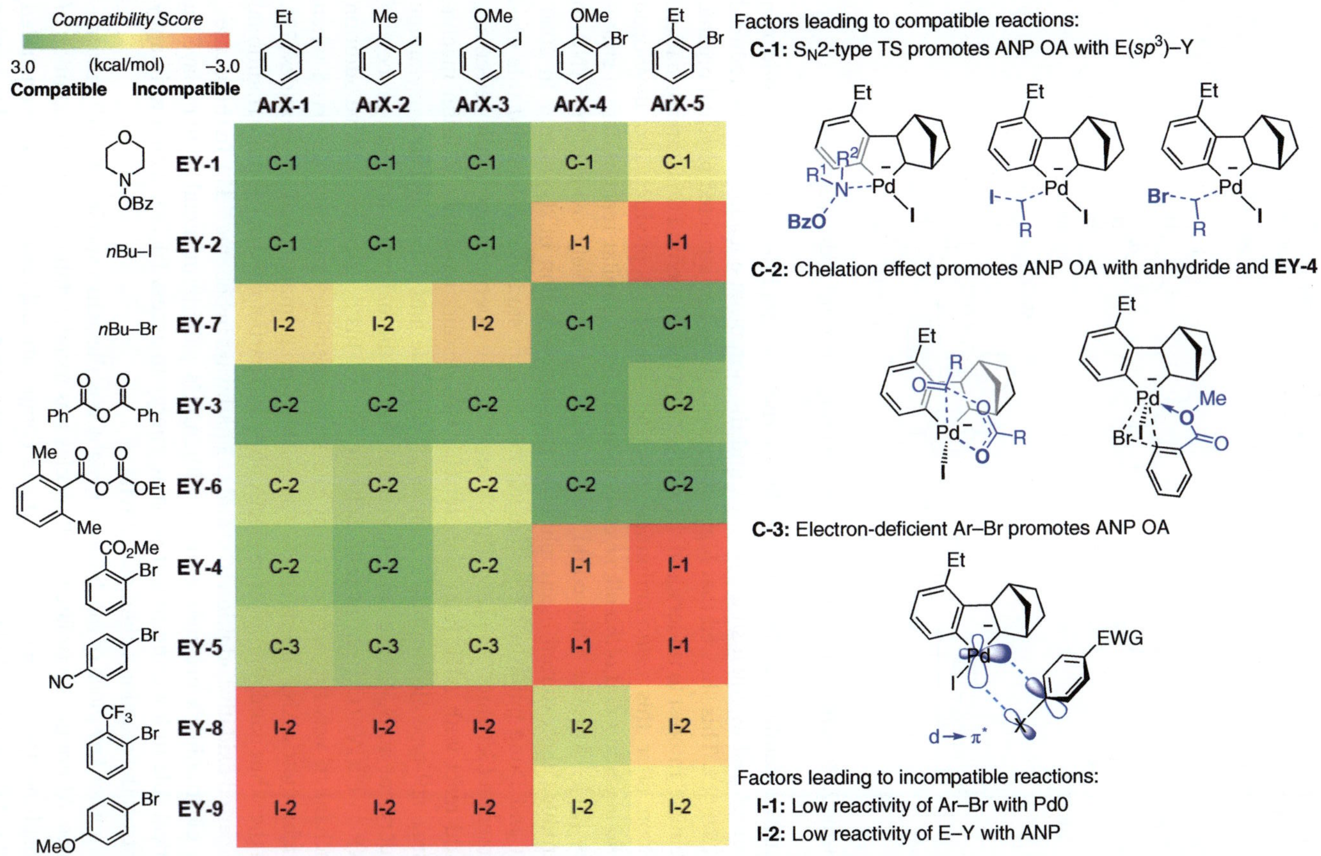

Scheme 7.17 Mechanism understanding of how electrophile selectively oxidizes the ANP intermediate to Pd(IV) in the presence of Pd(0).

center was believed to be the real catalytic active species which was responsible for the palladium(IV) intermediate formed during the whole catalytic cycle.

This key mechanism understanding has guided synthetic chemists to further develop new reagents to further expand the diversity of palladium/norbornene catalysis's electrophile scope.

For S_N-2-type oxidative addition, the bonding polarity and the leaving group's pKa clearly played a central role in determining the external electrophile's oxidizing ability. Based on this design principle, Dong and co-workers have developed *ortho*-thiolation and *ortho*-methoxylation (Scheme 7.18) [24]. For the *ortho*-thiolation reaction, they used an amide as leaving to control the nitrogen-sulfur bond's polarity, which proved to be highly tunable reagents for aryl iodide *ortho*-C—H bond thiolation reaction. They found the longer the N—S bond length it is, the higher reactivity was observed. This observation further supported the S_N-2-type oxidative addition mechanism. For *ortho*-methoxylation, the N—O bond's polarity has to be polarity-reversed for selective methoxylation instead of amination. To achieve that, the bis-sulfonyl group as a high electron-withdrawing group was installed on the nitrogen atom, but on the oxygen atom, only the electron-donating methyl group was tolerated. The DFT study also revealed that an S_N-2-type oxidative addition pathway was the key for this transformation.

7.3.2 Migratory Insertion vs *β*-carbon Elimination: Norbornene Modification

Since Prof. Catellani's original discovery, for simple aryl iodide without *ortho*-substitution, the palladium/norbornene catalysis always gave the bis-functionalization product. From a mechanism view, this observation proved to us that the *β*-carbon elimination would not happen until both *ortho*-positions of aryl halide were occupied [5g]. To achieve the mono-*ortho*-functionalization, Dong and co-workers systematically studied the structure relationship of substituted norbornene (Scheme 7.19) [25]. They hypothesized that the substitution at the bridge head position could generate steric repulsions with arene's *ortho*-substitution, which would make Pd(II) center much closer to the arene's *ipso*-carbon center. This would lower the transition-state energy for the *β*-carbon elimination steps. In sharp contrast, the standard norbornene would prefer further C—H activation without this key steric repulsion. Indeed, when the alkyl-substituted norbornene N2 was used, the *meta*-substituted aryl iodide gave the desired mono-functionalized *para*-substituted arenes. For less hindered *para*-substituted arene, 1,4-disubstituted norbornene was necessary for achieving mono-selectivity. This elegant work represents a nice example of how the minor conformational changes precisely tune the reactivity and selectivity of *β*-carbon elimination step, which was a far less-studied elementary step in the organometallic chemistry.

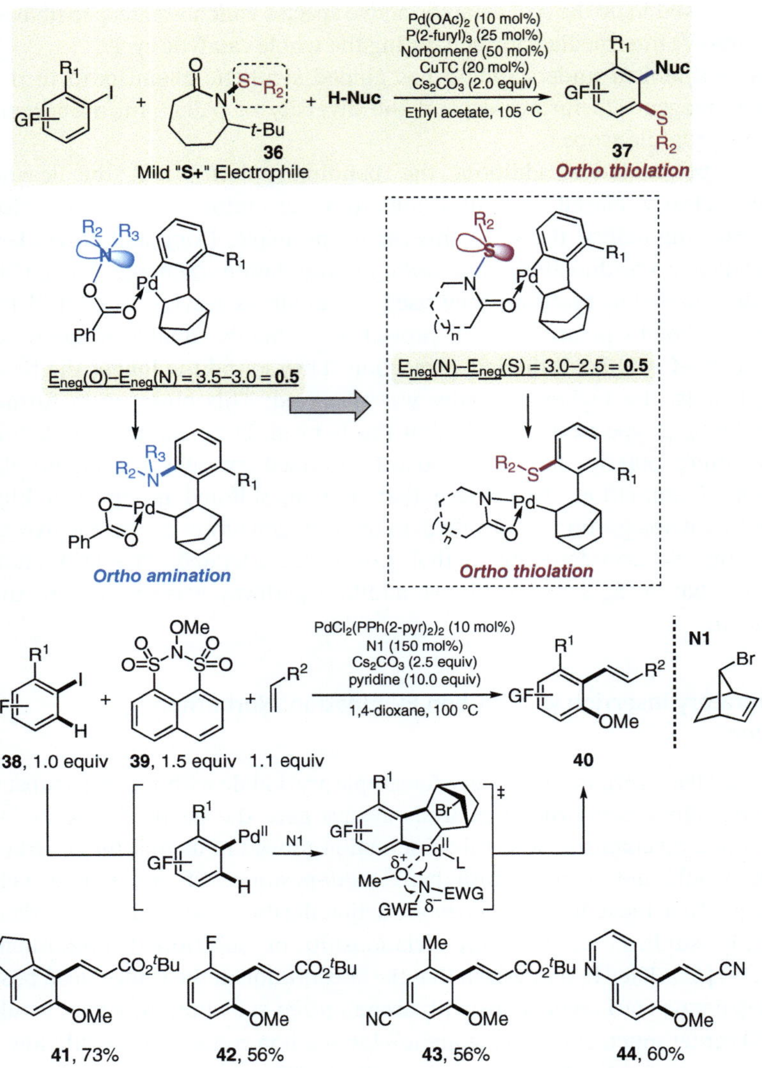

Scheme 7.18 *Ortho*-thiolation and *ortho*-methoxylation of aryl iodide.

7.4 Palladium(II)-initiated Palladium/Norbornene Catalysis

Compared with Pd(0)-initiated reaction, which functionalizes the *ortho*-position of aryl halide, Pd(II)-initiated reaction could selectively activate both *ortho*- and *meta*-positions of a simple arene. Although such Pd(II)/Pd(IV) cycle looks like more simplified than Pd(0)/Pd(II)/Pd(IV), Pd(II)-mediated C—H bond activation was much less robust than the Pd(0)-mediated oxidative addition into the aryl halide. Thus, such a process was not achieved until 2015 by the Yu and Dong group. Bach group had developed pioneering work to functionalize

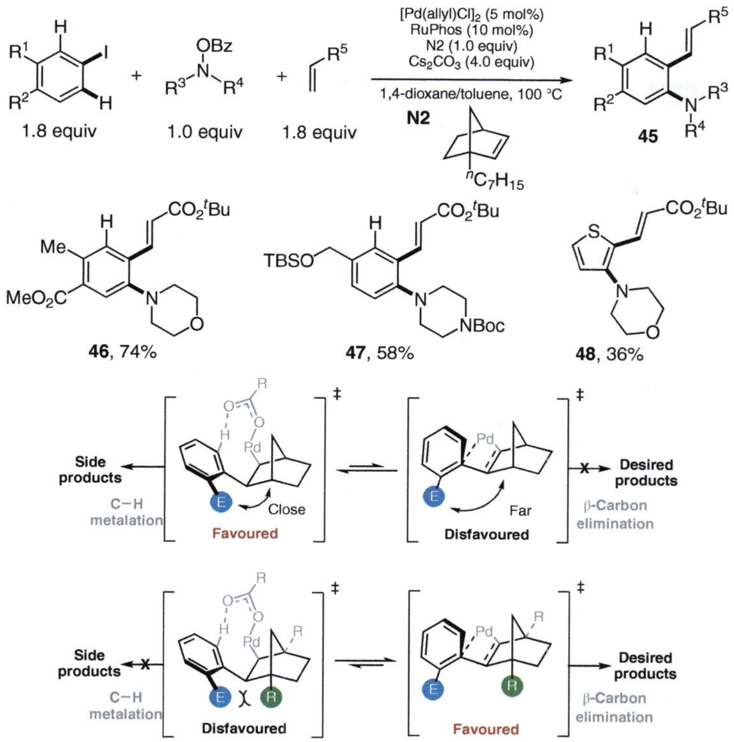

Scheme 7.19 *Ortho*-thiolation and *ortho*-methoxylation of aryl iodide.

the C-2 position of indole and pyrrole via the merger of Pd(II)-initiated N—H bond activation and palladium/norbornene catalysis.

7.4.1 N—H Bond-initiate Palladium/Norbornene Catalysis

The Bach group in 2011 reported an elegant work to achieve high C-2 selective alkylation of indole under mild basic conditions (Scheme 7.20) [26]. This transformation overrides the electronic nature of indole, in which the C-3 position was the nucleophilic and reactive position. Control experiments revealed that free indole's N—H was very critical to the success of this transformation. After a detailed physical organic chemistry study, the authors proposed the N—H bond activation pathway [27]. The Pd(II) center would deprotonate the acidic indole N—H and then the Pd—N bond would undergo syn-migration insertion with norbornene, which would guide Pd to activate the C-2 position. The sequential with a further oxidative addition with alkyl bromide will introduce the alkyl substitution at the C-2 position. This method proved to be highly practical for the synthesis of bioactive indole alkaloids [27a], due to its orthogonal regioselectivity to traditional indole alkylation reactions. The Bach group has successfully applied it to the synthesis of Goniomitine (Scheme 7.21).

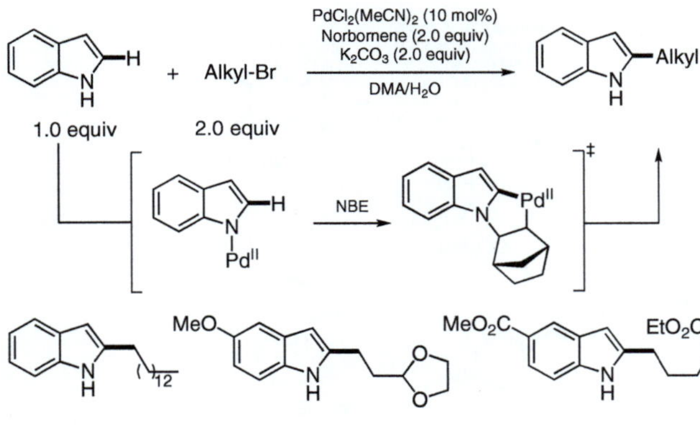

49, 67% **50**, 62% **51**, 87%

Scheme 7.20 C-2 selective alkylation of free indole.

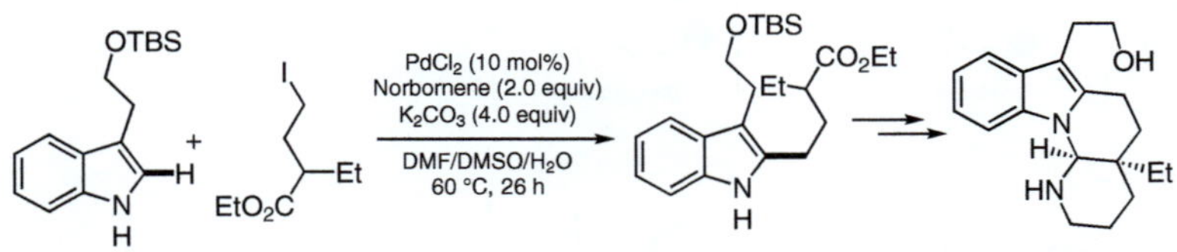

52, 1.0 equiv **53**, 4.0 equiv **54**, 73% **55**, (±)-Goniomitine

Scheme 7.21 Applying the Pd/norbornene to the total synthesis of Goniomitine.

The Bach group later expanded this type of chemistry to pyrrole through the same activating strategy (Scheme 7.22) [28]. However, for pyrrole-type substrates, one α-position of pyrrole was required to be blocked by the other functional groups. This was believed to facilitate the final β-carbon elimination step, which requires both α-positions of pyrrole be occupied. This observation was similar to the *ortho*-constraint observed for the Pd(0)/Pd(II)/Pd(IV) cycle.

The Bach group has further solved the reactivity issues related to the steric hindered three-substituted indole, and they directly alkylated the masked tryptophan to produce the unnatural amino acids in single chemical steps (Scheme 7.23) [29]. In 2017, Jiang and co-workers further expanded the electrophile scope from the primary alkyl halide to the aryl halide. However, the palladium acetate would also catalyze the indole *N*-arylation reaction or directed-C2 arylation, which complicated the reaction system and limited its real-world synthetic application [30]. Later in 2018, Liu group from SIOC later expanded this type of chemistry to the alkylatin with fluorinated alkyl iodide (Scheme 7.23) [31]. Trifluoroethyl iodide was not a good electrophile for the traditional two-electron oxidative addition, but ANP proved to be an excellent nucleophile to couple with the trifluoroethyl iodide for introducing the trifluoro ethyl group.

Ge and co-workers further developed this strategy for the functionalization of carbazoles (Scheme 7.24) [32]. Although the reaction mechanism and pathway were similar to Bach's pioneering work, this work represents the first examples of forming

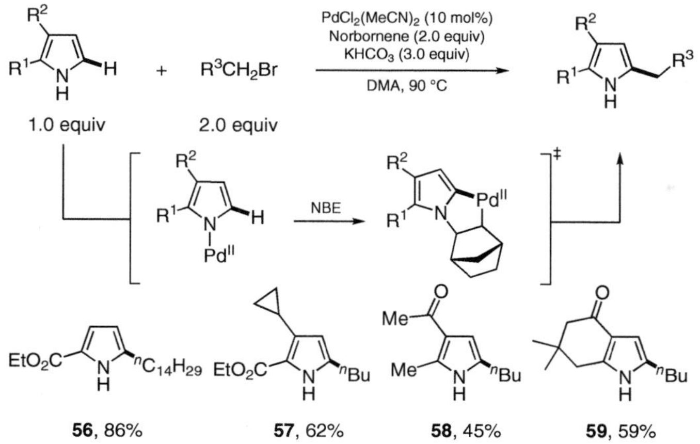

Scheme 7.22 C-2 selective alkylation of free pyrrole.

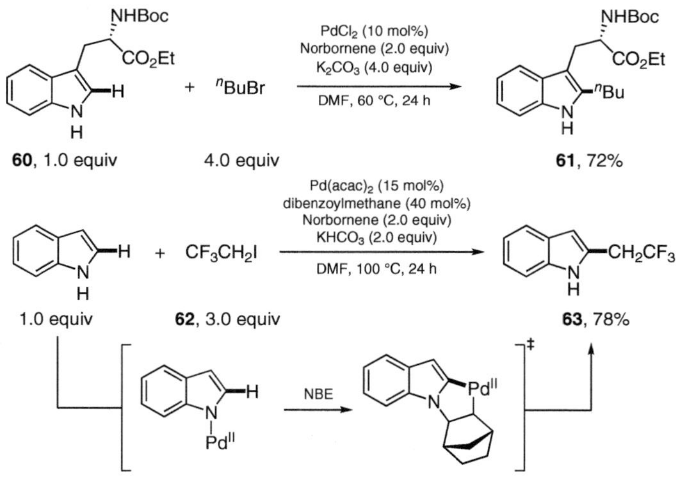

Scheme 7.23 C-2 selective alkylation of free indoles.

the six-membered aryl-norbornene–palladium cycle. Ge's work demonstrated that this six-membered ANP could also form Pd(IV) upon further oxidative addition and then β-carbon elimination to reform the N—Pd(II) bond. This new intermediate with a large ring size has provided a new chemical space to further expand the Pd/norbornene catalysis.

Beyond the simple primary alkyl halide, this method could also be used for glycosylation at the α position of pyrrole (Scheme 7.25) [33]. The reaction proved to be highly diastereoselective for mannose/rhamnose and ribose. The proposed mechanism involves an ANP intermediate to activate the chloro sugar to generate oxocarbenium, and then the palladium ate complex would selectively capture the corresponding oxocarbenium to generate the glycosylation product.

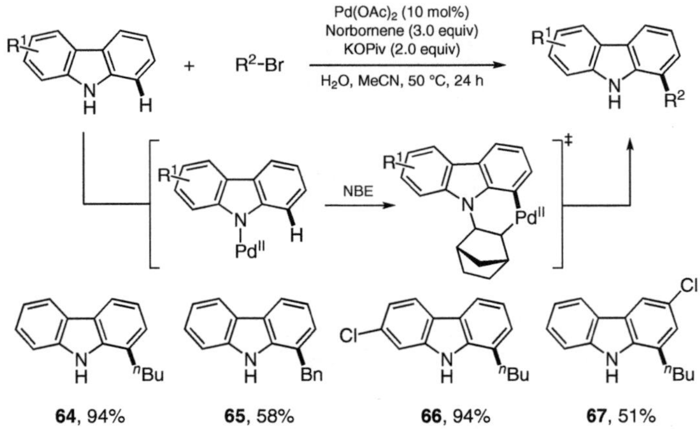

Scheme 7.24 Selective alkylation of carbazoles via six-membered ANP intermediates.

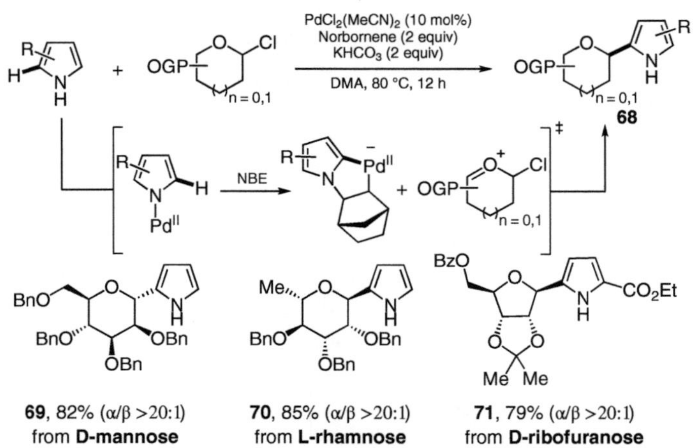

Scheme 7.25 Diastereoselective glycosylation of pyrroles via oxocarbenium intermediates.

7.4.2 C–H Bond-initiated Palladium/Norbornene Catalysis

7.4.2.1 Directed C–H Bond Activation

Directing group-assisted palladium-catalyzed arene *ortho*-C—H bond palladation was one of the major breakthroughs in the organometal chemistry in the past two decades. If we can merge it with palladium/norbornene cooperative catalysis, it would enable a highly *meta*-selective arene C—H bond activation. Although the design was simple, C—H palladation steps prefer neutral or acidic conditions and palladium/norbornene catalysis prefers basic conditions. How to merge these dramatically different conditions together has become a long-existing challenge.

In 2015, Hiyama and co-workers reported the merger of electron-rich alkyne-directed C—H palladation with Pd/norbornene catalysis (Scheme 7.26) [34].

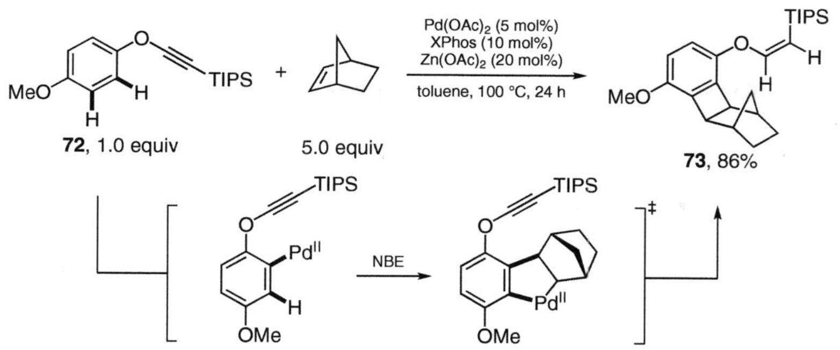

Scheme 7.26 Benzocyclobutane formation via the *meta*-selective C—H activation.

The alkyne served not only as the directing group but also as the oxidant that accepts two hydrides generated in this process. The reaction used zinc acetate as the base and did not add external oxidation, thus only forming a cyclobutane product.

During the same year, Dong and co-workers reported the simple benzyl-amine-directed *meta*-C—H activation (Scheme 7.27) [35]. Dimethyl benzyl amine proved to be the best directing group due to its labile bonding nature with a palladium catalyst. To merge the acidic condition for *ortho*-C—H palladation and the basic

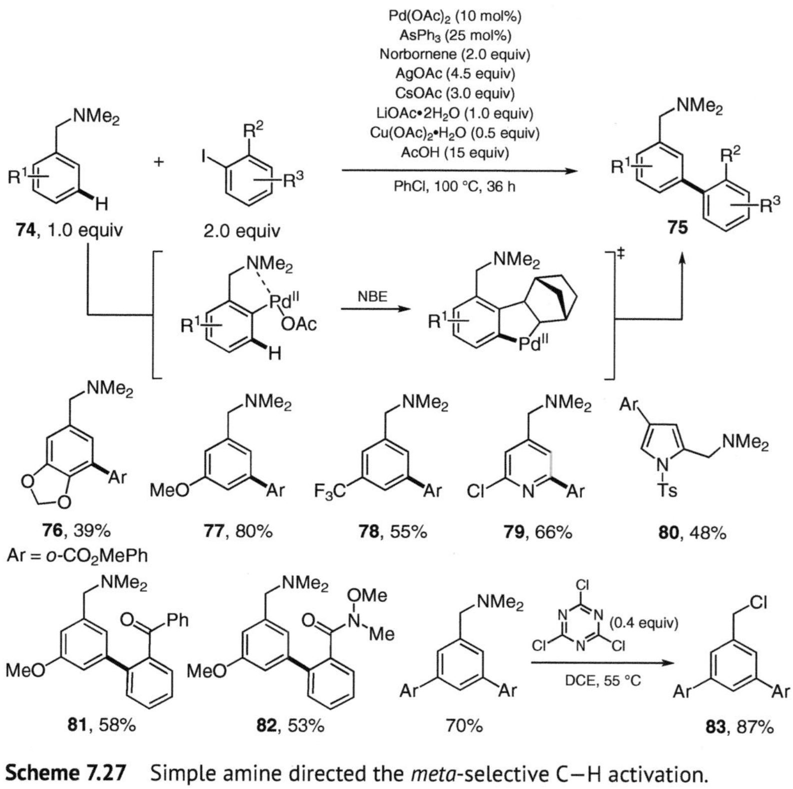

Scheme 7.27 Simple amine directed the *meta*-selective C—H activation.

condition for the *meta*-C—H activation steps, the authors employed acetate cocktails as buffer system to ensure both *ortho-/meta*-C—H bonds can be activated under the same catalytic cycle. It was worth noting that pyridine **79** and pyrrole **80** represented the first examples of directed *meta*-selective C—H activation demonstrated on the heterocycle for palladium catalysis. However, the arylation scope was limited to *ortho*-substituted electron-deficient aryl iodide, which limited its synthetic utility. Additionally, the dimethyl amine could be easily converted into corresponding benzyl chloride in high yield.

In 2015, Yu and co-workers reported engineered phenyl acetamide-directed *meta*-selective C—H bond alkylation and arylation reactions (Scheme 7.28) [36]. They have developed a novel pyridine ligand for this transformation, which greatly simplifies the reaction conditions. In this original report, they found that only the alkyl halide without the beta-hydrogens works smoothly. For the arylation reactions, the aryl halide scope was also restricted to those highly electron-deficient ones. For both Yu and Dong's *meta*-C—H functionalization work, they all showed that the reaction was 100% *meta*-selective, ignoring the substrate's electronic and steric bias. This clearly demonstrated the advantage of Pd/norbornene catalysis in the *meta*-C—H activation field.

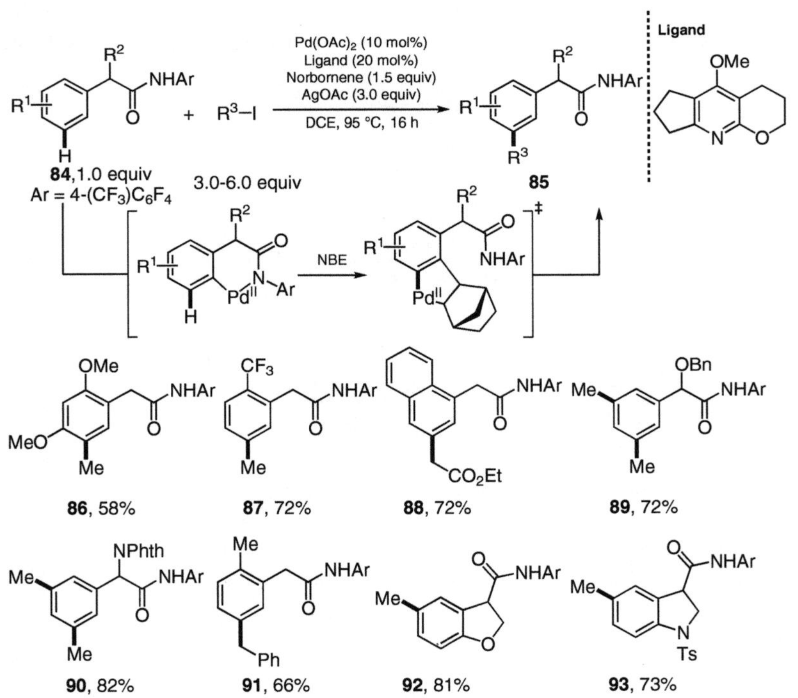

Scheme 7.28 Phenyl acetamide directed the *meta*-selective alkylation reactions.

Later, the Yu group solved the alkylation scope problem by norbornene and ligand engineering (Scheme 7.29) [37]. The major byproduct in the original report [36] was

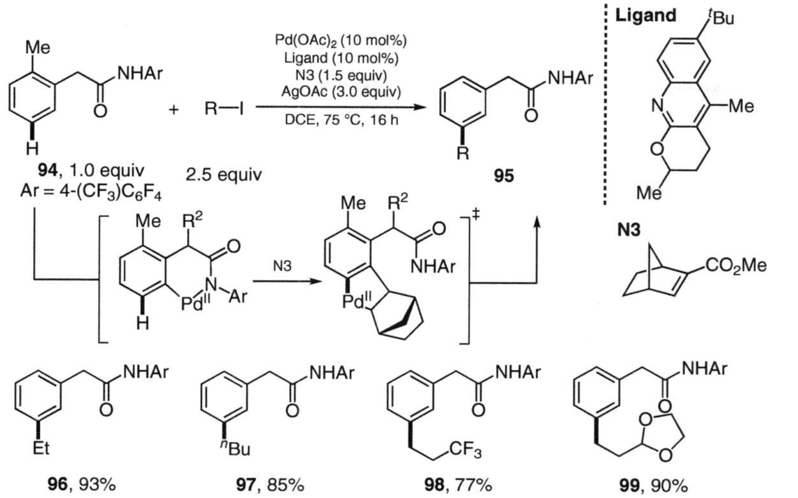

Scheme 7.29 Ester norbornene enabled the highly *meta*-selective alkylation reaction.

the benzocyclobutane, which was generated by Pd(II) reductive elimination from the ANP intermediate. To overcome this limitation, they installed an ester group at C2 position of the norbornene, which will prevent the benzocyclobutane formation via the corresponding quaternary carbon center formation. With this modified norbornene in hand, the alkylation scope was expanded to a large variety of primary alkyl iodides.

In 2016, the Yu group reported that when switching the directing group from the amide to covalent-linked pyridine as the directing group (Scheme 7.30) [38], they found that pyridone-type ligand would significantly improve the efficiency of the C—H activation steps. The combination of this directing group and ligand had enabled an extremely broad *meta*-arylation scope. The arene scope was far beyond the methyl 2-iodobenzoate. This pyridone ligand was believed to accelerate the C—H bond palladation step via a concerted metallation–deprotonation mechanism. This ligand later was widely adopted by other researchers for the challenging Pd/norbornene catalysis.

In the same year, Yu and co-workers also achieved the first *meta*-selective arene C—H bond amination and alkynylation (Scheme 7.31) [39]. These two transformations required using ester-substituted norbornene to avoid the benzocyclobutane formation. This work demonstrates that the Pd(II)/Pd(IV) cycle showed significant advantages over the Pd(0)/Pd(II)/Pd(IV) catalytic cycle. The Pd/norbornene co-catalyzed *ortho*-alkynylation of aryl iodide with corresponding alkynyl bromide was not a known transformation, due to the competing oxidative addition with Pd(0) catalyst. However, such conflicts in the catalytic cycle were nicely avoided by initiating this transformation with the Pd(II)-catalyzed C—H palladation step.

In 2016, Yu and co-workers reported the first *meta*-selective arene C—H chlorination of anilines and phenols (Scheme 7.32) [40]. They elegantly employed aryl chlorosulfate as a novel chlorinating reagent. This chlorination's regioselectivity

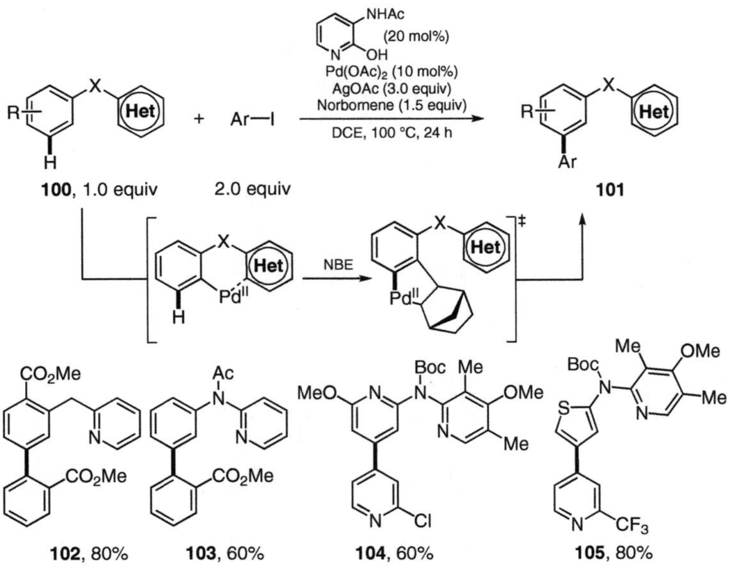

Scheme 7.30 Directing group enabled general *meta*-selective C—H bond arylation.

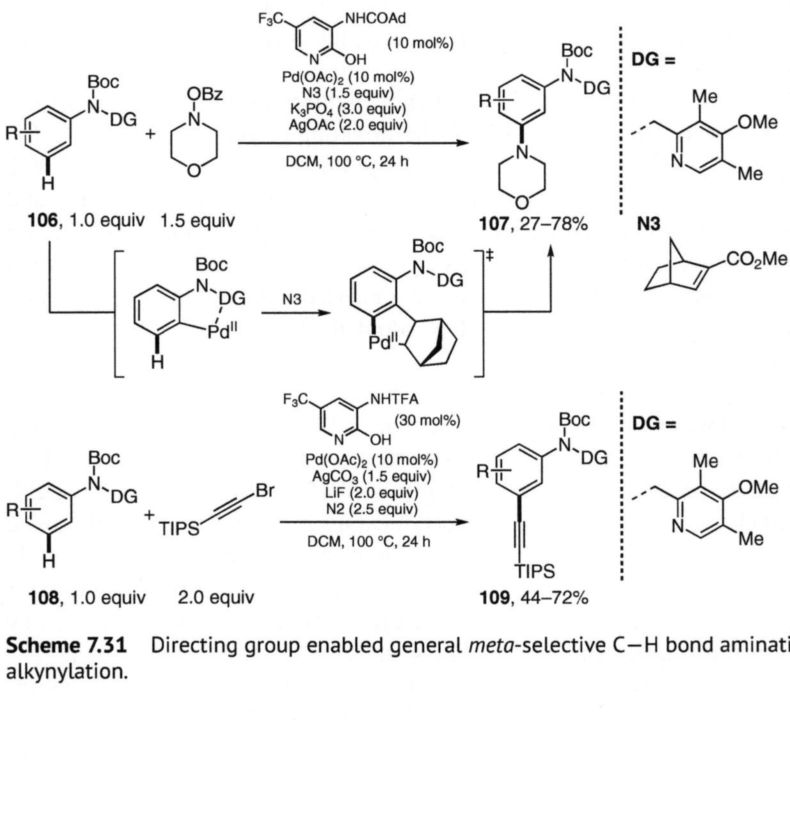

Scheme 7.31 Directing group enabled general *meta*-selective C—H bond amination and alkynylation.

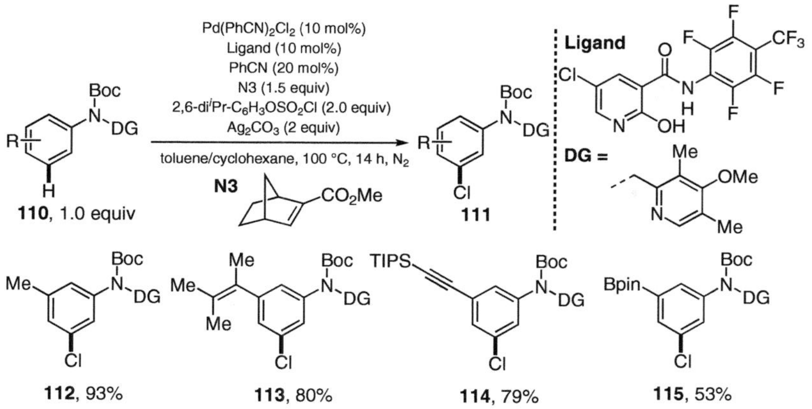

Scheme 7.32 Pd/norbornene-catalyzed *meta*-selective arene C—H chlorination of anilines and phenols.

proved to be highly *meta*-selective that overrides common para/*ortho*-selectivity for electrophilic aromatic halogenation reactions.

In 2017, Yu and co-workers further expanded this direct group strategy to common functional groups such as benzyl sulfonamide [41] and phenyl acetic acid [42] (Schemes 7.33 and 7.34). Both these transformations still required stoichiometric amount of ester-substituted norbornene to achieve the reaction generality for the aryl iodide scope.

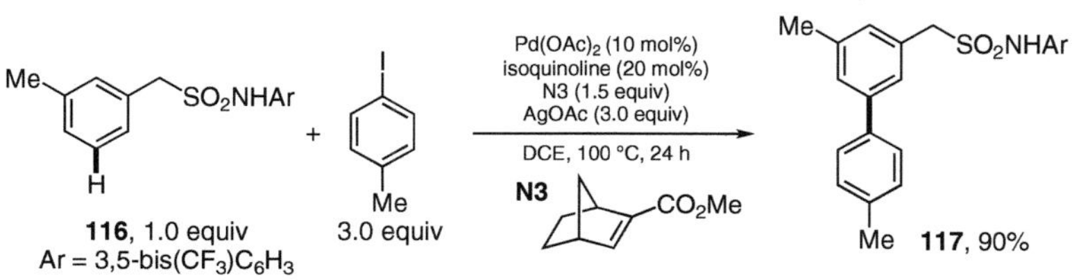

Scheme 7.33 Simple sulfonamide-directed *meta*-selective arene C—H bond arylation reaction.

In 2017, Shi and co-workers reported an interannular *meta*-selective C—H arylation of biaryl-2-trifluoroacetamides (Scheme 7.35) [43]. The authors had successfully isolated stoichiometric the key interannular biaryl *ortho*-position C—H palladation product **124**. This key observation demonstrates that a six-membered metallocycle was formed with a nitrogen atom directly bonded to a nitrogen atom that was successfully deficient by two arenes. The key intermediate **124** should further react with norbornene to couple with aryl iodide at the *meta*-position.

During the same period, the Yu group and other groups applied a very similar strategy to achieve highly *meta*-selective arylation reaction with a large variety of different directing groups including benzyl amine [44], benzyl alcohols [45], phenethylamines [46], and masked aromatic aldehydes [47]. The reaction

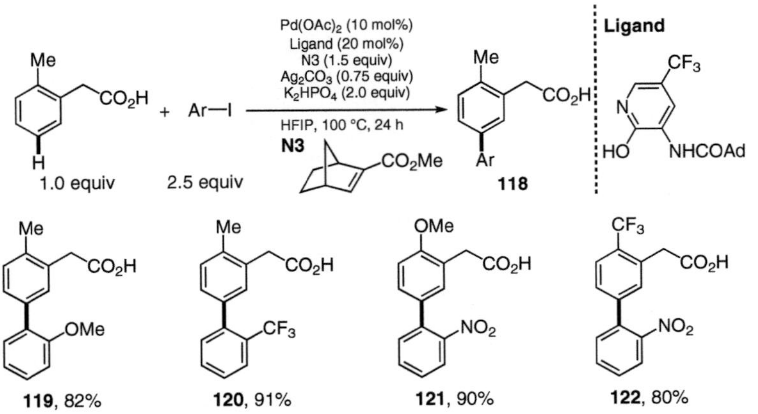

Scheme 7.34 Phenyl acetic-acid-directed *meta*-selective arene C–H bond arylation reaction.

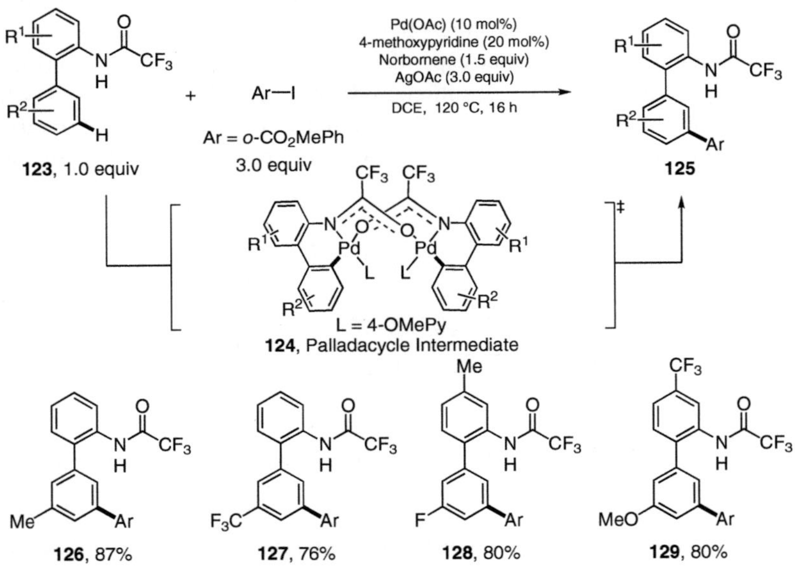

Scheme 7.35 Acetamide-directed interannular *meta*-arylation reaction.

mechanism and scope was also similar to previously discussed reaction outcomes. Thus, due to the limited space, we will not discuss those transformations in detail.

In 2018, the Yu group achieved a major breakthrough in enantioselective *meta*-C—H bond activation (Scheme 7.36) [48]. The reaction employed a racemic pyridone ligand to prompt C—H bond activation. The *ortho*-C—H activation would break the molecules' symmetry and generate the chiral benzylic center. The mechanism study revealed that the first *ortho*-C—H bond palladation steps were not highly enantioselective, and this step was highly reversible. Surprisingly, the

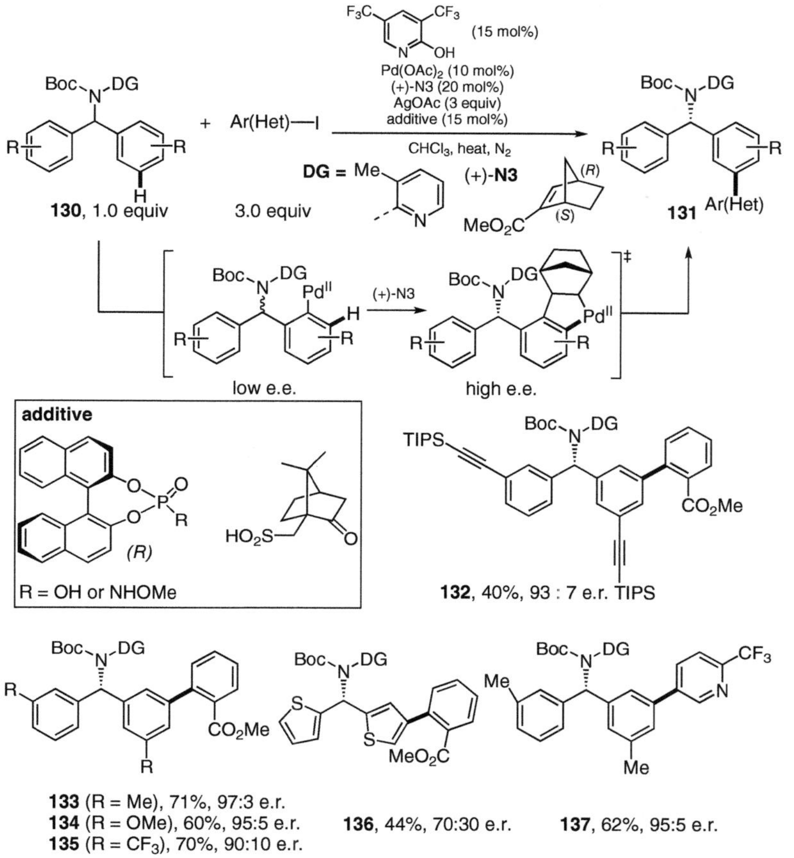

Scheme 7.36 Enantioselective remote *meta*-C—H bond arylation via chiral norbornene catalysis.

migration insertion with chiral norbornene was an enantiodetermining step and it was not reversible. The reaction scope was quite general and could differentiate two *meta*-substituted benzene with high enantioselectivity.

In 2020, the Yu group merged their template-directed C—H activation with palladium/norbornene catalysis to achieve highly selective C-6 arylation of quinolines (Scheme 7.37) [49]. This work proved the concept that norbornene was a universal transient mediator to translocate palladium from its original position to the corresponding *ortho*-position. This quinoline remote functionalization reaction set the foundation for developing other remote arene C—H bond functionalization strategies.

In 2020, Maiti and co-workers reported another big breakthrough in the field of Pd/norbornene co-catalysis. They have successfully merged the template-guided *meta*-selective C—H bond palladation reaction with Pd/norbornene co-catalysis to achieve highly *para*-selective arene C—H bond arylation reactions. This represents the first example of achieving the *para*-selective arene C—H bond activation

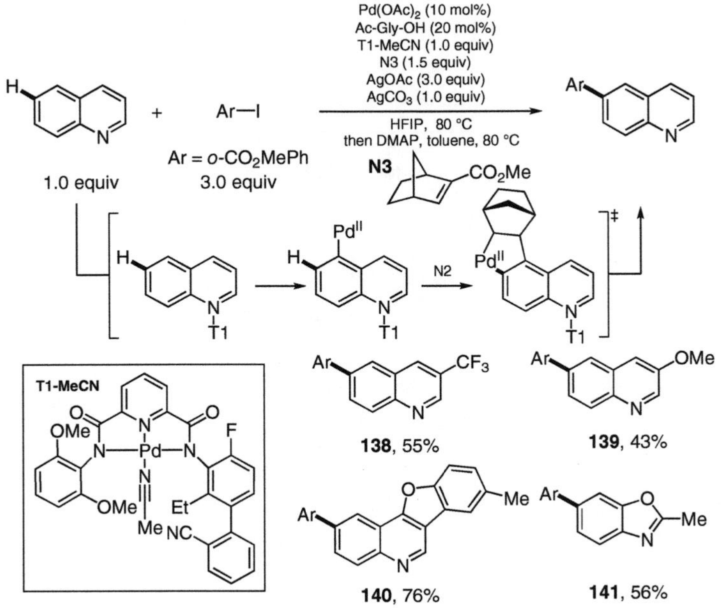

Scheme 7.37 Merger of template-directed C-5 quinoline activation with palladium/norbornene catalysis.

using Pd/norbornene catalysis. One major limitation of this work is the fact that substrates need to be blocked with both *ortho*-positions of the arene to achieve such desired *para*-selectivity. One major reason was that blocking the *ortho*-position would prevent the norbornene to translocate the palladium catalyst to the arene's *ortho*-position. Nevertheless, the reaction has a broad aryl iodide scope, and the directing template was quite easy to install and remove under mild reaction conditions (Scheme 7.38) [50].

In 2020, the Dong group merged their own *exo*-directing group strategies with their own alkene version palladium/norbornene catalysis (Scheme 7.39) [51]. The *exo*-directing group would enable C—H palladation at the closer position, and then norbornene would guide palladium catalyst to functionalize the distal position of alkene. This would provide extremely high control on the olefin's geometry and regioselectivity. The protocol introduces both arylation and alkylation with corresponding halides. For the olefin scope, both styrene and alkyl-substituted alkene could participate in this transformation.

In 2023, the Zhou group applied palladium/norbornene catalysis to synthesize the planar chiral ferrocenes (Scheme 7.40) [52]. The planar chirality was generated and determined by the *ortho*-C—H activation type. Then migratory insertion with bridge head substituted norbornene N2 would functionalize the *meta*-position of ferrocenes. The reaction gave excellent enantioselectivity for all the substrates. The difference between enantiodetermining steps compared to Yu group's earlier work [39] relied on the different electronic nature of ferrocenes and normal

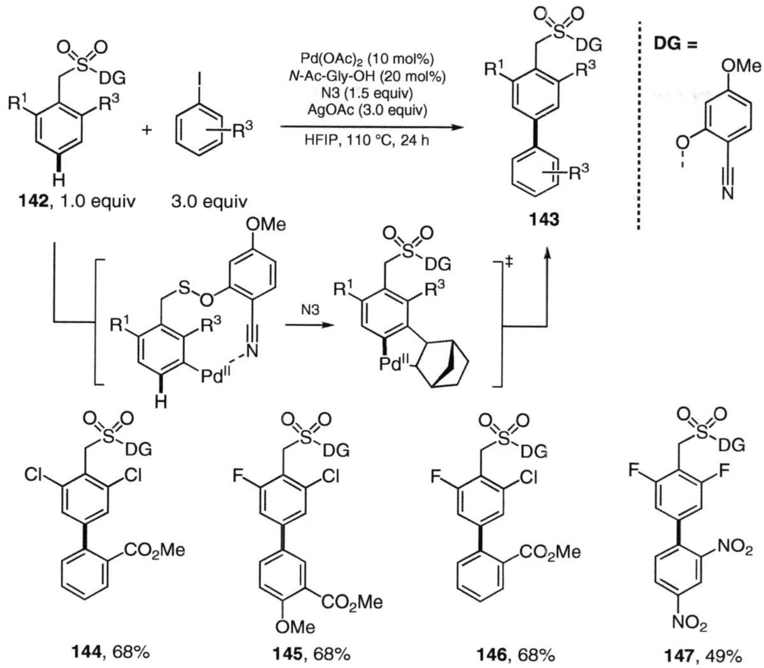

Scheme 7.38 Merger of template-directed *para*-C—H bond selective arylation.

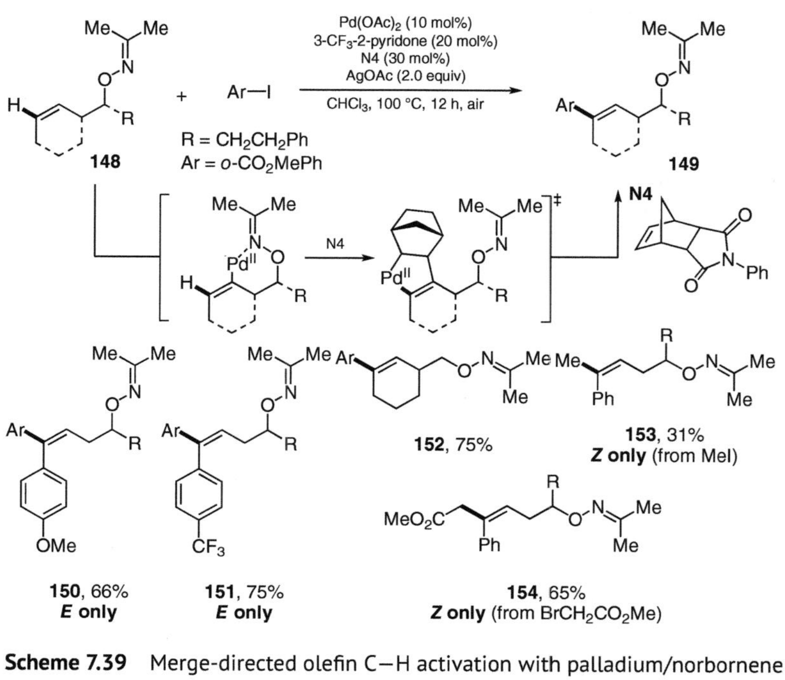

Scheme 7.39 Merge-directed olefin C—H activation with palladium/norbornene catalysis.

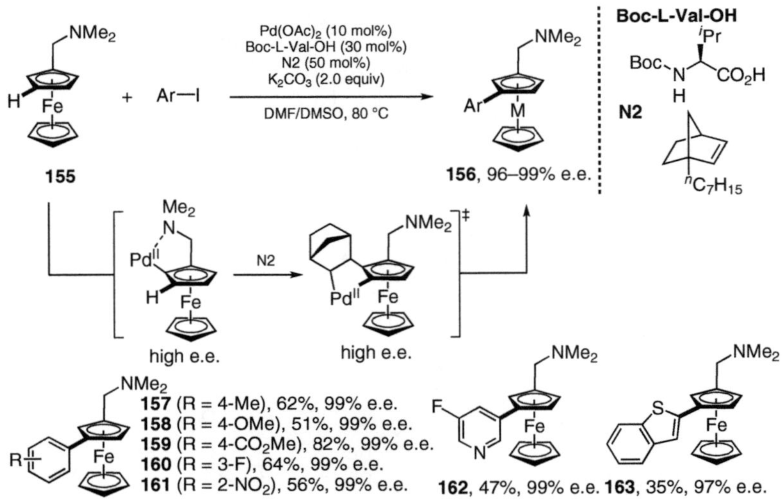

Scheme 7.40 Merger-directed enantioselective *ortho*-C–H activation with palladium/norbornene catalysis to generate planar chirality.

benzenes. Ferrocenes were much more electron-rich than benzenes, which made the C—H palladation easier and much less reversible, then triggered the change of enantiodetermining step.

7.4.2.2 Non-directed C–H Bond Activation

Similar to the directed C—H activation, the non-directed C—H bond activation could also be merged with palladium/norbornene catalysis. However, the first C—H palladation step becomes more challenging without the directing group's help, which usually limited the arene scope to those electron-rich arenes. However, this approach does not need any additional chemical steps to install and remove the directing group, so it attracts synthetic chemists to develop and apply these methods.

The first example was reported by the Yu group in 2019 [53], and they achieved the *meta*-selective C—H activation of anisole derivatives (Scheme 7.41). This elegant work fully takes advantage of palladium/norbornene catalysis to regio-convergently transfer both *ortho*- and *para*-C—H palladation intermediates to the *meta*-arylation product. The dual quinoline/ligand system was the key for the first C—H palladation steps, which were believed to be the rate-determined steps. The scope of aryl iodide was very broad, tolerating a large variety of different electronic and steric substitutions.

In the same year, Dong and co-workers reported palladium/norbornene catalyzed the regioselective difunctionalization of thiophene (Scheme 7.42) [54]. Compared with all previous Pd(II)/Pd(IV) cycle in which the Pd(II) center was always protonated off at the *ipso*-position, this work represents the first difunctionalization of two vicinal arene C—H bonds. However, the electrophile introduced at C3 position was still limited to 2-iodobenzoate, and the nucleophile introduced at C2 position

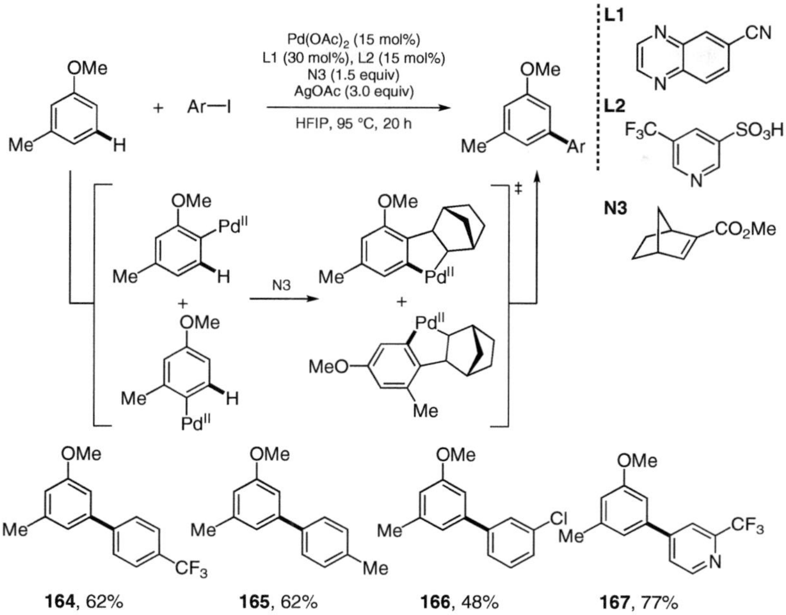

Scheme 7.41 *Meta*-arylation of anisole-type electron-rich arene.

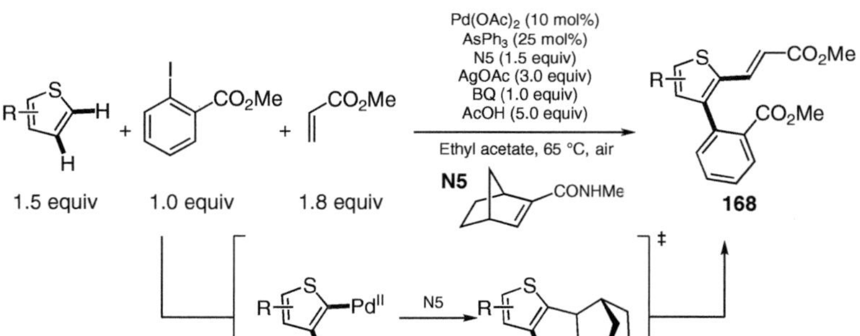

Scheme 7.42 Regioselective difunctionalization of thiophene.

was limited to acrylate derivatives. As a net oxidative reaction, the triphenyl arsenic proved to be a stable ligand to prevent palladium catalyst decomposition under air.

In 2020, the Yu group achieved the *meta*-selective C—H activation of fluoroarenes (Scheme 7.43) [55]. The first step non-directed C—H palladation step proved to be more challenging, which required two new different ligands combination. The aryl iodide scope was still limited to the electron-deficient arenes at this stage.

In 2021, Dong and co-workers achieved redox-neutral difunctionalization of five-membered heterocycles (Scheme 7.44) [56]. In this work, both aryl iodide and alkynyl bromide were coupled with an arene C—H bond with complete regioselectivity control. This demonstrates several mechanism insights: (i) The aryl

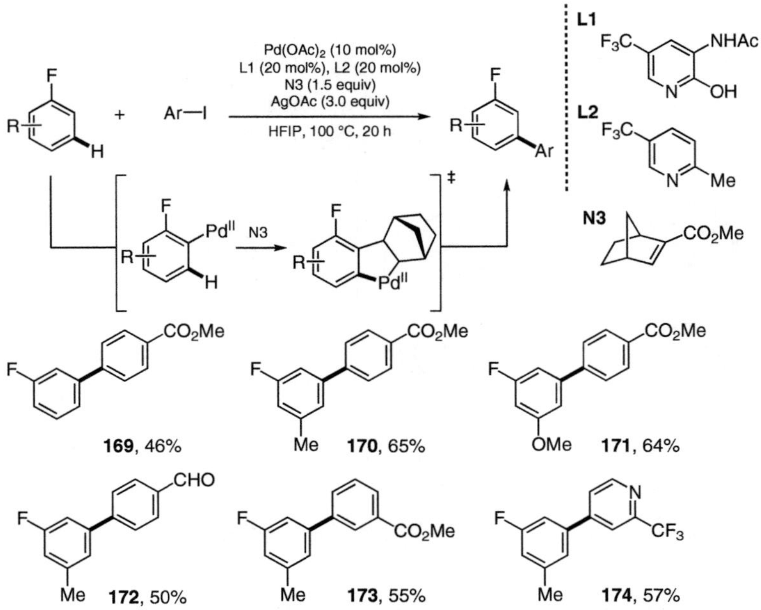

Scheme 7.43 *Meta*-arylation of mono-fluorine-substituted benzene.

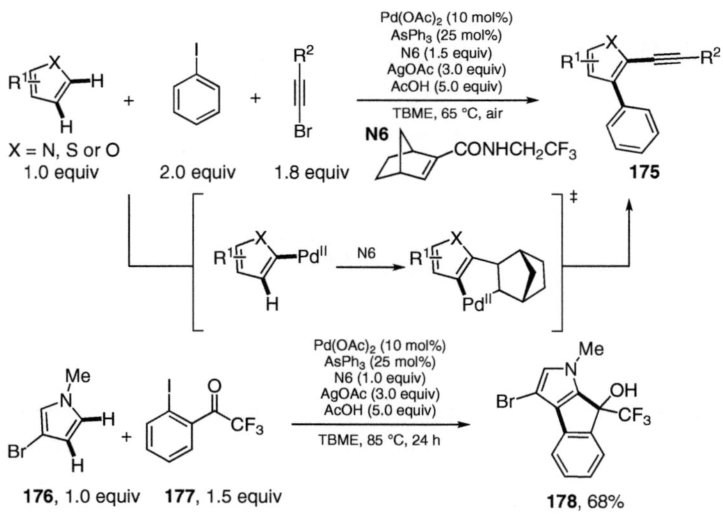

Scheme 7.44 Regioselective difunctionalization of pyrrole/furan/thiophenes.

iodide proved to be a better oxidant to selectively oxidize the ANP intermediate to the Pd(IV); (ii) alkynyl bromide selectively reacts with a simple aryl–Pd(II) intermediate other than ANP, which might undergo migratory insertion and then elimination pathway. Based on this mechanism insight, the author found that aryl–Pd(II) at the C2 position would also undergo 1,2-addition to aryl ketone to achieve intramolecular annulation to give the product **159**.

7.5 Summary and Conclusions

Prof. Catellani's early organometallic study and original work in 1997 have built all the mechanism insights for this complex Pd/norbornene catalytic cycle. Those mechanism insights had guided the synthetic chemists to further precisely tune the reactivity and selectivity of palladium/norbornene catalysis. Driven by the mechanism, people have solved the electrophile constraints, *ortho*-constraints, and aryl iodide constraints in the Pd(0)/Pd(II)/Pd(IV) cycle. Synthetic chemists have merged ortho C—H bond palladation with palladium/norbornene catalysis to achieve Pd(II)/Pd(IV) cycle, which smoothly solves the regioselective activation of the *meta*-C—H bond's issue. Although great progress has been made, the C—H bonds can be activated through palladium/norbornene catalysis were still limited to sp^2 C—H bonds. In the near future, this palladium/norbornene catalysis should be able to activate more inert sp^3 C—H bonds.

References

1 (a) Rouquet, G. and Chatani, N. (2013). Catalytic functionalization of C(sp^2)–H and C(sp^3)–H bonds by using bidentate directing groups. *Angew. Chem. Int. Ed.* 2013 (52): 11726–11743. (b) Lyons, T.W. and Sanford, M.S. (2010). Palladium-catalyzed ligand-directed C–H functionalization reactions. *Chem. Rev.* 110 (2): 1147–1169.

2 Daugulis, O., Roane, J., and Tran, L.D. (2015). Bidentate, monoanionic auxiliary-directed functionalization of carbon–hydrogen bonds. *Acc. Chem. Res.* 48 (4): 1053–1064.

3 Davies, H.M.L. and Morton, D. (2016). Recent advances in C—H functionalization. *J. Org. Chem.* 81 (2): 343–350.

4 Gandeepan, P. and Ackermann, L. (2017). Transient directing groups for transformative C—H activation by synergistic metal catalysis. *Chem* 4 (2): 1–24.

5 (a) Catellani, M. (2003). Catalytic multistep reactions via palladacycles. *Synlett* 34 (3): 0298–0313. (b) Catellani, M. (2005). Novel methods of aromatic functionalization using palladium and norbornene as a unique catalytic system. *Top. Organomet. Chem.* 14: 21–53. (c) Catellani, M., Motti, E., and Della Ca', N. (2008). Catalytic sequential reactions involving palladacycle-directed aryl coupling steps. *Acc. Chem. Res.* 41 (11): 1512–1522. (d) Martins, A., Mariampillai, B., and Lautens, M. (2010). Synthesis in the key of Catellani: norbornene-mediated

ortho C–H functionalization. *Top. Curr. Chem.* 292: 1–33. (e) Ye, J. and Lautens, M. (2015). Palladium-catalysed norbornene-mediated C—H functionalization of arenes. *Nat. Chem.* 7: 863–870. (f) Della Ca', N., Fontana, M., Motti, E. et al. (2016). Pd/norbornene: a winning combination for selective aromatic functionalization via C—H bond activation. *Acc. Chem. Res.* 49 (7): 1389–1400. (g) Wang, J. and Dong, G. (2019). Palladium/norbornene cooperative catalysis. *Chem. Rev.* 119 (12): 7478–7528. (h) Zhao, K., Ding, L., and Gu, Z. (2018). Development of new electrophiles in palladium/norbornenecatalyzed *ortho*-functionalization of aryl halides. *Synlett* 29: 129–140.

6 Horino, H., Arai, M., and Inoue, M. (1974). Reaction of norbornenes with phenylpalladium chloride. *Tetrahedron Lett.* 15 (8): 647–650.

7 Catellani, M. and Chiusoli, G.P. (1982). Catalytic activation of aromatic C—H bonds. *J. Organomet. Chem.* 239 (2): 35–37.

8 Bocelli, G., Catellani, M., Chiusoli, G.P. et al. (1984). Palladium-catalyzed C—C bond formation involving aromatic C—H activation. X-ray structure of 5-nitro -8-(2[2. 2.1]bicycloheptyl)-1,2,3,4,4a,8b-hexahydro-1,4-methanobiphenylene. *J. Organomet. Chem.* 265 (1): 9–11.

9 Catellani, M. and Chiusoli, G.P. (1985). Palladium-catalyzed synthesis of 1,2,3,4,4a,12b-hexahydro-1,4-methanotriphenylenes. *J. Organomet. Chem.* 286 (1): 13–16.

10 Catellani, M. and Chiusoli, G.P. (1988). Palladium-(II) and -(IV) complexes as intermediates in catalytic C—C bond-forming reactions. *J. Organomet. Chem.* 346 (1): 27–30.

11 (a) Catellani, M. and Mann, B.E. (1990). Conformational effects in elementary steps in catalytic reactions. Oxidative addition of 3-bromoprop-1-ene to a palladium(II) metallacyclic complex. *J. Organomet. Chem.* 390 (2): 251–255. (b) Bocelli, G., Catellani, M., and Ghelli, S. (1993). Regioselective ring opening of a palladium(IV) alkylaromatic metallacycle by benzyl group migration from palladium to the aromatic carbon and X-ray structure of the resulting palladium(II) complex. *J. Organomet. Chem.* 458 (1–2): 12–15.

12 Catellani, M. and Fagnola, M.C. (1995). Palladacycles as intermediates for selective dialkylation of arenes and subsequent fragmentation. *Angew. Chem. Int. Ed.* 33: 2421–2422.

13 Catellani, M., Frignani, F., and Rangoni, A. (1997). A complex catalytic cycle leading to a regioselective synthesis of o,o′-disubstituted vinylarenes. *Angew. Chem. Int. Ed.* 36: 119–122.

14 Chai, D.I., Thansandote, P., and Lautens, M. (2011). Mechanistic studies of Pd-catalyzed regioselective aryl C—H bond functionalization with strained alkenes: origin of regioselectivity. *Chem. Eur. J.* 17: 8175–8188.

15 (a) Maestri, G., Motti, E., Della Ca, N. et al. (2011). Of the *ortho* effect in palladium/norbornene-catalyzed reactions: a theoretical investigation. *J. Am. Chem. Soc.* 133 (22): 8574–8585. (b) Larraufie, M.-H., Maestri, G., Beaume, A. et al. (2011). Exception to the *ortho* effect in palladium/norbornene catalysis. *Angew. Chem. Int. Ed.* 50: 12253–12256.

16 Dong, Z. and Dong, G. (2013). *Ortho* vs ipso: site-selective Pd and norbornene-catalyzed arene C—H amination using aryl halides. *J. Am. Chem. Soc.* 135 (49): 18350–18353.

17 Catellani, M., Mealli, C., Motti, E. et al. (2002). Palladium–arene interactions in catalytic intermediates: an experimental and theoretical investigation of the soft rearrangement between η^1 and η^2 coordination modes. *J. Am. Chem. Soc.* 124 (16): 4336–4346.

18 Lautens, M. and Piguel, S. (2000). A new route to fused aromatic compounds by using a palladium-catalyzed alkylation–alkenylation sequence. *Angew. Chem. Int. Ed.* 39: 1045–1046.

19 Faccini, F., Motti, E., and Catellani, M. (2004). A new reaction sequence involving palladium-catalyzed unsymmetrical aryl coupling. *J. Am. Chem. Soc.* 126 (1): 78–79.

20 (a) Ca, N.D., Sassi, G., and Catellani, M. (2008). A direct palladium-catalyzed route to selectively substituted carbazoles through sequential C—C and C—N bond formation: synthesis of carbazomycin A. *Adv. Synth. Catal.* 350: 2179–2182. (b) Della Ca', N., Motti, E., and Catellani, M. (2008). Palladium-catalyzed synthesis of selectively substituted phenanthridine derivatives. *Adv. Synth. Catal.* 350: 2513–2516. (c) Motti, E., Della Ca', N., Xu, D. et al. (2012, 2012). Sequential Pd/norbornene-catalyzed process generates o-biaryl carbaldehydes or ketones via a redox reaction or 6H-dibenzopyrans by C—O ring closure. *Org. Lett.* 14 (22): 5792–5795.

21 (a) Dong, Z., Wang, J., Ren, Z., and Dong, G. (2015). *Ortho* C—H acylation of aryl iodides by palladium/norbornene catalysis. *Angew. Chem. Int. Ed.* 54: 12664–12668. (b) Wang, J., Zhang, L., Dong, Z. et al. (2016). Reagent-enabled *ortho*-alkoxycarbonylation of aryl iodides via palladium/norbornene catalysis. *Chem* 1 (4): 581–591.

22 Qi, X., Wang, J., Dong, Z. et al. (2020). Compatibility score for rational electrophile selection in Pd/NBE cooperative catalysis. *Chem* 6 (10): 2810–2825.

23 Xu, X. and Jiao, L. (2023). Stereochemistry of the reactions between palladacycle complexes and primary alkyl iodides. *Organometallics* 42 (7): 606–614.

24 (a) Li, R., Zhou, Y., Yoon, K.-Y. et al. (2019). Sulfenamide-enabled *ortho* thiolation of aryl iodides via palladium/norbornene cooperative catalysis. *Nat. Commun.* 10 (1): 3555. (b) Liu, X., Fu, Y., and Chen, Z. (2023). *Ortho*-C—H methoxylation of aryl halides enabled by a polarity-reversed N—O reagent. *Nat. Chem.* 15: 1391–1399.

25 Wang, J., Li, R., Dong, Z. et al. (2018). Complementary site-selectivity in arene functionalization enabled by overcoming the *ortho* constraint in palladium/norbornene catalysis. *Nat. Chem.* 10: 866–872.

26 Jiao, L. and Bach, T. (2011). Palladium-catalyzed direct 2-alkylation of indoles by norbornene-mediated regioselective cascade C—H activation. *J. Am. Chem. Soc.* 133 (33): 12990–12993.

27 (a) Jiao, L., Herdtweck, E., and Bach, T. (2012). Pd(II)-catalyzed regioselective 2-alkylation of indoles via a norbornene-mediated C—H activation: mechanism and applications. *J. Am. Chem. Soc.* 134 (35): 14563–14572.

(b) Jiao, L. and Bach, T. (2014). Regioselective direct C-H alkylation of NH indoles and pyrroles by a palladium/norbornene-cocatalyzed process. *Synthesis* 46: 35–41.

28 Jiao, L. and Bach, T. (2013). Palladium-catalyzed direct C—H alkylation of electron-deficient pyrrole derivatives. *Angew. Chem. Int. Ed.* 52: 6080–6083.

29 Potukuchi, H.K. and Bach, T. (2013). Selective C-2 alkylation of tryptophan by a Pd(II)/norbornene promoted C—H activation reaction. *J. Org. Chem.* 78 (23): 12263–12267.

30 Gao, Y., Zhu, W., Yin, L. et al. (2017). Palladium-catalyzed direct C2-arylation of free (N—H) indoles via norbornene-mediated regioselective C—H activation. *Tetrahedron Lett.* 58 (23): 2213–2216.

31 Zhang, H., Wang, H.-Y., Luo, Y. et al. (2018). Regioselective palladium-catalyzed C—H bond trifluoroethylation of indoles: exploration and mechanistic insight. *ACS Catal.* 8 (3): 2173–2180.

32 Elsaid, M., Ge, R., Liu, C. et al. (2023). Site-selective C—H functionalization of carbazoles. *Angew. Chem. Int. Ed.* 62: e202303110.

33 Chauhan, N.S. and Mandal, P.K. (2024). Palladium-catalysed norbornene-mediated regioselective direct C—H glycosylation of free (N—H) pyrrole derivatives. *Adv. Synth. Catal.* 366: 1291–1297.

34 Minami, Y., Kodama, T., and Hiyama, T. (2015). Dehydrogenative carbon–carbon bond formation using alkynyloxy moieties as hydrogen-accepting directing groups. *Angew. Chem. Int. Ed.* 54: 11813–11816.

35 Dong, Z., Wang, J., and Dong, G. (2015, 2015). Simple amine-directed meta-selective C—H arylation via Pd/norbornene catalysis. *J. Am. Chem. Soc.* 137 (18): 5887–5890.

36 Wang, X.-C., Gong, W., Fang, L.-Z. et al. (2015). Ligand-enabled meta-C—H activation using a transient mediator. *Nature* 519: 334–338.

37 Shen, P.-X., Wang, X.-C., Wang, P. et al. (2015). Ligand-enabled meta-C—H alkylation and arylation using a modified norbornene. *J. Am. Chem. Soc.* 137 (36): 11574–11577.

38 Wang, P., Farmer, M.E., Huo, X. et al. (2016). Ligand-promoted meta-C—H arylation of anilines, phenols, and heterocycles. *J. Am. Chem. Soc.* 138 (29): 9269–9276.

39 Wang, P., Li, G.-C., Jain, P. et al. (2016). Ligand-promoted meta-C—H amination and alkynylation. *J. Am. Chem. Soc.* 138 (29): 14092–14099.

40 Shi, H., Wang, P., and Suzuki, S. (2016). Ligand promoted meta-C—H chlorination of anilines and phenols. *J. Am. Chem. Soc.* 138 (29): 14876–14879.

41 Cheng, G., Wang, P., and Yu, J.-Q. (2017). meta-C–H arylation and alkylation of benzylsulfonamide enabled by a palladium(II)/isoquinoline catalyst. *Angew. Chem. Int. Ed.* 56: 8183–8186.

42 Li, G.-C., Wang, P., Farmer, M.E. et al. (2017). Ligand-enabled auxiliary-free meta-C—H arylation of phenylacetic acids. *Angew. Chem. Int. Ed.* 56: 6874–6877.

43 Ling, P.-X., Chen, K., and Shi, B.-F. (2017). Palladium-catalyzed interannular meta-C—H arylation. *Chem. Commun.* 53: 2166–2169.

44 (a) Wang, P., Farmer, M.E., and Yu, J.-Q. (2017). Ligand-promoted meta-C—H functionalization of benzylamines. *Angew. Chem. Int. Ed.* 56: 5125–5129. (b) Ding, Q., Ye, S., Cheng, G. et al. (2016). Ligand-enabled meta-selective C—H arylation of nosyl-protected phenethylamines, benzylamines, and 2-aryl anilines. *J. Am. Chem. Soc.* 139 (1): 417–425.

45 Li, Q. and Ferreira, E.M. (2017). Meta-selective C—H arylation of aromatic alcohols with a readily attachable and cleavable molecular scaffold. *Chem. Eur. J.* 23: 11519–11523.

46 (a) Han, J., Zhang, L., Zhu, Y. et al. (2016). Highly regioselective meta arylation of oxalyl amide-protected β-arylethylamine via the Catellani reaction. *Chem. Commun.* 52: 6903–6906. (b) Liu, J., Ding, Q., Fang, W. et al. (2018). Pd(II)/norbornene-catalyzed meta-C—H alkylation of nosyl-protected phenylalanines. *J. Org. Chem.* 83: 13211–13216.

47 Farmer, M.E., Wang, P., Shi, H. et al. (2018). Palladium-catalyzed meta-C—H functionalization of masked aromatic aldehydes. *ACS Catal.* 8 (8): 7362–7367.

48 Shi, H., Herron, A.N., Shao, Y. et al. (2018). Enantioselective remote meta-C—H arylation and alkylation via a chiral transient mediator. *Nature* 558: 581–585.

49 Shi, H., Lu, Y., Weng, J. et al. (2020). Differentiation and functionalization of remote C—H bonds in adjacent positions. *Nat. Chem.* 12: 399–404.

50 Dutta, U., Porey, S., Pimparkar, S. et al. (2020). *para*-selective arylation of arenes: a direct route to biaryls by norbornene relay palladation. *Angew. Chem. Int. Ed.* 59: 20831–20836.

51 Wu, Z., Fatuzzo, N., Dong, G. et al. (2020). Distal alkenyl C—H functionalization via the palladium/norbornene cooperative catalysis. *J. Am. Chem. Soc.* 142 (6): 2715–2720.

52 Zhou, L., Cheng, H.G., Li, L. et al. (2023). Synthesis of planar chiral ferrocenes via enantioselective remote C—H activation. *Nat. Chem.* 15: 815–823.

53 Liu, L.-Y., Qiao, J.X., Yeung, K.-S. et al. (2019). meta C—H arylation of electron-rich arenes: reversing the conventional site selectivity. *J. Am. Chem. Soc.* 141 (37): 14870–14877.

54 Li, R., Zhou, Y., Xu, X. et al. (2019). Direct vicinal difunctionalization of thiophenes enabled by the palladium/norbornene cooperative catalysis. *J. Am. Chem. Soc.* 141 (48): 18958–18963.

55 Liu, L.-Y., Qiao, J.X., Yeung, K.-S. et al. (2020). meta-selective C—H arylation of fluoroarenes and simple arenes. *Angew. Chem. Int. Ed.* 59: 13831–13835.

56 Li, R. and Dong, G. (2021). Redox-neutral vicinal difunctionalization of five-membered heteroarenes with dual electrophiles. *Angew. Chem. Int. Ed.* 60: 26184–26191.

8

Transient Directing Groups in C—H Bond Functionalization

Tsz-Kan Ma, Hannan M. Seyal, and James A. Bull

Department of Chemistry, Imperial College London, Molecular Sciences Research Hub, White City Campus, Wood Lane, London W12 0BZ, UK

8.1 Introduction

8.1.1 The Concept of Transient Directing Groups for C—H Functionalization

C—H functionalization has enabled new synthetic disconnections, leading to streamlined syntheses of complex organic molecules with improved efficiency and atom economy [1–7]. A common strategy for achieving excellent regioselectivity in C—H functionalization reactions is the use of covalently attached directing groups (DGs) (Figure 8.1a). These DGs contain Lewis basic sites that coordinate with the metal catalyst, positioning it near the target C—H bond of the substrate and facilitating C—H activation (Figure 8.1a). This forms an intermediate C—M bond, which subsequently productively converts into other carbon–carbon or carbon–heteroatom bonds through various mechanistic pathways, depending on the metal catalysts used (e.g. Pd, Ni, and Cu). Notable examples of DGs include Daugulis's bidentate 8-aminoquinoline and Yu's weakly coordinating acidic amides, which have demonstrated the versatility and breadth of this strategy across both aromatic and aliphatic substrates [8–10]. However, a significant limitation of this approach is the additional steps required for DG installation and removal. These steps not only necessitate extra pre-functionalization but also sometimes require harsh conditions for removal, limiting the applicability of a method in more complex structures and late-stage functionalization of bioactive compounds [11]. These two additional steps can offset the efficiency gains provided by a C—H functionalization approach, which has driven investigations into streamlined DG strategies [12, 13].

Transient C—H functionalization has emerged as an approach to address the limitations of powerful C—H functionalization synthetic methodology. Transient directing groups (TDGs) involve the in situ formation of a labile linkage that temporarily attaches a DG to facilitate C—H functionalization. The directing

Directed C–H Bond Functionalization: Concepts and Applications, First Edition.
Edited by Debabrata Maiti and Supriya Rej.

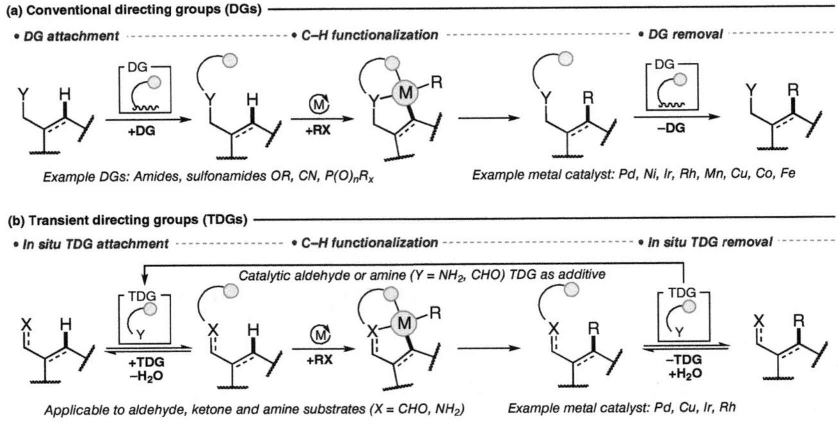

Figure 8.1 Conventional and transient directing groups for C—H functionalization. (a) Conventional directing groups and (b) the transient directing groups approach.

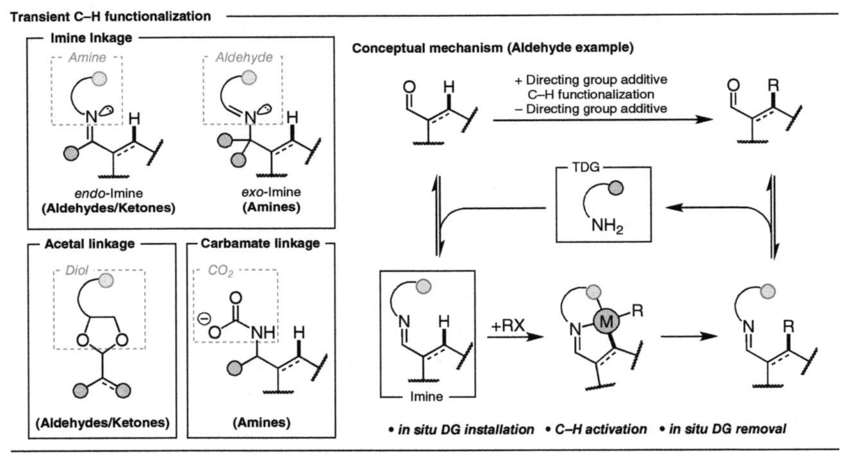

Figure 8.2 C—H functionalization with transient directing group overview.

motif of the resulting product is then readily hydrolyzed to regenerate the TDG additive, restoring the linking functionality for subsequent derivatization reactions (Figure 8.1b) [14–22]. While the majority of transient C—H functionalization reactions utilize dynamically formed imines as a catalytic DG, other acid-labile functional groups such as acetals or carbamic acids have been employed in a stoichiometric one-pot manner to direct C—H functionalization (Figure 8.2) [23–25]. The functionalization of phenol derivatives has also been investigated, with a phosphinite formed as the transient species to direct metalation [26–30].

The term "transient directing group" can refer to either the aldehyde or amine additive acting as an organocatalyst for imine formation or the imine formed from their condensation with the substrate. Since the resultant imine serves as the DG, it is more precise to refer to that as the TDG. Nonetheless, the additive is commonly referred to as the "TDG" also and we will do so here. Imines are readily formed by the condensation of the carbonyl groups of aldehydes or ketones with amines. The Lewis basic imine nitrogen is capable of coordinating with a variety of metal catalysts through the well-positioned non-bonding lone pairs of electrons to promote cyclometalation. A second binding motif, such as carboxylic acid, amide, or phenolate, can be incorporated to produce a bidentate ligand for C—H activation. The high tunability of these DGs makes them highly versatile for functionalizing both carbonyl and amine substrates. The major factor defining a TDG is that the imine is formed and cleaved in a highly dynamic and reversible environment as a transient species in the reaction. Therefore, no additional steps for the DG installation and cleavage are required. The dynamic nature of transient C—H functionalization enables the catalytic usage of the DG additive when the rates of the two catalytic cycles are compatible. Synchronizing the rates of the organocatalytic process for imine formation and cleavage with the metal-catalyzed C—H functionalization is key to a productive overall process. The lifetime of the TDG must be long enough for C—H functionalization to occur but short enough to be hydrolyzed, minimizing undesired side reactions such as difunctionalization and catalyst deactivation. Various strategies have been employed to achieve the optimal rate of TDG formation and C—H functionalization. For instance, fluorinated solvents have been used, in particular 1,1,1,3,3,3-hexafluoroisopropanol (HFIP) with water or acid additives, to control the imine equilibrium. Additionally, the use of 2-pyridone derivatives as ligand additives has proven to be a useful feature in unlocking new reactivities by lowering the energy barriers of key transition states.

This chapter will highlight selected examples of TDG-mediated C—H functionalization of aldehydes, ketones, and amines. The chapter is organized chronologically by substrate types and subdivided by metal catalyst and transformation. Most examples involve the use of imine DGs to functionalize complementary amine or aldehyde substrates. Other acid-labile DGs, such as acetals and carbamates, which can be installed and removed in one synthetic step, will also be discussed. For a comprehensive coverage of the literature on this topic, readers are encouraged to refer to recent review articles published in the field [14–21, 31–34].

8.1.2 Early Developments Using Stoichiometric Imine to Direct C—H Functionalization

Before a discussion on transient imines, this section discusses seminal studies using imine DGs. The earliest example of imine-directed C—H functionalization was reported by Murahashi in 1955. This work utilized an imine DG to achieve the carbonylation of aldimines using a dicobalt octacarbonyl catalyst under 100–200 atmospheric pressure of carbon monoxide at 220–230 °C (Figure 8.3a) [35]. Using palladium catalysis, early investigations into the C—H activation of pinacolone oxime by Shaw led to the isolation of a dimeric palladacycle (Figure 8.3b) [36]. Similarly, Dyke's study using *N*-benzyl imines also resulted in the isolation of dimeric palladacycles [37]. These pioneering studies established imines and

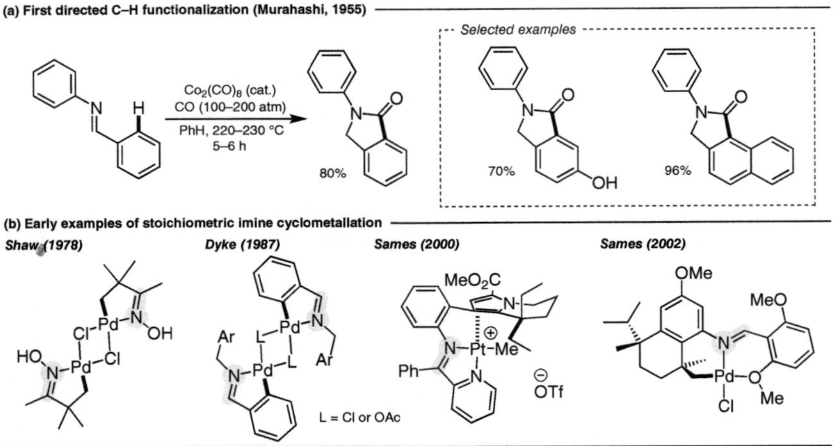

Figure 8.3 Early examples of imine-directed C−H functionalization. (a) The first example of imine-directed C−H functionalization and (b) additional imine-directed cyclometallation examples with Pd and Pt.

oximes as effective DGs for cyclometalation, although they did not further explore subsequent derivatization reactions. In the 2000s, Sames demonstrated the use of stoichiometric imines in platinum and palladium-mediated C—H activation as the key steps in the synthesis of the antibiotic Rhanzinilam and Teleocidin B4 [38, 39].

Building on these foundational studies, the application of imines and oximes as DGs for C—H functionalization significantly expanded. In 2009, Sanford reported the use of oximes and imines for C(sp^2)—H and C(sp^3)—H acetoxylation with $PhI(OAc)_2$ as the oxidant (Figure 8.4a) [40, 41]. Ji demonstrated the use of stoichiometric imines for C(sp^2)—H alkenylation of benzaldehydes with imine containing 2-((aminooxy)methyl)benzonitrile as an internal DG (Figure 8.4b) [42]. Moreover, stoichiometric imines have been utilized to direct metal-free C(sp^2)—H borylation (Figure 8.4c) [43].

Stoichiometric imine DGs have also been applied to amine functionalization, and Dong reported the use of 8-formylquinoline as a stoichiometric additive to form the iminoquinoline DG *in situ* with primary amines to direct C(sp^3)—H arylation with aryl iodonium salts as the electrophiles (Figure 8.4e) [44]. The exo-imine-type DG was applied to a broad range of aliphatic and aniline-derived amines to arylate the terminal methyl or methylene C—H bond. Additionally, Maiti demonstrated the possibility of using a pyrimidine-based imine to direct *meta*-C—H alkenylation of biaryl amines and aldehydes [45]. These works set the scene for the use of imines in a transient sense.

Recent developments have focused on utilizing transient imines as DGs. By leveraging the labile nature of the imine linker, *in situ* DG installation and cleavage can be achieved in one synthetic step, with the amine or aldehyde TDG additive acting as an organocatalyst. This concept was first demonstrated by Jun in 1997 on the rhodium-catalyzed aldehydic C—H hydroacylation of alkenes (Figure 8.5)

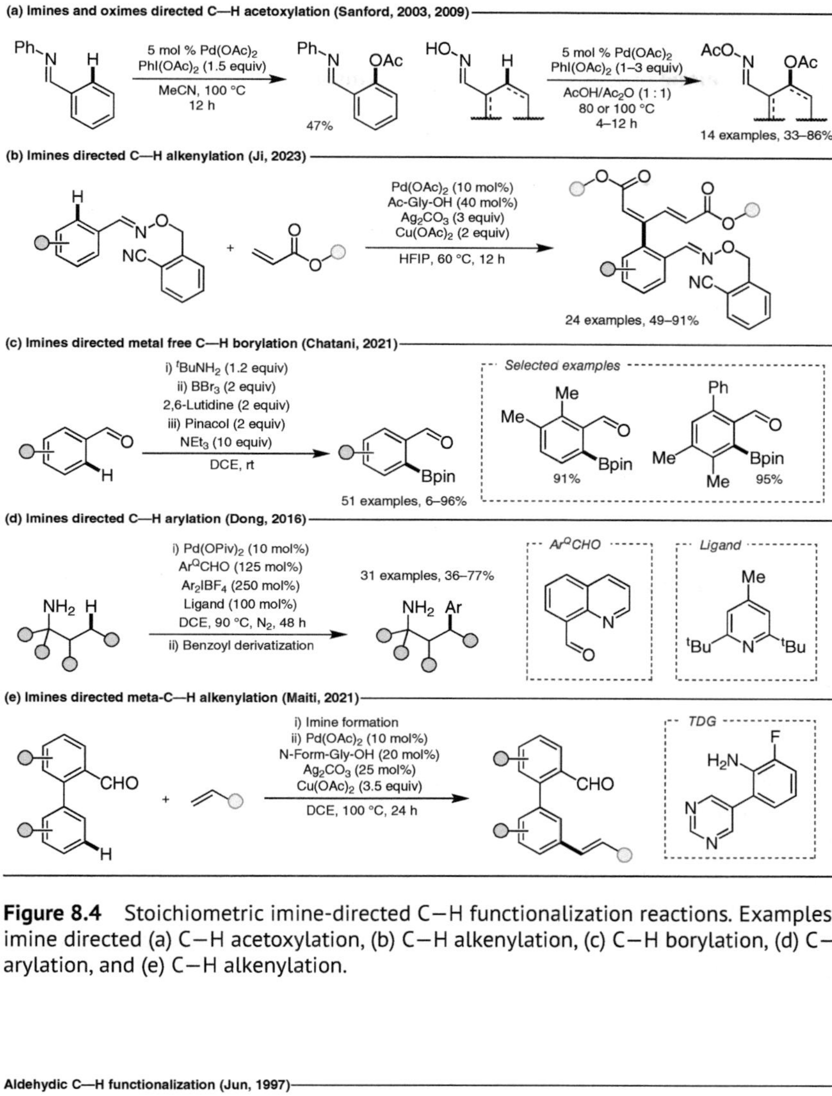

Figure 8.4 Stoichiometric imine-directed C—H functionalization reactions. Examples of imine directed (a) C—H acetoxylation, (b) C—H alkenylation, (c) C—H borylation, (d) C—H arylation, and (e) C—H alkenylation.

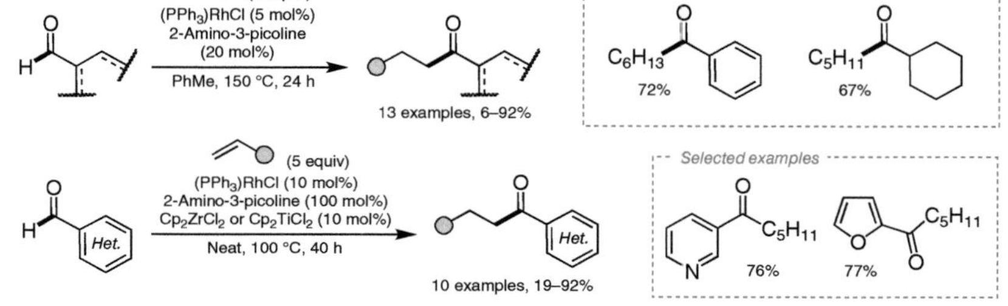

Figure 8.5 Pioneering work on transient directing-group-mediated C—H functionalization.

[46, 47]. In this process, 2-amino-3-picoline was used as an additive to form an imine intermediate with aldehydes, which then reacted with alkenes to produce the corresponding ketone products catalyzed by Wilkinson's catalyst.

A catalytic amount of the amine additive (20 mol%) was sufficient for both aliphatic and aromatic aldehydes. One equivalent of the amine additive was required for heteroaromatic substrates, along with a catalytic amount of bis(cyclopentadienyl)zirconium dichloride or bis(cyclopentadienyl) titanium dichloride to mask the heteroatom in these aldehydes to maintain the catalytic activity of the rhodium catalyst. However, it was the use of imines for the β-functionalization of aldehydes and γ-functionalization of amines that spurred the field. The next sections cover transient C—H functionalization of C—H bonds at sp^3 and sp^2 carbon centers.

8.2 Transient C(sp^3)—H Functionalization

8.2.1 C(sp^3)—H Functionalization of Aldehydes

Aldehydes are highly versatile building blocks, acting as accessible precursors to a range of functional groups. As such, there has been a concerted effort for the development of new methodologies that can functionalize aldehyde systems. The field has seen rapid development following Yu's influential work on palladium-catalyzed C(sp^3)—H functionalization of aldehydes and ketones, reported in 2016 (Figure 8.6) [48]. In this major breakthrough, an α-amino acid, glycine, was used as an amine additive to form imines with aldehyde substrates, directing C(sp^3)—H arylation with aryl and heteroaryl iodides. The reaction conditions included silver trifluoroacetate additives and used HFIP as a solvent with acetic acid (HFIP : AcOH, 3 : 1 or 9 : 1). HFIP was crucial for reactions with ketones and enantioselective reactions with aldehyde substrates and has since become a mainstay in C—H functionalization strategies, owing to unique solvent properties of high polarity, H-bond capabilities, and relative acidity [49–51].

In reactions with aldehyde substrates, Yu also added water either as a co-solvent or additive. In the absence of water, starting material degradation was observed. It

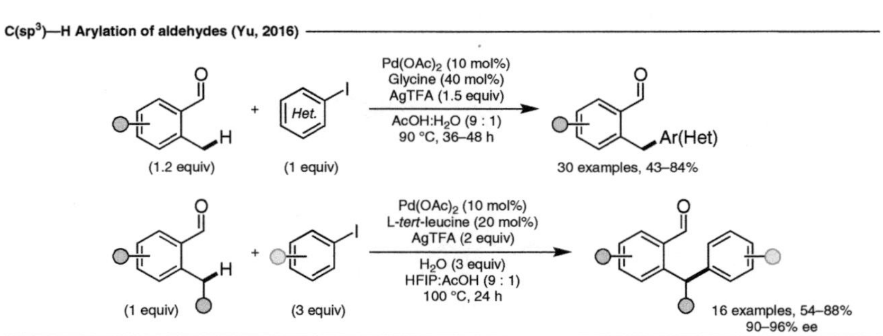

Figure 8.6 Pioneering work on transient C(sp^3)—H functionalization of aldehydes.

was proposed that this was the result of mismatched rates, comparing imine formation and subsequent coordination and C—H cleavage, leading to an accumulation and subsequent degradation of imine. The addition of water significantly reduced the level of starting material degradation. It was proposed that water reduced the concentration of imine within the reaction, resulting in better-matched rates when comparing the two steps – accounting for a decrease in starting material degradation and overall increase in yield toward the desired product.

Arylation was possible with a selection of electronically distinct aryl-coupling partners; however, reduced yields were obtained when using sterically hindered aryl iodides, e.g. 1-iodo-2-methylbenzene. Within the same report, L-*tert*-leucine was shown to form a successful chiral transient DG, allowing for enantioselective arylation in good yield and excellent *ee*'s. The origin of the observed enantioselectivity was proposed to be due to steric interactions between R groups of the substrate and R" groups of the TDG, accounted for by increased er as the steric bulk of the TDG was increased (L-valine, 85 : 15 vs L-*tert*-leucine, 98 : 2).

Ge and Li developed a C(sp^3)—H arylation process on secondary aliphatic aldehydes, which was especially notable given the potential for tautomerization (Figure 8.7) [52]. Through the use of β-amino acid, β-alanine, as a suitable TDG under mildly acidic conditions (AgTFA, AcOH, and HFIP), a selection of secondary aldehydes was successfully β-arylated in moderate to high yields.

A catalytic cycle was proposed to proceed as follows (Figure 8.8), now considered as broadly typical for palladium-catalyzed C—H functionalization. An imine forms between the aldehyde substrate and amine TDG. The resultant imine then coordinates to the Pd catalyst, here in a bidentate manner, though monodentate variations have also been described. The palladacycle orientates the metal catalyst proximal to a specific C—H bond (in this example β- to the imine, highlighted) and undergoes concerted metalation deprotonation to generate a Pd(II) fused-5,6-palladacycle. The palladacycle undergoes oxidative addition with the aryl-iodide-coupling partner, generating a Pd(IV) species. Reductive elimination affords the desired arylated product, and finally, ligand dissociation and imine hydrolysis recover both the TDG and Pd(II) catalyst, closing the cycle.

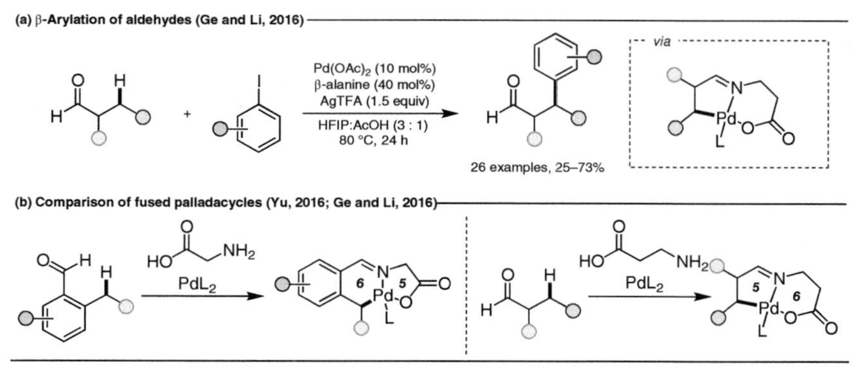

Figure 8.7 Transient C(sp^3)–H β-arylation of aliphatic aldehydes.

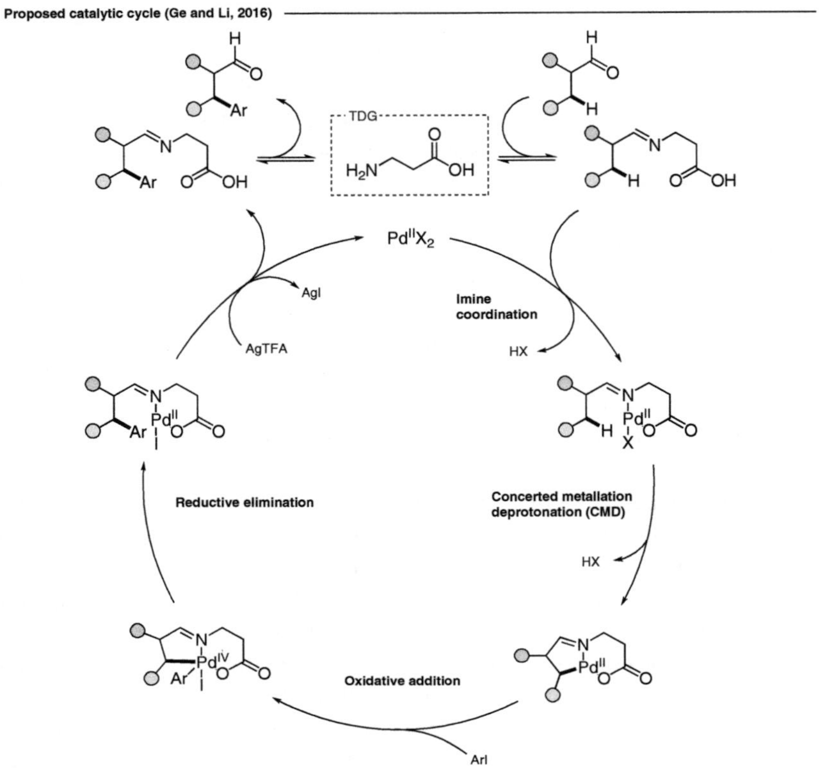

Figure 8.8 Proposed catalytic cycle of typical transient $C(sp^3)$–H arylations.

Compared with the work described by Yu, which used glycine to direct a formal γ-arylation of the benzylic $C(sp^3)$—H center, the use of β-amino acids, as shown by Ge and Li, optimizes the bite angle toward β-functionalization. This can likely be reasoned through the differences in the resultant palladacycles (Figure 8.9). The procedure was tolerant to a selection of aryl iodides, including *ortho*-substituted system, which saw no discernable decrease in yield compared with reactions with *para*-substituted analogs. Aldehyde scope was also investigated, with the reaction being tolerant of a range of acyclic and certain cyclic aldehydes. It is interesting to note, however, that desymmetrization of acyclic aldehydes gave approximately 1 : 1 mixtures of *mono-* and *di*-arylation. To provide some experimental evidence to support the proposed mechanism, an investigation into alternative reaction pathways was undertaken. Aliphatic aldehydes are known to form their corresponding Michael acceptors through a dehydrogenative pathway, which could foreseeably result in the desired arylation, without the need of a transient C—H functionalization approach. Reactions were carried out with isobutyraldehyde, producing a mixture of *mono-* and *di*-arylated products in the presence of a TDG but no products in its absence, exemplifying the need of the TDG (Figure 8.9a). This was contrasted by reactions with methacrylaldehyde, the corresponding Micheal acceptor, which gave

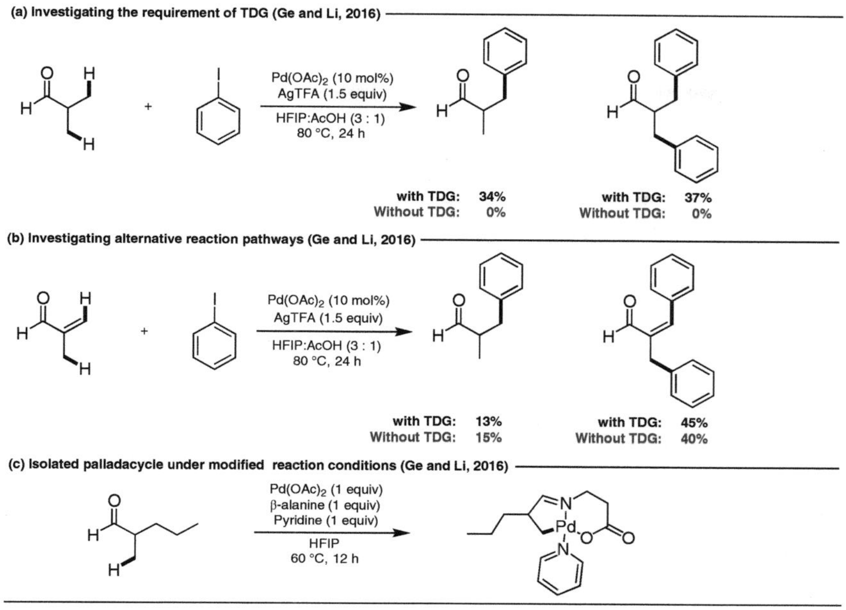

Figure 8.9 Mechanistic investigations of C(sp^3)−H arylation of secondary aldehydes. (a) control experiments showing the necessity of the TDG, (b) reaction with the Micheal acceptor, and (c) isolation of the palladacycle.

analogous mixtures of *mono-* and *di-*arylation in the presence or absence of the TDG, indicating that in the former case (reaction with isobutyraldehyde), *in situ* formation of methacrylaldehyde was unlikely to be formed (Figure 8.9b). Further supporting the proposed reaction mechanism, a fused [5,6]bispalladacycle, stabilized by pyridyl ligands, was isolated, effectively showing imine formation, coordination to the palladium catalyst, and C—H activation step (Figure 8.9c).

Bull described an in-depth study into the arylation of tertiary aliphatic aldehydes, highlighting key features in TDG design and use of additives, and gave interesting considerations regarding the proposed catalytic cycle (Figure 8.10a) [53]. A selection of bidentate and monodentate TDGs was screened with varying secondary binding sites, including amides, sulfonamides, and ethers. TDGs containing amide secondary binding sites showed overall increased reactivity, consequently favoring the formation of a mixture of *mono-*, *di-*, or *tri-*arylated products. Sulfonamide groups, on the other hand, displayed increased selectivity toward the desired monoarylation reaction, with aryliodides of different electronic properties behaving similarly. Ether groups as second coordinating sites were also successful, albeit, with reduced yields. Interestingly, an isolated palladacycle showed the formation of an asymmetric dimer, featuring both monodentate- and bidentate-binding modes of the TDG, with a bridging acetate ligand. The palladacycle was catalytically active, forming the monomer in solution. Following these reports, several different arylation procedures have been reported, expanding the aldehyde scope toward tertiary phenylacetaldehydes, secondary aliphatic aldehydes, and cyclic aldehydes. Most reports

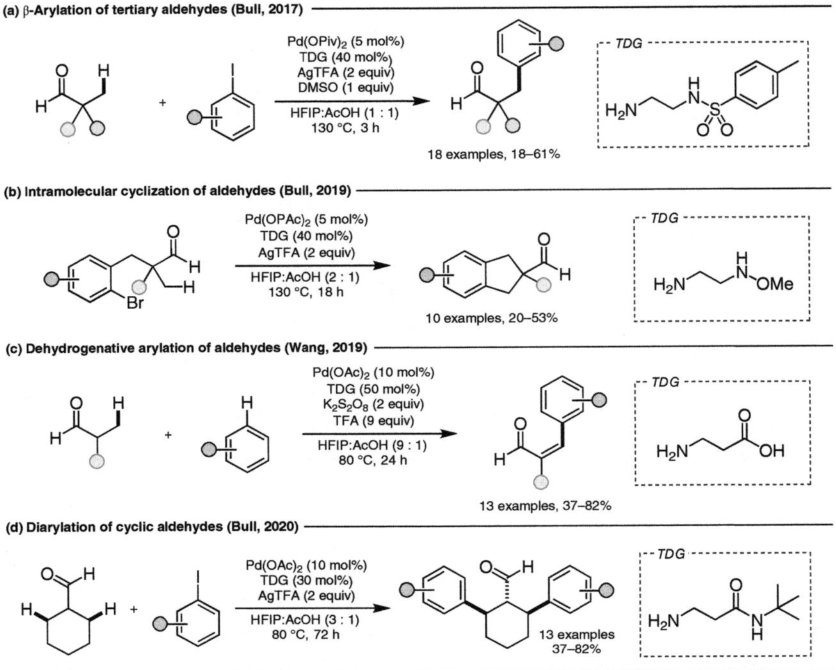

Figure 8.10 Notable examples in transient C(sp^3)—H functionalization of aldehydes. (a) Pd-catalyzed arylation of tertiary aldehydes, (b) intramolecular cyclization, (c) dehydrogenative C—H arylation, and (d) diarylation of cyclic aldehydes.

follow a similar procedure to the works described above, utilizing silver-based salts, HFIP : AcOH solvent mixtures, and β-alanine or amide derivatives thereof, as suitable TDGs [54–58].

In 2019, Bull reported intramolecular C—H functionalization of tertiary phenylacetaldehydes to form indanes, featuring an *ortho*-substituted arylbromide (Figure 8.10b) [59]. 2-Methoxyethyamine was selected as an optimal TDG, allowing for the selective formation of indane-carboxaldehyde in moderate yields. In the same year, Wang reported a cross-dehydrogenative coupling of secondary aldehydes and arenes, resulting in functionalized α,β-unsaturated aldehyde products with exclusive *E*-stereoselectivity (Figure 8.10c) [60]. In this method, the reaction involves α-palladation (Saegusa oxidation, with Pd(TFA$^+$) of the resulting enamine isomer after condensation of the aldehyde substrate with the TDG. The resulting palladacycle undergoes β-hydride elimination to furnish the key alkene intermediate, which can further undergo dehydrogenative arylation with potassium persulfate as the oxidant.

Most reports of C(sp^3)—H functionalization have focused on reactions with acyclic aliphatic aldehydes. It was only in 2020 that a report specifically on cyclic aldehydes was described, showcasing a selective diarylation procedure (Figure 8.10d) [61]. Bull described the use of amide derivatives of β-alanine as an optimal TDG toward the diarylation of cyclohexanecarboxaldehyde. Pyridone additives were included to promote reactivity. Pyridones are proposed to act as alternative ligands to palladium, acting as carboxylate surrogates to accelerate concerted-metalation-deprotonation (CMD) [62, 63]. Screening various pyridones found that those with electron-withdrawing groups were effective in increasing overall yields, though it was also notable that this was best seen where substituents were in the 5-position. The reaction was tolerant of a range of aryl-iodide-coupling partners, though *ortho*-substituted systems and heteroaryliodides were less successful. A further note is that the reaction proceeded with exclusive *trans*-diastereoselectivity, accounted for by the preference for a di-equatorial orientation of the substituents.

Following this, further methods using pyridones have been exemplified within the literature. However, a significant gap remains regarding the utility of transient C—H functionalization of C(sp^3)—H centers with other coupling partners, e.g. halogenation, carbonylation, silylation, or sulfonylation, as described with C(sp^2)—H centers (often with metal catalysts other than palladium).

8.2.2 C(sp^3)–H Functionalization of Ketones

Ketones are less electrophilic and display greater steric constraints compared with aldehydes. The imine formation and hydrolysis rates are reduced, and the resulting ketimines can form a mixture of *E* and *Z* isomers, potentially directing the metal catalyst in an inactive geometry. There have been fewer advancements in applying the transient C—H functionalization strategy to functionalize ketone substrates and all reported synthetic methodologies are currently limited to palladium catalysis.

In the 2016 seminal report by Yu, the C(sp^3)—H arylation of ketone was developed using glycine as the amine TDG additive in a 3 : 1 solvent mixture of HFIP : AcOH (Figure 8.11a) [48]. A wide range of aryl iodides with electron-donating or electron-withdrawing groups were well tolerated, as were various linear, α-branched, and cyclic ketone substrates. Ketones with cyclopropane, cyclobutene, or cyclohexane yielded the corresponding *syn* products with excellent diastereoselectivity. Additionally, Lin and Wei reported a similar transformation using

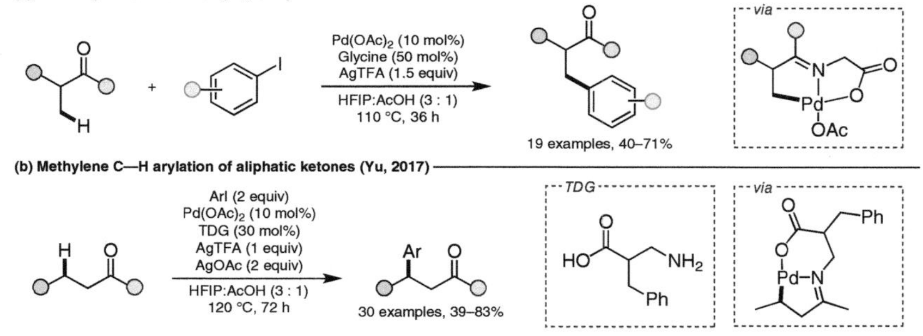

Figure 8.11 Early developments on Pd-catalyzed TDG-mediated C(sp^3)–H arylation of ketones. (a) Arylation of the methyl group of ketones, and (b) methylene C–H arylation of aliphatic ketones.

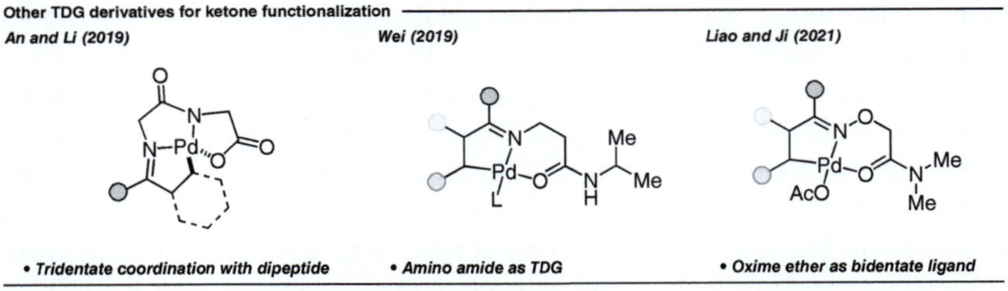

Figure 8.12 Other TDG derivatives for ketone functionalization.

2-amino-*N*-isopropyl-acetamide as the TDG, which formed an *N*, *N*-bidentate coordination with the palladium catalyst [56]. In the subsequent year, Yu further advanced this method to achieve the methylene $C(sp^3)$—H arylation of aliphatic ketones (Figure 8.11b) [64]. The use of α-benzyl β-alanine DG to form a six-membered palladacycle was found to be crucial. Additionally, a silver salt mixture was important to suppress the formation of undesired Ar–Ar homocoupling products.

Following these early reports, various other TDG derivatives were discovered (Figure 8.12). An and Li reported an improvement in functionalizing ketones using dipeptide TDG to form a tridentate ligand for palladium, effectively lowering the TDG loading to 20 mol%. A range of aliphatic ketones were arylated successfully [65]. On the other hand, Wei discovered the use of 3-amino-*N*-isopropylpropionamide as a more effective TDG than glycinamide to arylate the unactivated methyl and cyclic methylene motifs via the formation of [5,6]-bicyclopalladium intermediate [57]. Liao and Ji demonstrated the use of alkoxyamino TDG as a better nucleophile than the common TDG with amino group, forming a more stable oxime ether with ketones and acting as the bidentate ligand to form the corresponding [5, 6]-bicyclic palladium intermediate.

The use of electron-deficient pyridone ligands was also found to be important in unlocking additional ketone functionalization transformations. In 2022, Yu reported the first TDG-enabled γ-selective (hetero)arylation of aliphatic ketones (Figure 8.13a) [66]. 3-Nitro-5-(trifluoromethyl)-2-pyridinone was identified as the most effective electron-deficient 2-pyridone ligands in combination with lower loading of chloroacetic acid as an additive to accelerate the C—H cleavage step. Less than 10% of arylation was observed in the absence of the 2-pyridone ligand. The same 2-pyridone ligand and TDG system was also utilized by Sorensen in the synthesis of benzocyclobutenes via intramolecular methylene β-methylene-selective $C(sp^3)$—H arylation (Figure 8.13b) [67]. The use of 5-nitropyridone as an electron-deficient ligand was also found to be crucial in the selective arylation of tertiary C—H bonds of cyclobutylmethyl ketones to construct quaternary carbon centers (Figure 8.13c) [68].

The use of chiral TDGs is a promising strategy to enable enantioselective transformations. However, it is a challenging approach to employ due to the stereogenic center on the TDG being relatively far from the target C—H bond. In 2020, Yu successfully utilized D-valine combined with 5-nitropyridone as the X-type ligand

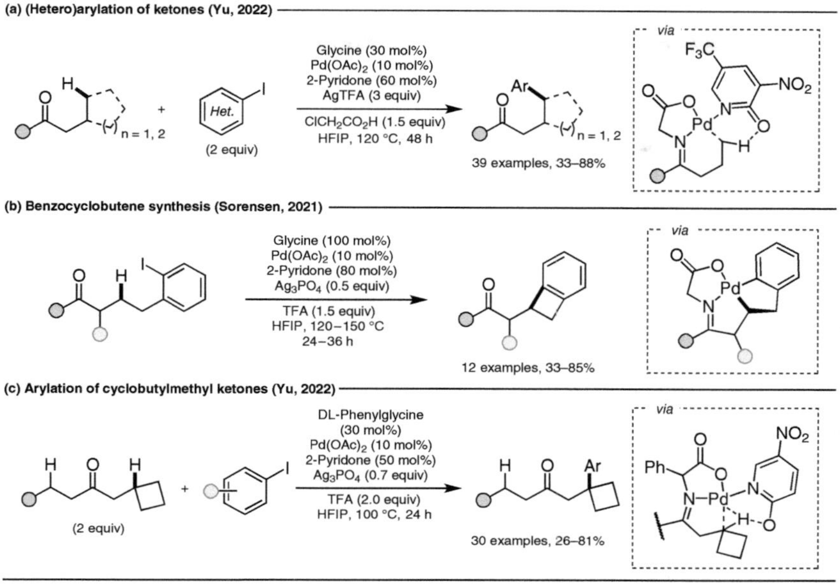

Figure 8.13 2-Pyridone-ligand-enabled C(sp^3)–H arylation of ketones. (a) Arylation of ketones, (b) synthesis of benzocyclobutene, and (c) arylation of cyclobutylmethyl ketones.

to achieve the enantioselective arylation of cyclobutyl ketones (Figure 8.14) [69]. The use of a silver salt was found to be crucial to this reaction during early optimization with D-cyclopentyl glycine as the chiral TDG. While most of the silver salts, such as AgTFA (14 : 86 er), AgOAc (14 : 86 er), and Ag_2CO_3 (10 : 90 er), gave moderate-to-good enantioselectivity to produce the same enantiomer as the major product, the use of Ag_3PO_4 resulted in a reversal of enantioselectivity (96 : 4 er). Further mechanistic studies with deuterium incorporation experiments indicated that the irreversible C—H cleavage step was rate limiting regardless of the identity of the silver salt. The origin of this reaction's enantioselectivity largely depends on the nature of the silver salt. When AgTFA was employed as an efficient halogen scavenger to promote the reductive elimination step, the chiral TDG was responsible for inducing asymmetric C—H palladation to control the enantioselectivity. When Ag_3PO_4 was used as the silver salt, the reductive elimination step became the rate-limiting step. As a result, the formation of the chiral Pd(IV) intermediate becomes the stereochemistry-determining step, resulting in the observed reversal of enantioselectivity.

8.2.3 C(sp^3)–H Functionalization of Amines

The functionalization of amine substrates is more challenging due to their high binding affinities to metal catalysts to form undesired metal complexes [70]. Competing α-oxidation of amines would lead to the formation of corresponding imine side products, further complicating the process [71]. As such, fewer advancements

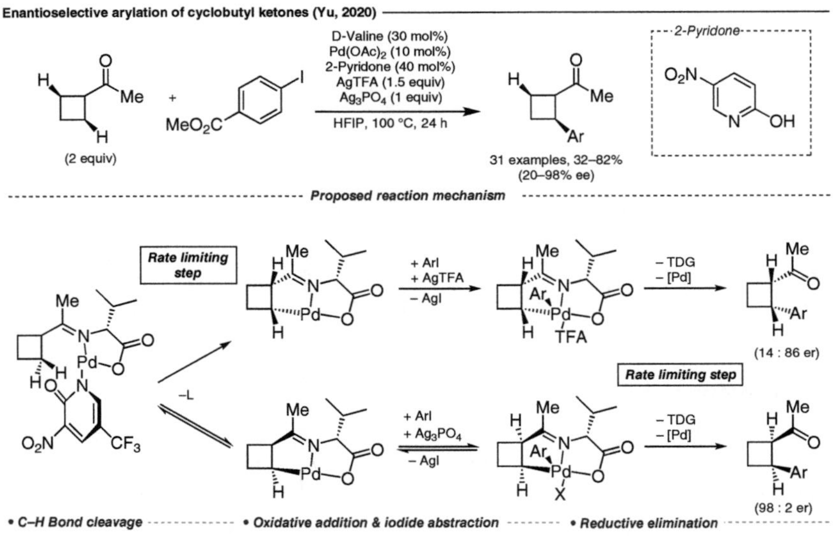

Figure 8.14 Enantioselective C(sp^3)—H arylation of cyclobutyl ketones.

have been made to directly functionalize amine substrates. In 2016, Yu reported the use of 2-hydroxynicotinaldehyde as a catalytic TDG for the arylation of free amines (Figure 8.15a) [72]. Water was added as an additive to facilitate imine hydrolysis after C—H activation, and the resulting arylated amine was Boc-protected for ease of purification. Methyl group, cyclic, and acyclic methylene C—H bonds can be arylated under optimal conditions. Amines with and without α-substituents were also well tolerated. A similar catalytic system was also employed by Kamenecka (Figure 8.15b), Liu, and Wang (Figure 8.15c) to functionalize amino esters [73, 74].

In addition to 2-hydroxynicotinaldehyde, other aldehyde derivatives have been discovered as competent TDGs for amine substrates (Figure 8.16). For example, Ge reported the use of glyoxylic acid as a catalytic TDG to achieve γ-arylation of primary alkylamines [75]. High-site selectivity was observed, favoring the γ-C(sp^3)—H bond of the methyl group. A range of functional groups such as alkoxyl, alkoxylcarbonyl, trifluoromethyl, and nitro groups were tolerated. On the other hand, Murakami and co-workers demonstrated the use of salicylaldehydes as TDG for γ-arylation. A bulky salicylaldehyde with *tert*-butyl groups at the 3- and 5-positions was identified as the optimal TDG [76]. Bull identified phenoxyacetaldehyde dimethyl acetal as the alkyl aldehyde TDG for the arylation of primary amines at the γ-position [77]. This commercially available acetal undergoes hydrolysis to the corresponding aldehyde under the reaction conditions to form imine with the amine substrates. Only a low loading of 15 mol% TDG was required. The secondary coordinating group at the α-position of the aldehyde could be modified, providing additional chelation to stabilize palladium intermediates, limiting imine isomerization, and increasing aldehyde electrophilicity to promote imine formation.

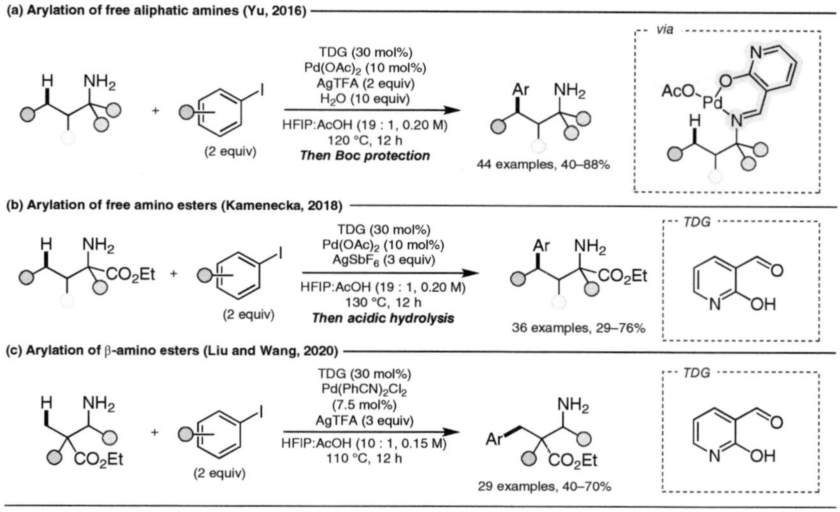

Figure 8.15 C(sp^3)–H arylation of amines with 2-hydroxynicotinaldehyde as TDG. (a) arylation of aliphatic amines, (b) arylation of α-amino esters, and (c) arylation of β-amino esters.

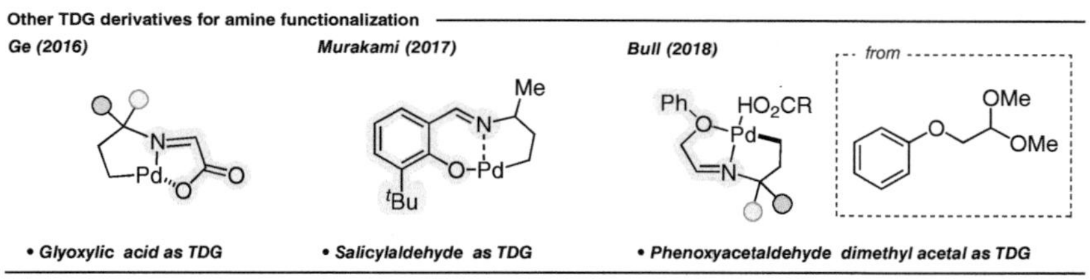

Figure 8.16 Other TDG derivatives for amine functionalization.

The competitive binding interactions between reaction components such as the TDG, amine substrates, and the resulting imine formed often lead to the formation of undesired metal complexes, further complicating the development of new methodologies to functionalize amine substrates with this strategy. To address these limitations, Yu introduced electron-deficient 2-pyridone ligands to enhance the γ-methylene arylation of alkyl amines to expand the scope to a wider range of heteroaryl iodides and bromides (Figure 8.17) [78]. 5-Trifluoromethyl- and 5-nitro-2-pyridone were identified as the suitable X-type ligands to act as an internal base to accelerate the C—H bond cleavage step. More importantly, it was found that TDGs that coordinate with Pd(II) to form a six-membered chelate are highly selective toward γ-C—H bonds, while TDGs that form a five-membered chelate with Pd(II) tend to activate δ-C—H bonds.

The transient C—H functionalization arylation strategy was utilized by Messaoudi and co-workers to functionalize aminosugars (Figure 8.18) [79]. 2-Hydroxynicotinaldehyde was identified as the optimal TDG and the use of electron-deficient 2-pyridone ligand was essential. Various aryl iodides with different functional

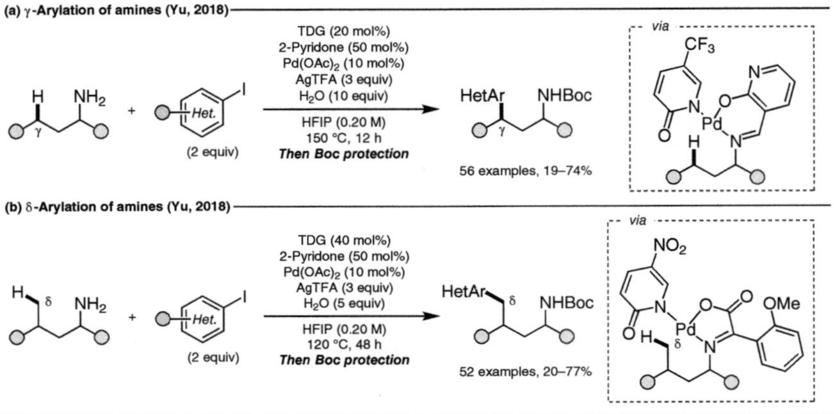

Figure 8.17 2-Pyridone-enabled Pd-catalyzed TDG-mediated C(sp^3)–H arylation of amines.

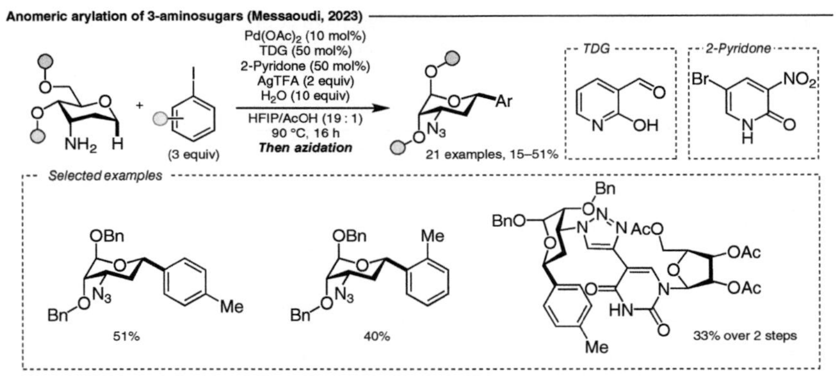

Figure 8.18 Anomeric C(sp^3)–H arylation of 3-aminosugars.

groups such as –OMe, –SMe, –Cl, and –OCF_3 were successful coupling partners to give the corresponding *C*-glycosides with exclusive α-anomeric stereoselectivity. The use of benzyl-protecting groups was found to be crucial for this reaction. Other protecting groups such as methyl, acetyl, naphthyl, or pivalyl gave the desired product in low yield. The developed protocol was used to synthesize a series of complex 3-triazolo-aryl *C*-glycosides after subsequent copper-catalyzed azide-alkyne cycloaddition (CuAAC reaction).

While most reported palladium-catalyzed transient C—H functionalization reactions focused on the arylation of different C—H bonds, other transformations have also been reported. For example, Yu developed the first example of C(sp^3)—H *γ*-methyl and methylene fluorination of cyclohexyl and linear aliphatic amines with 2-hydroxynicotinaldehyde as the TDG and *N*-fluorobenzenesulfonimide (NFSI) as the fluorinating reagent (Figure 8.19) [80]. Electron-deficient 3,5-disubstituted pyridone ligands were utilized to facilitate this reaction. Further computational and

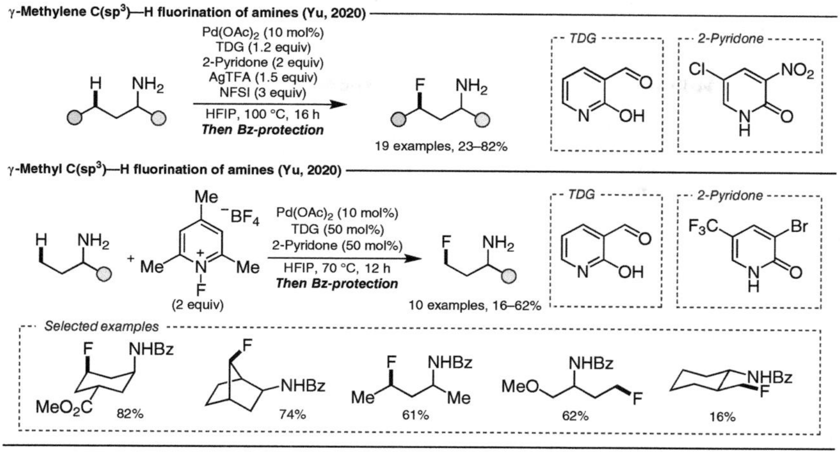

Figure 8.19 NFSI-mediated C(sp³)–H fluorination of amines.

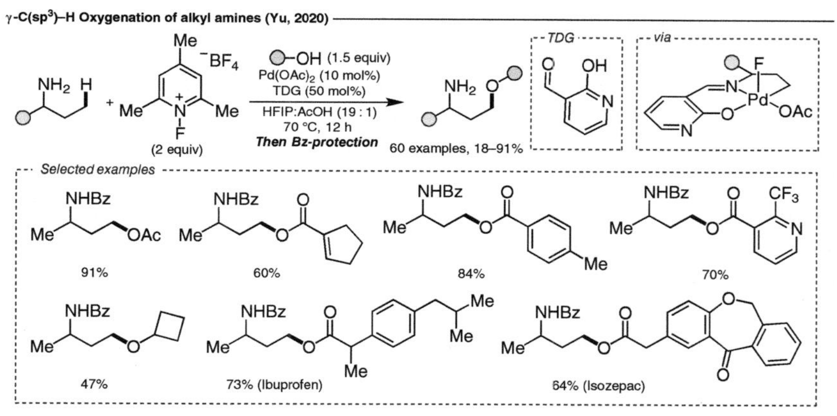

Figure 8.20 Palladium-catalyzed C(sp³)–H oxygenation of alkyl amines.

kinetic studies suggest that the pyridone ligand is involved in a bimetallic oxidative addition transition state (involving palladium and silver) for methylene fluorination while direct oxidative addition is preferred for methyl fluorination.

Based on the fluorination work, Yu subsequently developed γ-C(sp³)—H acyloxylation and alkoxylation of alkyl amines with 2-hydroxynicotinaldehyde as the TDG (Figure 8.20) [81]. *N*-Fluoro-2,4,6-trimethylpyridinium tetrafluoroborate was used as the [F⁺] bystanding oxidant, and the developed conditions are compatible with a wide range of amines with good functional group tolerance. Aromatic and aliphatic acids and primary, secondary, and tertiary alcohols are compatible nucleophiles for this oxygenation reaction. The use of natural product and drug molecule derivatives as coupling partners also demonstrated the synthetic utility of this reaction for late-stage functionalization.

8.3 Transient C(sp^2)–H Functionalization

8.3.1 C(sp^2)–H Functionalization of Aldehydes

Transient C—H functionalization on sp^2-carbon centers has been more extensively studied than C—H bonds at sp^3 carbon centers. Moreover, while methodology on C(sp^3)—H bonds relies heavily on palladium catalysis, a number of different transition metal catalysts have been used to effect a variety of transformations from C(sp^2)—H bonds. This section outlines key publications within the area, including different transition metal catalysts, different monodentate and bidentate TDGs, and the scope of transformations possible under the general strategy of transient C—H functionalization.

8.3.1.1 Palladium Catalysis

Palladium-catalyzed transient C(sp^2)—H functionalization was first described by Yu in 2017, in the *ortho*-arylation of benzaldehyde substrates (Figure 8.21a) [82]. α-Amino acid derivative, 2-aminoisobutyric acid, was used as an optimal TDG, with bulky *gem*-dimethyl groups providing the optimal bite angle to facilitate arylation. Notably, whereas arylation reactions on sp^3 centers often rely on the formation of fused [5, 6] or [5, 6] palladacycles, this case proceeds via a fused [5] palladacycle. The reaction was tolerant to a variety of aryliodide-coupling partners, including sterically hindered *ortho*-substituted arenes (55% yield when using methyl 2-iodobenzoate). Unfunctionalized iodopyridines and iodothiophenes were unsuitable coupling partners, due to competitive coordination and catalyst deactivation by the heteroatom, which remains a continuous challenge when working with coupling partners of this type. This was addressed through the use of 2-substituted pyridines and thiophenes, where coordination of the heteroatom to the catalyst was reduced, affording the desired products. Fused aromatic systems, e.g. iodoquinoline and iodoquinoxaline, were also tolerated, giving 33% and 40% yield, respectively, with 15 mol% $Pd(OAc)_2$. The scope of benzaldehydes for the arylation reaction showed good tolerance of electron-withdrawing and electron-donating groups in the *ortho*-, *meta*-, and *para*-positions, to afford the arylated aldehydes in excellent yield. *N*-Tosyl-3-formylindole reacted to give the 4-arylated indole in 82% yield.

In the same report, selective chlorination/bromination was also described, with NCS or NBS acting as chlorine or bromine sources, respectively (Figure 8.21b) [82]. Optimization studies showed that the reaction displayed increased success when moving to anthranilic acid (and derivatives) as the TDG, as well as a turn away from HFIP as a solvent, instead using DCE and reduced temperatures. Dihalogenation could also be promoted through the use of excess NCS/NBS and increasing the acidity of the system – using TFA as a co-solvent.

Monodentate aniline TDGs were later shown to be effective in transformations of this type, while Zhang reported an alternative chlorination methodology (via NCS) and a direct methoxylation reaction (Figure 8.21c) [83]. Notably, the TDGs feature one or more electron-withdrawing groups (trifluoromethyl). Pyridone ligands were also utilized, enabling the chlorination of benzaldehyde substrates in excellent

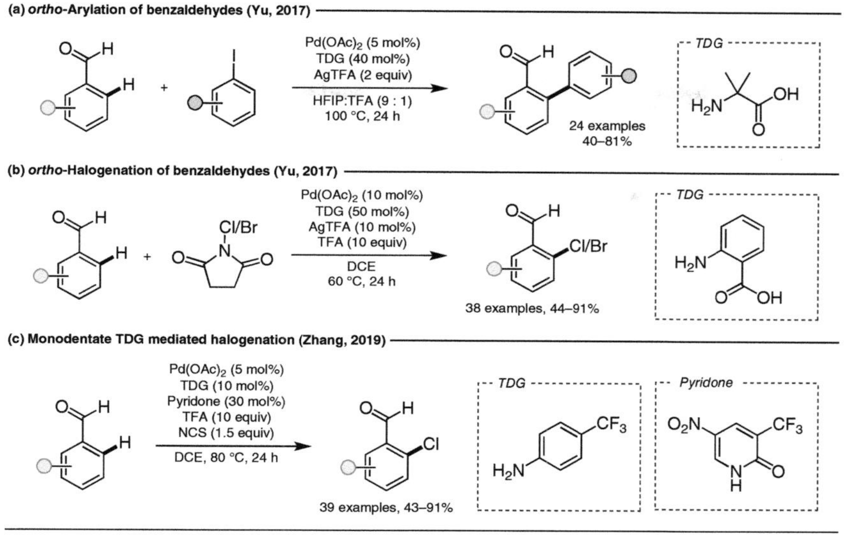

Figure 8.21 Pioneering examples of C(sp²)–H functionalization of aldehydes. (a) Arylation of benzaldehydes, (b) halogenation with bidentate directing group, and (c) halogenation with mono-dentate directing group.

yields. The reaction exhibited a broad functional group tolerance, including nitrile, ester, and sulfonamide functional groups on the aldehyde substrate. Additionally, a Celecoxib derivative could be chlorinated in a late-stage manner, with the imine DG overriding the effect of other coordinating functionalities on the substrate [84].

Sorensen highlighted the importance of TDG selection in influencing the result of reaction, reporting a transient C—H functionalization strategy to access fluorenones by reaction of benzaldehyde and aryl iodides (Figure 8.22) [85]. Compared with Yu's procedure, which used branched α-amino acids to promote arylation, the use of anthranilic acid in this case promoted cyclization. The reaction was proposed to

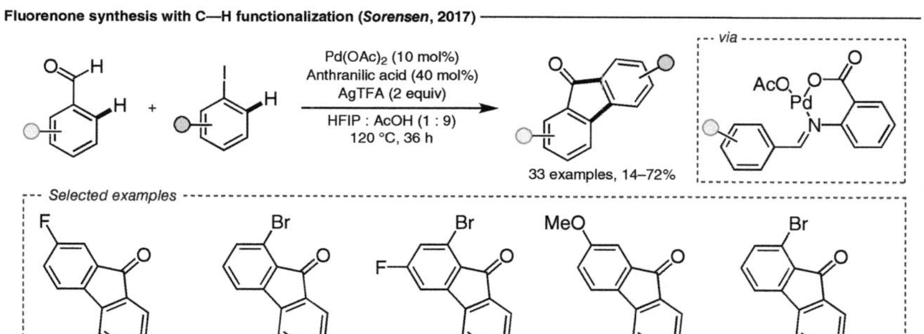

Figure 8.22 Fluorenone synthesis with dual C–H functionalization.

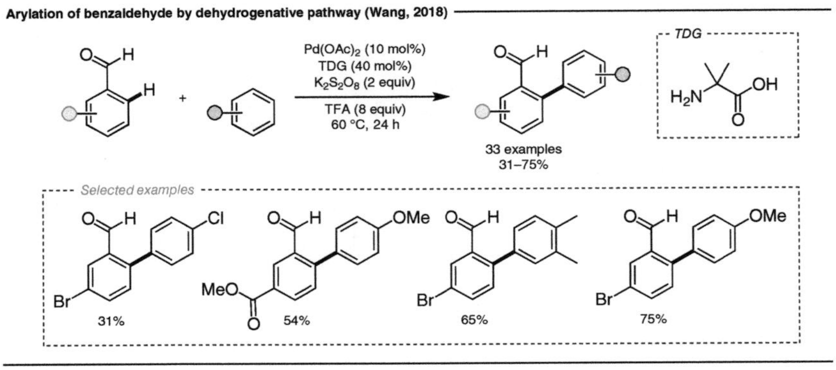

Figure 8.23 Pd-catalyzed dehydrogenative arylation of benzaldehyde.

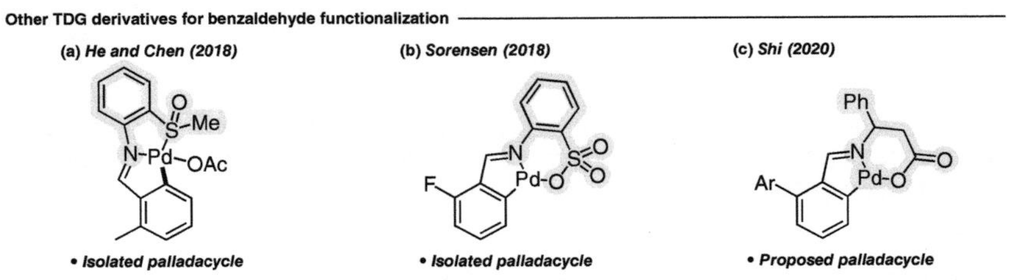

Figure 8.24 Other examples of bidentate TDG for C(sp^2)–H functionalization. (a) Sulfoxide-derived TDG, (b) orthanilic acid as TDG, and (c) the use of β-alanine derivatives for silylation.

proceed via two successive C—H functionalization reactions, accessing substituted fluorenones in low-to-moderate yields.

In 2018, Wang reported an arylation procedure operated by a dehydrogenative coupling mechanism using potassium persulfate as an oxidant. 2-Aminoisobutyric acid was retained as the optimal TDG for arylation; however, HFIP was not used as a solvent, with instead the arene being used as both reactant and solvent (Figure 8.23) [86].

He and Chen later reported on a further class of aniline-derived TDG, featuring an *ortho*-sulfoxide opposed to a carboxylic acid (Figure 8.24a) [87]. The sulfoxide was found to bind palladium as an L-type ligand through the sulfur lone pair, seen through isolated palladacycles. Biaryl products were obtained in moderate-to-high yields, with increased reactivity seen with electron-rich benzaldehydes as well as a selection of electronically distinct aryl-iodide-coupling partners.

Following these initial reports, which showcase the utility of C—H functionalization in coupling arenes, or halogens to C(sp^2)—H centers, a number of other transformations have been described, expanding on the scope of functionality that can be installed through transient C—H functionalization. This includes hydroxylation, fluorination, methylation, silylation, fluoroalkoxylation, alkoxycarbonylation,

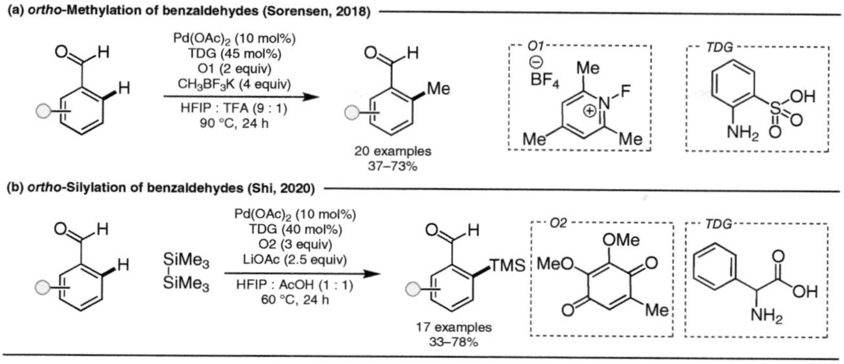

Figure 8.25 *ortho*-Functionalization of benzaldehydes. (a) Methylation and (b) silylation of benzaldehydes.

and diselenide formation. Selected examples that introduce different TDG motifs or additives or involve different mechanistic considerations are outlined below.

Orthanilic acid has been shown to be the optimal TDG of choice for some of these transformations: notably *ortho*-hydroxylation, fluorination, and methylation of benzaldehydes (Figure 8.24b) (Sorensen, 2018) [88]. The use of an electrophilic fluorine oxidant (O1) was key in all cases, responsible for forming a reactive a Pd(IV)-F species following cyclometalation, and allowing for subsequent ligand exchange with coupling partner of choice (Figure 8.25a). Of note was that the sulfonic acid imparted a significant increase in reactivity (+40% yield) in comparison with the carboxylic acid analog (anthranilic acid). β-Alanine derivatives are suitable DGs for silylation, as shown by Shi in 2020 (Figure 8.24c) [89]. A benzoquinone-type oxidant was used to promote the coupling (O2, Figure 8.25b). It is notable that the reaction scope only covered 2-aryl-substituted benzaldehyde substrates and saw exclusive *ortho*-silylation in yields of up to 78%. This allowed for the iterative use of multiple C—H functionalization strategies, e.g. atroposelective olefination, allylation, alkynylation, or naphthylation, followed by selective silylation. It should be noted, however, that only hexamethyldisilane could be used as a coupling partner, with other similar reagents (hexaphenyldisilane) being unreactive.

Zhou and Zhang show that copper bromide can be used as a suitable additive in accessing *ortho*-selenated benzaldehydes (Figure 8.26) [90]. Benzidine was used as the TDG, which is presumed to act as a monodentate DG with palladium. Copper bromide was included as an additive in the reaction with the report postulating three plausible pathways. In all three instances, imine formation and coordination to Pd, giving a Pd(II)palladacycle starts the cycle. In the primary cycle, oxidative addition to the diselenide gives a Pd(IV) species, followed by a reductive elimination to give *ortho*-selenated benzaldehyde products and a monoselenated Pd(II) complex. $Pd(OAc)_2$ is then expected to be regenerated by ligand exchange eliminating substituted-selenols, with copper bromide reacting with selenols to prevent catalyst poisoning. In an alternative pathway, the selenol byproducts directly undergo ligand exchange on the imine-coordinated Pd(II)palladacycle. The desired product is

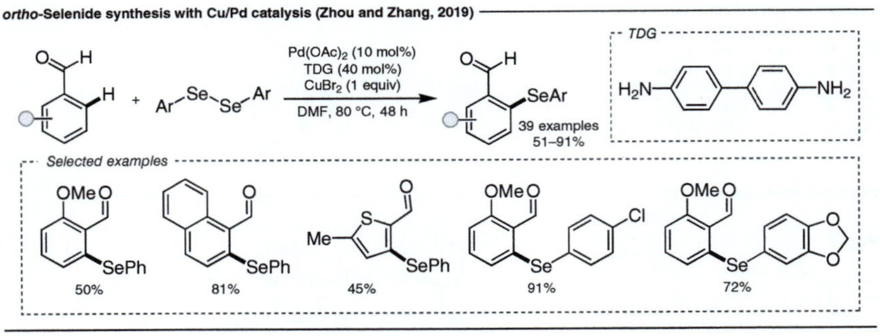

Figure 8.26 Pd-catalyzed C(sp^2)–H selenylation of benzaldehydes.

then formed by reductive elimination, with $CuBr_2$ needed to reoxidise Pd(0) back to Pd(II). Alternatively, under a third pathway, the monoselenated Pd(II) complex can coordinate to the imine of the starting material. High selectivity for monoselenation was observed in all cases, and the reaction functioned effectively with electron-rich and electron-poor substrates.

γ-Functionalization of C(sp^2)—H centers is far less common though there are notable examples that allow for atroposelective reactions and/or functionalization of heteroaromatic carboxaldehydes [91]. In 2017, Shi applied a chiral TDG strategy to functionalize the 2′-position of biaryl aldehydes (Figure 8.27a). Using chiral α-amino acid, L-tert-leucine as the TDG, axially chiral biaryls were obtained by a Heck-type coupling with electron-poor alkenes. Depending on the substrate, the reaction could proceed via dynamic kinetic resolution (DKR) or kinetic resolution (KR) pathways. Excellent *ee*'s of the olefinated products (95–99%) were observed across the substrate scope.

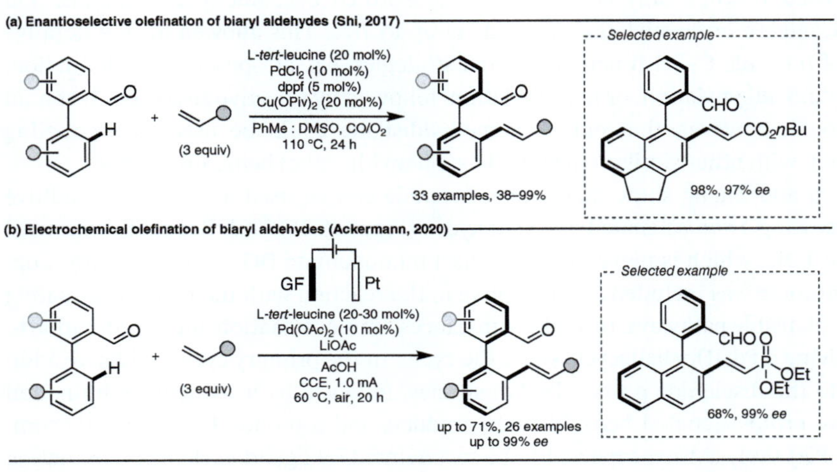

Figure 8.27 Atroposelective olefination by transient C(sp^2)–H functionalization. (a) Enantioselective olefination of biaryls and (b) electrochemical enantioselective olefination of biaryl aldehydes.

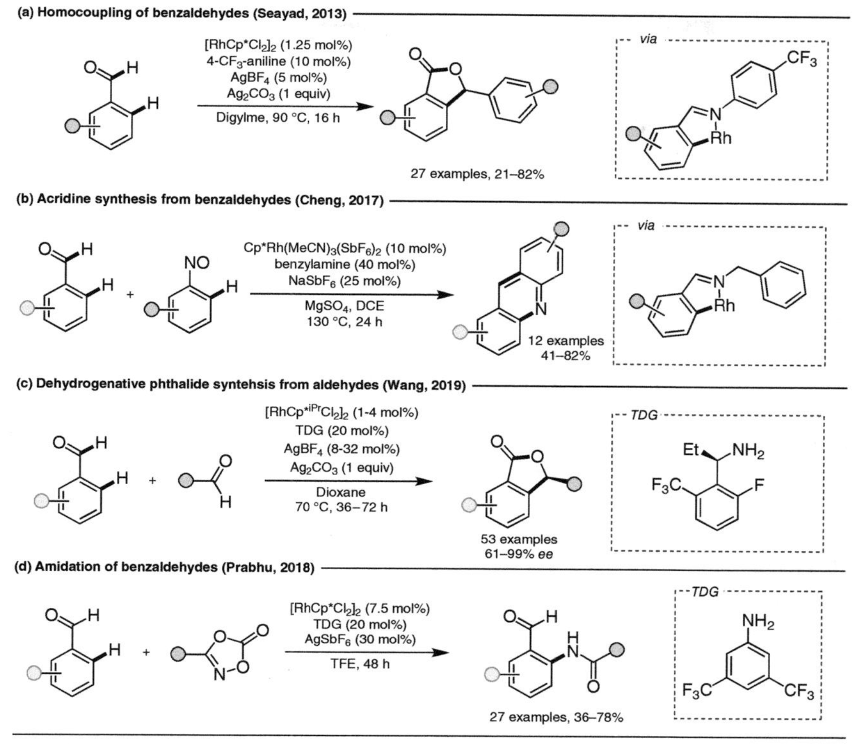

Figure 8.28 Rhodium-catalyzed C(sp^2)–H functionalization. (a) Homocoupling of benzaldehydes, (b) acridine synthesis, and (c) dehydrogenative phthalide synthesis.

Ackermann demonstrated the combination of an electrochemical approach with transient C—H functionalization as an alternative means to achieve an atroposelective olefination of biaryl systems (Figure 8.28b) [92]. Good yields and high *ee*'s were obtained in most cases, with several biaryl and *N*-naphthyl pyrroles being exemplified. Vinyl sulfones, phosphonates, amides, and ketones all proceeded in good yields and excellent enantioselectivity. Mechanistic studies supported the study, indicating that the CMD is rate-determining. Of note, chiral biaryl aldehydes products were also effectively transformed into highly enantioenriched helicene derivatives, highlighting the utility of the method in accessing products of this type via C—H functionalization.

8.3.1.2 Rhodium and Ruthenium Catalysis

Though palladium catalysis is by far the most common metal of choice within transient C(sp^2)—H functionalization, a number of different metals have been exemplified as suitable catalysts under transient C—H functionalization regimes. A number of reports have found success in the use of rhodium catalysis, in combination with TDG that can bind the metal center in a monodentate fashion. This has generally been applied in the synthesis of fused cyclic systems, e.g. acridines or phthalides.

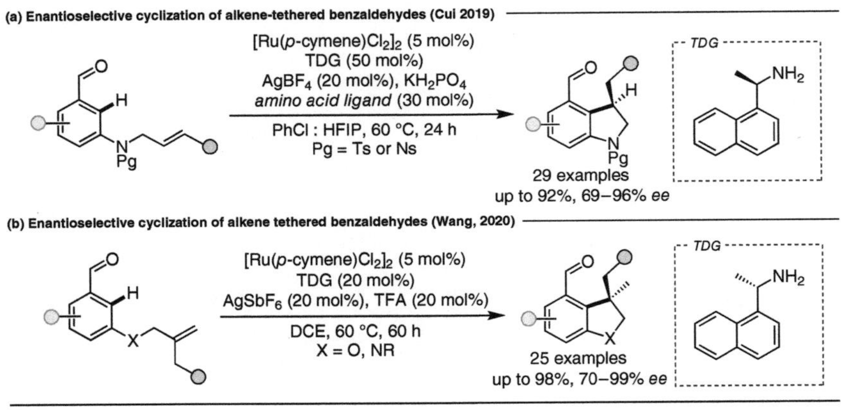

Figure 8.29 Ruthenium-catalyzed C(sp^2)—H functionalization. (a) Enantioselective indoline synthesis and, (b) enantioselective synthesis of indoline and dihydrobenzofuran.

Seayad and coworkers first reported the use of cationic rhodium catalysts in the presence of an aniline TDG to promote the dimerization of benzaldehydes (Figure 8.28a) [93]. The resultant phthalides were generated in moderate-to-high yields across a range of electronically diverse aldehydes. Benzylamine has been shown as a suitable TDG toward the synthesis of acridines (Cheng, 2017, Figure 8.28b) and enantioselective homo/heterocoupling to access phthalides (Wang, 2019, Figure 8.28c) [94, 95]. On the mechanism, it was proposed that these transformations proceed through a Rh(III)–Rh(I) catalytic cycle. Rh(III) is generated through *in situ* treatment of $[RhCp^*Cl_2]_2$ and Lewis acid $AgBF_4$, which is able to coordinate to the resultant imine, forming an *ortho*-rhodacycle. Association of a second aldehyde followed the addition of the Rh–C across the carbonyl group instigates nucleophilic attack of the new Rh-alkoxide to the electrophilic imine carbon. β-Hydride elimination and hydrolysis steps then afforded the observed product. On the other hand, Prabhu reported the application of rhodium catalysis with TDG in an amidative process on benzaldehydes (Figure 8.28d) [96].

Ruthenium catalysts have also been developed as suitable for transient C—H functionalization operations in similar transformations. Enantioselective methodologies described independently by Cui and Wang (2019 and 2020, respectively) are of particular note, allowing for the intramolecular hydroarylation of tethered alkenes for the synthesis of indolines (Cui, Figure 8.29a) and benzofurans/indolines (Wang, Figure 8.29b) [97, 98].

Both procedures used 1-(naphthalenyl)ethan-1-amine as a TDG and exhibited facial selectivity during the migratory insertion step. The *p*-cymene ligand on ruthenium restricts rotation of the DG C—N single bond fixing the orientation of the naphthyl and methyl groups. The smaller alkene then preferentially coordinates to the same side as the methyl group, rather than the naphthyl group, to avoid a steric clash, which leads to the stereochemical outcome observed.

Cui's work used an additional chiral amino-acid-derived ligand to further improve the enantioselectivity. Aldehydes bearing electron-donating and electron-withdrawing *ortho*-, *meta*-, and *para*-substituents (OMe, F, and CF_3) reacted effectively to give the corresponding indolines in good-to-excellent *ee*, up to 89% yield and 94% *ee* for the *meta*-methoxy example. More functionalized substituents on the tethered alkene were well tolerated, with silyl-protecting groups and esters compatible with the reaction.

Wang demonstrated under related conditions that dihydrobenzofurans could be synthesized from the corresponding oxygen-tethered alkenyl aldehyde. The amine transient DG was the sole source of enantioselectivity, achieving ring closure in yields and *ee*'s up to 98% for a range of substituted aldehydes bearing methoxy, nitro, aryl, and halide substituents.

8.3.1.3 Iridium and Cobalt Catalysis

Iridium and cobalt have been used in C—N bond-forming C—H functionalization reactions. For instance, Sorensen reported the alkylation of aromatic aldehydes with alkyltrifluoroborates as the coupling partner (Figure 8.30a) [99]. The developed process was compatible with a range of benzaldehydes, heterocyclic aldehydes as well as α,β-unsaturated aldehydes. Wu and Li developed a cobalt-catalyzed amidation process on benzaldehydes with dioxazolones, utilizing stoichiometric amount of *p*-chloroaniline as the TDG (Figure 8.30b) [100]. The use of cobalt catalyst avoids the need of expensive noble metals such as rhodium and iridium, and the developed conditions were able to produce a library of structurally diverse *ortho*-amidated benzaldehydes. A similar process was also reported by Sundararaju, featuring the use of a catalytic amount of aniline as TDG (Figure 8.30c) [100]. In both cases, a subsequent acid-mediated imine hydrolysis step was required.

8.3.1.4 Copper Catalysis

In 2022, Bull reported the first instance of copper-mediated transient C—H functionalization in a C(sp^2)—H sulfonylation through the use of β-alanine as TDG and sulfinate salts as a coupling partner (Figure 8.31) [101]. Mixed copper salts are proposed to act as metal catalyst, oxidant, and Lewis acid within the system, promoting reversible imine formation/cleavage. This enabled the synthesis of a variety of sulfones though an oxidative coupling of sulfinate salts and benzaldehydes. Product inhibition effects were observed, which were attributed to a relatively stable copper-ligated product imine forming and trapping the amino acid additive. The use of the aldehyde in excess alleviates product inhibition by shifting the reaction equilibrium to forming more starting material imine to promote the desired reaction.

8.3.2 C(sp^2)–H Functionalization of Ketones

8.3.2.1 Rhodium Catalysis

Building on previous work on rhodium-catalyzed aldehydic C—H hydroacylation of alkenes, [46] Jun reported catalytic alkylation of aromatic imines using Wilkinson's

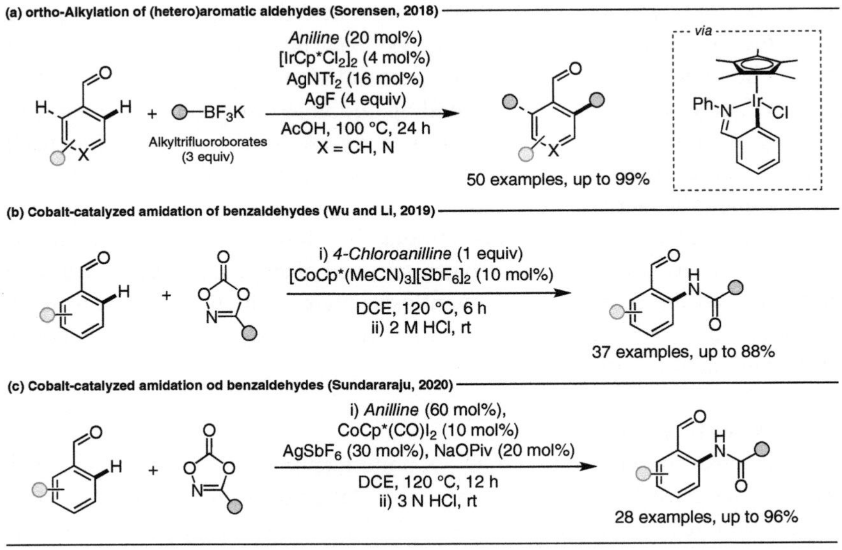

Figure 8.30 Ir- and Co-catalyzed C(sp^2)—H functionalization. (a) C—H alkylation, (b) and (c) amidation of benzaldehydes.

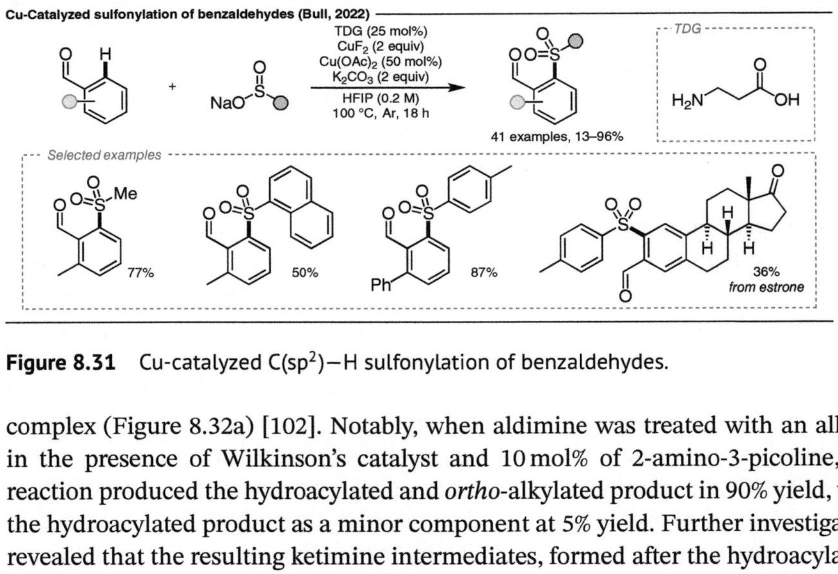

Figure 8.31 Cu-catalyzed C(sp^2)—H sulfonylation of benzaldehydes.

complex (Figure 8.32a) [102]. Notably, when aldimine was treated with an alkene in the presence of Wilkinson's catalyst and 10 mol% of 2-amino-3-picoline, the reaction produced the hydroacylated and *ortho*-alkylated product in 90% yield, with the hydroacylated product as a minor component at 5% yield. Further investigation revealed that the resulting ketimine intermediates, formed after the hydroacylation step, were the active species directing the *ortho*-alkylation. This study led to the development of a chelation-assisted Rh^{I}-catalyzed *ortho*-alkylation of aromatic ketimines or ketones with alkenes in 2002 (Figure 8.32b) [103]. A catalytic amount of benzylamines (50 mol%) was used as the amine additive to form the transient imine DG *in situ*, eliminating the need for prior preparation of ketimines from ketones. In 2018, Prabhu reported a Rh^{III}-catalyzed *ortho*-amidation of aromatic ketones with dioxazolone as the amidation reagent [96]. An aniline derivative

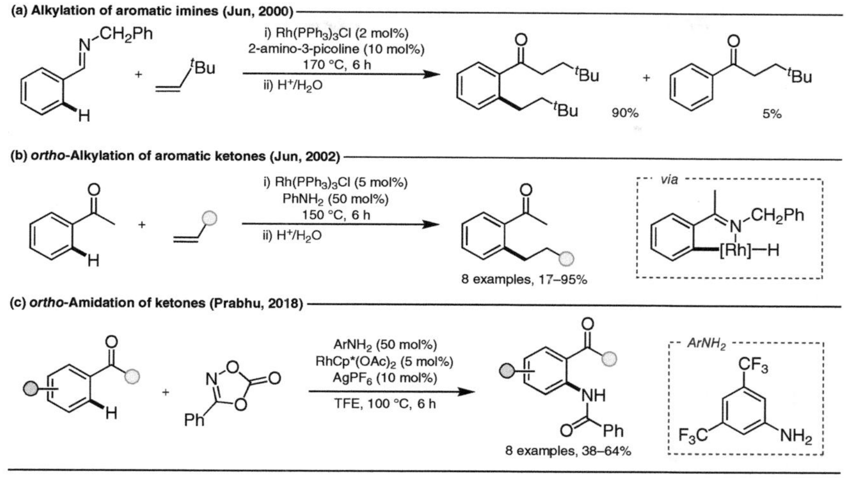

Figure 8.32 Rh-catalyzed TDG-mediated C–H functionalization. (a) Alkylation of imines, (b) alkylation of ketones, and (c) amidation of ketones.

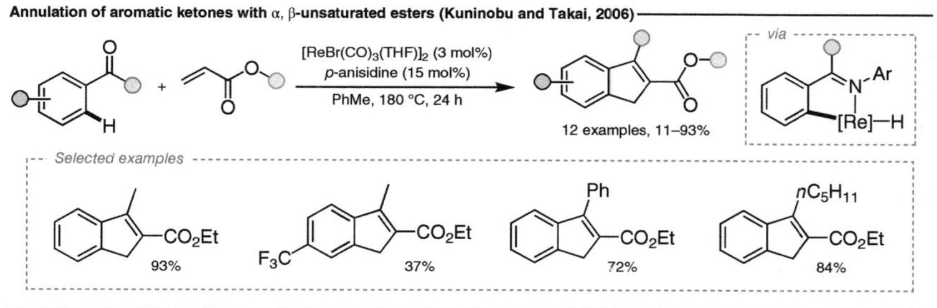

Figure 8.33 Re-catalyzed TDG-mediated C–H functionalization.

3,5-bis(trifluoromethyl)aniline was utilized as the amine additive to form the transient imine DG with ketone substrates in the presence of RhCp*$(OAc)_2$ and $AgPF_6$ as a chloride scavenger to direct the C—H amidation reaction.

8.3.2.2 Rhenium Catalysis

Kuninobu and Takai reported a rhenium-catalyzed annulation of aromatic ketones and α,β-unsaturated esters, mediated by a transient imine DG to produce indene derivatives (Figure 8.33) [104]. In this process, a ketimine is formed *in situ* by condensing acetophenone with *p*-anisidine, which then acts as the DG for the rhenium catalyst in C—H bond activation. The α,β-unsaturated ester inserts into the rhenium-carbon bond, and then the resulting alkyl-rhenium adds to the imine, which leads to the formation of the indene derivative and regenerates the *p*-anisidine.

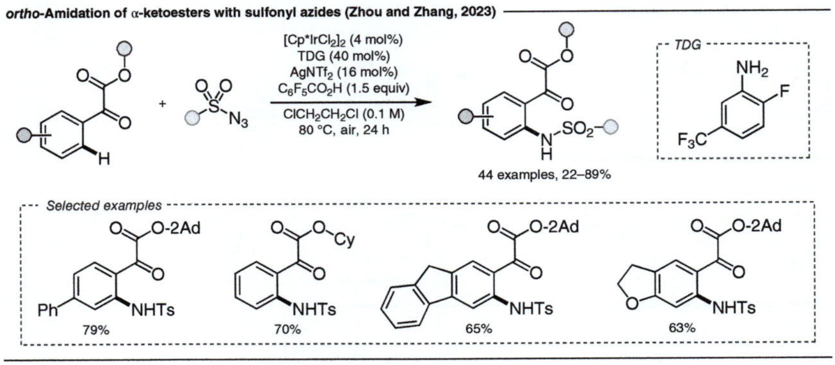

Figure 8.34 Ir-catalyzed TDG-mediated C–H functionalization.

8.3.2.3 Iridium Catalysis

In 2023, Zhou and Zhang developed an iridium-catalyzed *ortho*-amidation of α-ketoesters using organic sulfonyl azides as the amidating reagent (Figure 8.34) [105]. 2-Fluoro-5-(trifluoromethyl)aniline was identified as the optimal TDG for this process, with $[Cp^*IrCl_2]_2$ as the catalyst. The steric hindrance of the ester group significantly influenced the reaction outcomes, as increased steric bulk suppressed undesired aminolysis of the substrate with the TDG. The method was compatible with a wide range of electron-donating and electron-withdrawing functional groups on the aromatic ring, as well as fused ring systems. Other organic azides such as acyl azides and phosphoryl azides were also suitable amino sources for this C—H amidation process.

8.3.2.4 Palladium Catalysis

Jin developed the use of glycine as the TDG to direct the arylation of aromatic ketones with $Pd(PPh_3)_2Cl_2$ as the catalyst (Figure 8.35a) [106]. Later, Lin and Wei reported the use of 2-amino-*N*-isopropyl-acetamide as the TDG for the arylation of ketones, with applications in the synthesis of phenanthridinone alkaloids (Figure 8.35b). This study demonstrated that α-amino amides can provide more favorable *N*,*N*-bidentate coordination with the Pd catalyst compared with the *N*,*O*-bidentate coordination by α-amino acids [51]. In 2019, An and Li showed the use of H_2N-Gly-Gly-OH as the TDG for the arylation of ketones (Figure 8.35c) [65]. The tridentate coordination to the palladium catalyst further enhanced the reactivity of the ketone substrates, allowing the TDG loading to be lowered to 20 mol%. Additionally, Xu and Jin developed the enantioselective arylation of ferrocenyl ketones using *L*-*tert*-leucine as the chiral TDG, providing access to ferrocene derivatives with planar and stereogenic central chirality (Figure 8.35d) [107]. The use of PivOH instead of AcOH additives significantly improved both reactivity and enantioselectivity, and the developed conditions tolerated a wide range of aryl iodides.

Recent advances have further extended the transient $C(sp^2)$—H functionalization strategy to other palladium-catalyzed transformations. Lou and Xu developed a method to fluorinate the *ortho*-C—H bond of aromatic ketones using NFSI as

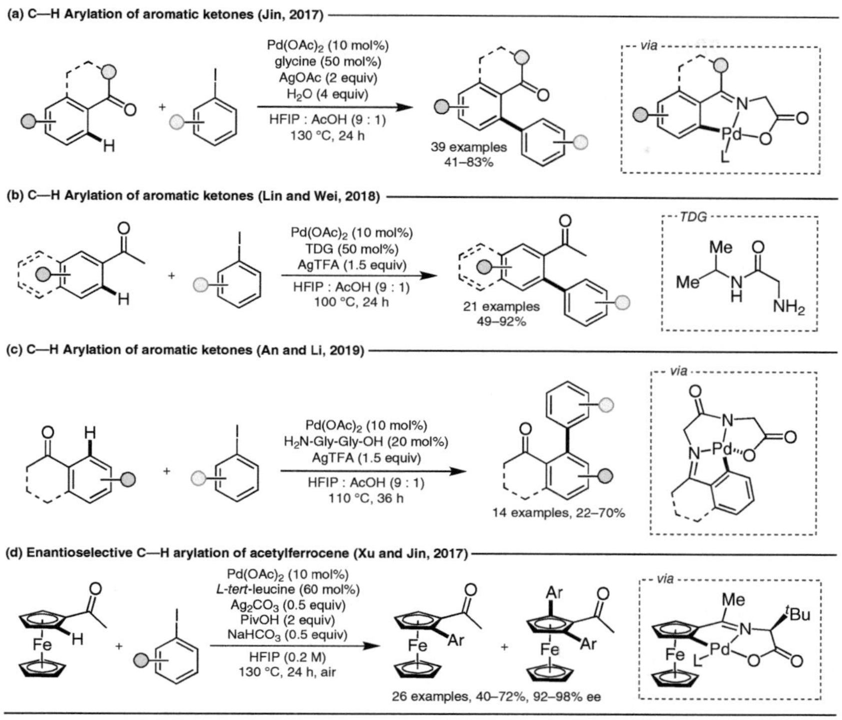

Figure 8.35 Pd-catalyzed TDG-mediated C(sp²)–H arylation. (a) Arylation of aromatic ketones, (b) synthesis of phenanthridinone alkaloids, (c) arylation of ketones with tridentate TDG, and (d) arylation of ferrocene.

the fluorinating agent (Figure 8.36a) [108]. The methodology was effective with both electron-donating and electron-withdrawing substituents on the substrate and was applied to the synthesis of anacetrapib, a cholesteryl ester transfer protein inhibitor developed by Merck for treating hypercholesterolemia. In 2022, Zhou and Zhang reported the palladium-catalyzed chlorination of aromatic α-ketoesters with *N*-chlorosuccinimide (NCS), mediated by a monodentate TDG with an aniline derivative (Figure 8.36b) [109]. This method was applied in a three-step synthesis of the anti-platelet agent clopidogrel. Additionally, they demonstrated the synthesis of 9-fluorenols from α-ketoesters via a one-pot sequential C—H arylation and intramolecular cyclization process (Figure 8.36c) [110]. Control experiments showed that the second cyclization step proceeded through a palladium-catalyzed C—H activation pathway rather than a silver-mediated intramolecular Friedel–Crafts reaction.

8.3.3 C(sp²)–H Functionalization of Amines

Following the previous work on the C(sp²)—H sulfonylation of benzaldehydes, the sulfonylation of benzylamines was developed by Bull (Figure 8.37) [101, 111]. Catalytic amounts of copper acetate with 2-hydroxynicotinaldehyde were an effective

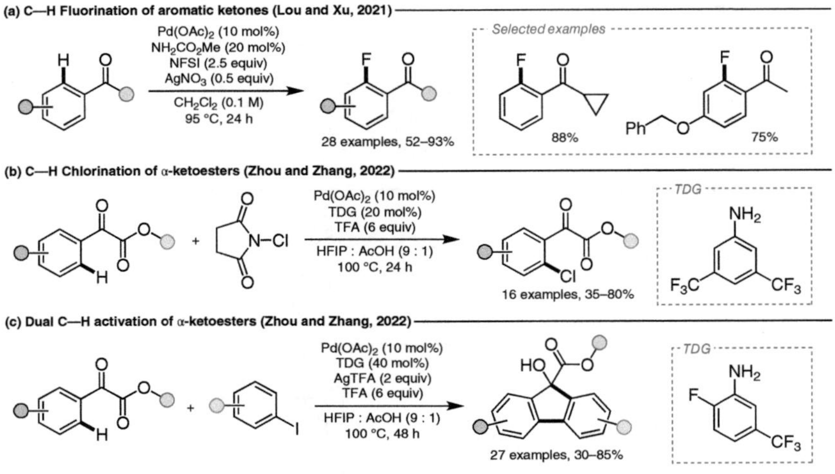

Figure 8.36 Other Pd-catalyzed TDG-mediated C(sp^2)–H transformations. (a) Fluorination of aromatic ketones, (b) chlorination of ketoesters and (c) dual C–H activation of ketoesters.

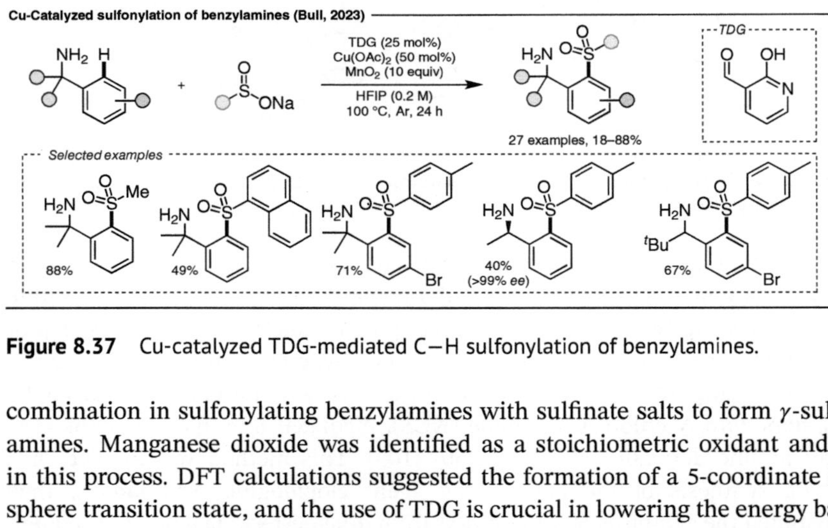

Figure 8.37 Cu-catalyzed TDG-mediated C–H sulfonylation of benzylamines.

combination in sulfonylating benzylamines with sulfinate salts to form γ-sulfonyl amines. Manganese dioxide was identified as a stoichiometric oxidant and base in this process. DFT calculations suggested the formation of a 5-coordinate inner sphere transition state, and the use of TDG is crucial in lowering the energy barrier for the C—H activation via the inner sphere CMD step involving a Wheland-type intermediate.

8.4 Conclusions and Outlook

The recent advancements in utilizing TDGs for C—H activation have marked a significant breakthrough in the field of C—H functionalization. This strategy offers a compelling alternative to conventional amide-linked DGs by enabling regioselective C(sp^2)—H and C(sp^3)—H bond activation. This strategy provides a unique advantage by allowing the formation of a DG with existing carbonyl and amine functionalities in the substrate to facilitate C—H functionalization with an organocatalyst for imine formation. DG cleavage post-functionalization also occurs in one pot. This has greatly improved the efficiency of various C—H to C—O, C—S, C—C, C—N, and C—X bond-forming processes using a variety of metal catalysts, including palladium, rhodium, ruthenium, iridium, and copper.

The increasing number of recent publications reflects the rapid development and high interest in this area. Advances in DG design and the use of external ligands have led to new transformations and reduced TDG loadings, making the processes more efficient. The use of ligand additives facilitates sluggish steps and ensures efficient rate matching of imine formation and C—H functionalization. Fluorinated solvents like HFIP, combined with acid additives such as acetic and chloroacetic acid, play a crucial role in maintaining the dynamic balance of imine concentration. Additionally, the incorporation of weakly coordinating acidic binding sites in TDGs enhances metal binding, improves functional group tolerance, and enables the late-stage functionalization of complex bioactive molecules.

Despite these advancements, significant opportunities for growth remain. Future research could focus on achieving more challenging regioselectivities, such as *meta*- and *para*-selective $C(sp^2)$—H bond activation, and expanding beyond the commonly studied palladium-catalyzed arylation to include a broader range of functionalization reactions for unactivated $C(sp^3)$—H bonds. The development of new TDGs will be essential to fully exploit the potential of transient C—H functionalization. Additionally, efforts to use base metal catalysts, more benign solvents, electrosynthetic methods, and milder reaction conditions will be key to making transient C—H activation more sustainable and broadly applicable.

References

1. Godula, K. and Sames, D. (2006). C—H bond functionalization in complex organic synthesis. *Science* 312 (5770): 67–72.
2. Wencel-Delord, J., Dröge, T., Liu, F., and Glorius, F. (2011). Towards mild metal-catalyzed C—H bond activation. *Chem. Soc. Rev.* 40 (9): 4740–4761.
3. Yamaguchi, J., Yamaguchi, A.D., and Itami, K. (2012). C—H bond functionalization: emerging synthetic tools for natural products and pharmaceuticals. *Angew. Chem. Int. Ed.* 51 (36): 8960–9009.
4. He, J., Wasa, M., Chan, K.S.L. et al. (2017). Palladium-catalyzed transformations of alkyl C—H bonds. *Chem. Rev.* 117 (13): 8754–8786.
5. Gandeepan, P., Müller, T., Zell, D. et al. (2019). 3d transition metals for C—H activation. *Chem. Rev.* 119 (4): 2192–2452.
6. St John-Campbell, S. and Bull, J.A. (2019). Base metal catalysis in directed $C(sp^3)$–H functionalisation. *Adv. Synth. Catal.* 361 (16): 3662–3682.
7. Maiti, D., Sebastian, A.T., Maji, S. et al. (2024). Palladium-catalyzed remote C—H functionalization: non-covalent interactions and reversibly bound templates. *Angew. Chem. Int. Ed.* 63 (44): e202410806.

8 Zaitsev, V.G., Shabashov, D., and Daugulis, O. (2005). Highly regioselective arylation of sp^3 C—H bonds catalyzed by palladium acetate. *J. Am. Chem. Soc.* 127 (38): 13154–13155.
9 Wasa, M., Chan, K.S.L., Zhang, X.-G. et al. (2012). Ligand-enabled methylene $C(sp^3)$—H bond activation with a Pd(II) catalyst. *J. Am. Chem. Soc.* 134 (45): 18570–18572.
10 Daugulis, O., Roane, J., and Tran, L.D. (2015). Bidentate, monoanionic auxiliary-directed functionalization of carbon–hydrogen bonds. *Acc. Chem. Res.* 48 (4): 1053–1064.
11 Guillemard, L., Kaplaneris, N., Ackermann, L., and Johansson, M.J. (2021). Late-stage C—H functionalization offers new opportunities in drug discovery. *Nat. Rev. Chem.* 5 (8): 522–545.
12 Rej, S., Ano, Y., and Chatani, N. (2020). Bidentate directing groups: an efficient tool in C—H bond functionalization chemistry for the expedient construction of C—C bonds. *Chem. Rev.* 120 (3): 1788–1887.
13 Das, J., Ali, W., and Maiti, D. (2023). The evolution of directing group strategies for $C(sp^3)$-H activation. *Trends Chem.* 5 (7): 551–560.
14 Zhao, Q., Poisson, T., Pannecoucke, X., and Besset, T. (2017). The transient directing group strategy: a new trend in transition-metal-catalyzed C—H bond functionalization. *Synthesis* 49 (21): 4808–4826.
15 St John-Campbell, S. and Bull, J.A. (2018). Transient imines as "next generation" directing groups for the catalytic functionalisation of C—H bonds in a single operation. *Org. Biomol. Chem.* 16 (25): 4582–4595.
16 Rasheed, O.K. and Sun, B. (2018). Advances in development of C—H activation/functionalization using a catalytic directing group. *ChemistrySelect* 3 (21): 5689–5708.
17 Gandeepan, P. and Ackermann, L. (2018). Transient directing groups for transformative C—H activation by synergistic metal catalysis. *Chem* 4 (2): 199–222.
18 Bhattacharya, T., Pimparkar, S., and Maiti, D. (2018). Combining transition metals and transient directing groups for C—H functionalizations. *RSC Adv.* 8 (35): 19456–19464.
19 Niu, B., Yang, K., Lawrence, B., and Ge, H. (2019). Transient ligand-enabled transition metal-catalyzed C–H functionalization. *ChemSusChem* 12 (13): 2955–2969.
20 Higham, J.I. and Bull, J.A. (2020). Transient imine directing groups for the C—H functionalisation of aldehydes, ketones and amines: an update 2018–2020. *Org. Biomol. Chem.* 18 (37): 7291–7315.
21 Amistadi-Revol, H., Liu, S., and Prévost, S. (2023). C—H functionalization of aldehydes and ketones with transient directing groups: recent developments. *Eur. J. Org. Chem.* 26 (31): e202300582.
22 Higham, J.I., Ma, T.-K., and Bull, J.A. When is an imine directing group a transient imine directing group in C–H functionalization? *Chem. Eur. J.* 30 (25): e202400345.
23 Kapoor, M., Liu, D., and Young, M.C. (2018). Carbon dioxide-mediated $C(sp^3)$-H arylation of amine substrates. *J. Am. Chem. Soc.* 140 (22): 6818–6822.

24 Kapoor, M., Chand-Thakuri, P., Maxwell, J. et al. (2019). Carbon dioxide-driven palladium-catalyzed C—H activation of amines: a unified approach for the arylation of aliphatic and aromatic primary and secondary amines. *Synlett* 30 (5): 519–524.

25 Xie, S., Li, S., Ma, W. et al. (2019). Chelation-directed remote *meta*-C—H functionalization of aromatic aldehydes and ketones. *Chem. Commun.* 55 (82): 12408–12411.

26 Lewis, L.N. (1985). Reexamination of the deuteration of phenol catalyzed by an orthometalated ruthenium complex. *Inorg. Chem.* 24 (25): 4433–4435.

27 Bedford, R.B., Coles, S.J., Hursthouse, M.B., and Limmert, M.E. (2003). The catalytic intermolecular orthoarylation of phenols. *Angew. Chem. Int. Ed.* 42 (1): 112–114.

28 Carrión, M.C. and Cole-Hamilton, D.J. (2006). Halide-free ethylation of phenol by multifunctional catalysis using phosphinite ligands. *Chem. Commun.* 43: 4527–4529.

29 Bedford, R.B., Betham, M., Caffyn, A.J.M. et al. (2008). Simple rhodium–chlorophosphine pre-catalysts for the ortho-arylation of phenols. *Chem. Commun.* 8: 990–992.

30 Yang, J.-F., Wang, R.-H., Wang, Y.-X. et al. (2016). Ligand-accelerated direct C–H arylation of BINOL: a rapid one-step synthesis of racemic 3,3′-diaryl BINOLs. *Angew. Chem. Int. Ed.* 55 (45): 14116–14120.

31 Yang, K., Song, M., Liu, H., and Ge, H. (2020). Palladium-catalyzed direct asymmetric C—H bond functionalization enabled by the directing group strategy. *Chem. Sci.* 11 (47): 12616–12632.

32 Liao, G., Zhang, T., Lin, Z.K., and Shi, B.F. (2020). Transition metal-catalyzed enantioselective C–H functionalization via chiral transient directing group strategies. *Angew. Chem. Int. Ed.* 59 (45): 19773–19786.

33 Bag, D., Verma, P.K., and Sawant, S.D. (2020). Chiral transient directing group strategies in asymmetric synthesis. *Chem. Asian J.* 15 (20): 3225–3238.

34 Ha, H., Lee, J., Park, M.H. et al. (2020). Transient directing group-assisted C—H bond functionalization of aliphatic amines: strategies for efficiency and site-selectivity. *Bull. Korean Chem. Soc.* 41 (6): 582–587.

35 Murahashi, S. (1955). Synthesis of phthalimidines from schiff bases and carbon monoxide. *J. Am. Chem. Soc.* 77 (23): 6403–6404.

36 Constable, A.G., McDonald, W.S., Sawkins, L.C., and Shaw, B.L. (1978). Palladation of dimethylhydrazones, oximes, and oxime *O*-allyl ethers: crystal structure of $[Pd_3(ON{=}CPr^iPh)_6]$. *J. Chem. Soc. Chem. Commun.* 23: 1061–1062.

37 Clark, P.W., Dyke, S.F., Smith, G., and Kennard, C.H.L. (1987). The cyclopalladation of benzylidenebenzylamines. *J. Organomet. Chem.* 330 (3): 447–460.

38 Johnson, J.A. and Sames, D. (2000). C—H bond activation of hydrocarbon segments in complex organic molecules: total synthesis of the antimitotic rhazinilam. *J. Am. Chem. Soc.* 122 (26): 6321–6322.

39 Dangel, B.D., Godula, K., Youn, S.W. et al. (2002). C—C bond formation via C—H bond activation: synthesis of the core of Teleocidin B4. *J. Am. Chem. Soc.* 124 (40): 11856–11857.

40 Dick, A.R., Hull, K.L., and Sanford, M.S. (2004). A highly selective catalytic method for the oxidative functionalization of C—H bonds. *J. Am. Chem. Soc.* 126 (8): 2300–2301.

41 Neufeldt, S.R. and Sanford, M.S. (2010). O-acetyl oximes as transformable directing groups for Pd-catalyzed C—H bond functionalization. *Org. Lett.* 12 (3): 532–535.

42 Wang, Y., Liu, X., Xu, X. et al. (2023). Palladium-catalyzed tandem diolefination reaction of benzaldehyde enabled by a remote directing group. *Org. Lett.* 25 (27): 4962–4967.

43 Rej, S. and Chatani, N. (2021). Transient imine as a directing group for the metal-free *o*-C—H borylation of benzaldehydes. *J. Am. Chem. Soc.* 143 (7): 2920–2929.

44 Xu, Y., Young, M.C., Wang, C. et al. (2016). Catalytic C(sp^3)–H arylation of free primary amines with an exo directing group generated in situ. *Angew. Chem. Int. Ed.* 55 (31): 9084–9087.

45 Bag, S., Jana, S., Pradhan, S. et al. (2021). Imine as a linchpin approach for meta-C—H functionalization. *Nat. Commun.* 12 (1): 1393.

46 Jun, C.-H., Lee, H., and Hong, J.-B. (1997). Chelation-assisted intermolecular hydroacylation: direct synthesis of ketone from aldehyde and 1-alkene. *J. Org. Chem.* 62 (5): 1200–1201.

47 Jun, C.-H., Lee, D.-Y., and Hong, J.-B. (1997). Hydroacylation of 1-alkene with heteroaromatic aldehyde by Rh(I) and additives. *Tetrahedron Lett.* 38 (38): 6673–6676.

48 Zhang, F.-L., Hong, K., Li, T.-J. et al. (2016). Functionalization of C(sp^3)–H bonds using a transient directing group. *Science* 351 (6270): 252–256.

49 Sinha, S.K., Bhattacharya, T., and Maiti, D. (2019). Role of hexafluoroisopropanol in C—H activation. *React. Chem. Eng.* 4 (2): 244–253.

50 Yu, C., Sanjosé-Orduna, J., Patureau, F.W., and Pérez-Temprano, M.H. (2020). Emerging unconventional organic solvents for C—H bond and related functionalization reactions. *Chem. Soc. Rev.* 49 (6): 1643–1652.

51 Bhattacharya, T., Ghosh, A., and Maiti, D. (2021). Hexafluoroisopropanol: the magical solvent for Pd-catalyzed C—H activation. *Chem. Sci.* 12 (11): 3857–3870.

52 Yang, K., Li, Q., Liu, Y. et al. (2016). Catalytic C—H arylation of aliphatic aldehydes enabled by a transient ligand. *J. Am. Chem. Soc.* 138 (39): 12775–12778.

53 St John-Campbell, S., White, A.J.P., and Bull, J.A. (2017). Single operation palladium catalysed C(sp^3)–H functionalisation of tertiary aldehydes: investigations into transient imine directing groups. *Chem. Sci.* 8 (7): 4840–4847.

54 Ma, F., Lei, M., and Hu, L. (2016). Acetohydrazone: a transient directing group for arylation of unactivated C(sp^3)-H bonds. *Org. Lett.* 18 (11): 2708–2711.

55 Park, H., Yoo, K., Jung, B., and Kim, M. (2018). Direct synthesis of anthracenes from *o*-tolualdehydes and aryl iodides through Pd(II)-catalyzed sp^3 C—H arylation and electrophilic aromatic cyclization. *Tetrahedron* 74 (16): 2048–2055.

56 Wang, J., Dong, C., Wu, L. et al. (2018). Palladium-catalyzed β-C–H arylation of ketones using amino amide as a transient directing group: applications to synthesis of phenanthridinone alkaloids. *Adv. Synth. Catal.* 360 (19): 3709–3715.

57 Dong, C., Wu, L., Yao, J., and Wei, K. (2019). Palladium-catalyzed β-C-H arylation of aliphatic aldehydes and ketones using amino amide as a transient directing group. *Org. Lett.* 21 (7): 2085–2089.

58 Gou, B.-B., Liu, H.-F., Chen, J., and Zhou, L. (2019). Palladium-catalyzed site-selective C(sp^3)–H arylation of phenylacetaldehydes. *Org. Lett.* 21 (17): 7084–7088.

59 St John-Campbell, S. and Bull, J.A. (2019). Intramolecular palladium(II)/(IV) catalysed C(sp^3)–H arylation of tertiary aldehydes using a transient imine directing group. *Chem. Commun.* 55 (62): 9172–9175.

60 Zhang, X.-L., Pan, G.-F., Zhu, X.-Q. et al. (2019). Dehydrogenative β-arylation of saturated aldehydes using transient directing groups. *Org. Lett.* 21 (8): 2731–2735.

61 St John-Campbell, S., White, A.J.P., and Bull, J.A. (2020). Methylene C(sp^3)–H β,β'-diarylation of cyclohexanecarbaldehydes promoted by a transient directing group and pyridone ligand. *Org. Lett.* 22 (5): 1807–1812.

62 Wang, P., Farmer, M.E., Huo, X. et al. (2016). Ligand-promoted meta-C—H arylation of anilines, phenols, and heterocycles. *J. Am. Chem. Soc.* 138 (29): 9269–9276.

63 Wang, P., Verma, P., Xia, G. et al. (2017). Ligand-accelerated non-directed C—H functionalization of arenes. *Nature* 551 (7681): 489–493.

64 Hong, K., Park, H., and Yu, J.-Q. (2017). Methylene C(sp^3)–H arylation of aliphatic ketones using a transient directing group. *ACS Catal.* 7 (10): 6938–6941.

65 Cheng, Y., Zheng, J., Tian, C. et al. (2019). Palladium-catalyzed C–H arylation of aliphatic and aromatic ketones using dipeptide transient directing groups. *Asian J. Org. Chem.* 8 (4): 526–531.

66 Li, Y.H., Ouyang, Y., Chekshin, N., and Yu, J.Q. (2022). Pd^{II}-catalyzed γ-C(sp^3)–H (hetero)arylation of ketones enabled by transient directing groups. *ACS Catal.* 12 (17): 10581–10586.

67 Provencher, P.A., Hoskin, J.F., Wong, J.J. et al. (2021). Pd(II)-catalyzed synthesis of benzocyclobutenes by β-methylene-selective C(sp^3)–H arylation with a transient directing group. *J. Am. Chem. Soc.* 143 (48): 20035–20041.

68 Cheng, J., Xiao, L., Qian, S. et al. (2022). Palladium(II)-catalyzed selective arylation of tertiary C–H bonds of cyclobutylmethyl ketones using transient directing groups. *Angew. Chem. Int. Ed.* 61 (18).

69 Xiao, L., Hong, K., Luo, F. et al. (2020). Pd^{II}-catalyzed enantioselective C(sp^3)–H arylation of cyclobutyl ketones using a chiral transient directing group. *Angew. Chem. Int. Ed.* 59 (24): 9594–9600.

70 Vicente, J., Saura-Llamas, I., Palin, M.G. et al. (1997). Orthometalation of primary amines. 4.1 orthopalladation of primary benzylamines and (2-phenylethyl)amine. *Organometallics* 16 (5): 826–833.

71 Hartwig, J.F., Richards, S., Barañano, D., and Paul, F. (1996). Influences on the relative rates for C–N bond-forming reductive elimination and β-hydrogen elimination of amides. A case study on the origins of competing reduction in the palladium-catalyzed amination of aryl halides. *J. Am. Chem. Soc.* 118 (15): 3626–3633.

72 Wu, Y., Chen, Y.-Q., Liu, T. et al. (2016). Pd-catalyzed γ-C(sp^3)–H arylation of free amines using a transient directing group. *J. Am. Chem. Soc.* 138 (44): 14554–14557.

73 Lin, H., Wang, C., Bannister, T.D., and Kamenecka, T.M. (2018). Site-selective γ-C(sp^3)–H and γ-C(sp^2)–H arylation of free amino esters promoted by a catalytic transient directing group. *Chem. Eur. J.* 24 (38): 9535–9541.

74 Wang, Z., Fu, Y., Zhang, Q. et al. (2020). Pd(II)-catalyzed direct γ-C(sp^3)-H arylation between free $β^2$-amino esters and $β^3$-amino esters and aryl iodides using a catalytic transient directing group. *J. Org. Chem.* 85 (12): 7683–7693.

75 Liu, Y. and Ge, H. (2017). Site-selective C—H arylation of primary aliphatic amines enabled by a catalytic transient directing group. *Nat. Chem.* 9 (1): 26–32.

76 Yada, A., Liao, W., Sato, Y., and Murakami, M. (2017). Buttressing salicylaldehydes: a multipurpose directing group for C(sp^3)–H bond activation. *Angew. Chem. Int. Ed.* 56 (4): 1073–1076.

77 St John-Campbell, S., Ou, A.K., and Bull, J.A. (2018). Palladium-catalyzed C(sp^3)–H arylation of primary amines using a catalytic alkyl acetal to form a transient directing group. *Chem. Eur. J.* 24 (67): 17838–17843.

78 Chen, Y.-Q., Wang, Z., Wu, Y. et al. (2018). Overcoming the limitations of γ- and δ-C—H arylation of amines through ligand development. *J. Am. Chem. Soc.* 140 (51): 17884–17894.

79 Ghouilem, J., Bazzi, S., Grimblat, N. et al. (2023). Transient imine as a directing group for the Pd-catalyzed anomeric C(sp^3)–H arylation of 3-aminosugars. *Chem. Commun.* 59: 2497–2500.

80 Chen, Y.-Q., Singh, S., Wu, Y. et al. (2020). Pd-catalyzed γ-C(sp^3)–H fluorination of free amines. *J. Am. Chem. Soc.* 142 (22): 9966–9974.

81 Chen, Y.-Q., Wu, Y., Wang, Z. et al. (2020). Transient directing group enabled Pd-catalyzed γ-C(sp^3)–H oxygenation of alkyl amines. *ACS Catal.* 10 (10): 5657–5662.

82 Liu, X.-H., Park, H., Hu, J.-H. et al. (2017). Diverse ortho-C(sp^2)–H functionalization of benzaldehydes using transient directing groups. *J. Am. Chem. Soc.* 139 (2): 888–896.

83 Li, F., Zhou, Y., Yang, H. et al. (2019). Monodentate transient directing group enabled Pd-catalyzed *ortho*-C—H methoxylation and chlorination of benzaldehydes. *Org. Lett.* 21 (10): 3692–3695.

84 Wang, Y.-F., Xu, W.-G., Sun, B. et al. (2019). Monodentate transient directing group assisted Pd-catalyzed direct dehydrogenative cross-coupling of benzaldehydes with arenes toward 9-fluorenones. *J. Org. Chem.* 84 (20): 13104–13111.

85 Chen, X.-Y., Ozturk, S., and Sorensen, E.J. (2017). Synthesis of fluorenones from benzaldehydes and aryl iodides: dual C—H functionalizations using a transient directing group. *Org. Lett.* 19 (5): 1140–1143.

86 Wang, D.-Y., Guo, S.-H., Pan, G.-F. et al. (2018). Direct dehydrogenative arylation of benzaldehydes with arenes using transient directing groups. *Org. Lett.* 20 (7): 1794–1797.

87 Mu, D., He, G., and Chen, G. (2018). Palladium-catalyzed ortho C–H arylation of benzaldehydes using ortho-sulfinyl aniline as transient auxiliary. *Chem. Asian J.* 13 (17): 2423–2426.

88 Chen, X.-Y. and Sorensen, E.J. (2018). Pd-catalyzed, ortho C—H methylation and fluorination of benzaldehydes using orthanilic acids as transient directing groups. *J. Am. Chem. Soc.* 140 (8): 2789–2792.

89 Wu, Y.-J., Yao, Q.-J., Chen, H.-M. et al. (2020). Palladium-catalyzed ortho-C-H silylation of biaryl aldehydes using a transient directing group. *Sci. China Chem.* 63 (7): 875–880.

90 Qiao, H., Sun, B., Yu, Q. et al. (2019). Palladium-catalyzed direct ortho-C—H selenylation of benzaldehydes using benzidine as a transient directing group. *Org. Lett.* 21 (17): 6914–6918.

91 Yao, Q.-J., Zhang, S., Zhan, B.-B., and Shi, B.-F. (2017). Atroposelective synthesis of axially chiral biaryls by palladium-catalyzed asymmetric C–H olefination enabled by a transient chiral auxiliary. *Angew. Chem. Int. Ed.* 56 (23): 6617–6621.

92 Dhawa, U., Tian, C., Wdowik, T. et al. (2020). Enantioselective pallada-electrocatalyzed C–H activation by transient directing groups: expedient access to helicenes. *Angew. Chem. Int. Ed.* 59 (32): 13451–13457.

93 Tan, P.W., Juwaini, N.A.B., and Seayad, J. (2013). Rhodium(III)-amine dual catalysis for the oxidative coupling of aldehydes by directed C—H activation: synthesis of phthalides. *Org. Lett.* 15 (20): 5166–5169.

94 Hu, W., Zheng, Q., Sun, S., and Cheng, J. (2017). Rh(III)-catalyzed bilateral cyclization of aldehydes with nitrosos toward unsymmetrical acridines proceeding with C—H functionalization enabled by a transient directing group. *Chem. Commun.* 53 (46): 6263–6266.

95 Li, G., Jiang, J., Xie, H., and Wang, J. (2019). Introducing the chiral transient directing group strategy to rhodium(III)-catalyzed asymmetric C–H activation. *Chem. Eur. J.* 25 (18): 4688–4694.

96 Hande, A.E., Ramesh, V.B., and Prabhu, K.R. (2018). Rh(III)-catalyzed ortho-C-(sp^2)–H amidation of ketones and aldehydes under synergistic ligand-accelerated catalysis. *Chem. Commun.* 54 (85): 12113–12116.

97 Li, Z.-Y., Lakmal, H.H.C., Qian, X. et al. (2019). Ruthenium-catalyzed enantioselective C—H functionalization: a practical access to optically active indoline derivatives. *J. Am. Chem. Soc.* 141 (40): 15730–15736.

98 Li, G., Liu, Q., Vasamsetty, L. et al. (2020). Ruthenium(II)-catalyzed asymmetric inert C–H bond activation assisted by a chiral transient directing group. *Angew. Chem. Int. Ed.* 59 (9): 3475–3479.

99 Chen, X.-Y. and Sorensen, E.J. (2018). Ir(III)-catalyzed ortho C—H alkylations of (hetero)aromatic aldehydes using alkyl boron reagents. *Chem. Sci.* 9 (48): 8951–8956.

100 Huang, J., Ding, J., Ding, T.-M. et al. (2019). Cobalt-catalyzed *ortho*-C(sp^2)–H amidation of benzaldehydes with dioxazolones using transient directing groups. *Org. Lett.* 21 (18): 7342–7345.

101 Higham, J.I. and Bull, J.A. (2022). Amine-catalyzed copper-mediated C–H sulfonylation of benzaldehydes via a transient imine directing group. *Angew. Chem. Int. Ed.* 61 (27): e202202933.

102 Jun, C.-H., Hong, J.-B., Kim, Y.-H., and Chung, K.-Y. (2000). The catalytic alkylation of aromatic imines by Wilkinson's complex: the domino reaction of hydroacylation and ortho-alkylation. *Angew. Chem. Int. Ed.* 39 (19): 3440–3442.

103 Jun, C.-H., Moon, C.W., Hong, J.-B. et al. (2002). Chelation-assisted RhI-catalyzed ortho-alkylation of aromatic ketimines or ketones with olefins. *Chem. Eur. J.* 8 (2): 485–492.

104 Kuninobu, Y., Nishina, Y., Shouho, M., and Takai, K. (2006). Rhenium- and aniline-catalyzed one-pot annulation of aromatic ketones and α,β-unsaturated esters initiated by C—H bond activation. *Angew. Chem. Int. Ed.* 45 (17): 2766–2768.

105 He, Y., Sun, B., Lu, X. et al. (2023). Iridium-catalyzed direct ortho-C—H amidation of α-ketoesters with sulfonyl azides using a transient directing group strategy. *J. Org. Chem.* 88 (7): 4345–4351.

106 Xu, J., Liu, Y., Wang, Y. et al. (2017). Pd-catalyzed direct ortho-C—H arylation of aromatic ketones enabled by a transient directing group. *Org. Lett.* 19 (7): 1562–1565.

107 Xu, J., Liu, Y., Zhang, J. et al. (2018). Palladium-catalyzed enantioselective C(sp^2)–H arylation of ferrocenyl ketones enabled by a chiral transient directing group. *Chem. Commun.* 54 (6): 689–692.

108 Wu, Q., Mao, Y.-J., Zhou, K. et al. (2021). Pd-catalysed direct C(sp^2)–H fluorination of aromatic ketones: concise access to anacetrapib. *Chem. Commun.* 57 (37): 4544–4547.

109 Wang, J., Wu, Y., Xu, W. et al. (2022). Monodentate transient directing group promoted Pd-catalyzed direct ortho-C—H arylation and chlorination of α-ketoesters for three-step synthesis of Cloidogrel racemate. *Tetrahedron* 123: 132980.

110 Xu, W., Zhang, Y., Wu, Y. et al. (2022). Direct assembly of diverse unsymmetrical tertiary 9-fluorenols via transient directing group-enabled palladium-catalyzed dual C—H bond activation of α-ketoesters. *J. Org. Chem.* 87 (16): 10807–10814.

111 Higham, J.I., Ma, T.-K., and Bull, J.A. (2023). Dual copper- and aldehyde-catalyzed transient C—H sulfonylation of benzylamines. *Org. Lett.* 25 (28): 5285–5290.

9

Redox Reactions in Ru(II)-Catalyzed C–H Activations

Suman Dana, Suman Ghosh, Mainak Koner, Nityananda Ballav, and Mahiuddin Baidya

Department of Chemistry, Indian Institute of Technology Madras, Chennai 600036, India

9.1 Introduction

In the past decades, transition-metal-catalyzed C—H bond activation reactions have transpired as a viable synthetic tool for the straightforward modification of organic molecules [1–8]. These transformations have revolutionized the conventional chemical synthesis, harvesting ubiquitous C—H bonds as a staple to incorporate new functionalities [1, 2, 4–6, 8–16]. They have found widespread application from the synthesis of high-value organic compounds from structurally simple manifolds to late-stage diversification of complex pharmaceuticals, peptides, and other biologically relevant compounds. These processes are advantageous as ubiquitous C—H bonds are harvested as a staple for the functionalization process. Thus, unlike typical functional group interconversion strategies, no pre-installation of functional groups is necessary and the overall waste generation is significantly reduced, making the synthetic process more environmentally friendly. Most of these transformations involve a redox cycle, which significantly influences the functionalization process. The genesis of such a transformation resembles the biosynthesis of terpenes, where in the oxidase phase different oxidative enzymes play a critical role in modifying the terpene's carbocyclic backbone [17–20]. Different oxidative enzymes catalyze the insertion of oxygen atoms in the terpene precursor, leading to the formation of alcohols, ketones, epoxides, and other oxygenated functional groups. Similarly, in transition-metal catalyzed C—H activations, external redox agents such as molecular oxygen, peroxides, silver or copper salts, and other mild oxidizing agents, play the key role to manipulate the oxidation state of the catalyst and help continuing the catalytic cycle. Continuous evolution in C—H activations has showcased the tremendous success of these approaches in catalyzing diverse C—C and C—heteroatom bond-forming reactions even in complex molecular settings. This method has appeared extremely effective even in industrial settings in the synthesis of functionalized aromatics, heterocycles, and pharmaceutical intermediates with high regio- and chemoselectivity [21].

Directed C–H Bond Functionalization: Concepts and Applications, First Edition.
Edited by Debabrata Maiti and Supriya Rej.

In this domain, ruthenium catalysts persisted as one of the most promising contenders in catalyzing these transformations [14, 15, 22–30]. Compared to other noble metal catalysts, ruthenium is inexpensive, bench-stable, and air- and moisture-tolerant. Further, the redox diversity of ruthenium catalysis exhibited unique features in comparison to other transition-metal catalysts. Ruthenium displays inimitable electronic properties and the ability to stabilize higher oxidation states [23–33]. Consequently, different redox processes can be attained in this catalytic regime, where Ru(II/0), Ru(II/IV), and Ru(II/III) are among the most notable redox cycles. This redox diversity offered several complementary C—H activation strategies in comparison to its noble metal congeners, which are solely limited to ruthenium catalysis. Most unique synthetic avenues were also achieved through Ru(II/III)-catalytic cycle, which helps in the direct functionalization of *meta*- and *para*-C—H bonds in the absence of sophisticated and designer external auxiliaries as the directing group [10]. These transformations are otherwise impossible through other transition-metal-catalyzed C—H activations. We have systematically described the redox features of ruthenium-catalyzed C—H activations in this chapter. Owing to the vastness of this area and limited space for the discussion, we have focused our discussion on the redox behavior of carboxylate-assisted Ru(II)-catalyzed C—H activations. Further, the area consists of enormous literature precedents, and thus, we have methodically summarized different catalytic manifolds with selected examples.

9.2 Background and Early Findings

Early findings in Ru(II)-catalyzed C—H activations were reported by Lewis and Smith, testifying the formation of cyclometalated Ru complex (Scheme 9.1a) [31]. Later, Murai *et al.* described the potential of Ru catalysis for a weak coordination-assisted hydroarylation of olefins (Scheme 9.1b) [32]. The reaction proceeded with a Ru(0)-active catalyst, which underwent oxidative addition to the *ortho*-C—H bond of acetophenone followed by the functionalization. This concept of Ru(0)-catalyzed C—H activation was further extended by Darses and coworkers using formate additives for the hydroarylation of diverse olefins (Scheme 9.1c) [33, 34]. Notably, Oi and Inoue described the early examples of Ru(II)-catalyzed arylation reactions in NMP solvent (Scheme 9.1d) [35]. Later, Ackermann group contributed significantly in this direction highlighting the important role of secondary phosphine oxide ligands in arylation reactions (Scheme 9.2a) [36–39]. The rate of the arylation was significantly influenced by the SPO ligands owing to a possible base assistance in the transition state. This analogy was further extended by the Ackermann group exploring the crucial carboxylate assistance in Ru(II)-catalyzed C—H activations, which later revolutionized this domain [14, 25, 40, 41].

The major operational mode of the Ru(II)-catalyzed C—H activations operates on three different modes. These are Ru(II/0), Ru(II/IV), and Ru(II/III) manifolds.

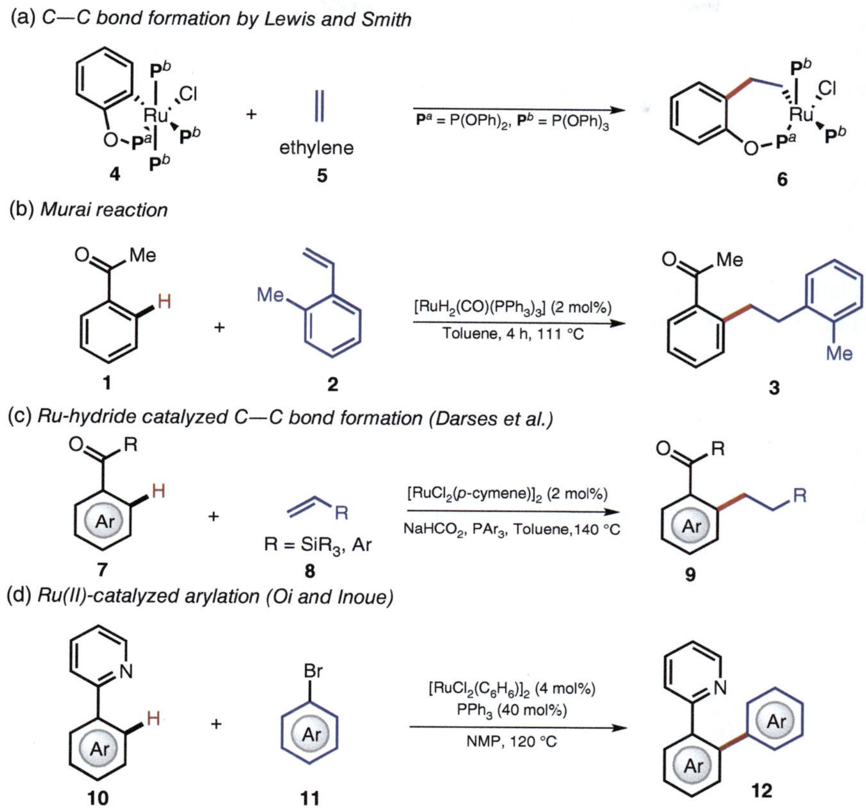

Scheme 9.1 Early findings in ruthenium catalyzed C—H activations.

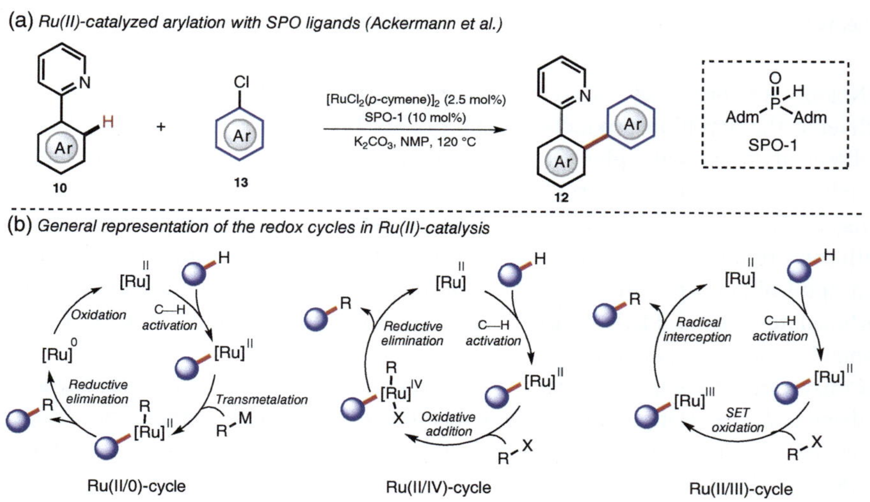

Scheme 9.2 Ruthenium(II)-catalyzed arylation reaction and general representation of the redox cycles.

The general mode of action of these catalytic systems is highlighted below (Scheme 9.2b). Notably, redox-neutral strategies are also well established in this platform, which is beyond the discussion of the current chapter.

9.3 Aromatic C–H Bond Activation Through Ru(II/IV)-catalyzed Reactions

Oxidative Ru(II/IV)-catalytic strategy is harnessed as one of the most popular strategies for the formation of C—C and C—heteroatom bonds among other catalytic manifolds reported in Ru(II)-catalyzed C—H activations [26, 29, 42]. The early crucial findings were established by the Ackermann group, where they screened more than 30 crucial additives to promote the arylation of *N*-aryl triazoles [40]. This thorough screening of additives inferred sterically hindered carboxylates as the best additive to promote the arylation at an optimal rate. The C—H bond activation step was reversible and various aryl electrophiles were compatible to this catalytic manifold (Scheme 9.3).

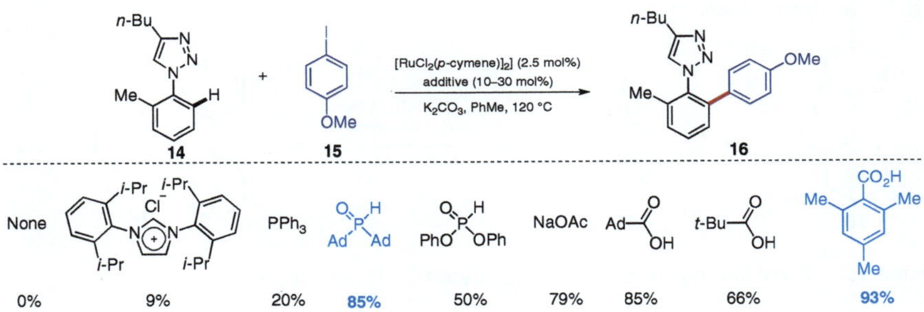

Scheme 9.3 Ruthenium(II)-catalyzed arylation with carboxylate additive.

Notably, aryl bromides and chlorides were susceptible to the arylation process and delivered the arylation products in high fidelity [41]. Later, they also established a direct arylation with phenols, in situ converting them into to aryl tosylates, which served as suitable aryl electrophiles under this catalytic manifold [43]. This Ru(II)-catalyzed arylation strategy was viable with a diverse range of substrates with non-removable and removable directing groups including heteroarenes, and applicable on preparative scale in different common organic solvents. Later, carboxylate-assisted Ru(II)-catalysis was applied to direct *ortho*-alkylation of C—H bonds (Scheme 9.4) [44–46].

Consistent efforts from different research groups later established carboxylate-assisted *ortho*-C—H arylations as a general protocol with different strong and weakly coordinating directing groups and typically used with a broad range of substrates (Scheme 9.5) [26].

Ackermann group further described arene-free ruthenium(II/IV)-catalyzed *ortho*-C—H arylation strategies with 2-arylpyridines [47–49]. While bench-stable

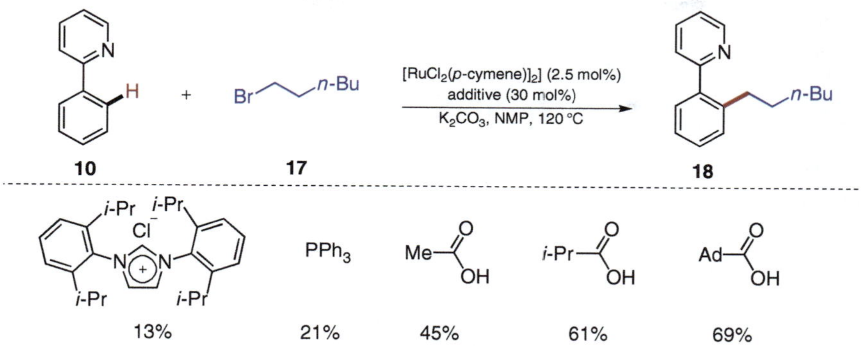

Scheme 9.4 Ruthenium(II)-catalyzed alkylation with carboxylate additive.

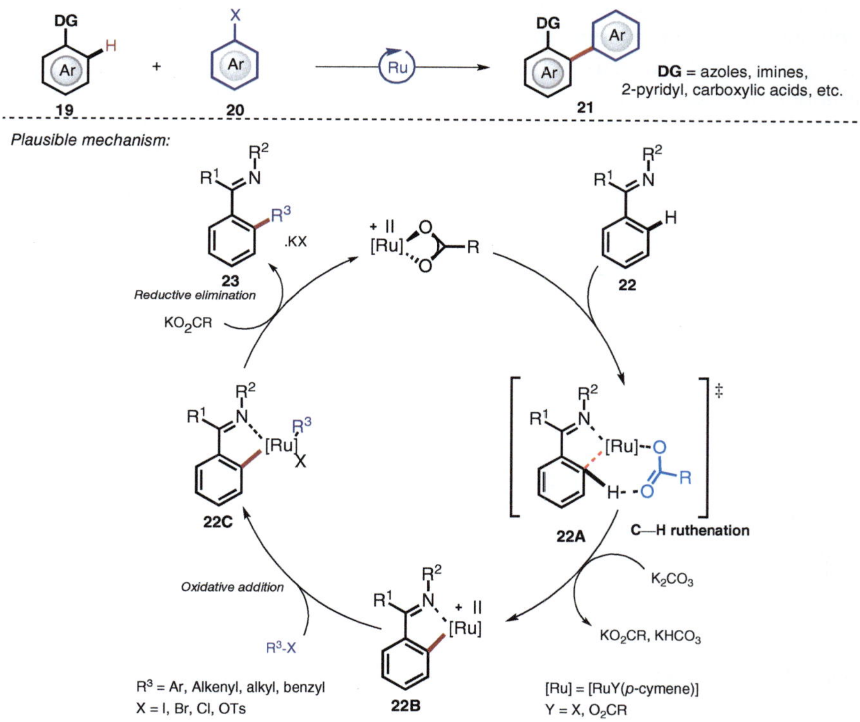

Scheme 9.5 General representation of ruthenium(II)-catalyzed arylation reactions.

$[Ru(p\text{-cymene})Cl_2]_2$ was used as the catalyst in this process, studies revealed the removal of *p*-cymene during the catalytic process. The arene ligand free Ru(II)-complex was found as the catalytically competent species, which was also supported by the DFT studies. The oxidative addition of aryl bromide to this Ru(II)-cyclometalated intermediate formed the short-lived Ru(IV)-species, which upon reductive elimination generated the *ortho*-arylated product (Scheme 9.6).

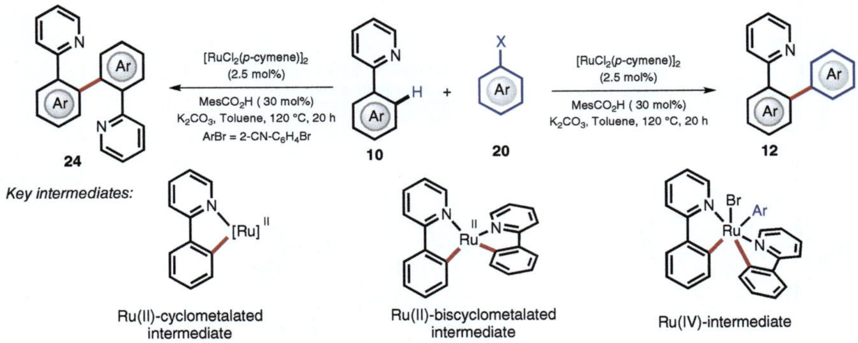

Scheme 9.6 Arene-ligand free ruthenium(II)-catalyzed C–H arylation.

Ackermann group recently reported an arylation of aryl triazoles and tetrazoles involving aryl dibenzothiophenium salts as the arylating agent [50]. Preliminary mechanistic studies revealed that the arylation was more susceptible with dibenzothiophenium salts compared to the aryl halides. The critical advantage of this process was the applicability of the current protocol in the late-stage functionalization of different complex active pharmaceutical ingredients (APIs) (Scheme 9.7).

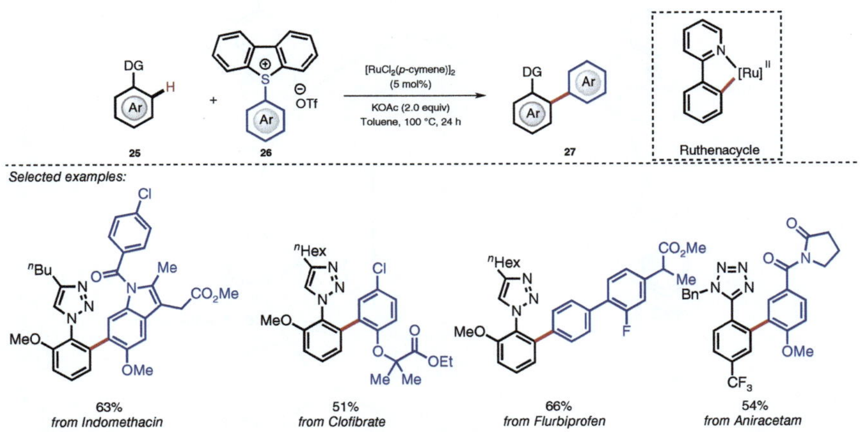

Scheme 9.7 Late-stage ruthenium(II)-catalyzed C–H arylation with dibenzothiophenium salts.

Larossa group harvested cyclometalated Ru(II)-complex as the catalyst to execute the late-stage C—H arylation of aryl pyridines [51]. Numerous biologically relevant aryl halides and triflates were accommodated under this protocol to avail their analogues. The catalytic process operated through a Ru(II/IV)-process, where the *bis*-cycloruthenated species was identified as the key intermediate (Scheme 9.8).

Cyclometalated Ru(II)-complex was used for the site-selective installation of methyl and d_3-methyl groups under mild conditions [52]. Different N-heterocycles and imines were used as the directing group in this process. Quaternary ammonium

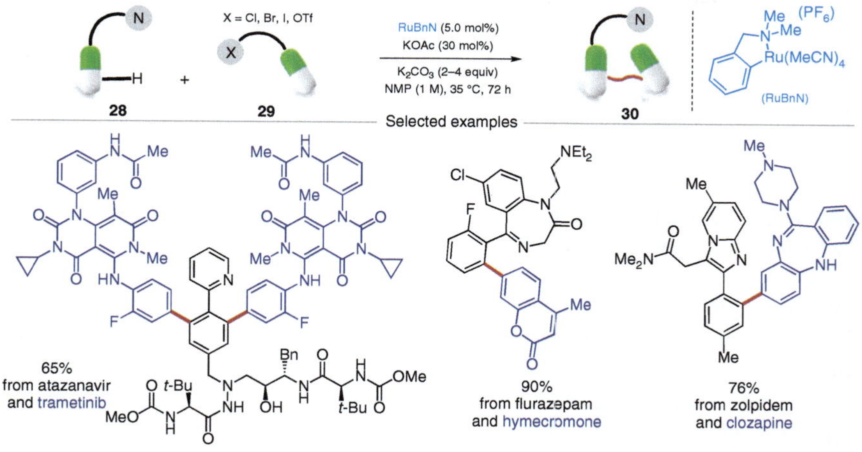

Scheme 9.8 Cyclometalated ruthenium(II)-complex catalyzed late-stage C—H arylation.

salt under the reaction conditions formed MeI as a crucial precursor for the methylation, which after oxidative addition formed transient Ru(IV)-species. Reductive elimination of this transient species generated the desired product (Scheme 9.9).

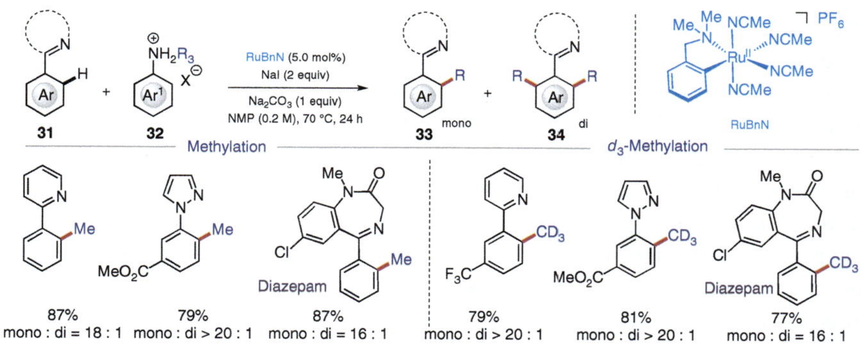

Scheme 9.9 Ruthenium-catalyzed monoselective C—H methylation and d^3-methylation of arenes.

This catalytic process was also applicable to secondary alkyl halides, which operated through a similar mechanistic scenario to generate the alkylation products in high yields [53]. The formation of *bis*-cyclometalated complex was critical for this process, which governed the *ortho*-selectivity instead of *meta*-functionalization (Scheme 9.10).

The Ru(II/IV)-catalytic manifold was also used for the *ortho*-C—H activation/annulation reaction with benzoic acids and sulfoxonium ylides [54]. The transformation proceeded through a base-assisted internal electrophilic substitution (BIES) type C—H activation, where the formation of Ru(IV)-cyclometalated intermediate was a key step. Here, sulfoxonium ylides served as the carbene

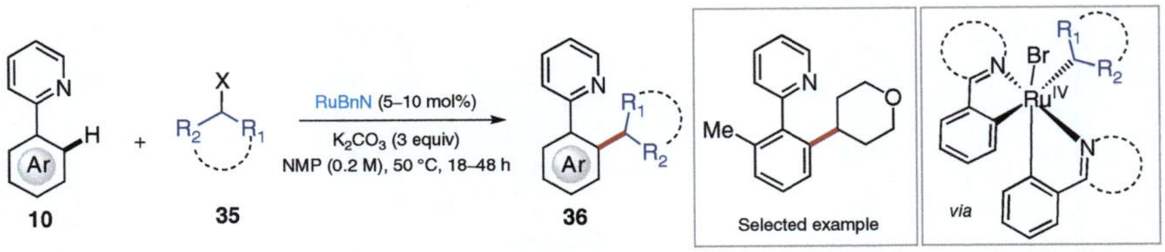

Scheme 9.10 Cyclometalated ruthenium(II)-catalyst enabled *ortho*-C–H alkylation with secondary alkyl bromides.

precursors, which after the release of DMSO generated the Ru(IV)-intermediate. Reductive elimination from which consequently generated the isocoumarins products with high diversity (Scheme 9.11).

The formation of C—heteroatom bond remains one of the most challenging transformations in the field of C—H activation. The strong binding affinity of the heteroatom ligands inhibits the reductive elimination for the formation of C—heteroatom bonds. However, the Ru(II/IV)-catalytic manifold served extremely well in this purpose. The cyclometalated Ru(II)-intermediate could realize an oxidatively induced reductive elimination to facilitate the reductive elimination process by forming short-lived Ru(IV)-intermediates. Different C—heteroatom bond-forming processes were reported utilizing this principle, selective examples of which are reported in the following section.

The formation of C—N bonds through strategic Ru(II)-catalyzed C—H bond activation is among the most popular approaches to incorporate amine functionalities in the molecule [28, 29, 42]. The groups of Chang, Ackermann, Sahoo, Jeganmohan, and others systematically explored this strategy over the years to synthesize diverse aminated arenes [28, 42, 55–62]. Different strongly and weakly coordinating directing groups were utilized for selective *ortho*-amidation reaction in combination with organic azides and sulfonyl azides [28, 29]. These transformations proceeded through the formation of cyclometalated Ru(II)-complex, which subsequently formed Ru(IV)-imido species (Scheme 9.12). Reductive elimination of this short-lived intermediate delivered the desired amidated product.

Sahoo group further used *N*-tosyloxyphthalimide as the reaction partner to promote selective *ortho*-imidation of removable sulfoximine directing group protected benzamides [56]. The catalytic process smoothly operated with diverse benzamides and delivered the desired product with high selectivity. The reaction proceeded through a Ru(II/IV)-pathway, where the reaction partner *N*-tosyloxyphthalimide acted as the oxidant to form the Ru(IV)-intermediate. The phthalimide protection was easily removed using hydrazine to obtain the free amine functionality (Scheme 9.13).

Yu group also contributed in this direction using *N*-benzoyloxymorpholine as the aminating agent [57]. This strategy functioned in a similar fashion, where the N—O bond served as the oxidant to generate Ru(IV)-intermediate (Scheme 9.14).

Recently, Chang group harvested dioxazolones as useful reagents for Ru(II)-catalyzed C—H amidation reactions [55]. Similar to prior examples, this strategy

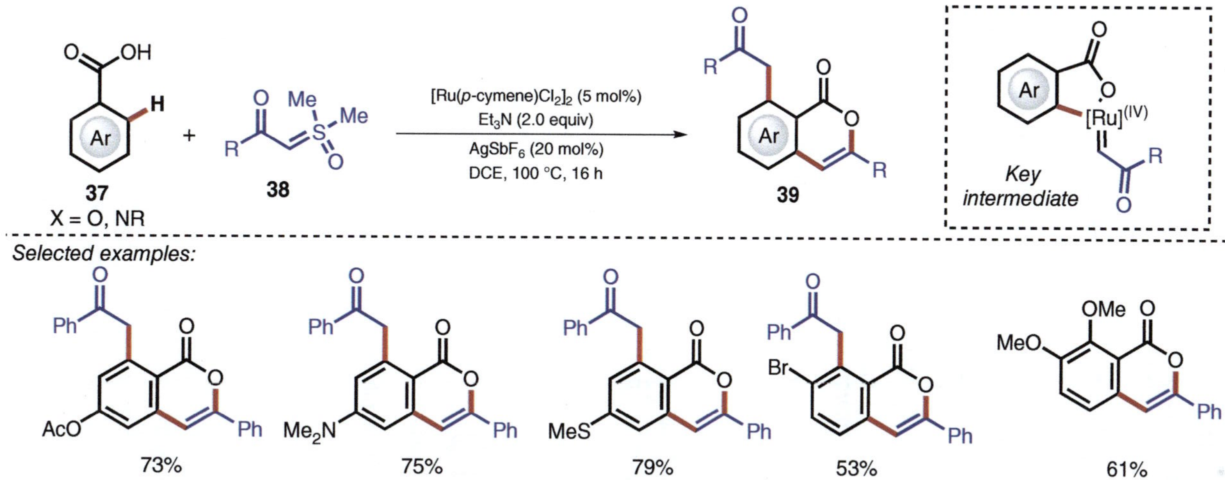

Scheme 9.11 Ruthenium(II)-catalyzed *ortho*-C–H alkylation/annulation with sulfur ylides.

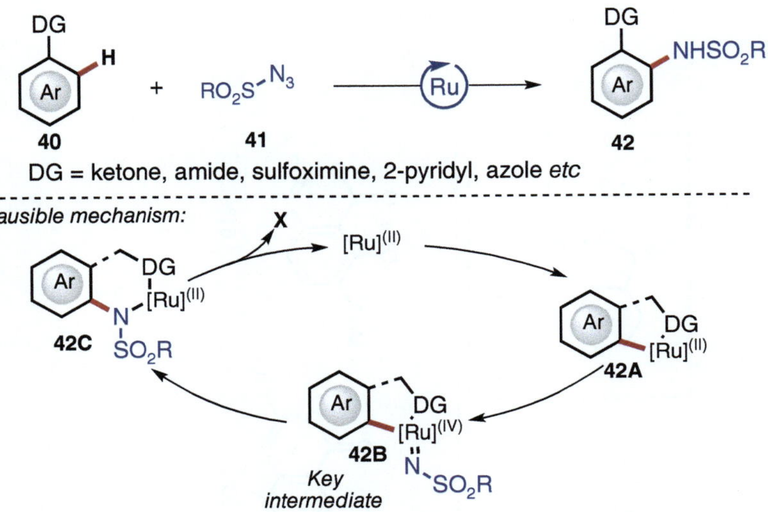

Scheme 9.12 Ruthenium(II)-catalyzed *ortho*-C–H amidation with sulfonyl azides.

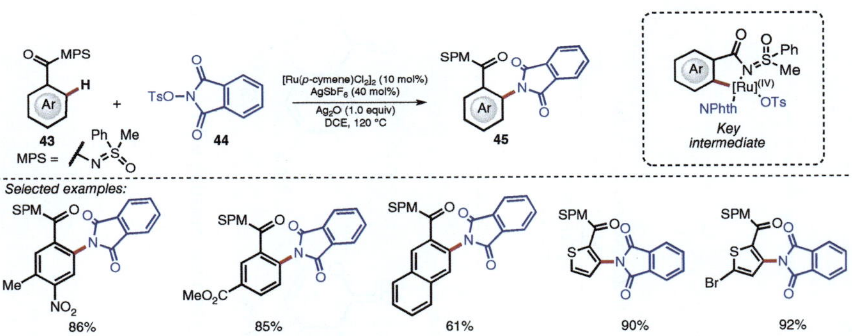

Scheme 9.13 Ruthenium(II)-catalyzed *ortho*-C–H imidation with *N*-tosyloxyphthalimide.

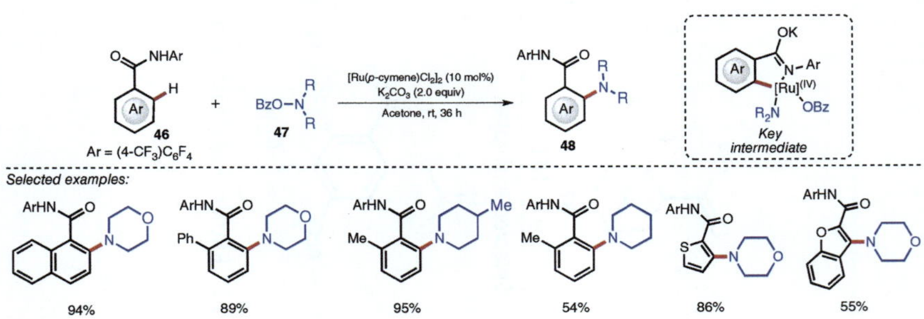

Scheme 9.14 Ruthenium(II)-catalyzed *ortho*-C–H amination with cyclic amine derivatives.

also operated with a Ru(II/IV)-catalytic cycle, where dioxazolone releases CO_2 gas to form an alleged Ru(IV)-imido intermediate. The reductive elimination of the Ru(IV)-imido species bestowed the desired product. They utilized these ideals for an iterative amidation of anilides accessing multi-amidated benzenes under very mild conditions (Scheme 9.15).

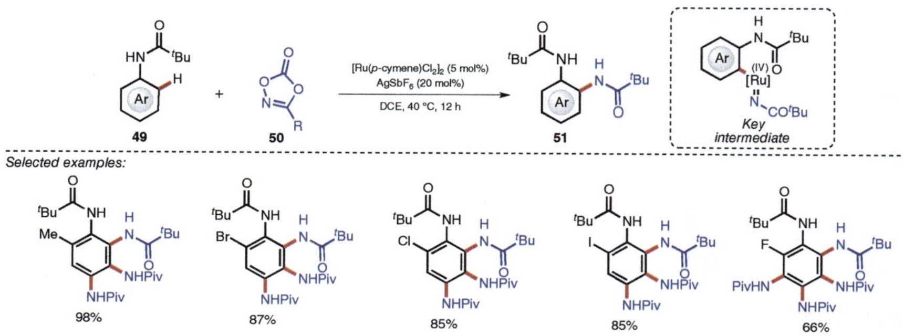

Scheme 9.15 Ruthenium(II)-catalyzed iterative *ortho*-C–H amidation with dioxazolones.

Carboxylate-assisted Ru(II)-catalysis has also been commonly used for direct C(sp^2)—O bond-forming reactions [28]. Various common strongly and weakly coordinating directing groups have been used for these C—H oxygenation reactions, where the catalytic process mostly relied on the Ru(II/IV)-manifold. Among the early examples, Ackermann and Rao reported *ortho*-hydroxylation with Ru(II)-catalysts in combination with strong oxidants such as $PhI(OAc)_2$, $K_2S_2O_8$, or HIO_3 [63–68]. A mixture of TFA and TFAA was used as the common solvent mixture in these transformations (Scheme 9.16a). This unique functionalization strategy was also fruitful for the direct functionalization of anisoles, where the *para*-C—H hydroxylation product was observed (Scheme 9.16b) [68].

Directed C—H hydroxylation reactions were also applicable to aryl acetamides and alkyl phenylacetyl esters through a distal weak *O*-coordination [69]. Similar to the prior reactions, this strategy also operated with a Ru(II/IV)-manifold, where $PhI(TFA)_2$ was used as the oxidant. The C—H activation process was rationalized to follow a CMD-type mechanism and the reaction was successful with different substituted arenes (Scheme 9.17).

The oxygenation process was also applicable to the selective benzoxylation of arenes, where the presence of strong oxidant such persulfate was necessary to form the Ru(IV)-intermediate [70–73]. The Ru(IV)-intermediate realized reductive elimination to deliver the oxygenated products. Different weakly and strongly coordinating groups were compatible to this catalytic process, accessing diverse aryl esters in good-to-excellent yields and high selectivity (Scheme 9.18).

This catalytic regime was also broadly applicable for the direct C—H chalcogenation of arenes with diverse substituted arenes [74–79]. Selenylation reactions mostly used diaryldiselenide and arylselenyl chloride as the electrophile, which reacted with the Ru(II)-cyclometalated complex to form the alleged transient

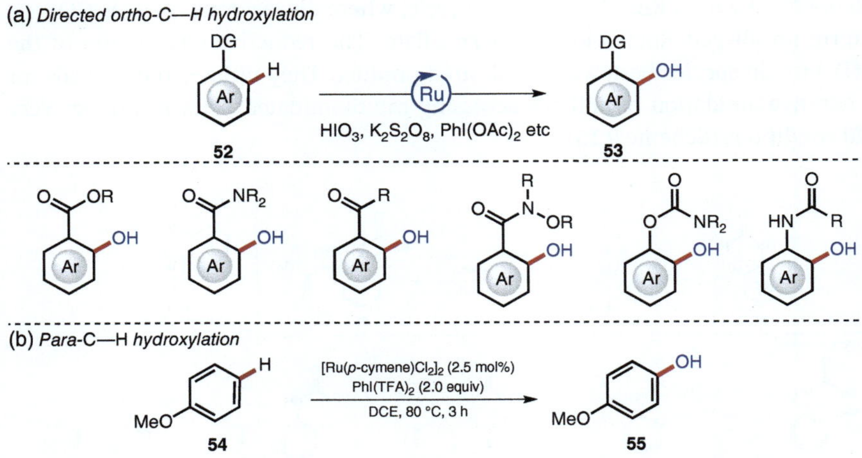

Scheme 9.16 Ruthenium(II)-catalyzed C–H hydroxylation.

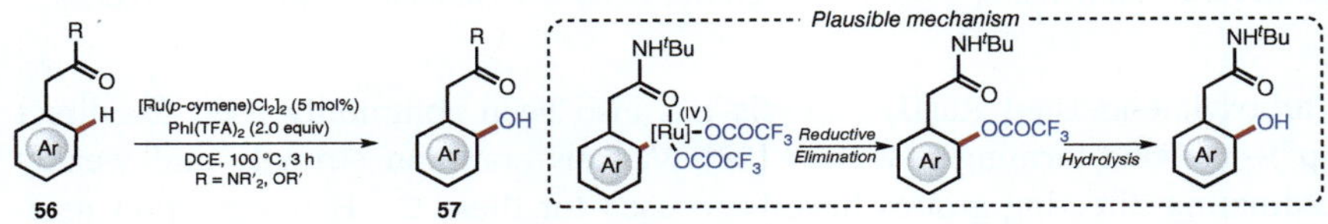

Scheme 9.17 Ruthenium(II)-catalyzed distal *ortho*-C–H hydroxylation.

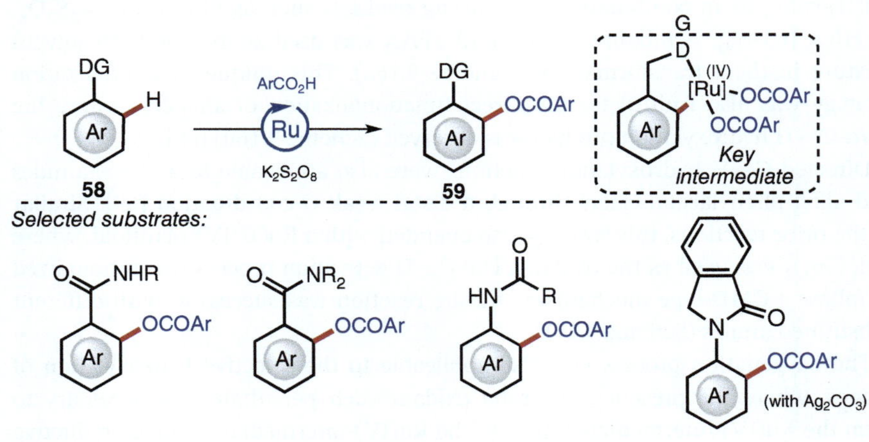

Scheme 9.18 Ruthenium(II)-catalyzed *ortho*-C–H aryloxylation.

Ru(IV)-complex. The reductive elimination of Ru(IV)-complex generated the desired chalcogenated products in good-to-excellent yields. Among notable contributions, Ackermann and co-workers reported the first *ortho*-selective selenylation of anilides under Ru(II)-catalysis [77]. The strategy was unequivocally applicable to various functionalized arenes, and additives $AgBF_4$ and $Cu(OTf)_2$ were

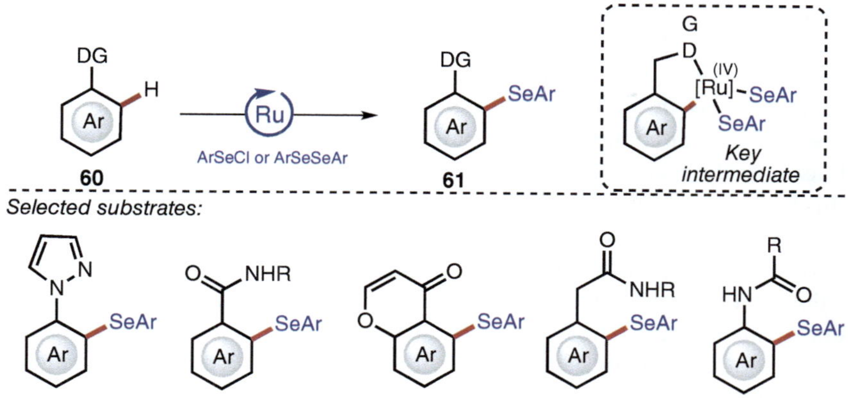

Scheme 9.19 Ruthenium(II)-catalyzed *ortho*-C—H selenylation.

crucial for the success of the transformation (Scheme 9.19). Later, Ma group delineated selenylation of aryl acetic amides and esters harnessing distal weak coordination [75].

Noteworthily, the Baidya group described the first *ortho*-selenylation and sulfenylation of benzoic acids [78]. In this case, the phosphine oxide additive was crucial to improve the reaction outcome. The products were easily manipulated to access diverse Chalcogenoxanthones in high yields (Scheme 9.20). These findings also represented a rare example of carboxylic acid directed *ortho*-C—H activation for C—heteroatom bond construction.

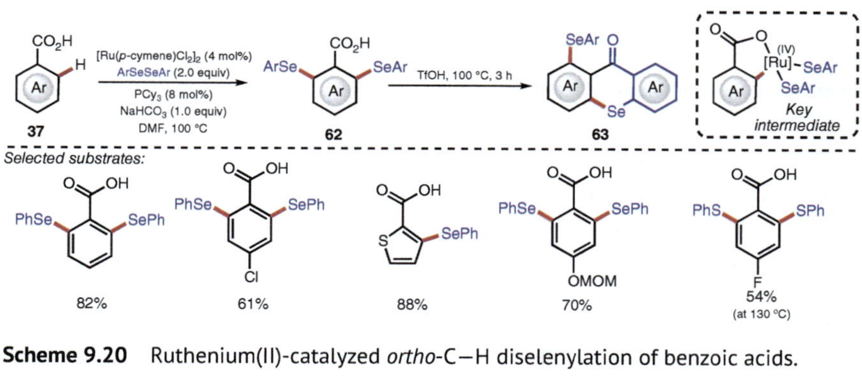

Scheme 9.20 Ruthenium(II)-catalyzed *ortho*-C—H diselenylation of benzoic acids.

9.4 Standard Ru(II/0)-catalyzed Reactions

In the regime of Ru(II)-catalyzed C—H activations, Ru(II/0)-catalytic process is among the most popular and efficient reaction manifolds [14, 26, 27]. This catalytic process has been harnessed for the installation of various common functional groups and annulation reactions involving the C—H bonds. One of the most common processes is the direct synthesis of vinyl arenes through *Heck*-type olefination reactions [27]. Vinyl arenes are important molecular scaffolds found in numerous natural products, pharmaceuticals, and organic materials [27]. In this context, Ru-catalyzed C—H bond activation reactions appeared as a suitable operative strategy to access this class of molecules under mild and operationally simple processes. Different common weakly and strongly coordinating directing groups such as 2-pyridyl, pyrazolyl, imine, ketones, amides, and nitriles have been used for this formal cross-dehydrogenative coupling reactions involving an arene and an olefin [27, 80–86]. All these transformations follow a common catalytic route, where the formation of cyclometalated Ru(II)-complex is the key, which upon migratory insertion with the olefin and β-hydride elimination form the desired olefinated product (Scheme 9.21). The Ru(0)-species formed after the β-hydride elimination is then reoxidized by an oxidant and the catalytic process is continued. Among the notable contributions, Yi group reported the ruthenium-hydride-catalyzed synthesis of vinyl arenes with ketones and amides with a specified Ru complex [86].

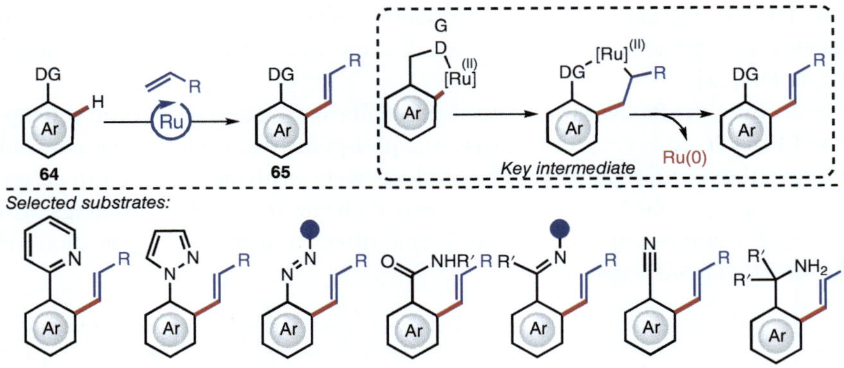

Scheme 9.21 Ruthenium(II)-catalyzed *ortho*-C–H olefination.

In 2011, groups of Satoh and Miura showcased olefination of heteroaromatic acid with acrylates in presence of Ru(II)-catalysis in DMF solvent, where $Cu(OAc)_2 \cdot H_2O$ was used as the terminal oxidant [87]. Further, Ackermann group delineated the olefination of aromatic carboxylic acids, where the olefination product further realized an intramolecular *oxa*-Michael addition to the activated olefin delivering phthalides as the product (Scheme 9.22) [88].

In the following period, Dixneuf and Bruneau reported an azole-directed olefination arenes [89], and Jeganmohan group described a similar process with weakly coordinating ketone directing group (Scheme 9.23) [85]. Consequently, this catalytic manifold was largely adopted by synthetic community to establish reactions with different other directing groups, where mostly the activated olefins were used as the reaction partners to obtain vinyl arenes [27].

In 2019, Baidya group described an interesting example of free amine-directed C—H activation/C—C cleavage cascade utilizing this catalytic manifold [90]. This cross-ring annulation was achieved with biphenyl amines and acrylonitrile. The cross-ring annulation process used a strongly coordinating amine directing

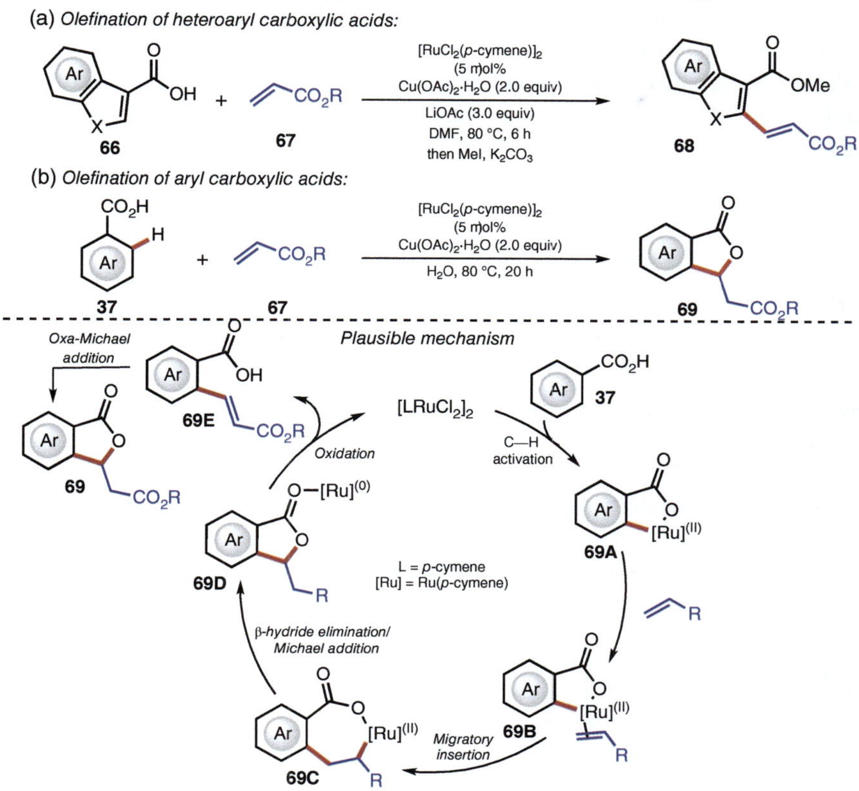

Scheme 9.22 Ruthenium(II)-catalyzed *ortho*-C—H olefination/annulation with benzoic acids.

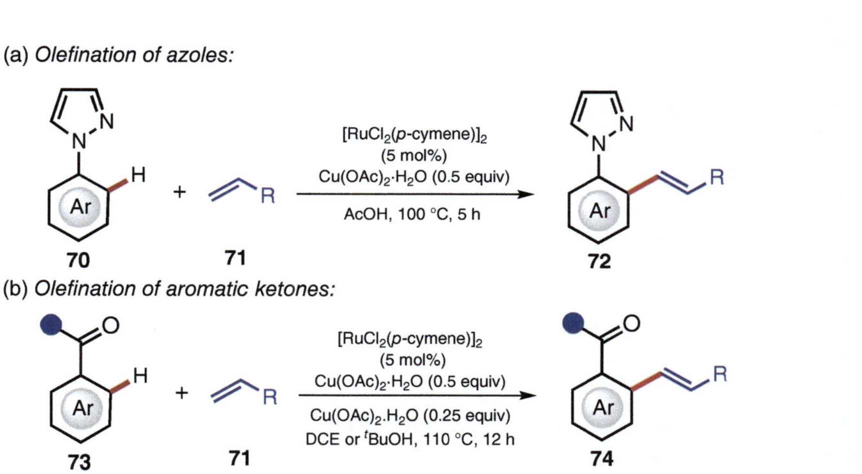

Scheme 9.23 Azole and ketone directed ruthenium(II)-catalyzed *ortho*-C—H olefination.

group in combination with a cationic Ru(II)-complex, which usually generated the cyclometalated ruthenacycle intermediate and followed by the formation of olefination product. The olefination product then underwent an *aza*-Michael addition with the free amine moiety and followed by realized the elimination of acetonitrile to construct the phenanthridines (Scheme 9.24).

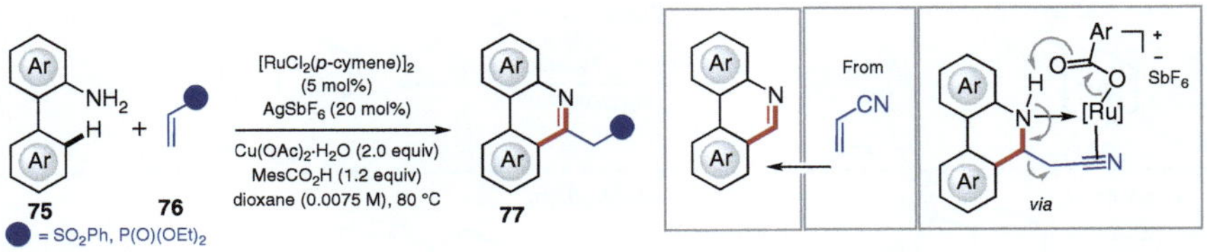

Scheme 9.24 Ruthenium(II)-catalyzed C–H activation/C–C activation cascade with biaryl amines.

Baidya group also demonstrated a site-selective sequential olefination of heteroaromatic carboxylic acids under aqueous conditions, which operated under mild conditions and used Cu(II)-salt as the terminal oxidant (Scheme 9.25) [91].

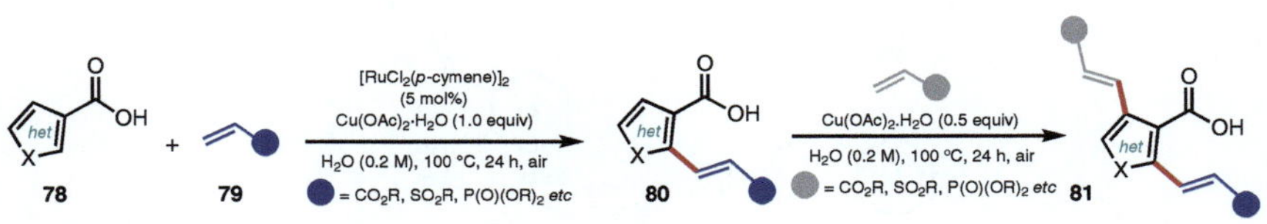

Scheme 9.25 Ruthenium(II)-catalyzed one-pot sequential C–H olefination of heteroarene carboxylic acids.

In 2021, Baidya group further described the use of electronically unbiased allyl sulfones as the olefin precursors for the olefination reaction [92]. This ambiphilic olefins served as useful synthetic handles to avail naphthalenes and lactams with aromatic ketones and amides, respectively (Scheme 9.26).

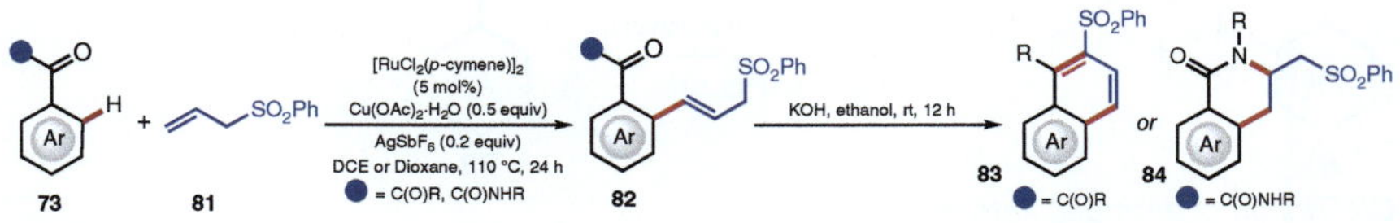

Scheme 9.26 Ruthenium(II)-catalyzed C–H olefination of arenes electronic unbiased olefins.

The combination of alkynes as the reaction partner in this catalytic manifold led to the formation of six-membered heterocyclic architectures [14]. The reactions are mostly compatible with aromatic carboxylic acids and amides, which form a similar ruthenacycle intermediate [93, 94]. Migratory insertion followed by reductive elimination leads to the formation of the annulated products (Scheme 9.27).

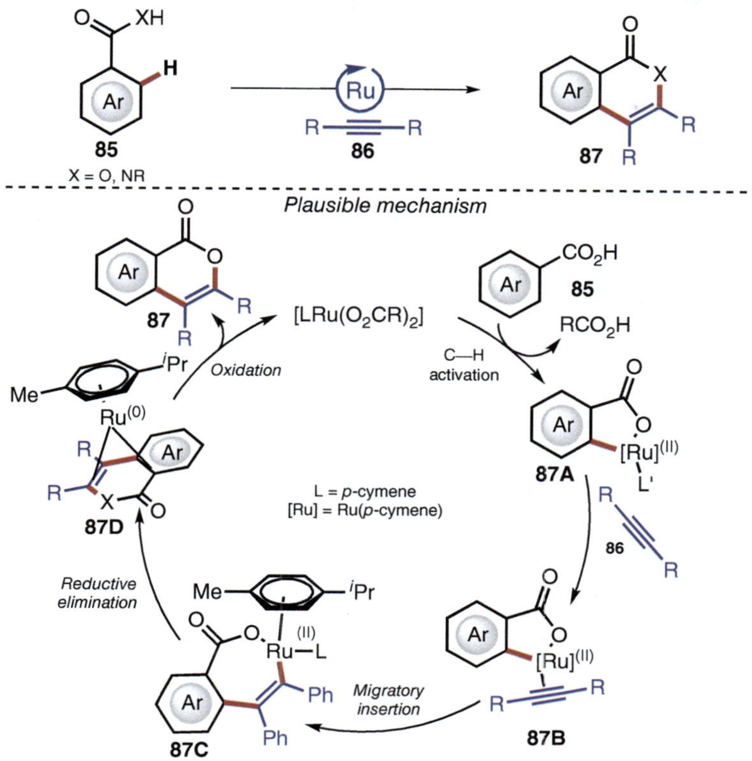

Scheme 9.27 Ruthenium(II)-catalyzed C—H activation/annulation with alkynes.

The Ru(II/0)-catalytic process was also applicable for the direct C—H arylation processes involving aryl boronic acids as the reaction partner [26]. In this case, the aryl boronic acid underwent transmetalation with ruthenacycle intermediate, reductive elimination of which bestowed the desired product and a Ru(0)-species [95, 96]. Oxidation of the Ru(0)-species regenerated the active catalyst to continue the catalytic cycle. This arylation process was successfully applied with various common directing groups (Scheme 9.28).

The Ru(II/0)-catalyzed direct C—H arylation process was also feasible with aryl silanes as the coupling partners [97–99]. In this case, fluoride salt additives were necessary to induce the transmetalation of the silane substrate with the ruthenacycle intermediate, which upon reductive elimination produced the desired olefination products. The arylation with aryl silanes was compatible with benzamides, thiobenzamides, and heteroaryl directing groups (Scheme 9.29).

A rare example of arene-arene cross-dehydrogenative coupling reaction was developed by the group of Baidya with aromatic carboxylic acids harnessing Ru(II/0)-catalytic manifold [100]. This catalytic strategy operated with an influence of non-covalent interaction with the soluble basic additive DBU, which stabilized the in situ generated carboxylate intermediates and led to the challenging

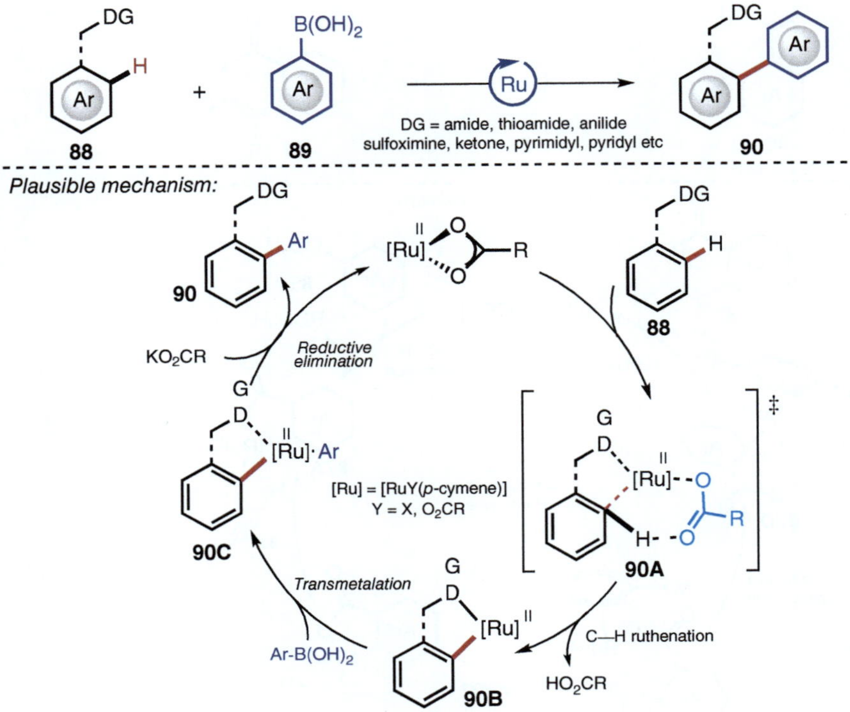

Scheme 9.28 Ruthenium(II)-catalyzed C–H arylation with aryl boronic acids.

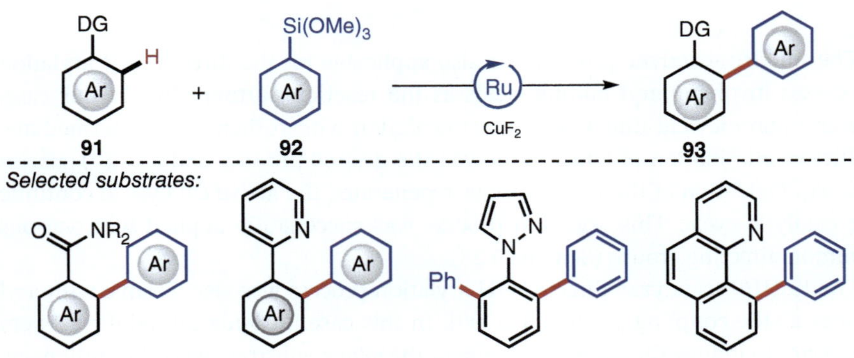

Scheme 9.29 Ruthenium(II)-catalyzed C–H arylation with aryl silanes.

cross-dehydrogenative coupling reaction. This hypothesis was also supported by several mechanistic experiments and DFT analysis. The reaction was successful both for homo- and hetero-dimerization of aromatic carboxylic acids, and the catalyst was regenerated using CuO oxidant (Scheme 9.30).

Ruthenium catalyzed C–H activation/annulation reactions have also been systematically utilized by the Sahoo group for sequential double-annulation

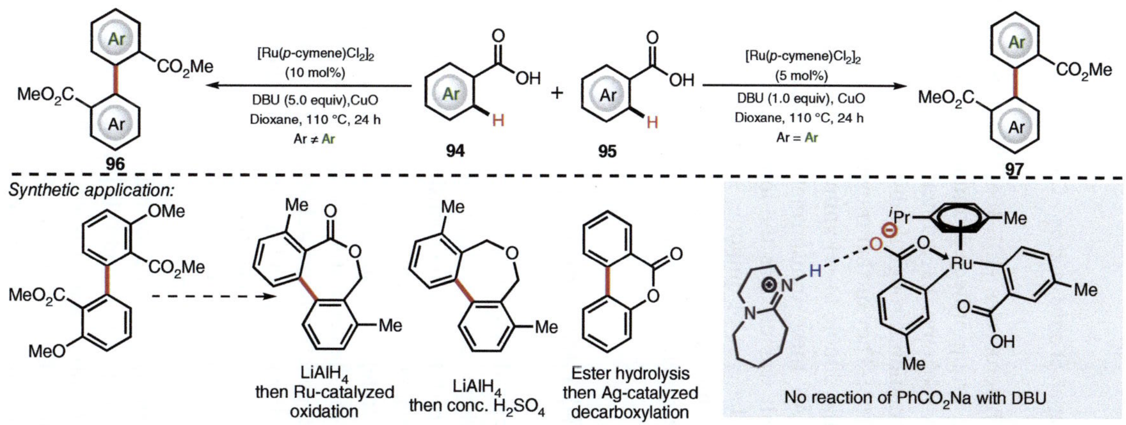

Scheme 9.30 Ruthenium(II)-catalyzed cross-dehydrogenative dimerization of benzoic acids.

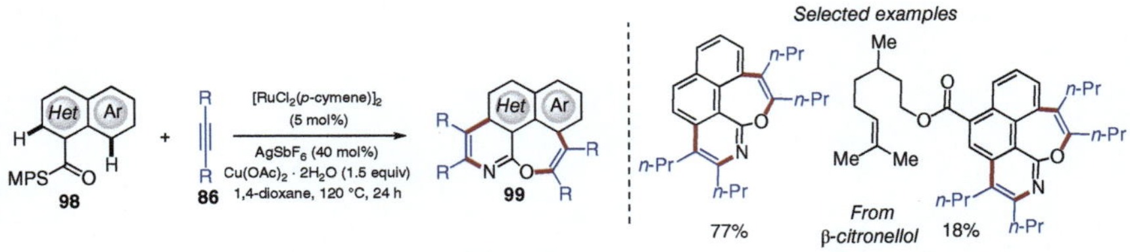

Scheme 9.31 Double annulation of *ortho-* and *peri*-C–H bonds of fused (hetero)arenes.

reaction forming diverse heterocycles [101–106]. They used methyl phenyl sulfoximine (MPS) directing group-protected amides as the precursor to initiate the diversification process (Scheme 9.31). Interestingly, they observed a unique polyaromatic structure formation through double *ortho-* and *peri*-C—H activation annulation with alkynes [102]. They further translated these findings in systematic fashion executing sequential double C—H annulation reactions with alkynes and quinone [101, 108–111]. All these transformations relied on Ru(II/0)-catalytic cycle, where the Cu(II)-salt was used as the terminal oxidant to regenerate the active Ru(II)-catalyst. In the case of quinone, it served as both oxidant and the reaction partner. Overall, this strategy showcased the power of Ru(II)-catalyzed C—H activations in the peripheral decoration of arenes in one pot (Scheme 9.32).

9.5 Aerobic Ru(II/0)-catalyzed Reactions

In recent trends in organic synthesis, devising green and sustainable catalytic transformations to forge high-value complex organic molecules is extremely significant. While the growth of bench-stable and inexpensive ruthenium(II)-catalyzed C—H activations displayed promising outcomes for various alkenylation, allylation, annulation reactions, all these strategies mostly relied on the use of external copper and silver salts as the oxidant for the catalyst regeneration [14, 15, 22–24, 26, 27, 107]. These metallic redox agents generate undesirable metallic waste. In contrast, molecular oxygen can be adopted as a green and sustainable oxidant for C—H activation reactions. Among notable contributions, Jiao and Huang groups independently delineated aerobic alkyne annulation strategies through C—H activations using expensive Pd and Rh catalysts [108, 109]. In 2015, Ackermann and coworkers demonstrated the first Ru(II)-catalyzed oxidative C—H activation/annulation using molecular oxygen as the sole oxidant [110]. The aerobic ruthenium catalysis operated under a mild condition and was compatible with different functionalized aromatic carboxylic acids and alkynes. Specifically, for heteroarenes, the functionalization occurred at the more acidic position, indicating the CMD pathway. Preliminary mechanistic studies revealed the formation of Ru(0)-sandwich complexes after the reductive elimination, which was susceptible to aerobic oxidation with molecular oxygen in acidic medium. The reaction between an alkyne with the ruthenacycle intermediate in the absence of an oxidant bestowed the sandwich Ru(0)-complex with isocoumarins as a neutral ligand. This study also provided

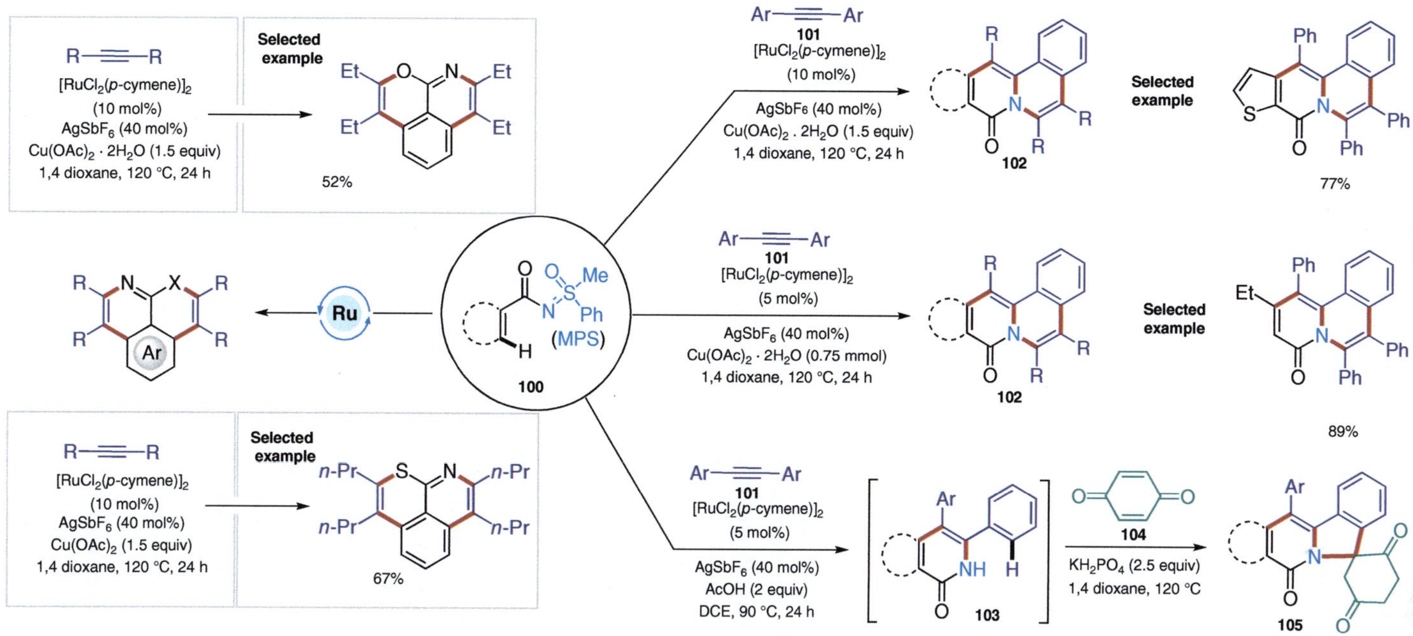

Scheme 9.32 Ruthenium(II)-catalyzed one-pot sequential double C—H annulation with arenes.

evidence for the facile reductive elimination from seven-membered metallacycle intermediate. The key Ru(0)-sandwich complex was verified by X-ray crystallography. The sandwich complex undergoes aerial reoxidation in the presence of the stoichiometric amount of AcOH to generate isocoumarins (Scheme 9.33b).

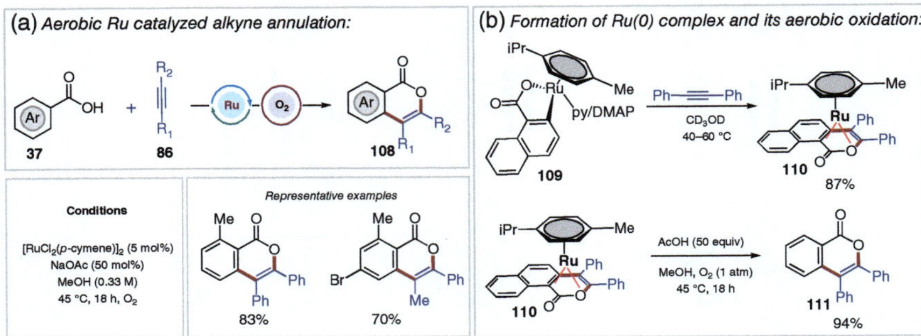

Scheme 9.33 Aerobic ruthenium catalysis for C–H/O–H annulation with alkynes.

Based on the experimental observation, a plausible reaction mechanism was proposed following an active Ruthenium(II)-bisacetate complex, which generated the ruthenacycle intermediate with the arene carboxylic acid after C—H activation. Next, the coordination and migratory insertion of the alkyne to the ruthenacycle intermediate **108A** produced the seven membered Ru(II)-intermediate **108C**. A subsequent reductive elimination of the ruthenacycle intermediate **108C** delivered the Ru(0)-sandwich complex **108D**. Isocoumarin products and active Ru(II)-catalyst were regenerated through an aerobic oxidation of **108D** and water was produced as the by-product from the reduction of molecular oxygen. These pioneering findings set the stage for aerobic C—H activation under Ru(II/0)-manifold (Scheme 9.34).

Despite the notable advancement with molecular oxygen as the terminal oxidant replacing the use of external metallic and non-metallic oxidants, the approach was primarily confined to the use of relatively toxic solvents compromising overall greenness of the system. Overcoming these limitations, in 2018, Ackermann group unveiled biomass-derived γ-valerolactone as a green solvent for the Ru(II)-catalyzed alkenylation of aromatic carboxylic acid with acrylates under aerobic conditions [111]. The olefinated product realized a subsequent *oxa*-Michael addition bestowing phthalides as the final products in the transformation. Similarly, Baidya group leveraged water as a green solvent for a formal cross-dehydrogenative coupling of aromatic carboxylic acid with olefins leading to diverse phthalides under mild conditions [112]. The catalytic protocol accommodated various alkene sources like vinyl phosphonate, vinyl sulfones, acrylates, and acrylonitrile as the reaction partners (Scheme 9.35).

Recently, photo-redox catalysis has also emerged as a sustainable transformative alternative for efficient redox manipulation in transition-metal catalysis [113, 114]. Generally, transition-metal-based polypyridyl complex and organic dyes act as photocatalysts to regulate the redox event in these transformations [115]. In this

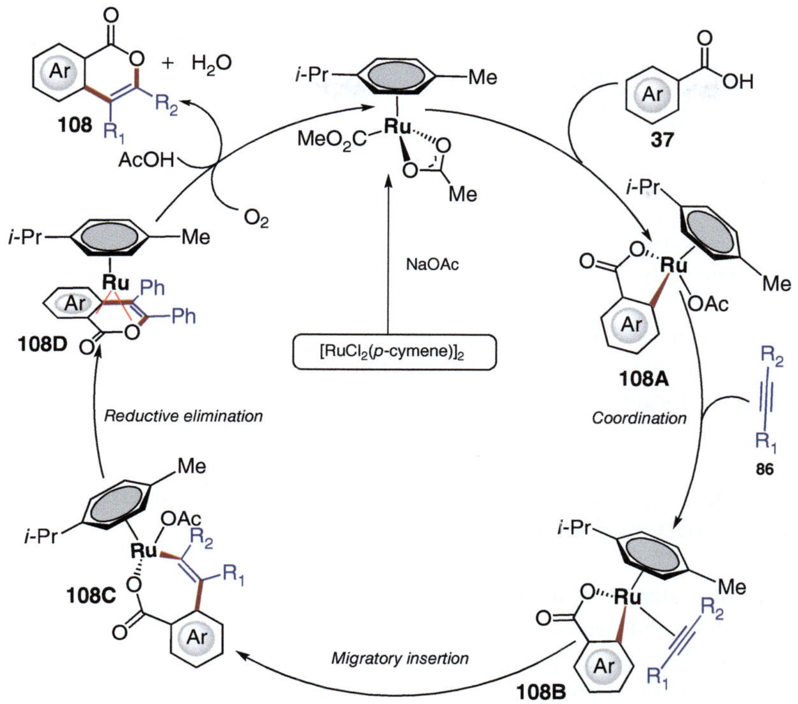

Scheme 9.34 Plausible mechanism of aerobic ruthenium(II)-catalyzed annulation.

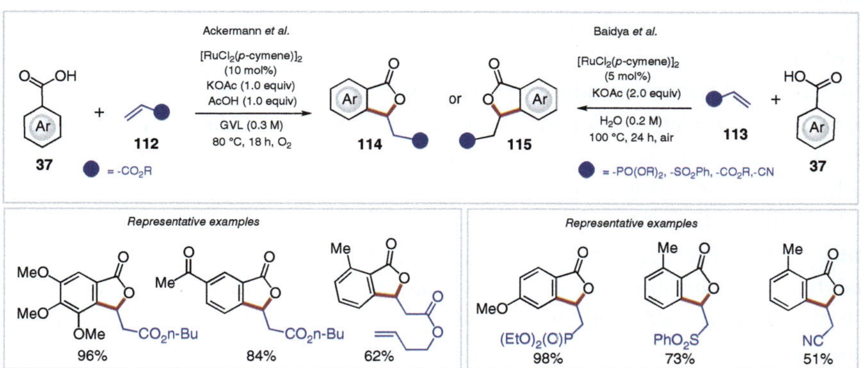

Scheme 9.35 Aerobic Ru(II)-catalyzed annulation of aromatic carboxylic acids with olefins.

direction, Baidya group reported Ru(II)-catalyzed olefination/annulation reaction with olefins using dye-sensitized metal-oxide semiconductor TiO_2 as the photocatalyst [116]. The dye on the surface of TiO_2 absorbs the visible light and goes to an excited electronic state. The excited dye molecule could donate one electron to the conduction band of semiconductor, which in turn generated superoxide anion radical from molecular oxygen. This superoxide anion radical acted efficiently as a

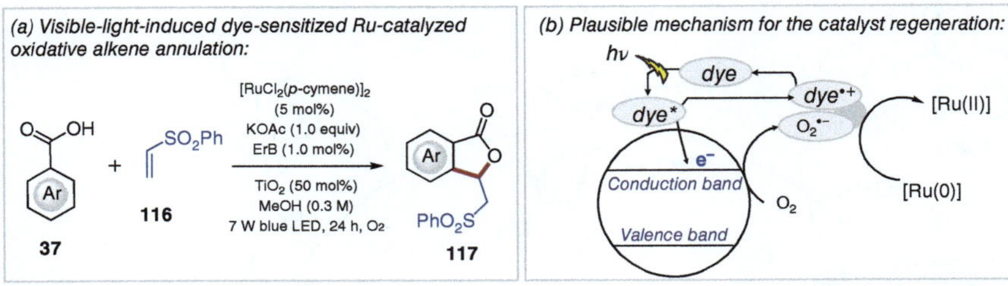

Scheme 9.36 Heterogeneous photocatalyst-mediated C–H activation with Ru(II)-catalysis.

terminal oxidant for the oxidation of Ru(0)-species to Ru(II)-species. The operationally simple strategy delivered different functionalized phthalides with the recovery of the photocatalyst (Scheme 9.36).

9.6 Ru(II)-catalyzed C–H Activations with the Directing Group as the Internal Oxidant

Transition-metal-catalyzed annulations of alkyne and alkene with arenes via C—H activations are atom- and step-economic tools for the straightforward construction of heterocycles [14, 15, 22–24, 107]. However, it is apparent from the preceding section that these transformations usually require external oxidants for the catalyst regeneration, which lead to the generation of undesirable waste. In this context, some noteworthy progress had been attained with the use of hydroxamic acid ester as a substrate, where the N—O bond of the *N*-methoxy benzamides can act as a terminal oxidant for the regeneration of the active catalyst, and thereby prevent the use of any external oxidant [107].

In 2011, Ackermann group employed hydroxamic acid derivatives and their esters to perform annulation with alkynes under Ru(II)-catalysis [117]. Abundant and non-toxic water was used as the optimized solvent, and sterically hindered KO_2CMes was used as the carboxylate-additive. The *in situ* generation of Ruthenium(II)-carboxylate species allowed the C—H activation/annulation strategy, where the cleavage of N—O bond regenerated the active catalyst after the catalytic cycle. The substrate scope was broad and general showcasing annulation of both aliphatic and aromatic alkyne toward the formation of isoquinolones (Scheme 9.37a).

The Ackermann group further unveiled a C—H/N—O bond activation via Ru(II)-catalysis employing alkynyl halides and hydroxamic acid ester in methanol for the synthesis of alkoxy isoquinolones with distal selectivity (Scheme 9.37b) [118]. Initially, ruthenium(II)-carboxylate catalytic species will be generated which effectuate the C—H cleavage to afford the metallacycle intermediate **122A** and subsequent selective coordination and insertion of the alkyne engendered the seven-membered metallacycle **122C**. Intermediate **122C** upon reductive elimination and a subsequent N—O bond cleavage forged the desired product (Scheme 9.38).

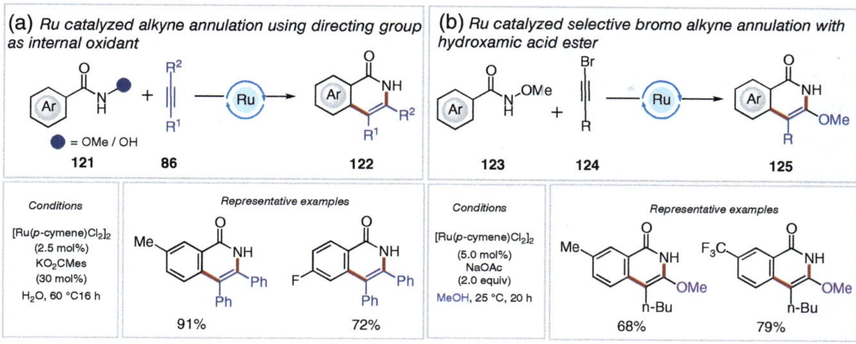

Scheme 9.37 Ru(II)-catalyzed annulation of alkynes with hydroxamic acid esters.

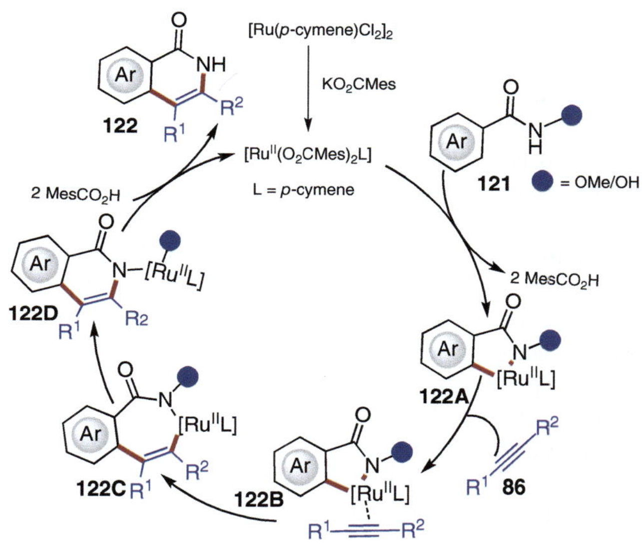

Scheme 9.38 Proposed reaction mechanism for the alkyne annulation.

In 2012, Jeganmohan group described a ruthenium(II)-catalyzed cyclization of ketoximes with terminal and internal alkynes affording diverse isoquinolones [119, 120]. The five-membered ruthenacycle intermediate was isolated, which reacted with the alkyne in the absence of any base constructing the isoquinolone product. In this case, NaOAc was used as a catalytic acetate source to initiate the *ortho*-C—H metalation, and the cleavage of N—OMe or N—OH bond helps in the regeneration of the active ruthenium species for the next cycle (Scheme 9.39).

In 2016, Chegondi and coworkers reported an annulation of aryl-substituted 2-acetylinic ketones with hydroxamic acid esters followed by a tandem cyclization under Ru(II)-catalysis with reverse regioselectivity [121]. The weak coordination of the carbonyl moiety present in the alkyne with the ruthenium was responsible for the reversal of the selectivity during the alkyne insertion. A consequent cascade reaction involving the carbonyl functionality gave the polycyclic products having indolizine core with high diversity with excellent functional group tolerance (Scheme 9.40a). While internal and terminal alkynes were used as a 2-carbon synthon for [4+2]-annulation under Ru catalysis to yield isoquinoline and isoquinolones, Liu group reported a versatile [4+1]-annulation strategy involving propargyl alcohols accessing isoindolinone derivatives (Scheme 9.40b) [122].

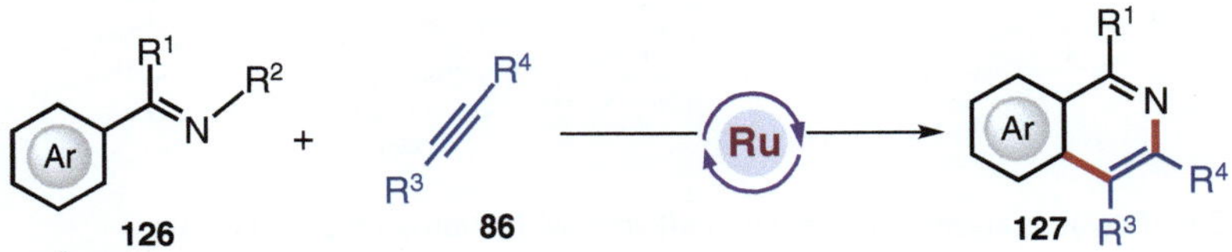

Scheme 9.39 Ru-catalyzed cyclization of ketoximes.

Mechanistically it was presumed that the N—O bond cleavage happens after the annulation to reoxidize the Ru(0)-species. However, Eycken and coworkers systematically defined an exclusive example of C—H/N—H annulation rection involving oxidizing directing groups, proving the catalytic sequence of the transformation [123]. Studies revealed that the N—O bond cleavage occurs before alkyne insertion as they showcased an intramolecular annulation employing alkyne tethered *N*-alkoxy benzamides via Ru(II)-catalysis. Detailed mechanistic investigation and DFT studies supported inverse annulation following a Ru(II)-Ru(IV)-Ru(II) pathway featuring N—O bond cleavage prior to the alkyne insertion (Scheme 9.40c).

Baidya group contributed significantly in this direction developing several [4+2]-annulation reactions between hydroxamic acid ester with olefins. First, enamides were consumed as the reaction partner under Ru(II)-catalysis to forge dihydroisoquinolinone-derived aminals scaffolds (Scheme 9.41a). The reaction showed a wide substrate scope tolerating various common functional groups [124]. The methodology was also successful with vinyl acetate to forge 3,4-Dihydroisoquinolones, where vinyl acetate acted as a formal acetylene precursor (Scheme 9.41a). Next, electronically unbiased olefins such as allyl amide and vinyl acetic acid amides were employed as the reaction partners with hydroxamic acid ester under the catalytic conditions to access [4+2]-annulated product in high yields [125, 126]. The reaction conditions were mild, operationally simple, and delivered a broad substrate scope including bioactive molecules with various functional groups (Scheme 9.41b). In all these transformations, fluorinated alcohol HFIP played a critical role in the product formation, as the transformation was futile in most of common organic solvents. The transformations were believed to be functioning through a Ru(II/IV)-cycle, similar to the mechanistic protocol as mentioned in Scheme 9.40d.

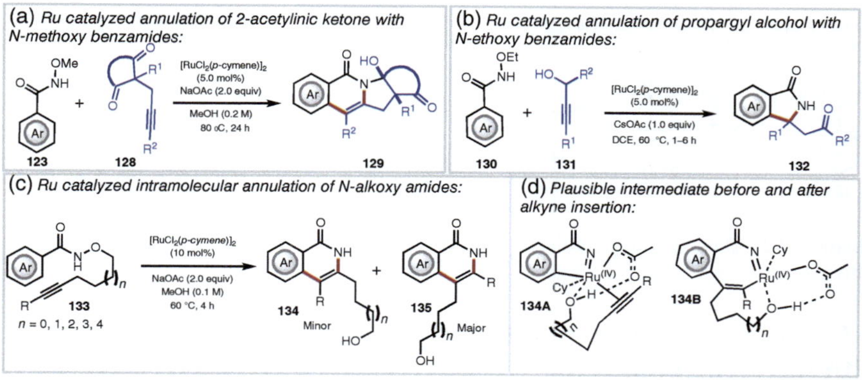

Scheme 9.40 Diverse Ru(II)-catalyzed C–H/N–H annulation with alkynes mediated by N–O bond cleavage.

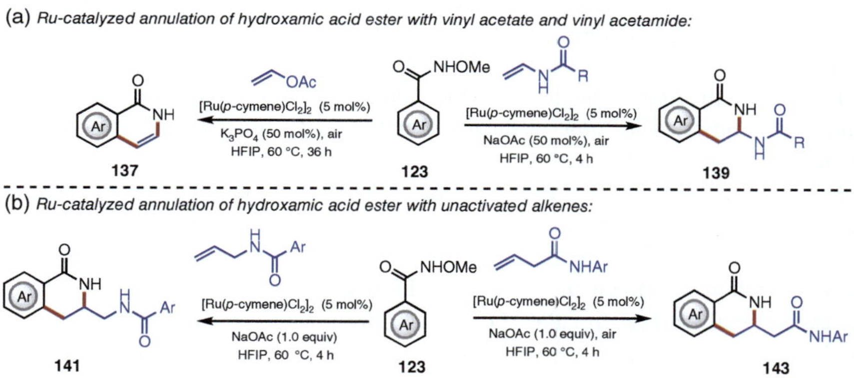

Scheme 9.41 Diverse Ru(II)-catalyzed C–H/N–H annulation with olefins mediated by N–O bond cleavage.

Allenes are considered as versatile building blocks in organic synthesis owing to the presence of two orthogonal bonds. Ackermann and coworkers developed a groundbreaking allenylation of hydroxamic acid esters with disubstituted silyl allene as a coupling partner under Ru(II)-catalysis [127]. The key to this reaction can be attributed to the use of ruthenium(II)-carboxylate species as an active catalyst. This methodology showcased a wide range of examples (Scheme 9.42a). Similarly, Baidya group employed hydroxamic acid ester with allenylphosphine oxides via Ru(II)-catalysis to synthesize substituted isoquinolinone frameworks with good-to-excellent yields [128]. The methodology was fruitful in delivering late-stage functionalization of various pharmacophores-derived molecules like tocopherol, probenecid, amino acids (Scheme 9.42b).

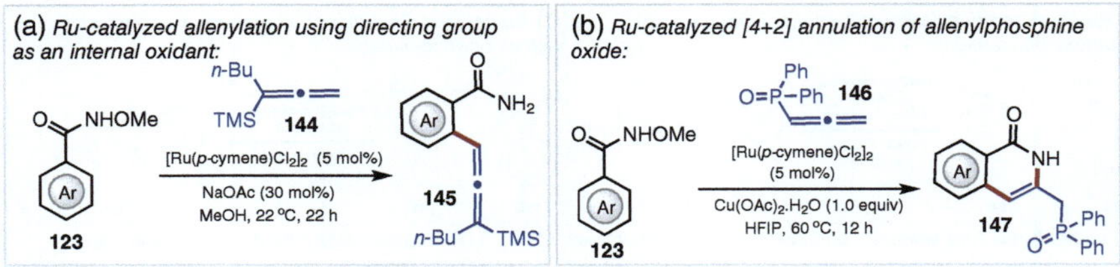

Scheme 9.42 Diverse Ru(II)-catalyzed C—H activation with allenes mediated by N—O bond cleavage.

9.7 Ru(II)-catalyzed *meta*- and *para*-C—H Activations with Ru(II/III)-manifold

In recent decades, substantial advancement has been achieved in ruthenium(II)-catalyzed *meta*- and *para*-selective C—H activation reactions [129, 130]. While most of the distal C—H activations rely on expensive palladium [131–133] or iridium [134–136] catalysts, complex templates [137–140], and ligand designs [141], significantly limiting their applicability and compromising the sustainability of C—H functionalization strategies, ruthenium(II)-catalysis reigns supreme in this regime avoiding most of these obstacles. Notably, *meta* and *para*-selective C—H functionalizations can be achieved under mild ruthenium catalysis via remote C—H functionalizations by *ortho*-C—H activation followed by cyclometallation. The cyclometalated intermediate generated from *ortho*-C—H activation induces a significant electronic bias at the arene ring, which governs the subsequent functionalization selectively at the *para*-position with respect to the Ru—C bond. This exceptional feature provides the opportunity for an exquisite *meta*-selective functionalization without the need of an external auxiliary to direct the selectivity. Furthermore, remote *para*-functionalization can be accomplished by tuning the electronic effect of cyclometalated complex with C—N bonds (Scheme 9.43a) [129, 130]. This concept originated from the pioneering work of Roper in 1994, where ruthenium complexes of benzene and toluene realized nitration selectively at the *para*-position relative to the ruthenium center, albeit with low yields and multiple by-products (Scheme 9.43b) [142].

In 1998, Coudret and coworkers confirmed that cyclometalated ruthenium-phenyl pyridine species could experience C—H halogenation at the *para*-position to the metal-carbon σ-bond. In this transformation, bromination was carried out using NBS, while a mixture of molecular iodine and $PhI(OAc)_2$ was employed as the iodinating agent (Scheme 9.43c) [143]. However, all these reactions were carried out using stoichiometric ruthenium complexes, and it was nearly a decade later when this concept was revisited and conceptualized in a more advanced catalytic context. In 2011, the Ackermann group, observed the formation of a *meta*-alkylated product as a by-product, while investigating *ortho*-selective alkylation with primary alkyl bromide (Scheme 9.43d) [144].

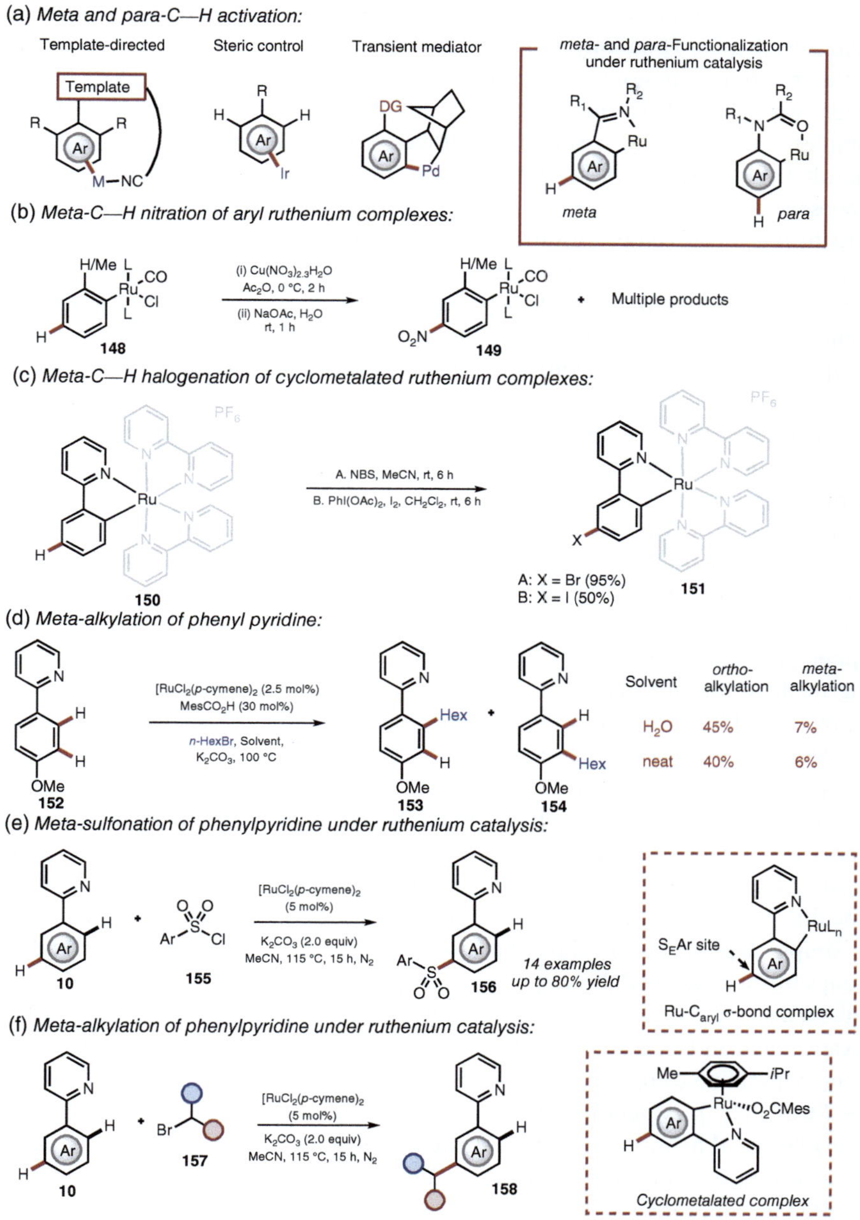

Scheme 9.43 Distal *meta*-C−H functionalization under ruthenium catalysis.

In the same year, Frost reported the first catalytic method for achieving *meta*-selective C—H sulfonylation of 2-aryl pyridines under ruthenium catalysis. In this reaction, *ortho*-selective product mixture was not observed and up to 80% yields was achieved. In the mechanistic studies, the synthesized cyclometalated complex was reacted with sulfonyl chloride, yielding the *meta*-selective product quantitatively. Based on these studies and prior stoichiometric reactions, the author proposed that the cyclometalated intermediate undergoes S_EAr C—H sulfonation *para* to the metal center, resulting in the net *meta*-sulfonated 2-phenylpyridine (Scheme 9.43e) [145]. In the same period, the Ackermann group independently developed a method for *meta*-alkylation using secondary alkyl halides as alkylating agents, which exclusively produced the *meta*-alkylation product (Scheme 9.43f) [146]. The sterically demanding 1,4,6-trimethyl(mesityl)benzoic acid additive played a crucial role in this transformation. Notably, the reaction was inhibited in the presence of the radical scavenger TEMPO, and the enantiopure alkyl halide was found to undergo racemization, indicating a homolytic cleavage of the carbon-halide bond.

Subsequently, efforts were made to understand the plausible mechanism of these transformations with a major contribution from the Ackermann and coworkers, sparking a surge of new methodologies in the field. In the following years, Ackermann [147] and Frost [148] reported the *meta*-alkylation reaction of 2-phenylpyridines using tertiary alkyl halides. The alleged involvement of a radical intermediate was apparent, which was intercepted with the cyclometalated arene via radical addition. The possibility of previously suggested S_EAr mechanism was ruled out (Scheme 9.43a). Consequently, a series of *meta*-functionalization reactions of 2-phenylpyridines were reported by Frost [149], Greaney [150], Huang [151], Zhang [152] group. The reaction was not only confined to the functionalization of 2-phenylpyridines as the sole substrate class. The Ackermann group significantly expanded its applicability for the *meta*-functionalization of anilines using a pyrimidine directing group, where one could achieve densely *meta*-functionalized aniline by simple deprotection of directing group [147]. Eventually, Ackermann and Li groups further extended the scope for the *meta*-functionalization of ketimine [153], diazobenzene [154, 155], and *o*-pyridyl phenol [156] (Scheme 9.44a). As a further advancement in this area, in 2020, Ackermann delineated the *meta*-alkylation of 2-phenylpyridine using the recyclable heterogeneous Ru(II)-hybrid catalyst. While the $RuCl_2PPh_3$(*p*-cymene) catalyst was found ineffective after the 1st run, this newly designed hybrid catalyst was efficient even after the 5th run, demonstrating the robustness of this heterogeneous catalyst (Scheme 9.44b) [157]. They further reported an innovative *meta*-selective alkylation reaction employing readily available amino-acid-derived Katritzky pyridinium salts as alkyl radical precursors. The process involves the generation of alkyl radicals through a deaminative cleavage of structurally diverse pyridinium salts (Scheme 9.44c) [158]. In this context, considering the significant biological relevance of C-aryl glycosides, they defined a groundbreaking *meta*-selective glycosylation reaction following this methodology. This method utilizes glycosyl bromides as versatile glycosylating agents, serving as glycoside synthons (Scheme 9.44d) [159].

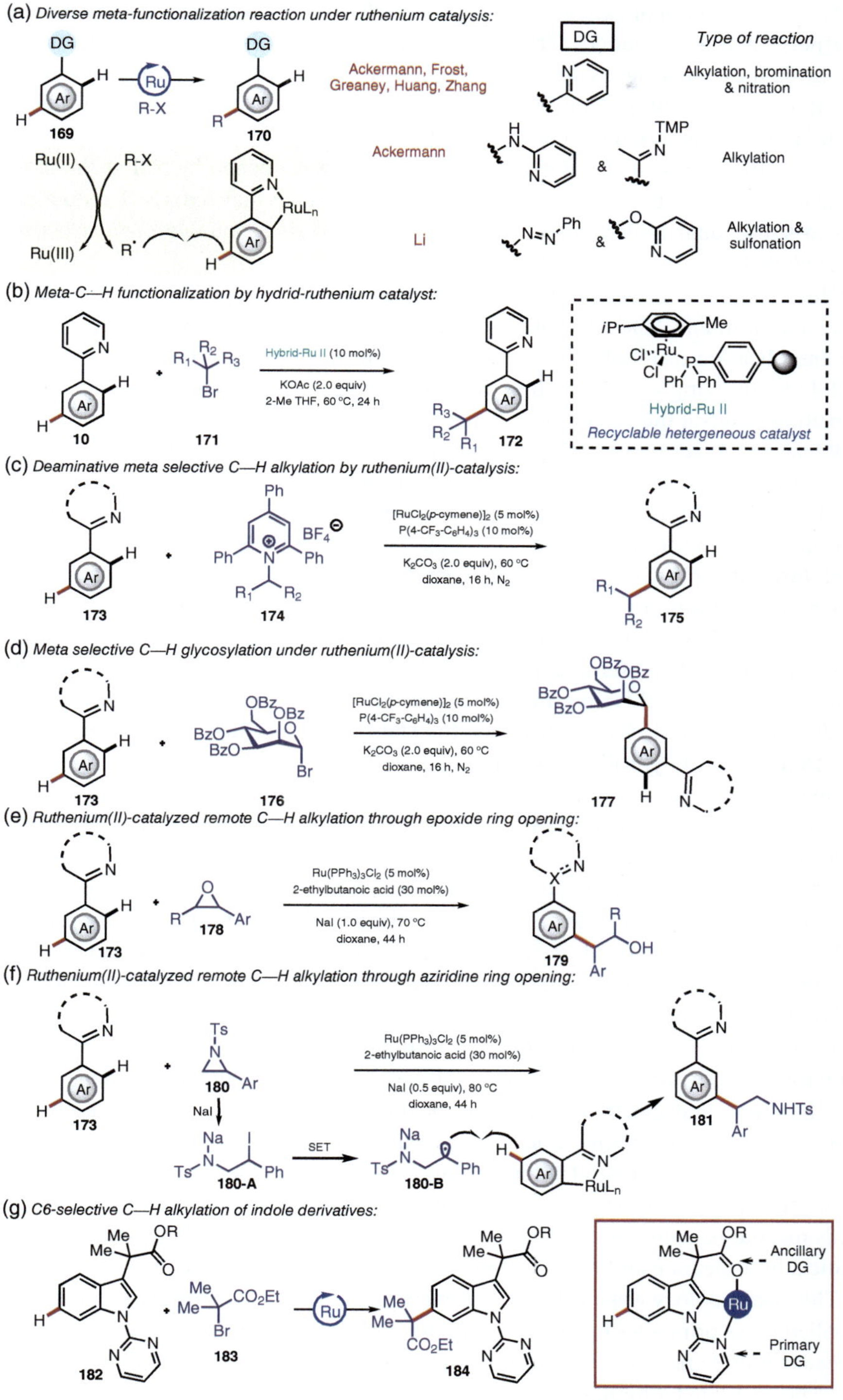

Scheme 9.44 Distal functionalization of arenes via Ruthenium catalysis.

The *meta*-functionalization was also achievable with other electrophilic coupling partners than alkyl halides. In this direction, Larrosa reported *meta*-alkylation of aromatics bearing *N*-directing groups using epoxide as an alkylating agent (Scheme 9.44e) [160]. Very recently, Wang and Ackermann testified a ruthenium(II)-catalyzed remote C—H alkylation through an aziridine ring opening reaction. In this transformation, the aziridine ring was opened by NaI at the benzylic position, forming 2-iodo-2-phenylethylamide, which realized a SET reduction and a subsequent radical quenching reaction, delivering valuable *meta*-selective phenylethylamine derivatives (Scheme 9.44f) [161].

This distal functionalization reaction strategy was also adopted for the functionalization of indole. The Frost group demonstrated the efficiency of a site-selective C6-modification of indole. They reported C6 alkylation on the benzenoid ring of indole by utilizing a primary directing group at the N1 position and an ancillary directing group at the C3 position (Scheme 9.44g) [162].

Recently, Ackermann group, through extensive experimental evidence and DFT studies, elucidated the possible reaction mechanism in detail (Scheme 9.45) [163]. In this proposed mechanism, initially, two molecules of the substrate coordinate with the Ru(II)-catalyst, accompanied by the dissociation of a ligand, and form the pre-C—H activation complex **185A**. Then a base-assisted electrophilic substitution (BIES) type C—H activation resulted in the cyclometalated intermediate **185B**. Although **185B** can realize a second C—H activation to form a *bis*-cyclometalated intermediate **185C**, DFT calculations revealed the catalytically inactive nature of this intermediate. Instead, the cyclometalated intermediate undergoes a ligand-to-ligand hydrogen transfer (LLHT) reaction to form intermediate **185D**. The catalytically active intermediate **185D** could initiate an inner-sphere single-electron transfer (ISET) with an alkyl bromide, generating an alkyl radical and a Ru(III)-intermediate, **185E**. The alkyl radical formed during this process is intercepted via either a radical coupling or a radical rebound pathway, depending on its steric and electronic properties. For tertiary alkyl radicals, the reaction preferentially follows the radical coupling pathway. This occurs at the *para*-position relative to the C—Ru bond, resulting in the formation of a carbene-like intermediate **185F**. The carbene-type intermediate subsequently experiences aromatization, followed by protodemetalation, and finally, the *meta*-functionalized product was delivered with high selectivity after decoordination. In contrast, secondary radicals can follow both a radical rebound and a radical coupling pathway, depending on the steric and electronic demands of the substrate and reaction conditions. For primary alkyl, methyl, and aryl radicals, the reaction exclusively proceeds through the radical rebound pathway. This is followed by reductive elimination and product deco-ordination, ultimately delivering the *ortho*-functionalized product. This mechanistic divergence emphasizes the critical influence of the nature of the radical intermediate and reaction conditions in navigating the selectivity.

This methodology was further advanced, when Liang and coworkers successfully applied these principles in a three-component reaction manifold, demonstrating its broader utility and versatility for the straightforward access to molecular complexity. They reported a highly challenging three-component *meta*-selective alkylation

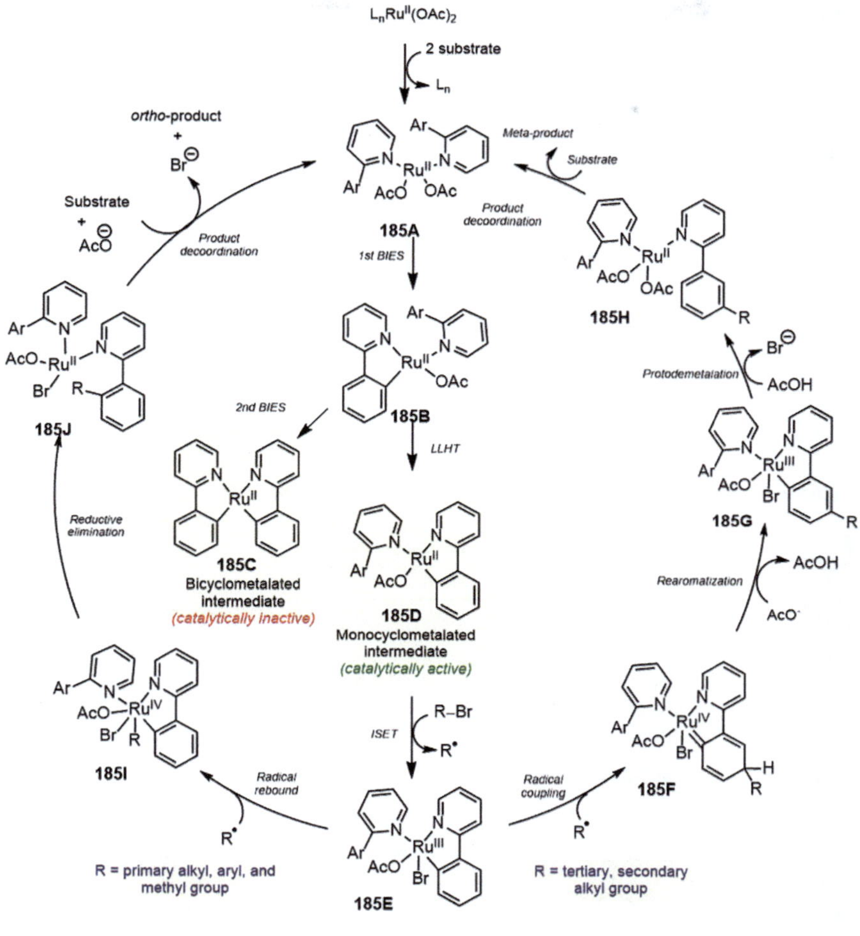

Scheme 9.45 Detailed reaction mechanism.

reaction involving fluoroalkyl halides, diverse alkenes, and phenylpyridines. According to the proposed reaction mechanism, the cycloruthenated intermediate initially undergoes single-electron transfer (SET) with the fluoroalkyl halide, generating a fluoroalkyl radical. This radical subsequently was intercepted by the olefin, forming a new carbon-centered radical. The newly formed radical then engages in C_{Ar}—H bond addition at the *para*-position relative to the C—Ru bond. Finally, aromatization and demetalation completed the catalytic cycle, yielding the desired product (Scheme 9.46a) [164].

In 2021, they extended this concept for an intramolecular variant of the preceding transformation, where a substrate bearing both alkyl halide and alkene functional groups first undergoes radical formation via C–halogen homolytic cleavage. An intramolecular radical addition the tethered olefin moiety generated a second radical species, which ultimately produced arylpyrrolidone derivatives (Scheme 9.46a) [165].

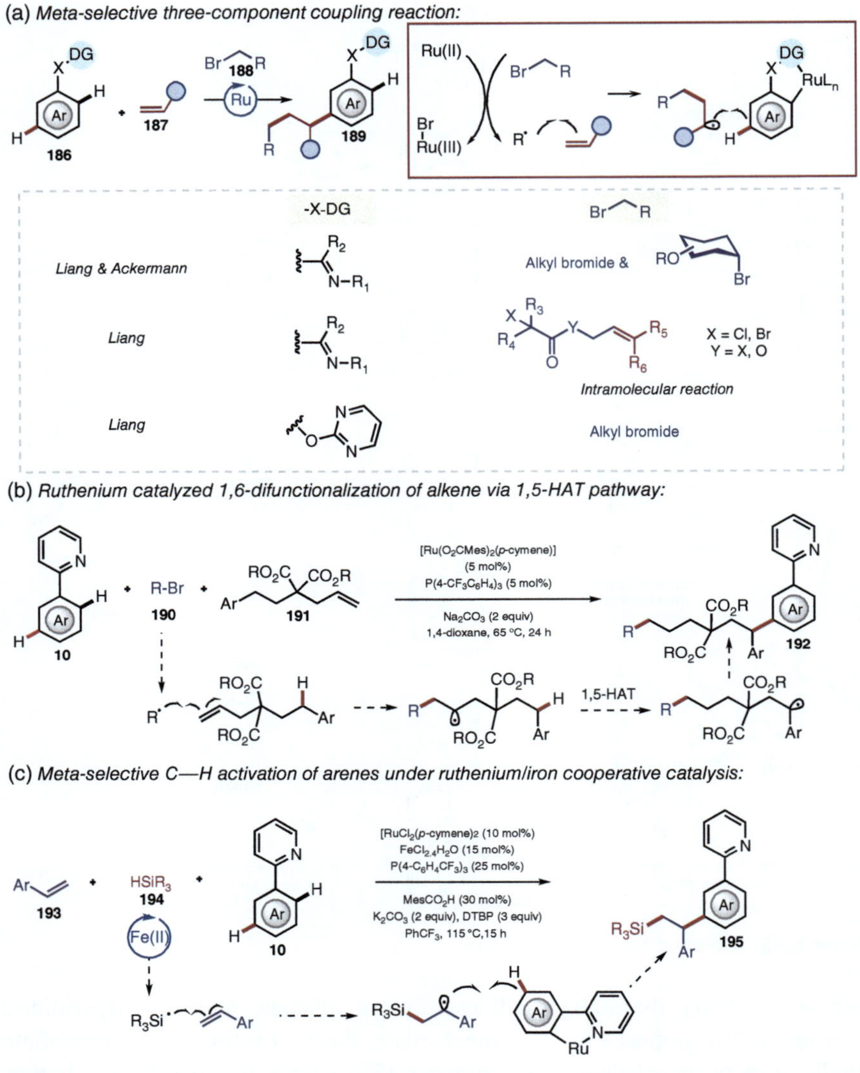

Scheme 9.46 Ruthenium-catalyzed meta functionalization of arenes.

The Ackermann group advanced this methodology by incorporating bioactive glycosyl bromides as the reaction partners, showcasing its applicability in complex molecular frameworks (Scheme 9.46a) [166]. Building on this progress, the Liang group also described a three-component *meta*-functionalization reaction for the functionalization of phenols, utilizing a pyrimidine directing group (Scheme 9.46a) [167]. This methodology was also successfully applied to the 1,6-difunctionalization of unactivated alkenes. In this reaction, the initial alkyl radical adds to the non-activated alkene, triggering a 1,5-hydrogen transfer (HAT) to generate a benzylic radical. This intermediate then undergoes radical coupling, leading to the formation of the *meta*-functionalized product. This extension underscores the remarkable versatility of the approach, enabling selective functionalization of otherwise inert alkenes and offering a robust strategy for molecular diversification (Scheme 9.46b) [168]. Very recently, the Hajra group reported a *meta*-selective silylarylation reaction under cooperative ruthenium and iron catalysis. In this transformation, a silyl radical was generated through Fe(II)-catalysis and then followed an addition to the olefin, forming a new carbon-centered radical. This newly formed radical intermediate subsequently led to the formation of the desired product (Scheme 9.46c) [169].

Recently, the power of this methodology was extended far beyond the realm of strong *N*-coordinating directing groups. In a captivating leap forward, Huang and Gooßen unveiled the striking efficacy of weakly coordinating carboxylic acid directing group in selective *meta*-alkylation reactions of carboxylic acid. The key to the success of this transformation relied on the utilization of a bidentate bipyridine ligand, which ensured the success of the reaction providing reasonable stability to the alleged reaction intermediates (Scheme 9.47) [170].

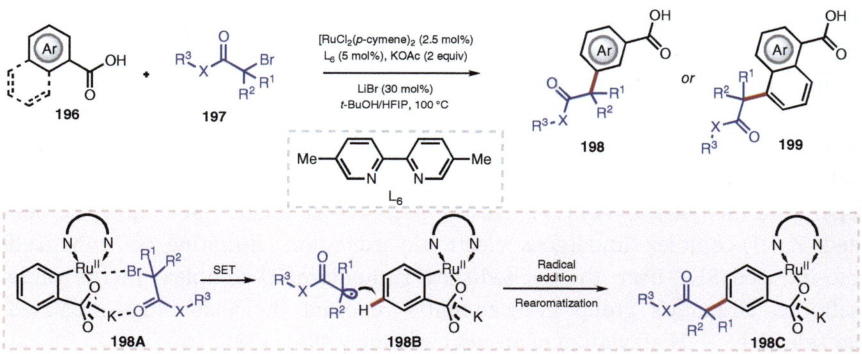

Scheme 9.47 Ligand enabled *meta*-C–H alkylation of aromatic carboxylic acids.

As highlighted in the preceding section, remote *para*-functionalization is elegantly managed by fine-tuning the electronic properties of the cyclometalated Ru(II)-complex. This strategy was successfully employed by Ackermann and Frost independently for the *para*-functionalization of anilines. Utilizing a pyrimidine directing group under ruthenium catalysis, they harnessed the electronic effects of the C—N bond to achieve precise *para*-selectivity (Scheme 9.48a) [171, 172]. Subsequently, a wide range of *para*-functionalization reactions were reported, employing various directing groups to expand the versatility of this approach (Scheme 9.48b) [173–176].

9.8 Ru(II)-catalyzed C–H Activations Under Photocatalysis

Since the pioneering work on *meta*-selective C—H activation by Ackermann and Frost, remarkable progress has been made in *meta*-C—H functionalization through

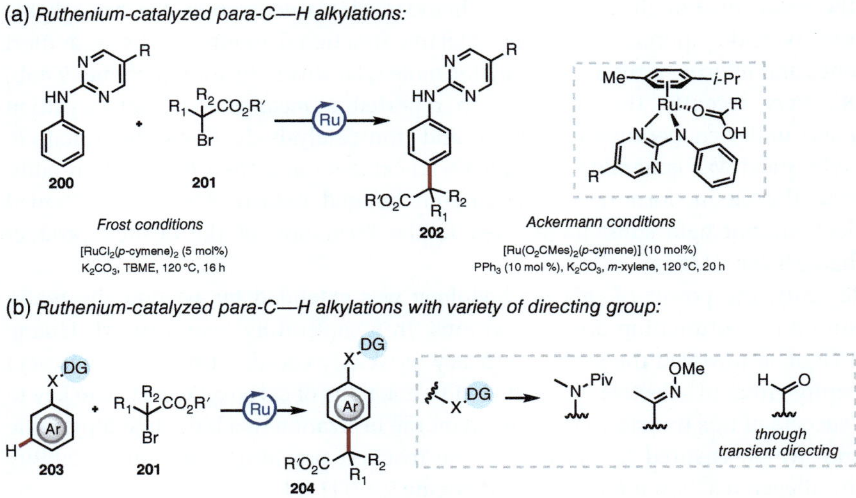

Scheme 9.48 Ru catalyzed *para*-functionalization of arenes.

chelation-assisted *ortho*-cycloruthenation. However, these strategies often require elevated reaction temperature, leading to diminished yields and reduced functional group tolerance. Overcoming these limitations, in 2019, Ackermann group reported visible-light-induced ruthenium(II)-catalyzed *meta*-alkylation reaction at room temperature (Scheme 9.49a) [177]. In the presence of visible light, the cyclometalated Ru(II)-complex undergoes electronic excitation, initiating an enhanced photo-induced SET from the excited-state ruthenium(II)-complex. In the same timeframe, Greaney's group independently reported the visible-light-mediated *meta*-selective C—H alkylation of arenes (Scheme 9.49a) [178].

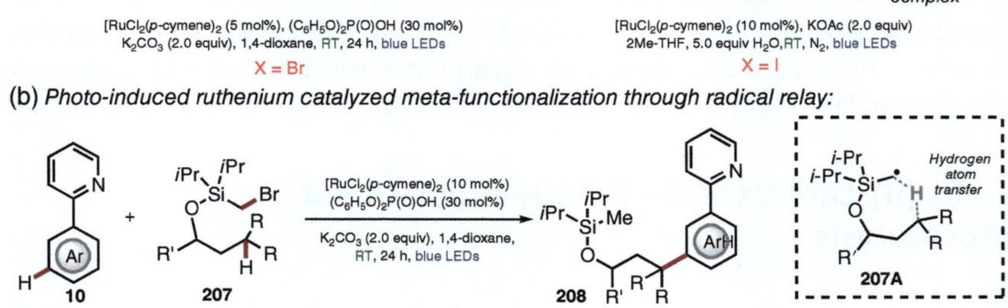

Scheme 9.49 Visible-light-mediated *meta*-C–H functionalization.

In 2022, the Ackermann group advanced this intriguing area of research by integrating hydrogen atom transfer (HAT) with *meta*-C—H activation of arenes, unveiling visible light-induced ruthenium-catalyzed transformations. Notably, their protocol eliminated the need for exogenous photocatalysts, as the ruthenium complex served a dual role: catalyzing both the *meta*-C—H activation and the *C*-centered *1,n*-HAT (where $n = 5$–7) processes. This dual functionality highlighted the growing sophistication in catalytic system design, pushing the boundaries of efficiency and selectivity in C—H activation (Scheme 9.49b) [179].

As discussed earlier, the radical rebound reaction was also feasible with an aryl radical, facilitating the formation of *ortho*-functionalized products. In this context, Ackermann and coworkers reported a significant breakthrough with the development of a visible-light-mediated *ortho*-C—H arylation reaction with aryl iodides (Scheme 9.50a) [180]. Greaney's group independently reported a similar visible-light driven *ortho*-C—H arylation methodology employing ruthenium(II)-catalysis (Scheme 9.50b) [178].

Building upon the success of *ortho*-C—H activation, Ackermann further expanded this approach to enable *ortho*-C—H benzylation and allylation reactions, broadening the scope and applicability of this visible-light-mediated approach (Scheme 9.50c) [181]. Very recently, they further extended the ruthenium(II)-catalyzed C—H arylation strategy for a polymerization reaction induced by visible-light irradiation under ambient temperature (Scheme 9.50d) [182]. All these transformations relied on the formation of ruthenacycle intermediate, which absorbs light in the visible region.

In addition to the remarkable advancements in photo-redox catalysis for ruthenium-catalyzed directed *meta*- and *ortho*-C—H functionalization, a few noteworthy examples showcasing *para*-selectivity have also been reported. In 2022, An and coworkers showcased an innovative visible-light-enabled ruthenium-catalyzed *para*-C—H difluoroalkylation of anilides (Scheme 9.50e) [183]. Recently, An, Li, and Wang's group further described a visible-light-mediated ruthenium(II)-catalyzed *para*-selective alkylation of unprotected anilines under mild and sustainable conditions (Scheme 9.50f) [184].

9.9 Ru(II)-catalyzed C—H Activations Under Electrocatalysis

Recent trends in organic synthesis have witnessed a tremendous shift toward the development of resource-economical synthetic strategies with decreased environmental hazards and increased energy-efficiency [185–187]. As discussed in the preceding section, typically, C—H activation reactions require stoichiometric amounts of metallic or non-metallic redox agents to manipulate the oxidation state of the transition-metal catalyst and facilitate the transformation, which generates substantial waste [4–17]. During the past decades, organic electrochemistry has regained momentum to enable the redox manipulation in transition-metal catalyzed C—H activation reactions under sustainable conditions [188–205].

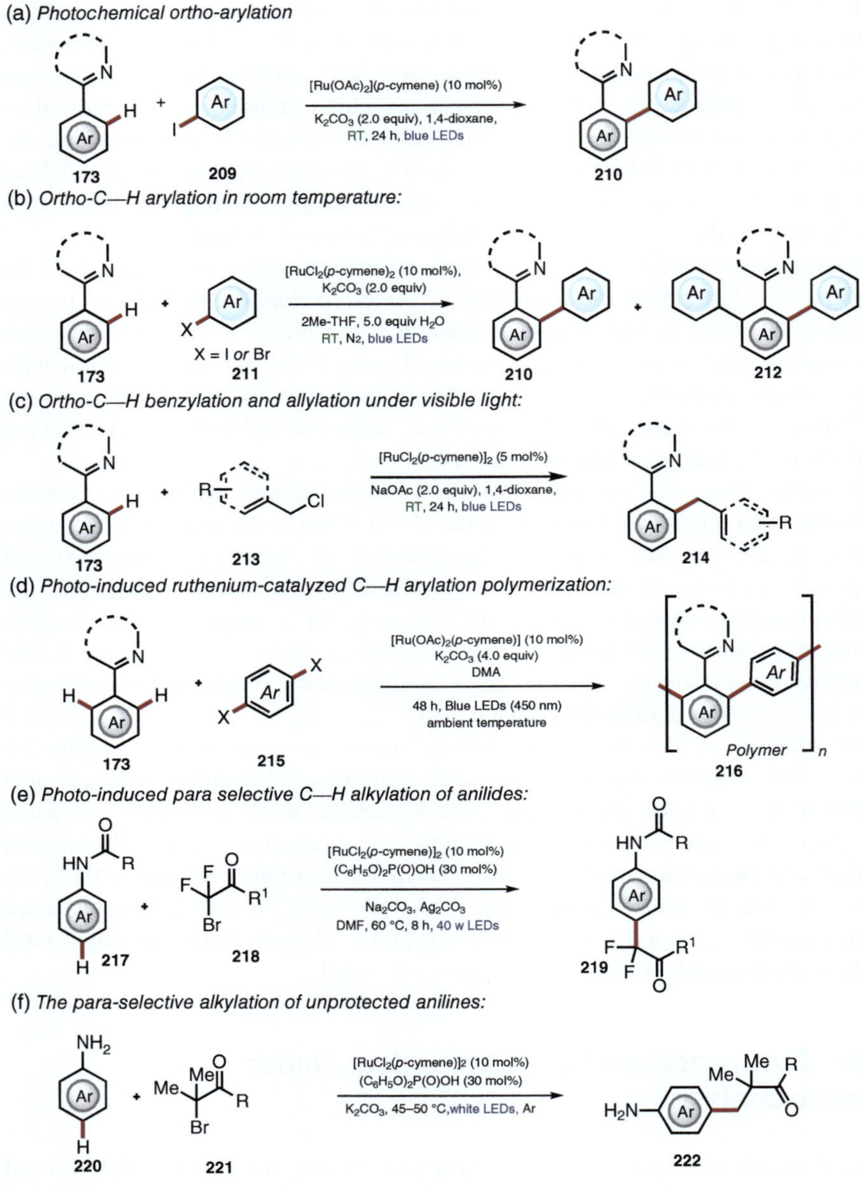

Scheme 9.50 Photo mediated *ortho* and *para*-C–H functionalization.

Here, electrons can be directly used in a traceless fashion to enable the redox manipulation. Notably, electrochemistry provides precise control over the redox potential necessary for a desired transformation, which is not trivial for typical chemical redox reagents. Thus, the merger of electrochemistry with C—H bond activations has revolutionized these processes addressing these inherent problems.

This is highly coveted, as an exceptional level of resource-economy could be obtained, accessing waste-free approaches using energy-efficient electricity as the green oxidant (Scheme 9.51). The only by-product of these processes is molecular hydrogen (H_2) through the hydrogen evolution reaction (HER) at the cathode, which is a useful by-product to enrich the green hydrogen economy [206–208].

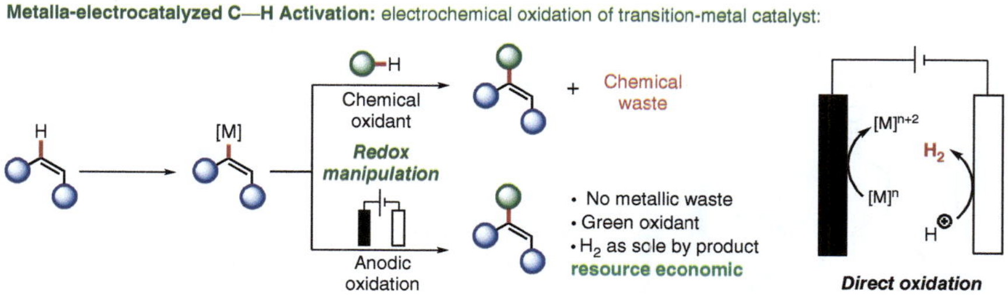

Scheme 9.51 General representation of metalla-electrocatalyzed C–H activation.

The conceptual merger of ruthenium(II)-catalyzed C—H activations under electrochemical conditions was first demonstrated by the Ackermann group through an annulation of alkynes with benzoic acids and benzamides (Scheme 9.52) [209]. The plausible reaction mechanism follows a regular formation of a cyclometalated intermediate **87A**, followed by the migratory insertion of the alkyne. Intermediate **87C**, thus formed, underwent reductive elimination to produce Ru(0)-intermediate **87D**, which was also isolated and confirmed. Anodic oxidation of this species delivered the product and regenerated the active Ru(II)-catalyst. Molecular hydrogen gas was obtained as the sole by-product in this process.

This approach was further applied for the synthesis of isoquinolines through a three-component reaction involving the annulation of alkynes (Scheme 9.53a) [210]. The reaction mechanism involved an oxidation-induced reductive elimination for the product formation and the direct oxidation of Ru(II)-catalyst at the anode helped in key oxidation steps. The insights of the oxidation-induced reductive elimination process were then systematically studied through imidazole-directed vinyl C—H bond activation/annulation reaction (Scheme 9.53a) [211]. In this study, the isolated Ru(II)-cyclometalated intermediate (**226B**) was exposed to electrochemical conditions, forming Ru(III)-intermediate (**226C**), which was isolated and confirmed. This Ru(III)-intermediate underwent rearrangement to fabricate the desired annulated products. Similarly, the annulation of alkynes was also realized with 1-naphthol and 1-naphthyl amine analogues (Scheme 9.53b) [212].

Recently, ruthenaelectro-catalysis was applied for the late-stage acyloxylation of tyrosine and oligopeptides (Scheme 9.54) [213]. A large variety of aromatic carboxylic acids were coupled through an oxidation-induced reductive elimination process, where plausibly an anodic oxidation-based generation of Ru(IV)-species was the key to success.

Recently, the principles of electro-oxidative ruthenium catalyzed C—H activations have also been translated for direct *ortho*- and *para*-selective C—H

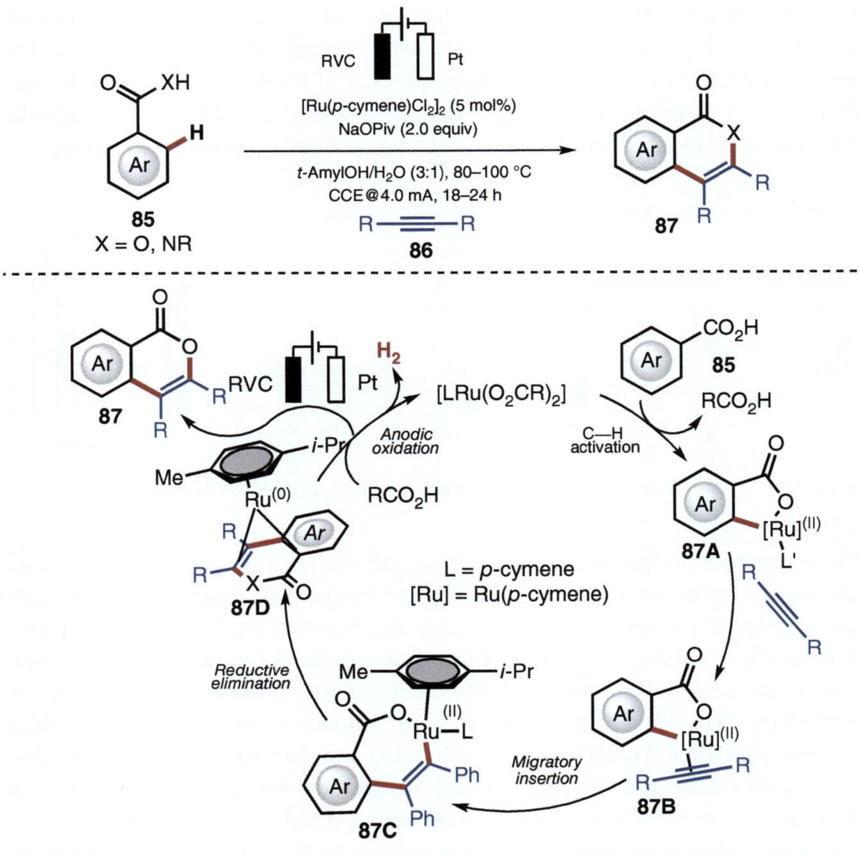

Scheme 9.52 Ru(II)-catalyzed annulation of alkynes with benzoic acids and benzamides.

phosphorylation reactions (Scheme 9.55) [214]. This cross-dehydrogenative strategy operated under mild electrochemical conditions, generating hydrogen as the sole by-product. Diverse *N*-heterocycles and arenes were accommodated under this catalytic protocol in combination with aryl/alkyl phosphine oxides, generating the phosphorylated products in good-to-excellent yields. Systematic DFT analysis suggested an unusual ruthenium(III/IV/II) catalytic manifold for the *ortho*-C—H phosphorylation. While the use of $RuCl_3{\cdot}3H_2O$ surprisingly bestowed a *para*-selective phosphorylation products in synthetically useful yields.

Ruthenaelectro catalysis was also extended for the first chelation-assisted *meta*-bromination of arenes by Ackermann and coworkers (Scheme 9.56) [215]. The transformation was catalyzed by inexpensive $RuCl_3{\cdot}nH_2O$ and used HBr as the bromide source. The C—H activation took place through an LLHT process and anodic oxidation of the Ru(II)-intermediate was responsible for the exclusive *meta*-bromination.

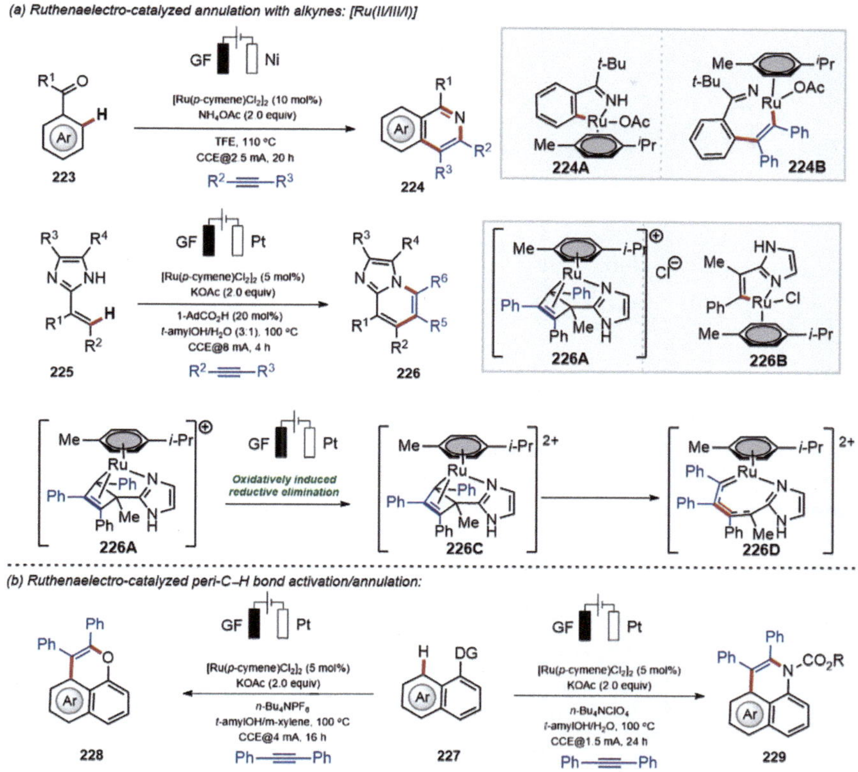

Scheme 9.53 Ruthenaelectrocatalyzed annulation of alkynes and *peri*-C—H bond activation.

Recently, Larossa group utilized ruthenaelectro catalysis for site-selective *ortho*-C—H arylation with aryl boron reagents under oxidative conditions, which was applicable with diverse heteroaromatic substrates (Scheme 9.57) [216]. This method empowered traditional chemical-oxidant-based arylation methods using electricity as the green oxidant.

9.10 Conclusion and Future Outlook

Ruthenium-catalyzed C—H activations have grown significantly as a powerful synthetic tool for the straightforward functionalization of arenes. Ruthenium catalysts showcase different modes of action due to the availability of possible oxidation states, leading to molecular diversification with high fidelity. Among notable redox modes, Ru(II/IV), Ru(II/0), and Ru(II/III) are the most popular

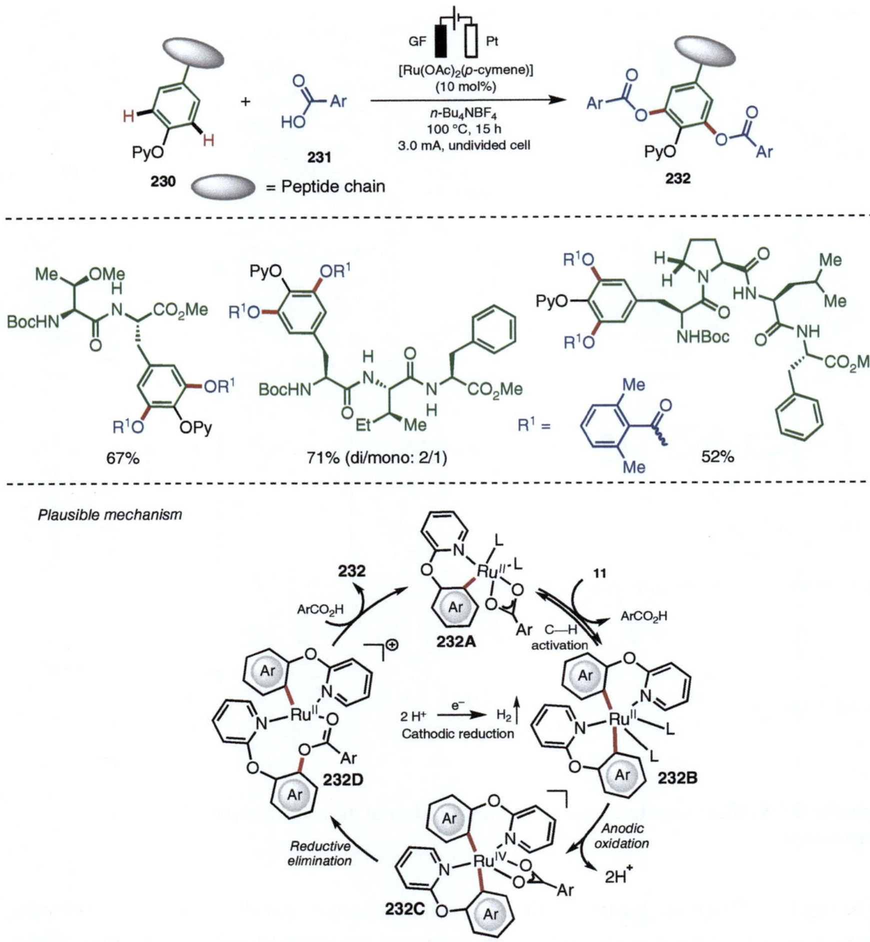

Scheme 9.54 Ruthenaelectrocatalyzed late-stage acyloxylation of tyrosine-containing peptides.

catalytic manifolds with distinct advantages for each cycle, which are harnessed for the site-selective *ortho*- and *meta*-C—H functionalization with a variety of strongly and weakly coordinating directing groups. The redox transformations are exploited under thermal, photochemical, and electrochemical conditions, where each of them showcases unique advantages. All these processes exhibited high functional group tolerance and many of them were particularly useful for the late-stage diversification of pharmaceuticals and complex bioactive molecules.

Despite the significant advancement in this area, ruthenium(II)-catalyzed C—H activations are still in their infancy compared to noble metal catalysis.

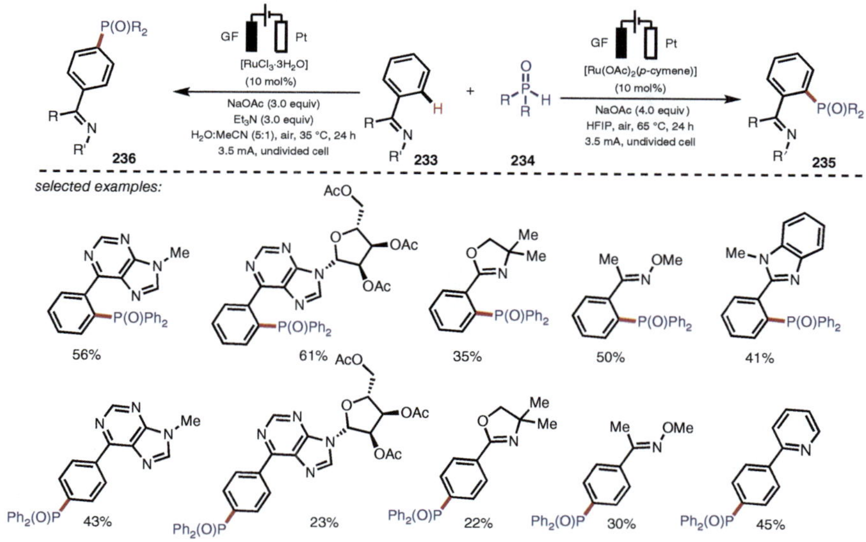

Scheme 9.55 Electro-oxidative ruthenium catalyzed *ortho*- and *para*-C—H phosphorylation.

There remains an immense synthetic space in this area, which necessitates continuous focus to expand this versatile regime of catalysis. While ruthenium catalysis has been extremely successful with a diversity of chelating groups, non-directed site-selective C—H activations are still elusive in this platform. Thus, significant focus is necessary for the catalyst and ligand development approaches to explore this area. While two-component reactions are well explored, multi-component variants of these transformations and cascade reactions relying on Ru(II)-catalyzed C—H activations to build rapid molecular complexity are still immature. Most of these redox reactions relied on the use of metallic or non-metallic oxidants. In contrast, recent developments harnessed visible light and electricity as sustainable sources to manipulate the critical redox events. However, these approaches still displayed limited utility, which calls for further advancement bring them into the mainstream. Also, ruthenium-catalyzed *meta*-selective reactions are mostly confined to C—C bond formation. Considering the high efficacy of ruthenium catalyzed distal C—H functionalization reactions, the exploration of C—heteroatom bond forming reactions is also need of the hour. Further, the use of heterogeneous ruthenium catalysis and hybrid catalytic systems is very little explored and only consists of sporadic examples. This form of catalysis is highly thought provoking and demands systematic exploration owing to its innate reusable and sustainable features. We expect that this summary of the advancement of ruthenium(II)-catalyzed C—H activations will act as a stimulus and fuel further development in this rapidly growing area of catalysis.

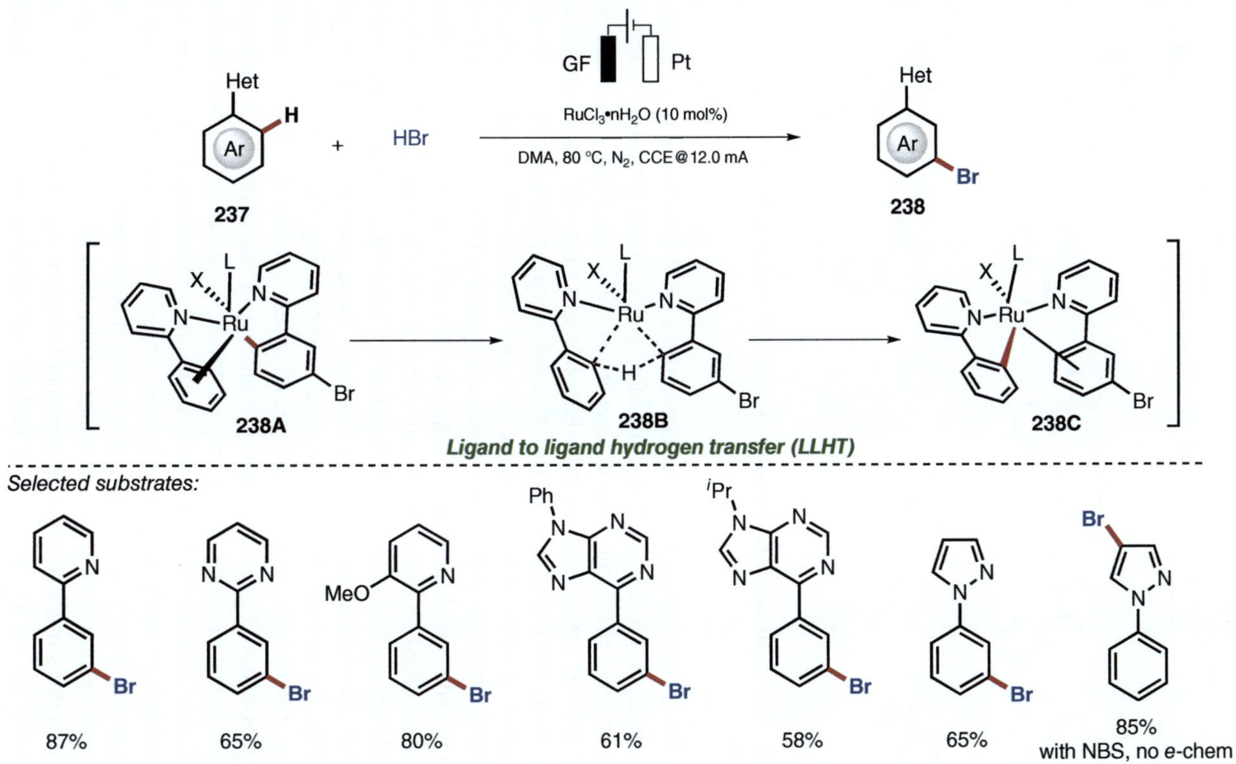

Scheme 9.56 Chelation-assisted *meta*-bromination of arenes.

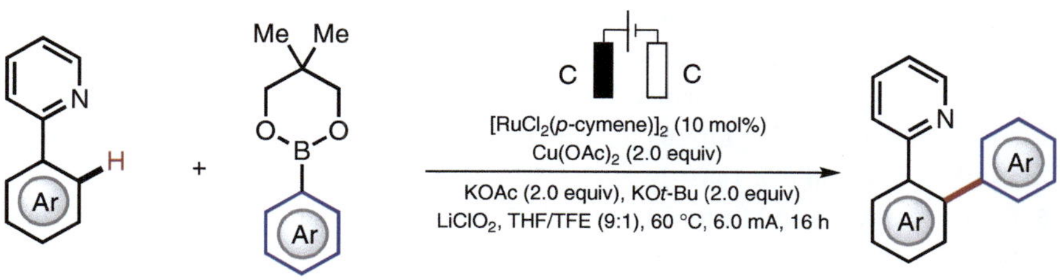

10 239 12

Scheme 9.57 Ruthenaelectrocatalyzed site-selective *ortho*-C–H arylation.

References

1 De Carvalho, R.L., Diogo, E.B.T., Homölle, S.L. et al. (2023). The crucial role of silver(I)-salts as additives in C–H activation reactions: overall analysis of their versatility and applicability. *Chem. Soc. Rev.* 52 (18): 6359–6378.

2 Kaplaneris, N. and Ackermann, L. (2022). Earth-abundant 3d transition metals on the rise in catalysis. *Beilstein J. Org. Chem.* 18: 86–88.

3 Liu, B., Romine, A.M., Rubel, C.Z. et al. (2021). Transition-metal-catalyzed, coordination-assisted functionalization of nonactivated C(sp^3)–H bonds. *Chem. Rev.* 121 (24): 14957–15074.

4 Dhawa, U., Kaplaneris, N., and Ackermann, L. (2021). Green strategies for transition metal-catalyzed C–H activation in molecular syntheses. *Org. Chem. Front.* 8 (17): 4886–4913.

5 Gandeepan, P., Müller, T., Zell, D. et al. (2019). 3d transition metals for C–H activation. *Chem. Rev.* 119 (4): 2192–2452.

6 Liu, W. and Ackermann, L. (2016). Manganese-catalyzed C–H activation. *ACS Catal.* 6 (6): 3743–3752.

7 Yang, X., Shan, G., Wang, L., and Rao, Y. (2016). Recent advances in transition metal (Pd, Ni)-catalyzed C(sp^3) H bond activation with bidentate directing groups. *Tetrahedron Lett.* 57 (8): 819–836.

8 Rouquet, G. and Chatani, N. (2013). Catalytic functionalization of C(sp^2)–H and C(sp^3)–H bonds by using bidentate directing groups. *Angew. Chem. Int. Ed.* 52 (45): 11726–11743.

9 Mei, R., Dhawa, U., Samanta, R.C. et al. (2020). Cobalt-catalyzed oxidative C–H activation: strategies and concepts. *ChemSusChem* 13 (13): 3306–3356.

10 Korvorapun, K., Samanta, R.C., Rogge, T., and Ackermann, L. (2021). Remote C–H functionalizations by ruthenium catalysis. *Synthesis* 53 (17): 2911–2946.

11 Li, X., Ouyang, W., Nie, J. et al. (2020). Recent development on Cp*Ir(III)-catalyzed C–H bond functionalization. *ChemCatChem* 12 (9): 2358–2384.

12 Li, S.S., Qin, L., and Dong, L. (2016). Rhodium-catalyzed C-C coupling reactions: via double C-H activation. *Org. Biomol. Chem.* 14 (20): 4554–4570.

13 Wencel-Delord, J., Dröge, T., Liu, F., and Glorius, F. (2011). Towards mild metal-catalyzed C–H bond activation. *Chem. Soc. Rev.* 40 (9): 4740.

14 Ackermann, L. (2014). Carboxylate-assisted ruthenium-catalyzed alkyne annulations by C-H/Het-H bond functionalizations. *Acc. Chem. Res.* 47 (2): 281–295.

15 De Sarkar, S., Liu, W., Kozhushkov, S.I., and Ackermann, L. (2014). Weakly coordinating directing groups for ruthenium(II)-catalyzed C—H activation. *Adv. Synth. Catal.* 356 (7): 1461–1479.

16 Davies, H.M.L. and Morton, D. (2016). Recent advances in C—H functionalization. *J. Org. Chem.* 81 (2): 343–350.

17 Baunach, M., Franke, J., and Hertweck, C. (2015). Terpenoid biosynthesis off the beaten track: unconventional cyclases and their impact on biomimetic synthesis. *Angew. Chem. Int. Ed.* 54 (9): 2604–2626.

18 Oldfield, E. and Lin, F.-Y. (2012). Terpene biosynthesis: modularity rules. *Angew. Chem. Int. Ed.* 51 (5): 1124–1137.

19 Ishihara, Y. and Baran, P.S. (2010). Two-phase terpene total synthesis: historical perspective and application to the Taxol® problem. *Synlett* 12: 1733–1745.

20 Bicas, J.L., Dionísio, A.P., and Pastore, G.M. (2009). Bio-oxidation of terpenes: an approach for the flavor industry. *Chem. Rev.* 109 (9): 4518–4531.

21 Dalton, T., Faber, T., and Glorius, F. (2021). C–H activation: toward sustainability and applications. *ACS Cent. Sci.* 7 (2): 245–261.

22 Gramage-Doria, R. and Bruneau, C. (2021). Ruthenium-catalyzed C–H bond functionalization in cascade and one-pot transformations. *Coord. Chem. Rev.* 428: 213602.

23 Duarah, G., Kaishap, P.P., Begum, T., and Gogoi, S. (2019). Recent advances in ruthenium(II)-catalyzed C–H bond activation and alkyne annulation reactions. *Adv. Synth. Catal.* 361 (4): 654–672.

24 Shan, C., Zhu, L., Qu, L.-B. et al. (2018). Mechanistic view of Ru-catalyzed C–H bond activation and functionalization: computational advances. *Chem. Soc. Rev.* 47 (20): 7552–7576.

25 Ackermann, L. (2011). Carboxylate-assisted transition-metal-catalyzed C–H bond functionalizations: mechanism and scope. *Chem. Rev.* 111 (3): 1315–1345.

26 Nareddy, P., Jordan, F., and Szostak, M. (2017). Recent developments in ruthenium-catalyzed C–H arylation: array of mechanistic manifolds. *ACS Catal.* 7 (9): 5721–5745.

27 Manikandan, R. and Jeganmohan, M. (2017). Recent advances in the ruthenium(II)-catalyzed chelation-assisted C–H olefination of substituted aromatics, alkenes and heteroaromatics with alkenes via the deprotonation pathway. *Chem. Commun.* 53 (64): 8931–8947.

28 Thirunavukkarasu, V.S., Kozhushkov, S.I., and Ackermann, L. (2014). C–H nitrogenation and oxygenation by ruthenium catalysis. *Chem. Commun.* 50 (1): 29–39.

29 Dana, S., Yadav, M.R., and Sahoo, A.K. (2016). Ruthenium-catalyzed C–N and C–O bond-forming processes from C–H bond functionalization. In: *C–H Bond Activation and Catalytic Functionalization I* (ed. P.H. Dixneuf and H. Doucet), 189–215. Cham: Springer International Publishing.

30 Dana, S. and Baidya, M. (2022). Ru-catalyzed C–H functionalization of benzoic acids. In: *Handbook of CH-Functionalization*, 1e, 1–53. Wiley.

31 Lewis, L.N. and Smith, J.F. (1986). Catalytic carbon-carbon bond formation via *ortho*-metalated complexes. *J. Am. Chem. Soc.* 108 (10): 2728–2735.

32 Murai, S., Kakiuchi, F., Sekine, S. et al. (1993). Efficient catalytic addition of aromatic carbon-hydrogen bonds to olefins. *Nature* 366 (6455): 529–531.

33 Martinez, R., Genet, J.-P., and Darses, S. (2008). Anti-Markovnikov hydroarylation of styrenes catalyzed by an in situ generated ruthenium complex. *Chem. Commun.* 33: 3855.

34 Martinez, R., Simon, M.-O., Chevalier, R. et al. (2009). C—C bond formation via C—H bond activation using an in situ-generated ruthenium catalyst. *J. Am. Chem. Soc.* 131 (22): 7887–7895.

35 Oi, S., Fukita, S., Hirata, N. et al. (2001). Ruthenium complex-catalyzed direct *ortho* arylation and alkenylation of 2-arylpyridines with organic halides. *Org. Lett.* 3 (16): 2579–2581.

36 Ackermann, L. (2010). Air-stable bifunctional HASPO preligands for metal-catalyzed cross-couplings and direct C–H bond arylations. *Isr. J. Chem.* 50 (5–6): 652–663.

37 Ackermann, L. (2006, 2006). Air- and moisture-stable secondary phosphine oxides as preligands in catalysis. *Synthesis* 10: 1557–1571.

38 Ackermann, L. (2005). Phosphine oxides as preligands in ruthenium-catalyzed arylations via C—H bond functionalization using aryl chlorides. *Org. Lett.* 7 (14): 3123–3125.

39 Ackermann, L. and Mulzer, M. (2008). Dehydrative direct arylations of arenes with phenols via ruthenium-catalyzed C—H and C—OH bond functionalizations. *Org. Lett.* 10 (21): 5043–5045.

40 Ackermann, L., Vicente, R., and Althammer, A. (2008). Assisted ruthenium-catalyzed C—H bond activation: carboxylic acids as cocatalysts for generally applicable direct arylations in apolar solvents. *Org. Lett.* 10 (11): 2299–2302.

41 Ackermann, L., Vicente, R., Potukuchi, H.K., and Pirovano, V. (2010). Mechanistic insight into direct arylations with ruthenium(II) carboxylate catalysts. *Org. Lett.* 12 (21): 5032–5035.

42 Shin, K., Kim, H., and Chang, S. (2015). Transition-metal-catalyzed C–N bond forming reactions using organic azides as the nitrogen source: a journey for the mild and versatile C–H amination. *Acc. Chem. Res.* 48 (4): 1040–1052.

43 Ackermann, L., Pospech, J., and Potukuchi, H.K. (2012). Well-defined ruthenium(II) carboxylate as catalyst for direct C–H/C–O bond arylations with phenols in water. *Org. Lett.* 14 (8): 2146–2149.

44 Ackermann, L. (2010). Metal-catalyzed direct alkylations of (hetero)arenes via C–H bond cleavages with unactivated alkyl halides. *Chem. Commun.* 46 (27): 4866–4877.

45 Ackermann, L. and Novák, P. (2009). Regioselective ruthenium-catalyzed direct benzylations of arenes through C—H bond cleavages. *Org. Lett.* 11 (21): 4966–4969.

46 Ackermann, L., Novák, P., Vicente, R., and Hofmann, N. (2009). Ruthenium-catalyzed regioselective direct alkylation of arenes with unactivated alkyl halides through C—H bond cleavage. *Angew. Chem. Int. Ed.* 48 (33): 6045–6048.

47 Ackermann, L., Althammer, A., and Born, R. (2008). [$RuCl_3(H_2O)_n$]-catalyzed direct arylations. *Tetrahedron* 64 (26): 6115–6124.

48 Ackermann, L., Althammer, A., and Born, R. (2007). [$RuCl_3(H_2O)_n$]-catalyzed direct arylations with bromides as electrophiles. *Synlett* 18: 2833–2836.

49 Rogge, T. and Ackermann, L. (2019). Arene-free ruthenium(II/IV)-catalyzed bifurcated arylation for oxidative C–H/C–H functionalizations. *Angew. Chem. Int. Ed.* 58 (44): 15640–15645.

50 Simon, H., Zangarelli, A., Bauch, T., and Ackermann, L. (2024). Ruthenium(II)-catalyzed late-stage incorporation of N-aryl triazoles and tetrazoles with sulfonium salts via C–H activation. *Angew. Chem. Int. Ed.* 63 (26): e202402060.

51 Simonetti, M., Cannas, D.M., Just-Baringo, X. et al. (2018). Cyclometallated ruthenium catalyst enables late-stage directed arylation of pharmaceuticals. *Nat. Chem.* 10 (7): 724–731.

52 Hogg, A., Wheatley, M., Domingo-Legarda, P. et al. (2022). Ruthenium-catalyzed monoselective C–H methylation and d_3-methylation of arenes. *JACS Au.* 2 (11): 2529–2538.

53 Wang, G.-W., Wheatley, M., Simonetti, M. et al. (2020). Cyclometalated ruthenium catalyst enables *ortho*-selective C–H alkylation with secondary alkyl bromides. *Chem* 6 (6): 1459–1468.

54 Liang, Y.-F., Yang, L., Rogge, T., and Ackermann, L. (2018). Ruthenium(IV) intermediates in C–H activation/annulation by weak O-coordination. *Chem. Eur. J.* 24 (62): 16548–16552.

55 Park, J., Lee, J., and Chang, S. (2017). Iterative C–H functionalization leading to multiple amidations of anilides. *Angew. Chem. Int. Ed.* 56 (15): 4256–4260.

56 Yadav, M.R., Shankar, M., Ramesh, E. et al. (2015). Ruthenium-catalyzed *ortho*-C–H mono- and di-imidation of arenes with N-tosyloxyphthalimide. *Org. Lett.* 17 (8): 1886–1889.

57 Shang, M., Zeng, S.-H., Sun, S.-Z. et al. (2013). Ru(II)-catalyzed *ortho*-C–H amination of arenes and heteroarenes at room temperature. *Org. Lett.* 15 (20): 5286–5289.

58 Thirunavukkarasu, V.S., Raghuvanshi, K., and Ackermann, L. (2013). Expedient C–H amidations of heteroaryl arenes catalyzed by versatile ruthenium(II) catalysts. *Org. Lett.* 15 (13): 3286–3289.

59 Zheng, Q.-Z., Liang, Y.-F., Qin, C., and Jiao, N. (2013). Ru(II)-catalyzed intermolecular C–H amidation of weakly coordinating ketones. *Chem. Commun.* 49 (50): 5654–5656.

60 Bhanuchandra, M., Yadav, M.R., Rit, R.K. et al. (2013). Ru(II)-catalyzed intermolecular *ortho*-C–H amidation of aromatic ketones with sulfonyl azides. *Chem. Commun.* 49 (45): 5225–5227.

61 Yadav, M.R., Rit, R.K., and Sahoo, A.K. (2013). Sulfoximine directed intermolecular o-C–H amidation of arenes with sulfonyl azides. *Org. Lett.* 15 (7): 1638–1641.

62 Kim, J., Kim, J., and Chang, S. (2013). Ruthenium-catalyzed direct C—H amidation of arenes including weakly coordinating aromatic ketones. *Chem. Eur. J.* 19 (23): 7328–7333.

63 Yang, Y., Lin, Y., and Rao, Y. (2012). Ruthenium(II)-catalyzed synthesis of hydroxylated arenes with ester as an effective directing group. *Org. Lett.* 14 (11): 2874–2877.

64 Thirunavukkarasu, V.S., Hubrich, J., and Ackermann, L. (2012). Ruthenium-catalyzed oxidative C(sp2)–H bond hydroxylation: site-selective C–O bond formation on benzamides. *Org. Lett.* 14 (16): 4210–4213.

65 Thirunavukkarasu, V.S. and Ackermann, L. (2012). Ruthenium-catalyzed C–H bond oxygenations with weakly coordinating ketones. *Org. Lett.* 14 (24): 6206–6209.

66 Yang, F. and Ackermann, L. (2013). Ruthenium-catalyzed C–H oxygenation on aryl weinreb amides. *Org. Lett.* 15 (3): 718–720.

67 Yang, X., Shan, G., and Rao, Y. (2013). Synthesis of 2-aminophenols and heterocycles by Ru-catalyzed C–H mono- and dihydroxylation. *Org. Lett.* 15 (10): 2334–2337.

68 Liu, W. and Ackermann, L. (2013). *Ortho*- and *para*-selective ruthenium-catalyzed C(sp2)–H oxygenations of phenol derivatives. *Org. Lett.* 15 (13): 3484–3486.

69 Bu, Q., Kuniyil, R., Shen, Z. et al. (2020). Insights into ruthenium(II/IV)-catalyzed distal C–H oxygenation by weak coordination. *Chem. Eur. J.* 26 (69): 16450–16454.

70 Yuan, Y.-C., Bruneau, C., Roisnel, T., and Gramage-Doria, R. (2019). Ruthenium(II)-catalysed selective C(sp2)–H bond benzoxylation of biologically appealing N-arylisoindolinones. *Org. Biomol. Chem.* 17 (32): 7517–7525.

71 More, N.Y., Padala, K., and Jeganmohan, M. (2017). Ruthenium-catalyzed C–H benzoxylation of tert-benzamides with aromatic acids by weak coordination. *J. Org. Chem.* 82 (23): 12691–12700.

72 Padala, K. and Jeganmohan, M. (2014). *Ortho*-benzoxylation of N-alkyl benzamides with aromatic acids catalyzed by ruthenium(II) complex. *Chem. Eur. J.* 20 (14): 4092–4097.

73 Padala, K. and Jeganmohan, M. (2013). Ruthenium-catalyzed oxidative *ortho*-benzoxylation of acetanilides with aromatic acids. *Chem. Commun.* 49 (83): 9651–9653.

74 Duan, B., Wu, Y., Gao, Y. et al. (2022). Regioselective peri -C–H selenylation of aromatic compounds with weakly coordinating ketone groups. *Chem. Commun.* 58 (82): 11555–11558.

75 Weng, Z., Fang, X., He, M. et al. (2019). Ruthenium catalyzed C–H selenylations of aryl acetic amides and esters via weak coordination. *Org. Lett.* 21 (16): 6310–6314.

76 Ma, W., Weng, Z., Fang, X. et al. (2019). Ruthenium-catalyzed C–H selenylations of benzamides. *Eur. J. Org. Chem.* 2019 (1): 41–45.

77 Ma, W., Weng, Z., Rogge, T. et al. (2018). Ruthenium(II)-catalyzed C–H chalcogenation of anilides. *Adv. Synth. Catal.* 360 (4): 704–710.

78 Mandal, A., Dana, S., Sahoo, H. et al. (2017). Ruthenium(II)-catalyzed *ortho*-C–H chalcogenation of benzoic acids via weak O-coordination: synthesis of chalcogenoxanthones. *Org. Lett.* 19 (9): 2430–2433.

79 Shu, S., Fan, Z., Yao, Q., and Zhang, A. (2016). Ru(II)-catalyzed direct C(sp^2)–H activation/selenylation of arenes with selenyl chlorides. *J. Org. Chem.* 81 (13): 5263–5269.

80 Reddy, M.C. and Jeganmohan, M. (2015). Ruthenium-catalyzed *ortho* alkenylation of aromatic nitriles with activated alkenes via C–H bond activation. *Chem. Commun.* 51 (53): 10738–10741.

81 Shome, S. and Singh, S.P. (2015). Solvent-free ruthenium(II)-catalyzed C–H activation: synthesis of alkenylarylpyrazole derivatives. *Eur. J. Org. Chem.* 27: 6025–6032.

82 Manikandan, R. and Jeganmohan, M. (2014). Ruthenium-catalyzed cyclization of anilides with substituted propiolates or acrylates: an efficient route to 2-quinolinones. *Org. Lett.* 16 (13): 3568–3571.

83 Reddy, M.C. and Jeganmohan, M. (2013). Ruthenium-catalyzed selective aerobic oxidative *ortho*-alkenylation of substituted phenols with alkenes through C–H bond activation. *Eur. J. Org. Chem.* 2013 (6): 1150–1157.

84 Padala, K. and Jeganmohan, M. (2012). Highly regio- and stereoselective ruthenium(II)-catalyzed direct *ortho*-alkenylation of aromatic and heteroaromatic aldehydes with activated alkenes under open atmosphere. *Org. Lett.* 14 (4): 1134–1137.

85 Padala, K. and Jeganmohan, M. (2011). Ruthenium-catalyzed *ortho*-alkenylation of aromatic ketones with alkenes by C–H bond activation. *Org. Lett.* 13 (23): 6144–6147.

86 Yi, C.S. and Lee, D.W. (2010). Intermolecular dehydrative coupling reaction of aryl ketones with cyclic alkenes catalyzed by a well-defined cationic ruthenium–hydride complex: a novel ketone olefination method via vinyl C–H bond activation. *Organometallics* 29 (8): 1883–1885.

87 Ueyama, T., Mochida, S., Fukutani, T. et al. (2011). Ruthenium-catalyzed oxidative vinylation of heteroarene carboxylic acids with alkenes via regioselective C–H bond cleavage. *Org. Lett.* 13 (4): 706–708.

88 Ackermann, L. and Pospech, J. (2011). Ruthenium-catalyzed oxidative C–H bond alkenylations in water: expedient synthesis of annulated lactones. *Org. Lett.* 13 (16): 4153–4155.

89 Arockiam, P.B., Fischmeister, C., Bruneau, C., and Dixneuf, P.H. (2011). Ruthenium diacetate-catalysed oxidative alkenylation of C–H bonds in air: synthesis of alkenyl N-arylpyrazoles. *Green Chem.* 13 (11): 3075–3078.

90 Chowdhury, D., Dana, S., Mandal, A., and Baidya, M. (2019). A ruthenium-catalyzed free amine directed (5+1) annulation of anilines with olefins: diverse synthesis of phenanthridine derivatives. *Chem. Commun.* 55 (79): 11908–11911.

91 Mandal, A., Bera, R., and Baidya, M. (2021). Regioselective C–H alkenylation and unsymmetrical bis-olefination of heteroarene carboxylic acids with ruthenium catalysis in water. *J. Org. Chem.* 86 (1): 62–73.

92 Dana, S., Giri, C.K., and Baidya, M. (2021). Ruthenium(II)-catalyzed regioselective C–H olefination of aromatic ketones and amides with allyl sulfones. *Org. Lett.* 23 (17): 6855–6860.

93 Reddy, M.C., Manikandan, R., and Jeganmohan, M. (2013). Ruthenium-catalyzed aerobic oxidative cyclization of aromatic and heteroaromatic nitriles with alkynes: a new route to isoquinolones. *Chem. Commun.* 49 (54): 6060–6062.

94 Chinnagolla, R.K. and Jeganmohan, M. (2012). Regioselective synthesis of isocoumarins by ruthenium-catalyzed aerobic oxidative cyclization of aromatic acids with alkynes. *Chem. Commun.* 48 (14): 2030–2032.

95 Chinnagolla, R.K. and Jeganmohan, M. (2012). Regioselective *ortho*-arylation and alkenylation of N-alkyl benzamides with boronic acids via ruthenium-catalyzed C–H bond activation: an easy route to fluorenones synthesis. *Org. Lett.* 14 (20): 5246–5249.

96 Zhang, J., Liu, Y., Jia, Q. et al. (2020). Ruthenium(II)-catalyzed C–H arylation of N,N-dialkyl thiobenzamides with boronic acids by sulfur coordination in 2-MeTHF. *Org. Lett.* 22 (17): 6884–6890.

97 Nareddy, P., Jordan, F., and Szostak, M. (2018). Ruthenium(II)-catalyzed direct C–H arylation of indoles with arylsilanes in water. *Org. Lett.* 20 (2): 341–344.

98 Nareddy, P., Jordan, F., and Szostak, M. (2017). Ruthenium(II)-catalyzed *ortho*-C–H arylation of diverse N-heterocycles with aryl silanes by exploiting solvent-controlled N-coordination. *Org. Biomol. Chem.* 15 (22): 4783–4788.

99 Nareddy, P., Jordan, F., and Szostak, M. (2017). Highly chemoselective ruthenium(II)-catalyzed direct arylation of cyclic and N,N-dialkyl benzamides with aryl silanes. *Chem. Sci.* 8 (4): 3204–3210.

100 Dana, S., Chowdhury, D., Mandal, A. et al. (2018). Ruthenium(II) catalysis/noncovalent interaction synergy for cross-dehydrogenative coupling of arene carboxylic acids. *ACS Catal.* 8 (11): 10173–10179.

101 Shankar, M., Saha, A., Ghosh, A. et al. (2021). Sulfur and nitrogen modulated one-pot double annulation of arenes. *J. Org. Chem.* 86 (21): 14942–14955.

102 Shankar, M., Rit, R.K., Sau, S. et al. (2020). Double annulation of *ortho*- and peri-C–H bonds of fused (hetero)arenes to unusual oxepino-pyridines. *Chem. Sci.* 11 (39): 10770–10777.

103 Shankar, M., Ghosh, K., Mukherjee, K. et al. (2018). One-pot unsymmetrical {[4 + 2] and [4 + 2]} double annulations of o/o'-C–H bonds of Arenes: access to unusual pyranoisoquinolines. *Org. Lett.* 20 (17): 5144–5148.

104 Mukherjee, K., Shankar, M., Ghosh, K., and Sahoo, A.K. (2018). An orchestrated unsymmetrical annulation episode of C(sp^2)–H bonds with alkynes and quinones: access to spiro-isoquinolones. *Org. Lett.* 20 (7): 1914–1918.

105 Shankar, M., Guntreddi, T., Ramesh, E., and Sahoo, A.K. (2017). Transformable sulfoximine assisted one-pot double annulation of vinylic C–H bonds with unactivated alkynes. *Org. Lett.* 19 (20): 5665–5668.

106 Shankar, M., Ghosh, K., Mukherjee, K. et al. (2016). Ru-catalyzed one-pot diannulation of heteroaryls: direct access to π-conjugated polycyclic amides. *Org. Lett.* 18 (24): 6416–6419.

107 Subhedar, D.D., Mishra, A.A., and Bhanage, B.M. (2019). N-methoxybenzamide: a versatile directing group for palladium-, rhodium- and ruthenium-catalyzed C–H bond activations. *Adv. Synth. Catal.* 361 (18): 4149–4195.

108 Shi, Z., Zhang, C., Li, S. et al. (2009). Indoles from simple anilines and alkynes: palladium-catalyzed C-H activation using dioxygen as the oxidant. *Angew. Chem. Int. Ed.* 48 (25): 4572–4576.

109 Zhang, G., Yang, L., Wang, Y. et al. (2013). An efficient Rh/O_2 catalytic system for oxidative C–H activation/annulation: evidence for Rh(I) to Rh(III) oxidation by molecular oxygen. *J. Am. Chem. Soc.* 135 (24): 8850–8853.

110 Warratz, S., Kornhaaß, C., Cajaraville, A. et al. (2015). Ruthenium(II)-catalyzed C-H activation/alkyne annulation by weak coordination with O_2 as the sole oxidant. *Angew. Chem. Int. Ed.* 54 (18): 5513–5517.

111 Bechtoldt, A., Baumert, M.E., Vaccaro, L., and Ackermann, L. (2018). Ruthenium(II) oxidase catalysis for C–H alkenylations in biomass-derived γ-valerolactone. *Green Chem.* 20 (2): 398–402.

112 Mandal, A., Garai, B., Dana, S. et al. (2020). Cross-dehydrogenative coupling/annulation of arene carboxylic acids and alkenes in water with ruthenium(II) catalyst and air. *Chem. Asian J.* 15 (23): 4009–4013.

113 Wang, C., Dixneuf, P.H., and Soule, J.-F. (2018). Photoredox catalysis for building C–C bonds from C(sp2)–H bonds. *Chem. Rev.* 118 (16): 7532–7585.

114 Nakajima, K., Miyake, Y., and Nishibayashi, Y. (2016). Synthetic utilization of α-aminoalkyl radicals and related species in visible light photoredox catalysis. *Acc. Chem. Res.* 49 (9): 1946–1956.

115 Hari, D.P. and König, B. (2014). Synthetic applications of eosin Y in photoredox catalysis. *Chem. Commun.* 50 (51): 6688–6699.

116 Dana, S., Dey, P., Patil, S.A., and Baidya, M. (2020). Enhancing Ru(II)-catalysis with visible-light-mediated dye-sensitized TiO_2 photocatalysis for oxidative C–H olefination of arene carboxylic acids at room temperature. *Chem. Asian J.* 15 (5): 564–567.

117 Ackermann, L. and Fenner, S. (2011). Ruthenium-catalyzed C-H/N-O bond functionalization: green isoquinolone syntheses in water. *Org. Lett.* 13 (24): 6548–6551.

118 Huang, H., Nakanowatari, S., and Ackermann, L. (2017). Selectivity control in ruthenium(II)-catalyzed C–H/N–O activation with alkynyl bromides. *Org. Lett.* 19 (17): 4620–4623.

119 Chinnagolla, R.K., Pimparkar, S., and Jeganmohan, M. (2012). Ruthenium-catalyzed highly regioselective cyclization of ketoximes with alkynesjia by C—H bond activation: a practical route to synthesize substituted isoquinolines. *Org. Lett.* 14 (12): 3032–3035.

120 Kornhaaß, C., Li, J., and Ackermann, L. (2012). Cationic ruthenium catalysts for alkyne annulations with oximes by C–H/N–O functionalizations. *J. Org. Chem.* 77 (20): 9190–9198.

121 Gollapelli, K.K., Kallepu, S., Govindappa, N. et al. (2016). Carbonyl-assisted reverse regioselective cascade annulation of 2-acetylenic ketones triggered by Ru-catalyzed C–H activation. *Chem. Sci.* 7 (7): 4748–4753.

122 Wu, X., Wang, B., Zhou, S. et al. (2017). Ruthenium-catalyzed redox-neutral [4 + 1] annulation of benzamides and propargyl alcohols via C–H bond activation. *ACS Catal.* 7 (4): 2494–2499.

123 Song, L., Zhang, X., Tang, X. et al. (2020). Ruthenium-catalyzed cascade C–H activation/annulation of N-alkoxybenzamides: reaction development and mechanistic insight. *Chem. Sci.* 11 (42): 11562–11569.

124 Dana, S., Sureshbabu, P., Giri, C.K., and Baidya, M. (2021). Ruthenium(II)-catalyzed C-H activation/annulation of aromatic hydroxamic acid esters with enamides leading to aminal motifs. *Eur. J. Org. Chem.* 9: 1385–1389.

125 Giri, C.K., Dana, S., and Baidya, M. (2022). Ruthenium(II)-catalyzed (4 + 2) annulative difunctionalization of non-conjugated alkenyl amides with hydroxamic acid esters. *Chem. Asian J.* 17 (23): e202200861.

126 Giri, C.K., Dana, S., and Baidya, M. (2021). Ruthenium(II)-catalyzed C–H activation and (4+2) annulation of aromatic hydroxamic acid esters with allylic amides. *Chem. Commun.* 57 (81): 10536–10539.

127 Nakanowatari, S. and Ackermann, L. (2015). Ruthenium(II)-catalyzed C–H functionalizations with allenes: versatile allenylations and allylations. *Chem. Eur. J.* 21 (45): 16246–16251.

128 Chowdhury, D., Koner, M., Ghosh, S., and Baidya, M. (2022). Regioselective annulation of allenylphosphine oxides with aromatic amides under ruthenium(II) catalysis. *Org. Lett.* 24 (20): 3604–3608.

129 Leitch, J.A. and Frost, C.G. (2017). Ruthenium-catalysed σ-activation for remote *meta*-selective C–H functionalisation. *Chem. Soc. Rev.* 46: 7145–7153.

130 Activation, B. and Korvorapun, K. (2021). Remote C–H functionalizations by ruthenium catalysis. *Synthesis* 53: A-X.

131 Zhang, Y., Shi, B., and Yu, J. (2009). Pd (II)-catalyzed olefination of electron-deficient arenes using 2, 6-dialkylpyridine ligands. *J. Am. Chem. Soc.* 131 (14): 5072–5074.

132 Ye, J. and Lautens, M. (2015). Palladium-catalysed norbornene-mediated C–H functionalization of arenes. *Nat. Chem.* 7 (11): 863–870.

133 Shi, H., Herron, A.N., Shao, Y. et al. (2018). Enantioselective remote *meta*-C–H arylation and alkylation via a chiral transient mediator. *Nature* 558 (7711): 581–585.

134 Cho, J.Y., Tse, M.K., Holmes, D. et al. (2002). Remarkably selective iridium catalysts for the elaboration of aromatic C–H bonds. *Science* 295 (5553): 305–308.

135 Mkhalid, I.A.I., Barnard, J.H., Marder, T.B. et al. (2010). C–H activation for the construction of C-B bonds. *Chem. Rev.* 110 (2): 890–931.

136 Mihai, M.T., Williams, B.D., and Phipps, R.J. (2019). *Para*-selective C–H borylation of common arene building blocks enabled by ion-pairing with a bulky countercation. *J. Am. Chem. Soc.* 141 (39): 15477–15482.

137 Leow, D., Li, G., Mei, T.S., and Yu, J.Q. (2012). Activation of remote *meta*-C–H bonds assisted by an end-on template. *Nature* 486 (7404): 518–522.

138 Lee, S., Lee, H., and Tan, K.L. (2013). *Meta*-selective C–H functionalization using a nitrile-based directing group and cleavable Si-tether. *J. Am. Chem. Soc.* 135 (50): 18778–18781.

139 Bag, S., Patra, T., Modak, A. et al. (2015). Remote *para*-C–H functionalization of arenes by a D-shaped biphenyl template-based assembly. *J. Am. Chem. Soc.* 137 (37): 11888–11891.

140 Bera, M., Maji, A., Sahoo, S.K., and Maiti, D. (2015). Palladium(II)-catalyzed *meta*-C–H olefination: constructing multisubstituted arenes through homo-diolefination and sequential hetero-diolefination. *Angew. Chem. Int. Ed.* 54 (29): 8515–8519.

141 Kuninobu, Y., Ida, H., Nishi, M., and Kanai, M. (2015). A *meta*-selective C–H borylation directed by a secondary interaction between ligand and substrate. *Nat. Chem.* 7 (9): 712–717.

142 Clark, G.R., Rickard, C.E.F., Roper, W.R. et al. (1996). Reactions of functionalised aryl groups σ-bound to ruthenium(II). *Inorganica Chim. Acta* 251 (1-2 PART II): 65–74.

143 Coudret, C., Fraysse, S., and Launay, J.P. (1998). Electrophilic functionalization of a cyclometallated ruthenium complex, an easy entry to new organometallic synthons. *Chem. Commun.* 6: 663–664.

144 Ackermann, L., Hofmann, N., and Vicente, R. (2011). Carboxylate-assisted ruthenium-catalyzed direct alkylations of ketimines. *Org. Lett.* 13 (7): 1875–1877.

145 Saidi, O., Marafie, J., Ledger, A.E.W. et al. (2011). Ruthenium-catalyzed *meta* sulfonation of 2-phenylpyridines. *J. Am. Chem. Soc.* 133 (48): 19298–19301.

146 Hofmann, N. and Ackermann, L. (2013). *Meta*-selective C–H bond alkylation with secondary alkyl halides. *J. Am. Chem. Soc.* 135 (15): 5877–5884.

147 Li, J., Warratz, S., Zell, D. et al. (2015). N-acyl amino acid ligands for ruthenium(II)-catalyzed *meta*-C–H tert-alkylation with removable auxiliaries. *J. Am. Chem. Soc.* 137 (43): 13894–13901.

148 Paterson, A.J., St John-Campbell, S., Mahon, M.F. et al. (2015). Catalytic *meta*-selective C–H functionalization to construct quaternary carbon centres. *Chem. Commun.* 51 (64): 12807–12810.

149 Marcé, P., Paterson, A.J., Mahon, M.F., and Frost, C.G. (2016). Mechanistic insight into ruthenium catalysed: *meta* -sulfonation of 2-phenylpyridine. *Cat. Sci. Technol.* 6 (19): 7068–7076.

150 Teskey, C.J., Lui, A.Y.W., and Greaney, M.F. (2015). Ruthenium-catalyzed *meta*-selective C-H bromination. *Angew. Chem. Int. Ed.* 54 (40): 11677–11680.

151 Yu, Q., Hu, L., Wang, Y. et al. (2015). Directed *meta*-selective bromination of arenes with ruthenium catalysts. *Angew. Chem. Int. Ed.* 54 (50): 15284–15288.

152 Fan, Z., Ni, J., and Zhang, A. (2016). *Meta*-selective C_{Ar}–H nitration of arenes through a $Ru_3(CO)_{12}$-catalyzed *ortho*-metalation strategy. *J. Am. Chem. Soc.* 138 (27): 8470–8475.

153 Li, J., Korvorapun, K., De Sarkar, S. et al. (2017). Ruthenium(II)-catalysed remote C–H alkylations as a versatile platform to *meta*-decorated arenes. *Nat. Commun.* 8 (Ii): 15430.

154 Li, G., Lv, X., Guo, K. et al. (2017). Ruthenium-catalyzed *meta*-selective C–H sulfonation of azoarenes with arylsulfonyl chlorides. *Org. Chem. Front.* 4 (6): 1145–1148.

155 Li, G., Ma, X., Jia, C. et al. (2017). Ruthenium-catalyzed *meta/ortho*-selective C–H alkylation of azoarenes using alkyl bromides. *Chem. Commun.* 53 (7): 1261–1264.

156 Li, G., Gao, P., Lv, X. et al. (2017). Synthesis of *m*-alkylphenols via a ruthenium-catalyzed C–H bond functionalization of phenol derivatives. *Org. Lett.* 19 (10): 2682–2685.

157 Choi, I., Müller, V., Wang, Y. et al. (2020). Recyclable ruthenium catalyst for distal *meta*-C–H activation. *Chem. Eur. J.* 26 (66): 15290–15297.

158 Wei, W., Yu, H., Zangarelli, A., and Ackermann, L. (2021). Deaminative *meta*-C–H alkylation by ruthenium(ii) catalysis. *Chem. Sci.* 12 (23): 8073–8078.

159 Wu, J., Kaplaneris, N., Pöhlmann, J. et al. (2022). Remote C–H glycosylation by ruthenium(II) catalysis: modular assembly of *meta*-C-aryl glycosides. *Angew. Chem. Int. Ed.* 61 (42): e202208620.

160 Bai, P.B., Durie, A., Wang, G.W., and Larrosa, I. (2024). Unlocking regioselective *meta*-alkylation with epoxides and oxetanes via dynamic kinetic catalyst control. *Nat. Commun.* 15 (1): 1–9.

161 Lan, H., Liu, Y., Ackermann, L. et al. (2024). Ruthenium(II)-catalyzed remote C–H alkylation of arenes using diverse *N*-directing groups through aziridine ring opening. *Org. Lett.* 26 (38): 4–9.

162 Leitch, J.A., McMullin, C.L., Mahon, M.F. et al. (2017). Remote C6-selective ruthenium-catalyzed C–H alkylation of indole derivatives via σ-activation. *ACS Catal.* 7 (4): 2616–2623.

163 Chen, X., Gülen, H.C., Wu, J. et al. (2023). Close-Shell reductive elimination versus open-shell radical coupling for site-selective ruthenium-catalyzed C–H activations by computation and experiments. *Angew. Chem. Int. Ed.* 62 (22): e202302021.

164 Wang, X.G., Li, Y., Liu, H.C. et al. (2019). Three-component ruthenium-catalyzed direct *meta*-selective C–H activation of arenes: a new approach to the alkylarylation of alkenes. *J. Am. Chem. Soc.* 141 (35): 13914–13922.

165 Gou, X.Y., Li, Y., Luan, Y.Y. et al. (2021). Ruthenium-catalyzed radical cyclization/ *meta*-selective C–H alkylation of arenes via σ-activation strategy. *ACS Catal.* 11 (7): 4263–4270.

166 Wu, J., Wei, W., Pöhlmann, J. et al. (2023). Domino *meta*-C–H ethyl glycosylation by ruthenium(II/III) catalysis: modular assembly of *meta*-C-alkyl glycosides. *Angew. Chem. Int. Ed.* 62 (20): 1–5.

167 Luan, Y.Y., Gou, X.Y., Shi, W.Y. et al. (2022). Three-component ruthenium-catalyzed *meta*-C–H alkylation of phenol derivatives. *Org. Lett.* 24 (5): 1136–1140.

168 Chen, S., Yuan, B., Wang, Y., and Ackermann, L. (2023). Ruthenium-catalyzed remote difunctionalization of nonactivated alkenes for double *meta*-C(sp^2)–H/C-6(sp3)–H functionalization. *Angew. Chem. Int. Ed.* 62 (26): 1–7.

169 Neogi, S., Bhunya, S., Kumar Ghosh, A. et al. (2024). Silylarylation of alkenes via meta-selective C–H activation of arenes under ruthenium/iron cooperative catalysis: mechanistic insights from combined experimental and computational studies. *ACS Catal.* 14 (7): 4510–4522.

170 Luo, X., Hou, P., Shen, J. et al. (2024). Ligand-enabled ruthenium-catalyzed *meta*-C–H alkylation of (hetero)aromatic carboxylic acids. *Nat. Commun.* 15 (1): 1–10.

171 Korvorapun, K., Kaplaneris, N., Rogge, T. et al. (2018). Sequential *meta*–/*ortho*-C-H functionalizations by one-pot ruthenium(II/III) catalysis. *ACS Catal.* 8 (2): 886–892.

172 Leitch, J.A., McMullin, C.L., Paterson, A.J. et al. (2017). Ruthenium-catalyzed *para*-selective C–H alkylation of aniline derivatives. *Angew. Chem. Int. Ed.* 56 (47): 15131–15135.

173 Wang, X.G., Li, Y., Zhang, L.L. et al. (2018). Ruthenium(ii)-catalyzed selective C–H difluoroalkylation of aniline derivatives with pyrimidyl auxiliaries. *Chem. Commun.* 54 (68): 9541–9544.

174 Yuan, C., Zhu, L., Chen, C. et al. (2018). Ruthenium(II)-enabled *para*-selective C–H difluoromethylation of anilides and their derivatives. *Nat. Commun.* 9 (1): 1–10.

175 Yuan, C., Zhu, L., Zeng, R. et al. (2018). Ruthenium(II)-catalyzed C–H difluoromethylation of ketoximes: tuning the regioselectivity from the *meta* to the *para* position. *Angew. Chem. Int. Ed.* 57 (5): 1277–1281.

176 Cheng, Y., He, Y., Zheng, J. et al. (2021). Ruthenium(II)-catalyzed *para*-selective C–H difluoroalkylation of aromatic aldehydes and ketones using transient directing groups. *Chin. Chem. Lett.* 32 (4): 1437–1441.

177 Gandeepan, P., Koeller, J., Korvorapun, K. et al. (2019). Visible-light-enabled ruthenium-catalyzed *meta*-C–H alkylation at room temperature. *Angew. Chem. Int. Ed.* 58 (29): 9820–9825.

178 Sagadevan, A., Charitou, A., Wang, F. et al. (2020). *Ortho* C-H arylation of arenes at room temperature using visible light ruthenium C-H activation. *Chem. Sci.* 11 (17): 4439–4443.

179 Wang, Y., Chen, S., Chen, X. et al. (2022). Photo-induced ruthenium-catalyzed double remote C(sp^2)–H / C(sp^3)–H functionalizations by radical relay. *Angew. Chem. Int. Ed.* 61 (32): 2–9.

180 Korvorapun, K., Struwe, J., Kuniyil, R. et al. (2020). Photo-induced ruthenium-catalyzed C–H arylations at ambient temperature. *Angew. Chem. Int. Ed.* 59 (41): 18103–18109.

181 Struwe, J., Korvorapun, K., Zangarelli, A., and Ackermann, L. (2021). Photo-induced ruthenium-catalyzed C–H benzylations and allylations at room temperature. *Chem. Eur. J.* 27 (65): 16237–16241.

182 Michiyuki, T., Maksso, I., and Ackermann, L. (2024). Photo-induced ruthenium-catalyzed C–H arylation polymerization at ambient temperature. *Angew. Chem. Int. Ed.* 63 (25): e202400845.

183 Cheng, Y., Zhang, X., An, G. et al. (2023). Visible-light-enabled ruthenium-catalyzed *para*-C–H difluoroalkylation of anilides. *Chin. Chem. Lett.* 34 (3): 3–7.

184 Lv, X., Cheng, Y., Zong, Y. et al. (2023). Visible-light-mediated ruthenium-catalyzed *para*-selective alkylation of unprotected anilines. *ACS Catal.* 13 (11): 7310–7321.

185 Trost, B. (1991). The atom economy—a search for synthetic efficiency. *Science* 254 (5037): 1471–1477.

186 Matlin, S.A., Mehta, G., Hopf, H., and Krief, A. (2016). One-world chemistry and systems thinking. *Nat. Chem.* 8 (5): 393–398.

187 Anastas, P.T. and Zimmerman, J.B. (2016). The molecular basis of sustainability. *Chem* 1 (1): 10–12.

188 Frontana-Uribe, B.A., Little, R.D., Ibanez, J.G. et al. (2010). Organic electrosynthesis: a promising green methodology in organic chemistry. *Green Chem.* 12 (12): 2099.

189 Horn, E.J., Rosen, B.R., and Baran, P.S. (2016). Synthetic organic electrochemistry: an enabling and innately sustainable method. *ACS Cent. Sci.* 2 (5): 302–308.

190 Sauermann, N., Meyer, T.H., Qiu, Y., and Ackermann, L. (2018). Electrocatalytic C–H activation. *ACS Catal.* 8 (8): 7086–7103.

191 Meyer, T.H., Finger, L.H., Gandeepan, P., and Ackermann, L. (2019). Resource economy by metallaelectrocatalysis: merging electrochemistry and C—H activation. *Trends Chem.* 1 (1): 63–76.

192 Gandeepan, P., Finger, L.H., Meyer, T.H., and Ackermann, L. (2020). 3d metallaelectrocatalysis for resource economical syntheses. *Chem. Soc. Rev.* 49 (13): 4254–4272.

193 Samanta, R.C., Meyer, T.H., Siewert, I., and Ackermann, L. (2020). Renewable resources for sustainable metallaelectro-catalysed C–H activation. *Chem. Sci.* 11 (33): 8657–8670.

194 Hilt, G. (2020). Basic strategies and types of applications in organic electrochemistry. *ChemElectroChem* 7 (2): 395–405.

195 Jiao, K.-J., Xing, Y.-K., Yang, Q.-L. et al. (2020). Site-selective C–H functionalization via synergistic use of electrochemistry and transition metal catalysis. *Acc. Chem. Res.* 53 (2): 300–310.

196 Heard, D.M. and Lennox, A.J.J. (2020). Electrode materials in modern organic electrochemistry. *Angew. Chem. Int. Ed.* 59 (43): 18866–18884.

197 Little, R.D. (2020). A perspective on organic electrochemistry. *J. Org. Chem.* 85 (21): 13375–13390.

198 Novaes, L.F.T., Liu, J., Shen, Y. et al. (2021). Electrocatalysis as an enabling technology for organic synthesis. *Chem. Soc. Rev.* 50 (14): 7941–8002.

199 Zhu, C., Ang, N.W.J., Meyer, T.H. et al. (2021). Organic electrochemistry: molecular syntheses with potential. *ACS Cent. Sci.* 7 (3): 415–431.

200 Samanta, R.C. and Ackermann, L. (2021). Evolution of earth-abundant 3 d-metallaelectro-catalyzed C–H activation: from chelation-assistance to C–H functionalization without directing groups. *Chem. Rec.* 21 (9): 2430–2441.

201 Rein, J., Zacate, S.B., Mao, K., and Lin, S. (2023). A tutorial on asymmetric electrocatalysis. *Chem. Soc. Rev.* 52 (23): 8106–8125.

202 Lodh, J., Paul, S., Sun, H. et al. (2023). Electrochemical organic reactions: a tutorial review. *Front. Chem.* 10: 956502.

203 Li, Y., Dana, S., and Ackermann, L. (2023). Recent advances in organic electrochemical functionalizations for specialty chemicals. *Curr. Opin. Electrochem.* 40: 101312.

204 Wang, Y., Dana, S., Long, H. et al. (2023). Electrochemical late-stage functionalization. *Chem. Rev.* 123 (19): 11269–11335.

205 Wang, Y., Zhao, R., and Ackermann, L. (2023). Electrochemical syntheses of polycyclic aromatic hydrocarbons (PAHs). *Adv. Mater.* 2300760.

206 Hyung Kweon, D., Jeon, I.-Y., and Baek, J.-B. (2021). Electrochemical catalysts for green hydrogen energy. *Adv. Energy Sustainability Res.* 2 (7): 2100019.

207 Boretti, A. (2021). There are hydrogen production pathways with better than green hydrogen economic and environmental costs. *Int. J. Hydrog. Energy* 46 (46): 23988–23995.

208 Chen, G., Li, X., and Feng, X. (2022). Upgrading organic compounds through the coupling of electrooxidation with hydrogen evolution. *Angew. Chem. Int. Ed.* 61 (42): e202209014.

209 Qiu, Y., Tian, C., Massignan, L. et al. (2018). Electrooxidative ruthenium-catalyzed C–H/O–H annulation by weak *O* -coordination. *Angew. Chem. Int. Ed.* 57 (20): 5818–5822.

210 Tan, X., Hou, X., Rogge, T., and Ackermann, L. (2021). Ruthenaelectro-catalyzed domino three-component alkyne annulation for expedient isoquinoline assembly. *Angew. Chem. Int. Ed.* 60 (9): 4619–4624.

211 Yang, L., Steinbock, R., Scheremetjew, A. et al. (2020). Azaruthena(II)-bicyclo[3.2.0]heptadiene: key intermediate for ruthenaelectro (II/III/I)-catalyzed alkyne annulations. *Angew. Chem. Int. Ed.* 59 (27): 11130–11135.

212 Mei, R., Koeller, J., and Ackermann, L. (2018). Electrochemical ruthenium-catalyzed alkyne annulations by C–H/Het–H activation of aryl carbamates or phenols in protic media. *Chem. Commun.* 54 (91): 12879–12882.

213 Hou, X., Kaplaneris, N., Yuan, B. et al. (2022). Ruthenaelectro-catalyzed C–H acyloxylation for late-stage tyrosine and oligopeptide diversification. *Chem. Sci.* 13 (12): 3461–3467.

214 Gou, X.-Y., Oliveira, J.C.A., Chen, S. et al. (2024). Ruthenaelectro-catalyzed C–H phosphorylation: *ortho* to *para* position-selectivity switch. *Chem. Sci.* https://doi.org/10.1039/d4sc06219a.

215 Wang, Y., Simon, H., Chen, X. et al. (2022). Distal ruthenaelectro-catalyzed *meta* -C–H bromination with aqueous HBr. *Angew. Chem. Int. Ed.* 61 (20): e202201595.

216 Kenyon, J. and Larrosa, I. (2024). Electrochemical ruthenium-catalysed directed C–H functionalization of arenes with boron reagents. *Adv. Synth. Catal.* 366 (4): 830–837.

10

Emerging Metal-free Directed C—H Functionalization

Rahul Bangari and Supriya Rej

Department of Biological and Synthetic Chemistry, Centre of Biomedical Research (CBMR), Sanjay Gandhi Post-Graduate Institute of Medical Sciences Campus, Raebareli Road, Lucknow, Uttar Pradesh 226014, India

10.1 Introduction

C—H bond functionalization offers a powerful tool in synthetic organic chemistry due to its atom efficiency and regioselective nature [1, 2]. The functionalization of a particular C—H bond is always a challenging task, since C—H bonds are abundant in organic compounds. In this regard, the directing strategy plays a crucial role, as it helps to target a selective C—H bond close proximal to it and maintain the regioselectivity [3, 4]. Notably, transition-metal catalytic-directed C—H functionalization strategies provide promising results with high site selectivity [5–7]. However, these protocols face major drawbacks, such as high cost, low impurity tolerance, requirement of precious metals, and harsh conditions [8, 9]. Most importantly, the leaching of metal catalysts can lead to the parts per million (ppm) level of metal contamination in the organic compound, which would be difficult to eliminate or detect. Hence, metal-catalytic protocols need a very stringent isolation procedure to confirm the hundred percent metal elimination from the final organic compound in a multi-step synthetic process or at least for synthesized medicinally relevant and drug compounds. Therefore, organic synthetic chemists primarily focus on developing metal-free C—H bond functionalization protocols, which would at least eliminate the chance of metal contamination in the product. Unlike metal-catalytic protocols, transition-metal-free C—H bond functionalization is indeed challenging. It has mostly been done by activating the C—H bond oxidatively (Scheme 10.1a) [10–15] or via an aromatic substitution process for an aromatic substrate (Scheme 10.1b) [16, 17]. In the former mentioned strategy, majorly C—N or C—O bond formation reactions were witnessed in the presence of an oxidant, whereas in latter cases, electrophilic aromatic substitution followed by proton abstraction is the step involved during the process. Details of

Directed C–H Bond Functionalization: Concepts and Applications, First Edition.
Edited by Debabrata Maiti and Supriya Rej.

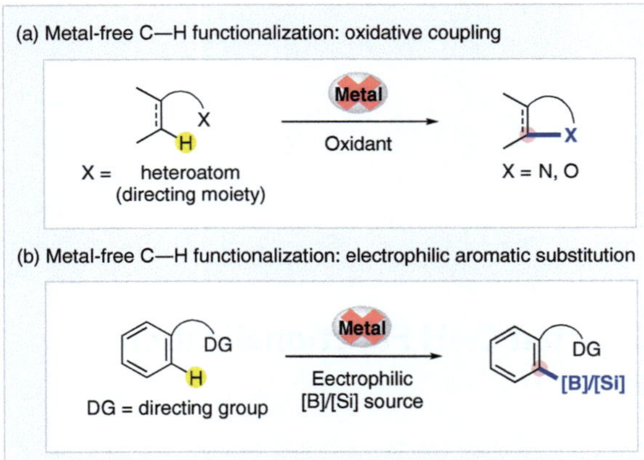

Scheme 10.1 Strategies for metal-free C–H bond functionalization via (a) radical pathways or (b) by electrophilic aromatic substitution.

these methods for regioselective metal-free C—H bond functionalization have been discussed in this chapter.

10.2 Metal-free Directed Oxidative C–H Functionalization

Oxidative C—H bond functionalization mainly forms a carbon–carbon or carbon–heteroatom bond directly by activating the C—H bond via a coupling reaction known as a cross-dehydrogenative (CDC) coupling reaction. An external oxidant is necessary for this reaction [15]. In this regard, various metal-based oxidants are well known; however, their uses are always questionable as stoichiometric metal salt generates as waste and chances of metal leaching are always there in this kind of reaction. Considering the recent focus on green synthesis, synthetic chemists are trying to adhere to the principle of green chemistry. This led to the use of metal-free oxidants such as DDQ, *tert*-butyl hydroperoxide, hydrogen peroxide, and even molecular oxygen. In oxidative C—H bond functionalization in the presence of an oxidant, a carbon-based free radical is generated by activating the C—H bond, which takes part in the CDC reaction. In some cases, a cationic intermediate can also form, which is subsequently attacked by nucleophilic heteroatom to form the carbon–heteroatom bond. In directed oxidative C—H bond functionalization at first heteroatom moiety (as shown in Scheme 10.2) gets activated by the oxidant leading to either radical or cationic species (**1** or **2**). Subsequently, the proximal C—H bond has been replaced via a radical pathway or by an electrophilic attack. Directed metal-free oxidative C—H bond functionalization majorly provides C—N and C—O bond formation, which are discussed section-wise below.

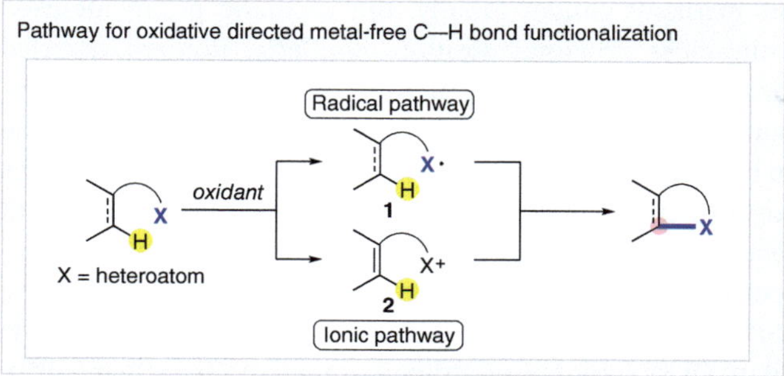

Scheme 10.2 Pathways for metal-free directed oxidative C–H bond functionalization.

10.3 Directed C–N Bond Formation

10.3.1 C(sp^2)–N Bond Formation

Directed oxidative C(sp^2)—N bond is formed mostly via the activation of nitrogen atoms by the metal-free oxidant, followed by an electrophilic attack by the proximal arene or alkenyl moiety (as shown in Scheme 10.2, bottom). In some cases, a radical pathway also followed, leading to the nitrogen-based radical. Later on, having an electrophilic nature N-based radical has been attacked by proximal arene moiety to form a C(sp^2)—N bond (as shown in Scheme 10.2, top). In 1990, Kikugwa et al. first reported an intramolecular directed C—N bond formation of *N*-methoxyamide derivatives using a hypervalent PhI(OCOCF$_3$)$_2$ in stoichiometric oxidant (Scheme 10.3a) [18]. The reaction led to the formation of indoline or tetrahydroquinoline derivatives (**4**). It was anticipated in the proposed mechanism that an electron-deficient nitrenium cation would initially generate from the

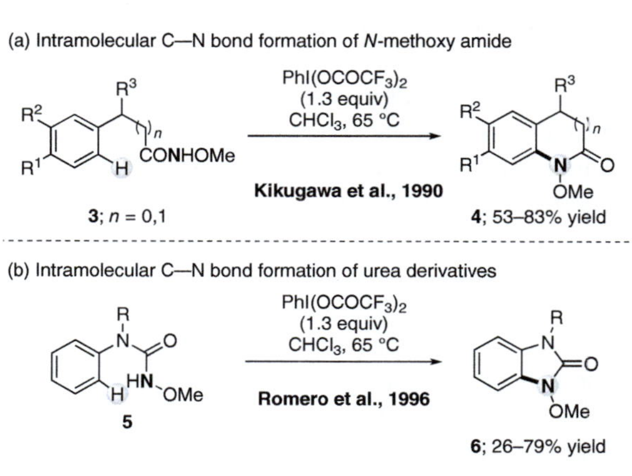

Scheme 10.3 Early reports on intramolecular C–N bond formation of *N*-methoxy amide derivatives.

reaction of oxidant and methoxy amide. Finally, the aromatic group attacked it intramolecularly via the close proximal position. Similarly, Romero reported that intramolecular oxidative C—N coupling of urea derivative **5** in the presence of a hypervalent iodine reagent leads to the formation of 2-benzimidazolinone **6** (Scheme 10.3b) [19]. In both of the reactions, *N*-methoxy-substituted substrate has been used as the neighboring methoxy group that can stabilize the electron-deficient nitrenium cation intermediate.

Following the same path, several different amide substrate applicabilities have been explored for directed intramolecular C(sp^2)—N bond formation using hypervalent iodine as an oxidant in stoichiometric ratio, summarized in Scheme 10.4. Importantly, this transformation offers access to a variety of medicinally or agrochemical significant heterocyclic scaffolds, such as Kikugawa reporting the synthesis of benzoxazolone **9** from *N*-acylaminophthalimides [20]; Fu reporting the oxidative amidation of *N*,*N*′-diarylurea derivatives to obtain *N*,*N*′-diaryl 2-benzimidazolinones **10** [21]; Zhang reporting quinolinediones **11** synthesis from 2,2-disubstituted-2-benzoylacetamides [22] and spirocyclopropyl oxindoles **12** synthesis from secondary cyclopropyl carboxamides [23]; Chang and Antonchick independently reporting the synthesis of carbazole (**15**) [24] or hyellazole (**16**) [25] scaffolds by intramolecular amidation of *N*-substituted amidobiphenyl or 2,6-diarylated acetanilide derivatives, respectively; Mal also reporting the carbazole (**17**) synthesis from biaryl sulfonamides using similar conditions [26]; 1,4-benzodiazepin-2-ones (**18**) [27], pyrrolo[2,1-*c*]-[1,4]-diazepin-5,11-diones (**19**) [28], indeno-1,4-diazepinones (**20**) [29], 1,4-benzodiazepine (**21**) [30], dibenzoxazepinone (**22**) [31], indoloquinolinones (**23** and **24**) [32, 33], quinolino[3,4-*b*]-quinoxalin-6(5*H*)-ones (**25**) [34], phenanthridinones (**26**) [35], and 2-bromo phenanthridinones (**27**) [36] that were synthesized similarly via intramolecular directed amination at the proximal position using a hypervalent iodine reagent.

Apart from the use of $PhI(OAc)_2$ or $PhI(OCOCF_3)_2$, oxone was also used successfully as an oxidant for the *in situ* generation of hypervalent iodine reagent in the presence of phenyl iodide; hence, C—H amination of aryl hydrazine dicarboxylates proceeds smoothly leading to benzo[*c*]cinnoline derivatives **28** (Scheme 10.4) [37]. In 2011, Antonchick introduced a catalytic system for intramolecular oxidative C—N amination of *N*-substituted aminobiphenyl derivatives to synthesize carbazoles **13** [38]. Notably, in the reaction mixture, a hypervalent iodine reagent is generated from the reaction of peracetic acid and 2,2′-diiodo-4,4′,6,6′-tetramethylbiphenyl (**14**), which is responsible for the oxidative amination reaction.

Besides the use of amide as a coupling part, enaminones were also used as substrates for intramolecular oxidative amination reactions using hypervalent iodine reagent leading to the formation of carbazolone derivatives **31** (Scheme 10.5) [39]. Likewise, Mal showed the synthesis of benzimidazole-fused phenanthridines **32** from intramolecular oxidative cyclization of benzimidazole units [40]. Authors anticipated that antiaromatic endocyclic nitrenium ions were involved as an intermediate. Benzimidazoles (**33**, **34**, and **35**) syntheses were achieved by Zhu and co-workers [41], Punniyamurthy and co-workers [42], and Long and

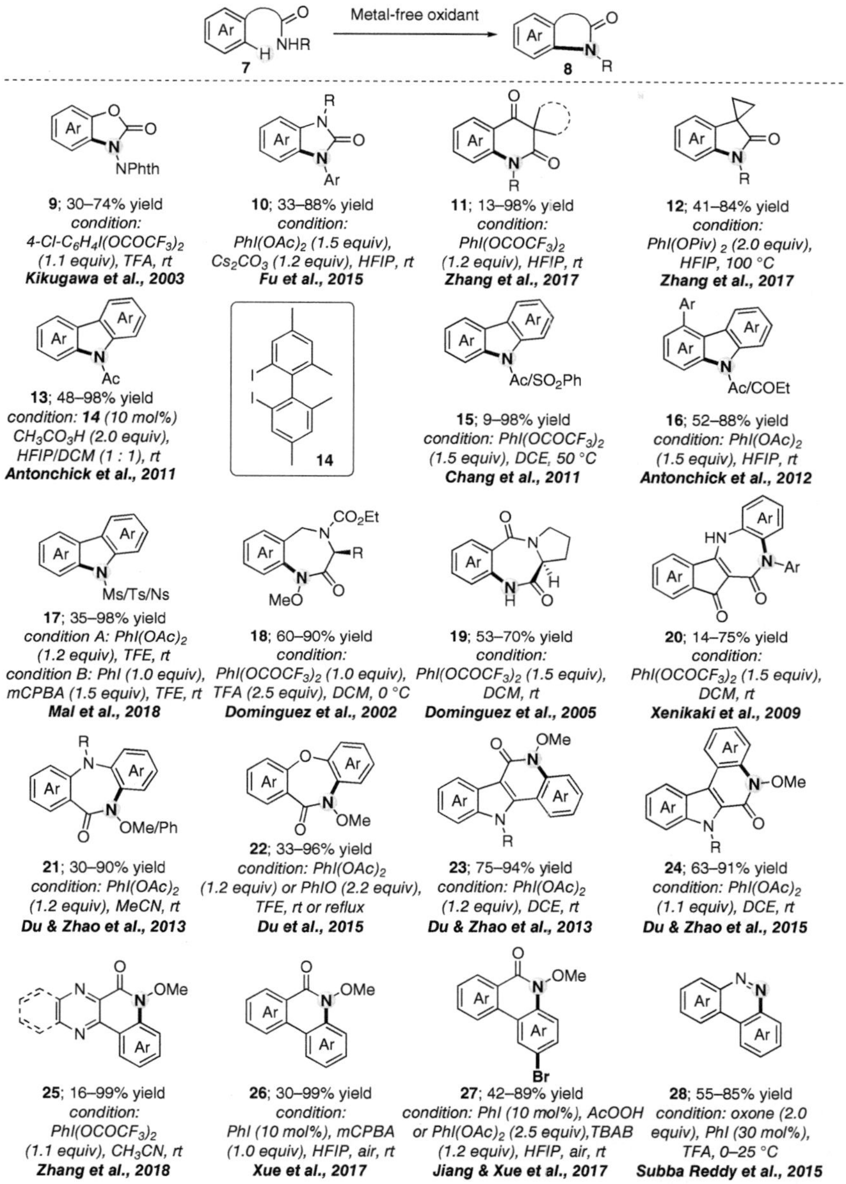

Scheme 10.4 Directed intramolecular C–H amination of various amide derivatives.

co-workers [43], independently using *N*-arylamidines, *N*′-aryl-*N*-tosyl/*N*-aryl/*N*-methylsulfonylamidines, and *N*′-aryl-*N*-alkyl derivatives. Zhang reported 2-amnio benzimidazole **36** synthesis from guanidine via C—H amination [44]. C—H cyclo amination of *N*-arylaminopyridines **37** and **38** was reported independently by Zhu and co-workers [45] and Das and co-workers [46] to construct

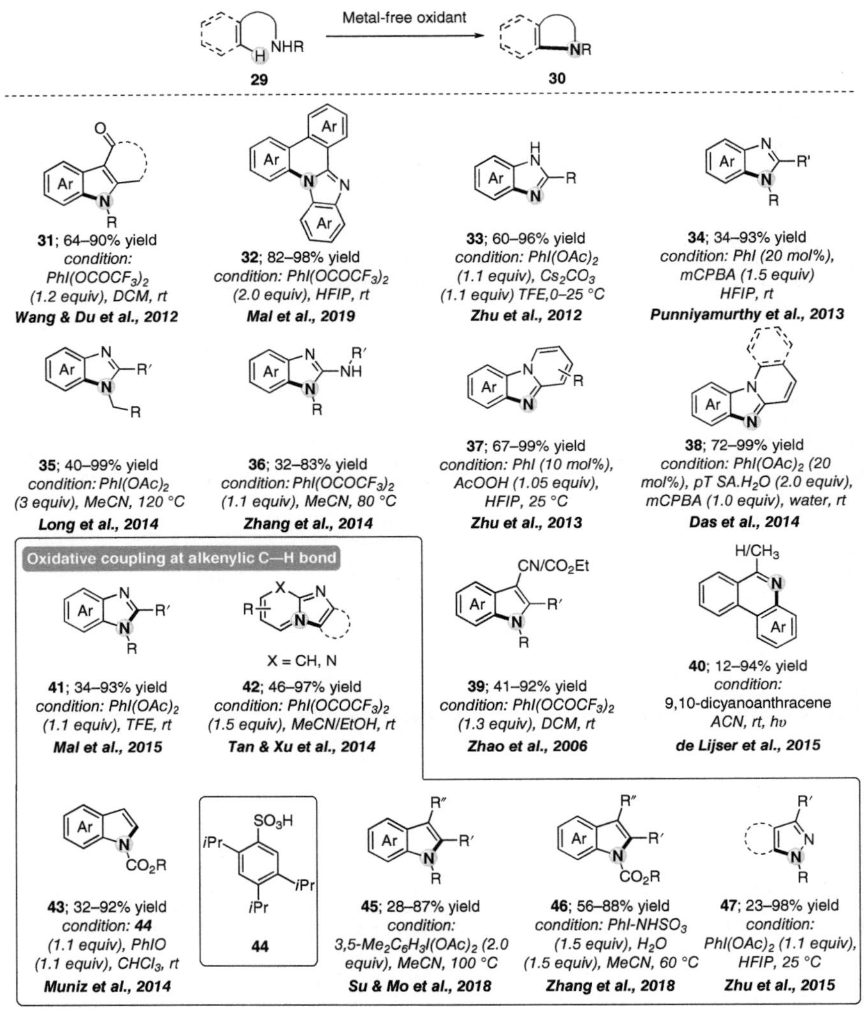

Scheme 10.5 Directed oxidative C–N coupling of amine and aryl/alkenyl C–H bond.

benzimidazole-fused heterocycles. Zhao reported a $PhI(OCOCF_3)_2$-mediated intramolecular cyclization to synthesized *N*-substituted Indole derivatives (**39**) [47]. Synthesis of phenanthridine **40** from 2′-arylbenzaldehyde oxime ethers was reported by de Lijser upon irradiation in the presence of 9,10-dicyanoanthracene [48]. The oxidative C—N bond formation is initiated by an electron transfer followed by the aryl ring's nucleophilic attack on the nitrogen center.

Oxidative coupling reactions of alkenyl C—H bond and nitrogen moiety were also studied to deliver significant heterocyclic moieties. Mal introduced a strategy of benzimidazole (**41**) synthesis from substituted-imine derivatives [49]. Tan and Xu reported the synthesis of imidazo[1,2-*a*]pyrimidines (**42**) from the cyclization of pyrimidinyl enamines [50]. Indoles (**43, 45,** and **46**) syntheses were described by Muniz and co-workers [51], Su/Mo and co-workers [52], and Zhang and co-workers [53] independently from 2-vinyl aniline and 2-alkenyl aniline via activating the alkenyl C—H bond. Zhu showcased an oxidative C—H cyclo amination of vinyl hydrazones to provide pyrazole **47** [54].

10.3.2 C(sp^3)–H Bond Formation

Oxidative C(sp^3)—N coupling mostly occurs through the radical pathway, as shown in Scheme 10.2a (top). At first, an oxidant initiates the reaction by generating a nitrogen-based radical, which activates the close proximal C(sp^3)—H bond, and a cyclized C—N couple product is generated. However, in some cases, ionic-based intermediates are also involved in delivering the oxidative couple product. In 2007, Kikugawa reported an intramolecular benzylic amination of *N*-acylaminophthalimides using PhI(OCOCF$_3$)$_2$ as an oxidant that led to the formation of aryl-substituted lactam derivatives (**50**, Scheme 10.6) [55]. It was proposed that *N*-phthalimido-*N*-acylnitrenium ions generated from *N*-acylaminophthalimides are responsible for activating a C(sp^3)—H bond to obtain the cyclized product. However, in the same year, Fan reported a similar transformation to attain pyrrolidines **51** from δ-C—H bond oxidation of sulfonamides, wherein

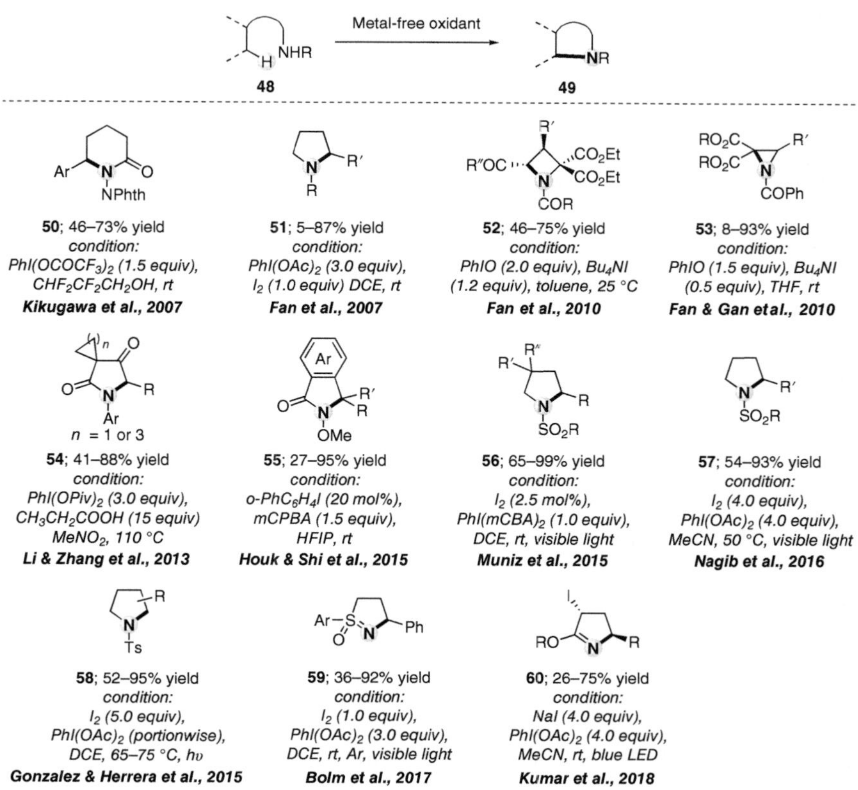

Scheme 10.6 Intramolecular oxidative C–N coupling via activating of C(sp^3)–H bond.

the author proposed that a radical mechanism was involved [56]. The hypervalent iodine reagent is used as a radical precursor to form sulfonamidyl radical, which eventually undergoes a 1,5-*H* shift to produce δ-carbon-based radical. Afterward, the carbon radical oxidized, and the formed carbocation was attacked by an intramolecular nitrogen atom to give the pyrrolidine derivatives.

Likewise, by using the hypervalent iodine-based reagent as an oxidant, several important saturated heterocyclic structural motifs were obtained, such as azetidine (**52**) [57], *N*-benzoyl aziridine (**53**) [58], tetramic acid (**54**) [59], γ-lactam (**55**, via activating of tertiary C—H bond) [60], 2-substituted *N*-sulfonated pyrrolidine (**56** or **57**) [61, 62], *N*-tosyl pyrrolidine (**58**) [63], dihydroisothiazole oxide (**59**) [64], and 4-Iodo-3,4-dihydropyrrole (**60**) [65]. In all of these cases, oxidation occurred at the proximal C—H position to produce the cyclized C–N-coupled heterocyclic derivatives.

10.4 Directed C—O Bond Formation

10.4.1 C(sp^2)—O Bond Formation

Oxidative C—O bond cyclization via the activation of proximal C—H bonds was also reported, and these too work similarly as described in the C—N bond formation section. In the case of an aromatic substrate (where the C—O bond is formed onto an aromatic C(sp^2)—H bond), an aromatic-based radical cation is expected to generate, which has been attacked by a close proximal nucleophilic oxygen center to yield the cyclized product. In 2004, Kita reported a chroman (**63**) synthesis from an aryl alkyl alcohol in the presence of oxidant $PhI(OCOCF_3)_2$ (Scheme 10.7) [66]. Later, Gu reported the directed oxidative C—O cyclization of 3-arylpropionic acids to yield 3,4-dihydrocoumarins (**64**), in which acid moiety acted as a nucleophile [67]. Du/Zhao and co-workers [68] and Martin and co-workers [69] also independently reported the coumarin derivative (**65** or **66**) synthesis via oxidative C—O cyclization using either stoichiometric hypervalent iodine reagent or a combination of catalytic aryl iodide and peracetic acid. Wirth reported a $PhI(OAc)_2$-mediated intramolecular cyclization of *ortho*-hydroxystilbenes to yield benzo or naphthofurans (**67**) [70]. The authors proposed that the iodine-based oxidant activates the double bond of stilbene to form an electrophilic three-membered iodonium species, which is further intramolecularly attacked by phenolic oxygen to deliver the cyclized product.

The oxygen atom of the carbonyl moiety can even participate in oxidative C—O bond cyclization, as shown in Scheme 10.8. Lupton and Hutt reported hypervalent iodine-catalyzed C—O cyclization to yield benzopyran derivatives (**71**) from the vinylogous ester bearing a benzyl group (**70**) [71]. Krishna demonstrated the synthesis of oxazolines (**73**) or oxazines (**74**) from *N*-allylamides (**72**) in the presence of the stoichiometric oxidant. Aryl-substituted *N*-allylamides provide the six-membered oxazines (**74**), whereas when no substitution is there, five-membered oxazolines (**73**) are formed [72]. Similarly, benzoxazole (**76**) [73] and oxazole (**78**, **81**, and **83**) [74–76] were synthesized by direct oxidative coupling of the C(sp^2)—H bond and an amide oxygen atom using hypervalent iodine reagent as an oxidant.

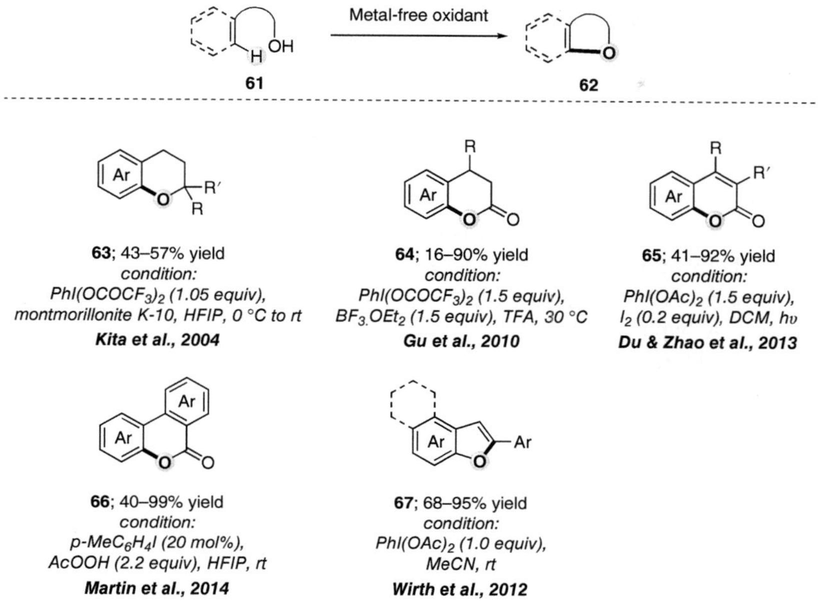

Scheme 10.7 Direct oxidative C—O cyclization via activation of the C(sp^2)—H bond.

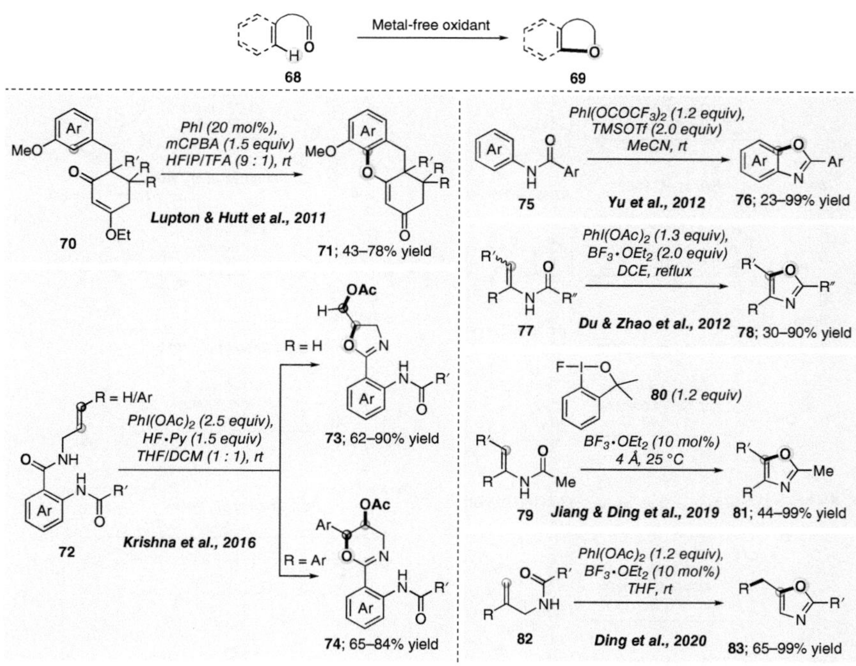

Scheme 10.8 Direct oxidative C—O cyclization of vinylogous ester or amide via activation of the C(sp^2)—H bond.

10.4.2 C(sp^3)–O Bond Formation

Structural editing and synthesizing oxygen-substituted saturated heterocycles were achieved via intramolecular oxidative C(sp^3)—O cyclization reaction. The list of discoveries in this field is summarized in Scheme 10.9. Fan reported a PhI(OAc)$_2$/Bu$_4$NI-mediated tandem acetoxylation followed by oxidative cyclization of *ortho*-acyl phenol (**86**) to offer α-acetoxy benzofurans (**87**) [77]. Tellitu and Dominguez showcased the facile construction of Furopyrimidinones (**89**) via allylic oxycarbonylation of 5-carbamoyl-substituted Biginelli adduct (**88**) [78]. Spirofurooxindoles (**91**) [79] and spiro-2,2′-benzo[*b*]furan-3,3′-ones (**93**) [80] were obtained by hypervalent iodine-mediated cascade oxidative C—O/C—C and C—O/C—O bond formation of 3-(2-hydroxyphenyl)-3-oxo-*N*-phenyl propanamides (**90**) and 3-hydroxy-1,3-bis(2-hydroxyphenyl)prop-2-en-1-ones (**92**), respectively. In 2014, Martin and co-workers [69] and Du/Zhao and co-workers [81] independently reported the intramolecular amination at the C(sp^3)—H adjacent to the nitrogen to construct benzoxazinone derivatives (**95** and **97**). Du even unlocked an asymmetric spirocyclization to deliver enantioselective spirofurooxindole **100** via cascade C—O and C—C bond formation reaction [82]. A chiral aryl iodine **99** was used in a catalytic amount that reacted *in situ* in the reaction with *m*CPBA to generate a chiral hypervalent iodine reagent; consequently, enantioselectivity was imposed in the final product.

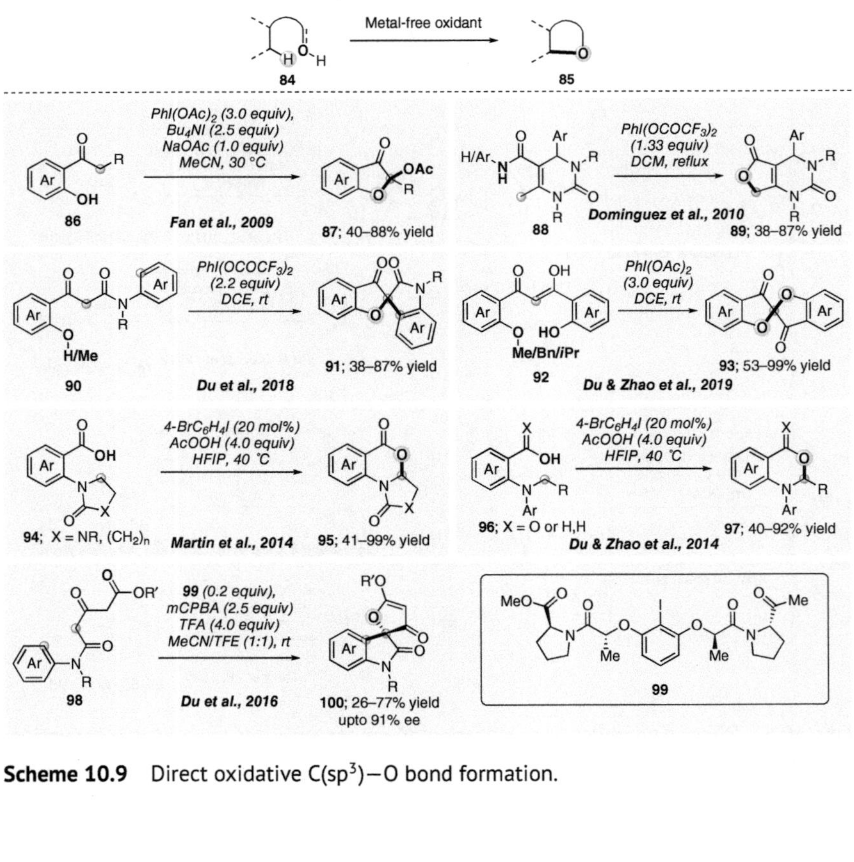

Scheme 10.9 Direct oxidative C(sp^3)–O bond formation.

10.5 Directed C–C Bond Formation

Apart from the C—heteroatom bond formation, directed oxidative coupling of the C—H/C—H bond is introduced for the formation of the C—C bond. In this regard, Wang and Ji showcased the construction of 6-alkyl phenanthridines (**102**) from 2-isocyanobiaryl (**101**) and 1,4-dioxane mediated by *tert*-butyl peroxybenzoate (TBPB) via a sequential C(sp^2)—H/C(sp^3)—H bond functionalization (Scheme 10.10a, top) [83]. In the same year, Cheng reported a similar transformation to construct phenanthridines derivative **103** using cycloalkane as a coupling partner and benzoyl peroxide (BPO) as an oxidant (Scheme 10.10a, bottom) [84]. An intramolecular CDC coupling of aryl aldehyde **104** was reported, leading to the formation of benzocyclic ketone **105** (Scheme 10.10b, top) [85]. The authors anticipated that an iodonium species generated from IPy_2BF_4/HBF_4 activated aldehyde **104**. A similar oxidative C—C bond formation reaction was reported by Du and Zhao via activating the aldehydic C—H bond/aryl C—H bond of 2-(*N*-arylamino)aldehyde (**106**) [86]. This coupling reaction provides biologically important acridone derivatives (**107**, Scheme 10.10b, bottom).

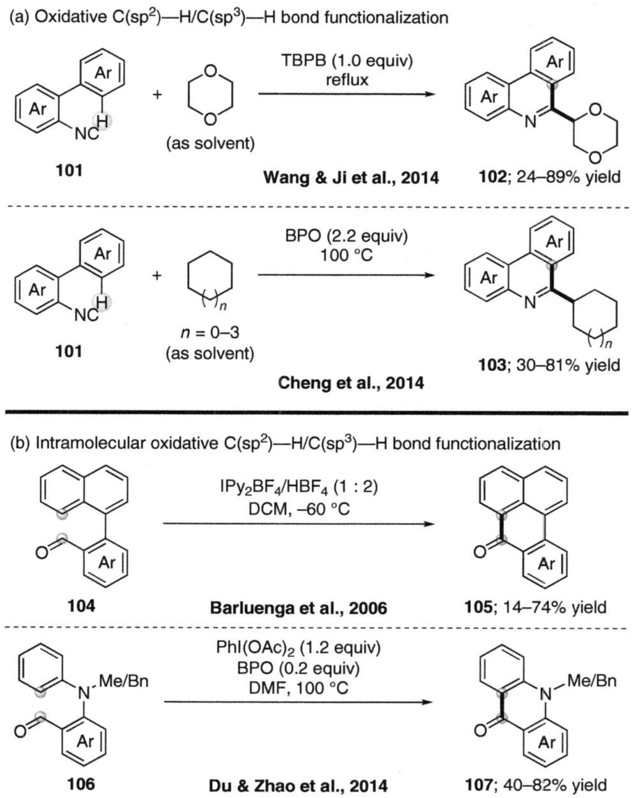

Scheme 10.10 Cyanide and aldehyde group-directed oxidative C–C bond formation.

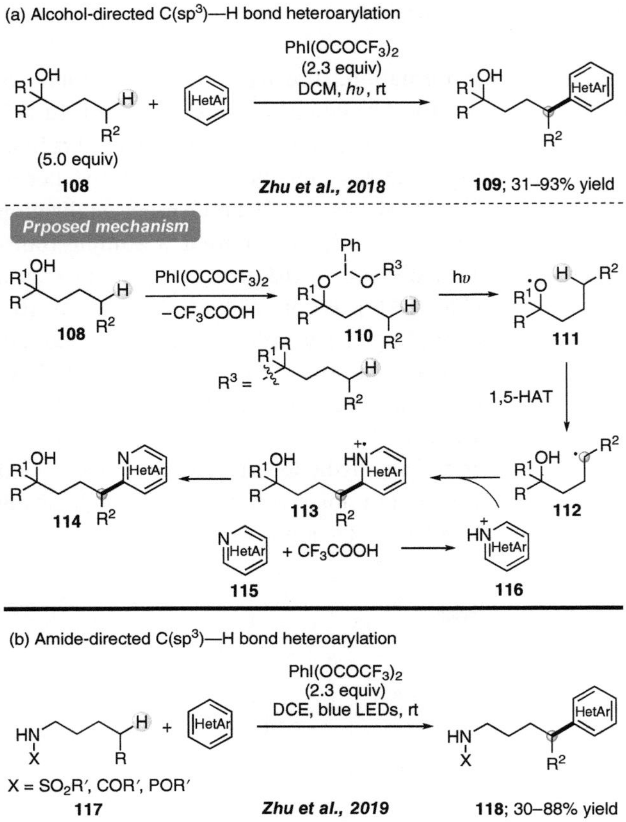

Scheme 10.11 Alcohol and amide-directed intermolecular C(sp^3)−H heteroarylation.

In 2018, Zhu reported an alcohol-directed intermolecular C(sp^3)—H heteroarylation of alcohol derivatives (**108**) under photochemical conditions by using $PhI(OCOCF_3)_2$ as an oxidant (Scheme 10.11a, top) [87]. The proposed mechanism suggested that an alkoxy radical (**111**) generated from an alcohol derivative prompted a regioselective 1,5-hydrogen atom transfer (HAT) and provided a carbon-based radical (**112**); subsequently, C—C coupling undergoes with the heteroarene. In the following year, the same group reported a similar C(sp^3)—H heteroarylation of amide derivatives (**117**) using the direction of NH moiety in the HAT step [88].

10.6 Directed C−H Borylation

10.6.1 C(sp^2)−H Borylation

Organoboron compounds are known as synthetic intermediates, since they have been used for many bond construction chemistry, medicinal compound modification, API chemical synthesis, and natural product synthesis [89, 90]. Transition-metal-catalyzed C—H borylation is a well-known tool to attain organoboron compounds, and many research groups already established several state-of-the-art strategies for this purpose [91–93]. However, precious metals (e.g. Rh, Ru, and Ir) are always required as catalysts, and if there is metal leaching, the contamination of metal cannot be ruled out. These are major setbacks for transition metal-catalyzed C—H borylation, especially for medicinally compound preparation. Therefore, strategies for metal-free C—H borylation are always a welcoming topic for researchers to work on. Considering C—H bond is abundant in organic compounds, it is also necessary to functionalize them selectively. Henceforth, the organic synthetic community came up with a new idea of transition-metal-free selective C—H borylation using a directing auxiliary to control the position of the substitution [17, 94–97]. In this regard, an electrophilic boron reagent is used, which first establishes coordination with the directing moiety and can direct the electrophilic boron center toward a certain position of arene or alkene. Hence, selective electrophilic borylation takes place, as depicted in Scheme 10.12.

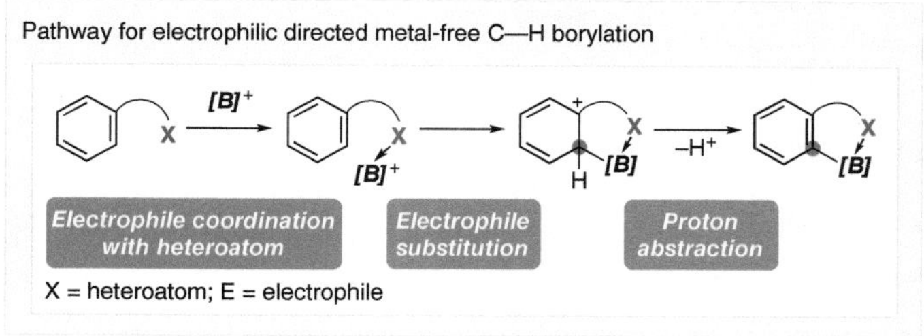

Scheme 10.12 Pathways of directed electrophilic substitution for C–B bond formation.

In 1958, Dewar first reported a -NH_2 group-directed C—H borylation of 2-amino biphenyl (**119**) via an electrophilic aromatic substitution reaction with electrophilic BCl_3 reagent in the presence of $AlCl_3$ (Scheme 10.13a) [98]. The same group also showcased that the direction of the -NH_2 group can be used for other substrates such as 2-amino styrene, 2-ethyl amino biphenyl, and 4-amino phenanthrene to obtain respective boracyclic compounds [99–101]. Secondary and tertiary amines were also used as directors for electrophilic C—H borylation [102–108]. Apart from amine derivatives, alcohol [109–112] or thiol [113] derivatives were also used as the directing auxiliary for obtaining selective electrophilic C—H borylation. As depicted in Scheme 10.13a, biphenyl 2-ol (**121**) can lead to **122** upon reaction with BCl_3 and $AlCl_3$ [109].

In 2009, Vedejs reported an electrophilic *ortho*-C—H borylation of *N*,*N*-dimethyl amine moiety (**123-125**) using the directing efficacy of amine moiety (Scheme 10.13b) [102, 103, 105]. In this reaction, hydroborane works as an electrophilic boron source and $[Ph_3C^+[B(C_6F_5)_4^-]]$ as a reaction mediator. Very recently, Chang and Park reported a pentafluoroarene-substituted borinic acid as a reagent for electrophilic

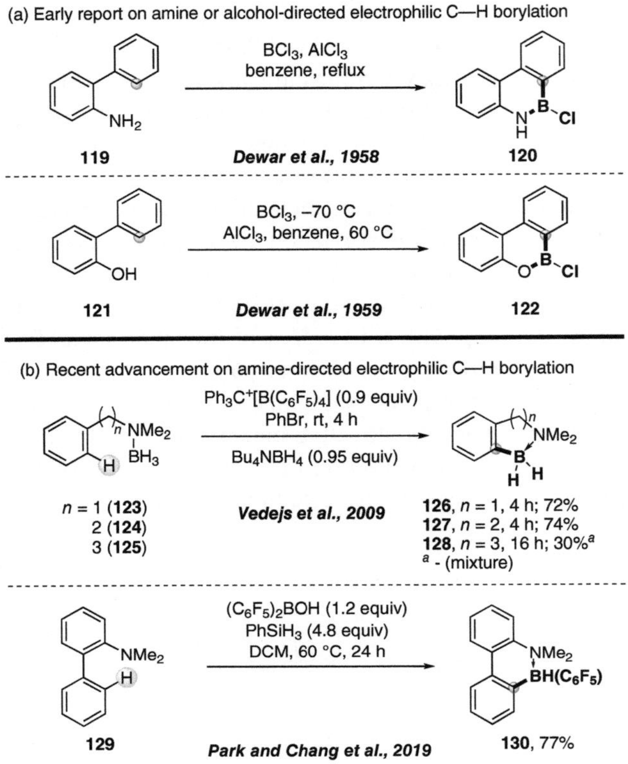

Scheme 10.13 (a) Early reports and (b) recent advancement on directed electrophilic borylation on arene.

C—H borylation of *N,N*-dimethyl amino bipenyl (**129**) (Scheme 10.13b) [107]. It was proposed that $HB(C_6F_5)_2$ is forming *in situ*, which is responsible for this electrophilic borylation.

Besides the functional group ($-NH_2$, –OH, and –SH) direction, heterocyclic moiety-directed electrophilic C—H borylation was studied extensively (Scheme 10.14). In a pioneering report, Letsinger and Maclean reported benzimidazole-directed *ortho*-C—H borylation leading to **133** and **134** using BCl_3 as an electrophilic boron reagent [114]. Ingleson showed that thiazole can be a good directing group for C7–H borylation of indole derivative (**135**) [115]. Recently, an electrophilic *ortho*-C–H borylation of azobenzene (**136**) was reported using BBr_3 [116]. Ingleson and Turner used the metal-free electrophilic C—H borylation method to synthesize donor-acceptor OLED materials (**137** and **138**) using benzothiadiazole as the directing moiety [117–120]. Quinoline moiety can also be used as an auxiliary to direct electrophilic C—H borylation (**139**) [114]. Murakami reported that pyridine-directed electrophilic C—H borylation of 2-phenyl pyridine followed by quenching with triphenyl aluminum yielded a *tetra*-coordinated triaryl borane system **140** [121]. Several other arene systems were borylated using pyridine auxiliary as a directing moiety, for instance, tetraphenyl-ethylene-substituted

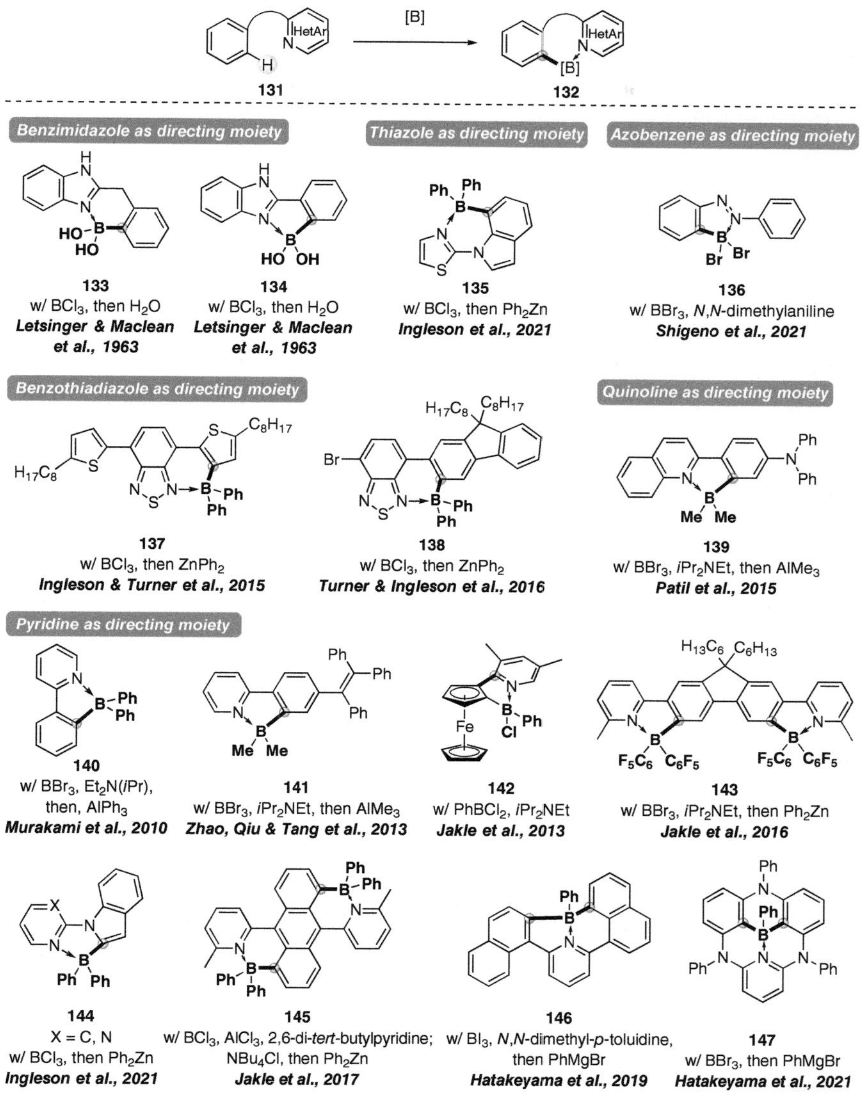

Scheme 10.14 Metal-free electrophilic C–H borylation directed by heterocycles.

phenyl pyridine (**141**) [122], pyridyl ferrocene (**142**) [123], ladder-type organoboron compound (**143**) [124], indole derivative (**144**) [115], anthracene moiety (**145**) [125], and polycyclic aromatic hydrocarbons (**146** & **147**) [126, 127]. Marder also reported an electrophilic C—H borylation of aza[5]helicines using the directing efficacy of pyridine to achieve bowl-shaped quasi-[7]circulenes [128].

Bourissou and co-workers [129] and Wang/Shi and co-workers [130] independently showed that an in-built phosphorus-based functional group could act as directing moiety to smoothly functionalize the electrophilic C—H borylation of arene moiety. Several phosphines containing arene moieties, such as naphthyl (**150**), biphenyl (**151**), *N*-phenylpyrrole (**152**), benzyl (**153**), and naphthylmethyl (**154**), were smoothly borylated upon reaction with BBr_3 in the presence of base (Scheme 10.15).

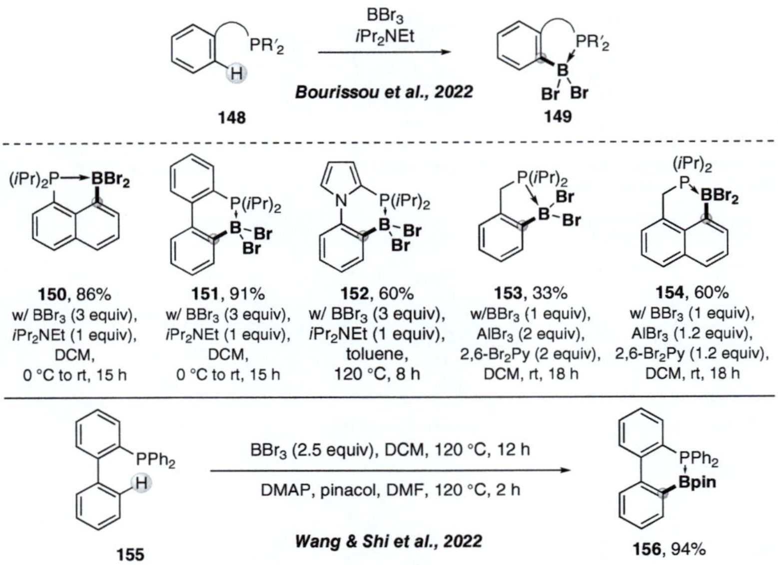

Scheme 10.15 Metal-free electrophilic C–H borylation of arene directed by phosphine moiety.

Considering the enormous progress of directing group-assisted C—H bond functionalization, it is always appreciable if a directing group can be removed or modified in a later stage. In this regard, the removable directing group plays a crucial role as the functional group can be retained after successfully removing the directing group in a post-functionalization stage. In 2012, Fu reported an electrophilic borylation of 2-phenoxypyridine to functionalize an *ortho*-C—H bond of phenol derivative (**159**; Scheme 10.16) [131]. Similarly, the pyridine group was installed in the thiophenol (**160)** [132], and aniline moiety (**162** and **163**) [133, 134], to borylated the respective *ortho*-C—H bond. Pyrimidine (**161**) [135], and amide groups (**164** and **165**) [136–139], worked perfectly as a directing auxiliary for *ortho*-C—H borylation of aniline derivative. A vinyl C—H bond was selectively borylated using the amide directing group and BBr_3 as an electrophilic boron reagent (**166**) [140]. Amide directing group is also efficient in C7–H (**167**), C5–H (**168**), and C4–H (**169**) borylation of indole derivatives [137, 138] as well as C2 borylation of pyrrole (**170**) [141]. Very recently, the thiobenzamide group was introduced as a directing group for *ortho*-selective borylation of arene (**171**) [142]. In 2013, Vedejs reported a P(III)-directed *ortho*-C—H borylation of phenol derivative (**172**→**174**) in which quenching by KHF_2 resulted the *in situ* cleavage of phosphorus directing group [143].

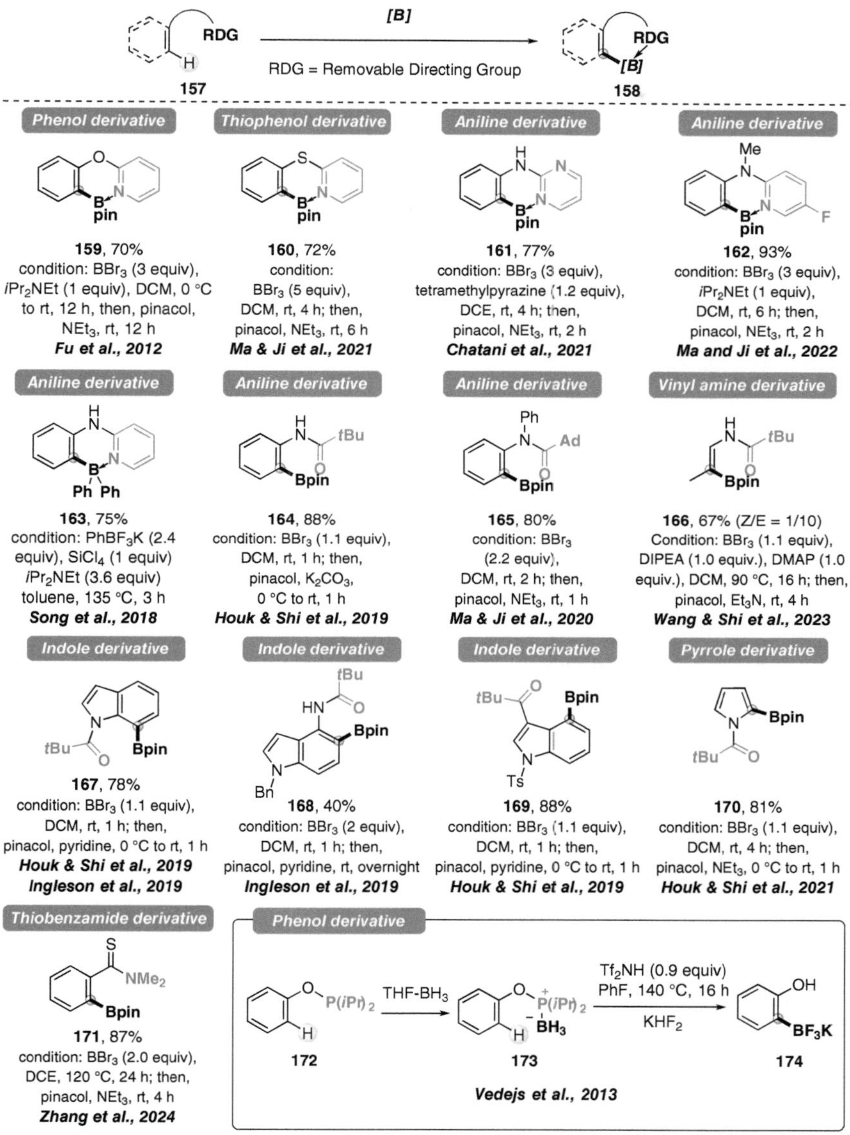

Scheme 10.16 Metal-free electrophilic C–H borylation directed by removable directing groups.

In a pioneering report, Chatani reported an electrophilic aromatic C–H borylation on electron-deficient benzaldehyde derivatives (benzaldehyde→**175**) by introducing a *tert*-butyl-group-substituted imine as a directing group. Imine directing group can be installed *in situ* in the reaction mixture and left at the end of the reaction. In this reaction, *in situ* prepared aldimine was reacted with BBr_3 in the presence of 2,6-lutidine leading to dibromo borane complex **176**, which

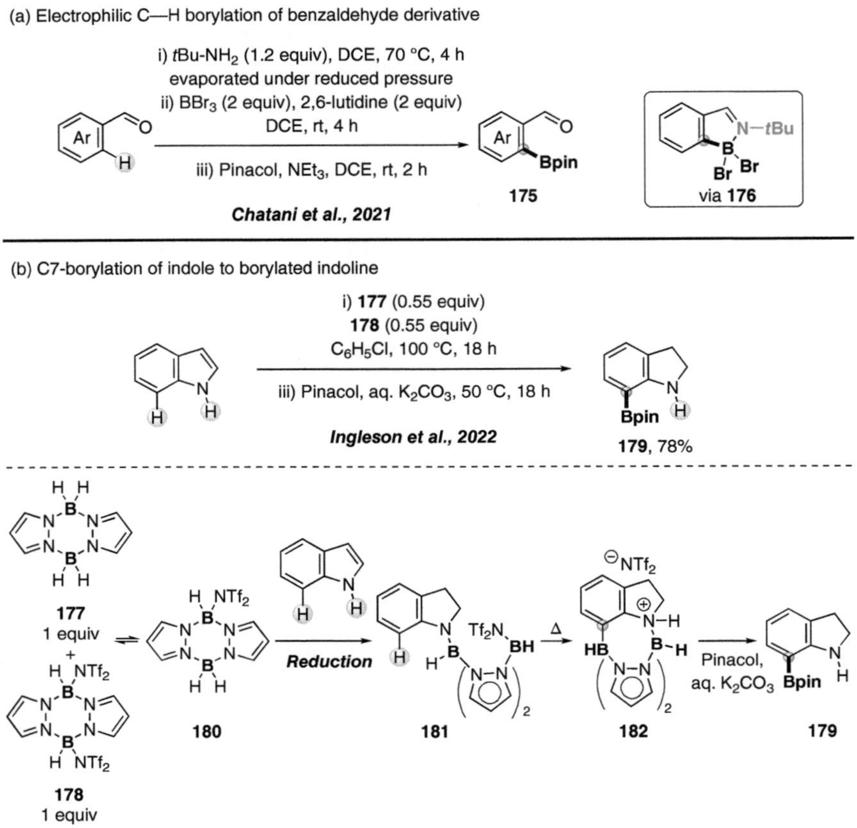

Scheme 10.17 Directed electrophilic borylation of benzaldehyde and indole.

was quenched by pinacol to afford borylated benzaldehyde **175** (Scheme 10.17a) [144, 145]. Ingleson also reported a transient borylation-directed borylation at the C7-position of indole to lead to borylated indoline **179** (Scheme 10.17b) [146]. The proposed mechanism suggested that at first, an *N*-borylation took place by the ditopic boron reagent **180**, which was *in situ* formed by the reactions of **177** and **178**. Subsequent reduction and C7-borylation took place to yield **179**.

10.6.2 C(sp^3)–H Borylation

In 2011, Vedejs discovered a revolutionary work of metal-free C(sp^3)—H borylation using an amine directing group [147]. A highly electrophilic boron cation was generated from amino-borane complex **183**, and simultaneously, intramolecular C(sp^2)—H and C(sp^3)—H borylation took place to lead to compound **184** (Scheme 10.18a) [105]. Very recently, Fontaine reported a pyridine-directed metal-free C(sp^3)—H borylation of amine derivative **187** (Scheme 10.18b) [148]. In fact, the author was able to selectively borylate the α-C—H (**188**) and β-C—H (**186**) bonds by switching the reaction condition.

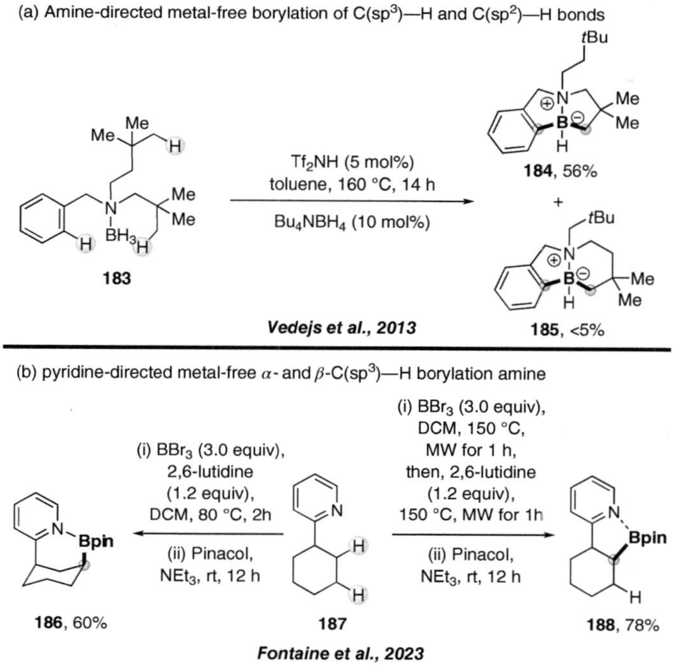

Scheme 10.18 Metal-free amine and pyridine-directed C(sp^3)–H borylation.

10.7 Directed C–H Silylation

Metal-free directed C(sp^2)—H silylation is mainly achieved intramolecularly. Primarily, the mechanism follows via the formation of a silyl radical (**190**, Scheme 10.19) or silyl cation (**199**, Scheme 10.20) and is followed by intramolecular aromatic substitution to form the C—Si coupled product [16]. Earlier, Kumada and coworkers found an oxidative protocol for the generation of a silyl-based radical from aryl alkyl silane **193**. This radical can undergo intramolecular coupling with the close proximal arene moiety, leading to the formation of the sila-cyclic product **194** (Scheme 10.19a) [149]. In this reaction, di-*tert*-butyl peroxide was used as an oxidant to form the silyl radical.

In 2015, Studer and co-workers [150] and Li and co-workers [151] independently reported the synthesis of 9-silafluorenes (**196** and **198**) via homolytic aromatic substitution reaction of biaryl silanes (**195** and **197**) in the presence of a radical initiator (Scheme 10.19b and c). Later, a visible-light-induced and Rose Bengal-catalyzed intramolecular radical silylation was showcased to obtain the same 9-silafluorene compound (**196**; Scheme 10.19d) [152].

Besides the radical pathway, cationic silicon can be generated in a reaction with the initiator (such as $B(C_6F_5)_3$ or $Ph_3C[B(C_6F_5)_3]$) and silane, which then leads to an electrophilic aromatic substitution at the closely situated arene to form the cyclized C—Si coupled product (Scheme 10.20). Kobayashi and Kawashima reported a

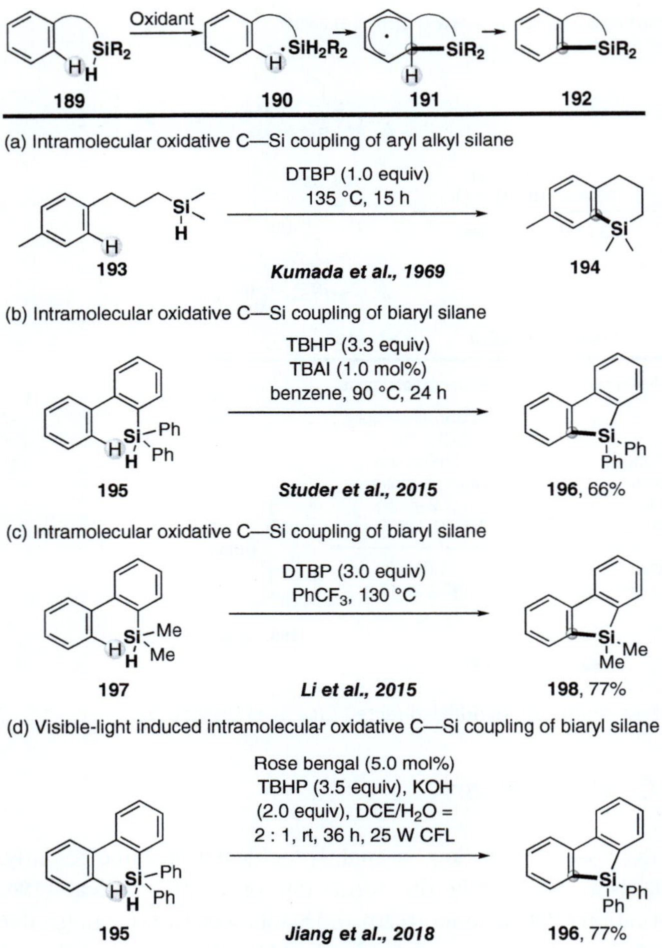

Scheme 10.19 Metal-free intramolecular oxidative C–H silylation via silyl radical formation.

cyclization reaction of biaryl silane **201** in the presence of the stoichiometric amount of trityl cation initiator leading to the formation of dibenzosilole derivative (**202**) via a sila-Friedel–Crafts reaction (Scheme 10.20a) [153]. Later, Ingleson showed that the same transformation can be achieved using a catalytic amount of $B(C_6F_5)_3$ as a catalyst (Scheme 10.20b) [154]. The reaction follows a similar mechanism via the formation of a silicon cation followed by intramolecular electrophilic aromatic substitution. Soon after, Oestreich designed a Ru-S-based catalyst (**204**) that can activate the silane substrate to generate a silicon cation. Subsequent cyclization eventually produced a similar dibenzosilole compound (**205**; Scheme 10.20c) [155]. The authors anticipated that the intramolecular electrophilic aromatic substitution undergoes via a sulfur-stabilized silicon cation intermediate.

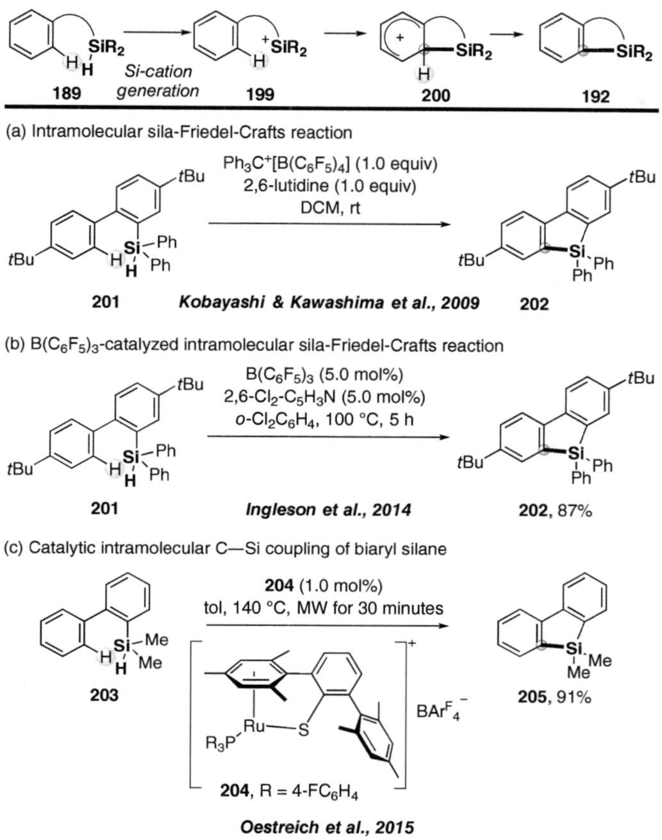

Scheme 10.20 Metal-free directed electrophilic C—H silylation via silicon cation formation.

10.8 Summary and Outlook

Metal-free transformation has always been a point of interest in synthetic chemistry due to its green nature, environmentally benign reaction condition, scalable synthesis, industrial application, and easy–to-handle protocols. Another point in metal-free transformation is that there is no chance of metal contamination in the final compound which again has enormous importance in biological synthesis, medicinal chemistry, and API chemical synthesis. Considering various metal-free transformation protocols, directed C—H bond functionalization has also been researched in considerable amount. Directed metal-free C—H bond functionalization is mainly achieved via oxidation chemistry or electrophilic aromatic substitution strategy. In this regard, to date, various significant bond formation reactions have been discovered such as C—C, C—N, C—O, C—B, and C—Si bond formation reactions. This chapter discusses each section in detail with their mechanistic point of view. Directed C—C, C—N, C—O, and C—Si coupling reactions are mainly obtained via intramolecular cyclization reactions. There are few pioneering reports in which metal-free functional group-directed protocols were introduced for intermolecular C—N coupling reactions. Notably, the directing group-assisted selective C—B bond formation strategy successfully constructed a library of organoborane compounds in a metal-free approach. Overall, we anticipate that this comprehensive chapter on metal-free directed C—H bond functionalization topic will largely help the synthetic community to understand the gaps in this field. We hope this field will flourish significantly in the near future considering its high demand in industry and API chemical synthesis.

Acknowledgments

S. Rej acknowledges financial support from ANRF (formerly known as SERB) for the Grant of Ramanujan Faculty (Fellowship No. RJF/2023/000038). Centre for Biomedical Research (CBMR) is greatly acknowledged for the provided infrastructure. R. Bangari expresses his thanks to CSIR New Delhi for the Junior Research Fellowship (File No. 09/0916(21013)/2025-EMR-I).

References

1 Xie, J. and Zhu, C. (2016). *Sustainable C(sp^3)-H Bond Functionalization*. Heidelberg: Springer.
2 Zhang, Y.-H., Shi, G.-F., and Yu, J.-Q. (2014). 3.23 Carbon–carbon σ-bond formation via C–H bond functionalization. In: *Comprehensive Organic Synthesis II*, 2e, 1101–1209. Amsterdam: Elsevier.
3 Rej, S., Ano, Y., and Chatani, N. (2020). Bidentate directing groups: an efficient tool in C–H bond functionalization chemistry for the expedient construction of C—C bonds. *Chem. Rev.* 120: 1788–1887.
4 Rej, S., Das, A., and Chatani, N. (2021). Strategic evolution in transition metal-catalyzed directed C—H bond activation and future directions. *Coord. Chem. Rev.* 431: 213683.
5 Gensch, T., Hopkinson, M.N., Glorius, F., and Wencel-Delord, J. (2016). Mild metal-catalyzed C–H activation: Examples and concepts. *Chem. Soc. Rev.* 45: 2900–2936.
6 Sambiagio, C., Schönbauer, D., Blieck, R. et al. (2018). A comprehensive overview of directing groups applied in metal-catalysed C–H functionalisation chemistry. *Chem. Soc. Rev.* 47: 6603–6743.
7 Gandeepan, P., Muller, T., Zell, D. et al. (2019). 3d transition metals for C–H activation. *Chem. Rev.* 119: 2192–2452.
8 Laird, T. (2012). Green chemistry is good process chemistry. *Org. Process. Res. Dev.* 16 (1): 1–2.

9 McElroy, C.R., Constantinou, A., Jones, L.C. et al. (2015). Towards a holistic approach to metrics for the 21st century pharmaceutical industry. *Green Chem.* 17 (5): 3111–3121.

10 Narayan, R., Matcha, K., and Antonchick, A.P. (2015). Metal-free oxidative C—C bond formation through C—H bond functionalization. *Chem. Eur. J.* 21 (42): 14678–14693.

11 Shamsabadi, A. and Chudasama, V. (2019). Recent advances in metal-free aerobic C–H activation. *Org. Biomol. Chem.* 17 (11): 2865–2872.

12 Shi, L., Li, T., and Mei, G.-J. (2022). Recent advances in transition-metal-free C–H functionalization of imidazo[1,2-*a*]pyridines. *Green Synth. Catal.* 3 (3): 227–242.

13 Li, Y.-N., Wang, B., Huang, Y.-K. et al. (2022). Recent advances in metal catalyst- and oxidant-free electrochemical C—H bond functionalization of nitrogen-containing heterocycles. *Front. Chem.* 10: 1–16.

14 Dahatonde, D.J., Bisht, S., and Batra, S. (2022). Metal-free oxidative C—H bond functionalization. In: *Handbook of C–H-Functionalization*, 1–207. Wiley.

15 Roy, S., Panja, S., Sahoo, S.R. et al. (2023). Enroute sustainability: metal free C—H bond functionalisation. *Chem. Soc. Rev.* 52 (7): 2391–2479.

16 Tyagi, A., Yadav, N., Khan, J. et al. (2021). Transition-metal-free C–H silylation: an emerging strategy. *Asian J. Org. Chem.* 10 (2): 334–354.

17 Rej, S. and Chatani, N. (2022). Regioselective transition-metal-free C(sp^2)–H borylation: a subject of practical and ongoing interest in synthetic organic chemistry. *Angew. Chem. Int. Ed.* 61: e202209539.

18 Kikugawa, Y. and Kawase, M. (1990). An electrophilic aromatic substitution by *N*-methoxyamides via hypervalent iodine intermediates. *Chem. Lett.* 581–582.

19 Romero, A.G., Darlington, W.H., Jacobsen, E.J., and Mickelson, J.W. (1996). Oxidative cyclization of acyclic ureas with bis(trifluoroacetoxy)iodobenzene to generate *N*-substituted 2-benzimidazolinones. *Tetrahedron Lett.* 37 (14): 2361–2364.

20 Kikugawa, Y., Nagashima, A., Sakamoto, T. et al. (2003). Intramolecular cyclization with nitrenium ions generated by treatment of *N*-acylaminophthalimides with hypervalent iodine compounds: Formation of lactams and spiro-fused lactams. *J. Org. Chem.* 68: 6739–6744.

21 Yu, J., Gao, C., Song, Z. et al. (2015). Metal-free oxidative C–H amidation of *N*,*N*′-diarylureas with PhI(OAc)2: synthesis of benzimidazol-2-one derivatives. *Eur. J. Org. Chem.* 5869–5875.

22 Zhang, Z., Gao, X., Li, Z. et al. (2017). PIFA-mediated oxidative cyclization of 1-aroyl-*N*-arylcyclopropane-1-carboxamides and their application in the synthesis of pyrrolo[3,2-*c*] quinolinones. *Org. Chem. Front.* 4: 404–408.

23 Zhang, Z., Zhang, Y., Huang, G., and Zhang, G. (2017). Organoiodine reagent-promoted intermolecular oxidative amination: synthesis of cyclopropyl spirooxindoles. *Org. Chem. Front.* 4: 1372–1375.

24 Cho, S.H., Yoon, J., and Chang, S. (2011). Intramolecular oxidative C–N bond formation for the synthesis of carbazoles: comparison of reactivity between the copper-catalyzed and metal-free conditions. *J. Am. Chem. Soc.* 133: 5996–6005.

25 Samanta, R., Kulikov, K., Strohmann, C., and Antonchick, A.P. (2012). Metal-free electrocyclization at ambient temperature: synthesis of 1-arylcarbazoles. *Synthese* 44 (15): 2325–2332.

26 Bal, A., Maiti, S., and Mal, P. (2018). Iodine(III)-enabled distal C–H functionalization of biarylsulfonanilides. *J. Org. Chem.* 83 (18): 11278–11287.

27 Herrero, M.T., Tellitu, I., Domínguez, E. et al. (2002). A novel and efficient iodine(III)-mediated access to 1,4-benzodiazepin-2-ones. *Tetrahedron Lett.* 43: 8273–8275.

28 Correa, A., Tellitu, I., Domínguez, E. et al. (2005). An efficient, PIFA mediated approach to benzo-, naphtho-, and heterocycle-fused pyrrolo[2,1-*c*][1,4] diazepines. An advantageous access to the antitumor antibiotic DC-81. *J. Org. Chem.* 70: 2256–2264.

29 Malamidou-Xenikaki, E., Spyroudis, S., Tsanakopoulou, M., and Hadjipavlou-Litina, D. (2009). A convenient approach to fused indeno-1,4-diazepinones through hypervalent iodine chemistry. *J. Org. Chem.* 74: 7315–7321.

30 Li, X., Yang, L., Zhang, X. et al. (2014). Construction of 1,4-benzodiazepine skeleton from 2-(arylamino)benzamides through $PhI(OAc)_2$-mediated oxidative C–N bond formation. *J. Org. Chem.* 79: 955–962.

31 Guo, X., Zhang-Negrerie, D., and Du, Y. (2015). Iodine(III)-mediated construction of dibenzoxazepinone skeleton from 2-(aryloxy)benzamides through oxidative C–N formation. *RSC Adv.* 3: 10715–10722.

32 Zhang, X., Zhang-Negrerie, D., Deng, J. et al. (2013). Synthesis of diversely substituted indoloquinolinones via Pd(II)/Cu(II)-mediated oxidative C—C bond formation and I(III)-mediated C—N bond formation. *J. Org. Chem.* 78: 12750–12759.

33 Li, Y., Zhang-Negrerie, D., Du, Y., and Zhao, K. (2015). A convenient synthesis of indoloquinolinones via 3-arylation of indole-2-carboxamides and PIDA-mediated C—N bond formation. *Tetrahedron* 71: 2927–2935.

34 Hu, C., Zhang, Z., Gao, W. et al. (2018). PIFA-promoted intramolecular oxidative C(aryl)–H amidation reaction: synthesis of quinolino[3,4-*b*] quinoxalin-6(5*H*)-ones. *Tetrahedron* 74: 665–671.

35 Liang, D., Yu, W., Nguyen, N. et al. (2017). Iodobenzene-catalyzed synthesis of phenanthridinones via oxidative C–H amidation. *J. Org. Chem.* 82: 3589–3596.

36 Liang, D., Sersen, D., Yang, C. et al. (2017). One-pot sequential reaction to 2-substituted-phenanthridinones from: *N*-methoxybenzamides. *Org. Biomol. Chem.* 15: 4390–4398.

37 Reddy, B.V.S., Reddy, C.R., Reddy, M.R. et al. (2015). Substrate directed C–H activation for the synthesis of benzo[*c*]cinnolines through a sequential C—C and C—N bond formation. *Org. Lett.* 17: 3730–3733.

38 Antonchick, A.P., Samanta, R., Kulikov, K., and Lategahn, J. (2011). Organocatalytic, oxidative, intramolecular C—H bond amination and metal-free cross-amination of unactivated arenes at ambient temperature. *Angew. Chem. Int. Ed.* 50 (37): 8605–8608.

39 Ban, X., Pan, Y., Lin, Y. et al. (2012). Synthesis of carbazolones and 3-acetylindoles via oxidative C—N bond formation through PIFA-mediated annulation of 2-aryl enaminones. *Org. Biomol. Chem.* 10 (18): 3606–3609.

40 Bera, S.K., Alam, M.T., and Mal, P. (2019). C–N coupling via antiaromatic endocyclic nitrenium ions. *J. Org. Chem.* 84 (18): 12009–12020.

41 Huang, J., He, Y., Wang, Y., and Zhu, Q. (2012). Synthesis of benzimidazoles by PIDA-promoted direct C(sp^2)-H imidation of *N*-arylamidines. *Chem. Eur. J.* 18 (44): 13964–13967.

42 Alla, S.K., Kumar, R.K., Sadhu, P., and Punniyamurthy, T. (2013). Iodobenzene catalyzed C–H amination of *N*-substituted amidines using *m*-chloroperbenzoic acid. *Org. Lett.* 15 (6): 1334–1337.

43 Lin, J.-P., Zhang, F.-H., and Long, Y.-Q. (2014). Solvent/oxidant-switchable synthesis of multisubstituted quinazolines and benzimidazoles via metal-free selective oxidative annulation of arylamidines. *Org. Lett.* 16 (11): 2822–2825.

44 Chi, Y., Zhang, W.-X., and Xi, Z. (2014). Oxidant-switchable selective synthesis of 2-aminobenzimidazoles via C–H amination/acetoxylation of guanidines. *Org. Lett.* 16 (24): 6274–6277.

45 He, Y., Huang, J., Liang, D. et al. (2013). C–H cycloamination of *N*-aryl-2-aminopyridines and *N*-arylamidines catalyzed by an in situ generated hypervalent iodine(III) reagent. *Chem. Commun.* 49 (66): 7352–7354.

46 Rao, D.N., Rasheed, S., Vishwakarma, R.A., and Das, P. (2014). Hypervalent iodine(iii) catalyzed oxidative C—N bond formation in water: synthesis of benzimidazole-fused heterocycles. *RSC Adv.* 4 (49): 25600–25604.

47 Du, Y., Liu, R., Linn, G., and Zhao, K. (2006). Synthesis of *N*-substituted indole derivatives via PIFA-mediated intramolecular cyclization. *Org. Lett.* 8 (26): 5919–5922.

48 Hofstra, J.L., Grassbaugh, B.R., Tran, Q.M. et al. (2015). Catalytic oxidative cyclization of 2′-arylbenzaldehyde oxime ethers under photoinduced electron transfer conditions. *J. Org. Chem.* 80 (1): 256–265.

49 Maiti, S. and Mal, P. (2015). Phenyliodine diacetate-mediated intramolecular C(sp^2)-H amidation for 1,2-disubstituted benzimidazole synthesis under metal-free conditions. *Adv. Synth. Catal.* 357 (7): 1416–1424.

50 Qian, G., Liu, B., Tan, Q. et al. (2014). (2014) Hypervalent iodine(III) promoted direct synthesis of imidazo[1,2-*a*] pyrimidines. *Eur. J. Org. Chem.* 22: 4837–4843.

51 Fra, L., Millán, A., Souto, J.A., and Muñiz, K. (2014). Indole synthesis based on a modified koser reagent. *Angew. Chem. Int. Ed.* 53 (28): 7349–7353.

52 Zhao, C.Y., Li, K., Pang, Y. et al. (2018). Iodine(III) reagent-mediated intramolecular amination of 2-alkenylanilines to prepare indoles. *Adv. Synth. Catal.* 360 (10): 1919–1925.

53 Xia, H.D., Zhang, Y.D., Wang, Y.H., and Zhang, C. (2018). Water-soluble hypervalent iodine(III) having an I—N bond. A reagent for the synthesis of indoles. *Org. Lett.* 20 (13): 4052–4056.

54 Liang, D. and Zhu, Q. (2015). A facile synthesis of pyrazoles through metal-free oxidative C(sp^2)–H cycloamination of vinyl hydrazones. *Asian J. Org. Chem.* 4 (1): 42–45.

55 Nagashima, A., Sakamoto, T., and Kikugawa, Y. (2007). Intramolecular benzylic cyclization with nitrenium ions generated from *N*-acylaminophthalimides using phenyliodine(III) Bis(trifluoroacetate): formation of phenyl substituted lactams. *Heterocycles* 74: 273–281.

56 Fan, R., Pu, D., Wen, F., and Wu, J. (2007). δ and α sp^3C—H bond oxidation of sulfonamides with $PhI(OAc)_2/I_2$ under metal-free conditions. *J. Org. Chem.* 72 (23): 8994–8997.

57 Ye, Y., Wang, H., and Fan, R. (2010). Stereoselective construction of highly functionalized azetidines via a [2 + 2]-cycloaddition. *Org. Lett.* 12 (12): 2802–2805.

58 Fan, R., Wang, H., Ye, Y., and Gan, J. (2010). $PhIO/Bu_4NI$ mediated oxidative cyclization of amidoalkylation adducts for the synthesis of *N*-benzoyl aziridines and oxazolines. *Tetrahedron Lett.* 51 (2): 453–456.

59 Mao, L., Li, Y., Xiong, T. et al. (2013). Synthesis of tetramic acid derivatives via intramolecular sp^3C–H amination mediated by hypervalent iodine(III) reagents/brønsted acids. *J. Org. Chem.* 78 (2): 733–737.

60 Zhu, C., Liang, Y., Hong, X. et al. (2015). Iodoarene-catalyzed stereospecific intramolecular sp^3C–H amination: reaction development and mechanistic insights. *J. Am. Chem. Soc.* 137 (24): 7564–7567.

61 Martínez, C. and Muñiz, K. (2015). An iodine-catalyzed Hofmann-Löffler reaction. *Angew. Chem. Int. Ed.* 54 (28): 8287–8291.

62 Wappes, E.A., Fosu, S.C., Chopko, T.C., and Nagib, D.A. (2016). Triiodide-mediated δ-amination of Secondary C—H bonds. *Angew. Chem. Int. Ed.* 55 (34): 9974–9978.

63 Paz, N.R., Rodríguez-Sosa, D., Valdés, H. et al. (2015). Chemoselective intramolecular functionalization of methyl groups in nonconstrained molecules promoted by *N*-iodosulfonamides. *Org. Lett.* 17 (10): 2370–2373.

64 Zhang, D., Wang, H., Cheng, H. et al. (2017). An iodine-mediated Hofmann-Löffler-Freytag reaction of sulfoximines leading to dihydroisothiazole oxides. *Adv. Synth. Catal.* 359 (24): 4274–4277.

65 Kumar, Y., Jaiswal, Y., and Kumar, A. (2018). Visible-light-mediated remote γ-C(sp^3)–H functionalization of alkylimidates: synthesis of 4-iodo-3,4-dihydropyrrole derivatives. *Org. Lett.* 20 (16): 4964–4969.

66 Hamamoto, H., Hata, K., Nambu, H. et al. (2004). A novel and direct synthesis of chroman derivatives using a hypervalent iodine(III) reagent. *Tetrahedron Lett.* 45 (11): 2293–2295.

67 Gu, Y. and Xue, K. (2010). Direct oxidative cyclization of 3-arylpropionic acids using PIFA or oxone: synthesis of 3,4-dihydrocoumarins. *Tetrahedron Lett.* 51 (1): 192–196.

68 Li, J., Chen, H., Zhang-Negrerie, D. et al. (2013). Synthesis of coumarins via $PIDA/I_2$-mediated oxidative cyclization of substituted phenylacrylic acids. *RSC Adv.* 3 (13): 4311–4320.

69 Wang, X., Gallardo-Donaire, J., and Martin, R. (2014). Mild ArI-catalyzed C(sp^2)–H or C(sp^3)–H functionalization/C–O formation: an intriguing catalyst-controlled selectivity switch. *Angew. Chem. Int. Ed.* 53 (41): 11084–11087.

70 Singh, F.V. and Wirth, T. (2012). Hypervalent iodine mediated oxidative cyclization of *o*-hydroxystilbenes into benzo- and naphthofurans. *Synthesis* 44 (8): 1171–1177.

71 Ngatimin, M., Frey, R., Andrews, C. et al. (2011). Iodobenzene catalysed synthesis of spirofurans and benzopyrans by oxidative cyclisation of vinylogous esters. *Chem. Commun.* 47 (42): 11778–11780.

72 Ranjith, J., Rajesh, N., Sridhar, B., and Radha Krishna, P. (2016). Intramolecular oxyacetoxylation of: *N*-allylamides: an expeditious synthesis of oxazolines and oxazines by using a $PhI(OAc)_2$/hydrogen fluoride-pyridine system. *Org. Biomol. Chem.* 14 (42): 10074–10079.

73 Yu, Z., Ma, L., and Yu, W. (2012). Phenyliodine bis(trifluoroacetate) mediated intramolecular oxidative coupling of electron-rich *N*-phenyl benzamides. *Synlett* 23 (10): 1534–1540.

74 Xu, K., Yang, R., Yang, S. et al. (2019). Hypervalent iodane mediated reactions of: *N*-acetyl enamines for the synthesis of oxazoles and imidazoles. *Org. Biomol. Chem.* 17 (40): 8977–8981.

75 Zheng, Y., Li, X., Ren, C. et al. (2012). Synthesis of oxazoles from enamides via phenyliodine diacetate-mediated intramolecular oxidative cyclization. *J. Org. Chem.* 77 (22): 10353–10361.

76 Xu, K., Yang, S., and Ding, Z. (2019). $PhI(OAc)_2$-mediated oxidative rearrangement of allylic amides: efficient synthesis of oxazoles and β-keto amides. *Org. Chem. Front.* 7 (1): 69–72.

77 Fan, R., Sun, Y., and Ye, Y. (2009). Iodine(III)-mediated tandem acetoxylation cyclization of *o*-acyl phenols for the facile construction of α-acetoxy benzofuranones. *Org. Lett.* 11 (22): 5174–5177.

78 Couto, I., Tellitu, I., and Domínguez, E. (2010). An intramolecular PIFA-mediated metal-free allylic oxycarbonylation reaction and its application to the preparation of furopyrimidinones. *J. Org. Chem.* 75 (22): 7954–7957.

79 Sun, D., Zhao, X., Zhang, B. et al. (2018). Synthesis of spirofurooxindoles via phenyliodine(III) bis(trifluoroacetate) (PIFA)-mediated cascade oxidative C—O and C—C bond formation. *Adv. Synth. Catal.* 360 (8): 1634–1638.

80 Xing, Q., Liang, H., Bao, M. et al. (2019). Metal-free synthesis of spiro-2,2′-benzo[*b*]furan-3,3′-ones via $PhI(OAc)_2$-mediated cascade spirocyclization. *Adv. Synth. Catal.* 361 (20): 4669–4673.

81 Zhang, N., Cheng, R., Zhang-Negrerie, D. et al. (2014). Hypervalent iodine-mediated oxygenation of *N*,*N*-diaryl tertiary amines: intramolecular functionalization of sp^3C—H bonds adjacent to nitrogen. *J. Org. Chem.* 79 (21): 10581–10587.

82 Cao, Y., Zhang, X., Lin, G. et al. (2016). Chiral aryliodine-mediated enantioselective organocatalytic spirocyclization: synthesis of spirofurooxindoles via cascade oxidative C—O and C—C bond formation. *Org. Lett.* 18 (21): 5580–5583.

83 Cao, J.-J., Zhu, T.-H., Wang, S.-Y. et al. (2014). *tert*-Butyl peroxybenzoate (TBPB)-mediated 2-isocyanobiaryl insertion with 1,4-dioxane: efficient synthesis of 6-alkyl phenanthridines via C(sp^3)–H/C(sp^2)–H bond functionalization. *Chem. Commun.* 50 (49): 6439–6442.

84 Sha, W., Yu, J.T., Jiang, Y. et al. (2014). The benzoyl peroxide-promoted functionalization of simple alkanes with 2-aryl phenyl isonitrile. *Chem. Commun.* 50 (65): 9179–9181.

85 Barluenga, J., Trincado, M., Rubio, E., and González, J.M. (2006). Direct intramolecular arylation of aldehydes promoted by reaction with IPy_2BF_4/HBF_4: Synthesis of benzocyclic ketones. *Angew. Chem. Int. Ed.* 45 (19): 3140–3143.

86 Zheng, Z., Dian, L., Yuan, Y. et al. (2014). $PhI(OAc)_2$-mediated intramolecular oxidative aryl-aldehyde Csp^2–Csp^2 bond formation: metal-free synthesis of acridone derivatives. *J. Org. Chem.* 79 (16): 7451–7458.

87 Wu, X., Zhang, H., Tang, N. et al. (2018). Metal-free alcohol-directed regioselective heteroarylation of remote unactivated C(sp^3)–H bonds. *Nat. Commun.* 9: 3343.

88 Tang, N., Wu, X., and Zhu, C. (2019). Practical, metal-free remote heteroarylation of amides: via unactivated C(sp^3)–H bond functionalization. *Chem. Sci.* 10 (28): 6915–6919.

89 Beletskaya, I.P., Alonso, F., and Tyurin, V. (2019). The Suzuki-Miyaura reaction after the Nobel prize. *Coord. Chem. Rev.* 385: 137–173.

90 Suzuki, A. (2011). Cross-coupling reactions of organoboranes: an easy way to construct C—C bonds (Nobel Lecture). *Angew. Chem. Int. Ed.* 50: 6722–6737.

91 Mkhalid, I.A.I., Barnard, J.H., Marder, T.B. et al. (2010). C–H activation for the construction of C—B bonds. *Chem. Rev.* 110: 890–931.

92 Ros, A., Fernández, R., and Lassaletta, J.M. (2014). Functional group directed C–H borylation. *Chem. Soc. Rev.* 43: 3229–3243.

93 Hartwig, J.F. (2011). Regioselectivity of the borylation of alkanes and arenes. *Chem. Soc. Rev.* 40: 1992–2002.

94 Ingleson, M.J. (2012). A perspective on direct electrophilic arene borylation. *Synlett* 23: 1411–1415.

95 Fontaine, F.-G. and Rochette, É. (2018). Ambiphilic molecules: from organometallic curiosity to metal-free catalysts. *Acc. Chem. Res.* 51: 454–464.

96 Iqbal, S.A., Pahl, J., Yuan, K., and Ingleson, M.J. (2020). Intramolecular (directed) electrophilic C–H borylation. *Chem. Soc. Rev.* 49: 4564–4591.

97 Li, Y. and Wu, X.-F. (2020). Direct C—H bond borylation of (hetero)arenes: evolution from noble metal to metal free. *Angew. Chem. Int. Ed.* 59: 1770–1774.

98 Dewar, M.J.S., Kubba, V.P., and Pettit, R. (1958). New heteroaromatic compounds part I. 9-Aza-10-boraphenanthrene. *J. Chem. Soc.* 3073–3076.

99 Dewar, M.J.S. and Dietz, R. (1959). New heteroaromatic compounds. Part III. 2,1-borazaro-naphthalene (1,2-dihydro-1-axa-2-boranaphthalene). *J. Chem. Soc.* 2728–2730.

100 Dewar, M.J.S., Kaneko, C., and Bhattacharjee, M.K. (1962). New heteroaromatic compounds. XVI. Compounds with heteroatoms at bridgeheads. *J. Am. Chem. Soc.* 84: 4884–4887.

101 Dewar, M.J.S. and Poesche, W.H. (1964). New heteroaromatic compounds. XXI. Some tetracyclic systems. *J. Org. Chem.* 29: 1757–1762.

102 Vedejs, E., Nguyen, T., Powell, D.R., and Schrimpf, M.R. (1996). Generation of reactive borenium ions in the 2,3-benzazaborolidine series. *Chem. Commun.* 4 (24): 2721–2722.

103 De Vries, T.S., Prokofjevs, A., Harvey, J.N., and Vedejs, E. (2009). Superelectrophilic intermediates in nitrogen-directed aromatic borylation. *J. Am. Chem. Soc.* 131: 14679–14687.

104 Hatakeyama, T., Hashimoto, S., Seki, S., and Nakamura, M. (2011). Synthesis of BN-fused polycyclic aromatics via tandem intramolecular electrophilic arene borylation. *J. Am. Chem. Soc.* 133 (46): 18614–18617.

105 Prokofjevs, A., Jermaks, J., Borovika, A. et al. (2013). Electrophilic C–H borylation and related reactions of B–H boron cations. *Organometallics* 32: 6701–6711.

106 Wang, X., Zhang, F., Schellhammer, K.S. et al. (2016). Synthesis of NBN-type zigzag-edged polycyclic aromatic hydrocarbons: 1,9-diaza-9a-boraphenalene as a structural motif. *J. Am. Chem. Soc.* 138 (36): 11606–11615.

107 Zhang, J., Jung, H., Kim, D. et al. (2019). Sequential C–H borylation and N-demethylation of 1,1′-biphenylamines: alternative route to polycyclic BN-heteroarenes. *Angew. Chem. Int. Ed.* 58: 7361–7365.

108 Yuan, K., Volland, D., Kirschner, S. et al. (2022). Enhanced N-directed electrophilic C–H borylation generates BN-[5]- and [6]helicenes with improved photophysical properties. *Chem. Sci.* 13 (4): 1136–1145.

109 Dewar, M.J.S. and Dietz, R. (1959). Aromatic boron-oxygen heterocycles. *Tetrahedron Lett.* 1: 21–23.

110 Zhou, Q.J., Worm, K., and Dolle, R.E. (2004). 10-hydroxy-10,9-boroxarophenanthrenes: versatile synthetic intermediates to 3,4-benzocoumarins and triaryls. *J. Org. Chem.* 69 (4): 5147–5149.

111 Numano, M., Nagami, N., Nakatsuka, S. et al. (2016). Synthesis of boronate-based benzo[*fg*]tetracene and benzo[*hi*]hexacene via demethylative direct borylation. *Chem. Eur. J.* 22 (33): 11574–11577.

112 Kang, J., Liu, J., and Chen, Z. (2024). Directing group controlled regioselective C–H borylation of 2-arylphenolic compounds at room temperature. *Org. Chem. Front.* 11 (15): 4249–4257.

113 Davis, F.A. and Dewar, M.J.S. (1968). New heteroaromatic compounds. XXX. A derivative of 10,9-borathiarophenanthrene. *J. Am. Chem. Soc.* 90: 3511–3515.

114 Shaikh, A.C., Ranade, D.S., Thorat, S. et al. (2015). Highly emissive organic solids with remarkably broad color tunability based on *N*,*C*-Chelate, four-coordinate organoborons. *Chem. Commun.* 51: 16115–16118.

115 Iqbal, S.A., Yuan, K., Cid, J. et al. (2021). Controlling selectivity in *N*-heterocycle directed borylation of indoles. *Org. Biomol. Chem.* 19: 2949–2958.

116 Shigeno, M., Imamatsu, M., Kai, Y. et al. (2021). Construction of 1,2,3-benzodiazaborole by electrophilic borylation of azobenzene and nucleophilic dialkylative cyclization. *Org. Lett.* 23: 8023–8027.

117 Crossley, D.L., Cade, I.A., Clark, E.R. et al. (2015). Enhancing electron affinity and tuning band gap in donor-acceptor organic semiconductors by benzothiadiazole directed C–H borylation. *Chem. Sci.* 6: 5144–5151.

118 Crossley, D.L., Vitorica-Yrezabal, I., Humphries, M.J. et al. (2016). Highly emissive far red/near-IR fluorophores based on borylated fluorene–benzothiadiazole donor-acceptor materials. *Chem. Eur. J.* 22: 12439–12448.

119 Crossley, D.L., Urbano, L., Neumann, R. et al. (2017). Post-polymerization C–H borylation of donor-acceptor materials gives highly efficient solid state near-infrared emitters for near-IR-OLEDs and effective biological imaging. *ACS Appl. Mater. Interfaces* 9: 28243–28249.

120 Dash, B.P., Hamilton, I., Tate, D.J. et al. (2019). Benzoselenadiazole and benzotriazole directed electrophilic C–H borylation of conjugated donor-acceptor materials. *J. Mater. Chem. C* 7: 718–724.

121 Ishida, N., Moriya, T., Goya, T., and Murakami, M. (2010). Synthesis of pyridine-borane complexes via electrophilic aromatic borylation. *J. Org. Chem.* 75: 8709–8712.

122 Zhao, Z., Chang, Z., He, B. et al. (2013). Aggregation-induced emission and efficient solid-state fluorescence from tetraphenylethene-based *N*,*C*–chelate four-coordinate organoborons. *Chem. Eur. J.* 19: 11512–11517.

123 Chen, J., Lalancette, R.A., and Jäkle, F. (2013). Stereoselective *ortho* borylation of pyridylferrocenes. *Organometallics* 32: 5843–5851.

124 Yusuf, M., Liu, K., Guo, F. et al. (2016). Luminescent organoboron ladder compounds via directed electrophilic aromatic C–H borylation. *Dalton Trans.* 45: 4580–4587.

125 Liu, K., Lalancette, R.A., and Jäkle, F. (2017). B-N lewis pair functionalization of anthracene: structural dynamics, optoelectronic properties, and O_2 sensitization. *J. Am. Chem. Soc.* 139: 18170–18173.

126 Oda, S., Abe, H., Yasuda, N., and Hatakeyama, T. (2019). Synthesis of tetracoordinate boron-fused benzoaceanthrylene analogs via tandem electrophilic C–H borylation. *Chem. Asian J.* 14 (10): 1657–1661.

127 Gotoh, H., Nakatsuka, S., Tanaka, H. et al. (2021). Syntheses and physical properties of cationic BN-embedded polycyclic aromatic hydrocarbons. *Angew. Chem. Int. Ed.* 60 (23): 12835–12840.

128 Zhang, X., Rauch, F., Niedens, J. et al. (2022). Electrophilic C–H borylation of aza[5]helicenes leading to bowl-shaped quasi-[7]circulenes with switchable dynamics. *J. Am. Chem. Soc.* 144 (48): 22316–22324.

129 Sadek, O., Le Gac, A., Hidalgo, N. et al. (2022). Metal-free phosphorus-directed borylation of C(sp^2)—H bonds. *Angew. Chem. Int. Ed.* 61: e202110102.

130 Lv, J., Zhang, X.-J., Wang, M. et al. (2022). BBr_3-mediated P(III)-directed C–H borylation of phosphines. *Chem. Eur. J.* 28: e202104100.

131 Niu, L., Yang, H., Wang, R., and Fu, H. (2012). Metal-free *ortho* C–H borylation of 2-phenoxypyridines under mild conditions. *Org. Lett.* 14: 2618–2621.

132 Wu, G., Pang, B., Wang, Y. et al. (2021). Metal-free *ortho*-selective C–H borylation of 2-phenylthiopyridines using BBr3. *J. Org. Chem.* 86 (8): 5933–5942.

133 Wu, G., Xu, X., Wang, S. et al. (2022). Metal-free directed C–H borylation of 2-(*N*-methylanilino)-5-fluoropyridines and 2-benzyl-5-fluoropyridines. *Chin. Chem. Lett.* 33 (4): 2005–2008.

134 Yang, K., Zhang, G., and Song, Q. (2018). Four-coordinate triarylborane synthesis: via cascade B-Cl/C–B cross-metathesis and C—H bond borylation. *Chem. Sci.* 9: 7666–7672.

135 Rej, S., Das, A., and Chatani, N. (2021). Pyrimidine-directed metal-free C–H borylation of 2-pyrimidylanilines: a useful process for tetra-coordinated triarylborane synthesis. *Chem. Sci.* 12: 11447–11454.

136 Wu, G., Fu, X., Wang, Y. et al. (2020). C–H borylation of diphenylamines through adamantane-1-carbonyl auxiliary by BBr_3. *Org. Lett.* 22: 7003–7007.

137 Lv, J., Chen, X., Xue, X.-S. et al. (2019). Metal-free directed sp^2-C–H borylation. *Nature* 575: 336–340.

138 Iqbal, S.A., Cid, J., Procter, R.J. et al. (2019). Acyl-directed *ortho*-borylation of anilines and C7 borylation of indoles using just BBr_3. *Angew. Chem. Int. Ed.* 58: 15381–15385.

139 Kothavale, S.S., Iqbal, S.A., Hanover, E.L. et al. (2023). Borylation-reduction-borylation for the formation of 1,4-azaborines. *Org. Lett.* 25 (49): 8912–8916.

140 Wang, T., Wang, Z.J., Wang, M. et al. (2023). Metal-free stereoconvergent C–H borylation of enamides. *Angew. Chem. Int. Ed.* 62 (44): e202313205.

141 Wang, Z.-J., Chen, X., Wu, L. et al. (2021). Metal-free directed C–H borylation of pyrroles. *Angew. Chem. Int. Ed.* 60: 8500–8504.

142 Lv, J., Liang, Y., Ouyang, Y., and Zhang, H. (2024). Metal-free *ortho* C–H borylation of thiobenzamides. *Org. Lett.* 26 (18): 3709–3714.

143 Cazorla, C., De Vries, T.S., and Vedejs, E. (2013). P-directed borylation of phenols. *Org. Lett.* 15: 984–987.

144 Yamazaki, K., Rej, S., Ano, Y., and Chatani, N. (2022). Origin of the enhanced reactivity in the *ortho* C–H borylation of benzaldehydes with BBr_3. *Org. Lett.* 24: 213–217.

145 Rej, S. and Chatani, N. (2021). Transient imine as a directing group for the metal-free *o*-C–H borylation of benzaldehydes. *J. Am. Chem. Soc.* 143: 2920–2929.

146 Millet, C.R.P., Noone, E., Schellbach, A.V. et al. (2023). Borylation directed borylation of *N*-alkyl anilines using iodine activated pyrazaboles. *Chem. Sci.* 14 (43): 12041–12048.

147 Prokofjevs, A. and Vedejs, E. (2011). *N*-directed aliphatic C–H borylation using borenium cation equivalents. *J. Am. Chem. Soc.* 133: 20056–20059.

148 Ashwathappa, P.K.S., Higashi, T., Desrosiers, V. et al. (2023). Metal-free directed site-selective Csp^3–H borylation of saturated cyclic amines. *Angew. Chem. Int. Ed.* 62: e202309295.

149 Sakurai, H., Hosomi, A., and Kumada, M. (1969). Silyl radicals V. Novel cyclization reactions involving homolytic aromatic silylation. *Tetrahedron Lett.* 22: 1757–1760.

150 Leifert, D. and Studer, A. (2015). 9-Silafluorenes via base-promoted homolytic aromatic substitution (BHAS) – the electron as a catalyst. *Org. Lett.* 17 (2): 386–389.

151 Xu, L., Zhang, S., and Li, P. (2015). Synthesis of silafluorenes and silaindenes via silyl radicals from arylhydrosilanes: intramolecular cyclization and intermolecular annulation with alkynes. *Org. Chem. Front.* 2 (5): 459–463.
152 Yang, C., Wang, J., Li, J. et al. (2018). Visible-light induced radical silylation for the synthesis of dibenzosiloles via dehydrogenative cyclization. *Adv. Synth. Catal.* 360: 3049–3054.
153 Furukawa, S., Kobayashi, J., and Kawashima, T. (2009). Development of a sila-Friedel-Crafts reaction and its application to the synthesis of dibenzosilole derivatives. *J. Am. Chem. Soc.* 131 (40): 14192–14193.
154 Curless, L.D. and Ingleson, M.J. (2014). $B(C_6F_5)_3$ – catalyzed synthesis of benzofused-siloles. *Organometallics* 33: 7241–7246.
155 Omann, L. and Oestreich, M. (2015). A catalytic SEAr approach to dibenzosiloles functionalized at both benzene cores. *Angew. Chem. Int. Ed.* 54 (35): 10276–10279.

11

Directed C(sp^3)−H Functionalization in Asymmetric Synthesis

Floris Buttard, Balu Ramesh, Javid Rzayev, and Tatiana Besset

Organic Chemistry Department, INSA Rouen Normandie, Univ Rouen Normandie, CNRS, Normandie Univ., Institut CARMeN, UMR 6064, INC3M FR 3038, F-76000 Rouen, France

11.1 Introduction

Transition-metal-catalyzed C—H bond activation strategies have opened a new paradigm for synthetic disconnections in organic synthesis by allowing the functionalization of C—H bonds that would be considered inert to more classical chemical transformations [1–5]. Besides undirected C—H bond activation approaches [6], which mainly rely on the steric and electronic properties of the substrates, the use of a directing group to bring the catalyst close to the targeted C—H bond position allows effective and regioselective transformations, even on the most challenging C(sp^3)—H positions [7–10]. The advances in this field over the last decades have opened the way to the possibility of controlling the stereoselectivity of such transformations, *via* the design of catalytic systems able to discriminate enantiotopic C—H bonds on prochiral substrates [11–16]. As the asymmetric functionalization of C(sp^3)—H bonds is of tremendous importance for the construction of chiral molecules, the approaches discussed in this chapter will likely have a high impact on the synthesis of key pharmaceutical ingredients and other high-interest molecules. An overview of known directed C(sp^3)—H bond functionalization strategies in asymmetric synthesis is provided, and this chapter is organized into three main sections: *transition-metal-catalyzed C(sp^3)—H bond activation reactions using chiral ligands* (Section 2), *transition-metal-catalyzed C(sp^3)—H bond activation reactions using chiral transient directing groups* (Section 3), and *supramolecular assembly-directed reactions* (Section 4).

11.2 Asymmetric Transition Metal-catalyzed C(sp^3)−H Bond Activation with Chiral Catalysts

One of the most successful approaches in achieving enantioselective transition-metal-catalyzed directed C(sp^3)—H activation has relied upon the use of a chiral

Directed C–H Bond Functionalization: Concepts and Applications, First Edition.
Edited by Debabrata Maiti and Supriya Rej.

catalyst. The ligand design in the strategies that will be discussed below has consequently been crucial in the development of new effective and selective C(sp^3)—H bond functionalization reactions. Most of these transformations use palladium catalysis, which will be discussed in the first Section (2.1), and several works on the use of rhodium, iridium, and cobalt catalysts, which have been reported for directed enantioselective C(sp^3)—H activation reactions, will be presented in the second Section (2.2).

11.2.1 Palladium Catalysis

11.2.1.1 Classical Directing Group-directed Asymmetric C(sp^3)–H Activation

First Examples and General Considerations The first example of enantioselective directed functionalization *via* the transition-metal-catalyzed activation of a C(sp^3)—H bond was reported in 2008 by the group of Yu [17]. In the presence of Pd(II) acetate, the chiral monoprotected cyclopropane amino acid ligand **L1a** and an alkyl boronic acid as a coupling partner, the 2-isopropylpyridine **1** was converted into the corresponding alkylated product **2** in a 38% yield and with a 37% enantiomeric excess (Scheme 11.1), the pyridine moiety acting as a directing group. Although the enantioinduction in this transformation was certainly modest, it represented the first groundbreaking example and proof of concept for asymmetric transition-metal-catalyzed C(sp^3)—H bond activation.

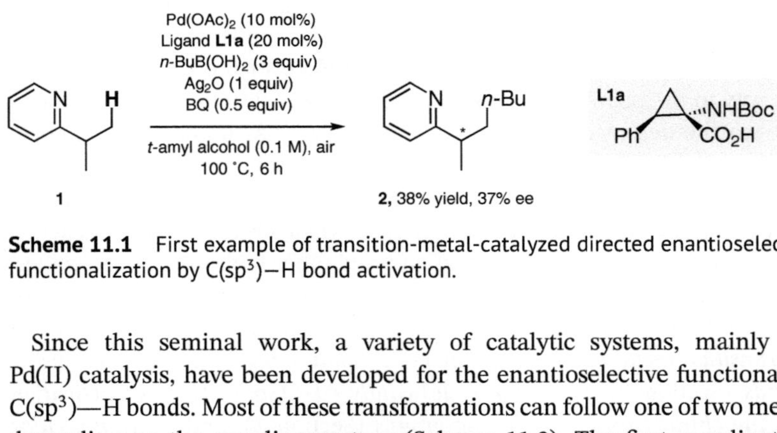

Scheme 11.1 First example of transition-metal-catalyzed directed enantioselective functionalization by C(sp^3)–H bond activation.

Since this seminal work, a variety of catalytic systems, mainly based on Pd(II) catalysis, have been developed for the enantioselective functionalization of C(sp^3)—H bonds. Most of these transformations can follow one of two mechanisms, depending on the coupling partner (Scheme 11.2). The first coordination of the palladium complex to the directing group on substrate **3** affords intermediate **int1**, which can undergo stereoselective C—H bond activation, most of the time through a concerted metalation-deprotonation (CMD) mechanism involving the chiral ligand **L*** (**TS1a** and **TS1b**), affording intermediate **int2** containing a C-Pd(II) bond. In most examples, this C—H bond metalation is the stereodetermining step of the reaction. The enantioselectivity of the transformation is often induced by the presence of destabilizing repulsive interactions between the substrate and the chiral ligand in the CMD transition structures (e.g., **TS1b**) for the metalation

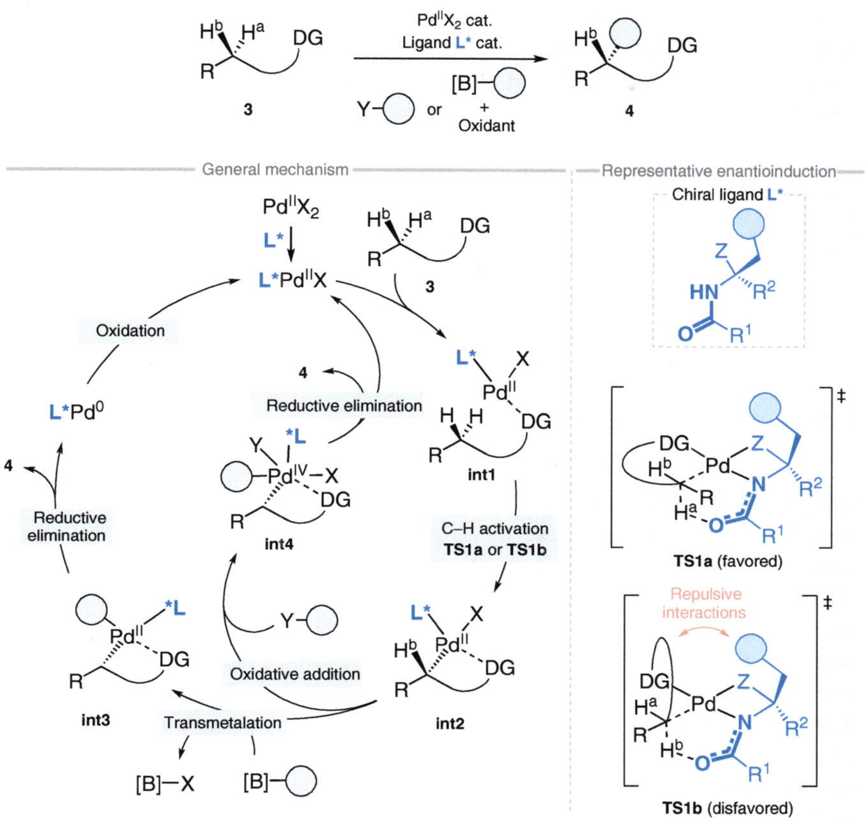

Scheme 11.2 Typical Pd-catalyzed enantioselective directed C(sp³)–H functionalization mechanism and enantioinduction.

of one of the enantiotropic hydrogens (H[b]), leading to an enantioselective C—H activation. The reaction may subsequently follow a Pd(II)/Pd(0) or a Pd(II)/Pd(IV) catalytic cycle. In the presence of boronic acid partners, transmetalation (toward **int3**) and reductive elimination afford product **4** along with a Pd(0) species that needs to be reoxidized into a Pd(II) complex to ensure catalytic turnover. In the presence of electrophilic partners, such as aryl halides, the oxidative addition on the Pd complex **int2** leads to a Pd(IV) species **int4**, which may undergo reductive elimination toward product **4** and regenerate the catalytically active Pd(II) complex.

In the seminal work from Yu and co-workers (Scheme 11.1) [17], the presence of a strong directing group, the pyridine moiety, can favor the non-enantioselective palladium-catalyzed C—H alkylation reaction, as the directing group itself provides a sufficient activation for the C—H bond cleavage. As such, strong directing groups can often outcompete the chiral ligand for the coordination to the metal catalyst and the C—H bond functionalization.

However, pyridine and quinoline moieties have recently been used as directing groups in enantioselective C(sp³)—H activation [18]. The use of a stoichiometric amount of trifluoroacetic acid led to a decrease in the coordinating ability of the pyridine group, and in the presence of the protected amino acid **L1b** as a chiral ligand to induce an enantioselective C—H bond cleavage, the products **5** were obtained in high yields and up to 86% enantiomeric excess (Scheme 11.3).

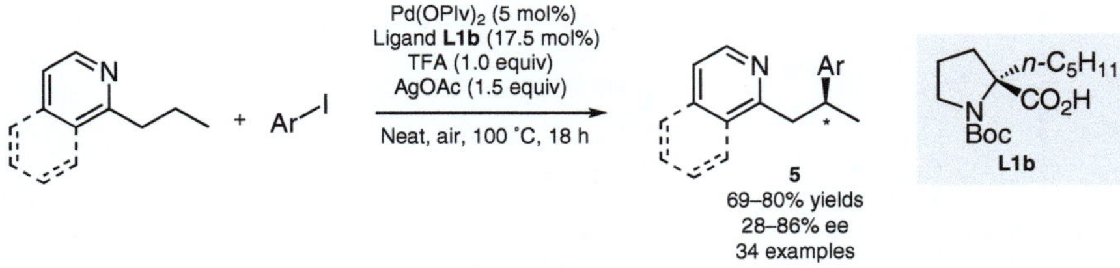

Scheme 11.3 Pyridine and quinoline as directing groups in enantioselective C(sp^3)–H arylation.

Polyfluorinated Aryl Amides as Directing Groups An alternative solution to avoid the competitive non-enantioselective C—H activation in the presence of the chiral ligand has been to use inherently weaker monodentate directing groups. The ligand design in these asymmetric transition-metal-catalyzed directed C(sp^3)—H activation approaches has been pivotal in developing efficient transformations, as the ligand must trigger the reactivity of the metal catalyst and ensure the enantiocontrol of the reaction by providing a chiral environment for the C—H activation step.

Based on their seminal results presented above and their previous works on the use of acidic *N*-arylamides as weakly coordinating directing groups for C—H bond functionalization reactions [19], the group of Yu developed several catalytic systems for enantioselective directed C—H bond functionalization. Using a Pd(II) catalyst and the chiral ligand **L2**, the enantioselective C—H arylation and alkylation of cyclopropylamide **6**, bearing a 4-cyano tetrafluoroaniline amide as a directing group, afforded products **7** in good yields and with 62–92% enantiomeric excesses (Scheme 11.4) [20]. Cyclopropyl and cyclobutyl moieties are indeed more reactive in C-H activation reactions, as their C(sp^3)—H bonds exhibit similar electronic properties as C(sp^2)—H bonds [21], and the rigid cyclic system allows for a more efficient C-H activation step [13].

Thus, cyclobutyl substrates were naturally employed in the study of new catalytic systems for enantioselective transition-metal-catalyzed C(sp^3)—H activation. Through a thorough ligand design, the group of Yu performed the enantioselective β-arylation (Scheme 11.5a) [22] and borylation (Scheme 11.5b) [23] of cyclobutyl amides **8** from aryl boronic esters and bis(pinacolato)diboron in the presence of *N*-protected α-amino-*O*-methylhydroxamic acid or acetyl-protected aminomethyl oxazoline ligands **L3** and **L4a**. The corresponding products **9** and **10** were obtained with high enantioselectivities. This approach has also been leveraged for the arylation and olefination of similar substrates in the presence of aryl and vinyl iodides [24] and for the desymmetrization of acyclic amides [23].

More recently, Colobert, Wencel-Delord, and co-workers developed a chiral sulfoxide ligand that enabled the smooth Pd-catalyzed arylation and alkynylation

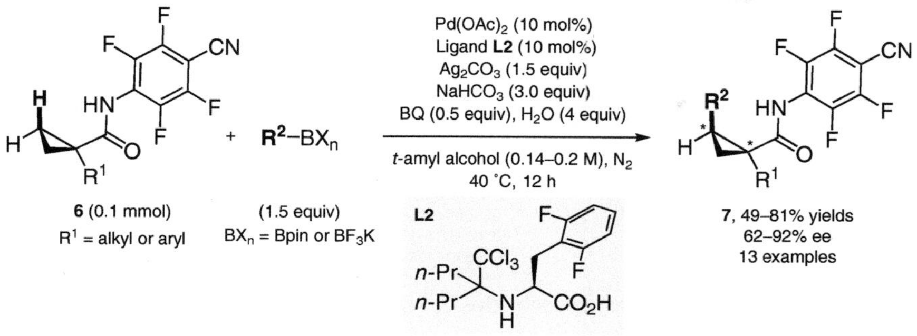

Scheme 11.4 Enantioselective directed C–H arylation and alkylation on cyclopropyl amides. BQ = 1,4-benzoquinone.

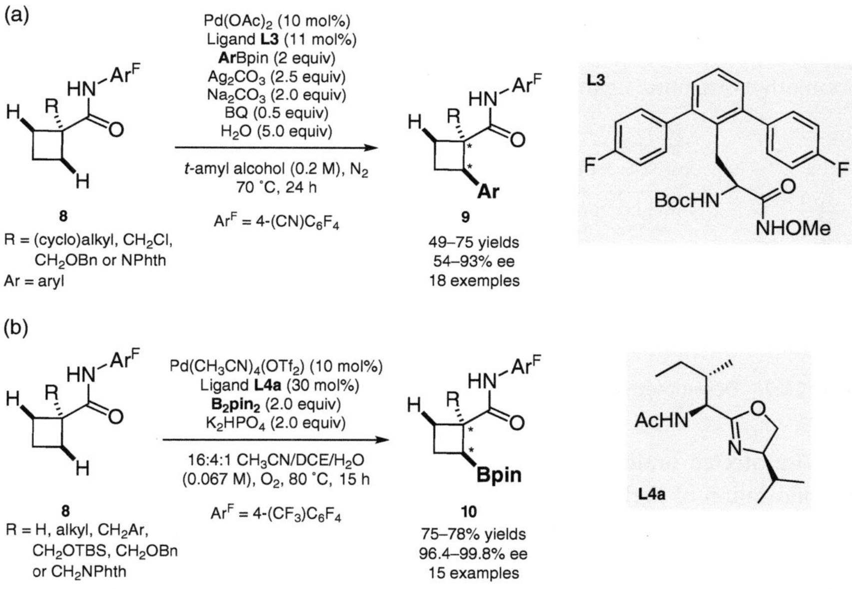

Scheme 11.5 Enantioselective functionalization of cyclobutyl amide substrates.

of cyclopropyl amides [25]. The group of Yu designed acetyl-protected aminoethyl quinoline ligands such as **L5a** to perform the arylation of cycloalkyl amides with larger rings, in particular, the cyclohexylamide **11** (Scheme 11.6) [26]. The reaction is effective for the arylation of cyclopentane and cycloheptane derivatives as well, and the stereoselectivity is greatly influenced by the substitution pattern of the aryl iodide, suggesting that the stereo-determining step is most likely the oxidative addition step in this transformation.

It is worth mentioning that the functionalization of C—H bonds on non-cyclic aliphatic amides appeared much more challenging. Indeed, the C(sp^3)—H bonds on this kind of methylene moieties are less reactive in C—H activation reactions, especially due to their higher heterolytic bond dissociation energies [21]. Nevertheless,

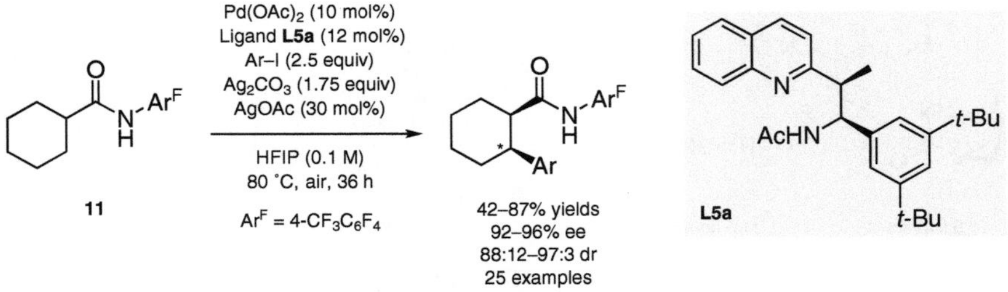

Scheme 11.6 Enantio- and diastereoselective C–H arylation of the cyclohexylamide **11**.

elegant approaches have been developed in order to functionalize such C(sp^3)—H bonds in an enantioselective fashion through the use of directing groups. Amides **12** bearing a chiral center at the α-position can be synthesized with high enantiomeric excesses *via* the selective C—C bond formation at the β-position, allowing for a desymmetrization of these two methyl groups in the presence of acetyl-protected aminomethyl oxazoline ligands **L4b** (Scheme 11.7) [27].

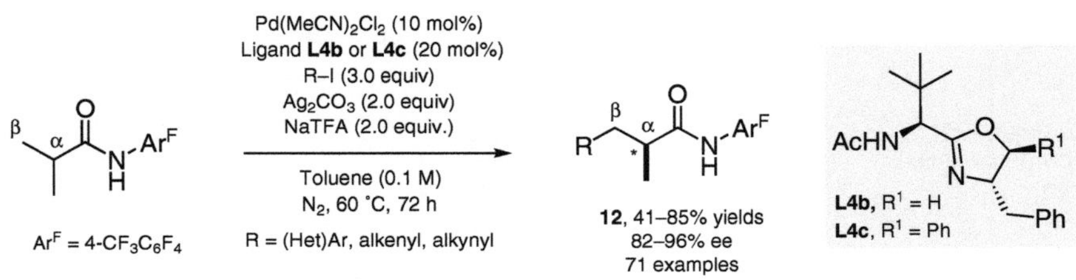

Scheme 11.7 Desymmetrization of isobutyric amides.

Acetyl-protected aminoethyl quinoline ligand **L5**, used for the stereoselective functionalization of cyclohexyl amides derivatives (Scheme 11.6) [26], was developed at first for the β-C–H arylation of linear substrates **13** (Scheme 11.8) [28]. Arylated products **14** were obtained with high efficiency and high enantioselectivity. The ligands **L5** can form a six-membered ring bis-chelated palladium complex (quinoline–Pd–NAc), which greatly accelerates the C—H activation event on the methylene group in the CMD process.

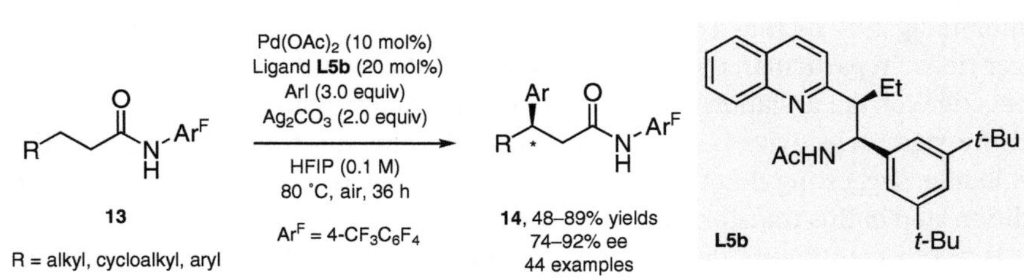

Scheme 11.8 Enantioselective C–H arylation.

Thioamides as Directing Groups Several strategies to perform the enantioselective transition-metal-catalyzed C(sp^3)—H activation on amine derivatives have also been developed, using thioamides as directing groups. These moieties were shown to act as suitable monodentate directing groups able to induce efficient enantioselective C—H bond activation in the presence of chiral ligands. The group of Yu reported the enantioselective Pd-catalyzed α-arylation of cyclic and acyclic thioamides **15** using the chiral phosphoric acid **L6a** as an anionic ligand (Scheme 11.9a) [29], furnishing the products **16** with high enantioselectivities. In the same vein, compounds **17** were arylated to form the products **18** with high yields and outstanding enantioselectivities in the presence of a hybrid palladium catalyst containing two different ligands: the phosphoramidite **L7a** and the Co(III) complex **L8**, which would presumably coordinate the palladium center with one of its carboxylate ligands (Scheme 11.9b) [30].

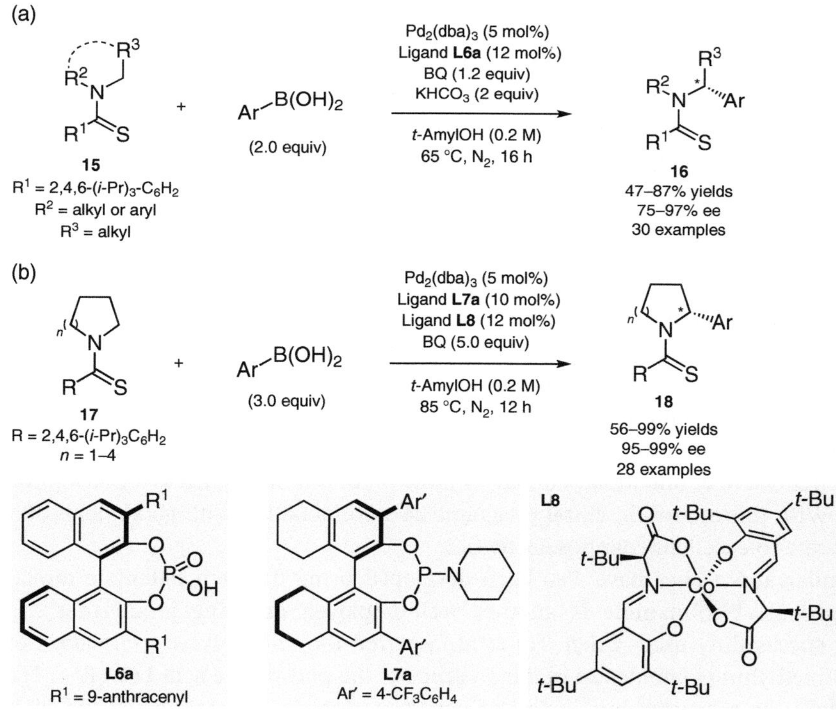

Scheme 11.9 Enantioselective arylation of thioamides.

Use of Bidentate Directing Groups In contrast with the previously discussed weakly coordinating monodentate directing groups, the use of bidentate directing groups such as those derived from 8-aminoquinoline was initially considered challenging due to the lack of accessible coordination sites on the metal center remaining for chiral ligands and the higher reactivity of the substrate itself in the absence of external ligand [31, 32]. As such, the use of strong directing groups, even in the association with a chiral ligand, tends to allow the C—H activation to occur without the ligand to induce the stereoselectivity and consequently lead to a non-enantioselective background reaction. Nevertheless, several strategies have been developed to stereoselectively functionalize compounds bearing strongly chelating directing groups in association with a chiral ligand for the palladium catalyst.

The use of acidic additives is common in transition-metal-catalyzed $C(sp^3)$—H activation reactions, as they are known to facilitate the cleavage of the C—H bond in the CMD mechanism [33]. Duan and co-workers hypothesized in 2015 that chiral acidic catalysts could be used to control the stereoselectivity of the C—H bond activation for the β-functionalization of 8-aminoquinoline amides **19** (Scheme 11.10a) [34]. The C2-symmetric chiral phosphoramidite **L9** was shown to be effective in inducing the enantioselective arylation of these substrates in the presence of iodide partners to afford the enantioenriched amides **20** with moderate-to-good enantioselectivities (26–80% ee). The groups of Chen, Peng, and He later demonstrated that this kind of enantioselective functionalization could be realized through a Pd(0)/Pd(II) mechanism, which was unprecedented for transition-metal-catalyzed $C(sp^3)$—H activation of derivatives using a bidentate directing group strategy (Scheme 11.10b) [35]. In the presence of phosphoramidite **L7b** and an additional ligand **L10**, products **21** were obtained with good enantioselectivities. The same group then showed that the use of the amide derived from 5-substituted 8-aminoquinoline as a directing group allowed the efficient Pd-catalyzed enantioselective arylation, alkenylation, and alkynylation of a wider range of compounds [36]. The synthesis of β-aryl-β-lactams *via* the Pd-catalyzed enantioselective intramolecular $C(sp^3)$—H amidation of 3-arylpropylamines **22** was reported in the presence of a difluoro-BINOL ligand **L11a** and an excess of aryl iodide **23** (Scheme 11.10c) [37]. The latter acted as an oxidant, able to generate a palladium(IV) intermediate after the C—H activation step, while limiting the C—C reductive elimination pathway in favor of an intramolecular C—N coupling [38]. With this approach, chiral β-lactams **24** were obtained with good yields and moderate-to-excellent enantioselectivities.

Similar approaches have also been developed using different bidentate directing groups. Picolinamide is another well-employed directing group that has been successfully used. Chen, He *et al.* reported the Pd-catalyzed γ-arylation of 1,1-dimethylpropylamides **25** in the presence of the phosphoric acid **L6b** (R^1 = H), to afford the corresponding enantioenriched products **26** in high yields and high enantioselectivities (Scheme 11.11a) [39]. A similar catalytic system was then used for the desymmetrization of bisaryl compounds **27** *via* a Pd-catalyzed atroposelective $C(sp^3)$—H arylation (Scheme 11.11b) [40].

Shi and coworkers have shown that the use of 2-pyridinylisopropyl auxiliaries as directing groups, previously designed in their group [41], enabled the Pd-catalyzed enantioselective β-arylation of compounds **28** in the presence of the phosphoric acid **L6d** (Scheme 11.12a) [42]. Their catalytic system was able to perform the C—H activation on methylene positions that are not electronically activated by the presence of an aryl group, and a large variety of arylated amides **29** were obtained with good

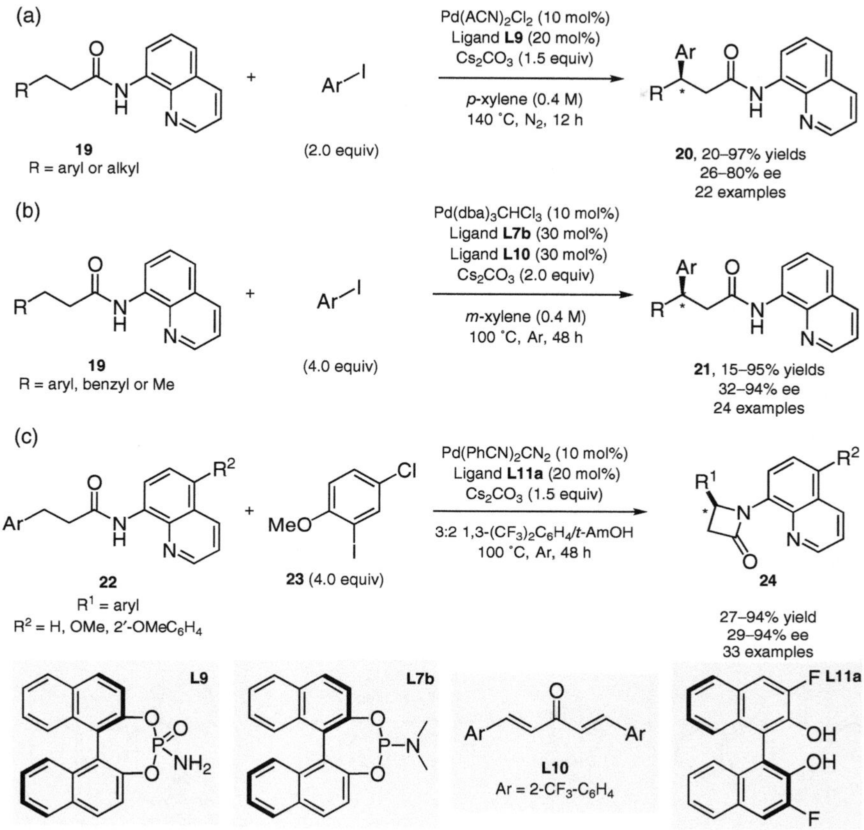

Scheme 11.10 Enantioselective 8-aminoquinoline-directed C–H functionalization.

yields and moderate-to-excellent enantiomeric excesses from aryl bromides as coupling partners. This strategy was later extended to the alkynylation of amides **30** in the presence of alkynyl bromides and the BINOL ligand **L11b** under palladium catalysis (Scheme 11.12b) [43].

11.2.1.2 Transition Metal-catalyzed C(sp^3)–H Activation Directed via the Oxidative Addition of Palladium on Aryl Halides or Pseudo Halides

The oxidative addition on a well-positioned aryl halide moiety of a Pd(0) catalyst can provide a Pd(II) intermediate with the metal positioned in proximity to a C(sp^3) center. In 2011, Kundig and co-workers used this strategy for the enantioselective methylene C(sp^3)—H functionalization of *N*-cycloalkyl-carbamates **32** in the presence of a chiral *N*-heterocyclic carbene ligand (formed from precursor **L12**, Scheme 11.13a) [44]. The reaction afforded the substituted fused indoline scaffolds **33** in good yields and enantioselectivities (up to 95% ee). Similarly, the group of Kagan developed an enantioselective method for the synthesis of 2-substituted chiral indolines **34** using the chiral diphosphine ligand **L13** and $Pd(OAc)_2$ as a catalyst, from various 2-halo-*N*-alkyl anilide **35** derivatives (Scheme 11.13b) [45]. Cramer and co-workers

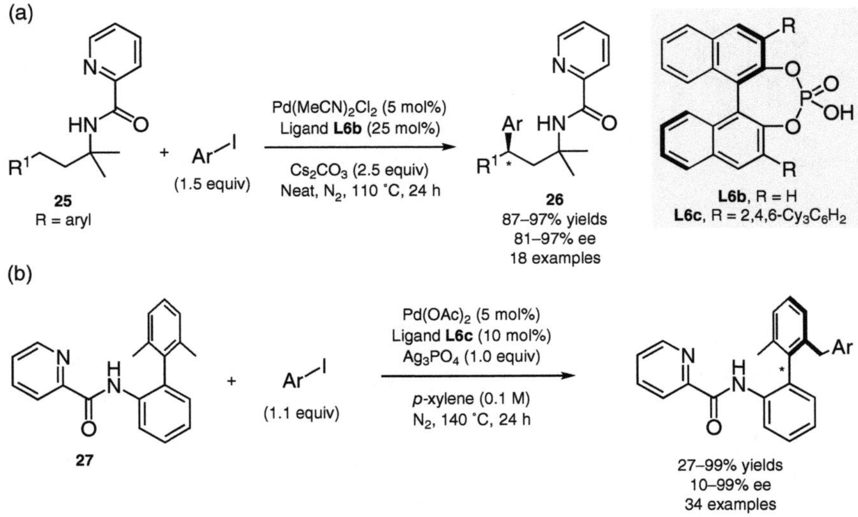

Scheme 11.11 Picolinamide-directed asymmetric $C(sp^3)$−H arylations.

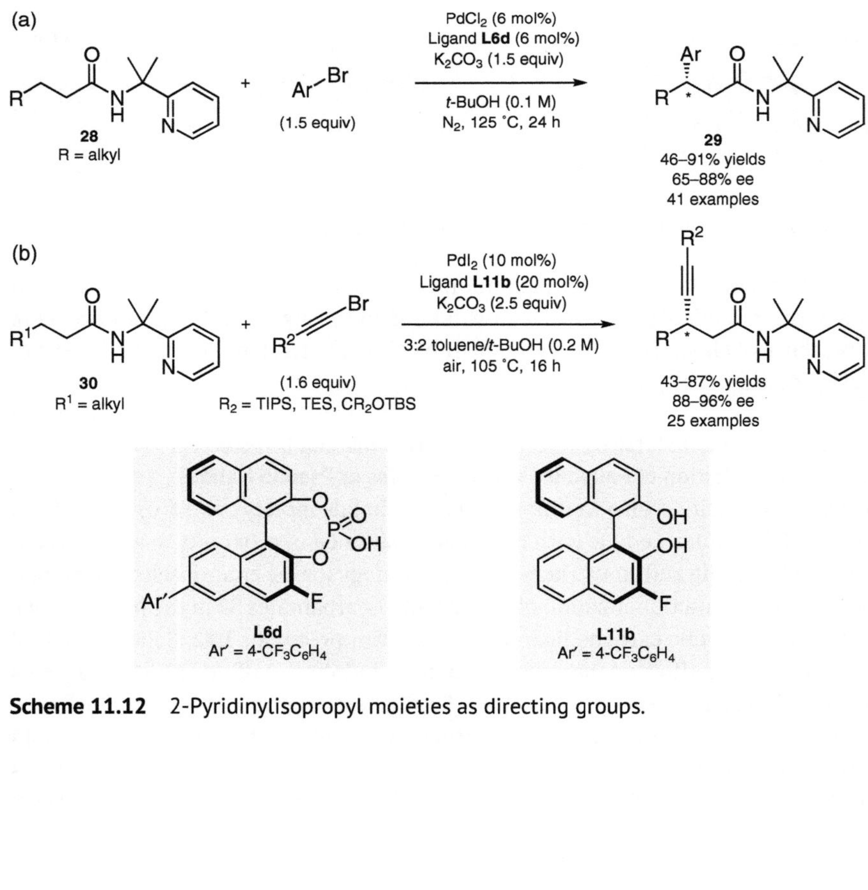

Scheme 11.12 2-Pyridinylisopropyl moieties as directing groups.

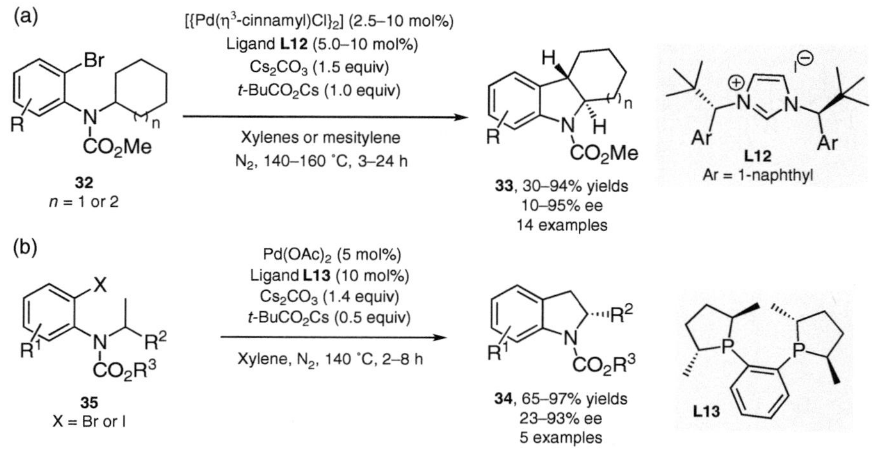

Scheme 11.13 Synthesis of fused indolines *via* enantioselective C(sp³)–H activation/cyclization.

used a similar approach on aryl triflates, providing various indolines in high yields and outstanding enantioselectivities with the use of a monodentate phosphine ligand in association with a sterically demanding carboxylic acid [46].

Later, Baudoin *et al.* developed the enantioselective intramolecular C(sp³)–H alkylation of aryl bromides **36** in the presence of the chiral phosphine ligand **L14**, allowing the synthesis of enantioenriched cyclopentane derivatives **37** in good-to-excellent yields and with high enantioselectivity (up to 80% ee, Scheme 11.14) [47].

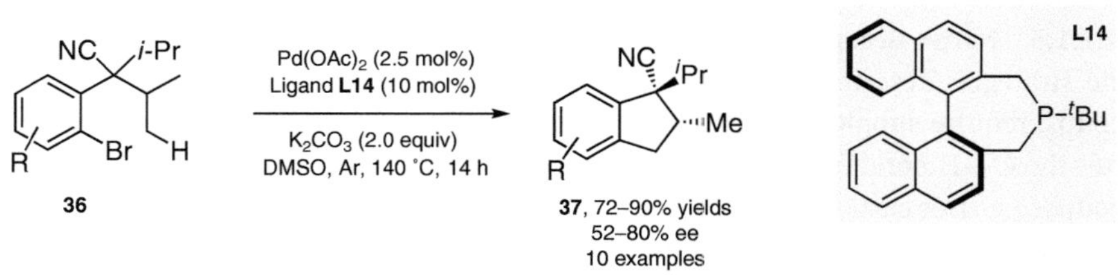

Scheme 11.14 Intramolecular C(sp³)–H arylation for the synthesis of substituted cyclopentanes.

The highly enantioselective synthesis of numerous functionalized bicyclo[*m.n.*1] skeletons **38** through the Pd-catalyzed desymmetrization of 1,3-diketones **39** was developed by Shi and co-workers (Scheme 11.15) [48]. For this transformation, the bidentate ferrocene ligand **L15** was key for the reactivity of the catalytic system and the high enantioselectivity of the reaction.

Shi and co-workers later designed a Pd(II)-catalyzed enantioselective synthesis of chiral bicyclic compounds **40** from bromoaryl derivatives **41** (Scheme 11.16) [49]. Using the strongly chelating 2-pyridinylisopropyl amide as a bidentate directing group, the enantioselective intramolecular arylation of a methylene C(sp³)–H bond at the β-position was achieved in the presence of phosphoric acid ligands **L6e** and **L6f**.

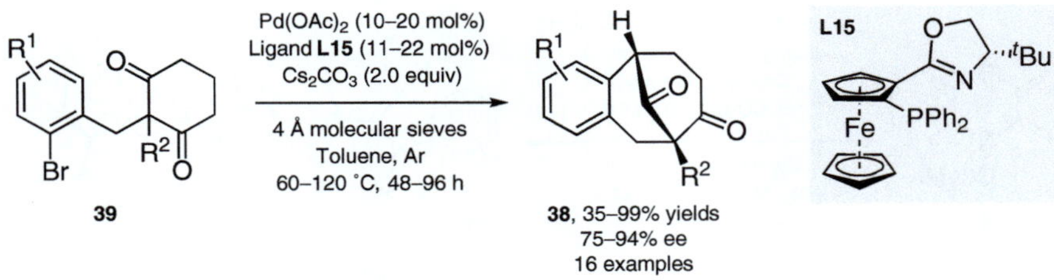

Scheme 11.15 Desymmetrization of 1,3-diketones to construct bicyclo[*m*.*n*.1] scaffolds.

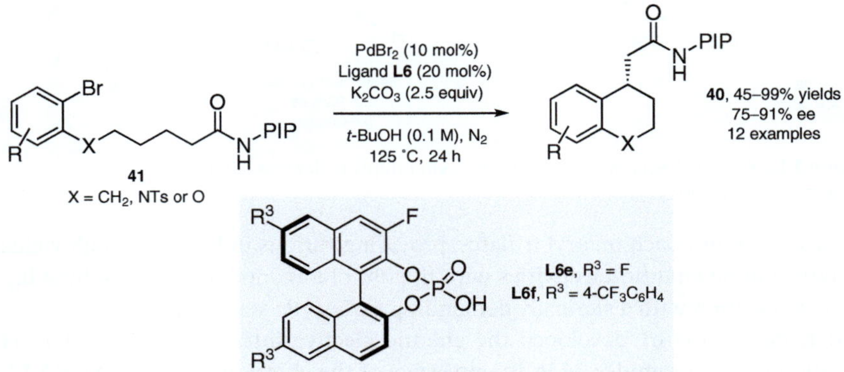

Scheme 11.16 Enantioselective synthesis of bicyclic scaffolds *via* an oxidative addition/C–H activation sequence.

11.2.1.3 Native Group-directed Asymmetric C(sp^3)–H Activation

The strategies described in the first Section (2.1.1), based on the use of directing groups, require supplementary steps to install and remove these functional groups after the C—H bond functionalization reaction. Hence, the use of native functional groups (e.g., free carboxylic acids, amines, or alcohols) as chelating auxiliaries for the transition-metal-catalyzed enantioselective C(sp^3)—H bond functionalization has attracted much attention due to its inherent step-economy compared with classical exogeneous directing groups.

In 2018, the group of Yu designed the first enantioselective directed transition-metal-catalyzed C(sp^3)—H activation using a native function as a directing group. The carboxylic acid moiety on substrates **42** is indeed able to act as a weakly coordinating group for the metal catalyst but usually suffers from a low reactivity for the C—H bond activation and a difficult control of the stereoselectivity due to the inherent flexibility of the resulting palladium-carboxylate (Scheme 11.17a) [50]. The authors designed the ethylenediamine-derived chiral ligand **L16a**, which enabled the enantioselective C(sp^3)—H bond arylation of various cyclopropyl carboxylic acids **42**, affording the corresponding enantioenriched products **43** in good yields and high enantioselectivities. This approach was extended to the Pd-catalyzed enantioselective arylation of 2-aminoisobutyric acids **44** *via* the desymmetrization of the geminal dimethyl groups. Yu and co-workers then leveraged this new

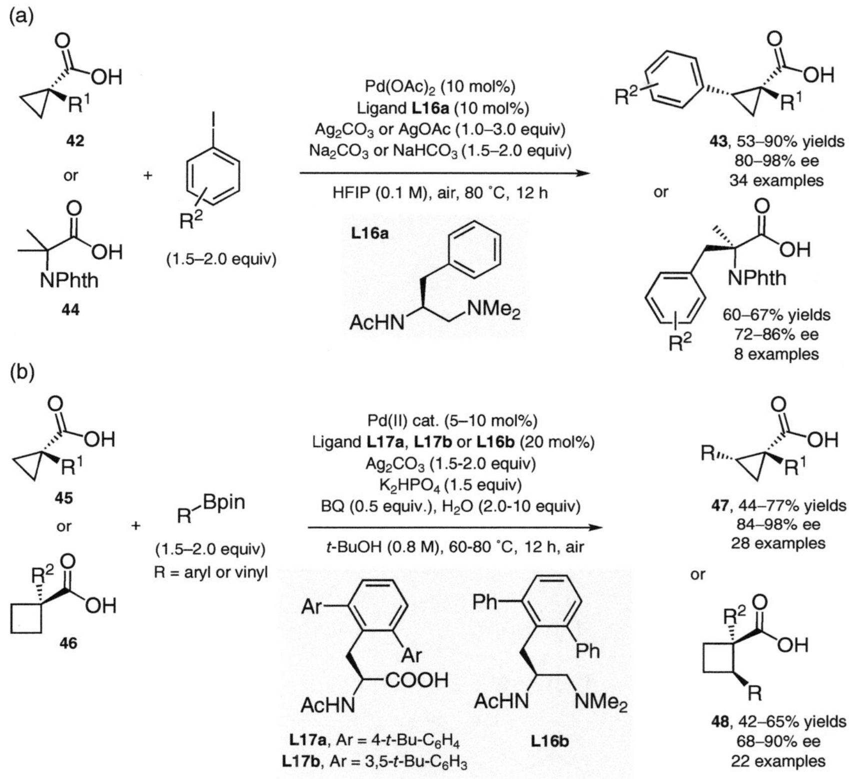

Scheme 11.17 Pd-catalyzed enantioselective C(sp^3)–H arylation of carboxylic acids.

approach for the enantioselective arylation and vinylation of cycloalkyl carboxylic acids **45** and **46** in the presence of monoprotected amino acids and aminoethyl amine ligands **L17a**, **L17b**, or **L16b**, using aryl and vinyl boronic ester as coupling partners through a Pd(II)/Pd(0) catalytic cycle (Scheme 11.17b) [51]. The chiral carboxylic acids **47** and **48** were obtained with moderate-to-good yields with high enantioselectivities (up to 90% ee).

The design of a new chiral oxazoline-pyridone ligand **L18a**, able to form six-membered ring metallacycles, allowed the site-selective and enantioselective γ-arylation of cycloalkyl carboxylic acids **49** to form the corresponding products **50** with high enantiomeric excesses (up to 99% ee) (Scheme 11.18) [52]. The same kind of ligand **L18b**, bearing a bulkier aryl substituent on the oxazoline moiety, was also shown to enable the more distal δ-C—H bond activation on substrates **51** to access the corresponding acids **52** with high selectivity.

Besides carboxylic acids, primary and secondary amines have also been used as native directing groups to perform transition-metal-catalyzed enantioselective C—H bond functionalization. Amines are challenging substrates in palladium-catalyzed reactions, especially due to their propensity to form stable and unreactive

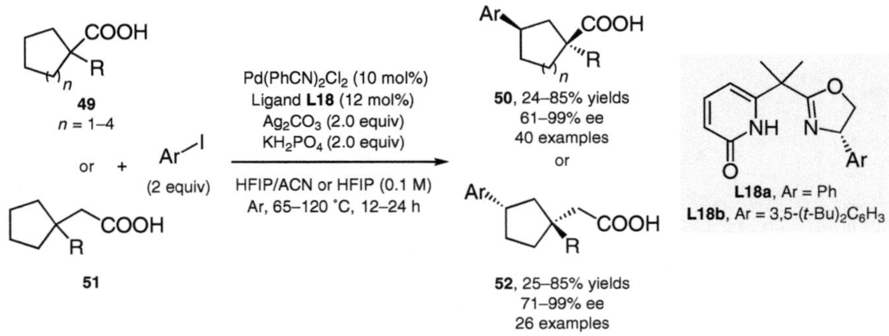

Scheme 11.18 Remote γ- and δ-C—H arylations of cycloalkyl carboxylic acids.

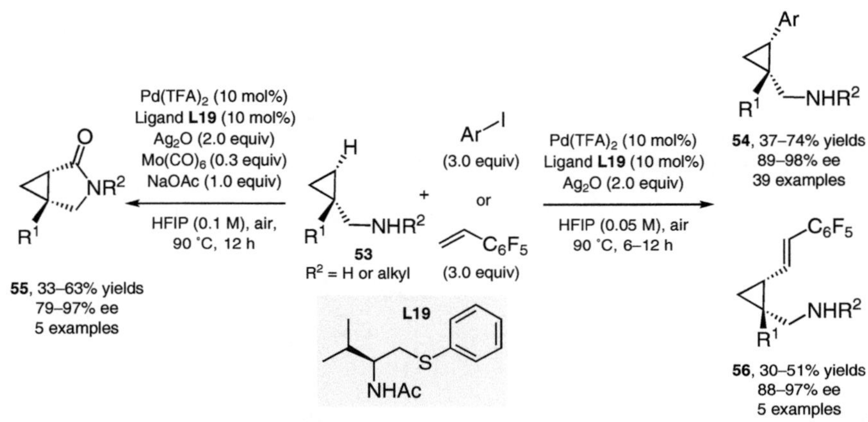

Scheme 11.19 Enantioselective C(sp^3)—H functionalization of cyclopropylmethylamine.

bis(amine)-Pd(II) complexes. To perform the C—H functionalization of cyclopropylamines **53**, the group of Yu designed the thioether ligand **L19** (Scheme 11.19) [53]. Its strong coordination ability and soft σ-donor character allowed the dissociation of the bis(amine)-Pd species and the formation of the reactive mono(amine)-Pd(II) complex that could undergo the desired cyclopalladation. The chiral amines **54** were obtained with excellent enantioselectivities in the presence of aryl iodides as coupling partners and were isolated as the corresponding carbamates after protection with benzoyl or *tert*-butoxycarbonyl groups. This approach was also successfully applied to the enantioselective C(sp^3)—H vinylation and carbonylation of amines **53** to afford a broad range of γ-lactams **55** and olefinated products **56** in good yield and with high enantioselectivity (up to 97% ee).

11.2.2 Use of Other Transition Metal Catalysts

All the previously discussed approaches used palladium(II) catalysts, which have been overwhelmingly privileged in enantioselective transition-metal-catalyzed C—H bond functionalization. Nevertheless, over the last decade, the reactivity of group 9 metals (such as rhodium, iridium, and cobalt) has also been leveraged in innovative synthetic methods for enantioselective C(sp^3)—H bond functionalization.

In 2011, Shibata and co-workers reported the first iridium-catalyzed asymmetric C(sp^3)—H activation approach, allowing for the enantioselective alkylation of 2-(ethylamino)pyridines **57** (Scheme 11.20) [54]. In the presence of an Ir(I) precursor, the BINAP ligand **L20** and olefines **58** as the coupling partners and the chiral amines **59** were obtained in good yields and high stereoselectivity. This development has represented one of the first highly enantioselective transition-metal-catalyzed C(sp^3)—H activation on an electronically unbiased methylene position along with Yu's work on palladium catalysis (Scheme 11.1) [17]. A similar catalytic system was used for the enantioselective C(sp^3)—H alkylation of 2-pyridinyl-substituted γ-butyrolactams [55].

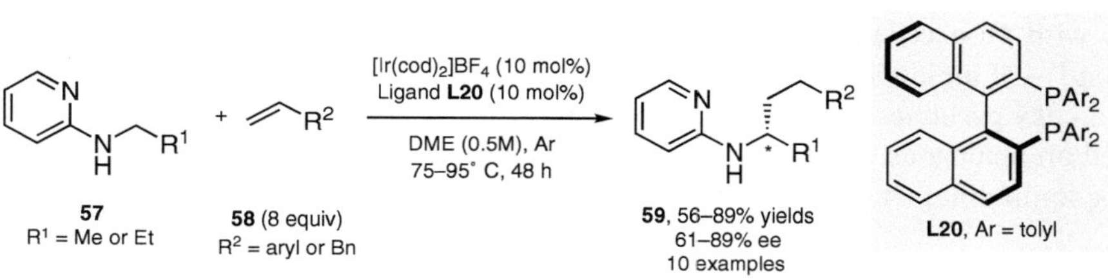

Scheme 11.20 Enantioselective alkylation of 2-(ethylamino)pyridines.

Iridium catalysis has subsequently been used for the asymmetric formation of C—B bonds. Xu and co-workers reported the enantioselective C(sp^3)—H borylation of cyclopropanecarboxamides **60** with (pinacolato)diboron by using an Ir(I) catalyst in the presence of the chiral bidentate borylated ligand **L21** (Scheme 11.21) [56]. The chiral boronic esters **61** were obtained in high yields and excellent enantioselectivities.

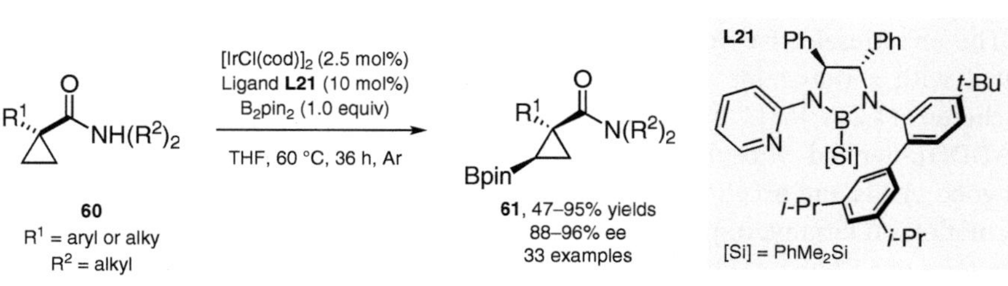

Scheme 11.21 Ir-catalyzed C(sp^3)–H borylation of cyclopropylamides.

In 2016, Glorius and co-workers developed the first rhodium(I)-catalyzed asymmetric C(sp^3)—H activation on quinolines **62** (Scheme 11.22) [57]. They used the N-heterocyclic carbene ligand obtained from the precursor **L22** to induce the selective C—H bond cleavage on the benzylic position. In the presence of aryl bromides, the corresponding triarylmethanes **63** were obtained with enantiomeric excesses ranging from 48 to 80%.

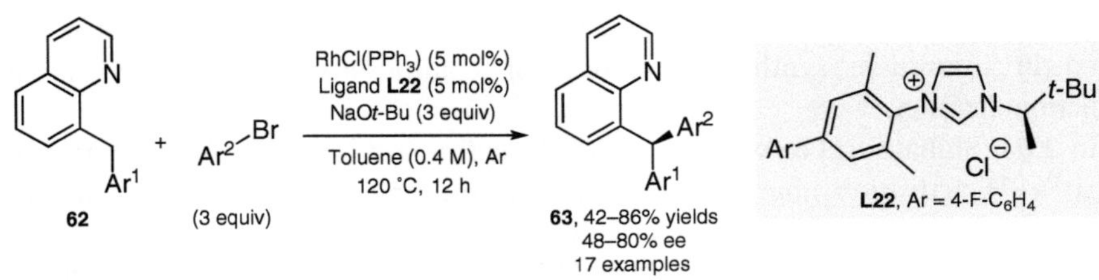

Scheme 11.22 Rh(I)-catalyzed enantioselective C(sp^3)–H bonds arylations.

Since this seminal report, various other strategies have been reported to perform rhodium-catalyzed C(sp^3)—H bond functionalization on 8-alkyl quinoline substrates. Matsunaga, Yoshino, and co-workers developed a highly efficient catalytic system for the amidation of alkylquinolines **64** with dioxazolones **65** in the presence of a chiral binaphthyl carboxylic acid ligand **L23**, affording the chiral amides **66** with an excellent enantioselectivity (up to 88% ee, Scheme 11.23) [58]. This C(sp^3)—H amidation approach was subsequently applied to the functionalization of 2-alkylpyridines through the desymmetrization of geminal methyl groups [59], and an analogous catalytic system was developed for the C(sp^3)—H alkylation of alkylquinolines **64** in the presence of activated olefinic partners [60].

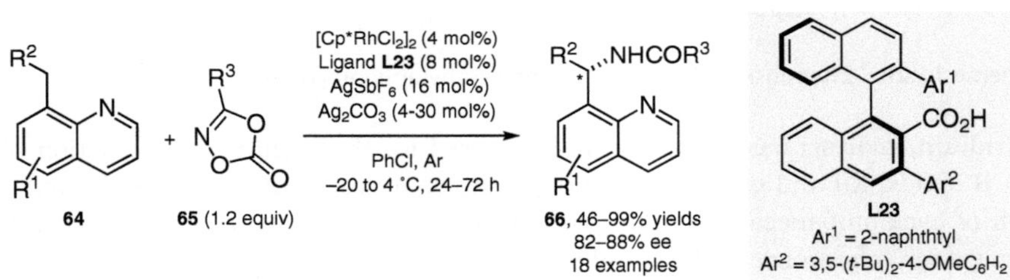

Scheme 11.23 Rh-catalyzed C(sp^3)–H amidation of 8-alkylquinolines.

The enantioselective arylation of various saturated aza-heterocycles **67**, substituted with a thioamide directing group, was reported by the group of Glorius (Scheme 11.24) [61]. In the presence of a rhodium(I) precatalyst and the TADDOL-derived phosphonite ligand **L24**, the heterocycles **68** were obtained in good yields and excellent enantioselectivities. Note that the site-selective transformation on tetrahydroquinoline derivatives is particularly challenging due to the presence of a C(sp^2)—H bond near the directing group.

In 2019, the cobalt(III)-catalyzed enantioselective C(sp^3)—H amidation of thioamides **69** with dioxazolones **70** was reported by the group of Matsunaga and Yoshino (Scheme 11.25) [62]. The use of the carboxylic acid ligand **L25**, involved in the CMD process, in association with a highly sterically demanding Co(III) catalyst was the key to ensuring a high enantioselective process. The products **71** bearing a chiral quaternary carbon center were accessed with overall high yields and excellent enantioselectivities through the desymmetrization of the geminal methyl groups of thioamides **69**. The scope of this transformation was then further extended to a larger variety of thioamides in the presence of a chiral 2-aryl ferrocene carboxylic acid ligand [63].

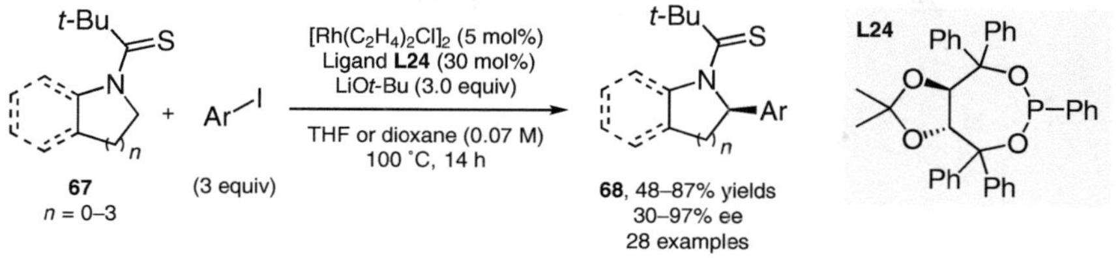

Scheme 11.24 Enantioselective arylation of aza-heterocycles.

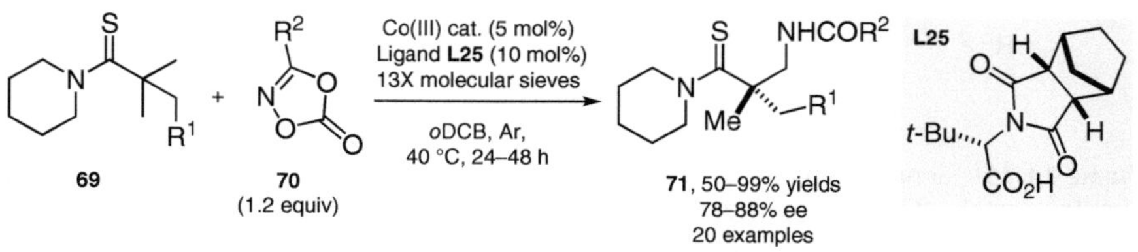

Scheme 11.25 Co-catalyzed C(sp^3)–H amidation of thioamides.

11.3 Chiral Transient Directing Groups for Asymmetric Transition Metal-catalyzed C(sp^3)–H Activation

To avoid the additional synthetic steps of installation and cleavage of the directing group or the use of achiral auxiliary [64], elegant strategies leveraging on the use of transient directing groups have recently emerged to achieve directed enantioselective transition-metal-catalyzed C(sp^3)—H bond activation. Unlike traditional directing groups, the transient directing groups are reversibly linked to the substrate, most of the time *via* the formation of an imine function between a carbonyl and an amine. The pro-directing group **cat1** (or organic catalyst) reacts with the substrate **72**, forming a transient directing group (**int4**, Scheme 11.26). Once the formation of the metallacycle and the C—H functionalization event is achieved (through **int5** and **int6**), the directing group is cleaved to form the product **73**, regenerating the transition-metal catalyst as well as the pro-directing group **cat1**. This strategy was pioneered by Jun and co-workers for rhodium-catalyzed C—H bond functionalization reactions [65] and has been successfully used in elegant developments enabling asymmetric C(sp^3)—H bonds activation reactions, which all use palladium catalysis [66–68].

The first example of such transformations was reported by the group of Yu in 2016 [69]. They showed that the bidentate imine **int7**, formed by the reversible condensation of the chiral amino acid L-*tert*-leucine on 2-alkylbenzaldehydes **74**, was an efficient directing group to perform the asymmetric palladium-catalyzed C(sp^3)—H bond activation on the benzylic position (Scheme 11.27). In the presence of aryl iodides, the chiral arylated benzaldehydes **75** were obtained in good yields and outstanding enantioselectivities (92–96% ee). In 2019, Zhou and co-workers showed that the Pd-catalyzed β-C(sp^3)—H arylation of phenyl acetaldehydes was achieved under similar reaction conditions, but with a much lower enantioselectivity (5–40% ee) [70].

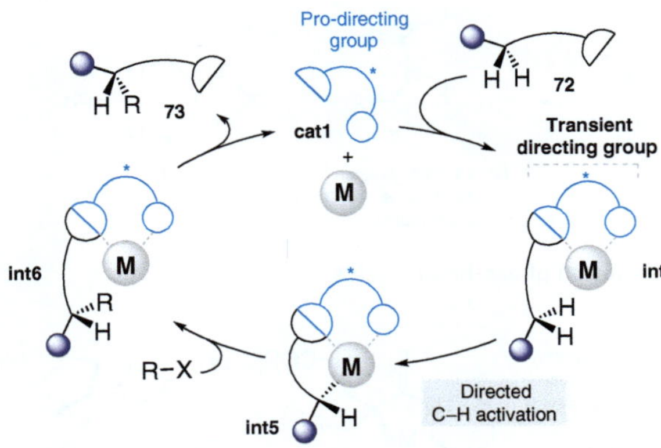

Scheme 11.26 General mechanism of chiral transient directing group in transition-metal-catalyzed C–H bond functionalization. M = transition metal.

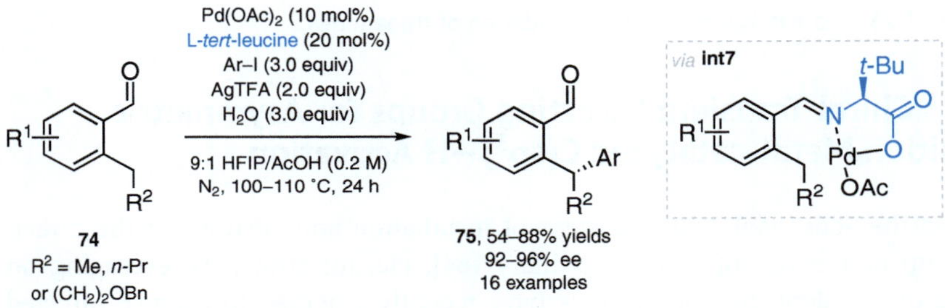

Scheme 11.27 Transient-group-directed enantioselective C(sp^3)–H arylation of benzaldehydes.

Yu and co-workers later extended this strategy to the C(sp^3)—H arylation of cyclobutyl ketones **76** (Scheme 11.28) [71]. Using *D*-Valine as the organocatalyst for the formation of the transient directing group on **int8**, in the presence of palladium(II) acetate and pyridone **L26** as an additional ligand, the β-arylated ketones **77** were obtained in good yields and excellent enantiomeric excesses up to 98%.

In 2018, the same group reported the enantioselective C(sp^3)—H bond fluorination of benzaldehydes using this transient directing group strategy [72]. This kind of transformation appeared particularly challenging due to a difficult reductive elimination towards the formation of the C(sp^3)–F bond. In their reaction conditions employing 1-fluoropyridinium tetrafluoroborate **78** as the fluorinated agent, the C(sp^3)—H fluorination on aldehydes **79** was achieved with good chemoselectivity for the C—F bond formation and high enantioselectivity in the presence of Pd(OAc)$_2$ and the chiral amino amide **cat2** as pro-directing group (Scheme 11.29). In this case, the use of an amino acid organocatalyst instead of the aminoamide **cat2** mainly afforded C(sp^3)–H acetoxylation products in acetic acid. The directing group formed by condensation of an amino acid on aldehydes **79** led to the formation of a five-coordinate Pd(IV) intermediate that renders the desired C(sp^3)—F bond

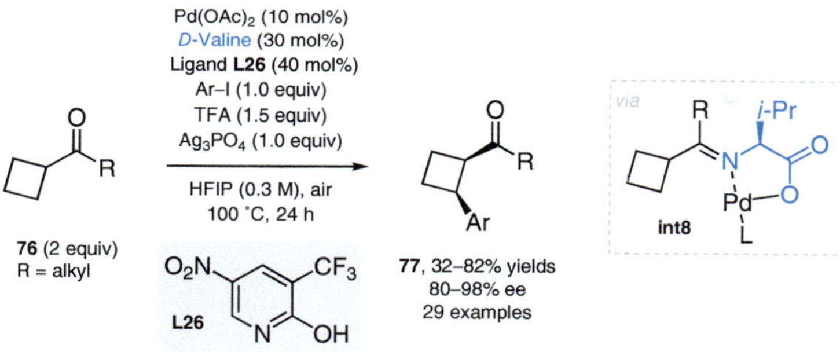

Scheme 11.28 Transient-group-directed β-arylation of cyclobutyl ketones.

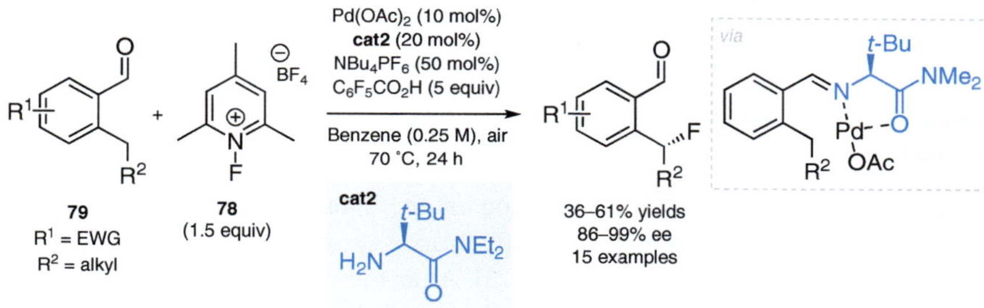

Scheme 11.29 Transient-group-directed enantioselective C(sp^3)–H fluorination.

formation pathway slower than the S$_N$2 C(sp^3)—O bond formation. Conversely, the aminoamide **cat2** allowed the formation of a cationic Pd(IV) species that favors the C(sp^3)–F reductive elimination.

11.4 Supramolecular Assembly-directed Hydrogen Atom Abstractions for Asymmetric C(sp^3)–H Bond Functionalization

Beyond transition-metal-catalyzed C(sp^3)—H bonds activation strategies, hydrogen atom abstraction approaches, related to the biological activity of cytochrome P450, have emerged to perform directed C—H bond functionalization reactions. The reactivity leveraged in these approaches relies on hydrogen atom abstraction by metal–oxo complexes or rhodium–nitrene species, with a reactive center that is positioned in proximity to the C—H bond to functionalize *via* supramolecular interactions. Consequently, several of these transformations are encompassed in the field of asymmetric directed C—H bond functionalization, as the supramolecular assembly, besides enabling site-selective reactions, also allows to perform enantioselective transformations.

In 1997, the group of Breslow developed several bio-inspired reactions for the regio- and stereoselective hydroxylation of steroid substrates. They designed complex catalysts containing a metalloporphyrin substituted with cyclodextrins,

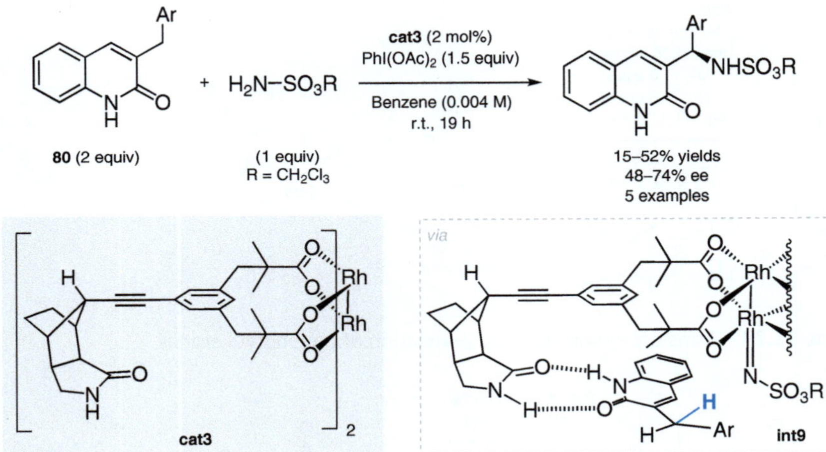

Scheme 11.30 Supramolecular assembly-directed enantioselective amination with rhodium nitrenes.

which ensured the molecular recognition of the substrate and the positioning of the targeted hydroxylation position in proximity of reactive oxo–manganese center *via* hydrophobic interactions [73–75]. After these pioneering works, several methods have been developed to perform C(sp^3)—H bond functionalizations using supramolecular interactions, such as hydrogen bonding, with simplified recognition modes. Bach and co-workers designed a chiral U-shaped H-bonding handle bonded to a bis-rhodium complex **cat3** for the enantioselective amination of a benzylic C(sp^3)—H bond on 2-quinolone substrates **80** (Scheme 11.30) [76]. H-bonding between the lactam moiety of **cat3** and the quinolone **80** was proposed to allow the presentation of one of the two enantiotropic hydrogens to the rhodium nitrene for C—H insertion in intermediate **int9**.

They then developed several analogous catalytic scaffolds based on similar H-bonding handles. Catalyst **cat4** bearing an Mn(III)-porphyrin complex allowed for the enantioselective exocyclic hydroxylation of quinolones **81** and the 4-hydroxylation of 3,4-dihydroquinolones in the presence of iodosobenzene (PhIO), *via* the formation of the oxo–manganese complex **int10**. (Scheme 11.31a) [77, 78]. Similarly, ligand **L27** was used in association with a silver complex to perform the enantioselective amination of 2-quinolones and 2-pyridones **82**, affording the corresponding chiral sulfonamides **83** in moderate-to-good yields with excellent enantioselectivities (Scheme 11.31b) [79].

11.5 Summary and Conclusions

In conclusion, numerous approaches for transition-metal-catalyzed asymmetric directed C(sp^3)—H bond functionalizations have been developed, allowing to access complex enantioenriched scaffolds in a single step from non-activated C—H bonds. Complementary strategies using well-designed chiral catalysts, transient

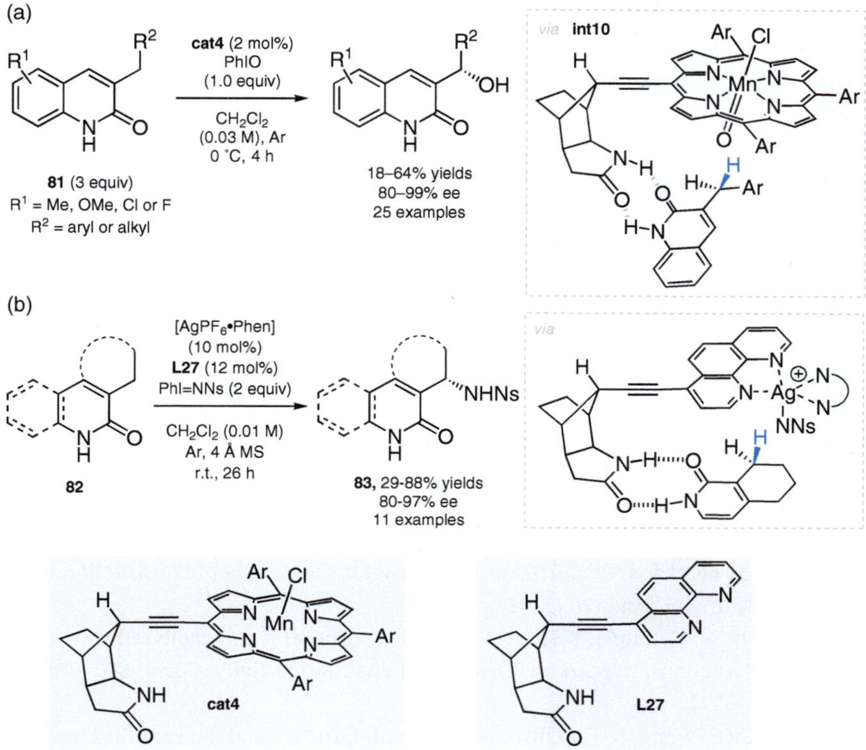

Scheme 11.31 Enantioselective C(sp^3)—H hydroxylation and amination directed by supramolecular assemblies.

directing groups, or supramolecular approaches have been designed, shedding light on promising alternatives to the use of chiral auxiliaries.

Nevertheless, the field is still in its infancy. Considering the importance of chiral molecules in organic chemistry and in compounds of interest, the developments discussed in this chapter are likely to precede many more breakthroughs in the future. Major advances in the selective functionalization of a wider range of derivatives and access to even more complex molecular scaffolds in a stereoselective fashion are expected.

References

1 Davies, H.M.L. and Morton, D. (2017). Collective approach to advancing C–H functionalization. *ACS Cent. Sci.* 3 (9): 936–943.

2 Gandeepan, P., Müller, T., Zell, D. et al. (2019). 3d transition metals for C–H activation. *Chem. Rev.* 119 (4): 2192–2452.

3 Rogge, T., Kaplaneris, N., Chatani, N. et al. (2021). C–H activation. *Nat. Rev. Methods Primers* 1 (1): 43.

4 Dalton, T., Faber, T., and Glorius, F. (2021). C–H activation: toward sustainability and applications. *ACS Cent. Sci.* 7 (2): 245–261.

5 Lam, N.Y.S., Wu, K., and Yu, J.-Q. (2021). Advancing the logic of chemical synthesis: C–H activation as strategic and tactical disconnections for C–C bond construction. *Angew. Chem. Int. Ed.* 60 (29): 15767–15790.

6 Kancherla, S., Jørgensen, K.B., and Fernández-Ibáñez, M.Á. (2019). Recent developments in palladium-catalysed non-directed C–H bond activation in arenes. *Synthesis* 51 (03): 643–663.

7 Lyons, T.W. and Sanford, M.S. (2010). Palladium-catalyzed ligand-directed C–H functionalization reactions. *Chem. Rev.* 110 (2): 1147–1169.

8 Jazzar, R., Hitce, J., Renaudat, A. et al. (2010). Functionalization of organic molecules by transition-metal-catalyzed C(sp^3)–H activation. *Chem. Eur. J.* 16 (9): 2654–2672.

9 Neufeldt, S.R. and Sanford, M.S. (2012). Controlling site selectivity in palladium-catalyzed C–H bond functionalization. *Acc. Chem. Res.* 45 (6): 936–946.

10 Rej, S., Ano, Y., and Chatani, N. (2020). Bidentate directing groups: an efficient tool in C–H bond functionalization chemistry for the expedient construction of C–C bonds. *Chem. Rev.* 120 (3): 1788–1887.

11 Giri, R., Shi, B.-F., Engle, K.M. et al. (2009). Transition metal-catalyzed C–H activation reactions: diastereoselectivity and enantioselectivity. *Chem. Soc. Rev.* 38 (11): 3242–3272.

12 Newton, C.G., Wang, S.-G., Oliveira, C.C. et al. (2017). Catalytic enantioselective transformations involving C–H bond cleavage by transition-metal complexes. *Chem. Rev.* 117 (13): 8908–8976.

13 Saint-Denis, T.G., Zhu, R.-Y., Chen, G. et al. (2018). Enantioselective C(sp^3)–H bond activation by chiral transition metal catalysts. *Science* 359 (6377): eaao4798.

14 Woźniak, Ł. and Cramer, N. (2019). Enantioselective CH bond functionalizations by 3d transition-metal catalysts. *Trends Chem.* 1 (5): 471–484.

15 Mas-Roselló, J., Herraiz, A.G., Audic, B. et al. (2021). Chiral cyclopentadienyl ligands: design, syntheses, and applications in asymmetric catalysis. *Angew. Chem. Int. Ed.* 60 (24): 13198–13224.

16 Zhang, Y., Zhang, J.-J., Lou, L. et al. (2024). Recent advances in Rh(I)-catalyzed enantioselective C–H functionalization. *Chem. Soc. Rev.* 53 (7): 3457–3484.

17 Shi, B.-F., Maugel, N., Zhang, Y.-H. et al. (2008). Pd^{II}-catalyzed enantioselective activation of C(sp^2)–H and C(sp^3)–H bonds using monoprotected amino acids as chiral ligands. *Angew. Chem. Int. Ed.* 47 (26): 4882–4886.

18 Li, H.-L., Yang, D.-F., Jing, H.-Q. et al. (2022). Palladium-catalyzed enantioselective C(sp^3)–H arylation of 2-propyl azaaryls enabled by an amino acid ligand. *Org. Lett.* 24 (6): 1286–1291.

19 Wasa, M., Engle, K.M., and Yu, J.-Q. (2009). Pd(0)/PR_3-catalyzed intermolecular arylation of sp^3 C–H bonds. *JACS* 131 (29): 9886–9887.

20 Wasa, M., Engle, K.M., Lin, D.W. et al. (2011). Pd(II)-catalyzed enantioselective C–H activation of cyclopropanes. *JACS* 133 (49): 19598–19601.

21 Bach, R.D. and Dmitrenko, O. (2004). Strain energy of small ring hydrocarbons. Influence of C–H bond dissociation energies. *JACS* 126 (13): 4444–4452.

22 Xiao, K.-J., Lin, D.W., Miura, M. et al. (2014). Palladium(II)-catalyzed enantioselective C(sp^3)–H activation using a chiral hydroxamic acid ligand. *JACS* 136 (22): 8138–8142.

23 He, J., Shao, Q., Wu, Q. et al. (2017). Pd(II)-catalyzed enantioselective C(sp^3)–H borylation. *JACS* 139 (9): 3344–3347.

24 Wu, Q.-F., Wang, X.-B., Shen, P.-X. et al. (2018). Enantioselective C–H arylation and vinylation of cyclobutyl carboxylic amides. *ACS Catal.* 8 (3): 2577–2581.

25 Jerhaoui, S., Djukic, J.-P., Wencel-Delord, J. et al. (2019). Asymmetric, nearly barrierless C(sp^3)–H activation promoted by easily-accessible *N*-protected aminosulfoxides as new chiral ligands. *ACS Catal.* 9 (3): 2532–2542.

26 Andrä, M.S., Schifferer, L., Pollok, C.H. et al. (2019). Enantio- and diastereoswitchable C–H arylation of methylene groups in cycloalkanes. *Chem. Eur. J.* 25 (36): 8503–8507.

27 Wu, Q.-F., Shen, P.-X., He, J. et al. (2017). Formation of α-chiral centers by asymmetric β-C(sp3)–H arylation, alkenylation, and alkynylation. *Science* 355 (6324): 499–503.

28 Chen, G., Gong, W., Zhuang, Z. et al. (2016). Ligand-accelerated enantioselective methylene C(sp^3)–H bond activation. *Science* 353 (6303): 1023–1027.

29 Jain, P., Verma, P., Xia, G. et al. (2017). Enantioselective amine α-functionalization *via* palladium-catalysed C–H arylation of thioamides. *Nat. Chem.* 9 (2): 140–144.

30 Jiang, H.-J., Zhong, X.-M., Yu, J. et al. (2019). Assembling a hybrid Pd catalyst from a chiral anionic Co^{III} complex and ligand for asymmetric C(sp^3)–H functionalization. *Angew. Chem. Int. Ed.* 58 (6): 1803–1807.

31 Rouquet, G. and Chatani, N. (2013). Catalytic functionalization of C(sp^2)–H and C(sp^3)–H bonds by using bidentate directing groups. *Angew. Chem. Int. Ed.* 52 (45): 11726–11743.

32 Daugulis, O., Roane, J., and Tran, L.D. (2015). Bidentate, monoanionic auxiliary-directed functionalization of carbon–hydrogen bonds. *Acc. Chem. Res.* 48 (4): 1053–1064.

33 Ackermann, L. (2011). Carboxylate-assisted transition-metal-catalyzed C–H bond functionalizations: mechanism and scope. *Chem. Rev.* 111 (3): 1315–1345.

34 Yan, S.-B., Zhang, S., and Duan, W.-L. (2015). Palladium-catalyzed asymmetric arylation of C(sp^3)–H bonds of aliphatic amides: controlling enantioselectivity using chiral phosphoric amides/acids. *Org. Lett.* 17 (10): 2458–2461.

35 Tong, H.-R., Zheng, S., Li, X. et al. (2018). Pd(0)-catalyzed bidentate auxiliary directed enantioselective benzylic C–H arylation of 3-arylpropanamides using the BINOL phosphoramidite ligand. *ACS Catal.* 8 (12): 11502–11512.

36 Huang, Y., Lv, X., Tong, H.-R. et al. (2024). Palladium-catalyzed enantioselective directed C(sp^3)–H functionalization using C5-substituted 8-aminoquinoline auxiliaries. *Org. Lett.* 26 (1): 94–99.

37 Tong, H.-R., Zheng, W., Lv, X. et al. (2020). Asymmetric synthesis of β-lactam *via* palladium-catalyzed enantioselective intramolecular C(sp^3)–H amidation. *ACS Catal.* 10 (1): 114–120.

38 Sun, W.-W., Cao, P., Mei, R.-Q. et al. (2014). Palladium-catalyzed unactivated C(sp^3)–H bond activation and intramolecular amination of carboxamides: a new approach to β-lactams. *Org. Lett.* 16 (2): 480–483.

39 Wang, H., Tong, H.-R., He, G. et al. (2016). An enantioselective bidentate auxiliary directed palladium-catalyzed benzylic C–H arylation of amines using a BINOL phosphate ligand. *Angew. Chem. Int. Ed.* 55 (49): 15387–15391.

40 Uchikura, T., Kato, S., Makino, Y. et al. (2023). Chiral phosphoric acid–palladium(II) complex catalyzed asymmetric desymmetrization of biaryl compounds by C(sp^3)–H activation. *JACS* 145 (29): 15906–15911.

41 Chen, F.-J., Zhao, S., Hu, F. et al. (2013). Pd(II)-catalyzed alkoxylation of unactivated C(sp^3)–H and C(sp^2)–H bonds using a removable directing group: efficient synthesis of alkyl ethers. *Chem. Sci.* 4 (11): 4187–4192.

42 Yan, S.-Y., Han, Y.-Q., Yao, Q.-J. et al. (2018). Palladium(II)-catalyzed enantioselective arylation of unbiased methylene C(sp^3)–H bonds enabled by a 2-pyridinylisopropyl auxiliary and chiral phosphoric acids. *Angew. Chem. Int. Ed.* 57 (29): 9093–9097.

43 Han, Y.-Q., Ding, Y., Zhou, T. et al. (2019). Pd(II)-catalyzed enantioselective alkynylation of unbiased methylene C(sp^3)–H bonds using 3,3′-fluorinated-BINOL as a chiral ligand. *JACS* 141 (11): 4558–4563.

44 Nakanishi, M., Katayev, D., Besnard, C. et al. (2011). Fused indolines by palladium-catalyzed asymmetric C–C coupling involving an unactivated methylene group. *Angew. Chem. Int. Ed.* 50 (32): 7438–7441.

45 Anas, S., Cordi, A., and Kagan, H.B. (2011). Enantioselective synthesis of 2-methyl indolines by palladium catalysed asymmetric C(sp^3)–H activation/cyclisation. *Chem. Commun.* 47 (41): 11483–11485.

46 Saget, T., Lemouzy, S.J., and Cramer, N. (2012). Chiral monodentate phosphines and bulky carboxylic acids: cooperative effects in palladium-catalyzed enantioselective C(sp^3)–H functionalization. *Angew. Chem. Int. Ed.* 51 (9): 2238–2242.

47 Martin, N., Pierre, C., Davi, M. et al. (2012). Diastereo- and enantioselective intramolecular C(sp^3)–H Arylation for the synthesis of fused cyclopentanes. *Chem. Eur. J.* 18 (15): 4480–4484.

48 Zhu, C., Wang, D., Zhao, Y. et al. (2017). Enantioselective palladium-catalyzed intramolecular α-arylative desymmetrization of 1,3-Diketones. *JACS* 139 (46): 16486–16489.

49 Han, Y.-Q., Zhang, Q., Yang, X. et al. (2021). Pd(II)-catalyzed enantioselective intramolecular arylation of unbiased C(sp3)–H bonds to construct chiral benzo-ring compounds. *Org. Lett.* 23 (1): 97–101.

50 Shen, P.-X., Hu, L., Shao, Q. et al. (2018). Pd(II)-catalyzed enantioselective C(sp^3)–H arylation of free carboxylic acids. *JACS* 140 (21): 6545–6549.

51 Hu, L., Shen, P.-X., Shao, Q. et al. (2019). Pd^{II}-catalyzed enantioselective C(sp^3)–H activation/cross-coupling reactions of free carboxylic acids. *Angew. Chem. Int. Ed.* 58 (7): 2134–2138.

52 Zhang, T., Zhang, Z.-Y., Kang, G. et al. (2024). Enantioselective remote methylene C–H (hetero)arylation of cycloalkane carboxylic acids. *Science* 384 (6697): 793–798.

53 Zhuang, Z. and Yu, J.-Q. (2020). Pd(II)-catalyzed enantioselective γ-C(sp^3)–H functionalizations of free cyclopropylmethylamines. *JACS* 142 (28): 12015–12019.

54 Pan, S., Endo, K., and Shibata, T. (2011). Ir(I)-catalyzed enantioselective secondary sp^3 C–H bond activation of 2-(alkylamino)pyridines with alkenes. *Org. Lett.* 13 (17): 4692–4695.

55 Tahara, Y., Michino, M., Ito, M. et al. (2015). Enantioselective sp^3 C–H alkylation of γ-butyrolactam by a chiral Ir(I) catalyst for the synthesis of 4-substituted γ-amino acids. *Chem. Commun.* 51 (93): 16660–16663.

56 Shi, Y., Gao, Q., and Xu, S. (2019). Chiral bidentate boryl ligand enabled iridium-catalyzed enantioselective C(sp^3)–H borylation of cyclopropanes. *JACS* 141 (27): 10599–10604.

57 Kim, J.H., Greßies, S., Boultadakis-Arapinis, M. et al. (2016). Rh(I)/NHC*-catalyzed site- and enantioselective functionalization of C(sp^3)–H bonds toward chiral triarylmethanes. *ACS Catal.* 6 (11): 7652–7656.

58 Fukagawa, S., Kojima, M., Yoshino, T. et al. (2019). Catalytic enantioselective methylene C(sp^3)–H amidation of 8-alkylquinolines using a Cp*Rh^{III}/chiral carboxylic acid system. *Angew. Chem. Int. Ed.* 58 (50): 18154–18158.

59 Kato, Y., Lin, L., Kojima, M. et al. (2021). Development of pseudo-C2-symmetric chiral binaphthyl monocarboxylic acids for enantioselective C(sp^3)–H functionalization reactions under Rh(III) catalysis. *ACS Catal.* 11 (7): 4271–4277.

60 Huang, L.-T., Fukagawa, S., Kojima, M. et al. (2020). Rhodium(III)/chiral carboxylic acid catalyzed enantioselective C(sp^3)–H alkylation of 8-ethylquinolines with α,β-unsaturated carbonyl compounds. *Org. Lett.* 22 (21): 8256–8260.

61 Greßies, S., Klauck, F.J.R., Kim, J.H. et al. (2018). Ligand-enabled enantioselective C –H activation of tetrahydroquinolines and saturated aza-heterocycles by Rh^I. *Angew. Chem. Int. Ed.* 57 (31): 9950–9954.

62 Fukagawa, S., Kato, Y., Tanaka, R. et al. (2019). Enantioselective C(sp^3)–H amidation of thioamides catalyzed by a $cobalt^{III}$/chiral carboxylic acid hybrid system. *Angew. Chem. Int. Ed.* 58 (4): 1153–1157.

63 Sekine, D., Ikeda, K., Fukagawa, S. et al. (2019). Chiral 2-aryl ferrocene carboxylic acids for the catalytic asymmetric C(sp^3)–H activation of thioamides. *Organometallics* 38 (20): 3921–3926.

64 Jerhaoui, S., Chahdoura, F., Rose, C. et al. (2016). Enantiopure sulfinyl aniline as a removable and recyclable chiral auxiliary for asymmetric C(sp^3)–H bond activation. *Chem. Eur. J.* 22 (48): 17397–17406.

65 Jun, C.-H., Lee, H., and Hong, J.-B. (1997). Chelation-assisted intermolecular hydroacylation: direct synthesis of ketone from aldehyde and 1-alkene. *J. Org. Chem.* 62 (5): 1200–1201.

66 Lapuh, M.I., Mazeh, S., and Besset, T. (2020). Chiral transient directing groups in transition-metal-catalyzed enantioselective C–H bond functionalization. *ACS Catal.* 10 (21): 12898–12919.

67 Bag, D., Verma, P.K., and Sawant, S.D. (2020). Chiral transient directing group strategies in asymmetric synthesis. *Chem. Asian J.* 15 (20): 3225–3238.

68 Liao, G., Zhang, T., Lin, Z.-K. et al. (2020). Transition metal-catalyzed enantioselective C–H functionalization via chiral transient directing group strategies. *Angew. Chem. Int. Ed.* 59 (45): 19773–19786.

69 Zhang, F.-L., Hong, K., Li, T.-J. et al. (2016). Functionalization of $C(sp^3)$–H bonds using a transient directing group. *Science* 351 (6270): 252–256.

70 Gou, B.-B., Liu, H.-F., Chen, J. et al. (2019). Palladium-catalyzed site-selective $C(sp^3)$–H arylation of phenylacetaldehydes. *Org. Lett.* 21 (17): 7084–7088.

71 Xiao, L.-J., Hong, K., Luo, F. et al. (2020). Pd^{II}-catalyzed enantioselective $C(sp^3)$–H arylation of cyclobutyl ketones using a chiral transient directing group. *Angew. Chem. Int. Ed.* 59 (24): 9594–9600.

72 Park, H., Verma, P., Hong, K. et al. (2018). Controlling Pd(IV) reductive elimination pathways enables Pd(II)-catalysed enantioselective $C(sp^3)$–H fluorination. *Nat. Chem.* 10 (7): 755–762.

73 Breslow, R., Zhang, X., and Huang, Y. (1997). Selective catalytic hydroxylation of a steroid by an artificial cytochrome P-450 enzyme. *JACS* 119 (19): 4535–4536.

74 Yang, J. and Breslow, R. (2000). Selective hydroxylation of a steroid at C-9 by an artificial cytochrome P-450. *Angew. Chem. Int. Ed.* 39 (15): 2692–2695.

75 Yang, J., Gabriele, B., Belvedere, S. et al. (2002). Catalytic oxidations of steroid substrates by artificial cytochrome P-450 enzymes. *J. Org. Chem.* 67 (15): 5057–5067.

76 Höke, T., Herdtweck, E., and Bach, T. (2013). Hydrogen-bond mediated regio- and enantioselectivity in a C–H amination reaction catalysed by a supramolecular Rh(II) complex. *Chem. Commun.* 49 (73): 8009–8011.

77 Burg, F., Gicquel, M., Breitenlechner, S. et al. (2018). Site- and enantioselective C–H oxygenation catalyzed by a chiral manganese porphyrin complex with a remote binding site. *Angew. Chem. Int. Ed.* 57 (11): 2953–2957.

78 Burg, F., Breitenlechner, S., Jandl, C. et al. (2020). Enantioselective oxygenation of exocyclic methylene groups by a manganese porphyrin catalyst with a chiral recognition site. *Chem. Sci.* 11 (8): 2121–2129.

79 Annapureddy, R.R., Jandl, C., and Bach, T. (2020). A chiral Phenanthroline ligand with a hydrogen-bonding site: application to the enantioselective amination of methylene groups. *JACS* 142 (16): 7374–7378.

12

Photoredox Catalysis in C—H Bond Functionalization

Sayak Ghosh, James Mortimer, and Patricia Z. Musacchio

The State University of New York at Buffalo, Department of Chemistry, Natural Sciences Complex, Buffalo, New York, 14260, USA

12.1 Introduction

In recent years, photoredox catalysis has become a "go-to" platform for the development of C—H functionalization methods, particularly for C(sp^3)—H bonds. The earliest modern reports of photoredox catalysis were demonstrated in the context of C—H functionalization. Nicewicz and MacMillan, in 2008, published a groundbreaking advance that paired visible-light-absorbing $Ru(bpy)_3Cl_2$ complexes with an imidazolidine organocatalyst to achieve an enantioselective α-alkylation of aldehydes using alkyl bromides [1] (Scheme 12.1). The transformation proceeds via single-electron transfer (SET) reduction of an activated alkyl bromide to generate a carbon-centered radical, which can add across the catalytic enamine species. The photocatalyst is turned over by oxidizing the resultant α-amino radical to an iminium, which furnishes an aldehyde product after hydrolysis. The platform was taken on to achieve C—H trifluoromethylation [2], cyanoalkylation [3], and benzylation [4] reactions. However, recent years have seen the dominance of a different, more general strategy for activating diverse types of C—H bonds – hydrogen-atom transfer (HAT). This chapter focuses on the use of HAT, where a hydrogenatom transfers from a given C—H bond to an open-shell species (X•), thus furnishing a carbon-centered radical and a neutral X—H species (Figure 12.1). Based on polarity-reversal catalysis developed by Brian Roberts, the current state of methods has witnessed significant advances for targeting electron-rich, hydridic C—H bonds with electrophilic radicals [5].

We choose to focus on redox pathways facilitated by traditional photoredox catalysts, but acknowledge that notable advances utilizing energy transfer mechanisms with transition metal complexes are also a blossoming field. For redox pathways, two types of SET cycles have been employed (Figure 12.2). Upon visible light excitation, a photocatalyst is promoted to an excited state, which can engage in oxidative or reductive SET to generate an array of electrophilic open-shell intermediates. Such species are often oxygen-, halogen-, sulfur-, or nitrogen-centered

Directed C–H Bond Functionalization: Concepts and Applications, First Edition.
Edited by Debabrata Maiti and Supriya Rej.

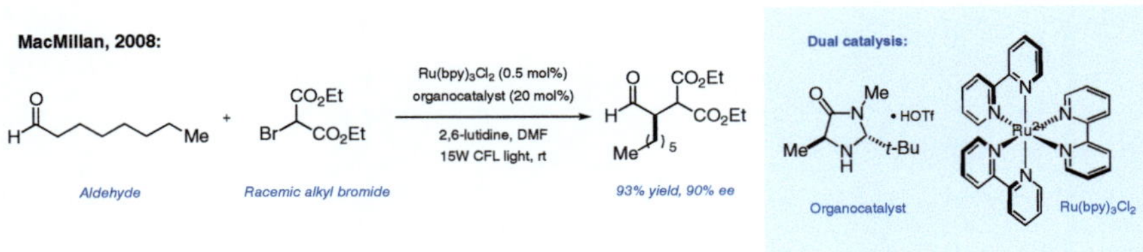

Scheme 12.1 Early example of C−H functionalization with photoredox catalysis.

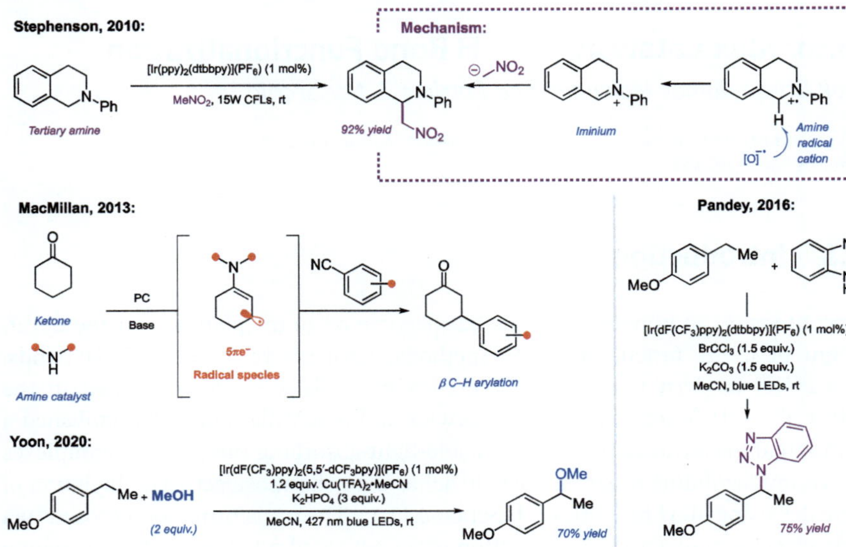

Scheme 12.2 Methods involving direct single-electron oxidation of reaction intermediates.

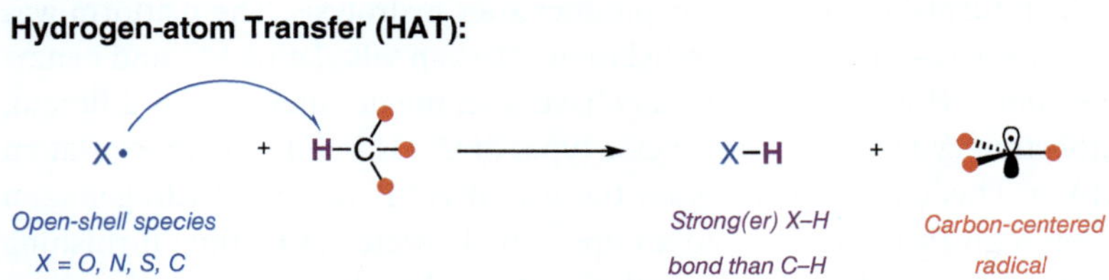

Figure 12.1 Hydrogen-atom transfer process.

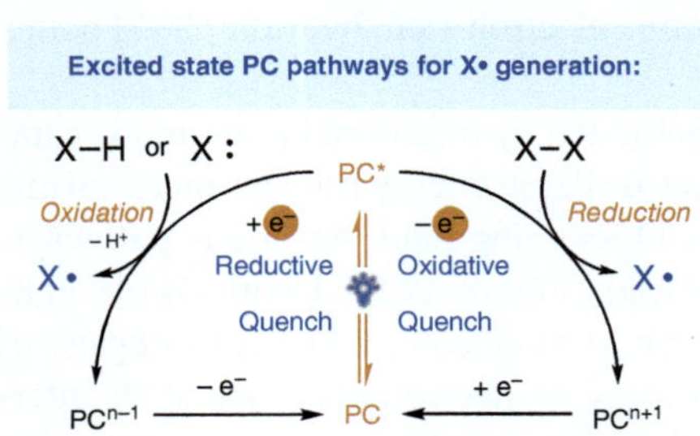

Figure 12.2 Photoredox catalysis electron transfer cycles.

radicals and are prone to facilitate homolytic cleavage of a given Csp^3—H bond, or HAT. This leaves behind a carbon-centered radical that is the key species in the bond-forming step. The Giese and Minisci reactions are common radical-coupling reactions that forge desirable C—C bonds. Oxidative and reductive radical-polar crossover events allow for the conversion of carbon radicals to their anionic and cationic equivalents so that they can engage with O^-, N^-, S^-, F^-, Br^-, and Cl^- nucleophiles, for both intermolecular couplings and intramolecular cyclizations. Metallaphotoredox catalysis allows for carbon-centered radicals to participate in classical cross-couplings with aryl halides and heteroatom reagents.

This fundamental photoredox cycle (excitation, electron transfer, and catalyst turnover) serves as the mechanistic core of the transformations discussed herein. Given the breadth of recent advances, this chapter does not attempt to be comprehensive [6–8]. Instead, it highlights distinct examples that exemplify key mechanistic strategies and reactivity trends, providing a platform for readers to explore broader developments in the field. The principles established here continue to guide the development of new photocatalytic platforms that exploit radical intermediates for efficient and selective C—H activation. Together, these examples demonstrate how photoredox catalysis enables the direct and selective conversion of inert C—H bonds into valuable synthetic intermediates, establishing it as a versatile tool in contemporary organic synthesis.

12.2 Direct Activation of Substrates via Oxidative SET

Oxidation of electron-rich moieties such as amines and aryl rings results in the generation of radical cation species. Adjacent C—H bonds to such electrophilic species are often activated to be more acidic than their neutral equivalents, and so facile deprotonation often follows to generate α-amino radicals and benzylic radicals, respectively (Scheme 12.2). In 2010, Stephenson and coworkers demonstrated that Ru or Ir photoredox catalysts could be used to perform C—H aza-Henry reactions on *N*-aryl tetrahydroisoquinolines [9]. The mechanism proceeds via direct oxidation of the tertiary amine, generating an amine radical cation. Subsequent HAT at the α-amino C—H of the transient species yields an iminium intermediate, which can further react with a nitromethane anion. The group extended the platform to include other nucleophiles, including indoles, which the team later used to inform a ring closing step in a total synthesis of (–)-pseudovincadifformine [10, 11]. Lastly, they demonstrated the effort on dimethylformamide structures to give C—H arylation products at α-amino sites [12]. Such efforts have set the foundations for modern publications like that from MacMillan and Doyle, who demonstrated that cyclic anilines were directly oxidized and deprotonated to give α-amino radicals, which could either undergo radical–radical coupling with dicyanobenzene motifs [13], or intercept nickel cross-coupling cycles [14], thus broadening the scope of aryl coupling partners.

Following the merger of photoredox catalysis with organocatalysis, MacMillan and his group went on to further introduce singly-occupied molecular orbital

(SOMO) catalysis and demonstrated that a 5πe⁻ β-enaminyl radical intermediate can be generated under photoredox conditions [15]. Oxidation of a catalytic enamine species, followed by deprotonation leads to the 5πe⁻ radical, which was leveraged to achieve first arylation, followed by alkylation [16], β-Mannich [17], and β-aldol reactions [18].

For benzylic C—H functionalization, electron-rich aryl rings can be oxidized by a photocatalyst to give aryl radical cations. Subsequent deprotonation of the benzylic C—H leads to a benzylic radical. Yoon coupled this strategy with Cu oxidants to generate benzylic carbocations by *in situ* oxidization of the benzylic radical, for an alkoxylation of benzylic C—H bonds where they were able to use abundant alcohol species [19]. Pandey has also activated benzylic C—H bonds toward reactivity with simple nucleophile partners using arene oxidation coupled with HAT [20], or by coupling with Selectfluor to give an intermediate benzylic fluoride, which is prone to nucleophilic displacement [21]. Although direct substrate oxidation provides site-selectivity in C—H functionalization designs, this strategy lacks the ability to activate other types of bonds, such as α-oxy or unactivated aliphatic C—H bonds. Thus, the majority of this chapter focuses on more general platforms.

12.3 Oxygen-centered Radicals

Considering oxygen's high electronegativity and the strength of the O—H bond [with bond dissociation energy (BDE), ~88–105 kcal/mol, oxygen-centered radicals present attractive opportunities as hydrogen-atom abstractors for the functionalization of electron-rich $C(sp^3)$—H bonds [22]. In addition to direct HAT processes, which are facilitated by light-absorbing compounds such as Eosin Y, the decatungstate anion, and various ketones (notably diaryl ketones and anthraquinones) under visible-light irradiation, the majority of oxygen radical-mediated transformations proceed via indirect HAT mechanisms [6]. In this context, the actual HAT species are generated *in situ* from specific oxygen-containing precursors. Oxidative and reductive quenching of the excited state of the photocatalyst by oxygen reagents is the most common way to generate these electrophilic radicals; however, energy transfer mechanisms are also utilized. Several examples of oxygen-containing precursors and radicals that facilitate "indirect" (HAT) processes are showcased in Figure 12.3:

12.3.1 Processes Utilizing Peroxide

The oxidative nature of peroxides allows them to induce a reductive quench of the photocatalyst's excited state; thus, a common mechanistic formula we will see in this section is: (i) SET reduction of the weak O—O bond by the excited state of the photocatalyst, followed by (ii) SET oxidation of a substrate to get to polar intermediates, thus enabling Giese reactions, Minisci additions, and carbocation formations.

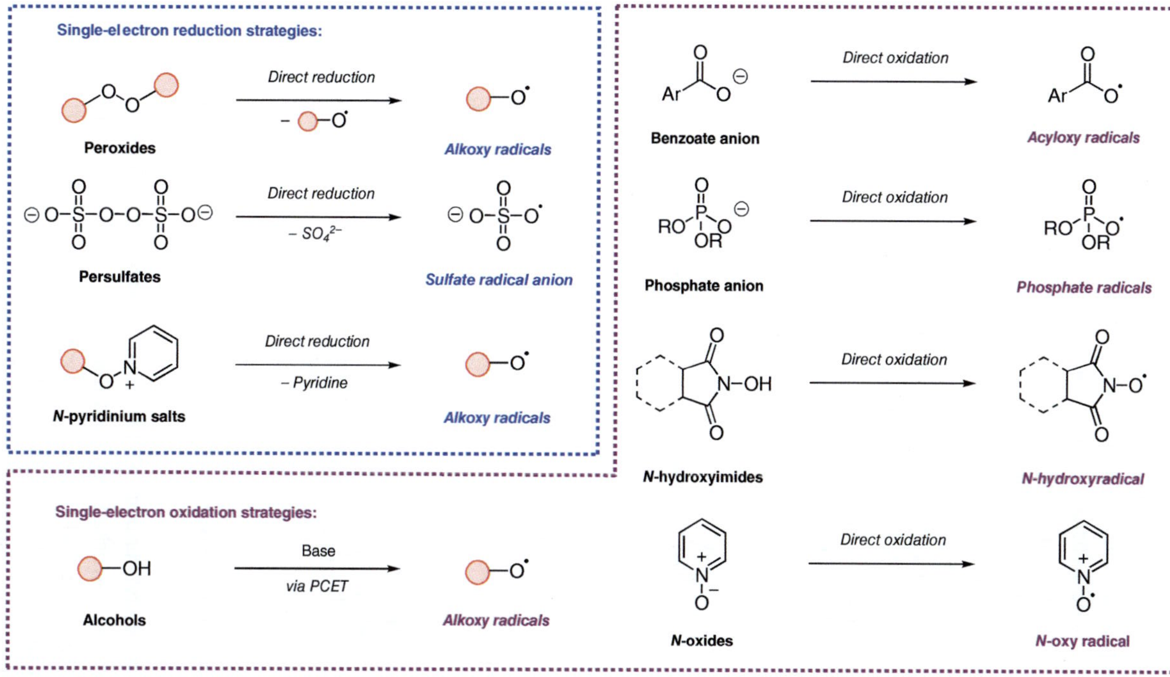

Figure 12.3 Oxygen-centered radical precursors and generation methods.

12.3.1.1 *Tert*-butyl Hydroperoxide (TBHP) as an Oxygen Radical Source

TBHP is widely regarded as a green oxidant, producing only water and *tert*-butanol as byproducts. When coupled with photoredox catalysis, it is commonly used as a single-electron oxidant to turn over photocatalyst cycles, but it can also be leveraged as an HAT reagent. In 2015, Wang proposed that the weak O—O bond of TBHP could be reduced, inducing fragmentation into hydroxide and *tert*-butoxy radical [23]. The electrophilic oxygen-centered radical can facilitate a hydrogen-atom abstraction from the hydride C—H bond on an aldehyde, thus generating acyl radicals. α,β-Epoxy ketones, spiroepoxy chromanones, and amides have all been prepared from aldehyde precursors through a photoredox catalysis/TBHP protocol [24, 25]. Shortly after, She and her colleagues demonstrated that α-oxy C—H bonds can also be engaged by *tert*-butoxy radicals. As shown in Scheme 12.3, the resultant α-oxy radical adds across the alkynoate to generate a vinyl radical, which then undergoes intramolecular cyclization, followed by oxidation to close the photocatalyst cycle. Other types of C—H bonds that have been targeted with *tert*-butoxy radicals include α-amino C(sp^3)—H bonds in amides and benzylic C—H bonds [26].

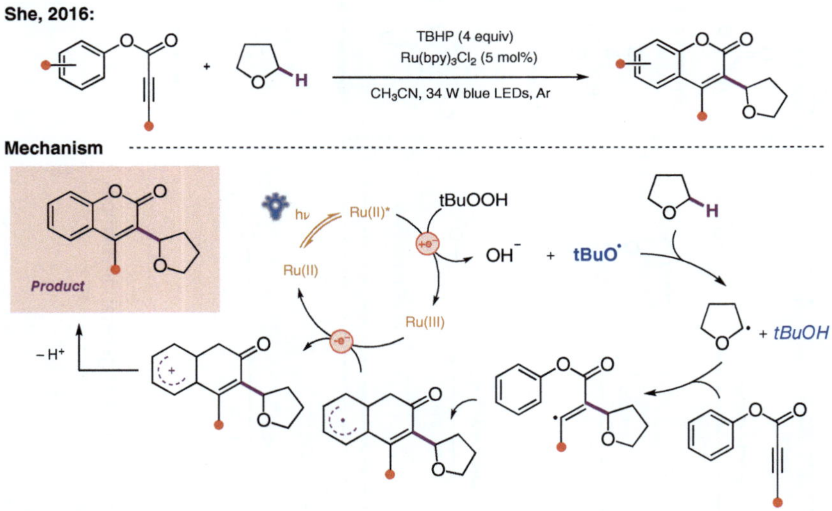

Scheme 12.3 Photocatalysis with tert-butyl hydroperoxide as the oxidant.

12.3.1.2 Di-*tert*-butyl Peroxide (DTBP) as an Oxygen Radical Source

In 2021, Stahl and coworkers reported the direct C(sp^3)—H methylation using an Ir-based photocatalyst and di-*tert*-butyl or dicumyl peroxide [27] (Scheme 12.4). Homolysis of the weak O—O bond in DTBP can be induced via energy transfer from the photocatalyst using visible light. The resulting alkoxyl radicals play a dual role in the reaction by facilitating the HAT process and generating a methyl radical via β-scission. Both carbon-centered radical species oxidatively couple to a nickel catalyst to forge the desired C—C bond, leading to a broad array of valuable methylated aliphatic products. Inspired by these results, Kramer combined photoredox

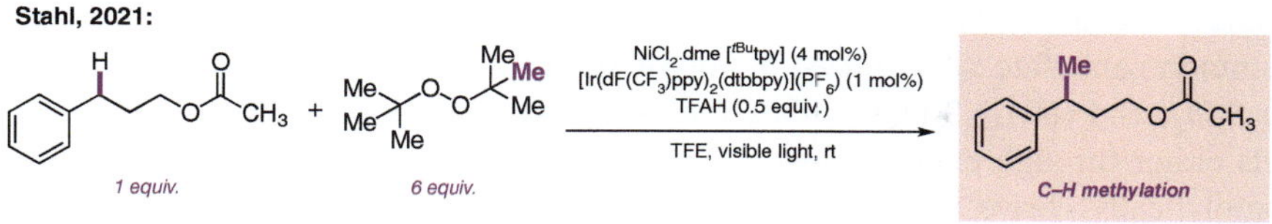

Scheme 12.4 Metallaphotoredox catalysis for a direct C(sp^3)–H methylation with peroxides.

catalysis/DTBP with copper catalysis to enable an enantioselective benzylic amidation [28]. Lastly, a three-component reaction coupling light hydrocarbons to alkenes and 1,3-dicarbonyl compounds was recently developed [29].

12.3.1.3 Dicumylperoxide (DCP) and Benzoyl Peroxide (BPO) as an Oxygen Radical Source

Merging dicumply peroxide with chiral phosphate anion catalysis, Phipps and coworkers are able to achieve an enantioselective Minisci reaction of alcohols to give enantiopure heterobenzylic alcohols [30]. The Minisci mechanism was also employed by DiRocco and Krska using BPO to perform an HAT at the α-oxy C—H of methanol to give hydroxymethylation of heteroarenes [31].

12.3.1.4 *Tert*-butyl Peroxybenzoate (TBPB) as an Oxygen Radical Source

In 2022, the Musacchio group reported on the wide-ranging applicability of the sequential (HAT + RPC) mechanism as a reliable platform for the catalytic formation of carbocations directly from C(sp^3)—H bonds under photocatalytic conditions [32] (Scheme 12.5). Photoredox catalysis was used to reductively fragment TBPB to give *tert*-butoxy radical and benzoate anion. After HAT of a benzylic C—H bond, the resultant benzylic radical is oxidized by the ground-state Ir(IV) of the photocatalyst to furnish a carbocation intermediate. Exogenous external nucleophiles, such as fluoride, alcohol, and azole sources, can then trap the carbocation, leading to C—H fluorination and etherification reactions. This work was conducted in conjunction with the Doyle group, whose system is highlighted in the carbon-centered radical section.

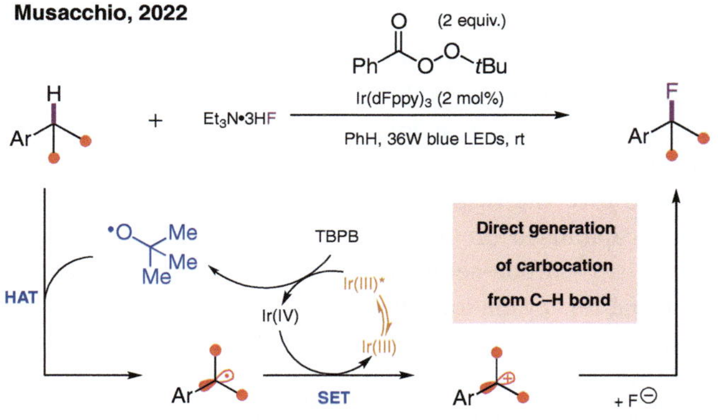

Scheme 12.5 Photoredox-catalyzed [HAT+RPC] nucleophilic fluorination.

12.3.1.5 Persulfate Salts as Oxygen Radical Source

Common persulfate salts utilized in photoredox catalysis methods include $K_2S_2O_8$, $Na_2S_2O_8$, and $(NH_4)_2S_2O_8$). Sulfate radical anions are generated from persulfate salts either through energy transfer from the excited state of the photocatalyst to homolytically cleave the O—O bond, or via single-electron reduction followed by heterolytic bond cleavage. In 2015, Macmillan and his coworker reported a groundbreaking photocatalytic cross-dehydrogenative coupling reactions for obtaining α-heteroarylation of ethers from common ether and light hydrocarbon synthons and heteroarenes [33] (Scheme 12.6). A sulfate radical anion produced by reductive SET between Ir(III)* and $Na_2S_2O_8$ serves as the HAT agent to generate α-oxy radicals. Addition of the radical to a protonated heteroarene, followed by oxidation and deprotonation to regenerate the aromatic ring, furnishes an array of alkylated heteroarene products.

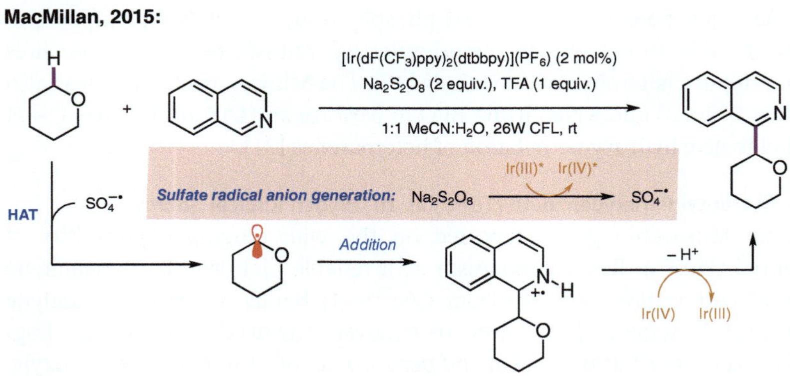

Scheme 12.6 Persuflate radical-mediated α-arylation of ethers.

12.3.2 Processes Using Pyridine *N*-Oxides and their Derivatives

N-Alkoxypyridinium salts have received some attention as reductive sources of alkoxy radicals. While Lakhdar and coworkers were the first to demonstrate that *N*-alkoxypyridinium salts could be activated with visible-light photoredox catalysis to furnish reactive alkoxy radicals, Hong and his team demonstrated that the photocatalyst/*N*-alkoxypyridinium combination was capable of achieving inter- and intramolecular C—H Minisci reactions [34, 35]. The N—O bond of the pyridinium salts can oxidize the excited state of a photoredox catalyst, thus triggering a fragmentation to give the respective alkoxy radical and a pyridine byproduct. The Zhu group capitalized on this C—H functionalization strategy and was able to trap the γ-oxy carbon-centered radical with copper catalysts to give distal arylation, azidation, cyanation, and thiocyanation products [36]. Notably, the *N*-alkoxypyridinium salts have allowed for the coupling of HAT with oxidative radical–polar crossover processes. Li and his team [37] developed a visible-light-mediated bifunctionalization of alkenes, where the alkoxy radical serves as both the HAT reagent, and then subsequently a nucleophile. In 2023, the Musacchio group reported a photocatalyzed benzylic C—H azolation reaction in which an *N*-methoxypyridinium salt was employed as an efficient reagent to facilitate C—H azolation at benzylic C(sp^3)—H bonds [38] (Scheme 12.7). Transient methoxy radicals generated after reduction of the *N*-methoxypyridinium reagent abstract a hydrogen atom from a range of benzylic C—H bonds. Oxidation by the Ir(IV) ground-state photocatalyst of the benzylic radical generates a benzylic carbocation, which is readily trapped by an array of azole motifs.

In 2022, Nicewicz and his coworkers developed a highly effective method for C—H alkylation reactions that used easily accessible pyridine *N*-oxides with an acridinium photoredox catalyst [39] (Scheme 12.8). The strongly oxidizing acridinium oxidizes the zwitterionic reagent to give highly electrophilic oxygen-centered radicals. The strength of the O—H bond allows for the abstraction of strong primary, secondary, and tertiary C—H bonds when electron-donating and electron-withdrawing moieties are present.

12.3.3 Processes Using Carboxylate Radicals

Carboxylate, or acyloxy, radicals are particularly attractive HAT reagents due to (i) high O—H BDE of 111 kcal/mol, and (ii) the ease in generating them via oxidative SET from readily accessible benzoate anions. In 2016, Glorius reported a highly selective tertiary C(sp^3)—H trifluoromethylthiolations employing catalytic amounts of both an Ir photocatalyst and sodium benzoate [40] (Scheme 12.9). Stern–Volmer experiments confirmed the reductive quenching of the Ir(III)* excited state by sodium benzoate. Regioselectivities for tertiary C—H sites were generally 19 : 1 or greater. In 2024, Yoshimi achieved alkylation of α-oxy C—H bonds utilizing 4-*tert*-butylbenzoic acid as the hydrogen atom transfer reagent, with 9,10-dicyanoanthracene (DCA) as the photocatalyst [41]. Lastly, Chen published a method that achieves tertiary C(sp^3)—H azidation with 1-azido-1,2-benziodoxol-3-(1H)-one (BI–N_3) and a Ru-based photocatalyst [42]. The authors propose a possible iodine radical intermediate as a viable HAT mediator. Alternatively, we propose that reduction of the I—O bond reduces the I(III) center to I(0) while fragmenting out an *o*-iodobenzoate radical.

12.3.4 Processes Utilizing Phosphate Radicals

In 2018, Nicewicz's and Alexanian's groups collaborated to demonstrate that strongly oxidizing excited acridinium catalysts can directly oxidize phosphate anions to phosphate radicals, providing a stereoelectronically different oxygen-centered radical for use as an HAT reagent. The team catalyzed the azidation of non-activated C(sp^3)—H bond by utilizing stoichiometric K_3PO_4 and a radical-compatible amination reagent – sulfonyl azide [43] (Scheme 12.10). Later, Kanai and Oisaki demonstrated that a catalytic amount of phosphate could be utilized to achieve the same transformation, but at α-heteroatom sites [44].

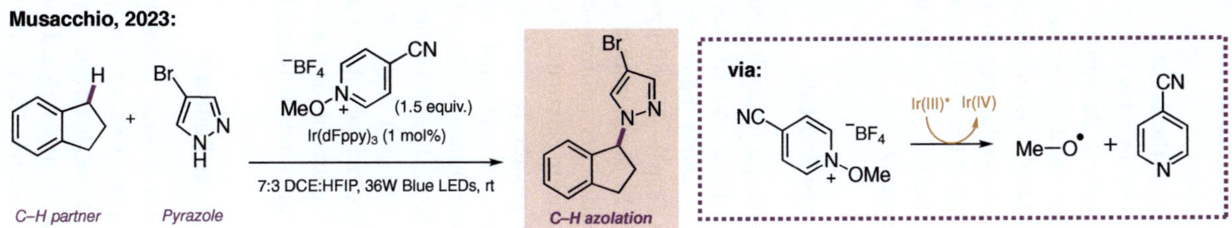

Scheme 12.7 *N*-Methoxypyridnium salts as reagents for benzylic C—H azolation.

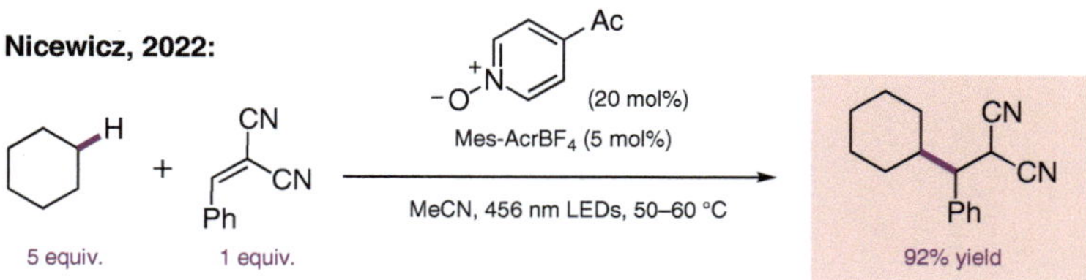

Scheme 12.8 Direct oxidation of pyridine N-oxides as an HAT species.

12.3.5 Direct and Indirect Use of Alcohols as Precursors

The holy grail of oxygen-centered radical methodology is to use alcohols as the direct sources of alkoxy radicals in a catalytic manner and without a stoichiometric oxidant. Zuo accomplished this feat by designing a ligand-to-metal charge transfer (LMCT) process with cerium salts and alcohol reagents to enable intermolecular couplings of synthons [45]. However, photoredox catalysis has been less general in achieving the same feat, largely due to the kinetic barrier involved in generating alkoxy radicals via oxidation. A recent advance in 2023 from the Wu group was able to access methoxy radicals directly from methanol using a thianthrene and an organic photocatalyst [46]. The team proposed the thianthrene radical cation, generated from oxidative SET with the photocatalyst, could oxidize methanol to achieve a Giese reaction with cyclic hydrocarbons. Prior to this work, there have been many other elegant strategies requiring prior functionalization of alcohols. In 2018, a photocatalyzed remote C(sp^3)—H heteroarylation reaction of tertiary alcohols was developed by Zhu (Scheme 12.11) [47]. α-Heteroaryl tertiary alcohols are designed to perform a successive HAT-migratory arylation to yield γ-heteroaryl ketones, with $K_2S_2O_8$ as the terminal oxidant. An intramolecular hydrogen bond between the alcohol and the heteroarene allows for an oxidative proton-coupled electron transfer (PCET) with the Ir(IV) catalyst, which generates an alkoxy radical under mild conditions. Next, 1,5-HAT furnishes a remote carbon-centered radical that will undergo a radical transposition of the heteroarene. The resultant ketyl radical is oxidized to furnish the neutral ketone product. A similar strategy was taken with cyanohydrins to achieve a distal C—H cyanation of cyanohydrins [48].

Another photoredox-catalyzed method for generating alkoxy radicals involves the use of perfluorinated hydroxybenziodoxole for the *in situ* production of an alcohol-based hypervalent iodine reagent, as published by Chen. A Ru photocatalyst reduces the weak I—O bond to give I(0) and an alkoxy radical to achieve a remote Minisci arylation [49]. Prior preparation of a benziodoxolone hypervalent iodine derived from commerical alcohols, was explored by Wallentin to give an intramolecular benzylic etherification [50] (Scheme 12.12).

Like alcohols, *N*-hydroxy substances have the potential to produce *N*-oxy radicals, which makes them effective HAT agents (Scheme 12.13). In 2011, Li and his colleagues used graphitic carbon nitride (g-C_3N_4) and *N*-hydroxyphthalimide (NHPI) for C(sp^3)—H oxygenation when exposed to photoirradiation [51]. Similarly, Gong and coworkers reported *N*-hydroxysuccinimide (NHS) as the HAT agent to formylate fluoroalkyl imines via the HAT of 1,3-dioxolane, with diacetyl performing either

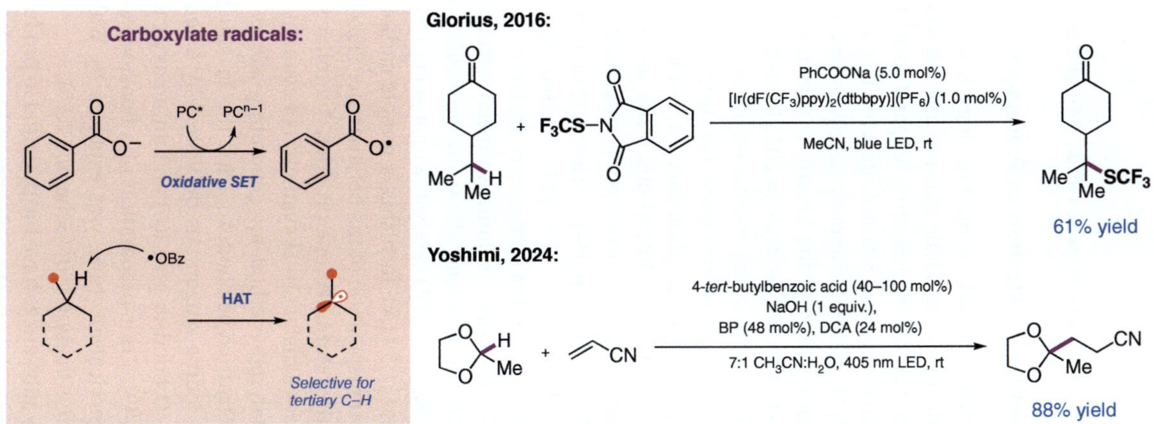

Scheme 12.9 Carboxylate radicals as HAT species for C—H bond functionalization.

Alexanian and Nicewicz, 2018:

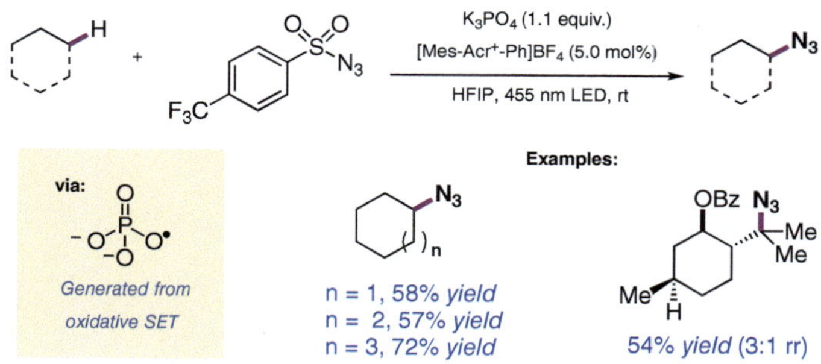

Scheme 12.10 Phosphate radicals from phosphate salts via direct SET oxidation.

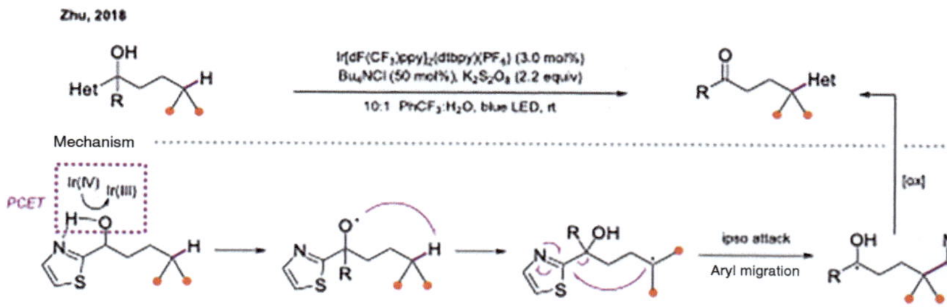

Scheme 12.11 Intramolecular hydrogen-bonding for PCET generation of alkoxy radicals.

Wallentin, 2020:

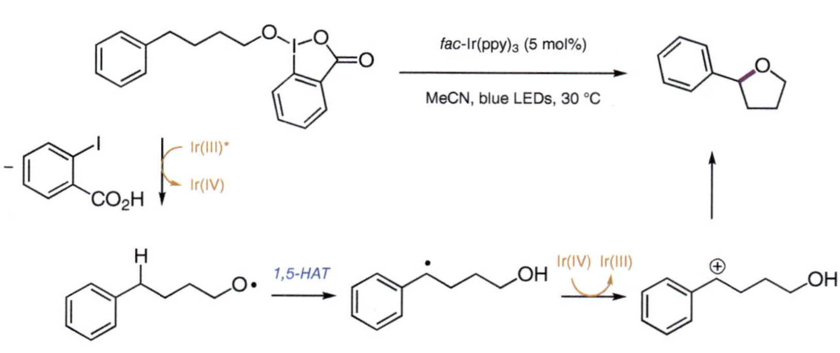

Scheme 12.12 Iodonium scaffolds for alkoxy radical generation.

direct HAT or PCET to generate phthalimidio-*N*-oxyl PINO [52]. In 2023, Jiang and his coworkers reported enantioselective protonation and visible-light-mediated photoredox catalytic intermolecular HAT as a novel approach to asymmetric olefin isomerization [53]. Using a dual catalytic system with a chiral phosphoric acid and a visible-light photosensitizer (DPZ), the HAT process is carried out with the aid of *N*-hydroxyimide.

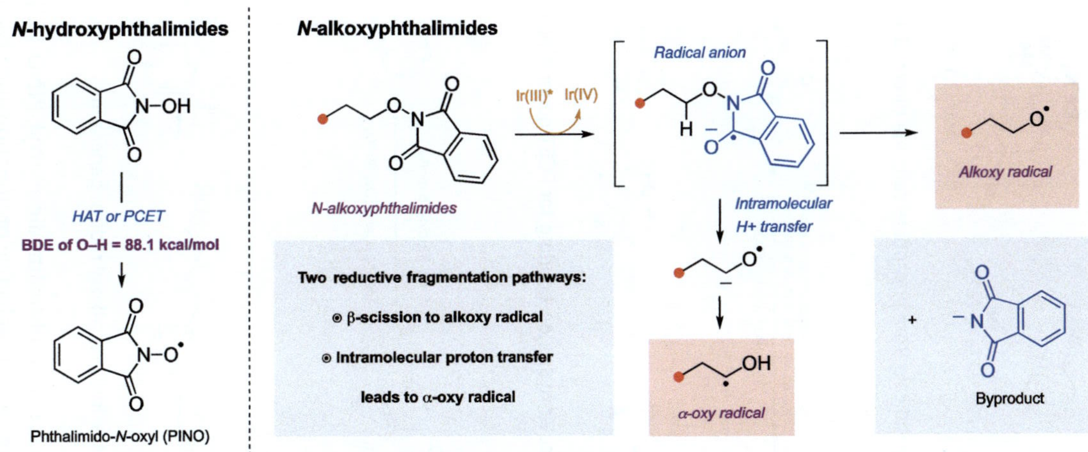

Scheme 12.13 Various strategies to generate *N*-oxy radicals from *N*-hydroxy substances.

N-Alkoxyphthalimides can also be a source of alkoxy radicals. Single-electron reduction of the phthalimide motif generates a radical anion, which can either undergo β-scission to give the alkoxy radical or undergo an intramolecular proton transfer, ultimately leading to an α-oxy radical. Examples of using the alkoxy radical for γ-functionalization via 1,5-HAT for C—H alkenylation and allylation have been demonstrated [54, 55]. Work by Chen and Duan explored the α-oxy radical intermediate, intercepting it with allyl sulfonates and imines, respectively [56].

In 2020, Yu and his team used oxyimino acids derived from pyruvic acid to generate alkoxy radicals via a photocatalyst-controlled decarboxylation and β-scission for remote δ-C—H halogenation with Selectfluor or ethyl trichloroacetate [57]. Common alcohols can be converted to *N*-alkoxyphthalimides and, in one pot, can be deprotected with hydrazine and condensed onto pyruvic acid to give oxyimino acid.

12.4 Decatungstate Catalysis

Among the most versatile and well-studied catalysts for sp^3 C—H activation is the decatungstate anion (DT), typically employed as tetrabutylammonium decatungstate (TBADT, $(n\text{-}Bu_4N)_4[W_{10}O_{32}]$) (Figure 12.4). This polyoxometalate photocatalyst enables efficient hydrogen atom transfer under near-UV irradiation and is uniquely capable of abstracting hydrogen atoms from strong aliphatic C—H bonds (BDE > 90 kcal mol^{-1}) under operationally simple conditions [58, 59]. Upon excitation with 365–390 nm light, TBADT enters a long-lived triplet excited state ($^3[W_{10}O_{32}]^{4-*}$), which acts as a highly electrophilic HAT species [60]. Abstraction of a hydrogen atom from a substrate C—H bond generates a carbon-centered radical, which can then engage in a range of downstream transformations, including Giese-type additions, Minisci alkylations, and metal-mediated cross-coupling. Notably, DT-mediated reactivity can proceed under redox-neutral, oxidative, or reductive conditions, depending on the overall catalytic design.

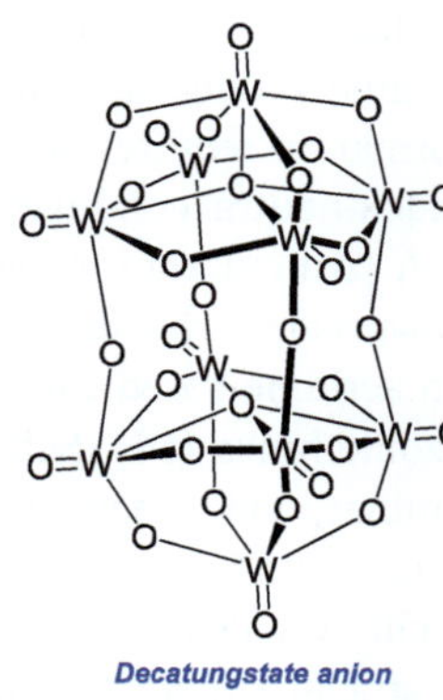

Figure 12.4 Structure of decatungstate anion.

The reactivity profile of TBADT spans a broad range *of sp*^{3} C—H bonds, with a strong preference for electron-rich sites (Scheme 12.14). Its excited state selectively abstracts hydrogen atoms from C—H bonds that are both sterically accessible and hydridic in nature. As such, secondary, benzylic, and α-heteroatom C—H bonds are most frequently functionalized [6]. While tertiary C—H bonds have lower BDEs and enhanced hydridicity, experimentally verified examples of product formation via tertiary-radical intermediates remain uncommon in the literature due to the steric hindrance imposed by the congested tertiary environment and the large metal-oxo species [61, 62].

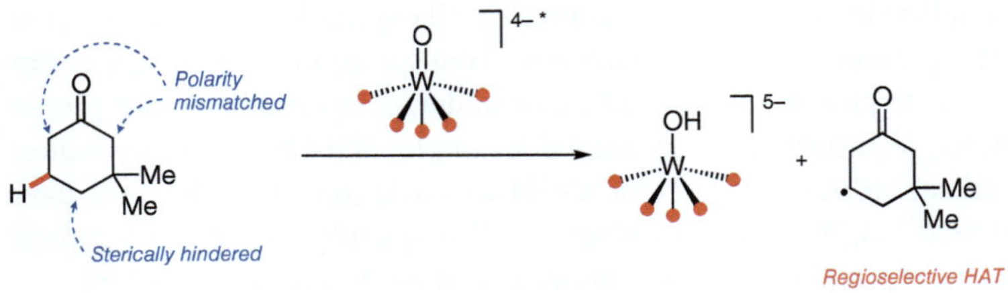

Scheme 12.14 Polarity matching for decatungstate HAT gives regioselective functionalization.

12.4.1 Application to Giese-Type Hydroalkylation and Olefin Addition

The use of decatungstate for radical alkylation of electron-deficient alkenes via sp^3 C—H activation has been foundational in expanding the scope of photoredox hydroalkylation. In a seminal study, Protti, Fagnoni, and coworkers demonstrated that photoexcited TBADT can abstract hydrogen atoms from simple alkanes, generating alkyl radicals that add across electron-poor olefins such as acrylates and maleates. A final HAT from TBADT–H• by the α-acyl radical furnishes both the desired product and regenerates the ground-state photocatalyst [63]. This transformation requires no external oxidants, highlighting TBADT's compatibility with redox-neutral conditions and functional group tolerance.

A particularly innovative example of DT catalysis was reported by Dilman and coworkers, who developed a dual photoredox system combining TBADT with an acridine-based catalyst to achieve the decarboxylative alkylation of imines and azomethines [64] (Scheme 12.15). In this three-component reaction, aldehydes, amines, and unactivated carboxylic acids are coupled to afford secondary amines under mild, visible-light-driven conditions. While the acridine photocatalyst facilitates PCET to generate alkyl radicals from carboxylic acids, TBADT plays a crucial supporting role by enabling turnover of the acridine catalyst through hydrogen atom abstraction from the intermediate Acr-H species. The authors demonstrate broad scope with respect to the aldehyde, amine, and acid components, including challenging substrates such as *N*-tosylimines, nitrones, and hydrazones.

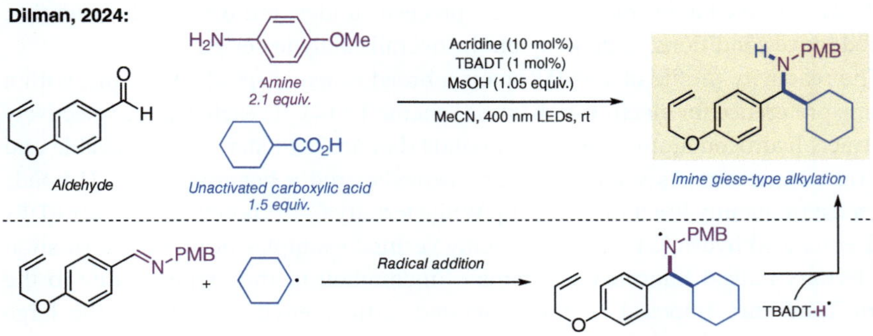

Scheme 12.15 Decatungstate-mediated Giese addition to an imine.

12.4.2 Application to Minisci Alkylation

In a seminal contribution to Minisci-type chemistry, Ravelli, Ryu, and coworkers demonstrated that TBADT enables cross-dehydrogenative coupling of heteroarenes with a wide array of C—H donors under mild photochemical conditions [65] (Scheme 12.16). Using sunlight or simulated solar light, TBADT abstracts a hydrogen atom from alkanes, ethers, amides, or aldehydes to generate alkyl radicals, which then add to protonated heteroarenes in the presence of a persulfate oxidant.

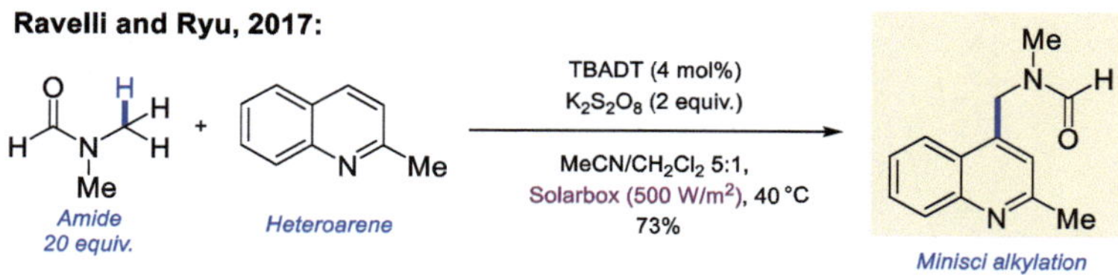

Scheme 12.16 Minisci addition of α-amidyl radicals.

12.4.3 Heteroatom Incorporation via Electrophilic Radical Trapping

In a distinct redox-neutral transformation, MacMillan's group also demonstrated that a decatungstate catalyst (NaDT) can be employed for the sulfinylation of aliphatic C(sp^3)—H bonds, affording alkyl sulfinates that serve as versatile intermediates for further derivatization [66] (Scheme 12.17). Aqueous SO_2 can trap out carbon-centered radicals derived from HAT processes, and reduction of the resulting sulfonyl radical gives the sulfinic acid product. These products can be directly converted into sulfones, sulfonamides, and sulfonyl fluorides in a modular fashion. A complementary advance was reported by Studer and coworkers, who developed a TBADT-catalyzed radical disulfuration strategy using tetrasulfides as electrophilic radical acceptors [67].

In addition to sulfur-based trapping strategies, decatungstate-catalyzed HAT has also been employed for C—H functionalization with other heteroatoms. Britton and coworkers reported a site-selective C(sp^3)—H fluorination using TBADT in conjunction with *N*-fluorobenzenesulfonimide (NFSI) as the fluorine source [68, 69] (Scheme 12.18). The alkyl radicals generated via TBADT-mediated HAT engage NFSI to furnish alkyl fluorides under mild conditions, enabling the fluorination of complex natural product scaffolds (such as Sclareolide) and bioactive molecules.

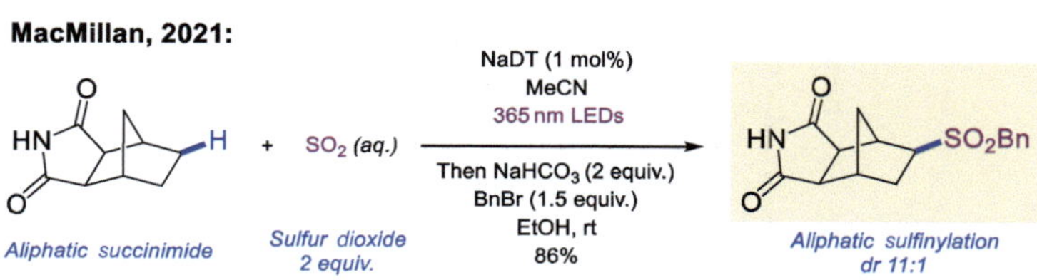

Scheme 12.17 Carbon-centered radical trapping with sulfur dioxide.

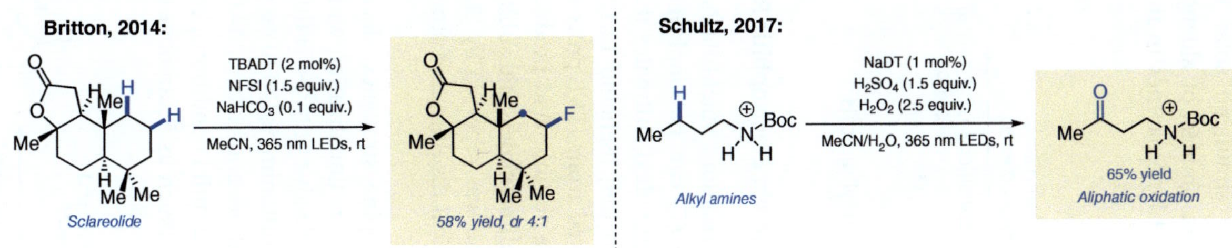

Scheme 12.18 Regioselective C—H functionalizations with decatungstate catalysis.

Separately, Schultz and coworkers demonstrated that decatungstate photocatalysis can facilitate remote C(sp^3)—H hydroxylation of amines in the presence of hydrogen peroxide, delivering functionalized ketones and lactams via selective C—O bond formation [70]. These studies underscore the potential of decatungstate to engage diverse heteroatom-based electrophiles for radical trapping, beyond sulfur-based species.

12.4.4 Application to Metallaphotoredox Cross-Coupling

TBADT can also be paired with secondary catalytic systems to enhance complexity and reactivity through metal-radical coupling. A transformation of this type was reported by MacMillan and coworkers in the context of organofluorine chemistry, where decatungstate was merged with copper catalysis to achieve direct C(sp^3)—H trifluoromethylation [71] (Scheme 12.19). In a dual-catalytic strategy, MacMillan and coworkers combined decatungstate-mediated HAT with copper-catalyzed trifluoromethylation to achieve direct C(sp^3)—H to C—CF_3 conversion using Togni's reagent II. Upon photochemical irradiation, TBADT abstracts a hydrogen atom from electron-rich or sterically accessible C—H bonds, generating alkyl radicals that are intercepted by a copper(II)-CF_3 species to form a transient Cu(III) intermediate. Reductive elimination furnishes the trifluoromethylated product. Site selectivity in this system reflects the polarity-matched nature of TBADT HAT. Protonated amines are selectively functionalized at β- or γ-positions rather than α-positions, as protonation reduces the hydridic character of the α-C—H bond. This effect enables regioselective functionalization of amine-containing scaffolds, including amino acid derivatives. Benzylic and unactivated aliphatic substrates are similarly functionalized at the most electron-rich and accessible sites, providing a predictable and broadly applicable platform for late-stage C—H trifluoromethylation.

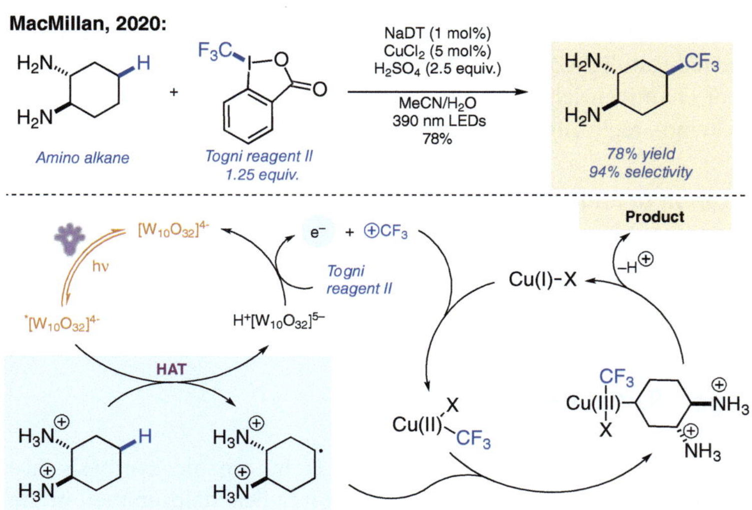

Scheme 12.19 Metallaphotoredox trifluoromethylation of alkanes with copper catalysis.

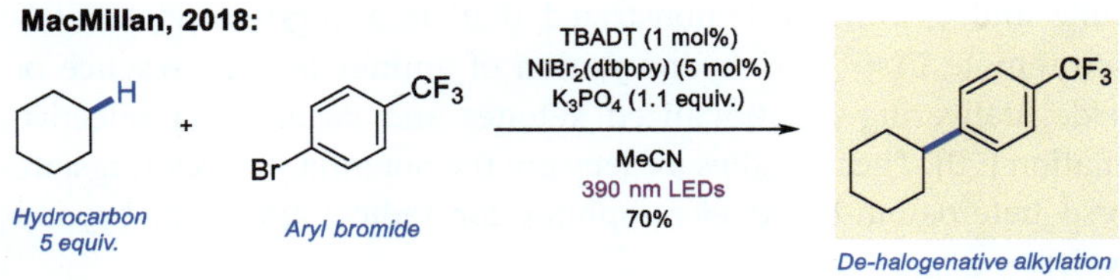

Scheme 12.20 Hydrocarbon arylation using a metallaphotoredox system.

A complementary approach was later disclosed by the same group, wherein TBADT was merged with nickel catalysis to enable direct arylation of strong aliphatic C(sp^3)—H bonds using aryl bromides as coupling partners [59] (Scheme 12.20). In this dual-catalytic system, carbon-centered radicals generated via TBADT-mediated HAT engage with Ni(0) to form nickel–alkyl intermediates that undergo oxidative addition and reductive elimination, delivering sp^3–sp^2 coupled products with high chemo- and site-selectivity. Following this advance, Noël and coworkers reported a rapid and scalable method for C(sp^3)—H acylation and arylation via decatungstate–nickel dual catalysis in flow [72]. In contrast to earlier batch protocols requiring many hours or even days, this continuous-flow approach achieves comparable or superior yields in only 5–15 minutes. Notably, the group was also able to push DT catalysis to perform HAT on methane, which has a strong C—H bond of 105 kcal/mol [58, 73, 74].

An illustrative example of redox-neutral C—H functionalization using decatungstate is the enantioselective three-component alkene dicarbofunctionalization developed by Nevado and coworkers, which enables the construction of vicinal C—C bonds with high enantioselectivity [75] (Scheme 12.21). This method interrupts the Giese mechanism by employing nickel catalysts to trap the α-acyl radical, thus leading to an enantioselective α-aryl esters via cross-coupling with aryl or alkenyl bromides. Importantly, the dual catalytic cycle avoids the use of prefunctionalized radical precursors or sacrificial oxidants. The synergy between the electrophilic excited-state TBADT and a chiral biimidazoline-nickel catalyst delivers a broad range of enantioenriched α-aryl/alkenyl carbonyls and 1,1-diarylalkanes with excellent chemo-, regio-, and stereoselectivity.

12.4.5 Application to Radical–Polar Crossover

In 2022, Noël and coworkers reported a decatungstate-catalyzed method for the oxidative coupling of electron-rich α-heteroatom C—H bonds with *N*-heterocycles using *tert*-butyl hydroperoxide (TBHP) as the terminal oxidant [76].

12.5 Halogen Radicals

The use of halogen radicals (*X*•; X = Cl, Br, I) in organic synthesis has a long-standing history, dating back to early studies on radical halogenation in the nineteenth century. One of the earliest documented uses of halogen radicals in photochemical C—H functionalization was Norrish and Porter's exploration of

Nevado, 2024;

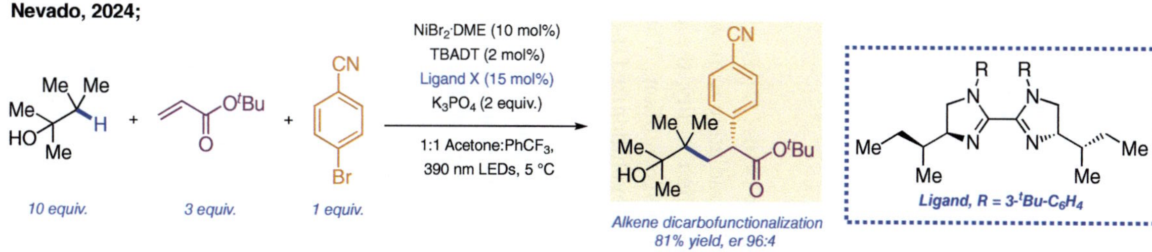

Scheme 12.21 Enantioselective alkene dicarbofunctionalization with metallaphotoredox catalysis.

Table 12.1 Halogen radical properties.

Halide	E^0 [X^-/X–] (V)[a]	BDE H-X (kcal/mol)	Reactivity	Applications
F^-	+3.30[b]	136	Extremely reactive	Limited in C—H funtionalization
Cl^-	+2.03	103	Very reactive	Aliphatic C(sp^3)—H bonds
Br^-	+1.60	88	Moderate reactivity	Tertiary, benzylic, activated C(sp^3)—H and formyl C(sp^2)—H aldehyde bonds
I^-	+1.06	71	Low reactivity	Limited in C—H funtionalization

a) Potential vs Saturated Calomel Electrode (SCE).
b) Estimated, rather than recorded.

halogen radical photogeneration in the 1950s [77]. During this period, radical-chain processes driven by ultraviolet (UV) irradiation were widely employed, particularly in industrial halogenation. Despite their utility, these approaches often suffered from poor selectivity and harsh reaction conditions. The advent of photoredox catalysis in the early twenty-first century revolutionized the field by enabling controlled generation of halogen radicals under mild visible-light conditions. Homolytic cleavage of an unactivated C(sp^3)—H bond typically demands substantial energy input, often exceeding 90 kcal/mol [59]. The reactivity of halogen radicals varies across the series from fluorine to iodine, influencing their applicability and the types of C—H bonds they can readily functionalize (Table 12.1).

While fluorine radicals are inaccessible, chlorine radicals (Cl•) are readily generated and exhibit strong hydrogen abstraction capabilities owing to their high BDE [78]. Chloride radicals have been successfully employed in photocatalytic C—H functionalization across various methodologies and are often ideal due to their high BDE potential, allowing them to target unactivated C—H positions. Many examples rely on LMCT processes for chloride radical generation with reports of SET-mediated chloride radical formation being scarce due to high oxidation potential of chloride anions [79]. Wu and colleagues demonstrated such an approach, employing a strongly oxidizing acridinium photocatalyst with catalytic amounts of HCl. Simple hydrocarbons like cyclooctane were allylated using allyl sulfones via a Giese addition [80]. However, their high reactivity similarly poses challenges for selectivity, requiring careful control of reaction conditions to guide productive functionalization. In a study by Russell and team, it was demonstrated that solvents capable of complexing with chlorine radicals, such as acetic acid, increased the selectivity of the chlorination process [81]. This complexation reduces the radicals' reactivity, allowing for more controlled and selective halogenation.

The use of photoredox catalysts has allowed for the activation of bromide anion to bromine radical via direct oxidative SET, allowing for the use of more atom-economical and less corrosive bromide salts as opposed to *N*-bromosuccinimide (NBS) or bromine (Br_2) [82, 83]. Although bromide is more easily oxidized than chloride, the resulting bromine radical is a weaker hydrogen atom abstractor than its chlorine counterpart. As a result, bromine radicals exhibit greater functional group tolerance but require more activated C—H sites for efficient abstraction such as [84] benzylic, allylic, α-amino or α-oxy positions, for engagement in Minisci reactions [85, 86], vinyl halides [87], or trapping by electrophiles such as CO_2 [83].

Iodine radicals (I•) are the most selective of the halogen series but are rarely employed for HAT due to the weak H—I bond [88]. Nonetheless, they serve important roles as intermediates in halogenation and coupling reactions where high reactivity is less critical. An additional advantage of halogen radical-mediated C—H functionalization lies in ease of purification: the HAT by-product is a strong acid that can be readily neutralized and removed from the reaction mixture.

Metallaphotoredox catalysis is the merger of photoredox catalysis with transition metal catalysis, with the two catalysts relying on each other to close both catalytic cycles either through SET or energy transfer processes. Nickel–halide complexes have been well studied in this context, particularly for establishing cross-coupling strategies between C—H partners and aryl/alkyl electrophiles through the intermediacy of a halogen radical. There are two primary pathways for halogen radical generation via Ni—X bond homolysis, (i) Direct oxidation of a Ni(II)—X to a Ni(III)—X species, which then undergoes ligand-to-metal-charge-transfer (LCMT) to generate the halogen radical, or (ii) Triplet energy transfer (^{3}EnT) of a Ni(II)—X complex to give an excited state species that can fragment out a halogen radical and Ni(I) (Figure 12.5). Initial reports from the Doyle group propose the SET-induced LMCT process to generate chlorine radicals, while the Molander group proposed ^{3}EnT for bromine radicals. Subsequent mechanistic studies by the Doyle team provided support for the latter mechanism [89–91]. In LMCT, absorption of a photon promotes charge transfer from the halide ligand to the metal within a metal halide complex, yielding a halogen radical and a reduced metal center.

The initial publication by Doyle and coworkers [92] reported coupling of aryl chlorides with ethereal solvents at the α-oxy positions (Scheme 12.22). Molander followed up with a similar report using aryl bromides and ethereal solvents. The Doyle group expanded the utility of the Cl• radical by demonstrating it can perform HAT at 1,3-dioxolane motifs to give benzaldehyde products after cross-coupling to give a ketal product, followed by hydrolysis. Use of this metallaphotoredox system to perform HAT at methylorthoformate produces methyl radicals, which they demonstrate can also cross-couple with aryl halides to give toluene products. Lastly, they also couple aliphatic hydrocarbons with acid chlorides to give a C—H acylation protocol. Murakami and colleagues demonstrated sp^3–sp^2 cross-coupling between benzylic and aldehyde C—H bonds in a photocatalytic system [93]. Notably, Paixão and König extend the halogen-radical-mediated C—H couplings to include alkyl halides [94]. Additionally, the Lu and Huo groups report on enantioselective C(sp^3)—H arylation of alkyl benzenes and protected amines, respectively, using chiral bis(imidazoline) ligands [95, 96].

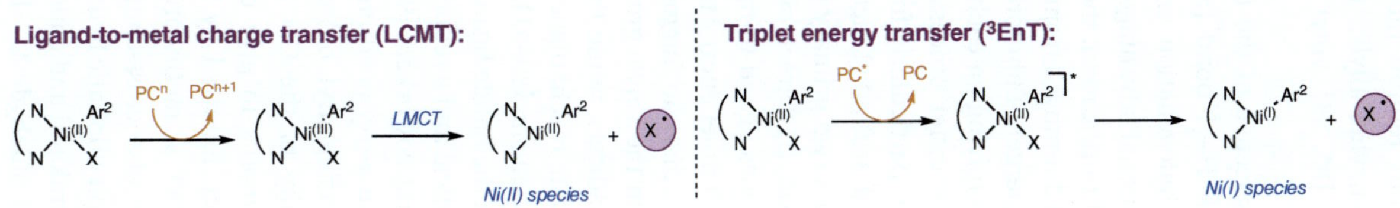

Figure 12.5 Metallaphotoredox methods for halogen radical generation.

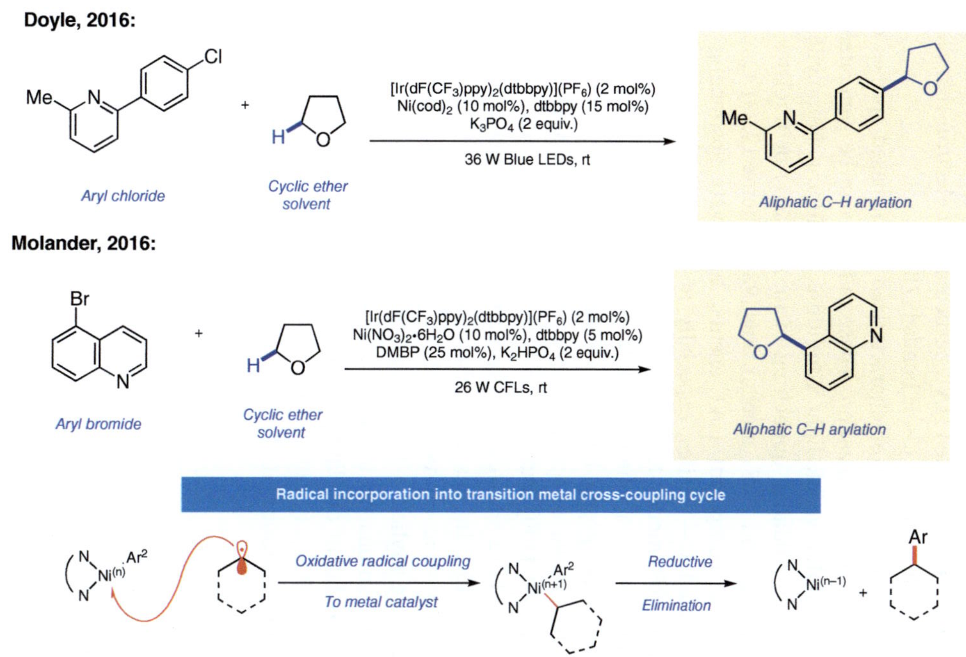

Scheme 12.22 Initial reports of C(sp^3)–H arylation with metallaphotoredox catalysis mediated by halogen radicals.

12.6 Thiyl Radicals

Sulfur radicals have emerged as versatile intermediates in photoredox-mediated C—H functionalisation due to their unique reactivity and ability to engage in selective HAT or radical addition processes. Among the various classes of sulfur radicals, thiyl radicals (RS•) are most widely used for direct C—H activation, while sulfonyl (RSO_2•) and thiocyanate-derived radicals have found more limited applications. The polar and often electrophilic nature of sulfur radicals enables them to engage in polarity-matched HAT, particularly at electron-rich sites such as benzylic, allylic, or α-heteroatom C—H bonds [97].

Thiyl radicals are the most well-characterized sulfur-based species employed in photoredox C—H functionalization. These radicals can be generated through various pathways under visible-light conditions, including homolytic cleavage of disulfides (RSSR), oxidative quenching of thiols (RSH) or thiolates (RS^-) by excited photocatalysts, and PCET mechanisms [98] (Scheme 12.23). Once formed, thiyl radicals act as competent HAT agents, selectively abstracting hydrogen atoms from activated C—H sites, especially at benzylic and allylic positions. These radicals exhibit moderate hydrogen atom transfer reactivity, with S—H bond dissociation energies in the range of 80–90 kcal/mol [99]. In a representative study by the MacMillan group, benzylic ethers were directly coupled with *N*-aryl imines (Schiff bases) via selective benzylic C—H activation enabled by thiyl radical-mediated HAT [100]. Building on this strategy, MacMillan and coworkers further demonstrated the versatility of thiyl radicals in the direct arylation of allylic sp^3 C—H bonds [101]. In this dual photoredox-organocatalytic system, triisopropylsilanethiol serves as a precursor to the thiyl radical, which mediates HAT to selectively activate allylic C—H bonds in unfunctionalized alkenes for coupling with arene radical anions. In a related study, they demonstrated that alcohols could also serve as latent alkylating agents, wherein thiyl radicals enable C—H abstraction to furnish α-oxy radicals for Minisci-type alkylation of heteroarenes under visible-light conditions [102].

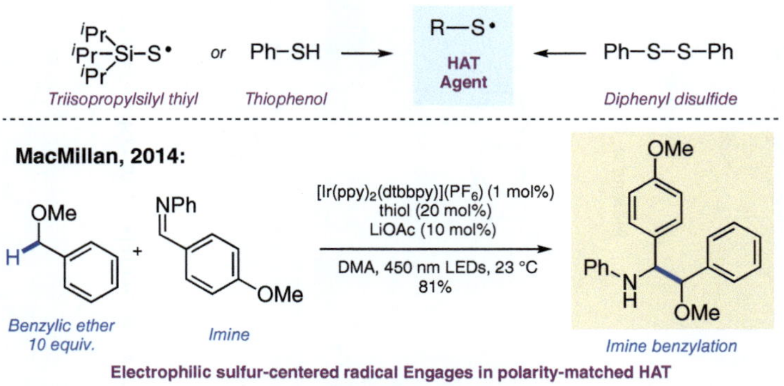

Scheme 12.23 Thiyl radical precursors and Schiff base coupling *via* HAT.

Further benzylic functionalization by thiyl radicals was demonstrated by König and coworkers, reporting selective photocarboxylation of benzylic C—H bonds

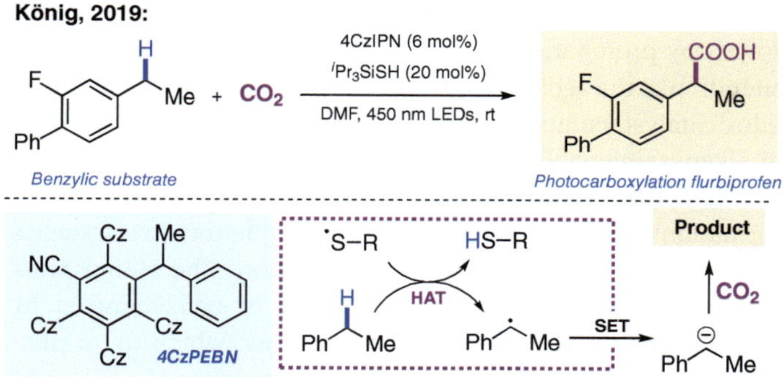

Scheme 12.24 Photocarboxylation of benzylic C–H bonds using thiyl radicals.

using CO_2 [103] (Scheme 12.24). Interestingly, the active photocatalyst was reported to be 4CzPEBN, with one cyano group in 4CzIPN substituted with benzylic substrate. Upon visible-light irradiation, the organic photocatalyst 4CzPEBN oxidizes a thiol substrate, which upon deprotonation yields the corresponding thiyl radical. This species then undergoes regioselective HAT at the benzylic position to generate a benzylic carbon-centered radical. Subsequent reduction by the photocatalyst in its reduced state, followed by CO_2 addition, affords the carboxylated product selectively at the benzylic position.

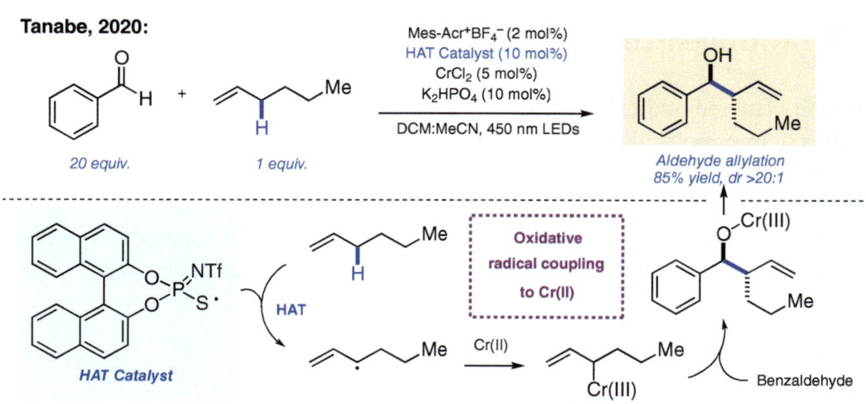

Scheme 12.25 Allylation of aldehydes using photochemical thiyl radical HAT and chromium catalysis.

Focusing on allylic C—H activation, Tanabe and coworkers developed a ternary hybrid catalytic system that enables direct allylation of aldehydes using simple unactivated alkenes, including industrial feedstocks such as 1-butene and 2-butene [104] (Scheme 12.25). This strategy combines photoredox catalysis, HAT, and chromium catalysis to deliver homoallylic alcohols under mild, visible-light-driven conditions. The mechanism begins with photoexcitation of an acridinium catalyst, which oxidizes a thiophosphoric imide HAT catalyst to generate a thiyl radical. This species abstracts an allylic hydrogen from the alkene, forming an allyl radical that is intercepted by a Cr(II) complex to yield an allylchromium(III) intermediate.

Subsequent nucleophilic addition to the aldehyde occurs via a six-membered transition state, followed by protonation to afford the homoallylic alcohol and a Cr(III) species. Regeneration of the active Cr(II) catalyst is achieved via reduction by the reduced photoredox catalyst, completing the catalytic cycle. This methodology expands the scope of alkenes amenable to allylation chemistry and showcases the utility of C—H activation strategies in convergent, atom-economical synthesis.

Beyond the transformations described above, several other photoredox strategies have exploited thiyl radicals for site-selective C—H activation. The Hamashima group have developed a regioselective benzylic arylation of benzylamines by combining SET and HAT catalysis using thiobenzoate anions, which upon photooxidation furnish thiyl radicals that selectively abstract *N*-benzylic C—H bonds, overriding the intrinsic *N*-methyl reactivity seen under purely SET-based conditions [105]. Using similar starting reagents, Chiba's group introduced a metal-free photocatalytic system based on polysulfide anions that facilitates SET and thiyl-radical-driven HAT in tandem, enabling α-tertiary amine synthesis through benzylic C—H functionalization of secondary benzylamines with cyanoarenes and ketones under mild, protecting-group-free conditions [106]. These diverse and increasingly sophisticated applications underscore the growing significance of thiyl radicals as versatile and tunable mediators in modern photoredox C—H functionalization.

12.7 Nitrogen-centered Radicals

Nitrogen-centered radicals (NCRs) have also been extensively utilized in photoinduced intermolecular HAT with much success as catalytic HAT reagents. Several examples of nitrogen-containing precursors and radicals that facilitate HAT processes are showcased in Figure 12.6. Like other radicals discussed in this chapter, generation strategies include both reductive and oxidative SET methods. A unique feature of NCR's is their ability to exist as charged aminium radical cations.

12.7.1 Processes Utilizing Aminyl Radicals

Initial instances of *N*-centered radicals (NCR) can be linked to the Hofmann–Löffler–Freytag (HLF) reaction, an intramolecular HAT reaction facilitated by aminyl radical intermediates. Nitrogen-halogen bonds have been recognized as precursors to δ-functionalized products during HLF reactions for over a century. Reductive SET, EnT, or the photolytic cleavage of N—X bonds can yield aminyl radicals. In 2015, Yu's group reported a photocatalytic method for conducting an HLF reaction to produce cyclic amides from *N*-chloroamides [107] (Scheme 12.26). When exposed to visible light, the photoreduction of the *N*-chloroalkyl amide, driven by photoexcited Ir*(III), leads to the cleavage of the N—Cl bond, thus producing the aminyl radical. Subsequent 1,5-HAT proceeds with its attached alkyl chain, forming a carbon-centered radical, which can undergo oxidation by

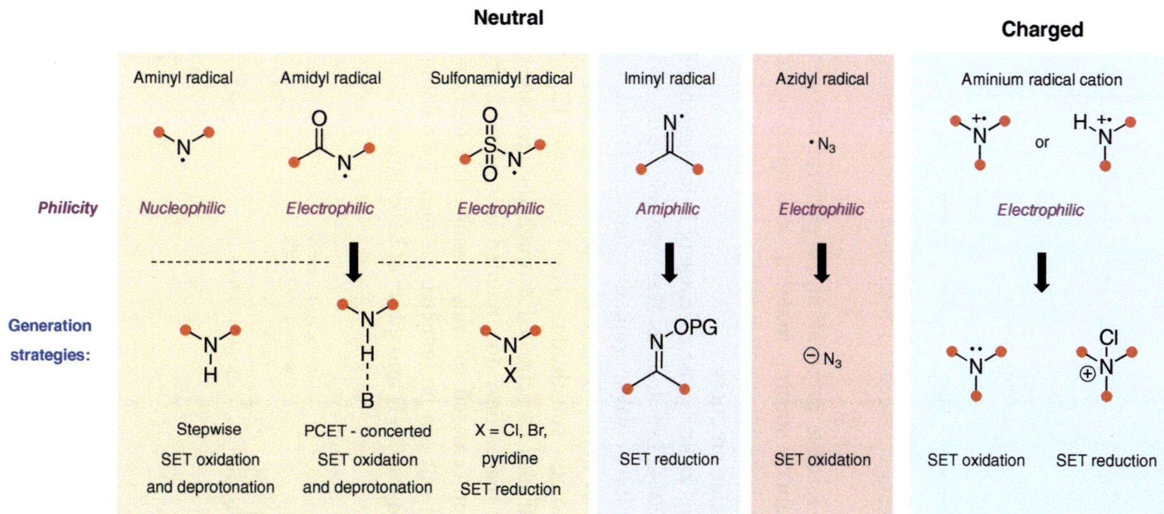

Figure 12.6 Nitrogen-centered radicals and their modes of generation.

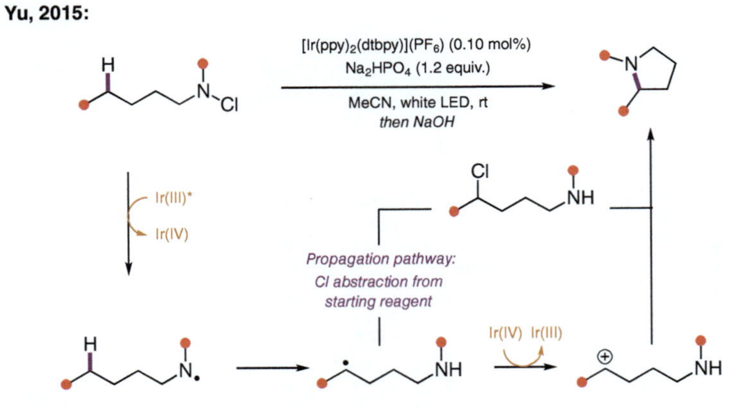

Scheme 12.26 Photoredox-catalyzed HLF reaction of N-chloroamides.

Ir(IV) to generate a reactive carbocation, or perform halogen abstraction with a different *N*-chloroamide starting material, leading to a 5-chloroalkyl amide. With the addition of a base, intramolecular S_N2 amidation occurs, resulting in the formation of cyclic amide products.

In addition to aminyl radicals, the cleavage of N-halogen bonds produces imidyl radicals as well. In 2014, Lee and his team revealed that iridium photocatalyst can facilitate the reaction between aromatic compounds and *N*-chlorophthalimide synthons to produce *N*-aryl products at room temperature through an imidyl radical-mediated aromatic substitution [108] (Scheme 12.27). The generation of imidyl radicals can also be achieved through the cleavage of the N—N bond [109, 110]. In one of the earliest documented instances of photoredox-mediated production of nitrogen-centered radicals through the cleavage of N—N bonds, an amino-pyridinium salt can be reduced, yielding 2,4,6-collidine and an imidyl radical, which leads to the C—H amination of heteroarenes [111].

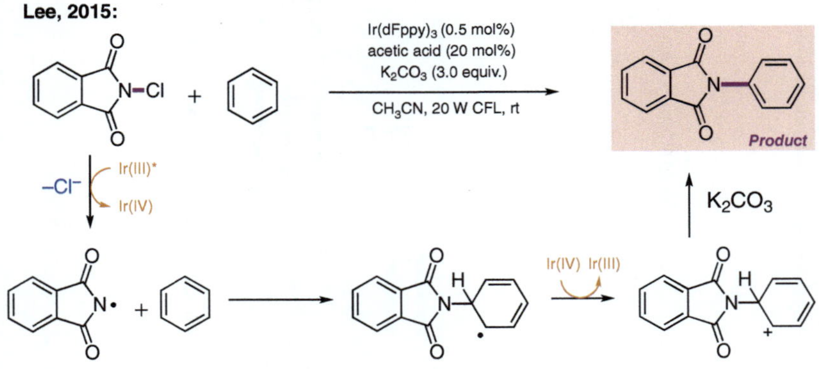

Scheme 12.27 Photoredox-catalyzed imidyl radical mediated aromatic substitution.

12.7.2 Processes Utilizing Iminyl Radicals

Iminyl radicals are significant synthetic intermediates and can be derived from the reduction of N—O bonds within *O*-aryl oximes, *O*-acyl oximes, ketoximes, aldoximes, and hydroxy acid- derived oximes. In 2017, Nevado and colleagues developed dual chiral Brønsted acid and photoredox catalysis to show the first instance of radical cyclization of *O*-acyl oximes and visible-light-mediated 1,5-HAT of iminyl radicals under mild and redox-neutral conditions [112] (Scheme 12.28). In 2018, Studer and Leonori published similar works using α-imino-oxy acids, or oximes with pendant carboxylic acids as a source of iminyl radicals [113, 114]. Studer employed the iminyl radical for a remote Giese reaction, while Leonori used them to furnish ring-opened products. In both reports, oxidative SET of the carboxylic acid triggers decarboxylation and β-scission, resulting in an iminyl radical.

12.7.3 Processes Utilizing Amidyl and Sulfonamidyl Radicals

Amidyl radicalshave become valuable synthetic intermediates for mediating HAT at strong C—H bonds. Traditionally, amidyl radicals were generated via reduction of a N—X bond. Knowles and colleagues showed that oxidative PCET could be established with photoredox catalysts to generate nitrogen-centered radicals directly from N—H bonds, thus avoiding dealing with unstable N—X adducts (Scheme 12.29). In this platform, an excited-state iridium oxidant and a weak phosphate base work together to simultaneously extract both a proton and an electron from the amide substrate, resulting in the formation of a reactive amidyl radical. Amide N—H bonds have a bond dissociation free energy (BDFE) of approximately 100 kcal/mol, making them strong HAT species [115, 116]. In back-to-back publications, Knowles and Rovis both demonstrated photoredox-catalyzed methods for using amide groups to mediate a Giese reaction at strong C—H bonds via the intermediacy of an amidyl radical [117, 118]. Meggers later rendered this remote Giese reaction enantioselective with chiral Lewis acid catalysis [119].

Oisaki and Kanai took a stepwise approach to generate catalytic sulfonamidyl radicals via a deprotonation and oxidation sequence for the C—H arylation of allylic hydrocarbons using 1,4-dicyanobenzene adducts [97]. Duan similarly reported a catalytic sulfonamide could be employed to facilitate a Giese reaction and Ni-mediated metallaphotoredox alkylation of the α-oxy position of cyclic ethers, using PCET [120]. Li and team developed a dual photoredox-Co catalyzed remote arylation of alkyl sulfonamides via a Minisci reaction [121]. Notably, Roizen has designed the use of sulfamate esters to achieve 1,6-HAT reactions for remote functionalization [122]. Reductive strategies to generate amidyl radicals with photoredox catalysis also existing, employing N—O motifs using *O*-aryl [123], *O*-benzoyl [124–126], and *O*-sulfonate ester [127] derivatives where the oxygen-motif is an anionic leaving group after a SET reduction.

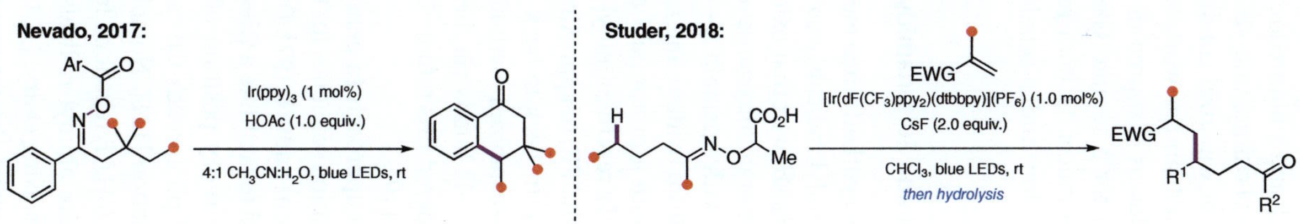

Scheme 12.28 Photoredox-catalyzed iminyl radical generation by N-O bond reduction.

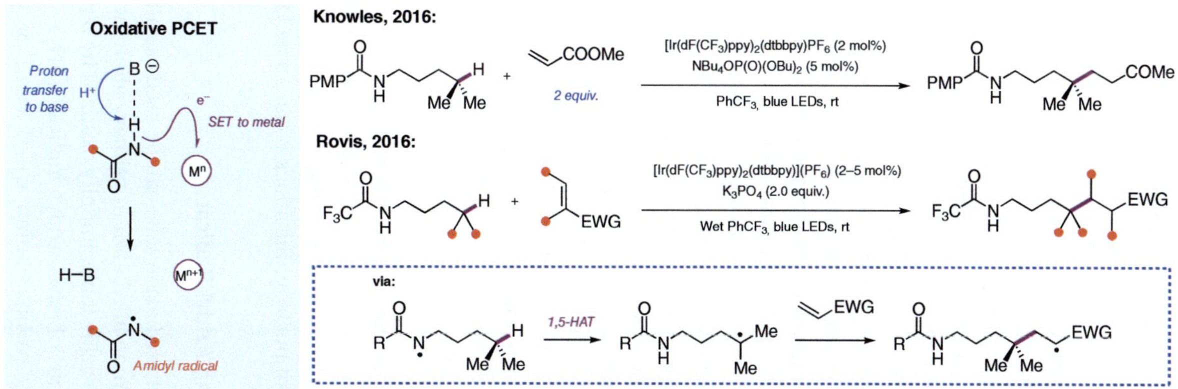

Scheme 12.29 Photoredox-catalyzed amidyl radical generation through oxidative PCET.

12.7.4 Processes Utilizing Nitrogen Radical Cations

Nitrogen radical cations have long been thought to be capable of HAT processes, however, most devolve into α-amino radicals through a very facile deprotonation of the adjacent C—H bonds. In 2015, MacMillan and coworkers discovered that quinuclidine structures can give nitrogen radical cations, which due to the bridged nature of the molecule, cannot convert into an α-amino radical (Scheme 12.30). Thus, the nitrogen radical cation is capable of being an effective HAT reagent of strong, hydridic C—H bonds. The report demonstrated a photoredox reaction that is facilitated by three different catalytic systems: an iridium-based photoredox catalyst, a HAT catalyst, and tetrabutylammonium phosphate (TBAP). The phosphate base functions as a hydrogen-bonding catalyst that engages the alcohol O—H, leading to a build of negative charge on the oxygen atom, which through hyperconjugation, can enhance the hydricity of any adjacent α-C—H bonds [128]. The quinuclidine radical cation, generated by reductive quenching of an excited state photocatalyst, can perform a site-selective HAT at the α-oxy C—H bond, thus generating an α-oxy radical of an unprotected alcohol. A Giese reaction, followed by intramolecular cyclization gave lactone products. Concurrently, in 2017, Macmillan and his coworkers reported C—H functionalization of aldehydes by using quinuclidine as a HAT reagent and developed a metallophoredox-mediated method for synthesizing ketones from aldehydes [129]. In 2016, they developed a triple-catalytic-mediated functionalization of α-amino and α-oxy C—H bond, using quinuclidine as a HAT catalyst [130]. In 2018, Wu and his coworkers reported organophotocatalyzed visible light-mediated difunctionalization of alkenes by using quinuclidine as an HAT catalyst [131]. Recently, in 2025, Huang and his coworkers [132] developed an iridium photoredox-catalyzed robust method for disulfidation of C—H bonds by using quinuclidine as an HAT catalyst. Martin and team employed quinuclidine derivatives as a way to functionalize adamantane structures [133].

In 2019, Leonori and his coworkers reported a C(sp^2)—H amination via intermediacy of an amine radical cation. The team employed NCS to *in-situ* chlorinate secondary amines, which can be reduced to give the active amine radical cation, which undergoes subsequent nucleophilic attack by an arene [134].

12.7.5 Processes Utilizing Azidyl Radical

In 2020, Cresswell demonstrated azide salts can be used as catalytic HAT reagents in the development of α-amino alkylation of primary aliphatic amines using electrophilic alkenes [135]. The azide ion is initially oxidized by photo-excited 4CzIPN, resulting in the formation of the azidyl radical ($\bullet N_3$). Site-selective HAT occurs from the relatively weak α-amino C—H bond of the primary amine (BDE = 89–91 kcal/mol), possibly enhanced by a polarity-matching effect between the electrophilic azidyl radical and the "hydridic" C—H bond.

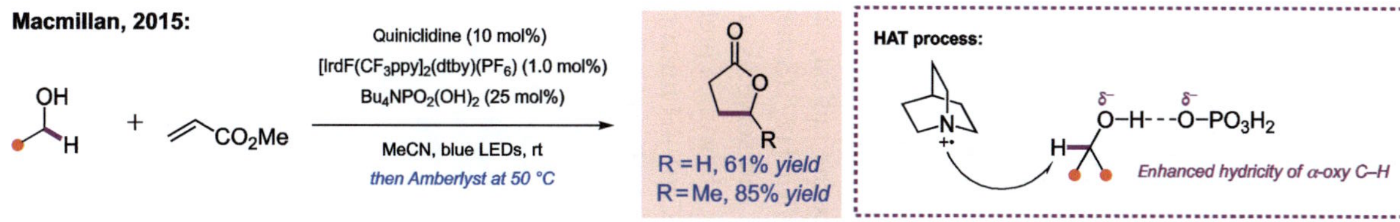

Scheme 12.30 Selective α-C—H alkylation of alcohols with quinuclidine radical cations.

12.8 Carbon-centered Radicals

Carbon-centered radicals are less common as a robust hydrogen-atom abstractor species for C—H bonds. Certain difficulties arise with the carbon radical-mediated HAT mechanism, such as the minor difference in BDE between reactive intermediates, slow kinetics during the HAT step, and undesired intermolecular HAT reactions. Recently, many groups have developed several HAT strategies that rely on a carbon-centered radical motif. In 2019, Studer and his coworkers announced a method for α-C—H alkylation and arylation of alkylboronic esters [136] (Scheme 12.31). Prior to the photoredox catalysis, an aryl-alkyl boronate complex is prepared from a boronic acid and organolithium reagent. A photocatalyst then activates trifluoromethyl iodide via single electron reduction to give an electrophilic trifluoromethyl radical (•CF_3). Subsequently, •CF_3 acts as a HAT reagent at the α-boryl C—H bond in the boronate complex, resulting in the formation of a radical anion intermediate, which is further oxidized by the photocatalyst. This generates a carbocation which reacts further via a 1,2 aryl or alkyl migration to afford α-substituted boronate products.

In 2021, Doyle reported a fluorination of Csp^3—H bonds with nucleophilic fluoride sources and an *N*-acetoxyphthalimide reagent. Reduction of the phthalimide motif leads to fragmentation of an acetate radical, which the authors propose can decarboxylate to generate a methyl radical [127] (Scheme 12.32). Mechanistic experiment support that the •CH_3 radical can serve as an efficient HAT reagent from both hydridic and acidic C—H bonds, showing a preference for acidic. Along with the fluorination, they showcased the Csp^3—H functionalisation with various other nucleophiles like alcohol, chloride, azide, thiol, and water.

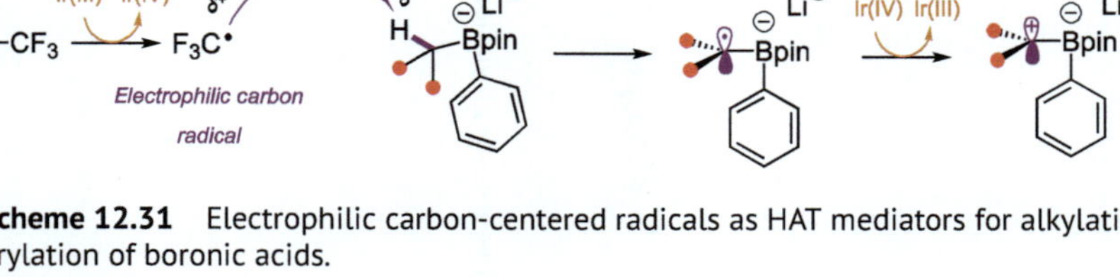

Scheme 12.31 Electrophilic carbon-centered radicals as HAT mediators for alkylation and arylation of boronic acids.

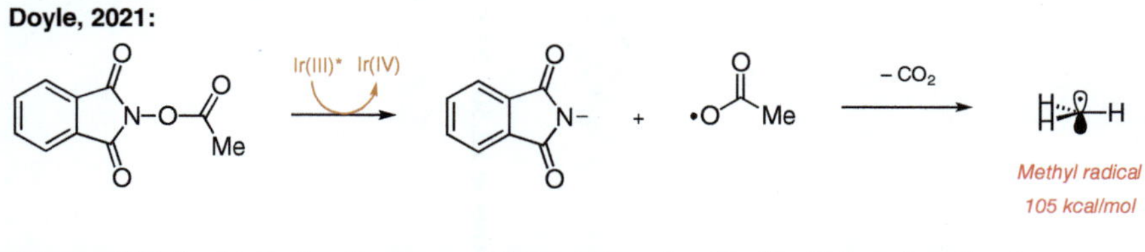

Scheme 12.32 Methyl radical as an HAT mediator for C(sp^3)–H functionalization.

Aryl Csp^2 radicals are also able to abstract hydrogen-atoms from strong $C(sp^3)$—H bonds given their higher bond energy of 113 kcal/mol [137]. Easily reducible C—X bonds such as aryl diazonium [138–140] or aryl iodides [141] are commonly coupled with strongly reducing photoredox catalysts to give aryl radicals capable of intramolecular HAT. The diazonium motif is either preformed or generated *in-situ*.

12.9 Summary and Conclusions

Although the birth of C—H functionalization started with C—H activation efforts mediated by traditional transition metal chemistry, the field has since expanded to encompass a variety of catalytic platforms. Among these, photoredox catalysis has become a widely adopted strategy – particularly for functionalizing $C(sp^3)$—H bonds – due to its versatility in reaction conditions and mechanisms. As highlighted within this chapter, the strength of photoredox catalysis lies in its capacity to provide diverse reaction conditions and mechanisms for those wishing to accomplish a C—H functionalization, allowing users to target an array of different C—H bond types, primarily via HAT processes. Early developments utilized carbon-centered radicals in Giese reactions for C—H alkylations and Minisci reactions to achieve C—H heteroarylation. Radical-polar crossover mechanisms have allowed for coupling of C—H bonds with abundant two-electron nucleophilic and electrophilic partners. And lastly, merging photoredox catalysis with transition metal catalysis has dramatically broadened the scope of HAT-enabled transformations, facilitating cross-coupling with organoboranes, alkyl and aryl halides, amines, alcohols, and others – and advancement that has accelerated photoredox adoption in industrial applications. The continued evolution of these approaches underscores the central role of photoredox catalysis in tackling the most challenging of transformations that face modern organic synthesis.

References

1 Nicewicz, D.A. and MacMillan, D.W.C. (2008). Merging photoredox catalysis with organocatalysis: The direct asymmetric alkylation of aldehydes. *Science* 322 (5898): 77–80.

2 Nagib, D.A., Scott, M.E., and MacMillan, D.W.C. (2009). Enantioselective α-trifluoromethylation of aldehydes via photoredox organocatalysis. *J. Am. Chem. Soc.* 131 (31): 10875–10877.

3 Welin, E.R., Warkentin, A.A., Conrad, J.C., and MacMillan, D.W.C. (2015). Enantioselective α-alkylation of aldehydes by photoredox organocatalysis: Rapid access to pharmacophore fragments from β-cyanoaldehydes. *Angew. Chem. Int. Ed.* 54 (33): 9668–9672.

4 Shih, H.-W., Vander Wal, M.N., Grange, R.L., and MacMillan, D.W.C. (2010). Enantioselective α-benzylation of aldehydes via photoredox organocatalysis. *J. Am. Chem. Soc.* 132 (39): 13600–13603.

5 Roberts, B.P. (1999). Polarity-reversal catalysis of hydrogen-atom abstraction reactions: concepts and applications in organic chemistry. *Chem. Soc. Rev.* 28 (1): 25–35.

6 Capaldo, L., Ravelli, D., and Fagnoni, M. (2022). Direct photocatalyzed hydrogen atom transfer (HAT) for aliphatic C–H bonds elaboration. *Chem. Rev.* 122 (2): 1875–1924.

7 Capaldo, L. and Ravelli, D. (2017). Hydrogen atom transfer (HAT): A versatile strategy for substrate activation in photocatalyzed organic synthesis. *Eur. J. Org. Chem.* 15: 2056–2071.

8 Bellotti, P., Huang, H.-M., Faber, T., and Glorius, F. (2023). Photocatalytic late-stage C–H functionalization. *Chem. Rev.* 123 (8): 4237–4352.

9 Condie, A.G., González-Gómez, J.C., and Stephenson, C.R.J. (2010). Visible-light photoredox catalysis: Aza-henry reactions via C–H functionalization. *J. Am. Chem. Soc.* 132 (5): 1464–1465.

10 Beatty, J.W. and Stephenson, C.R.J. (2015). Amine functionalization via oxidative photoredox catalysis: Methodology development and complex molecule synthesis. *Acc. Chem. Res.* 48 (5): 1474–1484.

11 Freeman, D.B., Furst, L., Condie, A.G., and Stephenson, C.R.J. (2012). Functionally diverse nucleophilic trapping of iminium intermediates generated utilizing visible light. *Org. Lett.* 14 (1): 94–97.

12 Dai, C., Meschini, F., Narayanam, J.M.R., and Stephenson, C.R.J. (2012). Friedel–crafts amidoalkylation via thermolysis and oxidative photocatalysis. *J. Org. Chem.* 77 (9): 4425–4431.

13 McNally, A., Prier, C.K., and MacMillan, D.W.C. (2011). Discovery of an α-amino C–H arylation reaction using the strategy of accelerated serendipity. *Science* 334 (6059): 1114–1117.

14 Zuo, Z., Ahneman, D.T., Chu, L. et al. (2014). Merging photoredox with nickel catalysis: Coupling of α-carboxyl sp^3-carbons with aryl halides. *Science* 345 (6195): 437–440.

15 Pirnot, M.T., Rankic, D.A., Martin, D.B., and MacMillan, D.W. (2013). Photoredox activation for the direct β-arylation of ketones and aldehydes. *Science* 339 (6127): 1593–1596.

16 Terrett, J.A., Clift, M.D., and MacMillan, D.W.C. (2014). Direct β-alkylation of aldehydes via photoredox organocatalysis. *J. Am. Chem. Soc.* 136 (19): 6858–6861.

17 Jeffrey, J.L., Petronijević, F.R., and MacMillan, D.W.C. (2015). Selective radical–radical cross-couplings: Design of a formal β-mannich reaction. *J. Am. Chem. Soc.* 137 (26): 8404–8407.

18 Petronijević, F.R., Nappi, M., and MacMillan, D.W.C. (2013). Direct β-functionalization of cyclic ketones with aryl ketones via the merger of photoredox and organocatalysis. *J. Am. Chem. Soc.* 135 (49): 18323–18326.

19 Lee, B.J., DeGlopper, K.S., and Yoon, T.P. (2020). Site-selective alkoxylation of benzylic C–H bonds by photoredox catalysis. *Angew. Chem. Int. Ed.* 59 (1): 197–202.

20 Pandey, G., Laha, R., and Singh, D. (2016). Benzylic C(sp^3)–H functionalization for C–N and C–O bond formation via visible light photoredox catalysis. *J. Org. Chem.* 81 (16): 7161–7171.

21 Sharma, K.S., Thadem, N., and Pandey, G. (2025). Visible-light-induced secondary benzylic C(sp^3)–H functionalization for nucleophilic substitution: An intermolecular C–X (C–N, C–C, and C–Br) bond forming reaction. *J. Org. Chem.* 90 (9): 3384–3390.

22 Chang, L., An, Q., Duan, L. et al. (2022). Alkoxy radicals see the light: New paradigms of photochemical synthesis. *Chem. Rev.* 122 (2): 2429–2486.

23 Li, J. and Wang, D.Z. (2015). Visible-light-promoted photoredox syntheses of α,β-epoxy ketones from styrenes and benzaldehydes under alkaline conditions. *Org. Lett.* 17 (21): 5260–5263.

24 Jung, S., Kim, J., and Hong, S. (2017). Cover picture: Visible light-promoted synthesis of spiroepoxy chromanone derivatives via a tandem oxidation/radical cyclization/epoxidation process (Adv. Synth. Catal. 22/2017). *Adv. Synth. Catal.* 359 (22): 3882–3882.

25 Iqbal, N. and Cho, E.J. (2016). Visible-light-mediated synthesis of amides from aldehydes and amines via in situ acid chloride formation. *J. Org. Chem.* 81 (5): 1905–1911.

26 Gaspa, S., Farina, A., Tilocca, M. et al. (2020). Visible-light photoredox-catalyzed amidation of benzylic alcohols. *J. Org. Chem.* 85 (18): 11679–11687.

27 Vasilopoulos, A., Krska, S.W., and Stahl, S.S. (2021). C(sp^3)–H methylation enabled by peroxide photosensitization and Ni-mediated radical coupling. *Science* 372 (6540): 398–403.

28 Chen, X., Lian, Z., and Kramer, S. (2023). Enantioselective intermolecular radical amidation and amination of benzylic C–H bonds via dual copper and photocatalysis. *Angew. Chem. Int. Ed.* 62 (13): e202217638.

29 Ouyang, X.-H., Li, Y., Song, R.-J. et al. (2019). Intermolecular dialkylation of alkenes with two distinct C(sp^3)—H bonds enabled by synergistic photoredox catalysis and iron catalysis. *Sci. Adv.* 5 (3): eaav9839.

30 Colgan, A.C., Proctor, R.S.J., Gibson, D.C. et al. (2022). Hydrogen atom transfer driven enantioselective minisci reaction of alcohols. *Angew. Chem. Int. Ed.* 61 (25): e202200266.

31 Huff, C.A., Cohen, R.D., Dykstra, K.D. et al. (2016). Photoredox-catalyzed hydroxymethylation of heteroaromatic bases. *J. Org. Chem.* 81 (16): 6980–6987.

32 Zhang, Y., Fitzpatrick, N.A., Das, M. et al. (2022). A photoredox-catalyzed approach for formal hydride abstraction to enable Csp^3–H functionalization with nucleophilic partners (F, C, O, N, and Br/Cl). *Chem Catal.* 2 (2): 292–308.

33 Jin, J. and MacMillan, D.W.C. (2015). Direct α-arylation of ethers through the combination of photoredox-mediated C–H functionalization and the minisci reaction. *Angew. Chem. Int. Ed.* 54 (5): 1565–1569.

34 Kim, I., Park, B., Kang, G. et al. (2018). Visible-light-induced pyridylation of remote C(sp^3)–H bonds by radical translocation of N-alkoxypyridinium salts. *Angew. Chem. Int. Ed.* 57 (47): 15517–15522.

35 Kim, I., Kang, G., Lee, K. et al. (2019). Site-selective functionalization of pyridinium derivatives via visible-light-driven photocatalysis with quinolinone. *J. Am. Chem. Soc.* 141 (23): 9239–9248.

36 Bao, X., Wang, Q., and Zhu, J. (2019). Remote C(sp^3)–H arylation and vinylation of N-alkoxypyridinium salts to δ-aryl and δ-vinyl alcohols. *Chem. Eur. J.* 25 (50): 11630–11634.

37 Zheng, M., Hou, J., Zhan, L.-W. et al. (2021). Visible-light-driven, metal-free divergent difunctionalization of alkenes using alkyl formates. *ACS Catal.* 11 (2): 542–553.

38 Das, M., Zamani, L., Bratcher, C., and Musacchio, P.Z. (2023). Azolation of benzylic C–H bonds via photoredox-catalyzed carbocation generation. *J. Am. Chem. Soc.* 145 (7): 3861–3868.

39 Schlegel, M., Qian, S., and Nicewicz, D.A. (2022). Aliphatic C–H functionalization using pyridine N-oxides as H-atom abstraction agents. *ACS Catal.* 12 (16): 10499–10505.

40 Mukherjee, S., Maji, B., Tlahuext-Aca, A., and Glorius, F. (2016). Visible-light-promoted activation of unactivated C(sp^3)–H bonds and their selective trifluoromethylthiolation. *J. Am. Chem. Soc.* 138 (50): 16200–16203.

41 Hirose, M., Sakaguchi, H., Hashimoto, R. et al. (2024). Benzoic acid serves as precursor of catalytic HAT reagent in a two-molecule photoredox system. *Chem. Eur. J.* 30 (55): e202402285.

42 Wang, Y., Li, G.-X., Yang, G. et al. (2016). A visible-light-promoted radical reaction system for azidation and halogenation of tertiary aliphatic C–H bonds. *Chem. Sci.* 7 (4): 2679–2683.

43 Margrey, K.A., Czaplyski, W.L., Nicewicz, D.A., and Alexanian, E.J. (2018). A general strategy for aliphatic C–H functionalization enabled by organic photoredox catalysis. *J. Am. Chem. Soc.* 140 (12): 4213–4217.

44 Wakaki, T., Sakai, K., Enomoto, T. et al. (2018). C(sp^3)–H cyanation promoted by visible-light photoredox/phosphate hybrid catalysis. *Chem. Eur. J.* 24 (32): 8051–8055.

45 An, Q., Xing, Y.-Y., Pu, R. et al. (2023). Identification of alkoxy radicals as hydrogen atom transfer agents in Ce-catalyzed C–H functionalization. *J. Am. Chem. Soc.* 145 (1): 359–376.

46 Zhang, X., Ning, S., Li, Y. et al. (2023). Hydrogen atom transfer for C(sp^3)–H functionalization enabled by photoredox/thianthrene/methanol synergistic catalysis. *ChemCatChem* 15 (11): e202300311.

47 Wu, X., Wang, M., Huan, L. et al. (2018). Tertiary-alcohol-directed functionalization of remote C(sp^3)–H bonds by sequential hydrogen atom and heteroaryl migrations. *Angew. Chem. Int. Ed.* 57 (6): 1640–1644.

48 Wang, M., Huan, L., and Zhu, C. (2019). Cyanohydrin-mediated cyanation of remote unactivated C(sp^3)–H bonds. *Org. Lett.* 21 (3): 821–825.

49 Li, G.-X., Hu, X., He, G., and Chen, G. (2019). Photoredox-mediated remote C(sp^3)–H heteroarylation of free alcohols. *Chem. Sci.* 10 (3): 688–693.

50 Rivero, A.R., Fodran, P., Ondrejková, A., and Wallentin, C.-J. (2020). Alcohol etherification via alkoxy radicals generated by visible-light photoredox catalysis. *Org. Lett.* 22 (21): 8436–8440.

51 Zhang, P., Wang, Y., Yao, J. et al. (2011). Visible-light-induced metal-free allylic oxidation utilizing a coupled photocatalytic system of g-C3N4 and N-hydroxy compounds. *Adv. Synth. Catal.* 353 (9): 1447–1451.

52 Yang, S., Zhu, S., Lu, D., and Gong, Y. (2019). Formylation of fluoroalkyl imines through visible-light-enabled H-atom transfer catalysis: Access to fluorinated α-amino aldehydes. *Org. Lett.* 21 (7): 2019–2024.

53 Liu, Y., Zhang, L., Zhang, Y. et al. (2023). Asymmetric olefin isomerization via photoredox catalytic hydrogen atom transfer and enantioselective protonation. *J. Am. Chem. Soc.* 145 (33): 18307–18315.

54 Zhang, J., Li, Y., Zhang, F. et al. (2016). Generation of alkoxyl radicals by photoredox catalysis enables selective C(sp^3)–H functionalization under mild reaction conditions. *Angew. Chem. Int. Ed.* 55 (5): 1872–1875.

55 Deng, Y., Nguyen, M.D., Zou, Y. et al. (2019). Generation of dithianyl and dioxolanyl radicals using photoredox catalysis: Application in the total synthesis of the danshenspiroketallactones via radical relay chemistry. *Org. Lett.* 21 (6): 1708–1712.

56 Wang, C., Yu, Y., Liu, W.-L., and Duan, W.-L. (2019). Site-tunable Csp^3–H bonds functionalization by visible-light-induced radical translocation of N-alkoxyphthalimides. *Org. Lett.* 21 (22): 9147–9152.

57 Herron, A.N., Liu, D., Xia, G., and Yu, J.-Q. (2020). δ-C–H mono- and dihalogenation of alcohols. *J. Am. Chem. Soc.* 142 (6): 2766–2770.

58 Laudadio, G., Deng, Y., van der Wal, K. et al. (2020). C(sp^3)–H functionalizations of light hydrocarbons using decatungstate photocatalysis in flow. *Science* 369 (6499): 92–96.

59 Perry, I.B., Brewer, T.F., Sarver, P.J. et al. (2018). Direct arylation of strong aliphatic C–H bonds. *Nature* 560 (7716): 70–75.

60 Hyodo, M., Iwano, H., Kasakado, T. et al. (2021). Using high-power UV-LED to accelerate a decatungstate-anion-catalyzed reaction: A model study for the quick oxidation of benzyl alcohol to benzoic acid using molecular oxygen. *Micromachines* 12 (11): 1307.

61 Ravelli, D., Protti, S., and Fagnoni, M. (2016). Decatungstate anion for photocatalyzed "window ledge" reactions. *Acc. Chem. Res.* 49 (10): 2232–2242.

62 Ravelli, D., Fagnoni, M., Fukuyama, T. et al. (2018). Site-selective C–H functionalization by decatungstate anion photocatalysis: synergistic control by polar and steric effects expands the reaction scope. *ACS Catal.* 8 (1): 701–713.

63 Protti, S., Ravelli, D., Fagnoni, M., and Albini, A. (2009). Solar light-driven photocatalyzed alkylations. Chemistry on the window ledge. *Chem. Commun.* 47: 7351–7353.

64 Rubanov, Z.M., Levin, V.V., and Dilman, A.D. (2024). Dual acridine/decatungstate photocatalysis for the decarboxylative radical addition of carboxylic acids to azomethines. *Org. Lett.* 26 (15): 3174–3178.

65 Quattrini, M.C., Fujii, S., Yamada, K. et al. (2017). Versatile cross-dehydrogenative coupling of heteroaromatics and hydrogen donors via decatungstate photocatalysis. *Chem. Commun.* 53 (15): 2335–2338.

66 Sarver, P.J., Bissonnette, N.B., and MacMillan, D.W. (2021). Decatungstate-catalyzed C(sp^3)–H sulfinylation: rapid access to diverse organosulfur functionality. *J. Am. Chem. Soc.* 143 (26): 9737–9743.

67 Zhang, J. and Studer, A. (2022). Decatungstate-catalyzed radical disulfuration through direct C-H functionalization for the preparation of unsymmetrical disulfides. *Nat. Commun.* 13 (1): 3886.

68 Yuan, Z. and Britton, R. (2023). Development and application of decatungstate catalyzed C–H 18F- and 19F-fluorination, fluoroalkylation and beyond. *Chem. Sci.* 14 (45): 12883–12897.

69 Halperin, S.D., Fan, H., Chang, S. et al. (2014). A convenient photocatalytic fluorination of unactivated C-H bonds. *Angew. Chem. Int. Ed.* 53 (18): 4690–4693.

70 Schultz, D.M., Lévesque, F., DiRocco, D.A. et al. (2017). Oxyfunctionalization of the remote C–H bonds of aliphatic amines by decatungstate photocatalysis. *Angew. Chem. Int. Ed.* 56 (48): 15274–15278.

71 Sarver, P.J., Bacauanu, V., Schultz, D.M. et al. (2020). The merger of decatungstate and copper catalysis to enable aliphatic C(sp^3)–H trifluoromethylation. *Nat. Chem.* 12 (5): 459–467.

72 Mazzarella, D., Pulcinella, A., Bovy, L. et al. (2021). Rapid and direct photocatalytic C(sp^3)–H acylation and arylation in flow. *Angew. Chem. Int. Ed.* 60 (39): 21277–21282.

73 Musgrave, C.B.I., Olsen, K., Liebov, N.S. et al. (2023). Partial oxidation of methane enabled by decatungstate photocatalysis coupled to free radical chemistry. *ACS Catal.* 13 (9): 6382–6395.

74 Tang, P., Zhu, Q., Wu, Z., and Ma, D. (2014). Methane activation: the past and future. *Energy Environ. Sci.* 7 (8): 2580–2591.

75 Hu, X., Cheng-Sánchez, I., Kong, W. et al. (2024). Nickel-catalysed enantioselective alkene dicarbofunctionalization enabled by photochemical aliphatic C–H bond activation. *Nat. Catal.* 7 (6): 655–665.

76 Wan, T., Capaldo, L., Laudadio, G. et al. (2021). Decatungstate-mediated C(sp^3)–H heteroarylation via radical-polar crossover in batch and flow. *Angew. Chem. Int. Ed.* 60 (33): 17893–17897.

77 Norrish, R.G.W. and Porter, G. (1950). Chemical reactions produced by very short light flashes. *Nature* 165.

78 Gonzalez, M.I., Gygi, D., Qin, Y. et al. (2022). Taming the chlorine radical: Enforcing steric control over chlorine-radical-mediated C–H activation. *J. Am. Chem. Soc.* 144 (3): 1464–1472.

79 Huang, C.-Y., Li, J., and Li, C.-J. (2022). Photocatalytic C(sp^3) radical generation via C–H, C–C, and C–X bond cleavage. *Chem. Sci.* 13 (19): 5465–5504.

80 Deng, H.-P., Zhou, Q., and Wu, J. (2018). Microtubing-reactor-assisted aliphatic C-H functionalization with HCl as a hydrogen-atom-transfer catalyst precursor in conjunction with an organic photoredox catalyst. *Angew. Chem. Int. Ed. Engl.* 57 (39): 12661–12665.

81 Russell, G.A., Ito, A., and Hendry, D.G. (1963). Solvent effects in the reactions of free radicals and atoms. VIII. The photochlorination of aralkyl hydrocarbons. *J. Am. Chem. Soc.* 85 (19): 2976–2983.

82 Armstrong, D.A., Huie, R.E., Koppenol, W.H. et al. (2015). Standard electrode potentials involving radicals in aqueous solution: Inorganic radicals (IUPAC Technical Report). *Pure Appl. Chem.* 87 (11–12): 1139–1150.

83 Saxena, B., Patel, I., and R., and Sharma, A. (2024). Visible light-induced bromine radical enhanced hydrogen atom transfer (HAT) reactions in organic synthesis. *RSC Sustain.* 2 (8): 2169–2189.

84 Yue, H., Zhu, C., Huang, L. et al. (2022). Advances in allylic and benzylic C–H bond functionalization enabled by metallaphotoredox catalysis. *Chem. Commun.* 58 (2): 171–184.

85 Wang, Z., Ji, X., Han, T. et al. (2019). LiBr-promoted photoredox minisci-type alkylations of quinolines with ethers. *Adv. Synth. Catal.* 361 (24): 5643–5647.

86 Wang, Z., Liu, Q., Ji, X. et al. (2020). Bromide-promoted visible-light-induced reductive minisci reaction with aldehydes. *ACS Catal.* 10 (1): 154–159.

87 Cao, C.-L., Zhang, G.-X., Xue, F., and Deng, H.-P. (2022). Photoinduced C–H monofluoroalkenylation with gem-difluoroalkenes through hydrogen atom transfer under batch and flow conditions. *Org. Chem. Front.* 9 (4): 959–965.

88 Darwent, B.dB. (1970). *Bond Dissociation Energies in Simple Molecules. National Bureau of Standards (U.S.), National Standard Reference Data System, NSRDS-NBS 31, U.S.* Government Printing Office.

89 Strieth-Kalthoff, F. and Glorius, F. (2020). Triplet energy transfer photocatalysis: Unlocking the next level. *Chem* 6 (8): 1888–1903.

90 Wang, Z., Yan, C.-X., Liu, R. et al. (2024). Photo-induced versatile aliphatic C–H functionalization via electron donor–acceptor complex. *Sci. Bull.* 69 (3): 345–353.

91 Cusumano, A.Q., Chaffin, B.C., and Doyle, A.G. (2024). Mechanism of Ni-catalyzed photochemical halogen atom-mediated C(sp^3)–H arylation. *J. Am. Chem. Soc.* 146 (22): 15331–15344.

92 Shields, B.J. and Doyle, A.G. (2016). Direct C(sp^3)–H cross coupling enabled by catalytic generation of chlorine radicals. *J. Am. Chem. Soc.* 138 (39): 12719–12722.

93 Kawasaki, T., Ishida, N., and Murakami, M. (2020). Dehydrogenative coupling of benzylic and aldehydic C–H bonds. *J. Am. Chem. Soc.* 142 (7): 3366–3370.

94 Santos, M.S., Corrêa, A.G., Paixão, M.W., and König, B. (2020). C(sp^3)–C(sp^3) Cross-coupling of alkyl bromides and ethers mediated by metal and visible light photoredox catalysis. *Adv. Synth. Catal.* 362 (12): 2367–2372.

95 Cheng, X., Lu, H., and Lu, Z. (2019). Enantioselective benzylic C–H arylation via photoredox and nickel dual catalysis. *Nat. Commun.* 10 (1): 3549.

96 Huan, L., Shu, X., Zu, W. et al. (2021). Asymmetric benzylic C(sp^3)-H acylation via dual nickel and photoredox catalysis. *Nat. Commun.* 12 (1): 3536.

97 Tanaka, H., Sakai, K., Kawamura, A. et al. (2018). Sulfonamides as new hydrogen atom transfer (HAT) catalysts for photoredox allylic and benzylic C–H arylations. *Chem. Commun.* 54 (26): 3215–3218.

98 Russo, C., Brunelli, F., Tron, G.C., and Giustiniano, M. (2023). Sulfur-centered radicals in visible light isocyanide photochemistry. *Eur. J. Org. Chem.* 26 (45): e202300743.

99 Benny, A., Nolan, M.D., and Scanlan, E.M. (2025). Harnessing radical mediated reactions of thioacids for organic synthesis. *Chem. Commun.* 61 (32): 5883–5898.

100 Hager, D. and MacMillan, D.W.C. (2014). Activation of C–H bonds via the merger of photoredox and organocatalysis: A coupling of benzylic ethers with schiff bases. *J. Am. Chem. Soc.* 136 (49): 16986–16989.

101 Cuthbertson, J.D. and MacMillan, D.W.C. (2015). The direct arylation of allylic sp^3 C–H bonds via organic and photoredox catalysis. *Nature* 519 (7541): 74–77.

102 Jin, J. and MacMillan, D.W.C. (2015). Alcohols as alkylating agents in heteroarene C–H functionalization. *Nature* 525 (7567): 87–90.

103 Meng, Q.-Y., Schirmer, T.E., Berger, A.L. et al. (2019). Photocarboxylation of benzylic C-H Bonds. *J. Am. Chem. Soc.* 141 (29): 11393–11397.

104 Tanabe, S., Mitsunuma, H., and Kanai, M. (2020). Catalytic allylation of aldehydes using unactivated alkenes. *J. Am. Chem. Soc.* 142 (28): 12374–12381.

105 Ide, T., Barham, J.P., Fujita, M. et al. (2018). Regio- and chemoselective Csp^3–H arylation of benzylamines by single electron transfer/hydrogen atom transfer synergistic catalysis. *Chem. Sci.* 9 (44): 8453–8460.

106 Li, H. and Chiba, S. (2022). Synthesis of α-tertiary amines by polysulfide anions photocatalysis via single-electron transfer and hydrogen atom transfer in relays. *Chem Catal.* 2 (5): 1128–1142.

107 Qin, Q. and Yu, S. (2015). Visible-light-promoted remote C(sp^3)–H amidation and chlorination. *Org. Lett.* 17 (8): 1894–1897.

108 Kim, H., Kim, T., Lee, D.G. et al. (2014). Nitrogen-centered radical-mediated C–H imidation of arenes and heteroarenes via visible light induced photocatalysis. *Chem. Commun.* 50 (66): 9273–9276.

109 Guin, J., Fröhlich, R., and Studer, A. (2008). Thiol-catalyzed stereoselective transfer hydroamination of olefins with N-aminated dihydropyridines. *Angew. Chem. Int. Ed.* 47 (4): 779–782.

110 Shin, S., Lee, S., Choi, W. et al. (2021). Visible-light-induced 1,3-aminopyridylation of [1.1.1]propellane with N-aminopyridinium salts. *Angew. Chem. Int. Ed.* 60 (14): 7873–7879.

111 Greulich, T.W., Daniliuc, C.G., and Studer, A. (2015). N-aminopyridinium salts as precursors for N-centered radicals – direct amidation of arenes and heteroarenes. *Org. Lett.* 17 (2): 254–257.

112 Shu, W. and Nevado, C. (2017). Visible-light-mediated remote aliphatic C–H functionalizations through a 1,5-hydrogen transfer cascade. *Angew. Chem. Int. Ed.* 56 (7): 1881–1884.

113 Jiang, H. and Studer, A. (2018). α-Aminoxy-acid-auxiliary-enabled intermolecular radical γ-C(sp^3)–H functionalization of ketones. *Angew. Chem. Int. Ed.* 57 (6): 1692–1696.

114 Dauncey, E.M., Morcillo, S.P., Douglas, J.J. et al. (2018). Photoinduced remote functionalisations by iminyl radical promoted C–C and C–H bond cleavage cascades. *Angew. Chem. Int. Ed.* 57 (3): 744–748.

115 Bordwell, F.G., Zhang, S., Zhang, X.-M., and Liu, W.-Z. (1995). Homolytic bond dissociation enthalpies of the acidic H-A bonds caused by proximate substituents in sets of methyl ketones, carboxylic esters, and carboxamides related to changes in ground state energies. *J. Am. Chem. Soc.* 117 (27): 7092–7096.

116 Bordwell, F.G., Harrelson, J.A. Jr., and Lynch, T.Y. (1990). Homolytic bond dissociation energies for the cleavage of .alpha.-nitrogen-hydrogen bonds in carboxamides, sulfonamides, and their derivatives. The question of synergism in nitrogen-centered radicals. *J. Org. Chem.* 55 (10): 3337–3341.

117 Choi, G.J., Zhu, Q., Miller, D.C. et al. (2016). Catalytic alkylation of remote C–H bonds enabled by proton-coupled electron transfer. *Nature* 539 (7628): 268–271.

118 Chu, J.C.K. and Rovis, T. (2016). Amide-directed photoredox-catalysed C–C bond formation at unactivated sp^3 C–H bonds. *Nature* 539 (7628): 272–275.

119 Yuan, W., Zhou, Z., Gong, L., and Meggers, E. (2017). Asymmetric alkylation of remote C(sp^3)–H bonds by combining proton-coupled electron transfer with chiral Lewis acid catalysis. *Chem. Commun.* 53 (64): 8964–8967.

120 Ma, Z.-Y., Li, M., Guo, L.-N. et al. (2021). Sulfonamide as photoinduced hydrogen-atom transfer catalyst for regioselective alkylation of C(sp^3)–H bonds adjacent to heteroatoms. *Org. Lett.* 23 (2): 474–479.

121 Wang, C., Chen, Z., Sun, J. et al. (2024). Sulfonamide-directed site-selective functionalization of unactivated C(sp^3)–H enabled by photocatalytic sequential electron/proton transfer. *Nat. Commun.* 15 (1): 5087.

122 Simons, R.T., Nandakumar, M., Kwon, K. et al. (2023). Directed photochemically mediated nickel-catalyzed (hetero)arylation of aliphatic C–H bonds. *J. Am. Chem. Soc.* 145 (7): 3882–3890.

123 Wu, K., Wang, L., Colón-Rodríguez, S. et al. (2019). Amidyl radical directed remote allylation of unactivated sp^3 C–H bonds by organic photoredox catalysis. *Angew. Chem. Int. Ed.* 58 (6): 1774–1778.

124 Chen, H., Fan, W., Yuan, X.-A., and Yu, S. (2019). Site-selective remote C(sp^3)–H heteroarylation of amides via organic photoredox catalysis. *Nat. Commun.* 10 (1): 4743.

125 Chen, H., Jin, W., and Yu, S. (2020). Enantioselective remote C(sp^3)–H cyanation via dual photoredox and copper catalysis. *Org. Lett.* 22 (15): 5910–5914.

126 Xu, K., Yang, J., Qin, H., and Liu, F. (2023). Amidyl radical directed γ-C(sp^3)–H functionalization with silyl enol ethers via photoredox catalysis. *Eur. J. Org. Chem.* 26 (31): e202300543.

127 Ruos, M.E., Kinney, R.G., Ring, O.T., and Doyle, A.G. (2023). A general photocatalytic strategy for nucleophilic amination of primary and secondary benzylic C–H bonds. *J. Am. Chem. Soc.* 145 (33): 18487–18496.

128 Jeffrey, J.L., Terrett, J.A., and MacMillan, D.W.C. (2015). O–H hydrogen bonding promotes H-atom transfer from α C–H bonds for C-alkylation of alcohols. *Science* 349 (6255): 1532–1536.

129 Zhang, X. and MacMillan, D.W.C. (2017). Direct aldehyde C–H arylation and alkylation via the combination of nickel, hydrogen atom transfer, and photoredox catalysis. *J. Am. Chem. Soc.* 139 (33): 11353–11356.

130 Shaw, M.H., Shurtleff, V.W., Terrett, J.A. et al. (2016). Native functionality in triple catalytic cross-coupling: sp^3 C–H bonds as latent nucleophiles. *Science* 352 (6291): 1304–1308.

131 Hou, J., Ee, A., Cao, H. et al. (2018). Visible-light-mediated metal-free difunctionalization of alkenes with CO_2 and silanes or C(sp^3)–H alkanes. *Angew. Chem. Int. Ed.* 57 (52): 17220–17224.

132 Wang, J., Gao, G., Shi, C. et al. (2025). Visible-light induced direct C(sp^3)–H bond disulfidation of saturated N-heterocycles through a hydrogen atom transfer (HAT) process. *Org. Chem. Front.* 12 (7): 2286–2291.

133 Yang, H.-B., Feceu, A., and Martin, D.B.C. (2019). Catalyst-controlled C–H functionalization of adamantanes using selective H-atom transfer. *ACS Catal.* 9 (6): 5708–5715.

134 Ruffoni, A., Juliá, F., Svejstrup, T.D. et al. (2019). Practical and regioselective amination of arenes using alkyl amines. *Nat. Chem.* 11 (5): 426–433.

135 Ryder, A.S.H., Cunningham, W.B., Ballantyne, G. et al. (2020). Photocatalytic α-tertiary amine synthesis via C–H alkylation of unmasked primary amines. *Angew. Chem. Int. Ed.* 59 (35): 14986–14991.

136 Wang, D., Mück-Lichtenfeld, C., and Studer, A. (2019). Hydrogen atom transfer induced boron retaining coupling of organoboronic esters and organolithium reagents. *J. Am. Chem. Soc.* 141 (36): 14126–14130.

137 Xing, L. and Zhang, Y. Recent advances in selective C-H bonds functionalization through aryl radical- mediated hydrogen atom transfer strategy. *Curr. Org. Chem.* 27 (9): 747–758.

138 Hollister, K.A., Conner, E.S., Spell, M.L. et al. (2015). Remote hydroxylation through radical translocation and polar crossover. *Angew. Chem. Int. Ed.* 54 (27): 7837–7841.

139 Du, S., Kimball, E.A., and Ragains, J.R. (2017). Visible-light-promoted remote C–H functionalization of o-diazoniaphenyl alkyl sulfones. *Org. Lett.* 19 (20): 5553–5556.

140 Huang, F.-Q., Dong, X., Qi, L.-W., and Zhang, B. (2016). Visible-light photocatalytic α-amino C(sp^3)–H activation through radical translocation: a novel and metal-free approach to α-alkoxybenzamides. *Tetrahedron Lett.* 57 (14): 1600–1604.

141 Chen, J.-Q., Wei, Y.-L., Xu, G.-Q. et al. (2016). Intramolecular 1,5-H transfer reaction of aryl iodides through visible-light photoredox catalysis: a concise method for the synthesis of natural product scaffolds. *Chem. Commun.* 52 (38): 6455–6458.

13

Dual Transition Metal/Photoredox Catalysis for Directed $C(sp^2)-H$ Activations

Akshay M. Nair and Martin Fañanás-Mastral

Centro Singular de Investigación en Química Biolóxica e Materiais Moleculares (CiQUS), Universidade de Santiago de Compostela, Santiago de Compostela 15782, Spain

13.1 Introduction

The past decade has seen the emanation of visible-light photocatalysis as a highly efficient platform for carrying out single-electron transformations [1]. These protocols rely on the use of photocatalysts, which are poor reductants or oxidants in the ground state, but upon photoexcitation become excellent redox mediators. Based on the nature of substrates and reaction conditions, the excited photocatalyst could act as either a single-electron oxidant or reductant. This allows the generation of radical intermediates bypassing the need of stochiometric amounts of radical initiators or UV irradiation. Additionally, these reactions proceed under mild and energy-efficient conditions typically employing LED lights as the energy source. All these features have made visible-light photocatalysis one of the most sustainable platforms to carry out radical reactions [2]. On the other hand, over the last six decades, transition-metal-catalyzed cross-couplings has become one of the main pillars of organic synthesis, finding widespread applications in both academia and industry [3]. These transformations involve the metal-catalyzed coupling of organometallic reagents with various electrophiles. They are highly reliable and allow the facile access to a wide range of complex organic molecules. A major drawback of these reactions is the prerequisite of stoichiometric amounts of organometallic reagents, which are not only expensive but also lead to substantial waste formation, potentially toxic in some instances (Scheme 13.1a). An interesting alternative to the stoichiometric use of organometallic reagents is the direct transition-metal-mediated functionalization of C—H bonds, which are ubiquitous in organic molecules. However, the presence of multiple chemically similar and non-biased C—H bonds in a molecule makes selective activation a major challenge (Scheme 13.1b). One of the most reliable approaches is the use of directing groups, which coordinate the metal and bring it to the proximity of specific C—H bonds leading to selective functionalization (Scheme 13.1c) [4]. The ingenious design

Directed C–H Bond Functionalization: Concepts and Applications, First Edition.
Edited by Debabrata Maiti and Supriya Rej.

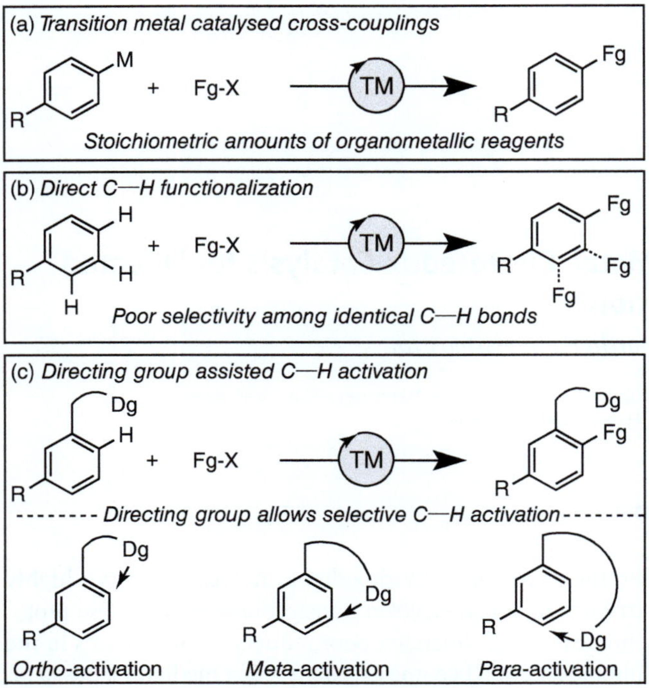

Scheme 13.1 Overview of directed C–H activations.

of directing groups has provided efficient tools to control regioselectivity among chemically similar C—H bonds, which include *ortho* vs *meta* vs *para* in arenes, β vs γ vs δ in aliphatic chains, or C-2 vs C-8 in quinolines. The judicious choice of directing groups, transition metals, and coupling partners has allowed a myriad of transformations ranging from aminations, oxygenations, arylations, alkenylation, alkylation, cascade cyclization, etc. As a result, directing group-assisted C—H activations is one of the most rapidly growing domains of organic synthesis allowing rapid access to molecular complexity from simple C—H precursors.

Despite all these advances, classical C—H activation still presents some drawbacks like the requirement of harsh reaction conditions and/or super stoichiometric amounts of additives (typically oxidants). Recent works on the merger of visible-light photocatalysis and directing group-assisted C—H activations have brought in an elegant approach to overcome these issues [5]. These processes leverage on the combination of the transition-metal catalyst's ability to activate C—H bonds with the superior redox properties of an excited photocatalyst. The scope of this chapter is to cover the state-of-the-art techniques in dual transition-metal/photoredox catalysis for $C(sp^2)$—H activation, with the aim of highlighting the main achievements and their key mechanistic features. We have divided the chapter into two sections based on the role of the photocatalyst being used either for transition-metal catalyst reoxidation or for substrate activation.

13.2 Photocatalysis for Transition Metal Catalyst Reoxidation

Many conventional directed C—H activation methods suffer from an economical and sustainable point of view due to the need of super stoichiometric amounts of oxidants, typically Cu or Ag salts (Scheme 13.2a) [6]. Oxidants play a crucial role in these reactions, allowing the formation of catalytically active high-valent metal species by oxidizing the low-valent complexes formed after reductive elimination. The search for viable alternatives has seen the development of protocols relying on the use of cheaper oxidants like persulfates, peroxides, or benzoquinone, which are still required in stoichiometric amounts leading to waste generation [7]. The ingenious design of oxidizing directing groups and coupling partners acting as internal oxidants (by N—N or N—O bond cleavage) has provided protocols that bypass the need of external oxidants. However, these protocols are limited to specific substrates and hence lack generality [8]. The use of oxygen as the terminal oxidant would be highly sustainable, and methods that allow the utilization of the oxidizing ability of oxygen in transition-metal-catalyzed C—H activation are of high interest [9].

Nevertheless, there are limited studies on this, because of the low redox potential of molecular oxygen when compared with the reduced metal species. Starting in early 2014, Rueping pioneered the merger of visible-light photocatalysis with transition-metal-catalyzed C—H activations using oxygen as the oxidant (Scheme 13.2b) [5a]. Herein, the low valent metal species generated after the C—C bond-forming event are oxidized by the excited photocatalyst, while molecular oxygen re-oxidizes the reduced photocatalyst. Oxygen is the stoichiometric oxidant and photocatalyst acts as an electron shuttle. Based on these initial studies, efficient and verdant dual catalytic C—H activations have been reported by Rueping and others involving a range of photocatalysts, transition-metal complexes, and substrate combinations.

The first example of a metallaphotoredox-mediated oxidative C—H activation described by Rueping was a dual rhodium/photocatalytic protocol for *ortho*-C—H olefination of *N,N*-dialkyl benzamides using $[Ru(bpy)_3](PF_6)_2$ as the photocatalyst at 80 °C under air (Scheme 13.3a) [10]. The photocatalytic transformation displayed a broad scope tolerating a range of primary, secondary, and Weinreb amides. The reaction also allowed a wide range of alkenes such as acrylates, vinyl silanes, or sulfones to be used as coupling partners (Scheme 13.3b). The proposed mechanism for this transformation is depicted in Scheme 13.3c. Initially, the carbonyl oxygen of the amide coordinates to Rh(III) and subsequent *ortho*-C—H activation leads to rhodacycle **B**. Coordination of the alkene and subsequent migratory insertion forms intermediate **C**, which upon β-hydride elimination leads to the olefinated product and Rh—H species **D**. The active Rh(III) complex **A**, likely after H-X elimination, is regenerated from **D** via oxidation by any of the following intermediates: (i) photoexcited Ru(II)* or (ii) the superoxide ion formed during the regeneration of the Ru(II) photocatalyst.

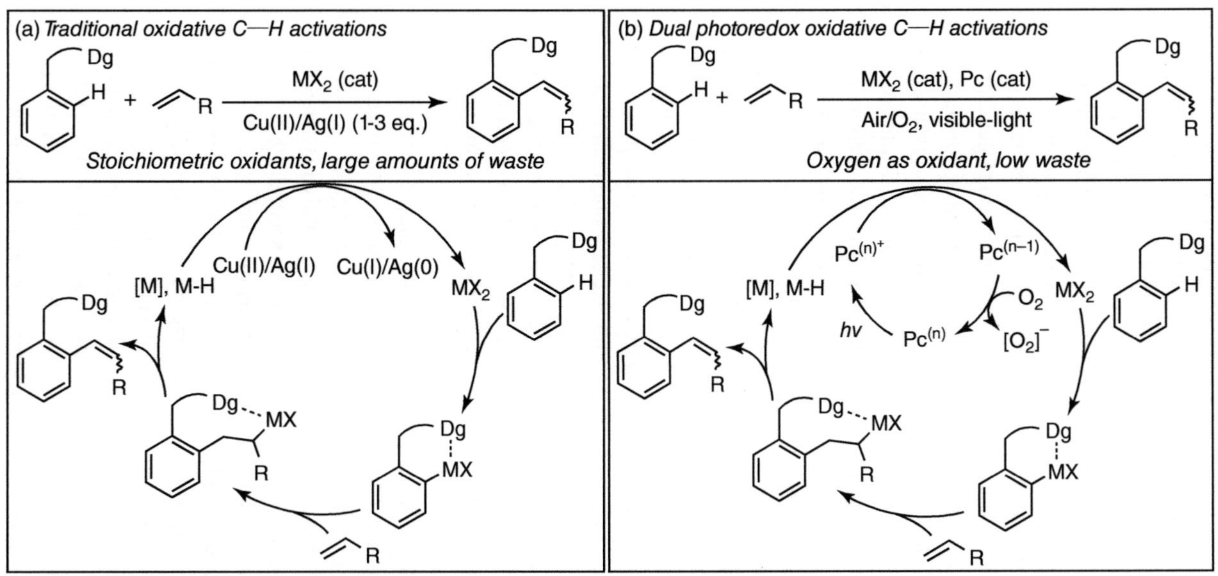

Scheme 13.2 Traditional and dual photocatalytic oxidative C—H activations.

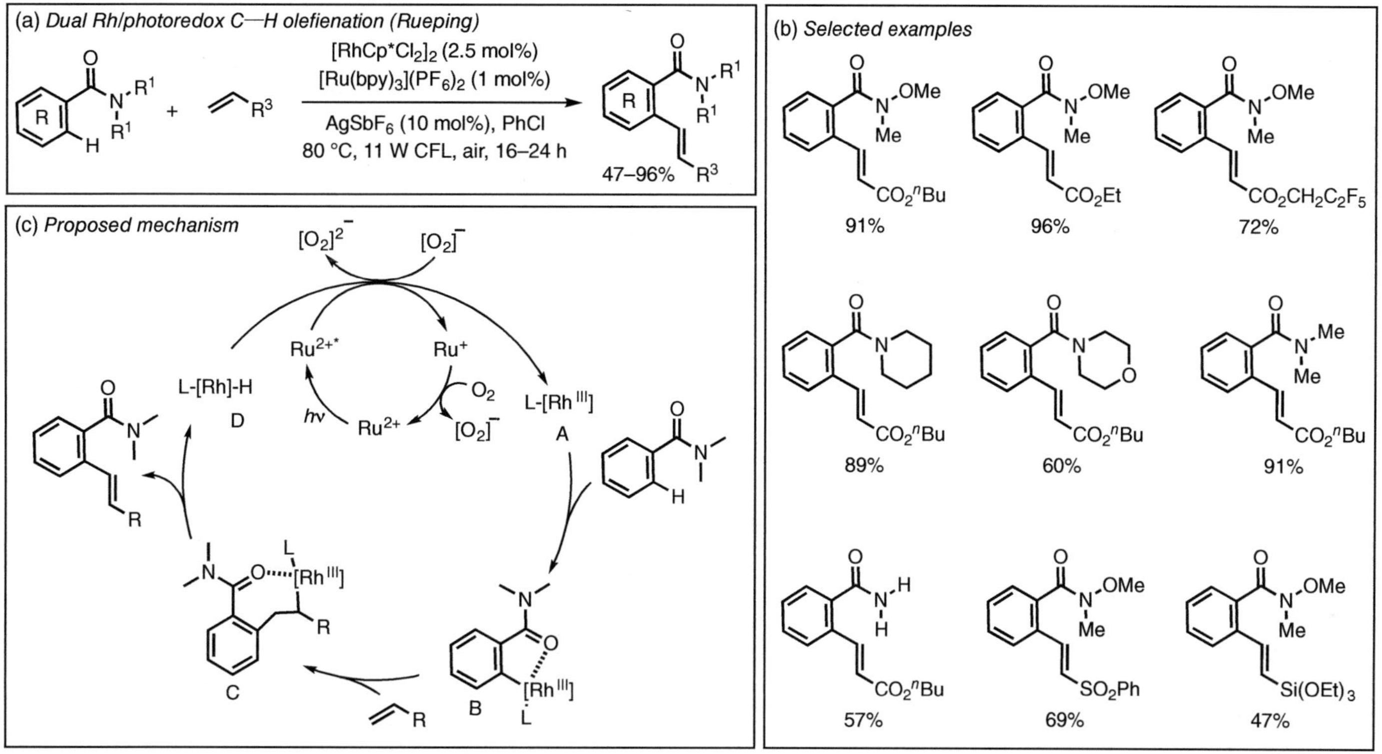

Scheme 13.3 Dual Rh/photocatalyzed alkenylation of amines and amides.

Having realized that the low valent Rh complexes could be efficiently reoxidized to the active Ru(III) via photocatalysis, the Rueping group then disclosed a similar dual rhodium photoredox protocol for carrying out (3 + 2) annulation of anilides with alkynes (Scheme 13.4a) [11]. The dual photocatalytic transformation delivered a range of diversely substituted indole derivatives in good-to-excellent yields (Scheme 13.4b). The reaction proceeds via a similar mechanism as explained above wherein the Rh(III) complex **A** is regenerated by the oxidative quenching of the excited Ru(II) photocatalyst (Scheme 13.4c).

In 2016, Rueping and co-workers reported the merger of photocatalysis and Ru-catalyzed C—H activation for carrying out a pyridine-directed *ortho*-olefination of phenyl ethers (Scheme 13.5a) [12]. The dual photocatalytic protocol relied on the use of $[Ir(ppy)_2(bpy)]PF_6$ as the photocatalyst under air. The reaction tolerated a range of substituents both on the alkene and the aryl ether. Notably, substrates containing labile groups such as aldehydes, ketones, esters, or nitriles were amenable for this transformation (Scheme 13.5b). In some cases, the authors observed hydroarylation products arising from the partial hydrogenation of the double bonds of the olefinated products. This was attributed to the consecutive olefin insertion into the in situ generated Ru—H complex, which was well documented previously [13]. The authors addressed these issues by using acid additives, which suppressed the formation of the hydroarylation products by converting the Ru—H complexes to Ru(0) by hydrogen evolution. Based on the previous literature and preliminary mechanistic studies, they proposed the reaction mechanism shown in Scheme 13.5c. Initially, the catalytically active Ru(II) complex **A** undergoes coordination and *ortho*-C—H activation with the phenyl pyridine ether to form complex **B**. The insertion of the olefin onto **B** leads to intermediate **C**, which upon β-hydride elimination delivers the olefinated product and Ru(II), intermediate **D**, which undergoes reductive elimination to form reduced Ru(0) complex **E**. The excited Ir(III)* photoredox catalyst regenerates the catalytically active Ru species via electron transfer forming Ir(II). The reduced Ir(II) is then oxidized to the active photocatalyst by atmospheric oxygen, generating a superoxide anion, which can also oxidize the reduced Ru complex **E**.

Sundararaju in 2018 reported a dual Co/photocatalytic (4 + 2) annulation of 8-aminoquinoline benzamides and internal alkynes using the inexpensive organic dye Na_2Eosin Y as the photocatalyst under O_2 atmosphere (Scheme 13.6a) [14a]. The reaction proceeds under ambient temperatures using white LED bulbs bypassing the need of stoichiometric amounts of metal oxidants. A range of structurally diverse benzamides and alkynes were well tolerated in this transformation affording the corresponding isoquinolones in good-to-excellent yields (Scheme 13.6b). A mechanism was proposed based on previous literature and preliminary mechanistic studies (Scheme 13.6c). Initially, Co(II) undergoes ligand exchange with the benzamide forming the active species **A**. The subsequent oxidation of **A** by the excited photocatalyst (reductive quenching) delivers Co(III) intermediate **B**, which undergoes C—H activation via concerted metalation deprotonation (CMD) to form cobaltacycle **C**. Coordination of the alkyne with **C** forms complex **D**, which upon migratory insertion, leads to the seven-membered

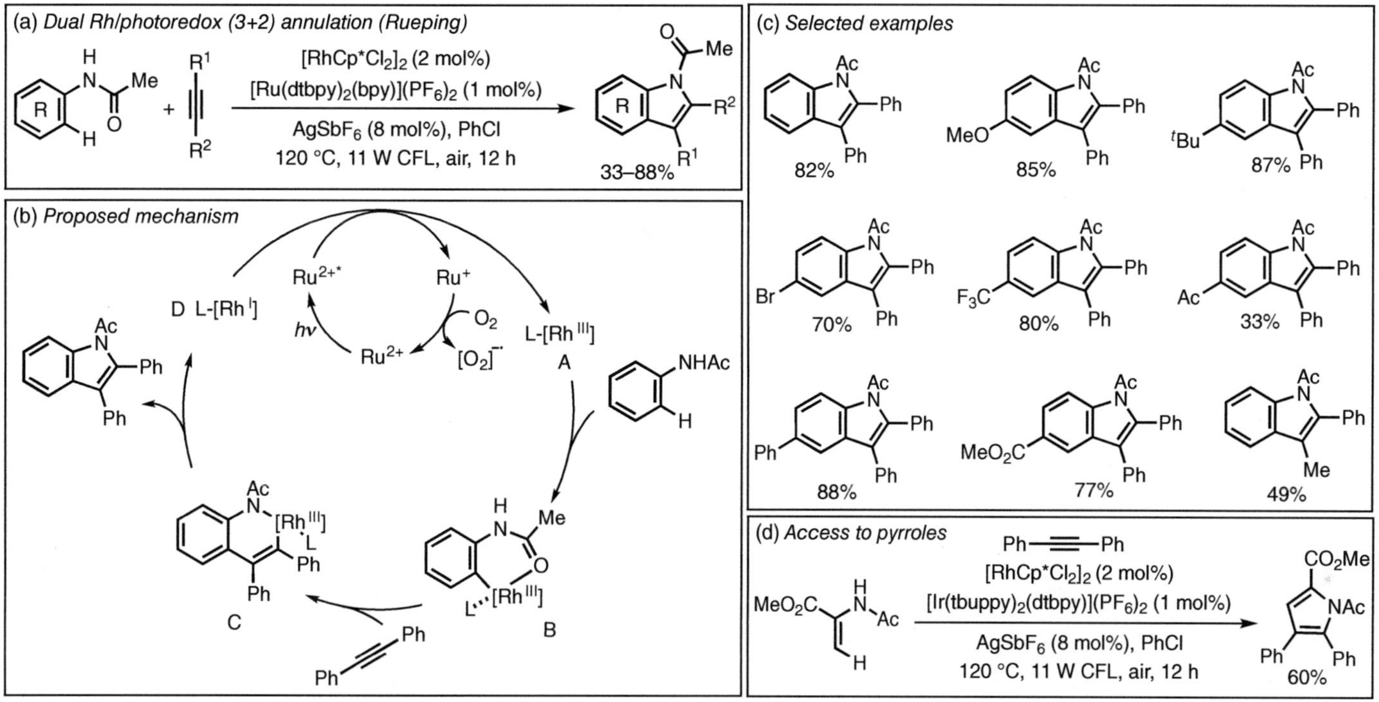

Scheme 13.4 Dual Rh/photocatalyzed annulation of anilides with alkynes.

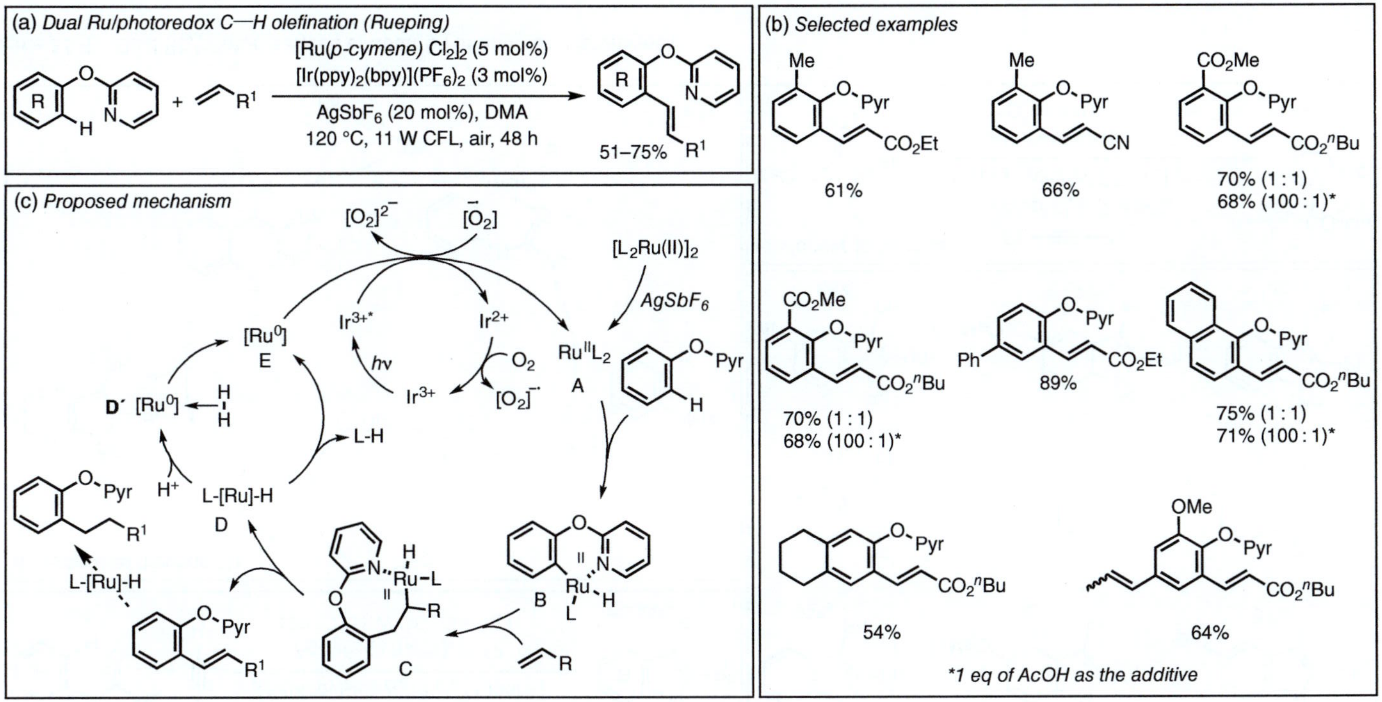

Scheme 13.5 Dual Ru/photocatalyzed pyridine-directed alkenylation of ethers.

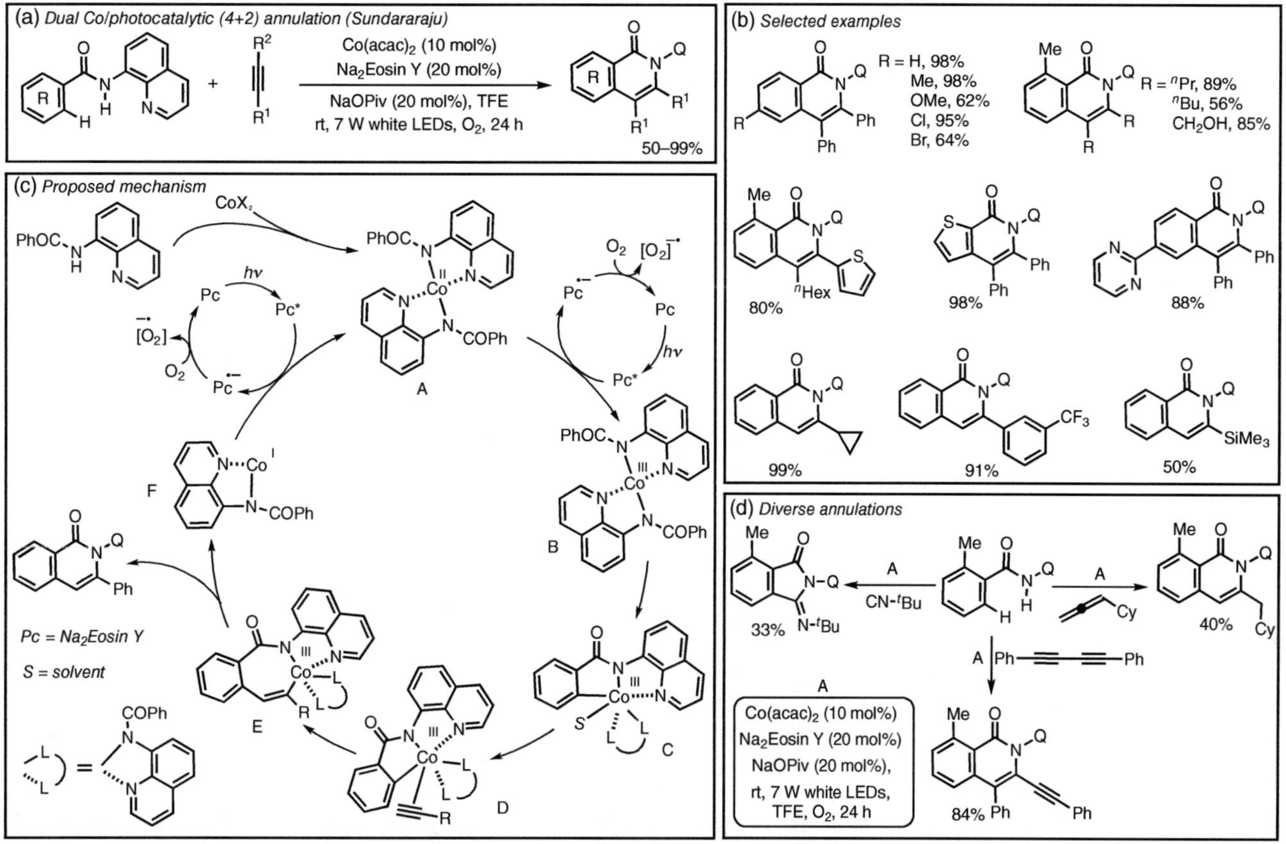

Scheme 13.6 Dual Co/photocatalyzed (4 + 2) annulation of benzamides with alkynes.

cobaltacycle **E**, which undergoes reductive elimination to deliver the isoquinolone product and Co(I) intermediate **F**. The photoexcited Na_2Eosin Y can then reoxidize **F** to the active complex **A**. The reduced Na_2Eosin Y is then converted into the active photocatalyst by oxidation with molecular oxygen. Under very similar conditions, the same group subsequently reported C—H alkynylations and (4 + 2) annulations using bromo alkynes instead of alkynes [14b].

Sundararaju further expanded their earlier dual catalytic system to carry out [4 + 2] annulations of benzamides with unactivated alkenes to access dihydroisoquinolones (Scheme 13.7a) [15]. They employed a dual catalytic system comprising $Co(acac)_2$ and Na_2Eosin Y in an oxygen atmosphere under irradiation by 7 W white LED bulbs. The Ghosh group in 2020 reported a dual cobalt photocatalytic protocol for carrying out (4 + 1) annulation of benzamides with maleimides to access isoindolone spirosuccinimides (Scheme 13.7b) [16]. The developed protocol relied on the use of $Co(acac)_2$ and Eosin Y at room temperature under irradiation by 3 W green LEDs in air bypassing the need of molecular oxygen. These reactions follow a similar mechanism as in Scheme 13.6c involving the key photocatalytic oxidation to form Co(III) species, which undergo C—H activation and then migratory insertion with the corresponding alkene.

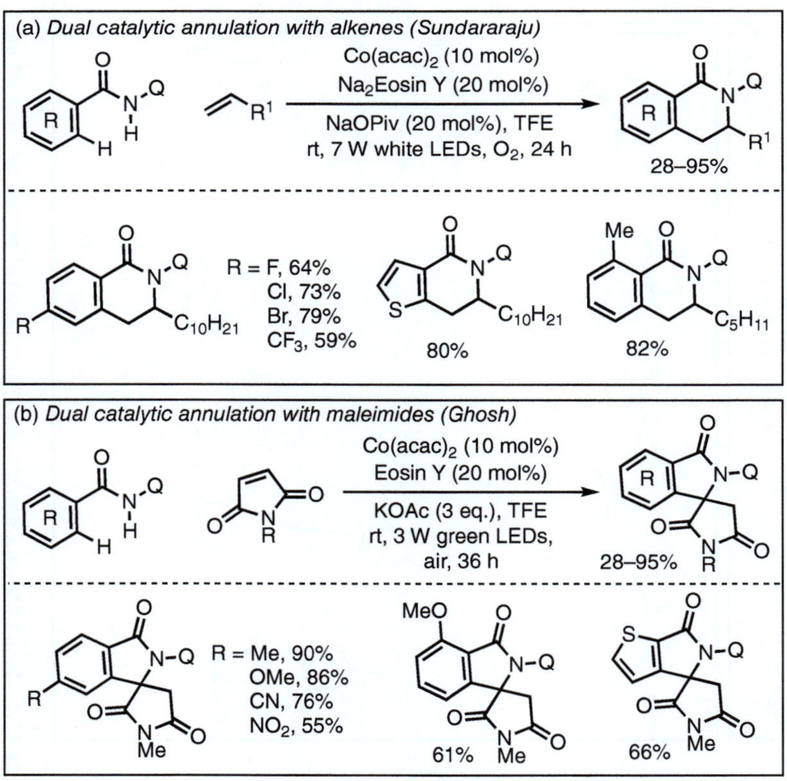

Scheme 13.7 Dual Co/photocatalyzed annulations of benzamides.

In 2021, Wu and co-workers extended the dual cobalt and photoredox strategy to a regiodivergent (4 + 2) annulation of 8-aminoquinoloine benzamides with allenes (Scheme 13.8a) [17]. The transformation involved the use of $Co(acac)_2$, Eosin Y, and irradiation using 15 W CFL at room temperature under an oxygen atmosphere. The reaction tolerated a range of substituted benzamides and electronically diverse allenes (Scheme 13.8b). They observed a regioselective dichotomy based on the nature of the allene. Electron-deficient allenyl esters delivered isoquinolones, while aryl allenes led to dihydroisoquinolones bearing an exocyclic double bond. The initial exchange, photocatalytic oxidation, and subsequent C—H activation form the cobaltacycle intermediate **C**, which then undergoes coordination with the allene to form **D** (Scheme 13.8c). Regioselective migratory insertion, depending on the stereo-electronic nature of the allene, leads to either seven-membered cobaltacycle intermediates **E** or **F**. The ensuing reductive elimination forms the cyclic products and the reduced Co(I) species, which is oxidized to the active Co(II) by the excited photocatalyst. Finally, the dihydroisoquinolone **G** isomerizes to form the isoquinolone product.

In 2021, Liu's group disclosed a dual cobalt photoredox protocol for concomitant dearomatization and (4 + 2) annulation of indoles with 8-aminoquinoline benzamides (Scheme 13.9) [18]. They utilized a dual catalytic system comprising $Co(OAc)_2$ and the photocatalyst $Ir(bt)_2(acac)$ in air under irradiation by 6 W blue LEDs, following a similar mechanism as discussed in Scheme 13.6c. A wide range of diversely substituted benzamides and indoles were found to be amenable under the reaction condition.

Very recently, a landmark work by the Sundararaju group provided an asymmetric version of the dearomatization, (4 + 2) annulation of indoles with benzamides previously demonstrated in a racemic fashion (Scheme 13.10a) [19]. A system based on Na_2EosinY as a photocatalyst and a chiral complex comprising $Co(OAc)_2 \cdot 4H_2O$ and chiral ligand (*S*)-Salox was used under air atmosphere using 7 W white LED bulb irradiation. The ingenious reaction design leverages the π-stacking of pyrimidyl indole with the chiral cobaltacycle generated after C—H activation to afford enantioselective access to tetrahydro-5*H*-indolo[2,3-c] isoquinolin-5-one derivatives. The reaction tolerates a wide range of substituted benzamides and indoles delivering the products in good yields with excellent enantiomeric excess of up to 99% (Scheme 13.10b).

Mechanistically, chiral intermediate A, generated from $Co(OAc)_2 \cdot 4H_2O$, (*S*)-Salox ligand, and benzamide, is oxidized by the excited photocatalyst upon reductive quenching to Co(III) intermediate **B**. Subsequently, the key enantiodetermining C—H activation leads to the diastereoselective formation of *mer*-[($\boldsymbol{S_L}$)-**ΔCo**]cobaltacycle **C**. The coordination of the π-bond of indole with the complex **C** and π–π stacking of the pyrimidine ring with the arene of the quinoline leads to the intermediate **D**, which upon migratory insertion forms the complex **E**. Subsequent reductive elimination generates Co(I) complex **G**, which upon product decoordination followed by photocatalyst-mediated oxidation and substrate coordination regenerates the active Co(II) intermediate **A**. Shi and Ackermann recently also reported similar dual Co/photocatalytic transformations for the enantioselective synthesis of tetrahydro-5*H*-indolo[2,3-c] isoquinolin-5-ones [20].

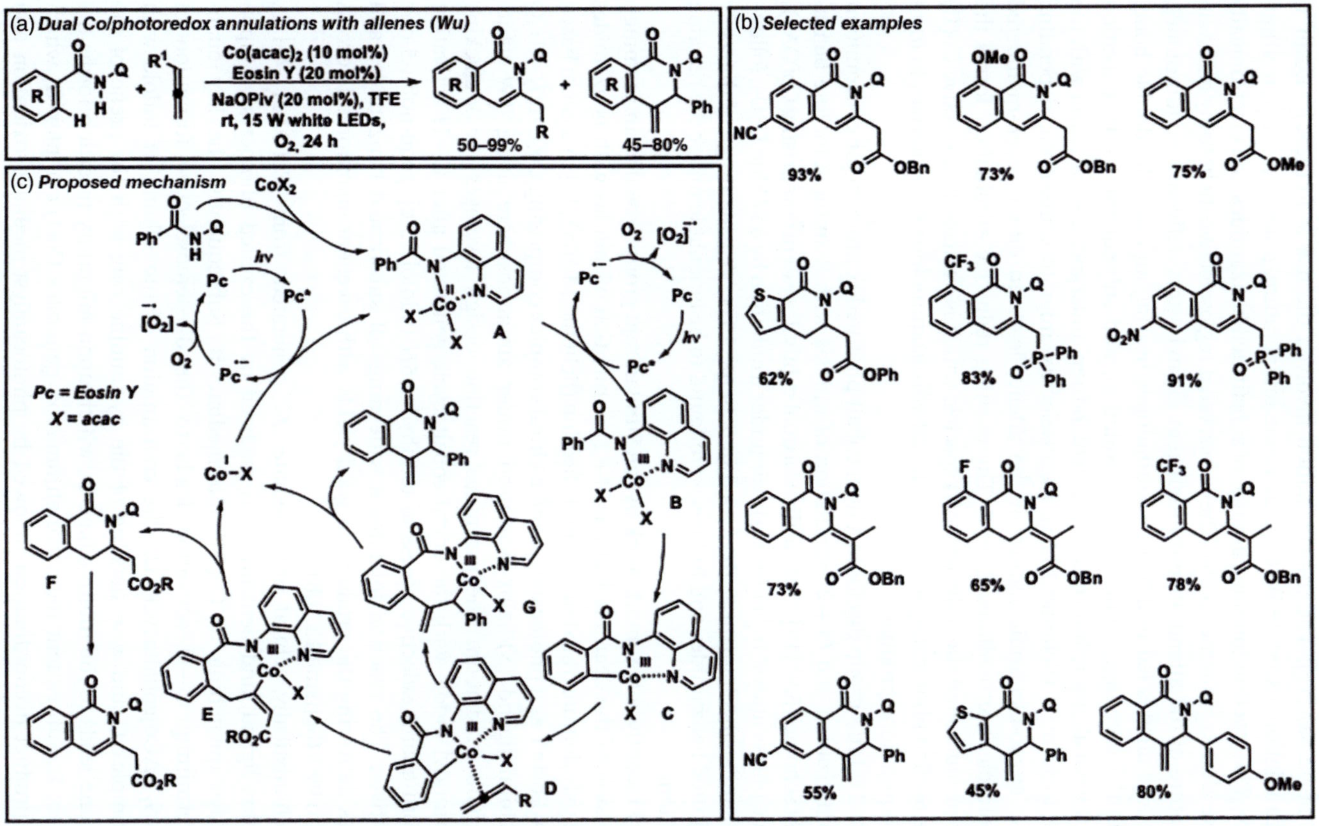

Scheme 13.8 Dual Co/photocatalyzed divergent (4 + 2) annulation of benzamides with allenes.

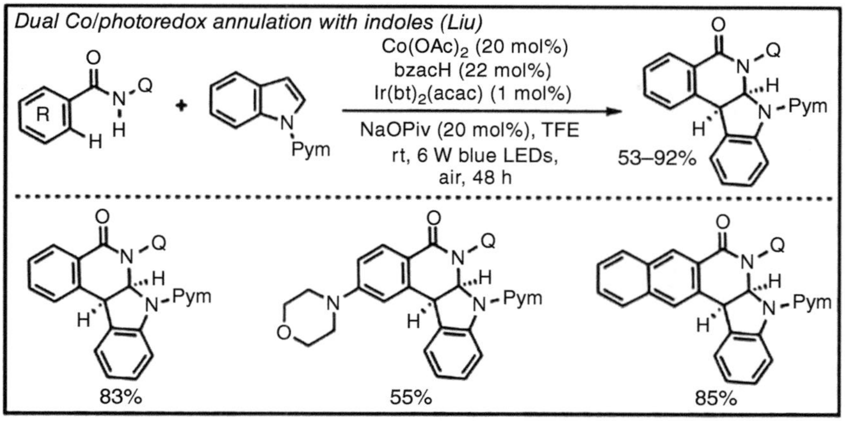

Scheme 13.9 Dual Co/photocatalyzed (4 + 2) annulation of benzamides with indoles.

Volla and co-workers reported in 2022 the use of a pyridine *N*-oxide directing group in dual cobalt/photoredox catalysis to carry a (4 + 1) annulation of benzamides and terminal alkynes (Scheme 13.11a) [21]. The reaction utilizes a combination of $Co(acac)_2$ and Na_2Eosin Y in DMSO under an oxygen atmosphere and irradiation by white LEDs. A range of substituted benzamides and terminal alkynes were tolerated under the reaction conditions affording the products in good-to-excellent yields (Scheme 13.11b).

The reaction follows a similar mechanism as discussed earlier involving ligand exchange, photocatalytic oxidation, and subsequent C—H activation forming the cobaltacycle **C** (Scheme 13.11c). Coordination of the alkyne to this species and further anion exchange leads to the intermediate **E**, which upon reductive elimination forms the internal alkyne **F**. The subsequent protonation of **F** results in the formation of the Co(I) species and the alkyne intermediate **G**, which undergoes a Co(II)-mediated intramolecular annulation to afford the cyclized products. Finally, the catalytically active Co(II) complex is regenerated by the photocatalytic oxidation of the in situ generated Co(I). Under slightly modified conditions, the authors were also able to carry out a pyridine *N*-oxide-directed alkoxylation of benzamides using alcohols (Scheme 13.11d).

In a seminal work, Maiti and co-workers reported a dual palladium/photoredox protocol for carrying out both non-directed and directed C—H olefinations of arenes in air bypassing the need of stoichiometric oxidants (Scheme 13.12a) [22]. Using a range of photocatalysts along with $Pd(OAc)_2$ and relying on an ingenious design of directing groups, they were able to selectively perform *ortho*-, *meta*-, or *para*-selective olefinations of diversely substituted phenyl rings (Scheme 13.12b). By using a chiral *N*-Ac-*L*-Ala ligand, they were also able to apply this dual palladium/photocatalytic system in a dynamic kinetic resolution of biaryls that led to the corresponding products with high enantioselectivity of up to 96 : 4 e.r. (Scheme 13.12d).

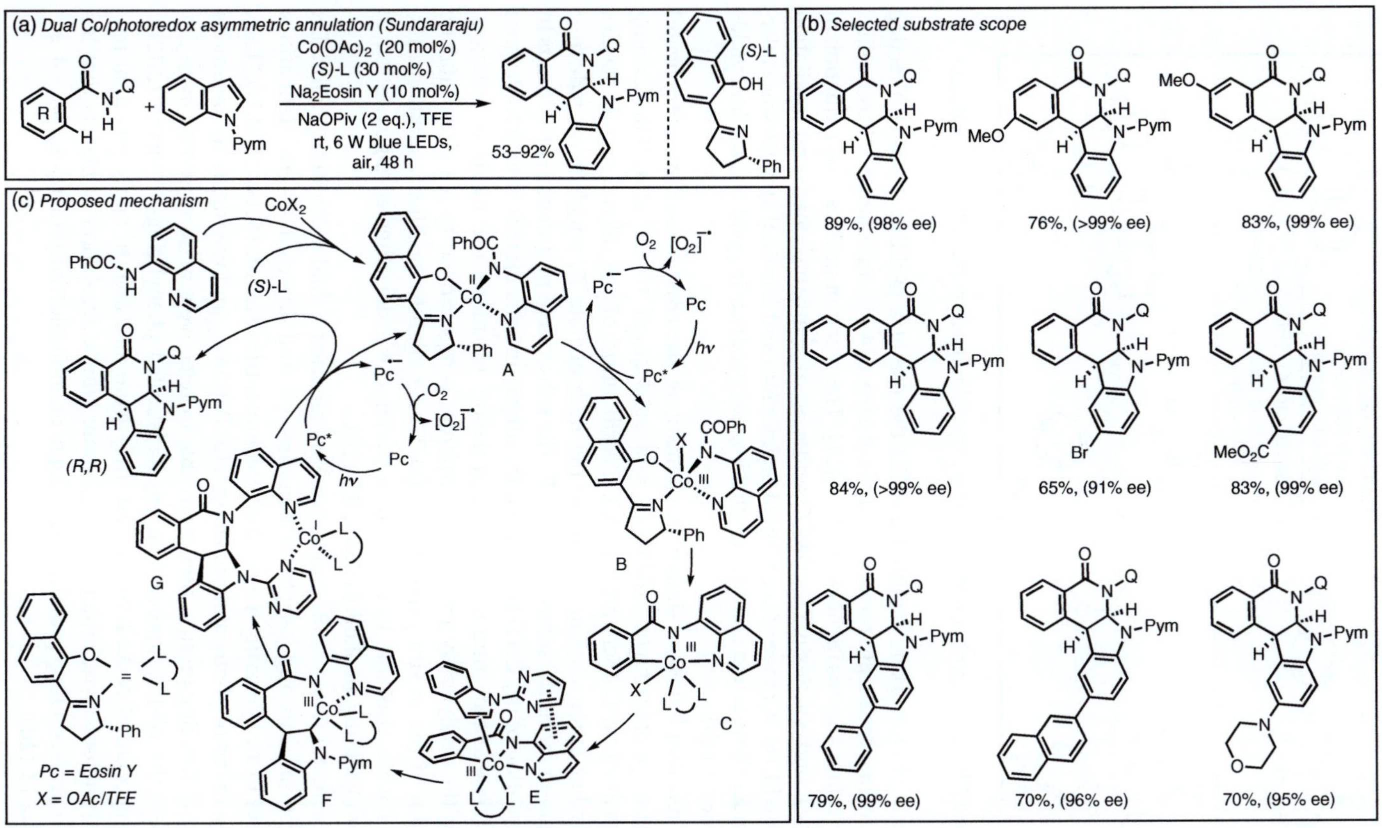

Scheme 13.10 Dual Co/photocatalyzed asymmetric (4 + 2) annulation of benzamides.

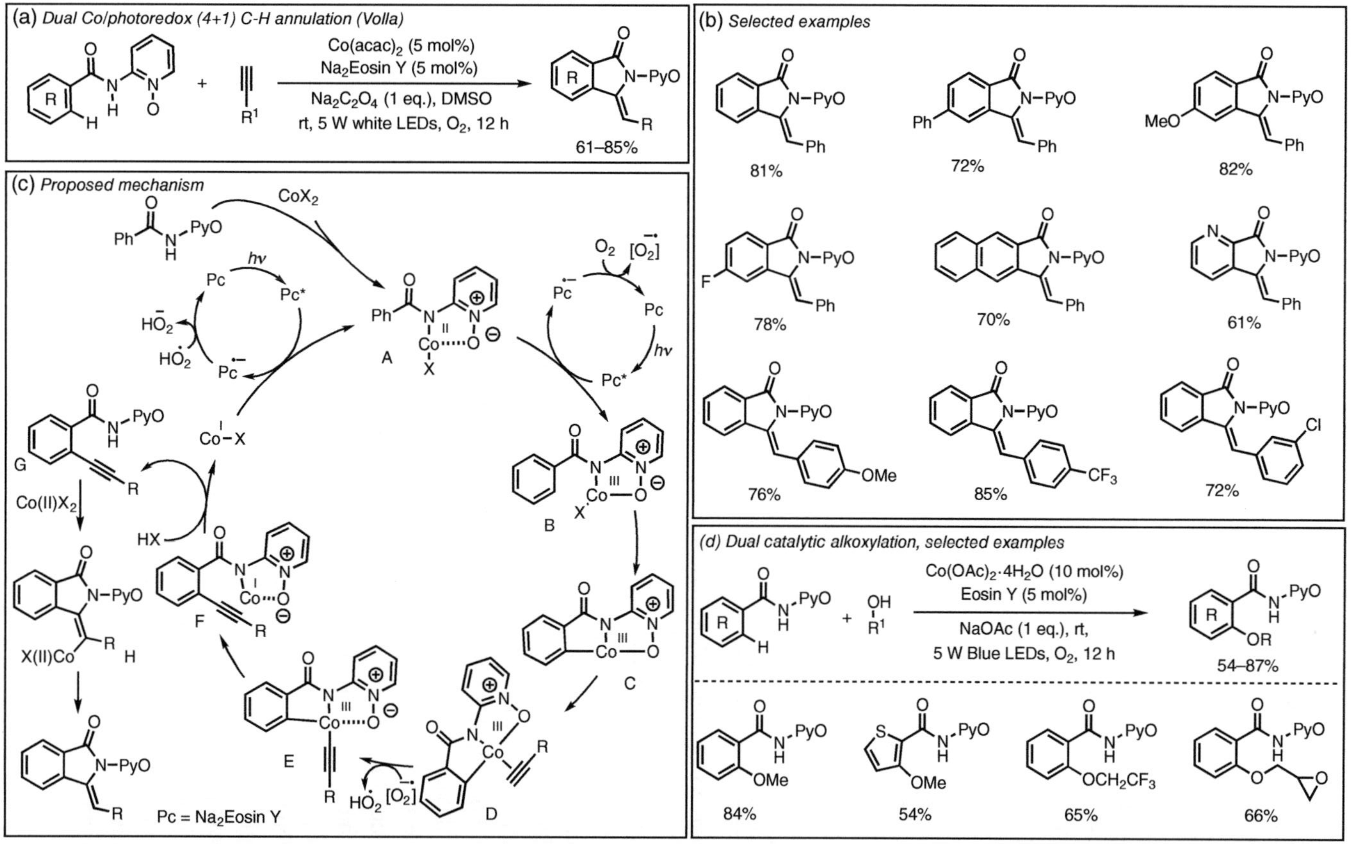

Scheme 13.11 Dual Co/photocatalyzed (4 + 1) annulation of benzamides with alkynes.

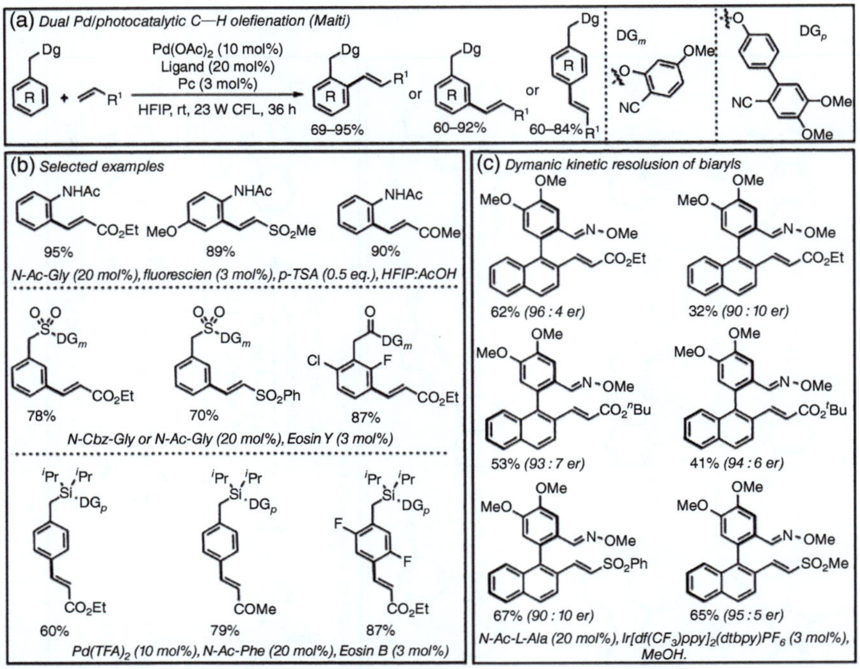

Scheme 13.12 Dual Pd/photocatalyzed C–H olefination.

Very recently, Volla and co-workers reported a dual copper photoredox-catalyzed alkynylation of aminoquinoline benzamides (Scheme 13.13a) [23]. They employed a dual catalytic combination of $Cu(OAc)_2.H_2O$ and $Ru(bpy)_3Cl_2$ in O_2 atmosphere under 5 W blue LED irradiation. The combination of copper catalyst and mild reaction conditions led to the isolation of direct alkynylation products, which was not possible in previous Cu-catalyzed transformations due to competing annulations [24]. The reaction tolerated a range of benzamides and alkynes even those tethered to complex natural products to afford the corresponding internal alkynes in good yields (Scheme 13.13b). Based on previous reports, a plausible mechanism was proposed (Scheme 13.13c). The initial ligand exchange, photocatalytic oxidation, and subsequent C—H activation generate the cupracyle **C**, which undergoes alkyne coordination and alkynylation to form the complex **E**. The ensuing reductive elimination and ligand exchange from **E** delivers the internal alkyne product and the reduced Cu(I) species **G**, which is oxidized by the excited photocatalyst completing the catalytic cycle. Under slightly modified conditions, the authors were also able to carry out diverse C—H functionalizations of benzamides such as amination, selenylation, sulfonylation, arylation, and nitration (Scheme 13.13d).

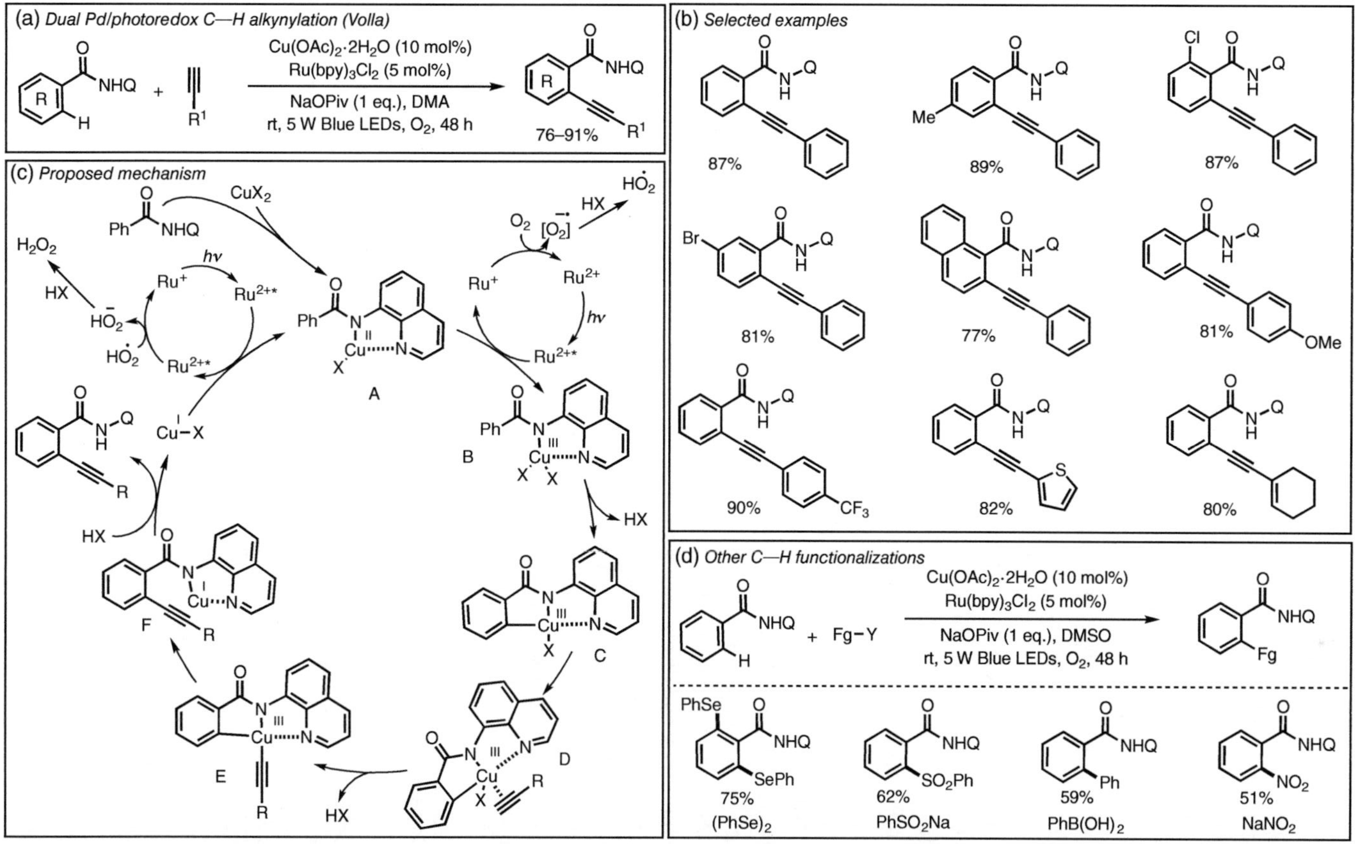

Scheme 13.13 Dual Cu/photocatalyzed alkynylation of benzamides with aryl acetylenes.

13.3 Photocatalysis for Coupling Partner Activation

In directed C—H activations, activation of coupling partners (oxidative addition) often requires high-energy demanding steps (Scheme 13.14a) [25]. In pioneering reports, Sanford and co-workers ingeniously addressed this issue by using photocatalysis for substrate activation in dual catalytic C—H arylations [26]. Ensuing studies by other groups have led to a further expansion of this strategy that has been applied for a range of functionalizations including arylation, acylation, and hydroxylation amination. These dual catalytic transformations involve a key generation of radical intermediates by single-electron transfer between the photocatalyst and one of the coupling partners (Scheme 13.14b). These radical intermediates then add on to the metal center of the C—H activated metallacycle and subsequent reductive elimination delivers the C—H functionalized products along with reduced metal species. Depending on the reaction conditions, the reduced metal complex is then reoxidized to the active catalyst by either the photocatalyst or an external oxidant. The mild conditions associated to this photocatalytic activation of one of the coupling partners allow these transformations to be carried out at room temperature in contrast to the high-temperature protocols involving oxidative additions.

In 2011, Sanford and co-workers reported the first example on the merger of photocatalysis and directed C—H activation for carrying out directed arylations using aryl diazonium salts [26a]. The reaction utilized a dual catalytic system of $Pd(OAc)_2$ and $Ru(bpy)_3Cl_2.6H_2O$ at room temperature (Scheme 13.15a). In a previous study, they had disclosed a similar arylation protocol using aryl diazonium salts, which required higher temperatures (100 °C) in acetic acid. The dual photocatalytic methodology was compatible with a range of diverse substrates including directing groups like pyridine, pyrimidine, pyrazole, amides, and oximes (Scheme 13.15b). The authors proposed a reaction mechanism wherein the initial directing group-assisted C—H activation by Pd(II) generates palladacycle **B** (Scheme 13.15c).

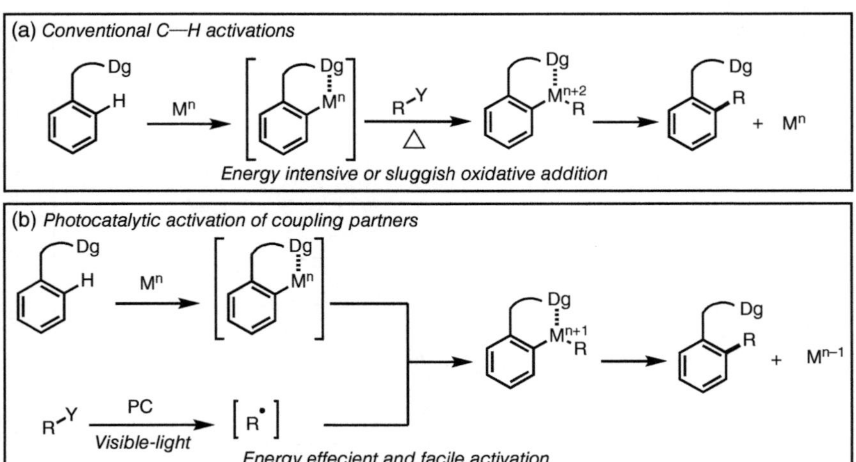

Scheme 13.14 Dual Cu/photocatalyzed alkynylation of benzamides with aryl acetylenes.

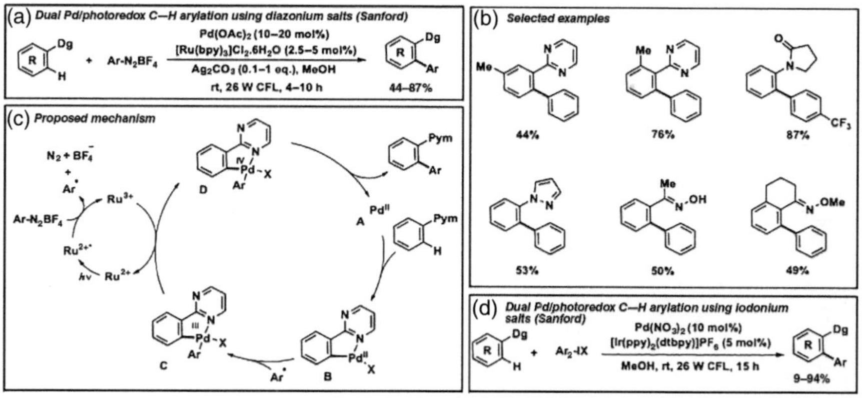

Scheme 13.15 Dual Cu/photocatalyzed alkynylation of benzamides with aryl acetylenes.

Synergistically, the excited Ru(II) photocatalyst undergoes oxidative quenching with the aryl diazonium salt to generate the key aryl radical intermediate and a Ru(III) complex. The aryl radical then adds onto palladacycle **B** to form the arylated Pd(III) complex **C**. The subsequent oxidation of **C** by Ru(III) generates the key Pd(IV) intermediate **D**, which upon reductive elimination delivers the arylated product and the catalytically active Pd(II) complex. Shortly after this, in 2012, the same group expanded this dual catalytic strategy for using diaryliodonium salts as the coupling partner (Scheme 13.15d) [26b]. The higher reduction potential of diaryliodonium salts as compared with diazonium salts demanded the use of a more reducing photocatalyst such as $Ir(ppy)_2(dtbpy)PF_6$.

These works by Sanford inspired other groups to develop similar dual catalytic arylations. Guo reported the arylation of 6-arylpurine derivatives (Scheme 13.16a) [27], while Spencer and Gopinath reported the arylation benzodiazepine cores (Scheme 13.16b) [28] of phenyl ureas (Scheme 13.16c), [29] respectively. These reports highlight the versatility of the dual Pd/Ru catalytic protocol to carry out late-stage modifications of medicinally important cores. Other photocatalysts were also employed in different arylation reactions. Balaraman group reported the dual catalytic arylation of anilides using a combination of palladium and inexpensive photocatalyst Eosin Y under additive-free conditions (Scheme 13.16d) [30]. Xu and co-workers documented the use of dihydroacridine ($AcrH_2$) as a photocatalyst for carrying out arylations of anilides and amides (Scheme 13.16e) [31]. The Gianetti group documented the use of helical carbenium ion (^{n}Pr-DMQA-BF_4) as a photocatalyst in dual catalytic C—H arylations under in red light irradiation (Scheme 13.16f) [32]. The use of low-energy red light makes this methodology milder and more benign as compared with other light sources.

Besides aryl diazonium or diaryliodonium salts, other types of photochemically activated substrates have been used in dual metal/photoredox mediated transformations. In 2015 Li, Wang et al. used a dual palladium/photocatalytic system to carry out decarboxylative C—H acylation of acetanilides using glyoxylic acids (Scheme 13.17a) [33]. They employed a catalytic combination of $Pd(OAc)_2$ and Eosin Y in an oxygen atmosphere under irradiation by green LEDs.

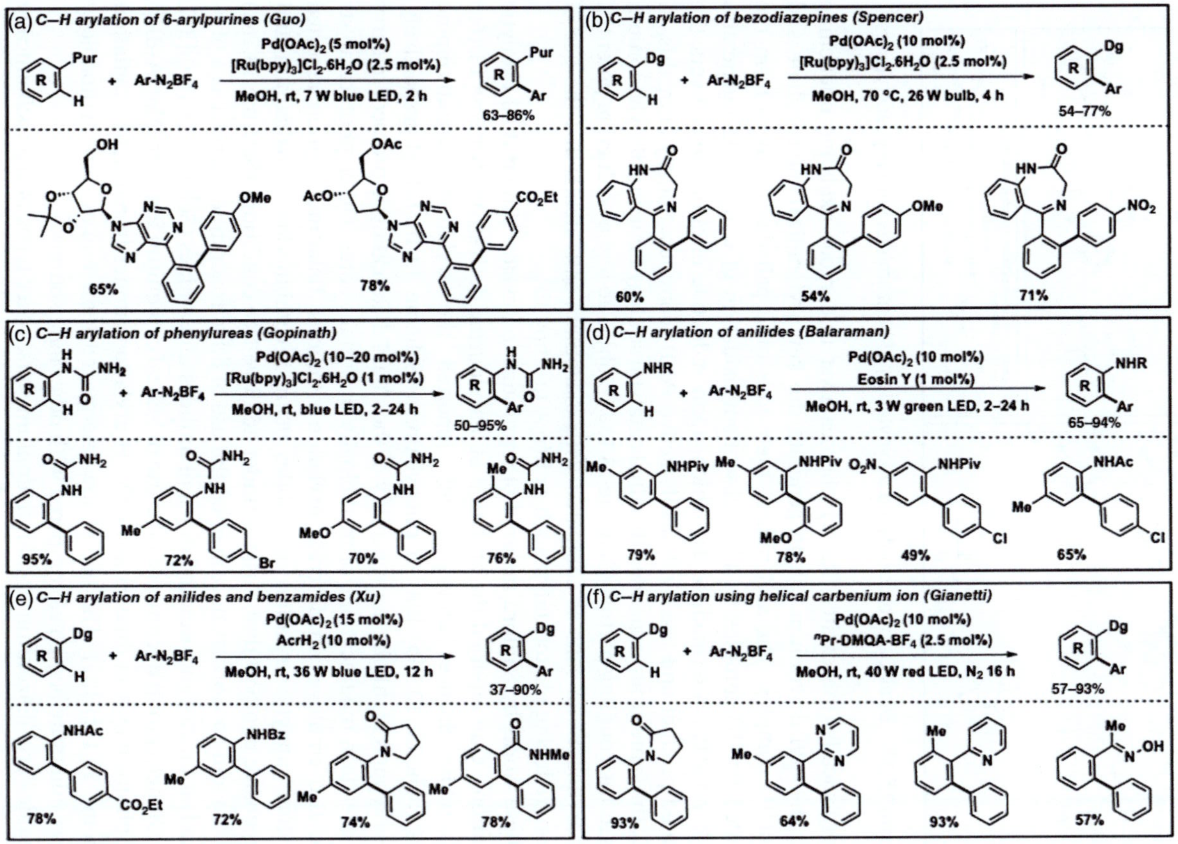

Scheme 13.16 Dual Pd/photocatalyzed C−H arylations.

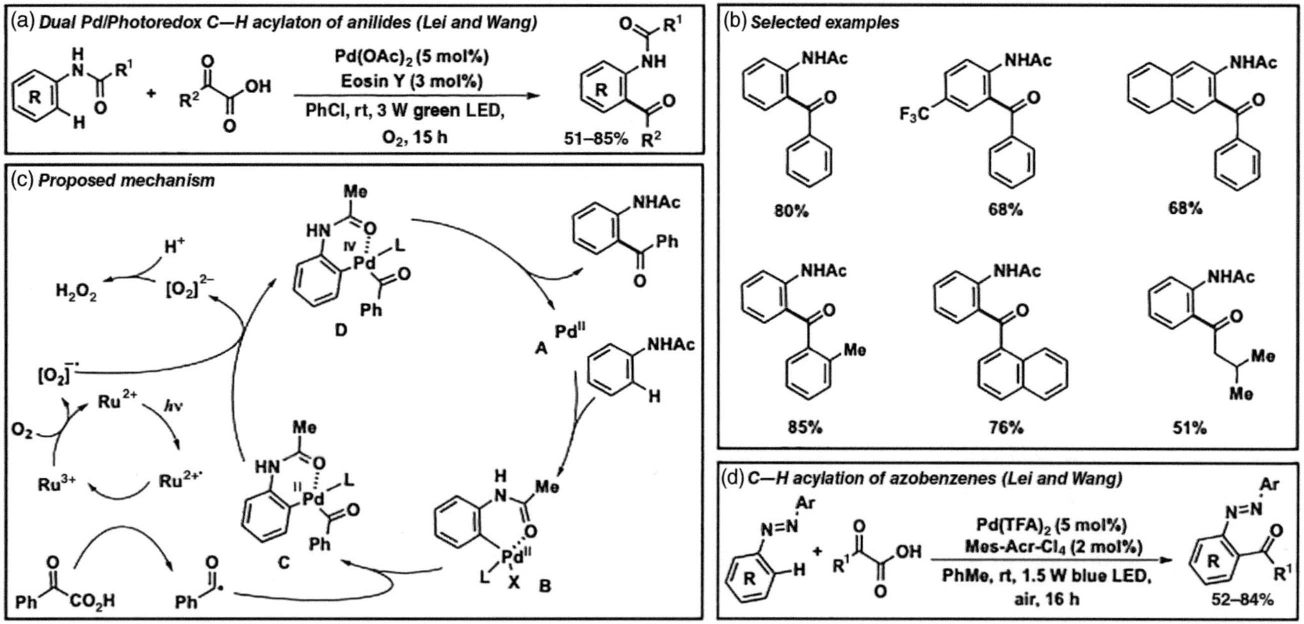

Scheme 13.17 Dual Pd/photocatalyzed decarboxylative C—H acylation.

The reaction proceeded at room temperature tolerating a range of acetanilides and glyoxylic acids (Scheme 13.17b). The proposed mechanism is initiated by an amide directed C—H activation by Pd(II) to form palladacycle intermediate **A** (Scheme 13.17c). Concomitantly, single-electron oxidation of the glyoxylic acid by excited photocatalyst and subsequent decarboxylation generates an acyl radical along with the reduced photocatalyst. The acyl radical adds on to **A** forming Pd(III) intermediate **B**. Then, molecular oxygen sequentially reoxidizes the reduced photocatalyst and the Pd(III) intermediate **B**. This not only closes the photocatalytic cycle but also generates the key Pd(IV) intermediate **C**, which undergoes reductive elimination to form the product and the active Pd(II) species **A**. The same group in 2016 reported a similar protocol for the acylation of azobenzenes using Mes-Acr-ClO_4 photocatalyst (Scheme 13.17d) [34].

The reports on decarboxylative acylation were then followed by other dual palladium photocatalytic acylations, which essentially follow a similar mechanism. Jain and co-workers reported a protocol for carrying out acylation of *N*-protected carbazoles using a combination of $Pd(OAc)_2$ and Eosin Y (Scheme 13.18a) [35]. Shortly after this, Gopinath reported a similar transformation using $[Ru(bpy)_3]Cl_2$ as the photocatalyst [36]. The Singh group used a combination of $Pd(OAc)_2$ and fluorescein dye to carry out decarboxylative acylation of phenylquinoxalin-2(1*H*)-ones (Scheme 13.18b) [37]. Jain disclosed a dual catalytic protocol for carrying out acylation of *N*-pheny-7-azaindoles (Scheme 13.18c) [38], while Zao et al. reported a similar protocol for carrying out acylation of *O*-methyl-ketoximes (Scheme 13.18d) [39].

Van der Eycken and Noël reported a dual palladium photocatalytic protocol for C-2 acylation of indoles using aldehydes in batch and flow (Scheme 13.19a) [40]. They utilized a combination of $Pd(OAc)_2$ and *fac*-$[Ir(ppy)_3]$ as catalysts and TBHP as the oxidant under blue LED irradiation. The reaction tolerated a variety of aldehydes and indole derivatives offering similar yields in both batch and flow conditions (Scheme 13.19b). The reaction follows a similar mechanism as discussed earlier, with the ^{t}BuO radicals formed by the oxidative quenching of the excited Ir(III) photocatalyst and ^{t}BuOOH undergoing HAT with the aldehyde to form an acyl radical (Scheme 13.19c). This acyl radical then adds onto the C—H activated palladacycle **B** to form Pd(III) complex **C**. The oxidation of complex **C** with the oxidized Ir(IV) generates the key Pd(IV) complex **D** and the active photocatalyst. The subsequent reductive elimination from **D** leads to the acylated products and the active Pd(II) species. Shortly after this, Jana and co-workers reported a similar strategy in batch using $Ru(bpy)_3Cl_2.6H_2O$ as the photocatalyst [41]. Very recently, the Xia group reported a similar protocol for the acylation of phenyl pyridines using phenanthraquinone as the HAT photocatalyst to generate acyl radicals from aldehydes (Scheme 13.19d) [42].

The Singh group in 2018 reported a dual palladium/photocatalytic protocol for the direct hydroxylation of 2-phenylpyridines and 2-phenylbenzothiazoles (Scheme 13.20a) [43]. They utilized a catalytic combination of 4-CzIPN and $Pd(OAc)_2$ in the presence of $BrCCl_3$ and oxygen. The reaction showcased a broad scope with good functional group tolerance (Scheme 13.20b). The proposed mechanism parallels those discussed earlier, with a key formation of a hydroxyl

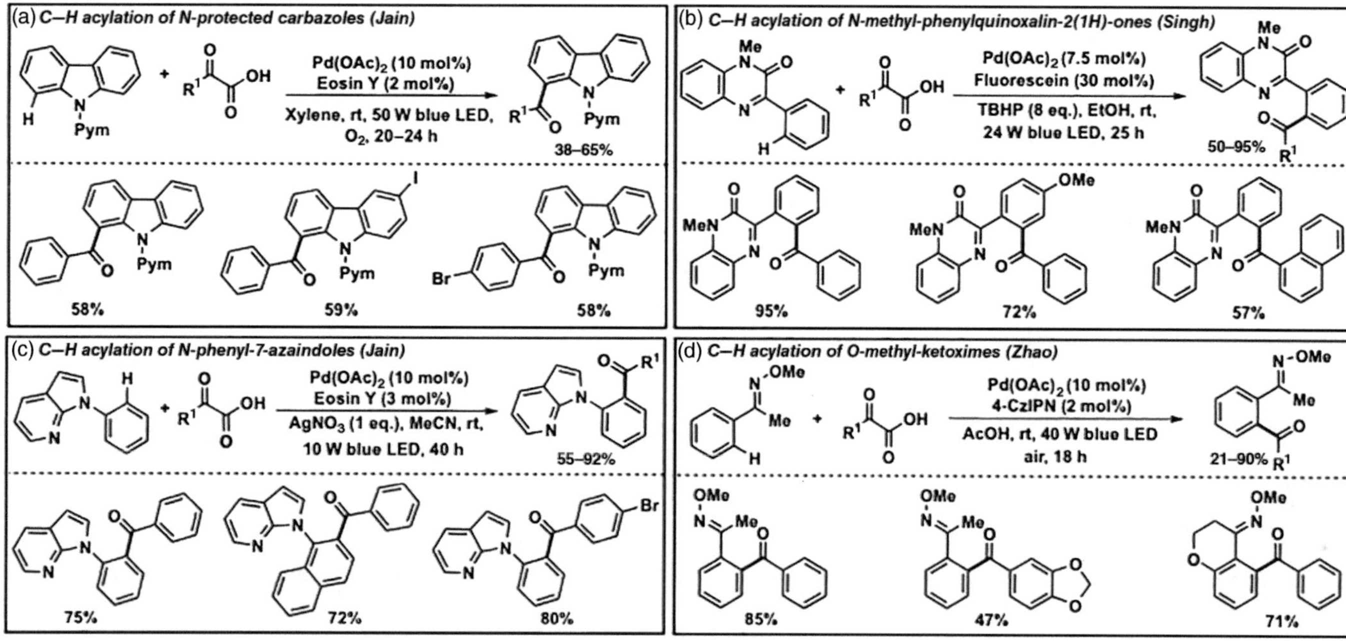

Scheme 13.18 Dual Pd/photocatalyzed decarboxylative C–H acylation.

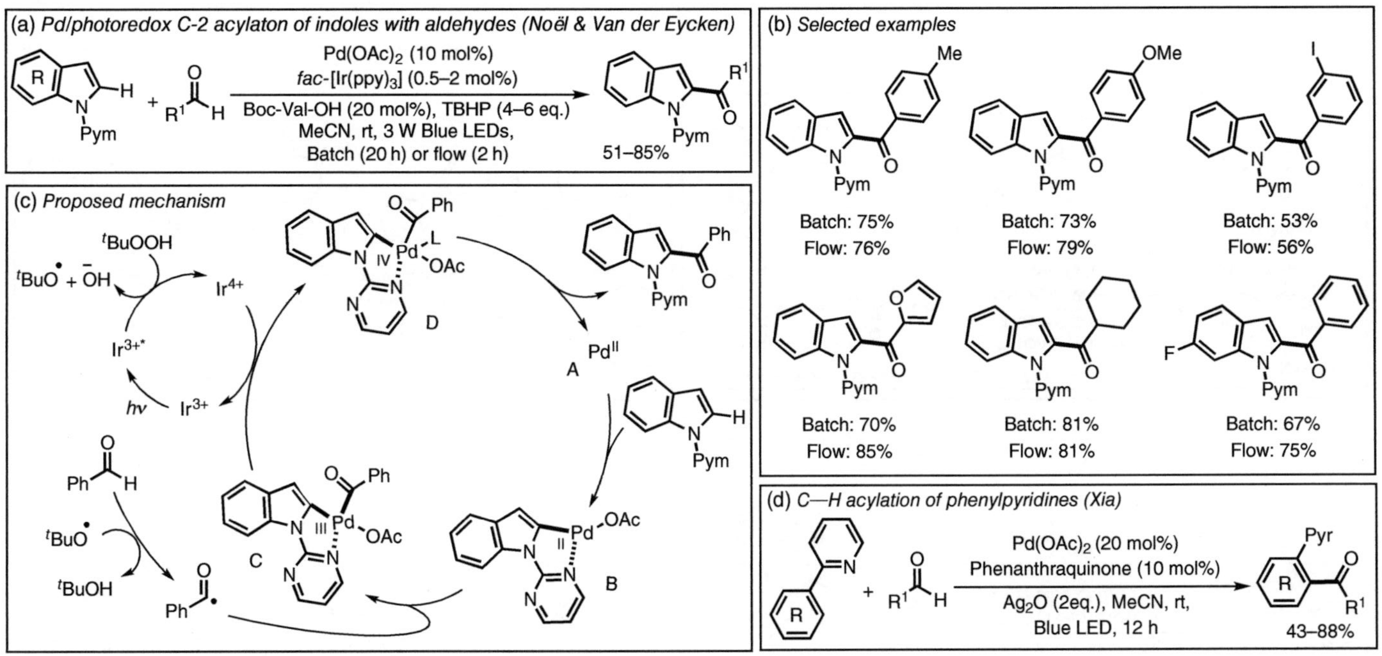

Scheme 13.19 Dual Pd/photocatalyzed C—H acylation using aldehydes.

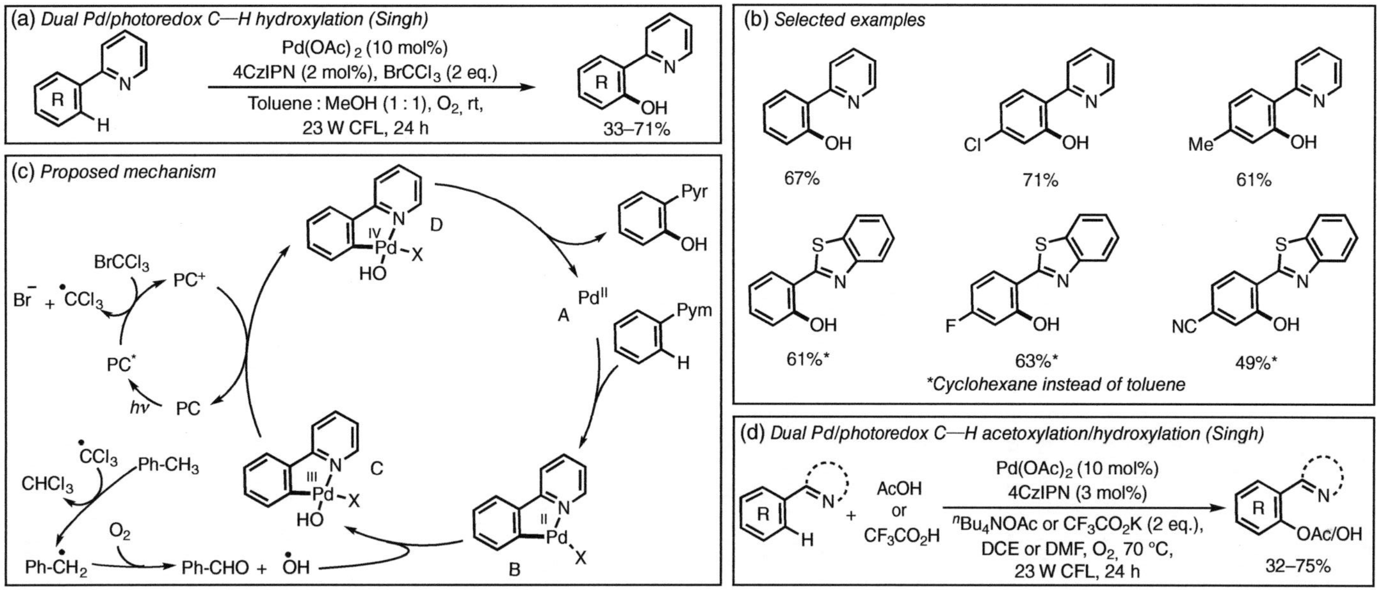

Scheme 13.20 Dual Pd/photocatalyzed C—H oxygenations.

radical as a coupling partner (Scheme 13.20c). First, the excited photocatalyst undergoes oxidative quenching with $BrCCl_3$ to generate CCl_3 radical, which then undergoes HAT with toluene to form benzyl radical (or with cyclohexane to form cyclohexyl radical). The carbon-centered radical combines with molecular oxygen to form benzaldehyde (or cyclohexanone) and the hydroxyl radical. The hydroxyl radical then adds onto the palladacycle **B** leading to intermediate **C**, which then undergoes single-electron oxidation by the oxidized photocatalyst to form the Pd(IV) species **D**. Subsequent reductive elimination from **D** delivers the product and the catalytically active Pd(II) species **A**. The same group later reported a dual catalytic protocol for carrying out directed C—H acetoxylation using acetic acid (Scheme 13.20d) [44]. They found that the use of trifluoro acetic acid leads to the corresponding hydroxyl products due to the rapid hydrolysis of the trifluoroacetate products under the reaction conditions.

In 2023, Maiti reported *meta*-acetoxylation of arenes by the merger of photocatalysis with palladium-catalyzed directing group-assisted *meta*-C—H activations using $PhI(OAc)_2$ (Scheme 13.21a) [45]. Using a combination of $Pd(OAc)_2$, Eosin Y, and a nitrile directing group capable of selectively activating the *meta*-position, the authors were able to carry out efficient acetoxylation of a range of arene derivatives including biphenyls (Scheme 13.21b). The reaction proceeds via a mechanism wherein the initial *meta*-directed C—H activation leads to the complex **B** (Scheme 13.21c). Simultaneously, the oxidative quenching of the photocatalyst by $PhI(OAc)_2$ generates the acetoxyl radical, which adds to **B** to form the Pd(III) complex **C**. The subsequent oxidation of **C** by the $EY^{\cdot+}$ generates the key Pd(IV) complex, which undergoes reductive elimination forming the acetoxylated products and the active catalyst **A**.

13.4 Summary and Conclusions

While photocatalysis and directed C—H activations have seen broad applications individually, their merger is a nascent field that has seen significant advancements in a short period of time. The use of photocatalysis for reoxidation of the reduced metal catalyst has led to protocols that bypass the use of stoichiometric oxidants and harsh reaction conditions. In this regard, photocatalytic reoxidation has been briefly explored with metals such as Rh, Ru, Pd, and Cu, while the same could be considered well explored in the case of Co. These efforts have given rise to green and sustainable C—H activations having better atom efficiency and process mass indices. On the other hand, protocols that allow the merger of photocatalytic radical formation and metal-catalyzed C—H activations have unlocked new paradigms related to excited-state catalysis. This has not only allowed access to previously inaccessible transformations but also led to facile activation of challenging coupling partners under mild energy-efficient conditions. These protocols rely mainly on the use of Pd as the metal catalyst leading to functionalizations such as arylation, acylation, or hydroxylation. Overall, the last couple of years have seen seminal advancements in directing group-assisted dual photocatalytic C—H activations, which include pioneering reports on asymmetric (4 + 2) annulations or facile and

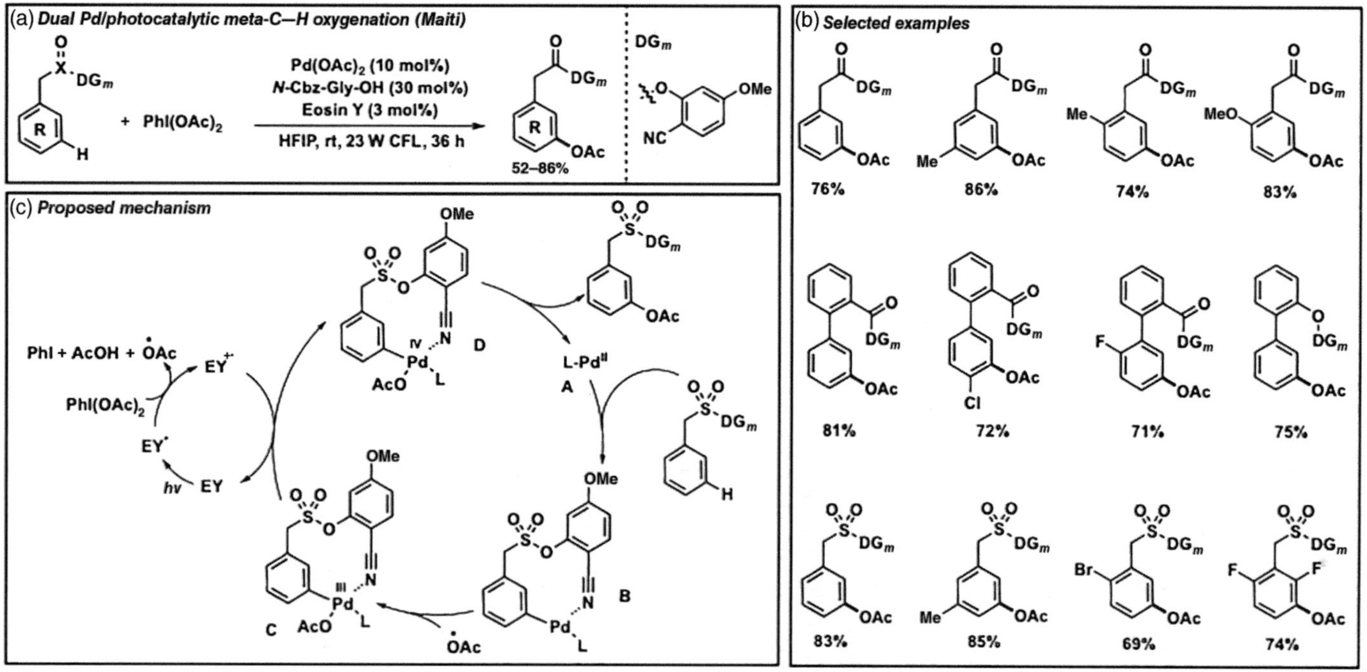

Scheme 13.21 Dual Pd/photocatalyzed *meta*-C—H oxygenations

selective functionalization on the different positions of the arene ring. Despite these advancements, there are significant prospects to look forward to in the future like the expansion of the chemical space of these reactions to other challenging functionalizations such as sulfonylation, halogenation, and alkylation. Also, the use of transient or innate directing groups would bring in further sustainability and step economy by preventing the need for the installation/removal of directing groups. New advancements in photocatalysis such as the use of recoverable heterogenous photocatalysts or catalysis under low-energy light irradiation could be applied to carry out more efficient directed C—H activations. Finally, the use of flow chemistry would allow large-scale reactions directly leading to industrial and pharmaceutical applications.

References

1 (a) Prier, C.K., Rankic, D.A., and MacMillan, D.W.C. (2013). Visible light photoredox catalysis with transition metal complexes: applications in organic synthesis. *Chem. Rev.* 113 (7): 5322–5363. (b) Narayanam, J.M.R. and Stephenson, C.R.J. (2011). Visible light photoredox catalysis: applications in organic synthesis. *Chem. Soc. Rev.* 40: 102–113.

2 Yoon, T.P., Ischay, M.A., and Du, J. (2010). Visible light photocatalysis as a greener approach to photochemical synthesis. *Nat. Chem.* 2: 527–532.

3 (a) de Meijere, A., Bräse, S., and Oestreich, M. (2013). *Transition-Metal- Catalyzed Cross-Coupling Reactions and More*. Weinheim, Germany: Wiley-VCH. (b) Seechurn, C.C.C.J., Kitching, M.O., Colacot, T.J. et al. (2012). Palladium-catalyzed cross-coupling: a historical contextual perspective to the 2010 Nobel prize. *Angew. Chem. Int. Ed.* 51 (21): 5062–5085. (c) Magano, J. and Dunetz, J.R. (2011). Large-scale applications of transition metal-catalyzed couplings for the synthesis of pharmaceuticals. *Chem. Rev.* 111 (3): 2177–2250.

4 (a) Docherty, J.H., Lister, T.M., McArthur, G. et al. (2023). Transition-metal-catalyzed C-H bond activation for the formation of C-C bonds in complex molecules. *Chem. Rev.* 123 (12): 7692–7760. (b) Sinha, S.K., Ghosh, P., Jain, S. et al. (2023). Transition-metal catalyzed C–H activation as a means of synthesizing complex natural products. *Chem. Soc. Rev.* 52: 7461–7503. (c) Chen, Z., Wang, B., Zhang, J. et al. (2015). *Org. Chem. Front.* 2: 1107–1295. (d) Rej, S., Das, A., and Chatani, N. (2021). *Coord. Chem. Rev.* 431: No. 213683.

5 (a) Fabrey, D.C. and Rueping, M. (2016). Merging visible light Photoredox catalysis with metal catalyzed C– H activations: on the role of oxygen and superoxide ions as oxidants. *Acc. Chem. Res.* 49 (9): 1969–1979. (b) Guillemard, L. and Wencel-Delord, J. (2020). When metal-catalyzed C–H functionalization meets visible-light photo-catalysis. *Beilstein J. Org. Chem.* 16: 1754–1804.

6 (a) Song, G., Wang, F., and Li, X. (2012). C–C, C–O and C–N bond formation via rhodium(iii)-catalyzed oxidative C–H activation. *Chem. Soc. Rev.* 41: 3651–3678. (b) Mei, R., Dhawa, U., and Samanta, R.C. (2020). Cobalt-catalyzed oxidative C–H activation: strategies and concepts. *ChemSusChem* 13 (13): 3306–3356.

7 Wencel-Delord, J., Dröge, T., Liu, F. et al. (2011). Towards mild metal-catalyzed C–H bond activation. *Chem. Soc. Rev.* 40: 4740–4761.

8 (a) Rakshit, S., Grohmann, C., Besset, T. et al. (2011). Rh(III)-catalyzed directed C–H olefination using an oxidizing directing group: mild, efficient, and versatile. *J. Am. Chem. Soc.* 133 (8): 2350–2353. (b) Huang, H., Ji, X., Wu, W. et al. (2015). Transition metal-catalyzed C–H functionalization of N-oxyenamine internal oxidants. *Chem. Soc. Rev.* 44: 1155–1171.

9 (a) Konnick, M.M. and Stahl, S.S. (2008). Reaction of molecular oxygen with a Pd^{II}-hydride to produce a Pd^{II}-Hydroperoxide: experimental evidence for an HX-reductive-elimination pathway. *J. Am. Chem. Soc.* 130 (17): 5753–5762. (b) Stahl, S.S. (2004). Palladium oxidase catalysis: selective oxidation of organic chemicals by direct dioxygen-coupled turnover. *Angew. Chem. Int. Ed.* 43 (26): 3400–3420.

10 Fabry, D.C., Zoller, J., Raja, S. et al. (2014). Combining rhodium and Photoredox catalysis for C-H Functionalizations of Arenes: oxidative heck reactions with visible light. *Angew. Chem. Int. Ed.* 53 (38): 10228–10231.

11 Kim, H.J., Fabry, D.C., Mader, S. et al. (2019). Photoredox/rhodium catalysis in C–H activation for the synthesis of nitrogen containing heterocycles. *Org. Chem. Front.* 6: 2319–2323.

12 Zoller, J., Fabry, D.C., Ronge, M.A. et al. (2014). Synthesis of indoles using visible light: Photoredox catalysis for palladium-catalyzed C-H activation. *Angew. Chem. Int. Ed.* 53 (48): 13264–13268.

13 Silantyev, G.A., Filippov, O.A., Tolstoy, P.M. et al. (2013). Hydrogen bonding and proton transfer to ruthenium hydride complex CpRuH(Dppe): metal and hydride dichotomy. *Inorg. Chem.* 52 (4): 1787–1797.

14 (a) Kalsi, D., Dutta, S., Barsu, N. et al. (2018). Room-temperature C–H bond functionalization by merging cobalt and photoredox catalysis. *ACS Catal.* 8 (9): 8115–8120. (b) Mandal, R., Barsu, N., Garai, B. et al. (2021). Room-temperature C-H bond alkynylation by merging cobalt and photocatalysis. *Chem. Commun.* 57: 12167–12170.

15 Kalsi, D., Barsu, N., Chakrabarti, S. et al. (2018). C–H and N–H bond annulation of aryl amides with unactivated olefins by merging cobalt(iii) and photoredox catalysis. *Chem. Commun.* 55: 11626–11629.

16 Sen, C., Sarvaiya, B., Sarkar, S. et al. (2020). Room-temperature synthesis of isoindolone spirosuccinimides: merger of visible-light photocatalysis and cobalt-catalyzed C–H activation. *J. Org. Chem.* 85 (23): 15287–15304.

17 Li, T., Li, J., Zhu, Z. et al. (2020). Metallaphotoredox-catalyzed C–H activation: regio-selective annulation of allenes with benzamide. *Org. Chem. Front.* 8: 928–935.

18 Ban, Y.-L., You, L., Wang, T. et al. (2021). Metallaphotoredox dearomatization of indoles by a benzamide-empowered [4 + 2] annulation: facile access to indolo[2,3-c]isoquinolin-5-ones. *ACS Catal.* 11 (9): 5054–5060.

19 Das, A., Kumaran, S., Ravi Sankar, H.S. et al. (2024). A dual cobalt-photoredox catalytic approach for asymmetric dearomatization of indoles with aryl amides via C–H activation. *Angew. Chem. Int. Ed.* 63 (40): e202406195.

20 (a) Teng, M.-Y., Liu, D.-Y., Mao, S.-Y. et al. (2024). Asymmetric Dearomatization of indoles through cobalt-catalyzed enantioselective C–H functionalization enabled by Photocatalysis. *Angew. Chem. Int. Ed.* 63 (40): e202407640. (b) Xu, Y., Lin, Y., Homölle, S.L. et al. (2024). Enantioselective cobaltaphotoredox-catalyzed C–H activation. *J. Am. Chem. Soc.* 146 (34): 24105–24113.

21 Kumar, S., Nair, A.M., and Volla, C.M.R. (2022). Dual photoredox and cobalt-catalysis for C(sp2)-H functionalization of benzamides. *Chem. Asian J.* 17 (19): e202200801.

22 Saha, A., Guin, S., Ali, W. et al. (2022). Photoinduced regioselective olefination of arenes at proximal and distal sites. *J. Am. Chem. Soc.* 144 (4): 1929–1940.

23 Nair, A.M., Rahaman, R., Patra, J. et al. (2024). Dual copper photoredox C–H alkynylation with arylacetylenes. *Org. Lett.* 26 (37): 7822–7827.

24 (a) Dong, J., Wang, F., and You, J. (2014). Copper-mediated tandem oxidative C(sp^2)–H/C(sp)–H Alkynylation and annulation of Arenes with terminal alkynes. *Org. Lett.* 16 (11): 2884–2887. (b) Tian, C., Dhawa, U., Scheremetjew, A. et al. (2019). Cupraelectro-catalyzed alkyne annulation: evidence for distinct C–H alkynylation and decarboxylative C–H/C–C manifolds. *ACS Catal.* 9 (9): 7690–7696.

25 (a) Daugulis, O., Do, H.-Q., and Shabashov, D. (2009). Palladium- and copper-catalyzed Arylation of carbon–hydrogen bonds. *Acc. Chem. Res.* 42 (8): 1074–1086. (b) Chen, X., Engle, K.M., Wang, D.-H. et al. (2009). Palladium(II)-catalyzed C–H activation/C–C cross-coupling reactions: versatility and practicality. *Angew. Chem. Int. Ed.* 48 (28): 5094–5115.

26 (a) Kalyani, D., McMurtrey, K.B., Neufeldt, S.R. et al. (2011). Room-temperature C–H arylation: merger of Pd-catalyzed C–H functionalization and visible-light photocatalysis. *J. Am. Chem. Soc.* 133 (46): 18566–18569. (b) Neufeldt, S.R. and Sanford, M.S. (2012). Combining transition metal catalysis with radical chemistry: dramatic acceleration of palladium-catalyzed C-H arylation with diaryliodonium salts. *Adv. Synth. Catal.* 354 (18): 3517–3522.

27 Liang, L., Xie, M.-S., Wang, H.-X. et al. (2017). Visible-light-mediated monoselective *ortho* C–H arylation of 6-arylpurine nucleosides with diazonium salts. *J. Org. Chem.* 82 (11): 5966–5973.

28 Khan, R., Boonseng, S., Kemmitt, P.D. et al. (2017). Combining Sanford arylations on benzodiazepines with the nuisance effect. *Adv. Synth. Catal.* 359 (18): 3261–3269.

29 Babu, S.S., Shahid, M., and Gopinath, P. (2020). Dual palladium–photoredox catalyzed chemoselective C–H arylation of phenylureas. *Chem. Commun.* 56: 5985–5988.

30 Sahoo, M.K., Midya, S.P., Landge, V.G. et al. (2017). A unified strategy for silver-, base-, and oxidant-free direct arylation of C–H bonds. *Green Chem.* 19: 2111–2117.

31 Jiang, J., Zhang, W.-M., Dai, J.-J. et al. (2017). Visible-light-promoted C–H arylation by merging palladium catalysis with organic photoredox catalysis. *J. Org. Chem.* 82 (7): 3622–3630.

32 Mei, L., Veleta, J.M., and Gianetti, T.L. (2020). Helical carbenium ion: a versatile organic photoredox catalyst for red-light-mediated reactions. *J. Am. Chem. Soc.* 142 (28): 12056–12061.

33 Zhou, C., Li, P., Zhu, X. et al. (2015). Merging photoredox with palladium catalysis: decarboxylative *ortho*-acylation of acetanilides with α-oxocarboxylic acids under mild reaction conditions. *Org. Lett.* 17 (24): 6198–6201.

34 Xu, N., Li, P., Xie, Z. et al. (2016). Merging visible-light photocatalysis and palladium catalysis for C−H acylation of azo- and azoxybenzenes with α-keto acids. *Chem. Eur. J.* 22 (7): 2236–2242.

35 Rajat; Jain, N. (2023). Photoredox/palladium dual catalysis for site-selective C−H arylation and acylation of N-protected carbazoles in visible light. *J. Org. Chem.* 88 (13): 8600–8608.

36 Shahid, M., Punnya, A.J., Babu, S.S. et al. (2017). Dual palladium- photoredox-mediated regioselective acylation of carbazoles and indolines. *J. Org. Chem.* 88 (19): 13686–13698.

37 Prince, M., Kumar, P. et al. (2024). Visible-light-driven regioselective decarboxylative acylation of N-methyl-3-phenylquinoxalin-2(1H)-one by dual palladium–photoredox catalysis through C–H activation. *ACS Omega* 9 (1): 651–657.

38 Rajput, S., Kaur, R., and Jain, N. (2022). Pd and photoredox dual catalysis assisted decarboxylative *ortho*-benzoylation of N-phenyl-7-azaindoles. *Org. Biomol. Chem.* 20: 1453–1461.

39 Wang, C., Yang, S., Huang, Z. et al. (2022). Palladium-catalyzed 4CzIPN-mediated decar- boxylative acylation reaction of O-methyl ketoximes with α-keto acids under visible light. *Molecules* 27 (6): 1980.

40 Sharma, U.K., Gemoets, H.P.L., Schröder, F. et al. (2017). Merger of visible-light photoredox catalysis and C–H activation for the room-temperature C-2 acylation of indoles in batch and flow. *ACS Catal.* 7 (6): 3818–3823.

41 Manna, M.K., Bairy, G., and Jana, R. Dual visible-light photoredox and palladium(ii) catalysis for dehydrogenative C2-acylation of indoles at room temperature. *Org. Biomol. Chem.* 15: 5899–5903.

42 Wang, H., Li, T., Hu, D. et al. (2017). Acylation of arenes with aldehydes through dual C-H activations by merging photocatalysis and palladium catalysis. *Org. Lett.* 23 (9): 3772–3776.

43 Shah, S.S., Paul, A., Bera, M. et al. (2018). Metallaphotoredox-mediated Csp2–H hydroxylation of arenes under aerobic conditions. *Org. Lett.* 20 (18): 5533–5536.

44 Shah, S.S., Shee, M., Singh, A.K. et al. (2020). Direct oxygenation of C–H bonds through photoredox and palladium catalysis. *J. Org. Chem.* 85 (5): 3426–3439.

45 Ali, W., Saha, A., Ge, H. et al. (2023). Photoinduced meta-selective C−H oxygenation of Arenes. *JACS Au.* 3 (6): 1790–1799.

14

Industrial and Flow Application of Directed C—H Bond Functionalization

Aritra Mukherjee, Rahul Bangari, and Supriya Rej

Centre of BioMedical Research (CBMR), Sanjay Gandhi Post-Graduate Institute of Medical Sciences Campus, Department of Biological and Synthetic Chemistry, Raebareli Road, Lucknow, 226014, India

14.1 Introduction

C—H bond functionalization has become one of the most explored fields in organic chemistry, especially in the last two decades, due to its wide prevalence in organic compounds [1–6]. C—H bonds' high prevalence in organic chemistry improves the scope for exploration and functionalization to obtain a library of compounds. In recent days, it has even decoded the selective functionalization of unbiased, distal, aliphatic, and even the least active C—H bonds with the use of directing auxiliaries [7, 8]. In this regard, transition-metal catalytic protocols provided a dynamic stage. It has even been explored for total synthesis, API chemical synthesis, and for the synthesis of drug-like molecules via the introduction of molecular complexity, which improves the scope for exploration of the medicinal field and the possibility of potential drug discovery. ACS GCI Pharmaceutical Roundtable in 2005 has acknowledged the C—H activation field as one of the useful and most prioritized reactions [9]. Considering the overwhelming use of C—H bond functionalization chemistry in organic synthesis, it has still been difficult to functionalize the C—H bond due to its inertness and high bond dissociation energy. Hence, highly reactive transition-metal catalysts, oxidants, harsh reaction conditions, and longer reaction times are required in many cases. Large-scale synthesis in an industrial setup became one of the notable drawbacks of C—H bond functionalization. To overcome this, recent developments have been performed using the continuous-flow setup. Since the introduction of flow synthesis in C—H bond functionalization chemistry has brought various advantages, for instance, safer reaction handling compared to batch process, safe handling of high-pressure conditions, control of accurate reaction temperature, efficient contact of chemicals and reagents in a different phase of the reaction, there is considerably high-surface-area-to-volume ratio for exposing the chemicals

Directed C–H Bond Functionalization: Concepts and Applications, First Edition.
Edited by Debabrata Maiti and Supriya Rej.

to the temperature or the light (in case of photochemical synthesis), and scalable property [10–14]. These multiple advantages of flow synthesis can help explore C—H bond functionalization chemistry further in industrial setups.

Considering the numerous advantages of continuous-flow reaction, it has widely been used in cross-coupling reactions and total synthesis [15–19]; nonetheless, it has scarcely been used in C—H bond functionalization chemistry. The introduction of continuous-flow reaction in C—H bond functionalization chemistry was made in the last decade, and this field is yet to grow widely. By taking into account the key theme of the book, we have summarized here the application of continuous-flow strategy in directed C—H bond functionalization reactions (Scheme 14.1). In this regard, few notable reviews and summarized articles are available discussing the significant application of continuous-flow reaction in C—H bond functionalization [20–24]; however, there are no focused summarized reports available covering directing strategy in C—H bond functionalization under continuous flow.

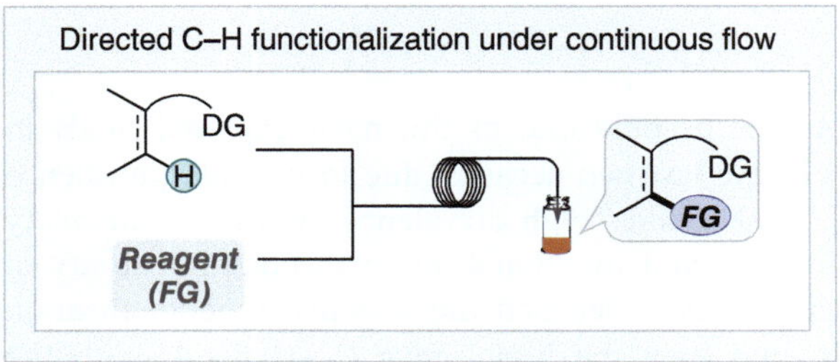

Scheme 14.1 Schematic representation of directed C–H bond functionalization under continuous flow.

14.2 Directed C(sp^2)–H Functionalization

In an early report, Reddy and Kappe reported homogeneous C—H bond functionalization in a continuous-flow system. A Cu-catalyzed oxidative C—O coupling of 2-carbonyl-substituted phenols or β-ketoesters with ethers via α-C—H bond activation was reported with *tert*-butyl hydroperoxide (TBHP) as an oxidant and using a two-feed high-temperature/pressure microreactor approach [25]. Soon after the introduction of the continuous-flow process in C—H bond functionalization, it has started to flourish. A notable number of studies have been reported on non-directed and directed C—H bond functionalization using the flow technique. Considering the scope of this chapter, we have focused on the direct C—H bond functionalization in which a directing group is involved.

14.2.1 Homogeneous Catalysis

In one of the early reports in 2013, Schnürch showcased a Ru-catalyzed directed *ortho*-arylation of 2-phenyl pyridine using aryl bromide in a continuous-flow process

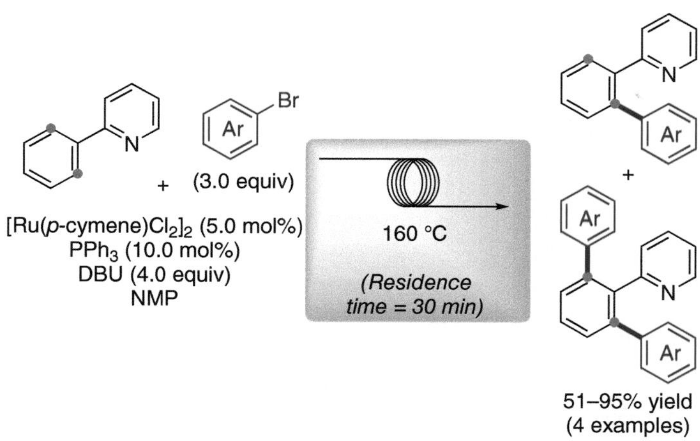

Scheme 14.2 Ru-catalyzed *ortho*-arylation of 2-phenyl pyridine under continuous flow.

(Scheme 14.2) [26]. Notably, the continuous-flow process showed efficiency over the batch process as the desired product can be obtained in very short residence time (30 minutes) albeit a mixture of mono- and di-arylation products.

In 2015, Gaunt reported a direct C—H amination via thermal decomposition of nitrene form from a nitro-substituted compound to yield the carbazole derivative **2** (Scheme 14.3), which is a key step for the concise and scalable total synthesis of Dictyodendrin B [27]. The process involves the use of dioxane at 180 °C in a continuous-flow process. Notably, over one gram of product **2** can be formed in 30 minutes. This process was adopted from Tokuyama's batch process of azide decomposition to yield the corresponding carbazole [28], and the efficacy of the flow process was realized over the batch reaction.

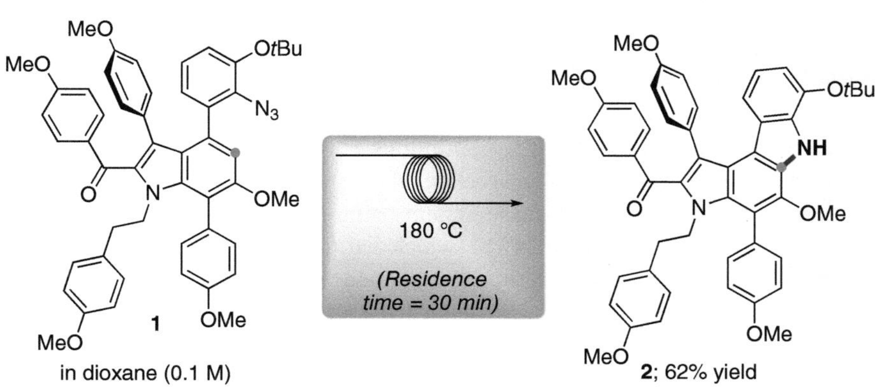

Scheme 14.3 Carbazole synthesis via C—H amination under continuous flow.

A useful and scalable Pd-catalyzed Catellani-type reaction [29–31] was designed under a continuous-flow process by Noël to afford *ortho*-disubstituted vinyl arenes,

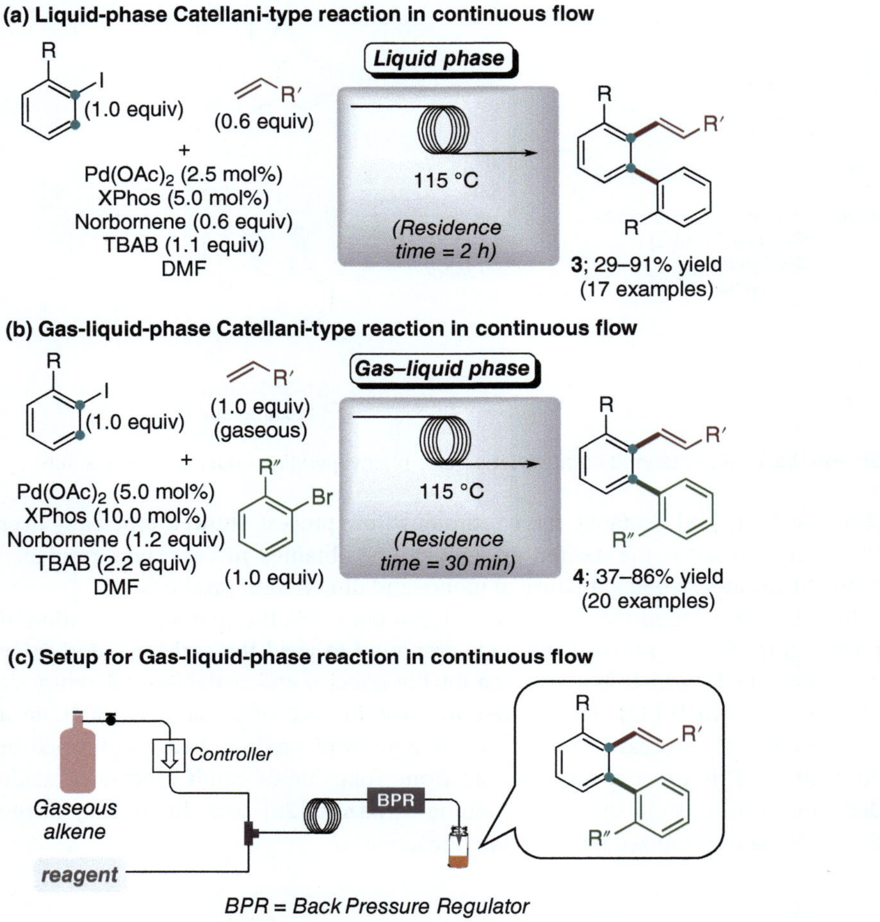

Scheme 14.4 Pd-catalyzed gas–liquid Catellani-type reaction under continuous flow.

as shown in Scheme 14.4 [32]. The continuous-flow process over batch was realized as the product transformation accelerated notably. The authors even expanded the process for the gas–liquid phase system using gaseous alkene (e.g. ethylene) as a coupling partner (Scheme 14.4b), which led to valuable styrene derivatives. The gas–liquid-phase reaction setup was attained by installing a mass flow controller (MFC) and a backpressure regulator (BPR) in the flow reactor, as shown in Scheme 14.4c.

A continuous-flow setup was designed for mild and scalable C2-acylation of *N*-pyrimidylindoles via dual photoredox/transition metal catalytic protocol using aldehyde as a coupling partner (Scheme 14.5a) [33]. The selective C—H bond functionalized product was obtained with the help of pyrimidine as a directing moiety. Importantly, micro-flow setup has reduced the reaction time significantly

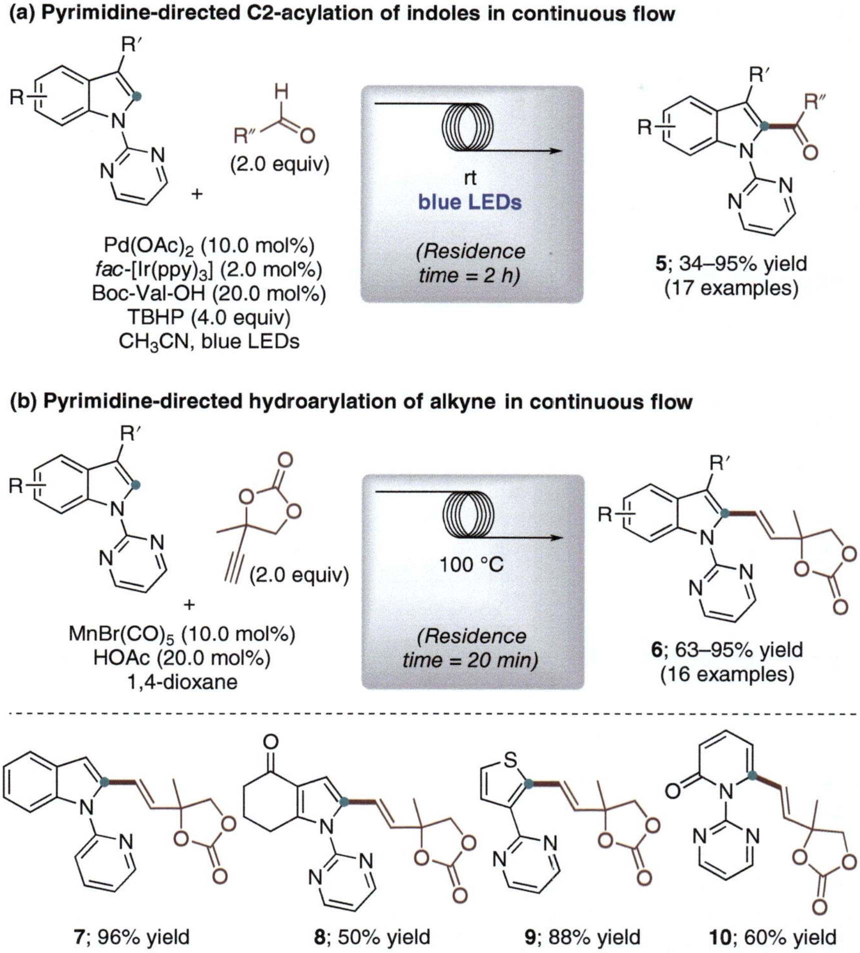

Scheme 14.5 Pyrimidine-directed C2-acylation/alkenylation of indole derivatives under continuous-flow setup.

from 20 to 2 hours. In a similar substrate, using the efficacy of pyrimidine directing group, a Mn-catalyzed protocol was reported for selective C2-alkenylation of *N*-pyrimidylindoles with propargylic alkynes under continuous-flow setup (Scheme 14.5b) [34].

Apart from indole derivatives, tetrahydro-1*H*-indole, thiophene, and 2-pyridone derivatives can also be smoothly alkenylated (Scheme 14.5b, bottom). The significance of the flow process was further emphasized as the reaction yielded the desired product within a very short time (residence time = 20 minutes) with a high yield.

A rhodaelectro-catalyzed C—H annulation of imidates and unsymmetrical alkynes was reported to yield isoquinoline derivatives under a continuous-flow

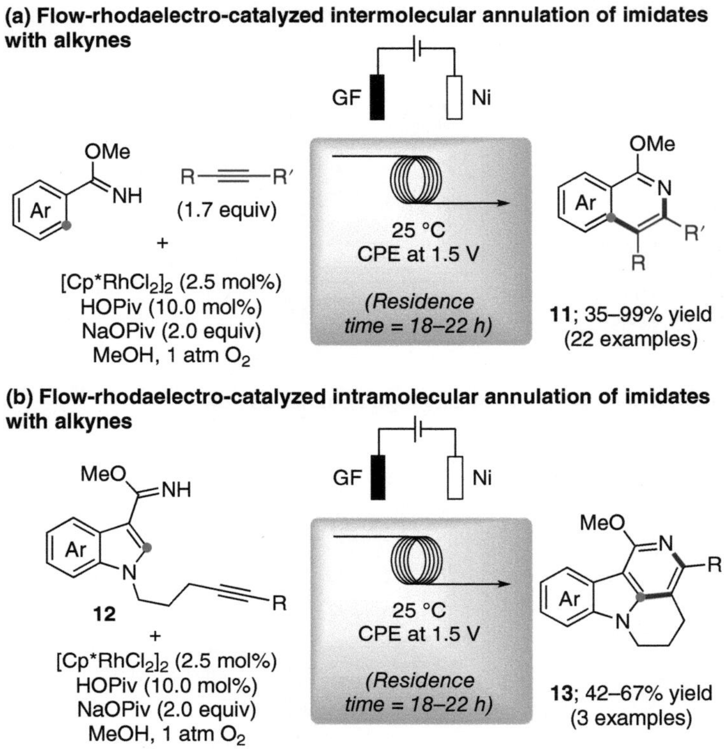

Scheme 14.6 Flow rhodaelectro-catalyzed inter- and intramolecular C–H/N–H annulation.

setup (Scheme 14.6). To ensure an effective turnover, a modular electro-flow cell was designed in combination with a porous graphite anode. Along with intermolecular C—H/N—H annulation, even an intramolecular annulation of **12** can be achieved in the designed setup (Scheme 14.6b) [35].

Ackermann has reported a sustainable $MnCl_2$-catalyzed C—H arylation of pyridine carboxamide derivatives using aryl magnesium bromide under a continuous-flow setup (Scheme 14.7a) [36]. The setup used a three-way connector to mix the reagent and flow reactor equipped with a back pressure regulator at 4 bar (Scheme 14.7a). Chatani also reported a Rh(I)/Rh(II)-catalyzed *ortho*-C—H alkylation of benzylamines with α,β-unsaturated carbonyl compounds and styrene derivatives under a continuous-flow setup (Scheme 14.7b) [37]. Comparison studies were conducted between batch and flow setup, and high efficiency of flow setup was realized as it provided very high product yields even in short residence time.

A Pd-catalyzed intramolecular direct arylation of 2-halobenzyl phenyl ether derivatives was designed in a continuous-flow capillary microreactor using ultrasonic irradiation to yield six-membered *O*-heterocycles (Scheme 14.8) [38]. The ether linkage can also be replaced by amine and amide to afford *N*-heterocycles.

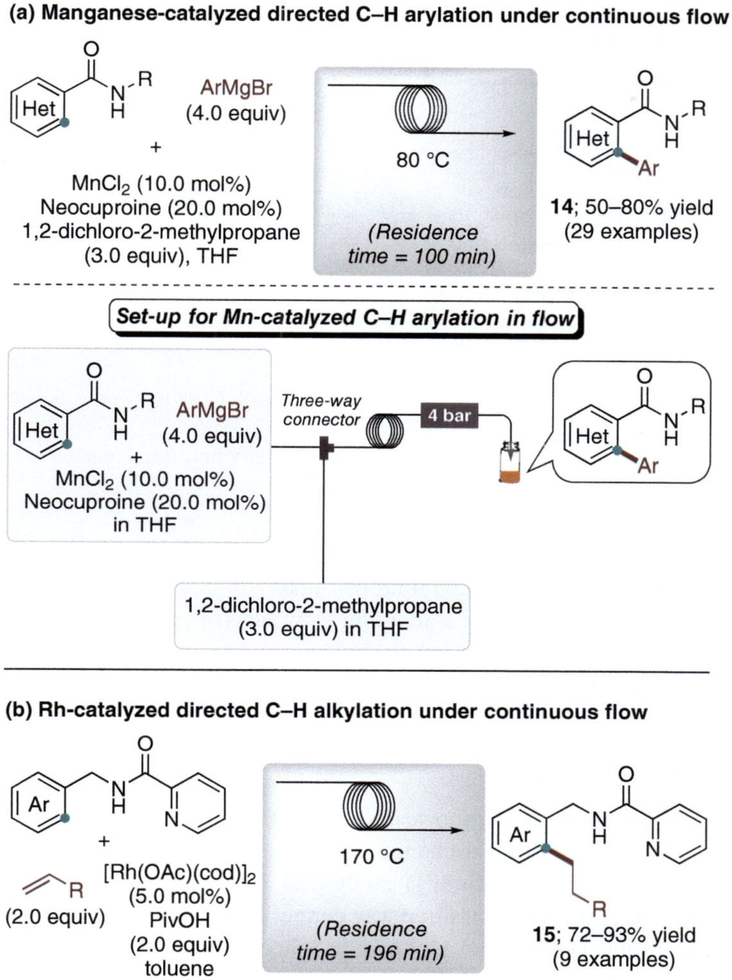

Scheme 14.7 (a) Mn-catalyzed C−H arylation of pyridine carboxamides and (b) Rh-catalyzed C−H alkylation of benzylamines under continuous flow.

The efficacy of the flow process was realized over batch as the comparison study shows that 90% conversion of starting material was obtained after 6 hours at 90 °C in the flow setup, whereas in batch setup to afford 90% conversion of starting material 24 hours of reaction time is required at 100 °C.

A flow electrochemical intramolecular C—H oxygenation of *N*-benzylamides was reported to synthesize 4*H*-1,3-benzoxazines with a very short residence time of 50 seconds (Scheme 14.9a) [39]. The reaction was designed in a transition-metal- and oxidant-free fashion and generates H_2 as the sole byproduct. The efficiency of flow reaction over batch was also witnessed in terms of reactivity. Wirth reported that a flow electrochemical process for dehydrogenative-directed annulation of *N*-aryl thioamides led to the synthesis of benzothiazole (Scheme 14.9b) [40]. Notably, the

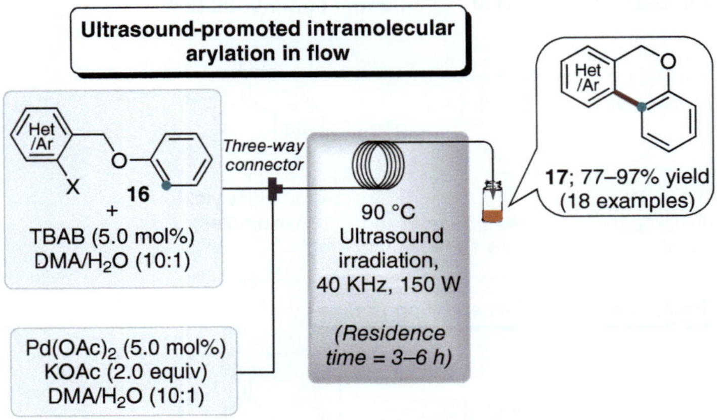

Scheme 14.8 Intramolecular arylation of aryl bromides in capillary flow microreactor using ultrasound irradiation.

described method works without the use of any catalyst or supporting electrolyte. Another dehydrogenative C—S cross-coupling of thioamide was reported under an electrochemical flow process using a modular flow electrolysis cell to afford benzofused *S*-heterocycles (Scheme 14.9c) [41]. The reaction offers convenient scalability and high efficiency.

14.2.2 Heterogeneous Catalysis

In one of the early reports, Noël showcased a continuous-flow process for *ortho*-directed hydrogen-deuterium exchange reaction of benzamide in flow microreactors using immobilized iridium(I) catalyst **24**. The catalyst was loaded in a packed-bed reactor and through a three-way connector substrate, and D_2 gas (equipped with a controller) was passed as shown in Scheme 14.10 [42].

In 2017, Vaccaro reported a heterogeneous catalyst system for directed *ortho*-alkenylation of acetanilide under flow process (Scheme 14.11a) [43]. The flow reactor was assembled using a heterogeneous catalyst (Pd/C) as a catalyst bed and biomass-derived γ-valerolactone (GVL) as a reaction medium. Importantly, the recovery/reuse and durability of the heterogeneous catalyst up to the 5th run with minimum leaching and high yield of the product showcased the efficiency of the process. The authors have also extended their work to a practical process of intramolecular direct C—H functionalization of 1,2,3-triazoles in a continuous-flow process using Pd/C as the heterogeneous catalyst and GVL as a reaction medium (Scheme 14.11b) [44]. Very recently, a significant discovery by Kobayashi showed an enantioselective version of C—H bond functionalization under continuous flow (Scheme 14.11c) [45]. In the report, the authors introduced a heterogeneous chiral Rh catalyst for stereoselective intramolecular hydroacylation of substituted 2-vinylbenzaldehyde. The developed process shows a high turnover number of >300 without any leaching of the catalyst.

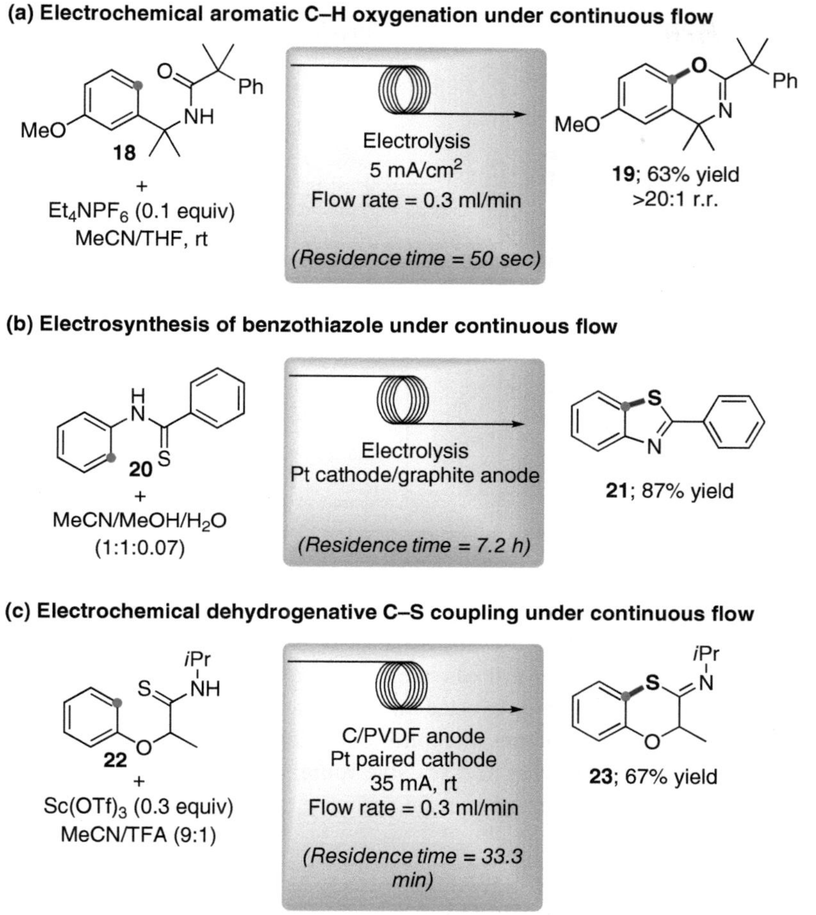

Scheme 14.9 Intramolecular C–O and C–S bond formation under electrochemical flow process.

14.3 Directed C(sp³)–H Functionalization

C(sp³)—H bond functionalization is always challenging compared to the C(sp²)—H bond due to the predictable less reactivity of them. Hence, C(sp³)—H bond functionalization is scarcely reported compared to the sp² version. In fact, directed C(sp³)—H bond functionalization under continuous-flow process is also fewer. One of the significant reports by Gaunt and Lapkin described a Pd-catalyzed continuous-flow synthesis of aziridines via C(sp³)—H activation of amine **28** (Scheme 14.12a) [46]. In the designed flow setup, the overall reaction time significantly reduced and increased the space–time yield. A Pd-catalyzed trifluoromethylthiolation of amides derived from the 8-aminoquinoline using *N*-[(trifluoromethyl)thio]phthalimide at C(sp³)—H bond was reported via a continuous-flow process with a residence

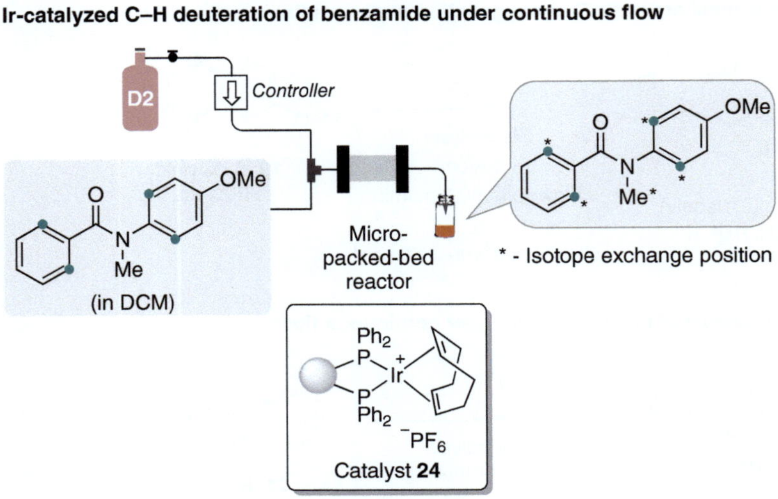

Scheme 14.10 Immobilized Ir-catalyzed C—H deuteration in a flow microreactor.

time of 20 minutes (Scheme 14.12b) [47]. Comparison studies suggested that the time of the reaction can be decreased by a factor of 100/200 in continuous-flow processes compared to the batch process, which further conveys the significance of flow synthesis. In 2021, Chatani reported an Ir(III)-catalyzed α-C—H amidation of 2-acylimidazoles with dioxazolones in a continuous-flow process (scheme 14.12c) [48]. The reaction provides an easy and scalable route for the synthesis of useful α-amino acid derivatives via a nitrene transfer pathway.

14.4 Summary and Conclusions

Considering the growing interest in flow chemical synthesis, it has become a key focus in industrial setups, aiming to achieve a large-scale, easy, reliable, and user-friendly synthetic process for target molecules. By taking into account the sustainability and the use of renewable energy in the synthetic paradigm, flow chemical synthesis took a pivotal role as it can offer several significant points, for instance, easy and safe-to-handle protocol, high-scale production, safe handling of high-pressure conditions, accurate reaction temperature control, high heat and mass transfer, ease of product isolation, reducing waste production, shortening the reaction time, reducing the reaction temperature, and having integrated multi-step synthesis. Along with the homogeneous phase reaction, flow reaction can also be planned for heterogeneous mediums such as gas–liquid or solid–liquid combinations, which further implies its efficacy over batch process. In a heterogeneous catalytic system, flow setup offers a tremendous advantage as effective recyclability of the catalyst can be done in several numbers of cycles. Although there are some disadvantages in flow setup such as clogging, solid formation inside the tube, and

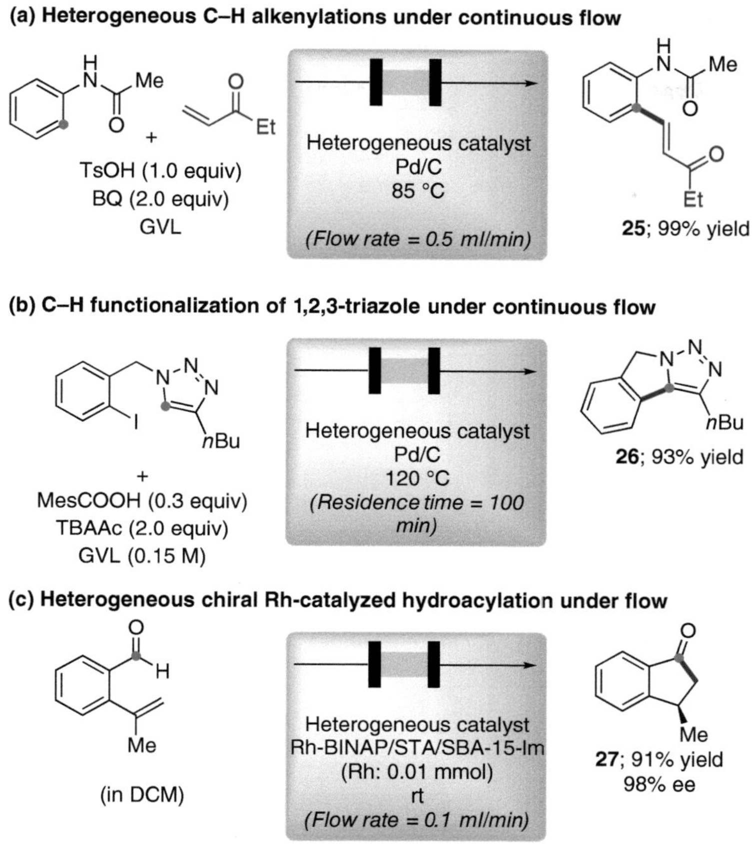

Scheme 14.11 Heterogeneous catalytic system for C–H bond functionalization under continuous flow.

leaching in heterogeneous cases, it still shows significant beneficial aspects over batch process. Considering the significance of flow synthesis in organic chemistry it has been overwhelmingly applied in cross-coupling reaction, acid–base reaction, catalyst-free synthesis, and in C—H bond functionalization chemistry. Although progress has been made in flow C—H bond functionalization, directed-C—H bond functionalization is needed to develop further. The application flow chemistry can provide a real breakthrough of directed-C—H bond functionalization chemistry by advancing in an industrial scale for kilogram-scale API synthesis. Hence, this can lead to the production of many potential drug candidates in large scale and can help for reducing the production cost, and thus would be helpful to mankind. In this chapter, we have summarized the directed C—H bond functionalization chemistry under continuous-flow techniques using homogeneous, heterogeneous, or catalyst-free conditions in which a directing group is involved for controlling the position selectivity. We anticipate that this chapter will be very useful to the synthetic community to understand the gaps in this field and proceed further.

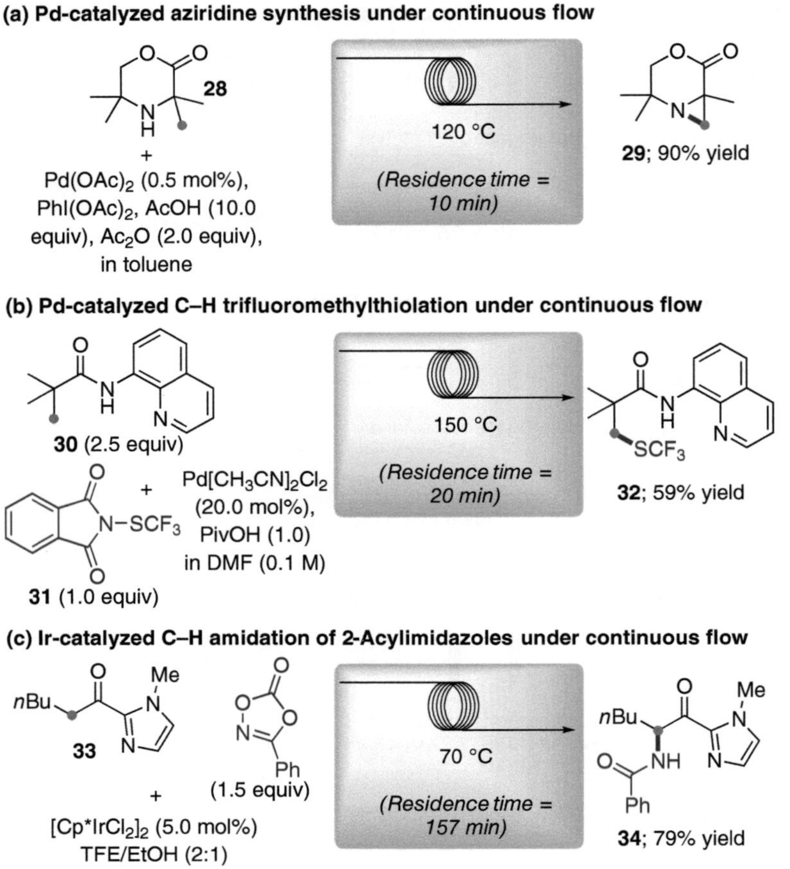

Scheme 14.12 Directed C(sp^3)–H bond functionalization under continuous flow.

We believe this is just the tip of the iceberg. The world of directed C—H bond functionalization under flow is brimming with untapped potential, and we're only scratching the surface of what's possible. With more breakthroughs on the horizon, the future of flow chemistry holds endless possibilities for innovation.

Acknowledgments

S. Rej acknowledges financial support from ANRF (formerly known as SERB) for the Grant of Ramanujan Faculty (Fellowship No. RJF/2023/000038). The Centre for Biomedical Research (CBMR) is greatly acknowledged for the provided infrastructure. R. Bangari expresses his thanks to CSIR New Delhi for the Junior Research Fellowship (File No. 09/0916[21013]/2025-EMR-I).

References

1 Yamaguchi, J., Yamaguchi, A.D., and Itami, K. (2012). C-H bond functionalization: emerging synthetic tools for natural products and pharmaceuticals. *Angew. Chem. Int. Ed.* 51: 8960–9009.

2 Gensch, T., Hopkinson, M.N., Glorius, F., and Wencel-Delord, J. (2016). Mild metal-catalyzed C-H activation: examples and concepts. *Chem. Soc. Rev.* 45: 2900–2936.

3 Abrams, D.J., Provencher, P.A., and Sorensen, E.J. (2018). Recent applications of C – H functionalization in complex natural product synthesis. *Chem. Soc. Rev.* 47: 8925–8967.

4 Gandeepan, P., Müller, T., Zell, D. et al. (2019). 3d Transition metals for C-H activation. *Chem. Rev.* 119: 2192–2452.

5 Dalton, T., Faber, T., and Glorius, F. (2021). C-H activation: toward sustainability and applications. *ACS Cent. Sci.* 7: 245–261.

6 Rej, S., Das, A., and Chatani, N. (2021). Strategic evolution in transition metal-catalyzed directed C–H bond activation and future directions. *Coord. Chem. Rev.* 431: 213683.

7 Sambiagio, C., Schönbauer, D., Blieck, R. et al. (2018). A comprehensive overview of directing groups applied in metal-catalysed C-H functionalisation chemistry. *Chem. Soc. Rev.* 47: 6603–6743.

8 Rej, S., Ano, Y., and Chatani, N. (2020). Bidentate directing groups: an efficient tool in C-H bond functionalization chemistry for the expedient construction of C-C bonds. *Chem. Rev.* 120: 1788–1887.

9 Constable, D.J.C., Dunn, P.J., Hayler, J.D. et al. (2007). Key green chemistry research areas—a perspective from pharmaceutical manufacturers. *Green Chem.* 9 (5): 411–420.

10 Vaccaro, L., Lanari, D., Marrocchi, A., and Strappaveccia, G. (2014). Flow approaches towards sustainability. *Green Chem.* 16 (8): 3680–3704.

11 Kobayashi, S. (2016). Flow "fine" synthesis: high yielding and selective organic synthesis by flow methods. *Chem. - An Asian J.* 11 (4): 425–436.

12 Akwi, F.M. and Watts, P. (2018). Continuous flow chemistry: where are we now? Recent applications, challenges and limitations. *Chem. Commun.* 54 (99): 13894–13928.

13 Baumann, M., Moody, T.S., Smyth, M., and Wharry, S. (2021). Evaluating the green credentials of flow chemistry towards industrial applications. *Synthesis (Stuttg).* 53 (21): 3963–3976.

14 Hone, C.A. and Kappe, C.O. (2021). Towards the standardization of flow chemistry protocols for organic reactions. *Chemistry-Methods* 1 (11): 454–467.

15 Baumann, M., Baxendale, I.R., and Ley, S.V. (2011). The flow synthesis of heterocycles for natural product and medicinal chemistry applications. *Mol. Divers.* 15 (3): 613–630.

16 Brandão, P., Pineiro, M., and Pinho e Melo, T.M.V.D. (2019). Flow chemistry: towards a more sustainable heterocyclic synthesis. *Eur. J. Org. Chem.* 2019 (43): 7188–7217.
17 Luque Navarro, P.M. and Lanari, D. (2020). Flow synthesis of biologically-relevant compound libraries. *Molecules* 25 (4): 909.
18 Jiao, J., Nie, W., Yu, T. et al. (2021). Multi-step continuous-flow organic synthesis: opportunities and challenges. *Chem. - A Eur. J.* 27 (15): 4817–4838.
19 Gambacorta, G., Sharley, J.S., and Baxendale, I.R. (2021). A comprehensive review of flow chemistry techniques tailored to the flavours and fragrances industries. *Beilstein J. Org. Chem.* 17: 1181–1312.
20 Echtermeyer, A., Amar, Y., Zakrzewski, J., and Lapkin, A. (2017). Self-optimisation and model-based design of experiments for developing a C-H activation flow process. *Beilstein J. Org. Chem.* 13: 150–163.
21 Laudadio, G. and Noël, T. (2017). Chapter 7 - flow chemistry perspective for CH bond functionalization. In: *Strategies for Palladium-Catalyzed Non-directed and Directed C-H Bond Functionalization* (ed. A.R. Kapdi and D.B.T.-S. Maiti) for P.-C.N.-D. and D.C.-H.B.F., 275–288. Elsevier.
22 Santoro, S., Ferlin, F., Ackermann, L., and Vaccaro, L. (2019). C-H functionalization reactions under flow conditions. *Chem. Soc. Rev.* 48 (10): 2767–2782.
23 Santoro, S., Ferlin, F., Ackermann, L., and Vaccaro, L. (2019). C–H functionalization reactions under flow conditions. *Chem. Soc. Rev.* 48: 2767–2782.
24 Govaerts, S., Nyuchev, A., and Noel, T. (2020). Pushing the boundaries of C–H bond functionalization chemistry using flow technology. *J. Flow Chem.* 10 (1): 13–71.
25 Kumar, G.S., Pieber, B., Reddy, K.R., and Kappe, C.O. (2012). Copper-catalyzed formation of C-O bonds by direct α-C-H bond activation of ethers using stoichiometric amounts of peroxide in batch and continuous-flow formats. *Chem. - A Eur. J.* 18 (20): 6124–6128.
26 Christakakou, M., Schön, M., Schnürch, M., and Mihovilovic, M.D. (2013). Arylation of pyridines via suzuki-miyaura cross-coupling and pyridine-directed C-H activation using a continuous-flow approach. *Synlett* 24 (18): 2411–2418.
27 Pitts, A.K., O'Hara, F., Snell, R.H., and Gaunt, M.J. (2015). A concise and scalable strategy for the total synthesis of dictyodendrin B based on sequential C-H functionalization. *Angew. Chem. Int. Ed.* 54 (18): 5541–5545.
28 Okano, K., Fujiwara, H., Noji, T. et al. (2010). Total synthesis of dictyodendrin A and B. *Angew. Chem. Int. Ed.* 49 (34): 5925–5929.
29 Wang, J. and Dong, G. (2019). Palladium/norbornene cooperative catalysis. *Chem. Rev.* 119 (12): 7478–7528.
30 Cheng, H.-G., Chen, S., Chen, R., and Zhou, Q. (2019). Palladium(II)-initiated catellani-type reactions. *Angew. Chem. Int. Ed.* 58: 5832–5844.
31 Dong, S. and Luan, X. (2021). Catellani reaction: an enabling technology for vicinal functionalization of aryl halides by palladium(0)/norbornene cooperative catalysis. *Chin. J. Chem.* 39 (6): 1690–1705.

32 Casnati, A., Gemoets, H.P.L., Motti, E. et al. (2018). Homogeneous and gas–liquid catellani-type reaction enabled by continuous-flow chemistry. *Chem. - A Eur. J.* 24 (53): 14079–14083.

33 Sharma, U.K., Gemoets, H.P.L., Schröder, F. et al. (2017). Merger of visible-light photoredox catalysis and C-H activation for the room-temperature C-2 acylation of indoles in batch and flow. *ACS Catal.* 7 (6): 3818–3823.

34 Wang, H., Pesciaioli, F., Oliveira, J.C.A. et al. (2017). Synergistic manganese(I) C–H activation catalysis in continuous flow: chemoselective hydroarylation. *Angew. Chem. Int. Ed.* 56 (47): 15063–15067.

35 Kong, W.-J., Finger, L.H., Messinis, A.M. et al. (2019). Flow rhodaelectro-catalyzed alkyne annulations by versatile C-H activation: mechanistic support for rhodium(III/IV). *J. Am. Chem. Soc.* 141 (43): 17198–17206.

36 Zhu, C., Oliveira, J.C.A., Shen, Z. et al. (2018). Manganese(II/III/I)-catalyzed C-H arylations in continuous flow. *ACS Catal.* 8 (5): 4402–4407.

37 Das, A. and Chatani, N. (2021). Rh(i)- And Rh(ii)-catalyzed C-H alkylation of benzylamines with alkenes and its application in flow chemistry. *Chem. Sci.* 12 (9): 3202–3209.

38 Zhang, L., Geng, M., Teng, P. et al. (2012). Ultrasound-promoted intramolecular direct arylation in a capillary flow microreactor. *Ultrason. Sonochem.* 19 (2): 250–256.

39 Xu, F., Qian, X.-Y., Li, Y.-J., and Xu, H.-C. (2017). Synthesis of 4H-1,3-benzoxazines via metal- and oxidizing reagent-free aromatic C–H oxygenation. *Org. Lett.* 19 (23): 6332–6335.

40 Folgueiras-Amador, A.A., Qian, X.-Y., Xu, H.-C., and Wirth, T. (2018). Catalyst- and supporting-electrolyte-free electrosynthesis of benzothiazoles and thiazolopyridines in continuous flow. *Chem. - A Eur. J.* 24 (2): 487–491.

41 Huang, C., Qian, X.-Y., and Xu, H.-C. (2019). Continuous-flow electrosynthesis of benzofused S-heterocycles by dehydrogenative C–S cross-coupling. *Angew. Chem. Int. Ed.* 58 (20): 6650–6653.

42 Habraken, E.R.M., Haspeslagh, P., Vliegen, M., and Noël, T. (2015). Iridium(I)-catalyzed ortho-directed hydrogen isotope exchange in continuous-flow reactors. *J. Flow Chem.* 5 (1): 2–5.

43 Ferlin, F., Santoro, S., Ackermann, L., and Vaccaro, L. (2017). Heterogeneous C-H alkenylations in continuous-flow: oxidative palladium-catalysis in a biomass-derived reaction medium. *Green Chem.* 19 (11): 2510–2514.

44 Ferlin, F., Luciani, L., Santoro, S. et al. (2018). A continuous flow approach for the C-H functionalization of 1,2,3-triazoles in γ-valerolactone as a biomass-derived medium. *Green Chem.* 20 (12): 2888–2893.

45 Saito, Y. and Kobayashi, S. (2024). Continuous-flow enantioselective hydroacylations under heterogeneous chiral rhodium catalysts. *Angew. Chem. Int. Ed.* 63 (1): e202313778.

46 Zakrzewski, J., Smalley, A.P., Kabeshov, M.A. et al. (2016). Continuous-flow synthesis and derivatization of aziridines through palladium-catalyzed C(sp3)–H activation. *Angew. Chem. Int. Ed.* 55 (31): 8878–8883.

47 Bouchard, A., Kairouz, V., Manneveau, M. et al. (2019). Continuous flow palladium-catalyzed trifluoromethylthiolation of C-H bonds. *J. Flow Chem.* 9 (1): 9–12.

48 Ohara, N., Das, A., Mahato, S.K., and Chatani, N. (2021). Synthesisof α-amino acid derivatives through the iridium-catalyzed α-C-H amidation of 2-acylimidazoles with dioxazolones under continuous-flo. *Chem. Lett.* 50 (9): 1722–1724.

15

Applications of Directed C—H Functionalization in Medicinal and Process Chemistry

Krishnamay Pal, Rajesh Sahu, and Anant R. Kapdi

Department of Chemistry, Institute of Chemical Technology, Nathalal Parekh Road, Matunga, Mumbai, 400019, India

15.1 Introduction

Catalysis has undergone a sea change in the past few decades, and the involvement of catalytic processes in academic research and industry has gone up significantly. Sustainability in catalysis has a lot of significance in the current era, and C—H bond functionalization strategy provides researchers with an opportunity to achieve this goal as the pre-functionalization step in the previous catalytic technologies (known to generate a significant amount of waste) can be avoided [1, 2]. The past decade has seen an upsurge in the number of protocols involving C—H bond functionalization of a wide variety of substrates starting from aromatics, heteroaromatics, aliphatic, olefins, and many more of synthetic and biological importance [3–5]. C—H bond functionalization strategies have also broken new grounds of achieving excellent selectivity, i.e. positional (*ortho*, *meta*, or *para* for aromatics; *delta* or *gamma* selectivity in alkanes), and demonstrated its potential with the synthesis of complex bio-active natural products [6–9]. In spite of these achievements in the field of C—H bond functionalization, there is a lack of application of such a powerful synthetic strategy to gram-scale or kilogram-scale synthesis of molecules of significance in medicinal chemistry or those of pharmaceutical relevance (APIs) [10, 11]. However, this scenario has changed dramatically in recent years with researchers applying C—H bond functionalization as a key synthetic step for achieving both the above objectives.

This chapter, therefore, details these achievements in two parts, namely (Figure 15.1),

(a) C—H functionalization in medicinal chemistry: Gram-scale synthesis of important bio-active molecules, pharmaceutical ingredients that have been identified to exhibit a wide spectrum of activity. These include Tie2 tyrosine kinase inhibitor, Angiotensin II receptor blocker, BRD3914, Zafirlukast (Accolate), Palomid 529, Febuxostat, Tryprostatin A, and Adiphenine.

Directed C–H Bond Functionalization: Concepts and Applications, First Edition.
Edited by Debabrata Maiti and Supriya Rej.

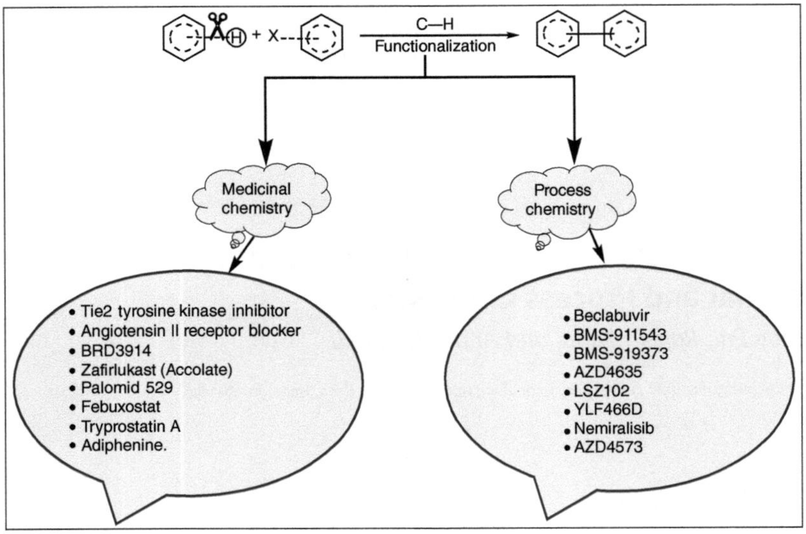

Figure 15.1 C–H Functionalization in medicinal and process chemistry.

(b) C—H functionalization in process chemistry: These include examples of pharmaceutical ingredients or the final drug candidate in kilogram or multikilogram scale using C—H functionalization as one of the key steps in achieving the required synthetic efficiency (Beclabuvir, BMS-911543, BMS-919373, AZD4635, LSZ102, YLF466D, Nemiralisib, and AZD4573).

15.2 C–H Functionalization in Medicinal Chemistry

Medicinal chemistry approaches have led to the identification of many potential drug candidates and the development of synthetic routes for these molecules. Researchers have employed many C—C bond-forming technologies in the past for developing synthetic methodologies for targeting such substrates [12–14], but the recent trend suggests the use of C—H bond functionalization as a more efficient strategy in achieving the desired goal of sustainability and cost-effectiveness [15]. The scale of synthesis, however, is restricted to grams as the purpose is to demonstrate the potential of the route to obtain the desired molecule in a lesser and more efficient way than the previously reported pathways. This section, therefore, comprises examples highlighting the potential of C—H bond functionalization strategy for the construction of several bio-active molecules.

15.2.1 Synthesis of Tie2 Tyrosine Kinase Inhibitor

Tie2 Tyrosine kinase Inhibitor is a drug that targets the Tie2 receptor found on the surface of endothelial cells and plays a crucial role in angiogenesis. In cancer, angiogenesis is known to become unregulated and forms unwanted blood vessels that can

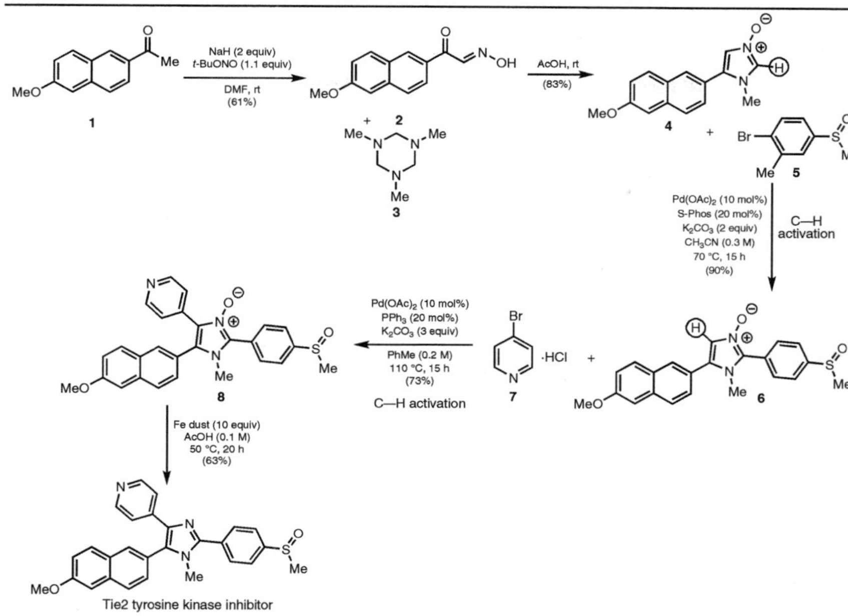

Scheme 15.1 Synthesis of Tie2 tyrosine kinase inhibitor.

lead to tumor growth. Tie2 inhibitor works by blocking the activity of Tie2, thereby inhibiting the angiogenesis and potentially slowing the growth and spread of the tumor [16]. The chemical synthesis of Tie2 tyrosine kinase inhibitor is a five-step process (Scheme 15.1) of substitution in the C2, C4, and C5 positions of the imidazole core structure [17]. The process was initiated by the treatment of ketone **1** with NaH and *t*-butylnitrile to give the α-ketoxime **2** in 61% yield followed by the further treatment of the obtained product **2** with 1,3,5-trimethyl-1,3,5-triazinane resulting in imidazole *N*-oxide in 83% yield and 51% yield over two steps. The second coupling partner bromoaryl sulfoxide was prepared separately via S_NAr reaction of 4-fluoro-2-methylbromobenzene with sodium methanethiolate with the formed sulfide subjected to further oxidation to the desired sulfoxide in 68% yield over two steps. The increased steric bulk of the aryl bromide coupling partner could play a crucial role in the outcome of the next step, which was the direct arylation between coupling partners **4** and **5**. A certain effect, observed as lower yield of the desired product, was obtained under the standard C2 arylation conditions. To circumvent this problem, the authors decided to employ S-Phos (20 mol%) as the ligand which in combination with $Pd(OAc)_2$ (10 mol%), K_2CO_3 (2 equiv) in CH_3CN as solvent for 15 hours at 70 °C provided the C—H arylation product in good yield. The next step of the synthesis involved another direct arylation strategy at the C4 position of imidazole **6** with the C—H activation reaction performed using 4-bromopyridine **7** as the coupling partner under the standard arylation condition of $Pd(OAc)_2$ (10 mol%), PPh_3(20 mol%), K_2CO_3 (3.0 equiv) in PhMe (0.2 M) as solvent for 15 hours at 110 °C. The triarylimidazole *N*-oxide thus formed was isolated in 73% yield. Later, selective reduction of the *N*-oxide moiety using Fe and acetic acid was performed (sulfoxide functionality was well tolerated) to give the Tie2 tyrosine kinase inhibitor an overall yield of 21%.

15.2.2 Synthesis of Angiotensin II Receptor Blocker

Angiotensin II receptor blockers (ARBs) are one of the most efficient antihypertensives due to their safety and high efficacy [18]. ARBs have a demand of over 100 tons/year worldwide for clinical treatment [19]. ARBs contain a biphenyltetrazole structural motif as the key and common building block. The previous synthetic routes have a significant drawback due to the requirement of a stoichiometric amount of expensive and/or hazardous organometallic reagents such as arylboronic acids and Grignard reagents to synthesize the biphenyltetrazole motif [20, 21]. C—H activation in the past decade has firmly developed as the major alternative to the cross-coupling processes as the dependency on pre-functionalization of the substrate is not there. Accordingly, to address the above problem, Seki and co-workers (Scheme 15.2) developed a C—H activation-based protocol [22], which was executed by the treatment of protected phenyltetrazole with aryl bromide in the presence of $RuCl_3{\cdot}xH_2O$ (Ru = 1.3 mol%) as an inexpensive catalyst with PPh_3 ligand and K_2CO_3 in NMP at 140 °C for 12 hours to obtain the desired biphenyl product in 81% yield. From the study, it was observed that the benzyl-protecting group has a crucial role in promoting the incorporation of the phenyl functionality at the ortho position via the *N*-coordinated Ru species. Optimization studies carried out on this process do reveal that the use of other phosphine ligands possessing electron-withdrawing groups or alkyl substituents as well as the employment of bidentate phosphine ligands resulted in a much lesser yield of the desired product. An inexpensive PPh_3 was observed to provide the most optimal result for the ortho-C—H arylation. The obtained biphenyl intermediate was further converted to the corresponding alcohol first using MeONa/MeOH at ambient temperature and then in situ brominated followed by condensed with L-Valine benzyl ester using Pd/C (0.06 equiv), HCO_2NH_4 (1.1 equiv) at 45 °C for six hours giving the desired ARB in 76% yield and an overall 55% yield.

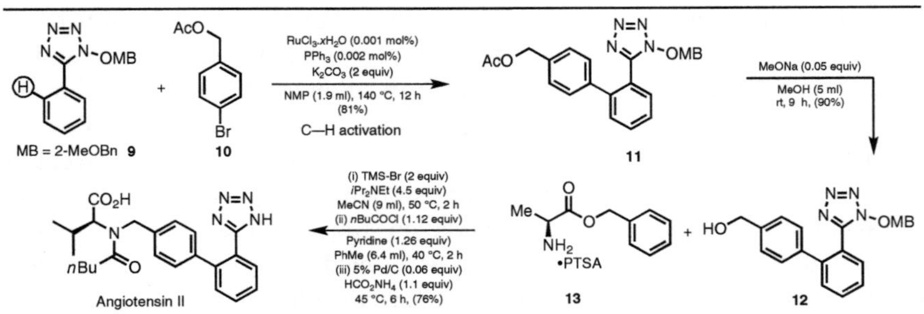

Scheme 15.2 Synthesis of Angiotensin II.

15.2.3 Synthesis of BRD3914

BRD3914 is an efficient antimalarial compound. It primarily works by inhibiting the plasmepsin X (an aspartyl protease), which is crucial for the egress and invasion stages of the Plasmodium lifecycle directly affecting the reproducibility and spreadability of the parasite [23]. Its specificity in targeting the parasite enzymes and the potential to counteract resistance highlight its therapeutic value in the ongoing battle against malaria. BRD3914, given its importance, was previously synthesized via a 14-step synthetic route in 2014. The extensive nature of the past synthetic routes does add up to a dramatic increase in the production cost, which is a key factor for the development of easily available and cheaper antimalarial therapeutics. Schreiber and co-workers in 2017 [24] identified an efficient synthetic route that employs a Pd-catalyzed, directed C(sp^3)—H arylation of azetidines at the C3 position to make it more cost effective (Scheme 15.3). The synthetic strategy initiates with TFA protection and the installation of 8-aminoquinoline on D-azetidine-2-carboxylic acid that was achieved by using 0.95 equiv of DIPEA, which helps to avoid epimerization at the α-position during the further HATU-mediated amide coupling providing the desired product **15** in 82% yield. Next, a one-pot sequential arylation via C—H activation followed by deprotection was performed that was initiated by the C—H arylation with 1-bromo-4-iodobenzene **16** using

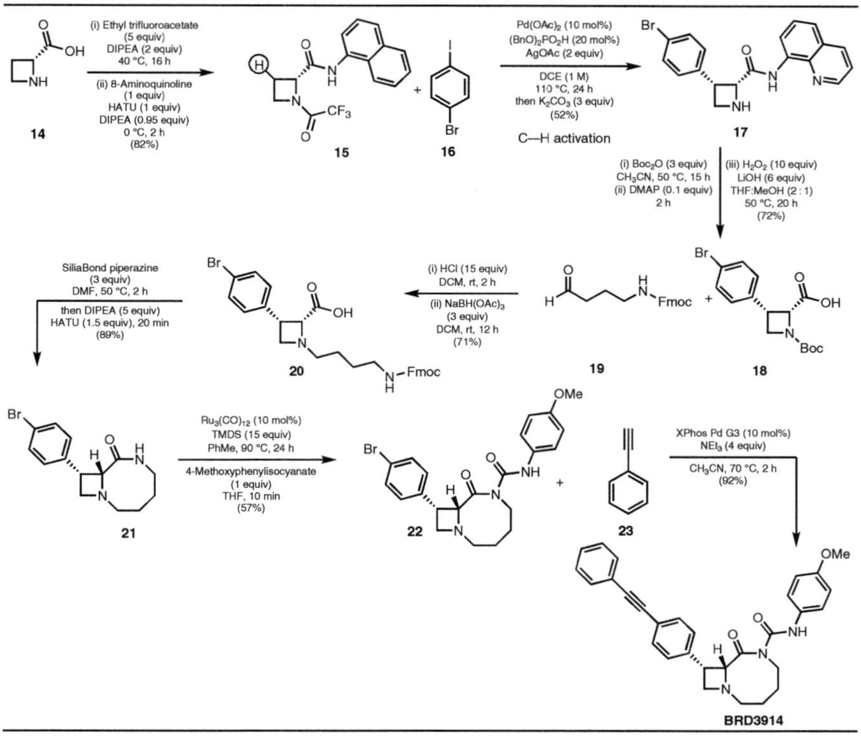

Scheme 15.3 Synthesis of BRD3914.

$Pd(OAc)_2$ (10 mol%) as a catalyst with $(BnO)_2PO_2H$ (20 mol%) as the ligand and AgOAc (2 equiv) in DCE at 110 °C for 24 hours, and on completion of the reaction, it was further subjected with the complete removal of protecting group using K_2CO_3 in MeOH:H_2O (9:1). Although the product obtained on completion of the above procedure was only 52%, it is certainly an advancement going away from the traditional synthetic processes. In the next step, the cleavage of the directing group with the retention of *cis*-stereochemistry afforded the synthetically useful carboxylic acid building block. The obtained carboxylic acid **18** was next subjected to deprotection using 4M HCl in 1,4-dioxane followed by the reductive amination using a bifunctional linker was carried out in a single flask resulting in a 71% yield of the desired product **20**. Furthermore, the removal of the Fmoc protecting group using SiliaBond piperazine and the sequential intramolecular amide bond formation led to the lactum **21** in 89% isolated yield. The reduction of lactum **21** under conditions proposed by Reeves and colleagues involving the use of $Ru_3(CO)_{12}$ and 1,1,3,3-tetramethyldisiloxane in PhMe at 90 °C for 24 hours proved to be effective in giving the diamine in good yield. The resulting diamine was trapped in situ with 4-methoxyphenyl isocyanate to afford the corresponding urea in 57% yield. In the final step, a Pd-catalyzed Heck alkynylation with phenylacetylene delivered BRD3914 in a 92% isolated yield and 10% overall yield.

15.2.4 Synthesis of Zafirlukast (Accolate)

Zafirlukast is a medication primarily used in the management of asthma. It is classified as a leukotriene receptor antagonist (LTRA), and it blocks the effect of leukotrienes, which are inflammatory chemicals the body is known to release in response to the external allergens or irritants [25]. Previously, Zafirlukast was synthesized from a prefabricated indole moiety, which exhibited lesser flexibility toward substitution in the benzene ring of indole. In 2018, the Chandrasekhar research group (Scheme 15.4) proposed a novel synthetic route for Zariflukast via the C—H activation of indoles using a transition-metal-free cyclization protocol. The synthetic methodology was initiated by performing a Sonogashira reaction of aryl bromide **24** using $PdCl_2(PPh_3)_2$ as the catalyst in the presence of CuI and Et_3N in dry DMF with trimethylsilyl acetylene (added slowly to the reaction mixture) at 80 °C for two hours. The TMS-protected alkynyl derivative without isolation was further subjected to deprotection of the TMS group by the addition of K_2CO_3 under heating conditions. A second Sonogashira reaction was then carried out under room-temperature conditions with $PdCl_2(PPh_3)_2$, Et_3N, CuI, and aryl iodide in DMF, and the reaction mixture was heated for three hours at 80 °C to provide biaryl acetylene **26** in 58% yield. Methyl 4-((2-(dimethyl amino)-5-nitrophenyl)ethynyl)-3-methoxybenzoate was now well set to undergo a C—H bond activation reaction leading to cyclization under transition-metal-free conditions by taking the substrate in dry DMF and adding $Na_2S_2O_8$ to give 3-methoxy-4-(1-methyl-5-nitro-1H-indole-3-carbonyl) benzoate **27** in 88% yield. Then, to a stirred solution of this cyclized intermediate **27** in MeOH was added, 5%

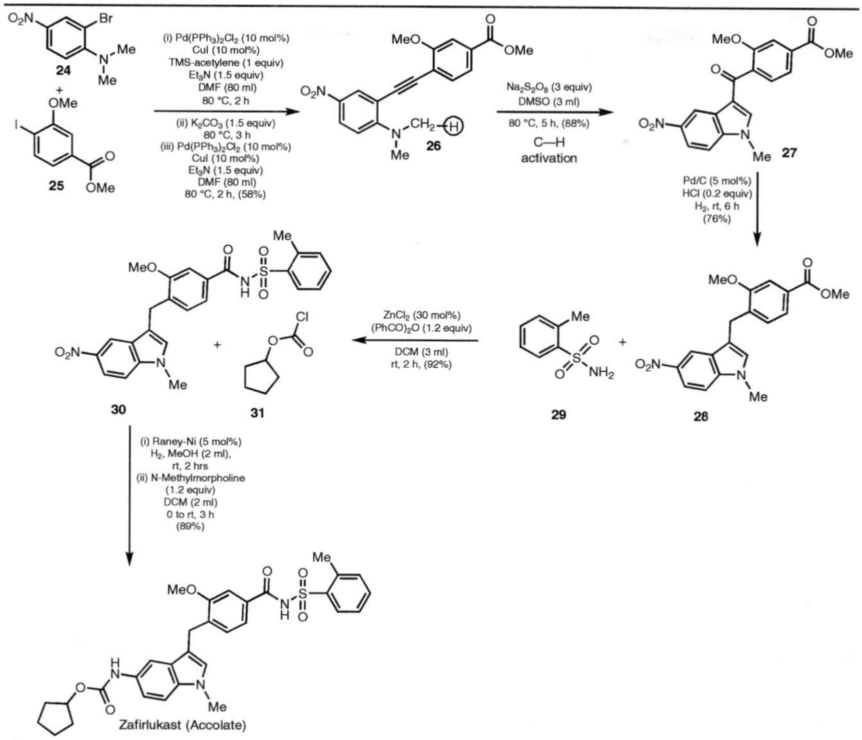

Scheme 15.4 Synthesis of Zafirlukast (Accolate).

Pd-C, and Conc. HCl under H_2 atmosphere and kept for stirring for six hours to give 3-methoxy-4-(1-methyl-5-nitro-1H-indole-3-carbonyl)benzoic acid. This benzoic acid intermediate underwent several transformations to finally provide Zafirlukast [26].

15.2.5 Synthesis of Palomid 529

Palomid 529 is a small-molecule inhibitor with potential therapeutic application in cancer treatment. It specifically targets the PI3K/Akt/mTOR signaling pathway, which is crucial in the regulation of cell growth, proliferation, and survival. Previously, it was synthesized in a 10-step long process in 2006 by D. Sherris with an overall yield of 3.9%, which makes it as costly as $1600/mol [27]. To synthesize it in a fewer-step route and better yield, Zhang and co-workers (Scheme 15.5) proposed an alternative dual chelation-assisted strategy via Ru(II)-catalyzed *ortho*-C—H cross-coupling of phenols with heteroaromatic amides [28]. The reaction sequence initiated by the treatment of 4-methoxy phenol **32** with 2-bromopyridine in the presence of CuI (10 mol%), picolinic acid (0.2 equiv) as the ligand, and K_3PO_4 (2 equiv) in DMSO for 24 hours at 90 °C to obtain **34** in 88% yield. Further, **34** was reacted with

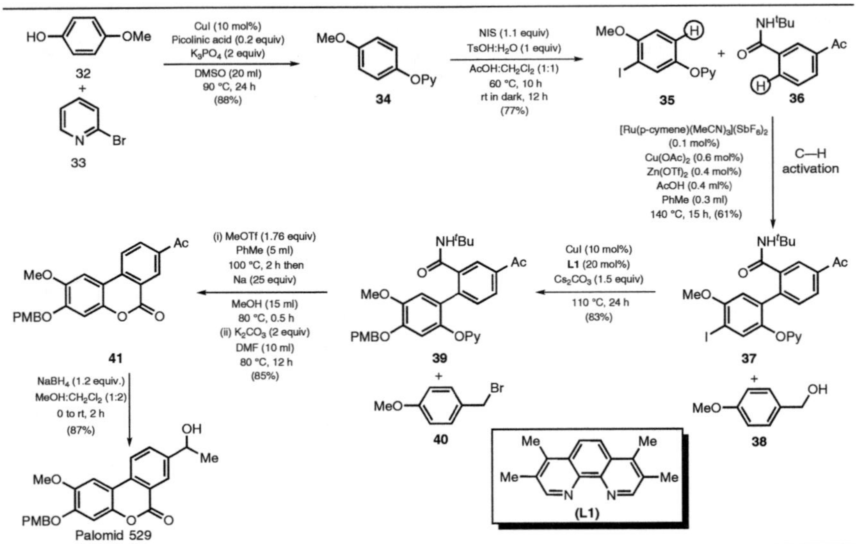

Scheme 15.5 Synthesis of Palomid 529.

N-iodosuccinimide for the installation of an iodo functionality at the *ortho*-position of the methoxy group giving **35** in 77% yield. Installation of the iodo functionality was crucial to execute the next reaction that involves a Ru(II)-catalyzed oxidative C—H/C—H cross-coupling reaction of **35** with 3-acetyl-*N*-(*tert*-butyl)benzamide using [Ru(p-cymene)(MeCN)$_3$](SbF$_6$)$_2$ as the catalyst in the presence of Cu(OAc)$_2$, Zn(OTf)$_2$, and AcOH in PhMe at 130 °C for 15 hours furnishing the biaryl **37** in 61% yield. The biaryl **37** thus obtained was further treated with 4-methoxybenzyl alcohol under the Cu-catalyzed Ullmann conditions to give an etherified product **39** in 83% yield. The resulting product **39** was then treated with 4-methoxybenzyl bromide to give **41**, and finally, the acetyl group of **41** was reduced using NaBH$_4$ providing Palomid 529 in 87% yield. With the help of this synthetic route starting from a much cheaper 4-methoxyphenol, Palomid 529 cost was brought down from $1600/mol to just $16/mol.

15.2.6 Synthesis of Febuxostat

Febuxostat is a medication primarily used to manage hyperuricemia (high uric acid level in the blood) in patients with gout by inhibiting xanthine oxidase and reducing uric acid product [29]. It serves as an important alternative for patients who are intolerant or unresponsive to allopurinol [30]. It was first developed by Teijin Pharma Ltd. and was approved as the first non-purine-type inhibitor of xanthine oxidase [31]. Previous synthetic methodologies starting from commercially available 4-cyanophenol febuxostat intermediate had allowed the synthesis to be achieved in four steps with an overall yield of 51% further requiring two additional steps to form Febuxostat [32]. However, a more efficient synthetic route was needed to make it commercially viable. As a result, in 2018, Komiyama and co-workers (Scheme 15.6) proposed a three-step synthesis process of febuxostat by using palladium and copper-catalyzed coupling reaction via C—H bond activation [33]. To achieve this, the authors first treated 5-bromo-2-fluorobenzonitrile with *t*-BuOK (1.05 equiv) and *t*-BuOH (1.1 equiv) in PhMe at ambient temperature for 2 hours to give **44**, which was then reacted with **45** in the presence of $PdCl_2$ (0.1 mol%), *t*-BuPCy (0.24 mol%), *i*-$PrCO_2H$(0.4 mol%), CuBr (1 mol%), and $KHCO_3$ (2.1 equiv) in PhMe at reflux for 15 hours to form the final intermediate **46** via C—H activated cross-coupling reaction in 91% yield. Finally, **46** was treated with 8N NaOH in THF and EtOH at 40–50 °C for five hours to produce febuxostat in 98% yield.

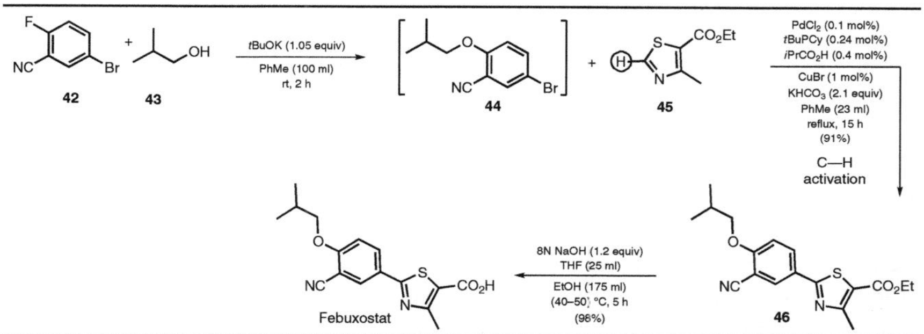

Scheme 15.6 Synthesis of Febuxostat.

15.2.7 Synthesis of Tryprostatin A

Tryprostatin A is a promising natural product with significant anticancer potential due to its ability to inhibit cell proliferation and disrupt microtubule function. It is isolated from the marine fungus *Aspergillus fumigatus*. It is a member of the alkaloid family of compounds. It completely inhibits the cell cycle progression of tsFT210 cells in the G2/M phase [34]. Previously, the total synthesis of Tryprostatin A has been done with an emphasis on the asymmetric preparation of the chiral 6-OMe-Trp moiety and a C2-selective prenylation of the indole ring [35]. In 2021, Yin, Qi, and Shi reported a total synthetic route of Tryprostatin A (Scheme 15.7) involving a key Pd(II)-catalyzed β-methyl C(sp^3)—H monoarylation step for the installation of the Trp skeleton from an inexpensive and readily available alanine [36]. The synthetic route is initiated by first the removal of the 8-aminoquinoline group of Trp **47** by the treatment with $BF_3{\cdot}Et_2O$ in methanol at 100 °C to provide Trp derivative **48** in 89% yield. Subsequently, the p-Ns group was removed from **48** using mercaptoacetic acid/K_2CO_3 mixture in DMF to obtain *N*-phthaloyl-6-OMeTrp **49** in 96% yield. For the prenylation reaction to be effective, *N*-phthaloyl-6-OMeTrp **49** was first converted to *N*-Boc-6-OMeTrp **36** in a two-step process with 95% yield. Prenylation step was next to be performed, which involved a Pd(II)-catalyzed C—H bond functionalization of the *N*-Boc-6-OMeTrp **50** with prenyl bromide (3.0 equiv) using $Pd(OAc)_2$ (10 mol%), PPh_3 (20 mol%), norbornene (5 equiv),

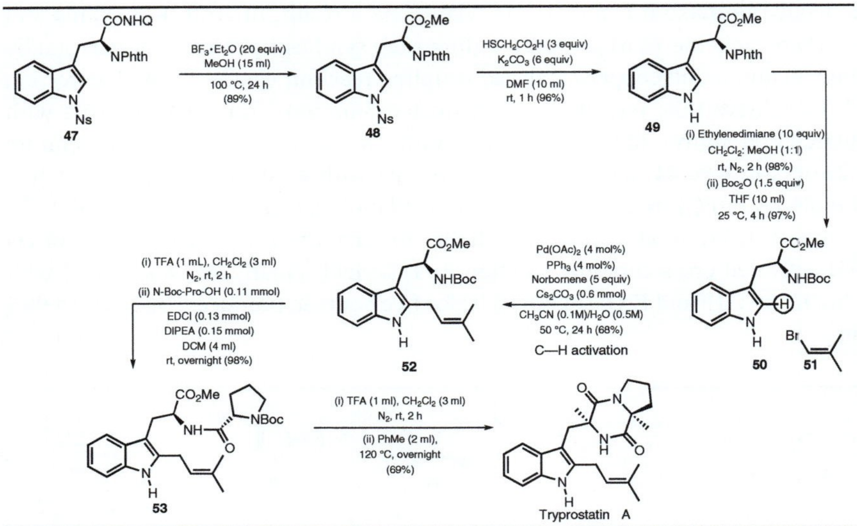

Scheme 15.7 Synthesis of Tryprostatin A.

and Cs_2CO_3 (3 equiv) in CH_3CN:H_2O (2:1 v/v) at 50 °C for 24 hours. The prenylated product **52** was obtained in 67% yield and was subsequently subjected to *N*-Boc deprotection and further coupling with *N*-Boc-Pro-OH leading to the formation of the dipeptide **53** in 98% yield. Finally, *N*-Boc deprotection of the resultant dipeptide **53** followed by an intramolecular cyclization step resulted in the formation of Tryprostatin A in 12 linear steps from alanine with a 25% overall yield.

15.2.8 Synthesis of Adiphenine

Adiphenine is known as Adiphenine hydrochloride or by its trade name Vagostyl. It is a synthetic anticholinergic and antispasmodic drug. Its ability to block muscarinic receptors leads to muscle relaxation and relief from spasms [37]. It was first synthesized by Daniel Bovet in the mid-twentieth century. In 2022, Navale and co-workers (Scheme 15.8) proposed an alternative synthetic route for the synthesis of adiphenine via scandium-catalyzed direct C—H bond functionalization [38]. In this process, first, C—H bond activation of benzene was done by using $Sc(OTf)_3$ (5 mol%) as a catalyst and coupled with α-phenyl-α-diazoacetate in DCE at 80 °C for four hours to give α,α-diarylacetate product **56** in 72% yield. Further, **56** was treated with $LiOH{\cdot}H_2O$ (2.5 equiv) in THF:MeOH:H_2O (1:1:1 v/v) at room temperature for two hours to obtain diphenylacetic acid **57** in 95% yield. Finally, diphenylacetic acid **57** was further reacted with Na_2CO_3 (5 equiv) and diethyl amine (20 equiv) in DCE at 80 °C for 16 hours to synthesize adiphenine in 85% yield.

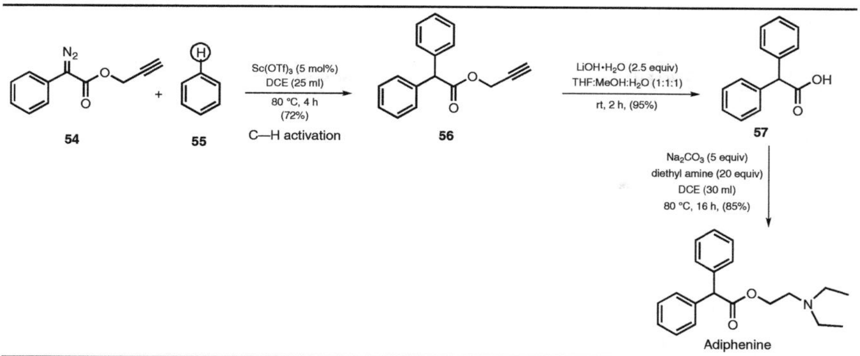

Scheme 15.8 Synthesis of Adiphenine.

15.3 C–H Functionalization in Process Chemistry

The previous section described the synthetic methodologies for achieving gram-scale synthesis of several bio-active molecules involving C—H bond functionalization as one of the key steps. Although it demonstrates the usefulness of C—H bond functionalization as a synthetic strategy for the construction of complex organic molecules without the requirement of pre-functionalization of substrates, it still limits the portrayal of the potential of C—H bond functionalization to larger-scale synthesis. Existing chemical processes for the large-scale synthesis of drug intermediates as well as active pharmaceutical ingredients (APIs) are marred with many processes-related issues, especially the requirement of protection–deprotection strategies, pre-functionalization of substrates, and usage of large quantities of solvents. All these aspects add significantly to the final cost of the API and sometimes become commercially less attractive. Process optimization (process chemistry) is an important area involving the identification of better alternative protocols for the existing synthetic technologies and improvement in the process conditions (parameters) to achieve improved output compared to the previously reported processes [39, 40]. C—H bond functionalization in recent years has emerged as a front runner for being the most preferred C—C bond [41–43] forming technology as it has demonstrated its potential to address the key issues of process chemistry detailed above. This section highlights some of the process chemistry examples of APIs or bio-active molecule synthesis on a kilogram or multikilogram scale utilizing C—H bond functionalization as an important synthetic strategy.

15.3.1 Kilogram-Scale Synthesis of Beclabuvir

Beclabuvir (BMS-791325) is a potential drug candidate for inhibiting HCV NS5B RNA polymerase [44]. HCV NS5B RNA polymerase is required for Hepatitis C virus replication, and a very selective inhibitor exists, as there is no analogous enzyme in human cells [45]. To meet the initial toxicological and clinical supply material demands, it was synthesized by the Gentles research group in 2014 [46], but a chemical route that would be amenable to rapid scale-up was required. To overcome this challenge, in 2018, the Bien group [47] proposed a kilogram-scale synthetic route for Beclabuvir via subsequent palladium-catalyzed intramolecular direct arylation (Scheme 15.9). Starting with *m*-anisaldehyde **58** (48.2 kg) and glacial acetic acid (325 l) in a 1000 l glass-lined reactor, bromine solution (Br_2, 77 kg in glacial acetic acid 12.3 l) was added to the reaction mixture (over 3.3 hours) at a maintained temperature of ≤29 °C to obtain the aldehyde **59** as an off-white solid in 86.9% (66.2 kg) yield. Furthermore, a 2000 l glass-lined reactor was charged with methyltriphenylphosphonium bromide (139.5 kg) in THF (416 l), and solid potassium *tert*-butoxide (45.9 kg) was added to the solution under a maintained temperature of 18–20 °C. To this was added a premixed solution of **59** (80 kg) and THF (200 l), and the resultant reaction mixture was stirred for three hours at 25 °C to obtain the desired product **60** (61.1 kg) with 69.9% yield. In a separate process, a 21.5 wt% solution of **60** in heptane (92.4 kg) was added to the Rh_2(S-DOSP)$_4$ (321.4 g) catalyst, and the solution was cooled to 15 °C. To this cold solution, **61** (169.6 kg) was added at 18–22 °C, and the reaction mixture was stirred for ~2 hours. On completion of the reaction, desired **62** (83.3 kg) was obtained after extraction with CH_2Cl_2 in 93.8% yield.

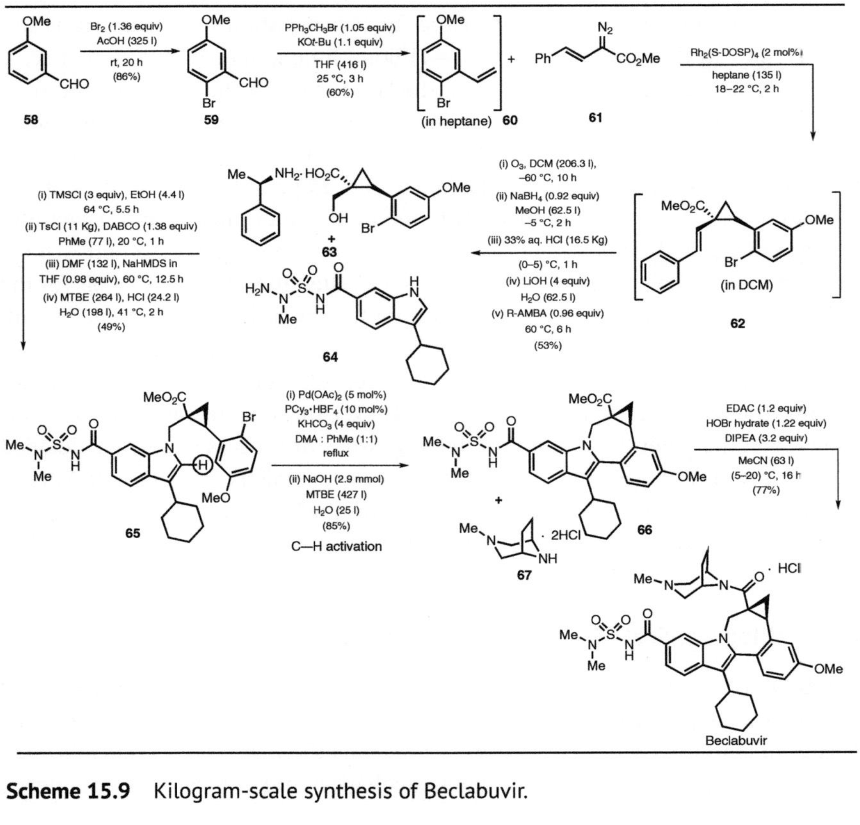

Scheme 15.9 Kilogram-scale synthesis of Beclabuvir.

The further process involved the solution of 25 kg of **62** in CH_2Cl_2 cooled to −60 °C and was treated with Ozone that was passed at 5.5 g/Nm^3 for 10 hours after which a solution of $NaBH_4$ (5.57 kg) in methanol was added slowly with the temperature maintained between −15 and −5 °C and kept stirring for 2 hours. The resultant mixture was added 33% aq. HCl (16.5 kg) and the solution cooled to 0–5 °C with stirring continued for one hour. The entire mixture was now added with water (62.5 l) and LiOH (6.25 kg), while after some time, MTBE (100 l) was charged to split the phases. Finally, the addition of isopropanol containing (*R*)-(+)-*α*-methylbenzylamine (7.5 kg) led to the formation of **63** (15.4 kg) as a gray solid in 56.5% yield.

The next phase of the synthetic route involved **63** (22 kg) taken in EtOH (193 l) to be added in a reactor containing 18.7 kg of TMS-Cl at 64 °C, and the reaction was continued for five hours. After the reaction, DABCO (8.8 kg) and a solution of p-toluenesulfonyl chloride (11 kg) in PhMe (77 l) were charged at 20 °C for one hour followed by the successive addition of DMF (132 l) and NaHMDS (47.5 kg) in THF at 60 °C. The reaction was eventually added MTBE (264 l), HCl (24.2 l), and H_2O (198 l) at 41 °C and on work-up gave **65** (16 kg) with a yield of 46.7%. A crucial juncture in the synthetic methodology is the C—H bond functionalization step that allows **65** to be subjected to a key C—H bond activation followed by cyclization. This was executed by first charging 42.8 kg of **65** into the reactor followed by the addition of potassium bicarbonate (26.1 kg), palladium(II) acetate (0.73 kg) as the catalyst, and tricyclohexylphosphonium tetrafluoroborate (2.4 kg) as the ligand in 1:1 DMA:PhMe solvent system. After the reaction stirred at reflux and the subsequent cooling of the solution to 40 °C, the addition of NaOH (2.9 mmol), MTBE (427 l), and H_2O (25 l) furnished 35.4 kg of **66** with 85% yield. In the final step, **66** (2.1 kg) was charged in a reactor containing CH_3CN (32.9 kg), and the solution was cooled to 17–23 °C and was slowly added with a bridged piperazine **67** (0.81 kg), EDAC (1.2 equiv), HOBr (1.22 equiv), and DIPEA (3.2 equiv) with the reaction temperature maintained between 5 and 20 °C. The reaction mixture after stirring for 16 hours provided Beclabuvir (1.654 kg) in 89.5% yield. This is the first known kilogram-scale strategy for Beclabuvir.

15.3.2 Commercial Synthesis of BMS-911543

BMS-911543 is a potent and selective inhibitor of Janus Kinase 2 (JAK2) for the treatment of different myeloproliferative disorders (MPDs) [48]. Functionally, BMS-911543 displayed significant anti-proliferative and pharmacodynamic effects dependent upon JAK2 signaling [49]. In 2015, the Fitzgerald group proposed a scalable synthetic route for the synthesis of BMS-911543 via Ni-mediated C—N bond formation [50]. This technique enabled the conversion of hydroxiamidine to tricyclic 2-aminopyridine via the Ullman–Goldberg–Buchwald cross-coupling reaction. This eight-step linear process was efficient (27–32% overall yield) to support the initial phases of clinical development. However, several aspects were later identified as non-ideal for the commercial use of this process. Based on those challenges, Fox et al. came up with an alternative long-term commercially viable synthetic route using Pd-catalyzed C—H functionalization providing better overall yield [51], an increased intermediate stability, and utilization of readily available reagents (Scheme 15.10). The synthetic procedure was initiated by performing ethylation of nitropyrrole **68** under phase-transfer conditions by reacting **68** with Et_2SO_4 (1.5 equiv), TBAB (2.0 equiv), and K_2CO_3 (1.5 equiv) in PhMe (737.6 l) at 65 °C for three hours to obtain **69** in quantitative yield. The next step is a key step involving a Pd-catalyzed C—H functionalization of compound **69**. The reaction was initiated by taking a mixture of potassium pivalate (84 kg) in PhMe (419.5 l), and the solution was degassed thoroughly. Compound **69** (404.5 kg) was then charged into the reactor followed by the addition of **70** (129.6 kg) and DIPEA (154.6 kg). After degasification of the mixture, Pd(Xantphos)Cl_2 (12.1 kg) was added as the catalyst, and the reaction was stirred at 95 °C for 13 hours to obtain **71** (83.6 kg) in 66% yield as a yellow solid. The success of the C—H functionalization step certainly has a major bearing on the efficiency of the overall synthetic protocol, and with this positive result in hand, the authors further went ahead with 62.7 kg of **71** that was dissolved in 6:1 THF:NMP solution at 20–30 °C, and 5% Pt/C (6.9 kg) was added to the mixture under H_2 atmosphere (0.15–0.30 MPa), and the reaction mixture was stirred for 40 hours at 35–45 °C. To this solution was added 10 mol% of NaOEt in EtOH (dropwise), and the mixture was allowed to react for 10 hours at 20–25 °C providing around 50 kg of **72** with 85% yield as an off-white solid. Further, the hydrolysis of **72** (7 g) was undertaken by adding *t*-AmOH (9.90 g), BippyPhos (98.8 mg), and Bu_4NOH (42 g), and the resultant solution was heated at 80 °C for 10 minutes. [PdCl(allyl)$_2$] (357 mg) in 105 ml of EtOH was added to the mixture and kept for 20 hours at 60 °C to obtain **73** (7.77 g) with 85% yield. Then, finally, in a suspension of **73** (50 g) and

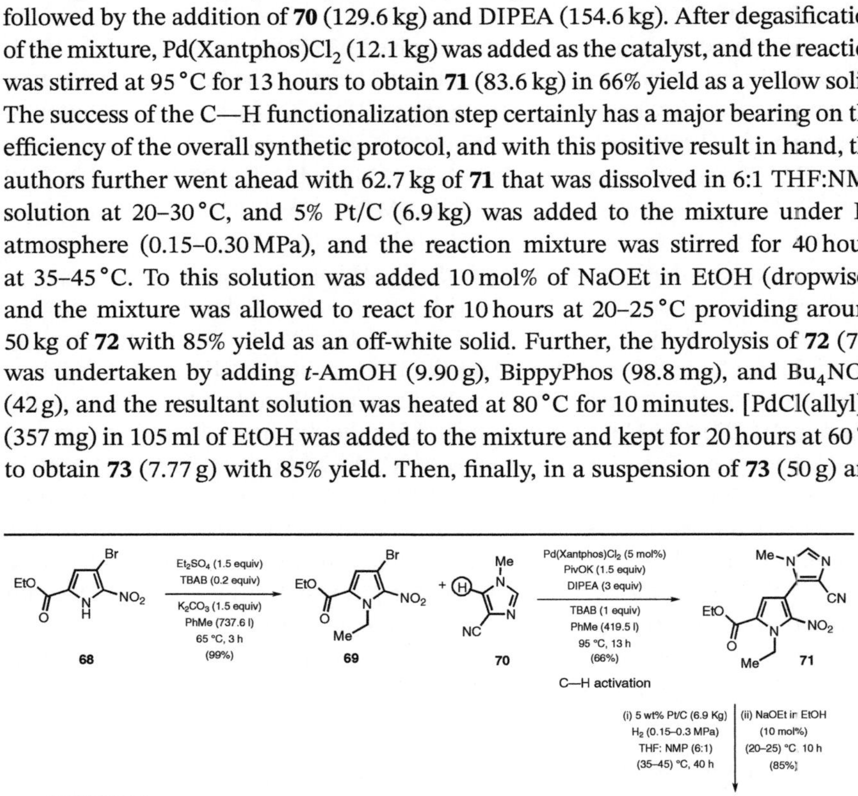

Scheme 15.10 Commercial synthesis of BMS-911543.

DCM (680 ml) at 20 °C, DPPCl (43.2 g) was added and kept for 2.5 hours and then charged with DIPEA (54.5 g) to form a brown solution, followed by the addition of **74** (23 g), DMAP (3.22 g), and DCM (50 ml) at room temperature over-night to finally obtain the BMS-911543 (52.7 g) with 92% yield.

15.3.3 Commercial Synthesis of BMS-919373

BMS-919373 is a selective inhibitor of the I_{Kur} (ultra-rapid delayed rectifier potassium) current, which is encoded by the KCNA5 gene and predominantly expressed in the atrium of the heart [52]. The synthetic route of BMS-919373 was centered around the derivatization of 2,4-dichloro-5-phenylquinazoline. However, in 2019, Wisniewski et al. (Scheme 15.11) proposed a synthetic route by using an aminomethylpyridine side chain at C-4 as a directing group for performing a selective C—H bond functionalization at C-5 [53].

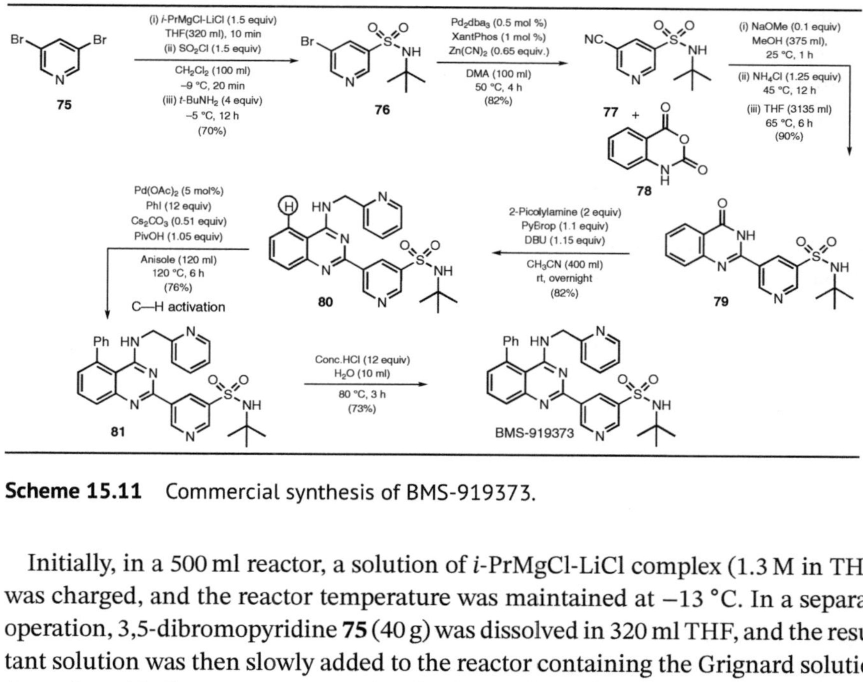

Scheme 15.11 Commercial synthesis of BMS-919373.

Initially, in a 500 ml reactor, a solution of *i*-PrMgCl-LiCl complex (1.3 M in THF) was charged, and the reactor temperature was maintained at −13 °C. In a separate operation, 3,5-dibromopyridine **75** (40 g) was dissolved in 320 ml THF, and the resultant solution was then slowly added to the reactor containing the Grignard solution dropwise with the temperature maintained at −8 °C. Subsequently, 1.5 equiv $SOCl_2$ in CH_2Cl_2 (100 ml) was added to the reaction mixture at a cooled temperature of −16 °C, and the temperature was further maintained below −9 °C for 20 minutes. Eventually, *t*-$BuNH_2$ (4.0 equiv) was added dropwise, and the reaction mixture was stirred for 12 hours at −5 °C to provide **76** (37.12 g) in 70% yield. Furthermore, 20 g of **76** was charged in a reactor along with Pd_2dba_3 (0.5 mol%), Xantphos (1.0 mol%), and $Zn(CN)_2$ (0.65 equiv) in 100 ml DMA, and the reaction mixture was stirred at 50 °C for four hours to obtain **77** (14.35 g) in 90% yield. **77** (25 g) thus obtained was further taken in a 1 l reactor and was charged with MeOH (297 g) followed by NaOMe

(25 wt%). The mixture was stirred at 25 °C for eight hours, and reaction contents were quenched by adding NH_4Cl (1.25 equiv) at 45 °C with the stirring continued for a further eight hours. MeOH was then removed from the reactor and replaced by THF to which 17.3 g of isatoic anhydride **78** was further added at 25 °C. The reactor was then heated with all the contents at 65 °C for six hours to yield **79** (32.5 g, 87% yield). For the introduction of the 2-(aminomethyl)pyridine functionality on the quinazoline moiety **79**, a slurry of **79** (20 g) in CH_3CN (20 ml/g) at 30 °C was treated with 1,8-diazabicyclo[5.4.0]undec-7-ene (1.15 equiv), Pybrop (1.1 equiv), and 2-(aminomethyl)pyridine (12.45 g) in a 1 l reactor. The reaction was stirred overnight to obtain **80** (23.04 g) as a white solid in 82% yield.

Finally, the stage was set to demonstrate the potential of the selective Pd-catalyzed C—H functionalization strategy for the synthesis of BMS-919373, by first taking iodobenzene (220 g) in a 250 ml reactor and was further charged with anisole (202.5 g), **80** (40.17 g), Cs_2CO_3 (14.24 g), pivalic acid (9.3 g), and $Pd(OAc)_2$ (963.5 mg) as the catalyst. The reaction mixture was stirred at 120 °C under N_2 for six hours to yield **81** (34.7 g) in 78%. To obtain BMS-919373, **81** (20 g) was placed in a 250 ml reactor and conc. HCl (12 equiv)/10 ml of water were added, and the resultant mixture was heated for three hours at 80 °C giving BMS-919373 as a white solid in 73% yield.

15.3.4 Multikilogram-Scale Preparation of AZD4635

AZD4635, also known as HTL1071, is an investigational small molecule that acts as an antagonist of the adenosine A_{2A} receptor ($A_{2A}R$) [54]. Previous reports have demonstrated the effectiveness of AZD4635 in inhibiting the $A_{2A}R$, thereby reversing adenosine-mediated immunosuppression. This inhibition enhances the anti-tumor immune response, primarily by reactivating CD8+ T cells and CD103+ dendritic cells (DCs) [55]. In 2012, Congreve et al. discovered the synthetic route of AZD4635 via a Suzuki–Miyaura coupling of aryl bromide with pinacol boronate ester [56]. Assessment of this process highlighted some areas for improvement. The modest 59% yield of the Suzuki–Miyaura coupling reaction, the requirement of a scavenger to remove the trace Pd metal impurity from API, the operability of the bromination step, and the lack of polymorph control in the recrystallization process provided researchers with an opportunity for process optimization. To address these challenges, Douglas et al. in 2019 (Scheme 15.12) proposed a multikilogram preparation strategy involving an Ir-catalyzed C—H borylation as the key step [57]. The process was initiated by charging a 100 l reactor with (1,5-cyclooctadiene)(methoxy)iridium(I) dimer (74.9 g), 4,4′-di-*tert*-butyl-2,2′-bipyridine (60.7 g), bis(pinacolato)diboron (8.03 kg), and *n*-heptane (35.5 kg) under N_2 condition, and to this was added 2-chloro-6-methylpyridine **82** (5.76 kg). The resultant reaction mixture was reacted at 80 °C for 18 hours to obtain **83** (8.2 kg) which will be employed as a coupling partner at a later stage of the synthesis. Ideally, the reaction proceeds through the well-known Ir-catalyzed C—H borylation mechanism and is a direct method for obtaining arylboronate esters rather than the requirement of installing a halide

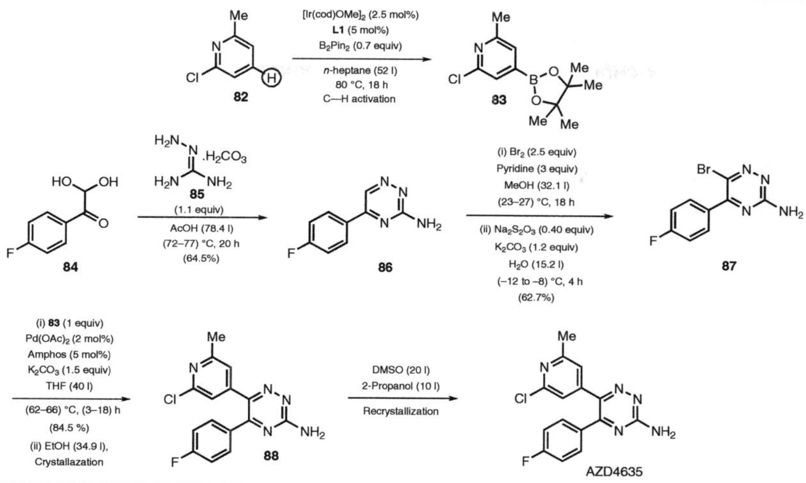

Scheme 15.12 Multikilogram-scale synthesis of AZD4635.

functionality followed by the classical Miyaura borylation. To synthesize the second coupling partner for reaction with **83** at a later stage, the authors charged 1-(4-fluorophenyl)-2,2-dihydroxyethanone **84** (6.85 kg), aminoguanidine bicarbonate **85** (6.03 kg), and acetic acid (86.2 kg) to a 100 l glass-lined reactor and maintained the temperature at 72–77 °C for 20 hours leading to the formation of **86** (4.94 kg) in 64.5% yield. In the bromination of **86**, 4.76 kg was then undertaken in a 100 l reactor to which methanol (30.2 kg), pyridine (5.94 kg), and bromine (10 kg) were subsequently added at 23–27 °C and stirred for 18 hours. The reaction was quenched with a mixture of $Na_2S_2O_3$ (1.58 kg) and K_2CO_3 (4.22 kg) in 15.2 l H_2O with the temperature maintained between −12 and −8 °C for four hours to obtain **87** (4.22 kg) as a white powder in 62.7% yield. The final step of the synthesis involved the coupling reaction between compounds **87** (4.37 kg) and **83** (4.11 kg) 100 l glass-lined reactor with potassium carbonate (3.36 kg) added as a base, 4-(di-*tert*-butylphosphino)-N,N-dimethylaniline (215 g) as the ligand, tetrahydrofuran (34.8 kg), water (17.5 kg), and palladium(II) acetate (72.8 g) as the catalyst. The reaction mixture was heated to 62–66 °C with stirring for 3–18 hours followed by crystallization using EtOH (34.9 l) to give **88** (4.32 kg) as a white solid in 84.5% yield. Finally, recrystallization of **88** (7.87 kg) using DMSO (20 l) and 2-propanol (10 l) allowed the final isolation of AZD4635 (6.51 kg) as a pure white powder in 82.7% yield.

15.3.5 Developed Synthetic Procedure of LSZ102

LSZ102 is a selective estrogen receptor (ER) degrader (SERD) with antineoplastic activity. It binds to the ER and degrades the receptor that helps for the treatment of Erα-positive breast cancer. The medicinal chemistry route for the synthesis of LSZ102 was assembled in an 11-step linear sequence with only 1% overall yield [58].

Several significant synthetic challenges were identified and were enlisted before developing an improved synthetic strategy: (i) several column chromatographic purification steps complicate the process and make it difficult for scale-up due to the large solvent usage, and (ii) preparation of a highly functionalized benzothiophene scaffold is a difficult proposition and inefficient leading to the use of undesirable reagents and solvents. To overcome these challenges, in 2020, the Baenziger group (Scheme 15.13) developed an efficient scale-up process using Pd-catalyzed C—H activation as one of the key steps [59].

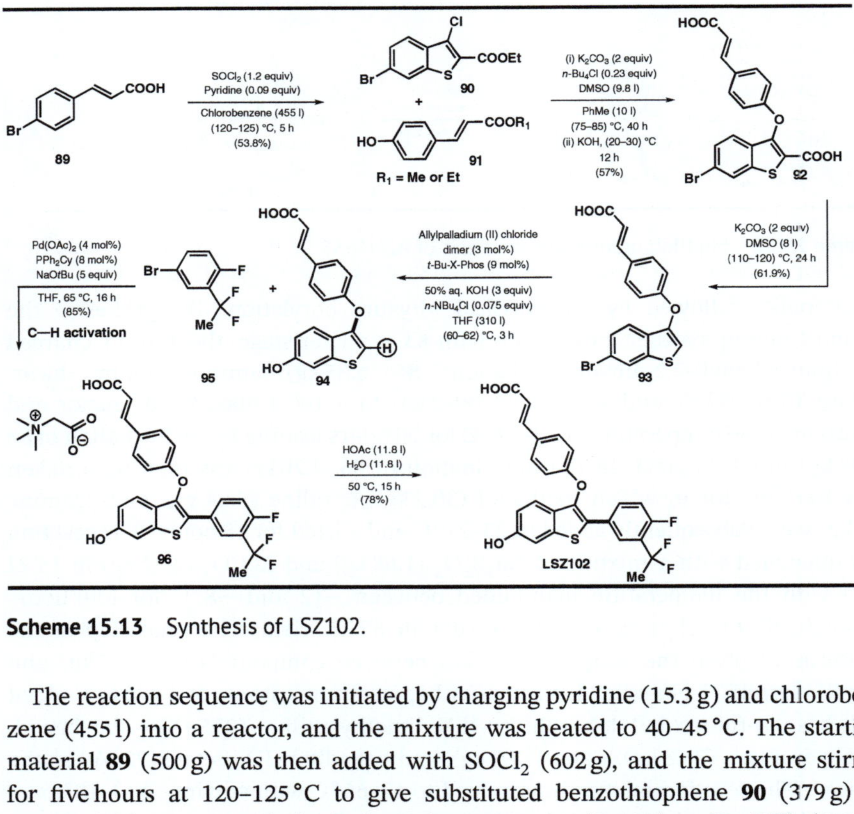

Scheme 15.13 Synthesis of LSZ102.

The reaction sequence was initiated by charging pyridine (15.3 g) and chlorobenzene (455 l) into a reactor, and the mixture was heated to 40–45 °C. The starting material **89** (500 g) was then added with $SOCl_2$ (602 g), and the mixture stirred for five hours at 120–125 °C to give substituted benzothiophene **90** (379 g) in 53.8% yield. Next, methyl-4-hydroxycinnamate **91** (2519 g), *n*-Bu_4NCl (600 g), and K_2CO_3 (2610 g) were suspended in DMSO (10.85 kg) and PhMe (8.66 kg) with benzothiophene **90** (3 kg), and the reaction mixture was stirred at 75–80 °C for 40 hours. Further treatment with KOH (1590 g) in water (3 l) results in the formation of diacid **92** (2252 g) as a brown solid in 57% yield. The obtained diacid **92** (1600 g) and K_2CO_3 (1324 g) were suspended in DMSO (8800 g) at 20–30 °C and kept for 24 hours at 110–120 °C with stirring to obtain bromobenzothiophene **93** (887 g) in 61.9% yield. A Pd-catalyzed hydroxylation reaction for the conversion of C—Br bond to C—OH was carried out on bromobenzothiophene **93** (108 kg) by treating it with tetrabutylammonium bromide (6.96 kg), allyl palladiumchloride dimer (0.329 kg), *t*Bu-X-Phos (1.22 kg), and 50% aq. KOH (3 equiv) in THF

(310 l) at 60–62 °C for three hours to obtain hydroxybenzothiophene **94** (75.9 kg) in 84.4% yield. The next step was the key step in the synthetic protocol with the hydroxybenzothiophene **94** (35 kg) being subjected to Pd-catalyzed C—H bond functionalization with aryl bromide **95** (34.8 kg) in degassed THF (704 kg) and $Pd(OAc)_2$ (1.01 kg) as catalyst in combination with cyclohexyldiphenylphosphine (2.41 kg) as the ligand in the presence of *t*-BuONa as the base in THF (20% m/m). This resulted in the formation of LSZ102 betaine cocrystal **96** (56.1 kg) in 85% yield, which was further subjected to a reaction with HOAc (11.8 l) and H_2O (11.8 l) at 50 °C for 15 hours, finally providing the crystalline final product LSZ102 (3.36 kg, 76% yield).

15.3.6 Developed Scalable Process of YLF466D

YLF466D is an allosteric activator of AMPK (AMP-activated protein kinase). AMPK plays a crucial role in cellular energy homeostasis, and its activation is associated with numerous beneficial effects on metabolism, including enhanced glucose uptake and fatty acid oxidation. Thus, AMPK is a potential therapeutic target for the treatment of myocardial ischemia–reperfusion injury (MIRI) [60]. As an AMPK activator, YLF466D is used in research to study these metabolic pathways and their implications in various diseases such as diabetes and obesity. In 2015, YLF466D was also investigated for antiplatelet activity and was found to be very much effective [61]. Given the importance of YLF466D, synthetic methodologies for its preparations were being developed with the likes of Yu and co-workers discovering a linear synthetic strategy [62]. However, it was observed that the methodology suffered from the formation of an isomerized byproduct as well as a dehalogenated byproduct as a part of the process leading to the requirement of multiple purification steps for obtaining the final product. One of the hydrolysis steps also was accompanied by partial double-bond isomerization, which complicated the synthetic procedure further. Thus, the possibility of employing multiple purifications, lower yields, and toxicity of the employed reagents makes the manufacturing of multikilograms of YLF466D, challenging. In 2021, Yin et al. (Scheme 15.14), however, developed a modified four-step linear synthetic procedure for the multikilogram-scale production of YLF466D keeping in mind the various problems associated with the previous processes. One of the key steps in the synthesis involved a Pd-catalyzed C—H bond functionalization [63].

The process was initiated by taking phenylpropiolic acid **97** (4500 g) in a 100 l reactor and was reacted with $PhNH_2$ (4300 g) and EDCI (7083 g) in CH_2Cl_2 (41.3 l) at 25 °C for 16 hours to obtain N,3-diphenylpropiolamide **98** (6626 g) as a white powder in 97.3% yield. Next, methyl 3-(bromomethyl)-benzoate **99** (6306 g) was reacted with N,3-diphenylpropiolamide **98** (5801 g) in the presence of K_2CO_3 (10.87 kg) in acetone (29.3 l) under reflux condition for 17 hours to provide methyl 3-((N,3-diphenylpropiolamido)methyl)benzoate **100** (9458 g, 97.6% yield) as an off-white solid. The next step was a key synthetic step involving a Pd-catalyzed C—H activation reaction between **100** (5003 g) and 1-chloro-4-iodobenzene **101** (4197 g) in the presence of $Pd(dppf)Cl_2{\cdot}CH_2Cl_2$ (335.0 g) as the catalyst and K_2CO_3

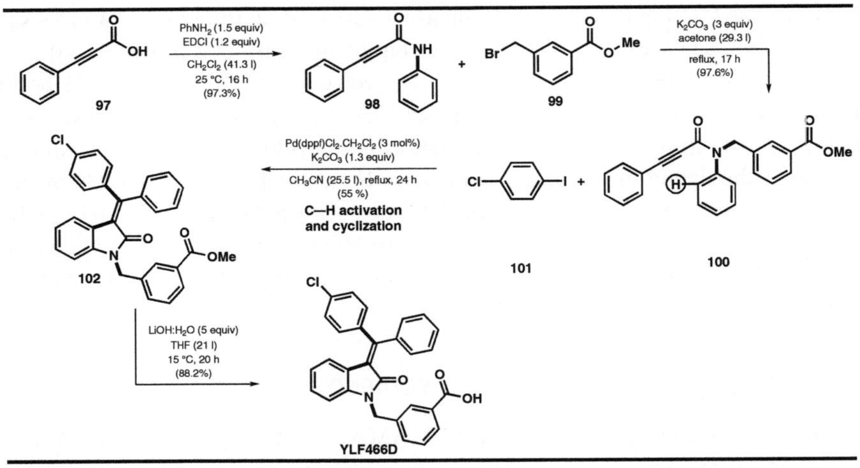

Scheme 15.14 Synthesis of YLF466D.

(2432 g) in CH_3CN (25.5 l) under reflux temperature for 24 hours leading to a tandem C—H bond functionalization followed by cyclization to obtain **102** (3612 g) as reddish-brown powder in 55% yield. Finally, the solution of **102** (4201 g) in THF (21 l) was treated with $LiOH{\cdot}H_2O$ (1836.0 g dissolved in 20.2 kg of water) that was added dropwise and allowed to react at 15 °C for 20 hours to achieve the final product YLF466D (3603 g) as an orange powder in 88.2% yield.

15.3.7 Multikilogram-Scale Synthetic Process of Nemiralisib

Nemiralisib (GSK2269557) is a potent and selective inhibitor of phosphoinositide 3-kinase delta (PI3Kδ) [64]. It is being developed by GlaxoSmithKline primarily for the treatment of respiratory diseases such as chronic obstructive pulmonary disease (COPD) and asthma [65]. PI3Kδ plays a crucial role in the inflammatory responses in the lungs, and inhibiting this enzyme can help reduce inflammation and improve lung function [66]. The initial scalable synthetic route was developed by Edney et al. in 2018 wherein they demonstrated the scalability up to 5 kg of the final product [67]. Despite extensive development work, this process had some significant issues to be resolved such as polymeric byproduct formation, lower purity, and costly intermediate synthesis. To find a shorter, cheaper, and more convergent route, Bream et al. in 2021 (Scheme 15.13) proposed a multikilogram scalable synthetic procedure via the introduction of an efficient Pd-catalyzed C—H functionalization step followed by an Ir-catalyzed C—H borylation [68].

In the first step of the process, $PdCl_2$ (0.74 kg) and XPhos (4.36 kg) were dissolved in CPME (10 l). 4-Chloro-1-(tetrahydro-2H-pyran-2-yl)-1H-indazole **104** (36 kg), (4-isopropylpiperazin-1-yl)(oxazol-5-yl)methanone **103** (34.78 kg), and potassium carbonate (35.64 kg) were then added to the reaction mixture followed by the addition of a solution of pivalic acid (9.294 kg) in CPME. The final mixture was then kept at reflux condition for five hours for the Pd-catalyzed C—H functionalization to be affected resulting in the formation of **105** (57.2 kg) in 89% yield. Ir-catalyzed borylation via C—H activation was the next step that was initiated by first introducing pinacolborane (40.8 kg) into the reactor already containing a stirred solution of (4-isopropylpiperazin-1-yl)(2-(1-(tetrahydro-2Hpyran-2-yl)-1H-indazol-4-yl)oxazol-5-yl)methanone **105** (54.00 kg), (1,5-cyclooctadiene)(methoxy)iridium-(I) dimer (0.864 kg), and 3,4,7,8-tetramethyl-1,10-phenanthroline (1.51 kg) in THF (243.2 kg) at 20 °C. The mixture was then heated at reflux condition for eight hours to obtain **106** (57.35 kg) in 83% yield. Furthermore, 4-chloroindole **107** (13.8 kg) was added to the stirred solution of **106** (50 kg), palladium acetate (0.2 kg), XPhos (0.65 kg), and potassium phosphate (46.4 kg) in MIBK (300 l)/water (250 l). After degassing, the reaction mixture was kept at 82 °C for 72 minutes until **108** (36.15 kg) was formed in 74% yield. Finally, **108** (34.75 kg) was treated with $AlCl_3$ (2.94 kg) and $LiAlH_4$ (22.6 kg) in THF (1 l) followed by the reaction with TMSCl (35.8 kg) in MeOH (244 l) at 50 °C for three hours after which NEt_3 (35.8 kg) was added at 20 °C resulting into the formation of the final product, Nemiralisib (24.8 kg) in 81% yield (Scheme 15.15).

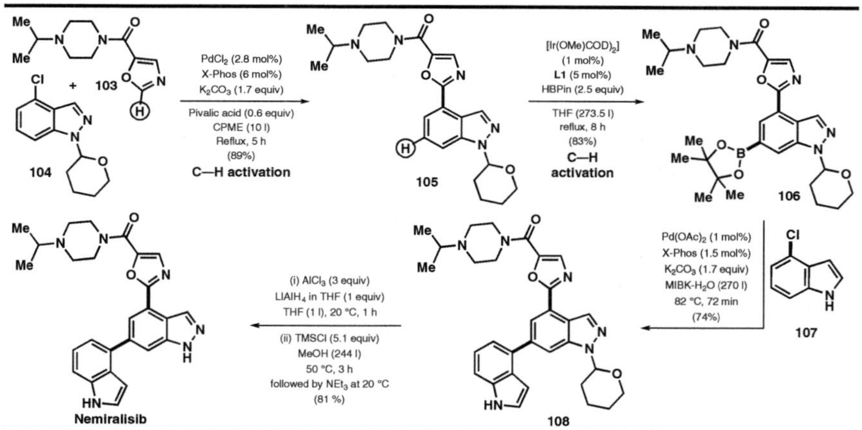

Scheme 15.15 Synthesis of Nemiralisib.

15.3.8 Kilogram-Scale Synthetic Process of AZD4573

AZD4573 is a potent and selective inhibitor of cyclin-dependent kinase 9 (CDK9). CDK9 plays a crucial role in regulating transcription elongation through the phosphorylation of RNA polymerase II [48]. The inhibition of CDK9 by AZD4573 leads to the downregulation of short-lived transcripts and labile proteins, including the anti-apoptotic protein MCL-1 [69]. This mechanism induces rapid apoptosis in cancer cells, making AZD4573 a promising therapeutic agent for hematologic malignancies. In 2020, Barlaam et al. introduced the first synthetic route of AZD4573 [70] following the medicinal chemistry route. Although the route was efficient in terms of rapid synthesis, for application toward the kilogram-scale supply of AZD4573, several other problems needed to be addressed as limitations regarding

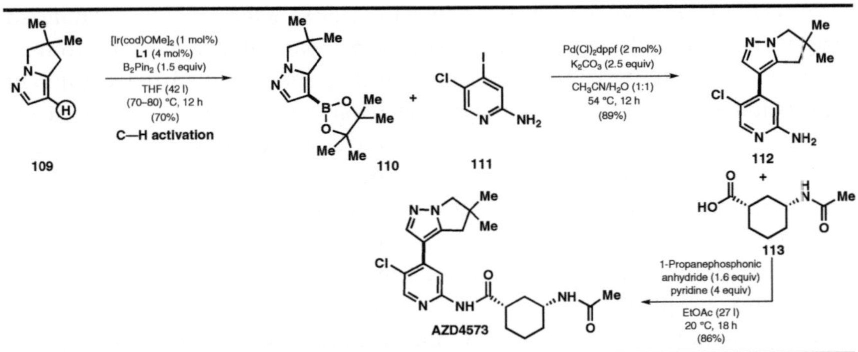

Scheme 15.16 Synthesis of AZD4573.

safety, solubility, and purification. In this regard, Karlsson et al. (Scheme 15.16) proposed an efficient multikilogram process for the commercial synthesis of AZD4573 in 2022 [71]. In the first step, they performed an Ir-catalyzed C—H borylation reaction by introducing 5,5-dimethyl-4,6-dihydropyrrolo[1,2-b]pyrazole **109** (4.1 kg) in an 80 l glass reactor with dry THF (42 l) added as a solvent. Bis(pinacolato) diboron (11.5 kg) and 3,4,7,8-tetramethyl-1,10-phenanthroline **L1** (0.29 kg) were added in the presence of bis[cyclooctadiene(methoxy)iridium] (0.20 kg), as an Ir-based catalyst, and the reaction mixture was allowed to react at 70–80 °C for 12 hours to furnish **110** (5.6 kg) as an off-white solid in 70% yield. In the next step, Suzuki–Miyaura cross-coupling between **110** (3.58 kg) and **111** (3.03 kg) in the presence of $PdCl_2$dppf (174 g), as a Pd-based catalyst, and K_2CO_3 (4.11 kg) in $CH_3CN:H_2O$ (1:1) was performed at 4 °C for 12 hours to provide **112** (2.76 kg) as a colorless solid in 88% yield. Furthermore, compound **112** (2.758 kg) and EtOAc (27 l) were charged to the reactor, followed by the addition of compound **113** (2.041 kg) and pyridine (3.39 l). 1-Propanephosphonic anhydride (10.01 kg) was then added to the reaction mixture slowly for 20 minutes with the temperature maintained below 23 °C. After the addition was completed, the mixture was stirred for 18 hours at 20 °C to finally obtain AZD4573 (3.96 kg) as colorless crystals in 86% yield and 99% purity by HPLC.

15.4 Conclusion

C—H bond functionalization has been existing as a useful synthetic C—C or C—X bond-forming technology for more than a decade and certainly has been explored extensively by researchers for the effective functionalization of a wide variety of arenes, heteroarenes, alkanes, alkenes, and many such substrates. Recent years have seen an upsurge in reports for the employment of C—H bond functionalization in gram-scale as well as kilogram-scale synthesis of many commercially and pharmaceutically important molecules. This demonstration of the potential of the C—H bond functionalization strategy to be identified as a key step in the synthesis of APIs and a sustainable alternative to the other existing C—C bond-forming technologies makes it a more attractive proposition for further applications. However, the true potential of this methodology has not been completely realized, which is evident from the number of examples discussed in this chapter. Studies pertaining to the portrayal of the usefulness of the reaction with a vis-à-vis comparison with previously employed reactions could be a good starting point. Current strategies only discuss the reduction in the number of steps rather than the overall impact of C—H bond functionalization on the reaction outcome. Green metrics could also be employed to further enhance its appeal and help promote its usage in industry.

References

1 Dutta, U., Maiti, S., Bhattacharya, T., and Maiti, D. (2021). Arene diversification through distal C(sp2)–H functionalization. *Science (80)*. 372 (6543): 5992.

2 Godula, K. and Sames, D. (2006). C—H bond functionalization in complex organic synthesis. *Science (80)*. 312 (5770): 67–72.

3 Hartwig, J.F. and Larsen, M.A. (2016). Undirected, homogeneous C—H bond functionalization: challenges and opportunities. *ACS Cent. Sci.* 2 (5): 281–292.

4 Guillemard, L., Kaplaneris, N., Ackermann, L., and Johansson, M.J. (2021). Late-stage C—H functionalization offers new opportunities in drug discovery. *Nat. Rev. Chem.* 5 (8): 522–545.

5 Li, J., De Sarkar, S., and Ackermann, L. (2015). *meta*- and *para*-Selective C—H functionalization by C—H activation. In: *Topics in Organometallic Chemistry* (ed. P.H. Dixneuf and H. Doucet), 1–41. Springer International Publishing.

6 Hirano, K. and Miura, M. (2013). C—H activation of heteroaromatics. *Sustain. Catal. Challenges Pract. Pharm. Fine Chem. Ind.* 233–267.

7 Dalton, T., Faber, T., and Glorius, F. (2021). C—H activation: toward sustainability and applications. *ACS Cent. Sci.* 7 (2): 245–261.

8 Sable, D.A., Gholap, A., Kommyreddy, S.P. et al. (2022). Heteroatom-assisted regio- and stereoselective palladium-catalyzed carboxylation of 9-allyl adenine. *J. Org. Chem.* 87 (19): 12574–12585.

9 Zhang, L. and Ritter, T. (2022). A perspective on late-stage aromatic C—H bond functionalization. *J. Am. Chem. Soc.* 144 (6): 2399–2414.

10 Cernak, T., Dykstra, K.D., Tyagarajan, S. et al. (2016). The medicinal chemist's toolbox for late stage functionalization of drug-like molecules. *Chem. Soc. Rev.* 45 (3): 546–576.

11 de Jesus, R., Hiesinger, K., and van Gemmeren, M. (2023). Preparative scale applications of C—H activation in medicinal chemistry. *Angew. Chem. Int. Ed. Engl.* 62 (45): e202306659.

12 Brown, D.G. and Boström, J. (2016). Analysis of past and present synthetic methodologies on medicinal chemistry: where have all the new reactions gone? *J. Med. Chem.* 59 (10): 4443–4458.

13 Buskes, M.J. and Blanco, M.-J. (2020). Impact of cross-coupling reactions in drug discovery and development. *Molecules* 25 (15).

14 Yan, X., Liu, X., Zhao, C., and Chen, G.-Q. (2023). Applications of synthetic biology in medical and pharmaceutical fields. *Signal Transduct. Target. Ther.* 8 (1): 199.

15 Kuhl, N., Hopkinson, M.N., Wencel-Delord, J., and Glorius, F. (2012). Beyond directing groups: transition-metal-catalyzed C—H activation of simple arenes. *Angew. Chem. Int. Ed. Engl.* 51 (41): 10236–10254.

16 Khan, K.A., Wu, F.T., Cruz-Munoz, W., and Kerbel, R.S. (2021). Ang2 inhibitors and Tie2 activators: potential therapeutics in perioperative treatment of early stage cancer. *EMBO Mol. Med.* 13 (7): e08253.

17 Campeau, L.-C., Stuart, D.R., Leclerc, J.-P. et al. (2009). Palladium-catalyzed direct arylation of azine and azole N-oxides: reaction development, scope and applications in synthesis. *J. Am. Chem. Soc.* 131 (9): 3291–3306.

18 Abraham, H.M.A., White, C.M., and White, W.B. (2015). The comparative efficacy and safety of the angiotensin receptor blockers in the management of hypertension and other cardiovascular diseases. *Drug Saf.* 38 (1): 33–54.

19 Li, E.C.K., Heran, B.S., and Wright, J.M. (2014). Angiotensin converting enzyme (ACE) inhibitors versus angiotensin receptor blockers for primary hypertension. *Cochrane Database Syst. Rev.* 8: CD009096.

20 Kubo, K., Kohara, Y., Imamiya, E. et al. (1993). Nonpeptide angiotensin II receptor antagonists. Synthesis and biological activity of benzimidazolecarboxylic acids. *J. Med. Chem.* 36 (15): 2182–2195.

21 Carini, D.J., Duncia, J.V., Aldrich, P.E. et al. (1991). Nonpeptide angiotensin II receptor antagonists: the discovery of a series of N-(biphenylylmethyl)imidazoles as potent, orally active antihypertensives. *J. Med. Chem.* 34 (8): 2525–2547.

22 Seki, M. (2011). Highly efficient catalytic system for C—H activation: a practical approach to angiotensin II receptor blockers. *ACS Catal.* 1 (6): 607–610.

23 Wells, T.N.C., van Huijsduijnen, R.H., and Van Voorhis, W.C. (2015). Malaria medicines: a glass half full? *Nat. Rev. Drug Discov.* 14 (6): 424–442.

24 Maetani, M., Zoller, J., Melillo, B. et al. (2017). Synthesis of a bicyclic azetidine with in vivo antimalarial activity enabled by stereospecific, directed C(sp(3))-H arylation. *J. Am. Chem. Soc.* 139 (32): 11300–11306.

25 Peters-Golden, M., Gleason, M.M., and Togias, A. (2006). Cysteinyl leukotrienes: multi-functional mediators in allergic rhinitis. *Clin. Exp. allergy J. Br. Soc. Allergy Clin. Immunol.* 36 (6): 689–703.

26 Paladugu, S., Mainkar, P.S., and Chandrasekhar, S. (2018). Synthesis of asthma drug zafirlukast (accolate) using intramolecular oxidative coupling via sp3 C—H bond activation. *ACS Omega* 3 (4): 4289–4294.

27 Xue, Q., Hopkins, B., Perruzzi, C. et al. (2008). Palomid 529, a novel small-molecule drug, is a TORC1/TORC2 inhibitor that reduces tumor growth, tumor angiogenesis, and vascular permeability. *Cancer Res.* 68 (22): 9551–9557.

28 Zhang, L., Zhu, L., Zhang, Y. et al. (2018). Experimental and theoretical studies on Ru(II)-catalyzed oxidative C—H/C—H coupling of phenols with aromatic amides using air as oxidant: scope, synthetic applications, and mechanistic insights. *ACS Catal.* 8 (9): 8324–8335.

29 Hu, M. and Tomlinson, B. (2008). Febuxostat in the management of hyperuricemia and chronic gout: a review. *Ther. Clin. Risk Manag.* 4 (6): 1209–1220.

30 Gao, L., Wang, B., Pan, Y. et al. (2021). Cardiovascular safety of febuxostat compared to allopurinol for the treatment of gout: a systematic and meta-analysis. *Clin. Cardiol.* 44 (7): 907–916.

31 Osada, Y., Tsuchimoto, M., Fukushima, H. et al. (1993). Hypouricemic effect of the novel xanthine oxidase inhibitor, TEI-6720, in rodents. *Eur. J. Pharmacol.* 241 (2–3): 183–188.

32 Li, L.X., Qiu, R., Wan, L.W. et al. (2015). Synthesis of the major metabolites of febuxostat. *Lett. Org. Chem.* 12 (3): 217–221.

33 Komiyama, M., Tsuchiya, H., Teramoto, M. et al. (2018). Process development of febuxostat using palladium- and copper-catalyzed C—H arylation. *Org. Process. Res. Dev.* 22 (9): 1306–1311.

34 Cui, C.B., Kakeya, H., Okada, G. et al. (1995). Tryprostatins A and B, novel mammalian cell cycle inhibitors produced by *Aspergillus fumigatus*. *J. Antibiot. (Tokyo)* 48 (11): 1382–1384.

35 Gan, T. and Cook, J.M. (1997). Enantiospecific total synthesis of tryprostatin A. *Tetrahedron Lett.* 38 (8): 1301–1304.

36 Yin, X.-S., Qi, W.-Y., and Shi, B.-F. (2021). Synthesis of tryptophan-containing 2,5-diketopiperazines via sequential C—H activation: total syntheses of tryprostatin A, maremycins A and B. *Chem. Sci.* 12 (39): 13137–13143.

37 Spitzmaul, G., Gumilar, F., Dilger, J.P., and Bouzat, C. (2009). The local anaesthetics proadifen and adiphenine inhibit nicotinic receptors by different molecular mechanisms. *Br. J. Pharmacol.* 157 (5): 804–817.

38 Navale, B.S., Laha, D., Banerjee, S. et al. (2022). Highly site-selective direct C—H bond functionalization of unactivated arenes with propargyl α-aryl-α-diazoacetates via scandium catalysis. *J. Org. Chem.* 87 (21): 13583–13597.

39 Zhang, T.Y. (2006). Process chemistry: the science, business, logic, and logistics. *Chem. Rev.* 106 (7): 2583–2595.

40 Meramo-Hurtado, S.I. and González-Delgado, Á.D. (2021). Process synthesis, analysis, and optimization methodologies toward chemical process sustainability. *Ind. Eng. Chem. Res.* 60 (11): 4193–4217.

41 Singh, K.S. (2019). Recent advances in C—H bond functionalization with ruthenium-based catalysts. *Catalysts* 9 (2): 173.

42 Chatterjee, I. (2023). C—H bond functionalization: recent discoveries and future directions. *Beilstein J. Org. Chem.* 19: 1568–1569.

43 Jana, R., Begam, H.M., and Dinda, E. (2021). The emergence of C—H functionalization strategy in medicinal chemistry and drug discovery. *Chem. Commun.* 57 (83): 10842–10866.

44 Rigat, K.L., Lu, H., Wang, Y.-K. et al. (2014). Mechanism of inhibition for BMS-791325, a novel non-nucleoside inhibitor of hepatitis C virus NS5B polymerase. *J. Biol. Chem.* 289 (48): 33456–33468.

45 Powdrill, M.H., Bernatchez, J.A., and Götte, M. (2010). Inhibitors of the hepatitis C virus RNA-dependent RNA polymerase NS5B. *Viruses* 2 (10): 2169–2195.

46 Gentles, R.G., Ding, M., Bender, J.A. et al. (2014). Discovery and preclinical characterization of the cyclopropylindolobenzazepine BMS-791325, a potent allosteric inhibitor of the hepatitis C virus NS5B polymerase. *J. Med. Chem.* 57 (5): 1855–1879.

47 Bien, J., Davulcu, A., DelMonte, A.J. et al. (2018). The first kilogram synthesis of beclabuvir, an HCV NS5B polymerase inhibitor. *Org. Process. Res. Dev.* 22 (10): 1393–1408.

48 Vardiman, J.W., Thiele, J., Arber, D.A. et al. (2009). The 2008 revision of the World Health Organization (WHO) classification of myeloid neoplasms and acute leukemia: rationale and important changes. *Blood* 114 (5): 937–951.

49 Purandare, A.V., McDevitt, T.M., Wan, H. et al. (2012). Characterization of BMS-911543, a functionally selective small-molecule inhibitor of JAK2. *Leukemia* 26 (2): 280–288.

50 Fitzgerald, M.A., Soltani, O., Wei, C. et al. (2015). Ni-catalyzed C—H functionalization in the formation of a complex heterocycle: synthesis of the potent JAK2 inhibitor BMS-911543. *J. Org. Chem.* 80 (12): 6001–6011.

51 Fox, R.J., Cuniere, N.L., Bakrania, L. et al. (2019). C—H arylation in the formation of a complex pyrrolopyridine, the commercial synthesis of the potent JAK2 inhibitor, BMS-911543. *J. Org. Chem.* 84 (8): 4661–4669.

52 Finlay, H.J., Savage, S.A., Stevens, J.M. et al. (2020). The discovery and chemical development of BMS-919373: A Selective I_{Kur} inhibitor for the potential treatment of atrial fibrillation. In: *Complete Accounts of Integrated Drug Discovery and Development: Recent Examples from the Pharmaceutical Industry Volume 3*, vol. 1369, 2–43.

53 Wisniewski, S.R., Stevens, J.M., Yu, M. et al. (2019). Utilizing native directing groups: synthesis of a selective $\mathrm{I_{Kur}}$ inhibitor, BMS-919373, via a regioselective C—H arylation. *J. Org. Chem.* 84 (8): 4704–4714.

54 Borodovsky, A., Wang, Y., Ye, M. et al. (2017). Abstract 5580: preclinical pharmacodynamics and antitumor activity of AZD4635, a novel adenosine 2A receptor inhibitor that reverses adenosine mediated T cell suppression. *Cancer Res.* 77: 5580.

55 Borodovsky, A., Barbon, C.M., Wang, Y. et al. (2020). Small molecule AZD4635 inhibitor of A(2A)R signaling rescues immune cell function including CD103(+) dendritic cells enhancing anti-tumor immunity. *J. Immunother. Cancer* 8 (2).

56 Congreve, M., Andrews, S.P., Doré, A.S. et al. (2012). Discovery of 1,2,4-triazine derivatives as adenosine A2A antagonists using structure based drug design. *J. Med. Chem.* 55 (5): 1898–1903.

57 Douglas, J.J., Adams, B.W.V., Benson, H. et al. (2019). Multikilogram-scale preparation of AZD4635 via C—H borylation and bromination: the corrosion of tantalum by a bromine/methanol mixture. *Org. Process. Res. Dev.* 23 (1): 62–68.

58 Tria, G.S., Abrams, T., Baird, J. et al. (2018). Discovery of LSZ102, a potent, orally bioavailable selective estrogen receptor degrader (SERD) for the treatment of estrogen receptor positive breast cancer. *J. Med. Chem.* 61 (7): 2837–2864.

59 Baenziger, M., Baierl, M., Devanathan, K. et al. (2020). Synthesis development of the selective estrogen receptor degrader (SERD) LSZ102 from a suzuki coupling to a C—H activation strategy. *Org. Process. Res. Dev.* 24 (8): 1405–1419.

60 Paiva, M.A., Rutter-Locher, Z., Gonçalves, L.M. et al. (2011). Enhancing AMPK activation during ischemia protects the diabetic heart against reperfusion injury. *Am. J. Physiol. Heart Circ. Physiol.* 300 (6): 2123–2134.

61 Liu, Y., Park, J.-M., Oh, S.-J. et al. (2015). Antiplatelet effect of a newly developed AMP-activated protein kinase activator YLF-466D. *Eur. J. Pharmacol.* 760: 81–87.

62 Yu, L.-F., Li, Y.-Y., Su, M.-B. et al. (2013). Development of novel alkene oxindole derivatives as orally efficacious AMP-activated protein kinase activators. *ACS Med. Chem. Lett.* 4 (5): 475–480.

63 Yin, J., Zhan, D., Ma, H. et al. (2021). Optimization and development of a scalable palladium-catalyzed C—H activation process for the geometry-selective preparation of kilograms of YLF466D, a potent AMP-activated protein kinase activator. *Org. Process. Res. Dev.* 25 (10): 2260–2269.

64 Begg, M., Amour, A., Jarvis, E. et al. (2023). An open label trial of nemiralisib, an inhaled PI3 kinase delta inhibitor for the treatment of activated PI3 kinase delta syndrome. *Pulm. Pharmacol. Ther.* 79: 102201.

65 Ameriks, M.K. and Venable, J.D. (2009). Small molecule inhibitors of phosphoinositide 3-kinase (PI3K) delta and gamma. *Curr. Top. Med. Chem.* 9 (8): 738–753.

66 Ali, K., Bilancio, A., Thomas, M. et al. (2004). Essential role for the p110delta phosphoinositide 3-kinase in the allergic response. *Nature* 431: 1007–1011.

67 Edney, D., Hulcoop, D.G., Leahy, J.H. et al. (2018). Development of flexible and scalable routes to two phosphatidinylinositol-3-kinase delta inhibitors via a common intermediate approach. *Org. Process. Res. Dev.* 22 (3): 368–376.

68 Bream, R.N., Clark, H., Edney, D. et al. (2021). Application of C—H functionalization in the development of a concise and convergent route to the phosphatidylinositol-3-kinase delta inhibitor nemiralisib. *Org. Process. Res. Dev.* 25 (3): 529–540.

69 Booher, R.N., Hatch, H., Dolinski, B.M. et al. (2014). MCL1 and BCL-xL levels in solid tumors are predictive of dinaciclib-induced apoptosis. *PLoS One* 9 (10): e108371.

70 Barlaam, B., Casella, R., Cidado, J. et al. (2020). Discovery of AZD4573, a potent and selective inhibitor of CDK9 that enables short duration of target engagement for the treatment of hematological malignancies. *J. Med. Chem.* 63 (24): 15564–15590.

71 Karlsson, S., Benson, H., Cook, C. et al. (2022). From milligram to kilogram manufacture of AZD4573: making it possible by application of enzyme-, iridium-, and palladium-catalyzed key transformations. *Org. Process. Res. Dev.* 26 (3): 601–615.

Index

Directed C–H Bond Functionalization: Concepts and Applications, First Edition.
Edited by Debabrata Maiti and Supriya Rej.

d

e

q

r

s

t

v

w

y

z